CHEMICAL ZOOLOGY

Volume XI

MAMMALIA

Contributors to This Volume

R. C. ALOIA

T. W. GOODWIN

DWAIN D. HAGERMAN

R. WESLEY LEID, JR.

MELVIN L. MOSS

LETTY MOSS-SALENTIJN

E. T. PENGELLEY

F. J. REITHEL

BRADLEY T. SCHEER

SANDER SHAPIRO

A. R. TAMMAR

JEFFREY F. WILLIAMS

CHEMICAL ZOOLOGY

Edited by **MARCEL FLORKIN**
LABORATOIRES DE BIOCHIMIE
INSTITUT LÉON FREDERICQ
LIÈGE, BELGIUM

BRADLEY T. SCHEER
PROFESSOR EMERITUS
DEPARTMENT OF BIOLOGY
UNIVERSITY OF OREGON
EUGENE, OREGON

Volume XI

MAMMALIA

ACADEMIC PRESS   New York   San Francisco   London   1979
A Subsidiary of Harcourt Brace Jovanovich, Publishers

ACADEMIC PRESS, INC.
111 Fifth Avenue, New York, New York 10003

United Kingdom Edition published by
ACADEMIC PRESS, INC. (LONDON) LTD.
24/28 Oval Road, London NW1 7DX

Library of Congress Cataloging in Publication Data

Florkin, Marcel.
    Chemical zoology.

    Includes bibliographies.
    CONTENTS:--v. 1. Protozoa.--v. 2. Porifera,
Coelenterata, and Platyhelminthes.--[etc.] --v. 11. Mam-
malia.
    1. Biological chemistry. I. Scheer, Bradley Titus,
Date joint author. II. Kidder, George Wallace. ed.
III. Title. [DNLM: 1. Amphibia. 2. Reptiles.
M1 H276 v. 9 1974 / QL641 A526 1974]
QP514.F528   591.1'92   67--23158
ISBN 0--12--261041--5 (v. 11)

PRINTED IN THE UNITED STATES OF AMERICA

79 80 81 82   9 8 7 6 5 4 3 2 1

Contents

List of Contributors ix

Preface xi

Contents of Other Volumes xiii

Chapter 1. **Lipid Composition of Cellular Membranes of Hibernating Mammals**

R. C. ALOIA AND E. T. PENGELLEY

  I. Introduction 1
  II. Scope of the Review 4
  III. Background: Low Temperatures and Membrane Phospholipids 6
  IV. Fatty Acids of Membrane Phospholipids of Hibernators 10
  V. Phospholipids of the Membranes of Active and Hibernating Squirrels 22
  VI. Discussion 33
    References 42

Chapter 2. **Membrane Structure and Function during Hibernation**

R. C. ALOIA

  I. Introduction 49
  II. Lipid Alterations Unrelated to Low Temperature Membrane Function ... 51
  III. Nonspecific Temperature Effects and Lipid Metabolism during Hibernation 52
  IV. Membrane Compartmentation and Relationship to the Lipids of
     Hibernating Mammals 53
  V. Lipid Alterations and Low Temperature Membrane Function 67
  VI. Conclusion and Future Research 69
    References 71

Chapter 3. **Mineral Metabolism and Bone**

MELVIN L. MOSS AND LETTY MOSS-SALENTIJN

  I. Introduction 77
  II. General Theory of Mineral Metabolism 78

III. Structural Aspects of Mineral Metabolism 80
IV. Calcification Processes of Bone 86
V. Bone Resorption 88
VI. Regulation 89
VII. Mode of Action of Mineral Homeostatic Agents 92
References 98

Chapter 4. **Mammalian Endocrines**

Bradley T. Scheer

I. Introduction 103
II. Pituitary 105
III. Adrenal Glands 118
IV. Sex Hormones 123
V. Thyroid Gland 128
VI. Parathyroid Glands 138
VII. Endocrine Pancreas 139
VIII. Pineal Body 148
IX. Kidney 152
References 154

Chapter 5. **Biochemical Aspects of Viviparity**

Sander Shapiro and Dwain D. Hagerman

I. Introduction 159
II. Placental Peptide Hormones 161
III. Placental Steroidogenesis 173
IV. Placental Enzymes 177
V. The Uterine Environment of the Fetus 185
VI. Summary 188
References 189

Chapter 6. **Lactation**

F. J. Reithel

I. Introduction 199
II. Mammary Gland Structure and Its Changes 200
III. Milk 202
IV. Hormonal Control of Growth and Function 204
V. Regulation of Genetic Expression 211

VI. Secretion of Milk 216
VII. The Process of Milk Ejection 219
VIII. Selected Chemical Features Involved in Synthesis 219
IX. Regression and Involution 223
X. Epilogue 224
References 225

Chapter 7. **Helminth Parasites and the Host Inflammatory System**

R. WESLEY LEID, JR. AND JEFFREY F. WILLIAMS

I. Introduction 229
II. Pathways to Inflammation 230
III. Tissue Reactions to Helminth Parasites 243
IV. Parasites and the Pathways to Inflammation 247
V. Conclusion 265
References 266

Chapter 8. **Pigments of Mammals**

T. W. GOODWIN

I. Introduction 273
II. Structural Colors 273
III. Red Colors 274
IV. Melanins 274
V. Miscellaneous Yellow to Black Pigments: Fuscins 280
VI. Pigments That Play Little or No Part in External Coloration 280
References 284

Chapter 9. **Bile Salts in Mammalia**

A. R. TAMMAR

I. Introduction 287
II. Mammalian Bile Salts 290
III. Possible Significance of Mammalian Bile Salt Differences 299
References 301

Author Index 303

Subject Index 329

List of Contributors

Numbers in parentheses indicate the pages on which the authors' contributions begin.

R. C. ALOIA (1, 49), Department of Anesthesiology, Loma Linda University School of Medicine, Loma Linda, California 92354

T. W. GOODWIN (273), Department of Biochemistry, The University of Liverpool, Liverpool L69 3BX, England

DWAIN D. HAGERMAN (159), Departments of Obstetrics and Gynecology and Biochemistry, University of Colorado School of Medicine, Denver, Colorado 80262

R. WESLEY LEID, JR. (229), Department of Pathology, Veterinary Clinical Center, Michigan State University, East Lansing, Michigan 48824

MELVIN L. MOSS (77), Department of Anatomy, College of Physicians and Surgeons, Columbia University, New York, New York 10032

LETTY MOSS-SALENTIJN (77), Department of Anatomy, College of Physicians and Surgeons, Division of Orofacial Growth and Development, School of Dental and Oral Surgery, Columbia University, New York, New York 10032

E. T. PENGELLEY (1), Biology Department, University of California, Riverside, California 92521

F. J. REITHEL (199), Department of Chemistry, University of Oregon, College of Arts and Sciences, Eugene, Oregon 97403

BRADLEY T. SCHEER* (103), Department of Biology, University of Oregon, Eugene, Oregon 97403

SANDER SHAPIRO (159), Department of Gynecology and Obstetrics, Center for Health Sciences, University of Wisconsin, Madison, Wisconsin 53706

* Present address: 1905 Mission Ridge Road, Santa Barbara, California 93103

List of Contributors

A. R. TAMMAR (287), Biochemistry and Chemistry Department, Guy's Hospital Medical School, London SE1 9RT, England

JEFFREY F. WILLIAMS (229), Department of Microbiology and Public Health, Veterinary Clinical Center, College of Veterinary Medicine, Michigan State University, East Lansing, Michigan 48824

Preface

In the Preface to the first volume of this treatise we emphasized, more than a decade ago, the growing importance to zoologists of biochemical approaches to classic problems and of new problems of a biochemical nature. To chemists we pointed out the opportunities for investigation of a variety of animal forms in the discovery of new substances and in the broadening of understanding of the biological importance of known substances. The contents of the first ten volumes of this treatise have abundantly confirmed these emphases.

We have been encouraged and heartened by the favorable reception accorded these volumes, and we anticipated that, for the class Mammalia, two volumes would be required to maintain the breadth and depth of coverage which have characterized the treatment of other animal taxa in previous volumes. Because of a number of different reasons we were unable to obtain the chapters planned, and as a result the contents of this volume are limited to those topics for which we could obtain contributions.

We are grateful to those authors who have taken the time and effort to collaborate with us. The standards of their contributions are of the same high quality that has characterized the best in previous volumes. Because of the difficulties noted and the fact that we have now retired from active academic life, we have decided to terminate the treatise with this volume. Our thanks are due the staff of Academic Press who have produced volumes of highest quality and lasting value from the materials supplied them, and the authors and associate editors who have made this treatise possible by their collaboration.

MARCEL FLORKIN
BRADLEY T. SCHEER

Contents of Other Volumes

Volume I: PROTOZOA

Systematics of the Phylum Protozoa
  John O. Corliss

Chemical Aspects of Ecology
  E. Fauré-Fremiet

Carbohydrates and Respiration
  John F. Ryley

Nitrogen: Distribution, Nutrition, and Metabolism
  George W. Kidder

Lipid Composition, Nutrition, and Metabolism
  Virginia C. Dewey

Growth Factors in Protozoa
  Daniel M. Lilly

Transport Phenomena in Protozoa
  Robert L. Conner

Digestion
  Miklós Müller

The Chemistry of Protozoan Cilia and Flagella
  Frank M. Child

Protozoan Development
  Earl D. Hanson

Nucleic Acids of Protozoa
  Manley Mandel

Carbohydrate Accumulation in the Protist—A Biochemical Model of
    Differentiation
  Richard G. Pannbacker and Barbara E. Wright

Chemical Genetics of Protozoa
  Sally Lyman Allen

Chemistry of Parasitism among Some Protozoa
  B. M. Honigberg

AUTHOR INDEX—SUBJECT INDEX

Volume II

Section I: PORIFERA

The Sponges, or Porifera
  Paul Brien

Skeletal Structures of Porifera
  M. Florkin

Pigments of Porifera
  T. M. Goodwin

Nutrition and Digestion
  Raymond Rasmont

Composition and Intermediary Metabolism—Porifera
  C. S. Hammen and Marcel Florkin

Chemical Aspects of Hibernation
  Raymond Rasmont

Section II: COELENTERATA, CTENOPHORA

Introduction to Coelenterates
  J. Bouillon

Pigments of Coelenterata
  T. W. Goodwin

Chemical Perspectives on the Feeding Response, Digestion, and Nutrition of Selected Coelenterates
  Howard M. Lenhoff

Intermediary Metabolism—Coelenterata
  C. S. Hammen

The Chemistry of Luminescence in Coelenterates
  Frank H. Johnson and Osamu Shimomura

Coelenterata: Chemical Aspects of Ecology: Pharmacology and Toxicology
  C. E. Lane

Section III: PLATYHELMINTHES, MESOZOA

Introduction to Platyhelminthes
Bradley T. Scheer and E. Ruffin Jones

Nutrition and Digestion
J. B. Jennings

Intermediary Metabolism of Flatworms
Clark P. Read

Platyhelminthes: Respiratory Metabolism
Winona B. Vernberg

Growth Development and Culture Methods: Parasitic Platyhelminths
J. A. Clegg and J. D. Smyth

Chemical Aspects of the Ecology of Platyhelminths
Calvin W. Schwabe and Araxie Kilejian

Responses of Trematodes to Pharmacological Agents
Ernest Bueding

The Mesozoa
Bayard H. McConnaughey

AUTHOR INDEX—SUBJECT INDEX

Volume III

Section I: ECHINODERMATA

General Characteristics of the Echinoderms
Georges Ubaghs

Ionic Patterns
Shirley E. Freeman and W. P. Freeman

Feeding, Digestion, and Nutrition in Echinodermata
John Carruthers Ferguson

Carbohydrates and Carbohydrate Metabolism of Echinoderms
Philip Doezema

Lipid Metabolism
U. H. M. Fagerlund

Pigments in Echinodermata
T. W. Goodwin

Fertilization and Development
  Tryggve Gustafson

Pharmacology of Echinoderms
  Ragnar Fänge

Section II: NEMATODA AND ACANTHOCEPHALA

The Systematics and Biology of Some Parasitic Nematodes
  M. B. Chitwood

The Biology of the Acanthocephala
  Ivan Pratt

Skeletal Structures and Integument of Acanthocephala and Nematoda
  Alan F. Bird and Jean Bird

Culture Methods and Nutrition of Nematodes and Acanthocephala
  Morton Rothstein and W. L. Nicholas

Carbohydrate and Energy Metabolism of Nematodes and Acanthocephala
  Howard J. Saz

Lipid Components and Metabolism of Acanthocephala and Nematoda
  Donald Fairbairn

Nitrogenous Components and Their Metabolism: Acanthocephala and
  Nematoda
  W. P. Rogers

Osmotic and Ionic Regulation in Nematodes
  Elizabeth J. Arthur and Richard C. Sanborn

Chemical Aspects of Growth and Development
  W. P. Rogers and R. I. Sommerville

The Pigments of Nematoda and Acanthocephala
  Malcolm H. Smith

Pharmacology of Nematoda
  J. Del Castillo

Chemistry of Nematodes in Relation to Serological Diagnosis
  José Oliver-Gonzáles

Chemical Ecology of Acanthocephala and Nematoda
  Alan F. Bird and H. R. Wallace

Gastrotricha, Kinorhyncha, Rotatoria, Kamptozoa, Nematomorpha, Nemertina, Priapuloidea
  Ragnar Fänge

AUTHOR INDEX—SUBJECT INDEX

Volume IV: ANNELIDA, ECHIURA, AND SIPUNCULA

Systematics and Phylogeny: Annelida, Echiura, Sipuncula
  R. B. Clark

Nutrition and Digestion
  Charles Jeuniaux

Respiration and Energy Metabolism in Annelids
  R. Phillips Dales

Respiratory Proteins and Oxygen Transport
  Marcel Florkin

Carbohydrates and Carbohydrate Metabolism: Annelida, Sipunculida, Echiurida
  Bradley T. Scheer

Nitrogen Metabolism
  Marcel Florkin

Guanidine Compounds and Phosphagens
  Nguyen van Thoai and Yvonne Robin

Annelida, Echiurida, and Sipunculida—Lipid Components and Metabolism
  Manfred L. Karnovsky

Inorganic Components and Metabolism; Ionic and Osmotic Regulation: Annelida, Sipuncula, and Echiura
  Larry C. Oglesby

Pigments of Annelida, Echiuroidea, Sipunculoidea, Priapulidea, and Phoronidea
  G. Y. Kennedy

Growth and Development
  A. E. Needham

Endocrines and Pharmacology of Annelida, Echiuroidea, Sipunculoidea
  Maurice Durchon

Luminescence in Annelids
  Milton J. Cormier

AUTHOR INDEX—SUBJECT INDEX

Volume V: ARTHROPODA Part A

Arthropods: Introduction
  S. M. Manton

Arthropod Nutrition
  R. H. Dadd

Digestion in Crustacea
  P. B. van Weel

Digestion in Insects
  R. H. Dadd

Carbohydrate Metabolism in Crustaceans
  Lyle Hohnke and Bradley T. Scheer

Metabolism of Carbohydrates in Insects
  Stanley Friedman

Nitrogenous Constituents and Nitrogen Metabolism in Arthropods
  E. Schoffeniels and R. Gilles

Lipid Metabolism and Transport in Arthropods
  Lawrence I. Gilbert and John D. O'Connor

Osmoregulation in Aquatic Arthropods
  E. Schoffeniels and R. Gilles

Osmoregulation in Terrestrial Arthropods
  Michael J. Berridge

Chemistry of Growth and Development in Crustaceans
  Larry H. Yamaoka and Bradley T. Scheer

Chemical Aspects of Growth and Development in Insects
  Colette L'Hélias

AUTHOR INDEX—SUBJECT INDEX

Volume VI: ARTHROPODA Part B

The Integument of Arthropoda
  R. H. Hackman

Hemolymph—Arthropoda
  Charles Jeuniaux

Blood Respiratory Pigments—Arthropoda
  James R. Redmond

Hemolymph Coagulation in Arthropods
  C. Grégoire

Respiration and Energy Metabolism in Crustacea
  Kenneth A. Munday and P. C. Poat

Oxidative Metabolism of Insecta
  Richard G. Hansford and Bertram Sacktor

Excretion—Arthropoda
  J. A. Riegel

Pigments—Arthropoda
  T. W. Goodwin

Endocrines of Arthropods
  František Sehnal

Chemical Ecology—Crustacea
  F. John Vernberg and Winona B. Vernberg

Toxicology and Pharmacology—Arthropoda
  M. Pavan and M. Valcurone Dazzini

AUTHOR INDEX—SUBJECT INDEX

Volume VII: MOLLUSCA

The Molluscan Framework
  Charles R. Stasek

Structure of the Molluscan Shell
  C. Grégoire

Shell Formation in Mollusks
  Karl M. Wilbur

Byssus Fiber—Mollusca
  E. M. Mercer

Chemical Embryology of Mollusca
  C. P. Raven

Pigments of Mollusca
  T. W. Goodwin

Respiratory Proteins in Mollusks
  F. Ghiretti and A. Ghiretti-Magaldi

Carbohydrates and Carbohydrate Metabolism in Mollusca
  Esther M. Goudsmit

Lipid and Sterol Components and Metabolism in Mollusca
  P. A. Voogt

Nitrogen Metabolism in Mollusks
  Marcel Florkin and S. Bricteux-Grégoire

Endocrinology of Mollusca
  Micheline Martoja

Ionoregulation and Osmoregulation in Mollusca
  E. Schoffeniels and R. Gilles

Aspects of Molluscan Pharmacology
  Robert Endean

Biochemical Ecology of Mollusca
  R. Gilles

AUTHOR INDEX—SUBJECT INDEX

Volume VIII: DEUTEROSTOMIANS, CYCLOSTOMES, AND FISHES

Section I: DEUTEROSTOMIANS

Introduction to the Morphology, Phylogenesis, and Systematics of
    Lower Deuterostomia
  Jean E. A. Godeaux

Biochemistry of Primitive Deuterostomians
  E. J. W. Barrington

Section II: VERTEBRATES (CRANIATA)

General Characteristics and Evolution of Craniata or Vertebrates
Paul Brien

A. Cyclostomes

Osmotic and Ionic Regulation in Cyclostomes
James D. Robertson

Endocrinology of the Cyclostomata
Sture Falkmer, Norman W. Thomas, and Lennart Boquist

B. Fishes

Biochemical Embryology of Fishes
A. A. Neyfakh and N. B. Abramova

The Muscle Proteins of Fishes
H. Tsuyuki

Plasma Proteins in Fishes
Robert E. Feeney and W. Duane Brown

Respiratory Function of Blood in Fishes
Gordon C. Grigg

Nitrogen Metabolism in Fishes
R. L. Watts and D. C. Watts

Respiratory Metabolism and Ecology of Fishes
Robert W. Morris

Electrolyte Metabolism of Teleosts—Including Calcified Tissues
Warren R. Fleming

Pigments of Fishes
George F. Crozier

Endocrinology of Fishes
*I. Chester Jones, J. N. Ball, I. W. Henderson,
T. Sandor, and B. I. Baker*

Bile Salts in Fishes
A. R. Tammar

AUTHOR INDEX—SUBJECT INDEX

Volume IX

Section I: AMPHIBIA

Biochemistry of Amphibian Development
  H. Denis

Gastric Secretion in Amphibia
  Gabriel M. Makhlouf and Warren S. Rehm

Salt Balance and Osmoregulation in Salientian Amphibians
  Bradley T. Scheer, Marus W. Mumbach, and Allan R. Thompson

Bile Salts of Amphibia
  A. R. Tammar

Amphibian Hemoglobins
  Bolling Sullivan

Endocrinology of Amphibia
  Wilfred Hanke

Venoms of Amphibia
  Gerhard G. Habermehl

Section II: REPTILIA

Plasma Proteins of Reptilia
  Herbert C. Dessauer

Intermediary Metabolism of Reptiles
  R. A. Coulson and Thomas Hernandez

Digestion in Reptiles
  G. Dandifosse

Water and Mineral Metabolism in Reptilia
  William H. Dantzler and W. N. Holmes

Bile Salts in Reptilia
  A. R. Tammar

Seasonal Variations in Reptiles
  M. Gilles-Baillien

Reptilian Hemoglobins
  Bolling Sullivan

Endocrinology of Reptilia—The Pituitary System
Paul Licht

Venoms of Reptiles
Findlay E. Russell and Arnold F. Brodie

AUTHOR INDEX—SUBJECT INDEX

Volume X: AVES

Introduction
Donald S. Farner

Plasma and Egg White Proteins
R. G. Board and D. J. Hornsey

Chemical Embryology
B. M. Freeman

Feather Keratins
Alan H. Brush

Avian Pigmentation
Alan H. Brush

Uropygial Gland Secretions and Feather Waxes
Jürgen Jacob

Avian Endocrinology
Albert H. Meier and Blaine R. Ferrell

Calcium Metabolism in Birds
Shmuel Hurwitz

Energy: Expenditures and Intake
Larry L. Wolf and F. Reed Hainsworth

Respiratory Proteins in Birds
A. G. Schnek, C. Paul, and C. Vandecasserie

AUTHOR INDEX—SUBJECT INDEX

CHAPTER 1

Lipid Composition of Cellular Membranes of Hibernating Mammals

R. C. Aloia and E. T. Pengelley

I.	Introduction	1
II.	Scope of the Review	4
III.	Background: Low Temperatures and Membrane Phospholipids	6
IV.	Fatty Acids of Membrane Phospholipids of Hibernators	10
	A. Animals	10
	B. Methodology	10
	C. Membrane Fatty Acids and Hibernator Tissue	14
V.	Phospholipids of the Membranes of Active and Hibernating Squirrels	22
	A. Review of Phospholipid Metabolism	22
	B. Brain Phospholipids of Mammals Capable of Hibernation	24
	C. Liver Phospholipids of Mammals Capable of Hibernation	26
	D. Kidney Phospholipids of Mammals Capable of Hibernation	27
	E. Lung Phospholipids of Mammals Capable of Hibernation	30
	F. Skeletal Muscle Phospholipids of Mammals Capable of Hibernation	30
VI.	Discussion	33
	A. Brain Phospholipids	33
	B. Liver Phospholipids	37
	C. Kidney Phospholipids	39
	D. Lung and Skeletal Muscle Phospholipids	41
	References	42

I. Introduction

Homeothermic organisms have a tightly regulated zone of normal body temperature; a normothermic range (39° ± 2°C), within which they function without metabolic stress. Their euthermic temperature zone over which survival is possible for a prolonged time with varying degrees of energy expenditure, is only a few degrees either side of their normothermic range (Hensel *et al.*, 1973). Heterothermic or hibernating mammals are unique in that their euthermic temperature range is approximately 0°–38°C. Their normothermic range is similar to the euthermic limits since at both extremes of this range they function in a normal physiological condition, the active or the hibernating state (Swan, 1975). However, their "coenothermic" or usually active

1

CHEMICAL ZOOLOGY, VOL. XI
Copyright © 1979 by Academic Press, Inc.
All rights of reproduction in any form reserved.
ISBN 0-12-261040-5

body temperature range is toward the upper end of their euthermic zone. The golden mantled ground squirrel, *Citellus lateralis,* has been observed to function within a coenothermic range of approximately 35°–39°C (R. C. Aloia, unpublished observations). In this parameter, heterotherms differ from poikilotherms which can remain active throughout their euthermic range.

Some hibernating mammals become torpid on a regular, daily basis while others, such as the bat, can remain in a dormant state for periods of several weeks (Johansson, 1967; Menaker, 1962). Many hibernating species such as the ground squirrel borrow into the soil for approximately 2 to 3 feet where temperature fluctuations during the 4 or 5 winter months remain rather constant (Gates, 1962) and moderate protection from predators is ensured. The physiological adaptation of hibernation in many cases is a mechanism allowing conservation of energy when food is scarce and sustained homeothermy would be metabolically costly (Hudson, 1967). For example, woodchucks (*Marmota monax*) maintained in a cold room 6°C and deprived of food will usually hibernate while those at the same temperature but fed, usually will not (Davis, 1967). Lack of water, high ambient temperatures, and other factors requiring costly energy expenditures can also be demonstrated to lead to the induction of torpidation (Mrovsovsky, 1976; Swan, 1975).

The actual environmental stimulus, or *Zeitgeber,* which initiates the preparation and entrance into the torpid state seems to vary with the species. Photoperiod and temperature are the most common, but certainly others exist (Pengelley and Asmundson, 1974; Brenner and Lyle, 1974; Kramm, 1973). The primary purpose of the *Zeitgeber* seems to be to entrain the animal's internal oscillatory clock or heterothermic cycle (periods of the activity and hibernation) to the appropriate time of the year. Without such *Zeitgebers* the cyclic phenomenon of hibernation would "free run" with a periodicity of approximately a year, but usually slightly less (Pengelley and Asmundson, 1974). In the golden mantled ground squirrel, *Citellus lateralis,* the hibernating cycle has been shown to be a true endogenous circannual rhythm since it was demonstrable in animals which were (a) born in captivity and hence received no initial entrainment from the environment, (b) kept on a constant light and temperature regimen, and (c) maintained for a period of 3 years.

In the several months preceding the hibernating season the ground squirrel consumes large quantities of food and more than doubles its body weight (final weight about 300–350 gm). Most of the weight gained is in the form of brown and white adipose tissue which is used as an energy source during the depressed metabolic state of hibernation and during arousal. It is estimated that 90% of the weight loss during the hibernating season occurs during the periodic arousals (Kayser, 1953). The mean weight loss per day during hibernation in the marmot is 2.2–3.8 gm, while during arousal the weight loss is 13.7–25.5 gm (Bailey and Davis, 1965).

During the yearly phase of torpidation the ground squirrel establishes a pattern of hibernation consisting of periodic bouts of depressed metabolism for durations of approximately 8–12 days, under laboratory conditions (Pengelley, 1967). The initial bout of hibernation is signaled by a resistance phase in which cyclic interactions occur between excitatory and inhibitory mechanisms or sympathetic–parasympathetic systems (South *et al.*, 1972). The heart rate, metabolic rate (Lyman, 1958) and respiratory rate (Landau, 1956) decline prior to any reduction in body temperature (T_b), thus indicating that the entry is not simply a reversion to poikilothermy, but a metabolically regulated process. Skipped beats and long periods of asystole occur under parasympathetic influence as a primary means of slowing the heart (Lyman and O'Brien, 1963). After several preliminary reductions, the T_b finally falls to within 3–4 degrees above ambient, while the heart rate and metabolic rate are at their nadir. After the T_b declines, increased sympathetic tone is established and rapid increases in heart and metabolic rate can occur upon disturbance (Lyman and O'Brien, 1963).

The length of time spent at the lower body temperature prior to arousal for *Citellus lateralis* is initially short, usually a day or two. However, over a period of several weeks, the length of time spent in the depressed metabolic state increases to a period of approximately 8–12 days (Pengelley, 1967). At the end of this period the animal will arouse to the normothermic body temperature, void itself and possibly consume food. However, if food is removed, the animal will usually remain in subsequent bouts of hibernation for periods longer than 8–12 days. Toward the end of the hibernating season, the length of the hibernating bouts gradually diminishes until the squirrel finally emerges and does not reenter the hibernating state until the following fall.

During the arousal process, which requires about 2–3 hours in the ground squirrel, the sympathetic nervous system is activated (Lyman and O'Brien, 1963). There is an increase in heart rate and metabolic rate and a powerful vasoconstriction of the posteriorly directed blood vessels. Consequently, the posterior half of the animal remains essentially at ambient temperature while the thoracic muscle mass warms up to approximately 35°C. Initially, body heat is provided by brown fat and skeletal muscle via nonshivering thermogenesis (Himms-Hagen et al., 1975). This process involves either an uncoupling of oxidative phosphorylation (Lindberg et al., 1976) or active Na$^+$ + K$^+$-ATPase activity with consequent enhanced ATP hydrolysis and heat production (Horwitz and Eaton, 1975). From approximately 15°C (T_b) onward, shivering thermogenesis appears to supplant the nonshivering process and the temperature of the thoracic region increases primarily through muscular contraction. At this point, vasodilation occurs and the temperature of the hindquarters of the organism rapidly increases by the circulation of warm blood (Lyman and O'Brien, 1960).

II. Scope of the Review

During the period of deep hibernation, the T_b of the golden mantled ground squirrel, *Citellus lateralis,* is approximately 2°–4°C, since the laboratory cold boxes are set at 2° ± 1°C. The metabolic rate and heart rate of such an animal is reduced by over 98% (Dawe and Spurrier, 1975; Johansson, 1967). In nonhibernating mammals the heart would normally enter asystole or ventricular fibrillation as body temperatures approached 15°–20°C (Johansson, 1969; Illanes and Marshall, 1964; Hirvonen, 1956). Action potentials recorded with intracellular electrodes from rabbit auricles ceased at approximately 15°C while those from ground squirrel atria could be recorded down to approximately 6°C (lowest temperature attempted; Marshall and Willis, 1962). Nervous system excitability of nonhibernating mammals is also severely reduced below about 15°–20°C (South et al., 1972) and these organisms could not survive prolonged exposures (Hensel et al., 1973). However, the tissues of the torpidants or hibernators are apparently adapted and can remain in a low temperature state for extended periods.

More than a hundred years ago, the observation was made that the homeothermic heart ceased to beat at approximately 17°C, whereas the hibernators heart only stopped below 0°C (Tait, 1922; Martin and Applegarth, 1890; Horvath, 1876). The suggestion was proffered that the degree of saturation of the depot fat may be responsible for the continued automaticity of the hibernator's heart (Tait, 1922). Indeed, numerous studies have confirmed a relationship between temperature and total lipid saturation. Hedgehog and marmot fat was shown to remain fluid at −17°C while that of nonhibernators solidified at 15°C (Dill and Forbs, 1941). Hamster depot fat was also shown to become more unsaturated just prior to hibernation (Fawcett and Lyman, 1954).

Obligate homeotherms indigenous to cold climates such as the caribou and arctic herring gull are adapted in such a manner that, although distal body temperatures reach <10°C, muscular and nervous system membranes continue to function (Swan, 1975). Because of the poorly insulated foot and lower leg regions, the distal end of the sciatic nerve was found capable of conduction at 0°C while the proximal end of the nerve was blocked at 8°C. Chemical analysis of tissue lipids (total lipids) reveals a higher degree of unsaturation in the distal than the proximal regions of the extremities. Numerous other studies have demonstrated a relationship between low temperature and increased unsaturation of depot fat, plasma lipid or total lipid extract of organs and tissues from hibernating mammals (Esher et al., 1973; Krulin and Sealander, 1972; Plattner et al., 1972; Paulsrud and Dryer, 1968; Suomalainen and Saarikoskic, 1967; Williams and Plattner, 1967; Wells et al., 1965; Kodoma and Pace, 1964; Fraenkel and Hof, 1940). However, since biological membranes contain no depot fat or triglyceride (Rouser et al., 1968) and triglyceride biosynthetic mechanisms differ from those of phospholipids (Packter, 1973), such relationships of low temperature and fatty acid unsaturation remain purely speculative when evaluating the capacity of the hibernator's cell membranes to remain functional at temperatures approaching 0°C. The reader is therefore referred to the recent papers mentioned above for triglyceride metabolism and utilization during hibernation.

The cellular membranes of hibernating hibernators are capable of continued function at temperatures approaching 0°C. Structural integrity, enzyme activity, ionic gradients, and transport mechanisms are sustained in many organs during hibernation. There is ample evidence that these membrane functions are not sustained, or at least not to the same degree, in nonhibernating mammals at low body temperatures (Bowler and Duncan, 1968; Gruener and Avi-Dor, 1966; Swanson, 1966). The primary emphasis of this review therefore will focus upon the molecular mechanisms of low temperature membrane acclimation in mammals which are capable of hibernation. As such, emphasis will be given to phospholipid metabolism and alterations in steady state levels of phospholipid classes which are commensurate with hibernation and which might possibly account for low temperature membrane function. There are several recent reviews covering various aspects of this topic (Cannon and Polnaszek, 1976; Aloia and Rouser, 1975; Willis et al., 1972; Willis, 1967) and for this reason, as well as to provide an indepth analysis of recent results, only those papers pertaining to membrane lipids and hibernation published in 1975 and thereafter will be reviewed.

III. Background: Low Temperatures and Membrane Phospholipids

Figure 1 illustrates several typical Arrhenius plots: the log of enzyme activity or other measurable function plotted against the reciprocal of the absolute temperature. Enzyme activities from membranes of nonhibernating mammals usually undergo Arrhenius plot discontinuities as temperature is reduced (Raison, 1972). Theoretical, physicochemical considerations have suggested that the Arrhenius discontinuity is indicative of a phase shift from one dominant reaction to another (Kumamoto *et al.*, 1971), or from one enzyme conformational form to another (Massey *et al.*, 1966). At the phase transition temperature, both mutually exclusive isothermal reactions are in equalibrium. Below the break temperature there is usually a reduction in enzyme activity and an increase in the activation energy (E_a) of the

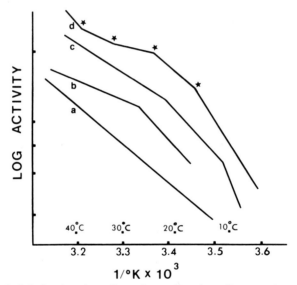

FIG. 1. Typical Arrhenius plots of membrane functions. Curve a. A straight-line Arrhenius plot typical of functions of the membranes from mammals during their hibernating state. Curve b. An Arrhenius plot exhibiting one discontinuity. Curve c. An Arrhenius plot exhibiting two discontinuities typical of most transitions of membrane lipids from the liquid-crystalline to the crystalline state. The upper break temperature corresponds to the beginning of the transition and the lower break corresponds to the end of the transition when the lipids presumably enter a solid state. Curve d. An Arrhenius plot exhibiting four discontinuities typical of the transitions observed in LM fibroblasts (Wisnieski *et al.*, 1974a,b). The break temperatures correspond to approximately 37, 32, 22, and 15°C; the first and third being assigned to the beginning and end of the lipid transition of the inner monolayer of the membrane, and the second and fourth to similar phase shifts of the outer monolayer.

process, which thus becomes energetically less feasible (see Chapter 2 by Aloia).

The reduction in enzyme activity and increased activation energy has usually been considered to result from a phase transition of the membrane lipids from a liquid-crystalline or a fluid state to a crystalline or solid state as temperature decreases (Hulbert *et al.*, 1976; Hazel and Schuster, 1976; Raison, 1973). This phenomenon has been demonstrated in both *in vivo* and *in vitro* membrane and bilayer systems (Chapman, 1976; Melchior and Steim, 1976; Fox, 1975). Conversely, as the temperature is raised, lipids apparently pass from an all-trans configuration in the solid state in which the acyl chains are closely packed and usually positioned at an angle to the normal surface of the membrane, to a predominantly *gauche* configuration at which time the acyl chains become mobile and the membrane bilayer undergoes a lateral expansion and a transverse thinning (Melchior and Steim, 1976; Sturtevant, 1974). X-Ray analysis detects the all-*trans*, hexagonally packed lipids as a sharp 4.15 Å band which becomes more diffuse, centering around 4.6 Å, as hexagonal close packing is abolished and the fatty acids enter a liquid-crystalline, mobile state at higher temperatures (Yellin and Levin, 1977; Levine, 1976). Differential scanning calorimetry can detect the fluid to solid or liquid-crystalline to crystalline (LC–C) phase transition of biological membranes and lipids extracted from the membranes, as well as the degree of cooperativity of the various lipids (Steim, 1974; Reinert and Steim, 1970). As such, it is estimated that approximately 70–80% of the membrane lipids exist in bilayer form.

Arrhenius discontinuities have also been detected for membrane enzyme activities and for both isolated membranes and lipids extracted from such membranes using physical techniques such as electron paramagnetic and nuclear magnetic resonance (Horwitz, 1972; Melhorn and Keith, 1972). In one such study, Arrhenius breaks were detected at 19°C at three different levels of cellular organization, the rate of the heart beat, succinate oxidation from mitochondria isolated from the heart of the same species and for spin-labeled heart mitochondrial membranes (McMurchie *et al.*, 1973). Arrhenius breaks have been detected in erythrocyte membranes (Zimmer and Schirmer, 1974; Lacko *et al.*, 1973; Wood and Beutler, 1967), liver microsomal membranes (Zakim and Vesey, 1976; McMurchie *et al.*, 1973), numerous enzymes of the mitochondrial electron transport chain (Watson *et al.*, 1975; Lee and Gear, 1974; Lenaz *et al.*, 1972) and for many other membrane systems (see Chapter 2 for details).

Although most membrane phase transitions have been demonstrated to be due to lipids, there are several which have been shown to result primarily from conformational changes of the membrane proteins (Vessey and Zakim, 1974; Inesi *et al.*, 1973; Eletr *et al.*, 1973; Eletr and Inesi, 1972). For example, liver UDP-glucuronyl transferse, a microsomal enzyme that requires lipid for activity, undergoes two Arrhenius discontinuities at approximately 19° and 32°C. Both can be detected with lipophilic spin probes, although after heat denaturation and sonic disruption, only the 19°C break is observed. Phospholipase A_2 treatment destroys the 19°C break and it cannot be detected when measuring enzyme activity. However, both temperature breaks are absent when measuring the lipid phase fluidity with lipophilic spin probes (Eletr *et al.*, 1973). It was concluded from these data that the lower 19°C Arrhenius discontinuity was due to a lipid phase transition since it remained after heat denaturation and was removed with phospholipase A_2 when enzyme activity was measured. Conversely, the 32°C break was probably due to protein conformational changes since it was removed with heat treatment and was not with phospholipase A_2 when enzyme activity was assayed. Both temperature breaks were no longer evident after phospholipase A_2 when measuring the physical state of the lipids with spin probes since the phospholipase disrupted the normal lipid structure and protein–lipid interaction (Eletr *et al.*, 1973). One important point to notice is that the 32°C-protein-induced Arrhenius break is transmitted to the lipid phase of the membrane since it is detectable with lipophilic probes. Similar studies indicate that the upper 35°C-Arrhenius discontinuity observed with steady state Ca^{2+} transport by sarcoplasmic reticular Ca^{2+}-ATPase is also due to a protein conformational change (Inesi *et al.*, 1973).

In the early phase of the recent surge in membrane temperature research (ca. 1970), a single Arrhenius discontinuity was demonstrable for enzyme activity and other membrane functions, and was interpreted to be the result of a lipid solidification below the LC–C temperature (Fig. 1, curve b; Lyons, 1972; Fox and Tsukugoshi, 1972). However, it was soon realized that sharp transition temperatures (variance of a degree or two) were characteristic of only pure lipid bilayer systems. Both differential scanning calorimetry and electron spin resonance have demonstrated a single, sharp transition temperature (T_C) below which the lipids enter the solid state, only when pure lipid systems or binary systems composed of acyl chains differing by only 2 carbon units were analyzed (Chapman and Urbina, 1974; Shimshick and McConnell, 1973; Linden *et al.*, 1973). This form of LC–C transition is referred to as eutectic crystallization. However, when binary lipid systems differ by <2 carbon atoms, monotectic crystallization is observed and two distinct phase transition temperatures are demonstrable with differential scanning calorimetry (Van Dijck *et al.*, 1976). In real biological membranes with a heterogeneous complex mixture of lipids a single Arrhenius discontinuity could either represent the beginning, the end, or some point within the broad temperature range through which the different species of lipid solidify (Fig. 1, curve c). Indeed in binary mixtures of lipids lateral phase separations were detected by electron spin resonance, in which the higher melting point lipid of the mixture would begin to phase separate lateral to the surface of the membrane and form isolated regions of gel state (solidified) phospholipid (Raison and McMurchie, 1974; Shimshick and McConnell, 1973). Similar lateral phase separations (LPS) can also be detected in binary phosphatidylcholine systems by freeze-fracture electron microscopy (Grant *et al.*, 1974).

Consequently, more recent studies have reported two-phase transition temperatures, T_h and T_l which corresponded to the upper and lower, or the beginning and end of the temperature range over which the lipids of the system solidify (Shimshick and McConnell, 1973; Tsukugoshi and Fox, 1973). With more data points taken at smaller temperature intervals, the second phase transition can be detected. Recently, reports have appeared which demonstrate phase transitions at four distinct temperatures (Fig. 1, curve d). Phase transitions of approximately 37°, 32°, 22°, and 15°C have been detected when measuring $Na^+ + K^+$-ATPase, and α-aminoisobutyrate transport in LM fibroblast cells; the first and third temperature being assigned to the inner bilayer half, the second and fourth assigned to the outer half of the bilayer (Wisnieski *et al.*, 1974a,b).

Although Arrhenius plot discontinuities can be detected for most mammalian cell membrane functions, many Arrhenius plots derived from membrane functions of hibernating mammals are straight lines (Fig. 1, curve a) with a constant activation energy over the entire temperature range examined (however, see Chapter 2; Raison, 1972, 1973; Lyons, 1972; Raison *et al.*, 1971a; Raison and Lyons, 1971). Consequently, continued membrane–enzyme function during hibernation has been proposed (Lyons, 1972). Indeed, residual enzyme activity has been demonstrated in tissues of hibernating hibernators at temperatures approaching 0°C, whereas enzyme activity is inhibited in tissues of active hibernators and of nonhibernators (Goldman and Albers, 1975; Goldman and Willis, 1973a; Bowler and Duncan, 1969; Willis and Li, 1969; Bowler and Duncan, 1968). Similar straight-line

Arrhenius plots and continued enzyme activity are also demonstrable with poikilothermic organisms adapted to low temperatures (Hazel and Prosser, 1974). Perhaps the most thoroughly documented change in membrane lipids of poikilothermic organisms associated with reduced temperatures has been the increase in unsaturated fatty acids (Leslie and Buckley, 1976; Driedzic and Roots, 1975; Thorpe and Ratledge, 1973; Caldwell and Vernberg, 1970; Kreps *et al.*, 1969; Morris and Schneider, 1969; Baranska and Wlodawer, 1969; Roots, 1968; Knipprath and Mead, 1966; Farkas and Herodek, 1964). Although an increase in fatty acid unsaturation of depot fat, triglycerides, and total lipid extracts has been shown to occur in hibernating hibernators (see above), few studies have examined the fatty acid composition of the membrane phospholipids of hibernating mammals to determine if modifications occur at reduced temperatures.

IV. Fatty Acids of Membrane Phospholipids of Hibernators

A. ANIMALS

The hibernating or heterothermic species employed in our studies discussed in this chapter and Chapter 2 is *Citellus lateralis*, the golden mantled ground squirrel. The animals are trapped locally, brought to the laboratory and housed in individual cages at ambient temperature on a 12/12 light–dark cycle with food and water *ad libitum*. The term "active" animal is used to designate an animal which was sacrificed in the summer during July, August, or September at a body temperature of 37°C, whereas, the term "hibernating" designates an animal sacrificed during the third to the seventh day of an established hibernating cycle (>8 day bouts in hibernation) at a rectal temperature of 3°–5°C, during January, February, or March. These considerations are important since seasonal variation in enzyme activity has been demonstrated (Fang and Willis, 1974). All animals were sacrificed by decapitation and organs removed and frozen in liquid nitrogen in the following order and time: Brain, 45 seconds; liver, $1\frac{1}{2}$ min; lung, $2\frac{1}{4}$ min; kidney $3\frac{1}{2}$ min; skeletal muscle $4\frac{1}{2}$ min. All organs are stored at $-70°C$ until used.

B. METHODOLOGY

Detailed methodology concerning extraction and analysis of phospholipids from tissues has been previously published (Aloia, 1978c; Rouser *et al.*, 1969a). Therefore, only a few cautionary points will be mentioned. After tissue lipids have been extracted with cloroform–

methanol (2:1) they must be purified to remove nonlipid contaminants (proteins, amino acids) and gangliosides. Several of the currently used procedures involve aqueous phase washes with water and various salt solutions (Bligh and Dyer, 1959; Folch et al., 1957). However, these procedures vary from sample to sample and have been shown to occasionally result in inadequate washing if not done carefully (Rouser et al., 1972). Multiple washes with KCl, for example, result in some of the acidic phospholipids and sulfatide partitioning into the upper aqueous phase (Rouser et al., 1972). Radiolabeled sulfatide continued to shift into the upper phase with as many as ten consecutive washes with KCl (R. C. Aloia and G. Rouser, unpublished observations.) Gangliosides appear to partition into the lower phase with Ca^{2+} washes and into the upper phase with Na^+, K^+, and Mg^{2+} washes (Kruger and Mendler, 1970). Lastly, acidic phospholipids can become linked through Ca^{2+} to inorganic phosphate and pull the latter into the lower phase (Cotmore et al., 1971). Consequently, the most universal and most efficient procedure for purifying lipid extracts seems to be the Sephadex G-25 column chromatographic procedure of Rouser (Rouser et al., 1969a).

After the lipid extracts are purified, membrane phospholipids must then be separated from neutral lipids. This can easily be accomplished on the appropriate silicic acid column, by one-dimensional thin layer chromatography (1D-TLC,), or by two-dimensional thin layer chromatography (2D-TLC) if each phospholipid class is to be analyzed (Rouser et al., 1969a). Figure 2 illustrates a typical 2D-thin layer chromatogram of liver phospholipids demonstrating the complete separation of all phospholipid classes, even the minor components, by means of one of the solvent systems of Rouser (Rouser et al., 1969b). If phospholipid classes are to be analyzed, plates are sprayed with sulfuric acid/formaldehyde (97:3), charred at 180°C for 45 min and lipid spots identified by comparison of known standards. Each spot is aspirated into a test tube, digested with perchloric acid to release phosphorus, and color reagents added. After heating to 100°C to develop a color, the molar quantity of phosphorus is determined by spectrophotometric analysis (see Aloia, 1978c, for details). If fatty acids are to be analyzed, the plates must be sprayed with water or other appropriate agent (0.01% Rhodamine 6G) and viewed under ultraviolet light to identify lipid spots, and allowed to dry in a N_2 atmosphere. Each dry lipid is then aspirated into a tube and either derivatized along with silica gel or eluted with chloroform–methanol from a small pipet (Pasteur) containing a glass wool plug. There are many methods utilized to render the fatty acids more volatile and suitable for gas–

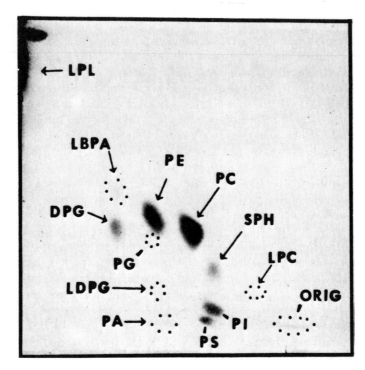

FIG. 2. Two-dimensional thin layer chromatogram of liver phospholipids. Individual phospholipid classes were separated by two-dimensional thin layer chromatography after removal of nonlipid contaminants by Sephadex column chromatography. Solvent systems used for first dimension: chloroform/methanol/ammonia, 65:25:5; for second dimension: chloroform/acetone/methanol/acetic acid/water, 3:4:1:1:0.5. Spots difficult to reproduce photographically are outlined. LPL, less polar lipid; DPG, diphosphatidylglycerol; PE, phosphatidylethanolamine; PC, phosphatidylcholine; PG, phosphatidylglycerol; LDPG, lysophosphatidylglycerol; LPC, lysophosphatidylcholine; LBPA, lysobisphosphatidic acid; PI, phosphatidylinositol; PS, phosphatidylserine; PA, phosphatidic acid; SPH, sphingomyelin.

liquid chromatographic analysis. We prefer boron trifluoride (BF_3; 14%) in methanol (Morrison and Smith, 1964). Caution is required so that sphingomyelin is not incompletely methylated (requires 60 min, 110°C) and other phospholipids are not overheated so that double bonds are lost (Rouser *et al.*, 1970). Additionally, a high temperature (110°C) is required if silica gel is present with the fatty acid (Feldman *et al.*, 1965).

After methylation and extraction into an organic solvent such as hexane or heptane, fatty acid methyl esters are ready to be injected into the gas chromatograph. Fatty acid methyl esters are best separated on polar stationary phases of polyester, such as diethyleneglycol succinate (DEGS), or, on the newer cyanosilicones, such as SP 2340 from Supelco, Inc., or Silar 10C from Applied Sciences. The senior author prefers the cyanosilicone columns because of the longer column life and higher temperature maximum (275°C). Normally, the fatty acid mixture of the phospholipid is resolved into its component species and carefully identified by chromatography of known standards. The quantity of each fatty acid is usually expressed as "area %," or the area of one peak divided by the area of all peaks. A modern gas chromatograph determines area and area % values by digital integration. However, since fatty acids respond differently in the chromatographic detector than hydrocarbons, a more quantitative method of analysis of precise changes in fatty acid composition is by weight or mole % values.

One of the most universal detectors for fatty acid methyl esters is that of flame ionization. Use of this detector is based on the assumption that the ionization of each hydrocarbon methylene unit produces a maximum molar response/mole of carbon. However, with fatty acids, the presence of oxygen atoms distorts the proportion of active carbon atoms from about 67% for C_{12} to approximately 75% for C_{24} (Ackman, 1972). An example of this can be seen in Table I which illustrates typical retention times, area %, and mole % values for three fatty acid methyl esters, 18, 20 and 22 carbons long. It can be seen that the area % value underestimates the shorter chain hydrocarbon and overestimates the longer chain hydrocarbon. When the calibration factors (moles/area) are multiplied by the absolute area values, the molar, and consequently, mole % values can be derived. These values correspond to the amount originally added to the mixture.

The programming procedure to determine mole % values requires tremendous care since the precise weight of the purified fatty acid or quantitative fatty acid mixture must be determined. Quantitative standards of fatty acid methyl ester (FAME) mixtures or purified

TABLE I

FATTY ACID METHYL ESTER STANDARDS, AREA % VERSUS MOLE %[a]

Carbon number	Retention time	Area	Area %	Calibration factor	Mole %
18:0	10.75	657,000	33.51	1	36.69
20:0	15.27	654,000	33.36	2	33.02
22:0	20.23	649,400	33.13	4	30.29

[a] R. C. Aloia, unpublished observations.

FAME's of only one fatty acid which can be ordered from the major suppliers are not weighed precisely. Purity and percent composition of a mixture are guaranteed, but the actual weight of the standard supplied in the vial is only approximate. Therefore, calibration factors based on weight or mole % cannot be calculated. To perform this calibration, the quantitative FAME mixture must be weighed on a microbalance (e.g., Perkin Elmer AD-2). If the acyl chains are longer than 16 carbons, an aliquot dried on a heating block (ca. 75°C) and weighed on a micro balance will be accurate, and molar quantities of the standard can be calculated. Conversely, the standard can be prepared in a larger volume (e.g., 10 ml) and an aliquot similarly weighed. FAME's less than 16 carbons will volatize and inaccurate weights will be obtained. If shorter-chained methyl esters are in the standard vial, difference weighing in a closed container must be undertaken and the sample must be transferred as quickly as possible to a measured tube.

C. Membrane Fatty Acids and Hibernator Tissue

There are many studies relating the hibernating physiological state and changes in lipid unsaturation. However, the vast majority of these are concerned with neutral lipids and the reader is referred to the references mentioned earlier. These studies can tell us little about the possible mechanisms of adaptation of the cell membranes to low temperatures since biological membranes contain no triglycerides (Rouser et al., 1968). There have been relatively few studies focusing upon the degree of unsaturation of the phospholipid fatty acid composition of hibernating animals. However, the adaptive value of an increase in unsaturated fatty acids in biological membranes can be seen from the following.

Numerous reports have shown that reactivating a delipidated membrane-bound enzyme (which requires a lipid for its activity) with unsaturated fatty acids results in an increase in enzyme function. In one study, reactivating $Na^+ + K^+$-ATPase with lipids containing unsaturated fatty acids at a temperature above their liquid-crystalline (LC–C) phase transition, results in approximately a tenfold increase in activity compared to reactivation with a lipid whose acyl chains are below their LC–C temperature (Kimelberg and Papahadjopoulos, 1972). Bovine brain $Na^+ + K^+$-ATPase activity has been shown to exhibit an Arrhenius discontinuity at 18°C while frog kidney cortex ATPase undergoes a phase transition at 10°C (Tanaka and Teruya, 1973). However, delipidating the bovine brain enzyme and reactivating it with a more unsaturated lipid, similar in composition to lipids of the frog kidney cortex, can lower the Arrhenius break temperature from 18° to 10°C (Tanaka and Teruya, 1973). A similar enhancement in activity can be demonstrated for succinic dehydrogenase from fish epaxial muscle. Reactivation of the enzyme with lipid from 5°C-acclimated fish results in greater activity than when lipids from 25°C-acclimated fish are used (Hazel, 1972). Additionally, induction of β-glucoside and β-galactoside transport in *Escherichia coli* is severely reduced at temperatures below the LC–C phase transition temperature (T_c) of the predominant lipids in the cell membrane. It can be demonstrated that membrane function and fluidity, as determined by measuring glycoside transport and the spectral characteristics of the spin label TEMPO (2,2,6,6-tetramethylpiperidine-1-oxyl), are sustained at successively lower temperatures when fatty acids of increasing unsaturation are included in the growth medium. For example, the beginning (T_h, high) and the end (T_1, low) of the LC–C phase transition in *E. coli* are 38° and 31°C when cells are cultured on elaidic acid $(18 : 1_t)$, 32° and 15°C when cultured on oleic acid $(18 : 1_c)$, and 28° and 8°C when cultured on linoleic acid $(18 : 2_{cc};$ Linden *et al.*, 1973; Tsukugoshi and Fox, 1973). Analysis of the membrane demonstrates that the fatty acids in the culture medium represented the majority of those in the membrane. Without such alterations in membrane fatty acid composition, permeability abnormalities would occur (Fox, 1975; McElhaney *et al.*, 1973). The above evidence appears to substantiate the fact that higher levels of unsaturated fatty acids in cellular membranes will enhance the capacity to function at low temperatures.

To the knowledge of the senior author there have only been two thorough examinations of the fatty acid composition of individual phospholipid classes of an organ or subcellular organelle of hibernating species (Aloia, 1978a; Goldman, 1975). In one study (Goldman, 1975), phospholipid classes of brain microsomes from the syrian hamster (*Mesocricetus auratus*) were separated by 2D-thin layer chromatography using the solvent system of Rouser (Rouser *et al.*, 1970). A summary of these results is presented in Table II. It can be seen that the monoenoic fatty acids of 16 and 18 carbons increased in several phospholipid classes in the brain microsomes of hibernating hamsters. Linoleic acid (18 : 2) and linolenic acid (18 : 3) increased significantly in phosphatidylcholine, whereas oleic acid (18 : 1) and arachidonic acid (20 : 4) increased significantly in phosphatidylethanolamine and phosphatidylserine during hibernation. The Unsaturation Index (UI; % composition × number of double bonds) was shown to increase in hibernation only in phosphatidyl-

TABLE II

FATTY ACID COMPOSITION OF BRAIN MICROSOME PHOSPHOLIPIDS FROM ACTIVE AND HIBERNATING HAMSTERS[a]

Carbon number	PC		PE		PS		PI		SPH		Plasmalogen PE (fatty aldehyde)	
	A	H	A	H	A	H	A	H	A	H	A	H
16:0	55.8[b]	48.7	10.7	10.2	2.6	3.0	9.2	8.9	5.8	6.6	31.1	25.7[c]
16:1	2.2	3.3[c]	1.3	1.5	0.6	0.6	1.7	1.5	2.2	1.9	3.0	3.6
18:1	25.5	30.4	11.4	13.7[c]	11.2	14.8[c]	7.1	10.0[c]	3.0	3.6	14.4	27.2[c]
18:2	0.7	1.2[c]	0.8	0.8	T[d]	T	T	T	0.5	0.5	—	—
18:3	0.6	1.1[c]	0.6	0.7	0.5	0.6	—	—	—	—	—	—
20:4	5.0	5.4	15.4	20.3[c]	1.4	2.1[c]	32.1	32.1	2.1	5.2	—	—
22:6	—	—	29.5	27.2	32.6	32.0	—	—	—	—	—	—
UI[e]	49	61	268	276	223	225	140	141	16	28		

[a] Data modified from Goldman (1975). A, active; H, hibernating.
[b] Values expressed as area %.
[c] Significant difference between active and hibernating hamsters.
[d] T, trace amount.
[e] Unsaturation Index (% composition × number of double bonds) determined from complete data in original paper.

choline and sphingomyelin (49 to 61 and 16 to 28, respectively). In phosphatidylethanolamine, phosphatidylserine and phosphatidyl-inositol, although the UI is high, it is similar in both physiological states being 268 versus 276, 223 versus 225, and 140 versus 141, re-spectively, for active and hibernating conditions. On the other hand, ethanolamine-plasmalogen exhibited a significant increase in oleic acid from 14 to 28% in the hibernating state.

Table III presents the fatty acid class composition of brain phospho-lipids from the active and hibernating squirrel, *Citellus lateralis*. In-dividual phospholipid classes are separated by 2D-thin layer chromatography using the solvent system of Rouser (Aloia, 1978c; Rouser *et al.*, 1970). There appears to be a significant increase in linoleic acid (18:2) during hibernation in four of the five lipid classes reported, PE, PC, PI and PS. Oleic acid (18:1) and docosahexaenoic acid (22:6) do not appear to be altered during hibernation and arachidonic acid (20:4) is increased only in phosphatidyl-ethanolamine. Hexadecenoic acid (16:1) increases significantly in PC, PI and PS and exhibits the same tendency in PE, although not significantly. Additionally, there appears to be a reduction in the predominant saturated species, 16:0 and 18:0 (palmitic and stearic acid) during the hibernating state in both PE and PC. The UI in-creases slightly in all phospholipid classes except sphingomyelin; the average increase being 12.7%.

An increased percentage of unsaturated fatty acids is also seen in brain tissue from poikilotherms adapted to low temperatures (Horiuchi, 1977; Leslie and Bucklie, 1976; Driedzic and Roots, 1975; Kreps *et al.*, 1969; Morris and Schneider, 1969; Roots, 1968). In-creased quantities of unsaturated fatty acids have been demonstrated in many organs besides brain (Caldwell and Vernberg, 1970; Baranska and Wlodawer, 1969; Knipprath and Mead, 1966; Farkas and Herodek, 1964), as well as in plant species and bacteria (Kates and Paradis, 1973; Sinensky, 1974, respectively). An increased level of unsaturation in cellular membranes of organisms under low tempera-ture stress has been demonstrated so frequently that it is considered synonomous with increased membrane fluidity. The enzyme $Na^+ + K^+$-ATPase of the hamster brain has been shown to undergo a transformation to a low temperature-tolerant species during cold ex-posure prior to hibernation (Goldman and Albers, 1975; Bowler and Duncan, 1969). Although its activity is drastically reduced at 5°C, the enzyme from the brain of hibernating hamsters is approximately twice as active as that derived from an active hamster brain at that tempera-ture (Goldman and Albers, 1975). This sustained enzymatic activity

TABLE III

Fatty Acid Composition of Whole Brain from Active and Hibernating Squirrels

Carbon number	PE[a,e] A[b]	PE[a,e] H[b]	PC A	PC H	PI A	PI H	PS A	PS H	SPH A	SPH H
16:0	14.3 ± 1.5[c]	13.9 ± 0.4	46.6 ± 1.2	41.9 ± 0.7[j]	4.8 ± 1.3	3.4 ± 0.9	33.2 ± 2.5	33.8 ± 1.4	9.5 ± 2.5	11.5 ± 2.2
16:1	1.6 ± 0.4	2.13 ± 0.1	2.7 ± 0.2	4.3 ± 0.5[j]	0.5 ± 0.02	0.8 ± 0.1[a]	0.7 ± 0.1	1.0 ± 0.1[i]	—	—
18:0	18.0 ± 1.9	13.6 ± 0.8[g]	8.9 ± 0.7	7.6 ± 0.2[j]	37.5 ± 3.1	32.9 ± 2.1	18.5 ± 3.2	14.7 ± 1.0	36.9 ± 2.6	35.6 ± 1.5
18:1	25.1 ± 2.1	24.9 ± 2.5	28.7 ± 0.6	29.6 ± 0.8	29.8 ± 1.7	28.2 ± 2.8	10.7 ± 1.2	10.2 ± 2.0	T[k]	T[k]
18:2	1.81 ± 0.6	3.1 ± 0.2[g]	1.1 ± 0.2	4.2 ± 0.4[j]	0.7 ± 0.1	2.1 ± 0.1[j]	0.5 ± 0.1	1.8 ± 0.8[h]	T[k]	T[k]
18:3	3.7 ± 0.6	2.6 ± 0.4[h]	1.4 ± 0.1	1.2 ± 0.1	2.5 ± 0.1	2.4 ± 0.2	1.1 ± .2	1.6 ± 0.9	T[k]	T[k]
20:4	9.7 ± 1.5	13.1 ± 1.0[i]	4.2 ± 0.7	3.9 ± 0.2	3.9 ± 1.7	6.0 ± 1.4	T[k]	T[k]	T[k]	T[k]
22:6	8.4 ± 2.2	9.0 ± 1.3	2.5 ± 0.7	3.3 ± 0.5	12.4 ± 1.7	13.2 ± 4.0	2.8 ± 0.1	2.6 ± 1.4		
24:0	5.0 ± 0.8	4.9 ± 0.3	T[k]	T[k]	3.4 ± 0.7	3.7 ± 0.8	1.2 ± 0.3	1.0 ± 0.5	1.8 ± 0.4	2.0 ± 0.2
24:1	1.6 ± 0.3	1.4 ± 0.2	T[k]	T[k]	0.9 ± 0.2	0.9 ± 0.2	1.9 ± 0.3	1.9 ± 0.4	25.1 ± 0.6	28.3 ± 3.2
UI[d]	142	159 (Δ12%)	71	82 (Δ15%)	135	150 (Δ11%)	43	51 (Δ19%)	58	56

[a] See Fig. 2 for abbreviations.
[b] A, active animal; H, hibernating animal.
[c] Area % values
[d] UI = Unsaturated Index: Area % × number of double bonds; values determined from raw data, see original publication, Aloia, 1978a.
[e] PE values are for fatty acids only. Aldehyde fraction was removed.
[f] $P < 0.01$.
[g] $P < 0.02$.
[h] $P < 0.05$.
[i] $P < 0.025$.
[j] $P < 0.001$.
[k] $< 0.5\%$.

could well be due primarily to the increased quantities of unsaturated fatty acids (Tables II and III; Aloia, 1978a; Goldman, 1975).

Recent studies have indicated that specific metabolic alterations during hibernation may account for an increase in unsaturated fatty acids similar to that found in some of the brain phospholipids. Chemical analysis has indicated that thyroid hormone secretion is drastically reduced in the fall prior to hibernation (Hulbert and Hudson, 1976). It was proffered that the absence of thyroid hormone during hibernation was responsible for the increase in the relative proportion of unsaturated fatty acids in hibernating species which in turn accounted for the straight-line Arrhenius plots derived from mitochondrial succinate oxidation from hibernating animals (Hulbert and Hudson, 1976). Recent studies corroborate this possibility by demonstrating that thyroxine-treated rats experience a decrease in unsaturation. Specifically, in heart mitochondria, thyroxine induced a reduction in linoleic ($18:2$; 28 to 24%) and an elevation in stearic acid ($18:0$; 21 to 23%), while in heart microsomes a reduction in linoleic (24 to 21%) and arachidonic acid ($20:4$; 18 to 14%), and an elevation of palmitic ($16:0$; 15 to 17%) and stearic acid (21 to 27%) (Steffen and Plattner, 1976). Furthermore, an elegant study by Hulbert *et al.* (1976), demonstrates an increase in unsaturated fatty acids from 61 to 71% (area %) and a decrease in the onset of the LC–C phase transition from 23° to 10°C in thyroidectomized rats. Additionally, within 12 hours after the administration of thyroxine, the phase transition temperature increased and fatty acid unsaturation decreased significantly toward normal values (Hulbert *et al.*, 1976). Consequently, possible quiescence of the thyroid in hibernating species may have an effect on increasing the content of unsaturated fatty acids in some membranes of hibernating animals and provide the necessary enhancement in membrane fluidity to allow continued membrane function at reduced temperatures.

However, several recent fatty acid analyses seem to demonstrate that the content of unsaturated fatty acids declines in heart and liver tissue and mitochondria during hibernation. For example, although oleic acid and arachidonic acid increase from 18.1 to 28.7% and 9.5 to 11.4%, respectively, in liver mitochondria from hibernating squirrels, the content of docosahexaenoic acid ($22:6$) is reduced from 5.8 to 1.0% (Lerner *et al.*, 1972). Consequently, the Unsaturation Index actually decreases from 142 to 128. Table IV illustrates selective, summary data from a similar study (Plattner *et al.*, 1976). It is readily apparent from an examination of the table that the degree of unsaturation of the membrane lipids, expressed as the Unsaturation Index (% composition × number of double bonds), decreases in the liver mitochondria from

TABLE IV

FATTY ACID COMPOSITION OF MITOCHONDRIA FROM HEART AND LIVER OF GROUND SQUIRRELS (*Citellus tridecemlineatus*)[a]

Carbon number	Liver				Heart			
	Young animals		Adults		Young animals		Adults	
	A[b]	H	A	H	A	H	A	H
16:0	17.8%[c]	22.9	18.0	26.1[d]	17.7	14.6	2.0	1.9
17:0	0.14	0.14	0.05	1.1[d]	0.37	0.58	0.43	0.83
18:0	36.5	22.0[d]	35.6	24.2[d]	14.0	17.4[d]	14.2	11.4
18:1	30.1	43.1	31.6	35.6	11.5	16.4[d]	15.3	17.2
18:2	4.4	2.8[d]	6.0	3.7[d]	22.7	19.1[d]	24.1	25.4
20:4	—	—	1.7	0.35[d]	5.5	5.3	6.6	7.7
22:4	0.63	ND	ND	ND	19.6	15.1	14.9	15.0
UI[e]	62.4	57.6	59.2	53.0	221.4	190.3	208.6	209.0

[a] Data modified from Plattner *et al.* (1976) (see original reference for complete details).

[b] A, active; H, hibernating squirrels.

[c] Area % values.

[d] Comparison between values from active and hibernating animals are significantly different.

[e] Unsaturation Index: % occurrence × number of double bonds.

both young and old hibernating squirrels (Table IV). There does not appear to be a tendency to increase the quantity of unsaturated fatty acids. There is no significant difference in oleic acid (18:1) in liver mitochondria from active and hibernating (young and old) animals, yet it increases significantly in heart mitochondria from young hibernators. Linoleic acid (18:2) decreases during hibernation in liver mitochondria from young and old animals and in heart mitochondria from young animals. Stearic acid (18:0) decreases in the liver mitochondria from hibernating (young and old) squirrels, while it increases in heart mitochondria from young hibernating squirrels. There appear to be no significant differences in heart mitochondrial fatty acid values from adult active and hibernating squirrels (Table IV).

Similar observations were made in an investigation of the membrane phospholipids from the hearts of active and hibernating squirrels *Citellus lateralis* (R. C. Aloia, unpublished observations; Table V). It is apparent that the Unsaturation Index either changes very little or decreases slightly in the hearts of adult hibernating animals. There

TABLE V

FATTY ACID COMPOSITION OF TOTAL PHOSPHOLIPID FRACTION OF HEART MUSCLE FROM GROUND SQUIRRELS (Citellus lateralis)[a]

Carbon number	A_1[b]	A_2	A_3	A_4	$A\bar{x}$	$H\bar{x}$	H_1	H_2	H_3
	(3)[c]	(3)[c]	(3)[c]	(3)[c]	(12)[c]	(12)[c]	(2)[c]	(5)[c]	(5)[c]
16:0	19.89[d]	20.43	19.87	19.03	19.80	15.39	14.14	17.72	14.31
16:1	0.62	0.66	0.64	0.61	0.63	0.87	0.56	1.04	1.02
18:0	17.05	15.77	17.86	17.06	16.94	21.25	20.80	22.13	20.84
18:1	12.07	9.96	9.03	11.04	10.53	11.29	11.43	11.36	11.07
18:2	23.07	26.25	24.34	25.72	24.96	27.43	28.59	26.33	27.38
18:3	0.78	0.75	0.80	0.74	0.77	0.50	0.66	0.26	0.59
20:4	10.50	9.01	8.59	10.33	9.61	10.93	11.23	8.62	12.93
22:1	ND	ND	ND	ND	—	0.92	ND	ND	0.92
24:0	0.52	0.39	0.39	0.38	0.42	0.20	0.07	0.16	0.36
24:1	1.10	0.59	0.63	0.61	0.73	0.40	0.67	—	0.51
22:5	2.33	2.17	2.33	1.92	2.18	1.33	1.30	1.41	1.29
22:6	11.62	14.02	15.33	12.56	13.43	10.35	10.50	10.97	9.58
UI[e]	185.6	197.0	200.6	192.2	193.8	181.7	186.2	173.2	185.7

[a] R. C. Aloia, unpublished data (phospholipid fraction separated from total lipids by 1D-TLC in hexane/ether/acetic acid; 70:30:1).
[b] A, active; H, hibernating animals.
[c] Number of animals/pool group.
[d] Mole % values.
[e] UI = Unsaturation Index; % occurrence × number of double bonds.

appears to be a definite reduction of the long-chain polyunsaturated fatty acids, docosahexa- (22 : 6) and docosapentaenoic acid (22 : 5), and only a slight increase in oleic (18 : 1), linoleic (18 : 2), and arachidonic acid (20 : 4) in the hibernating state. In regard to enhancing membrane disorder and the degree of membrane fluidity by increasing the number of double bonds during hibernation, it would appear that these results are inconclusive. Additionally, the quantity of the high melting point stearic acid (18 : 0) seems to increase in the hibernating state. Consequently, perhaps the bulk lipid phase of the membranes from heart and liver tissue of hibernators does not require an increase in unsaturation to continue functioning at low temperatures.

V. Phospholipids of the Membranes of Active and Hibernating Squirrels

A. Review of Phospholipid Metabolism

Membrane phospholipid metabolism centers around phosphatidic acid (PA) which is the pivotal or key molecule in the biosynthetic scheme (Fig. 3). The phospholipids are either derived from cytidine diphosphate (CDP)-diacylglycerol, (1,2-diacyl-*sn*-glycerol-3-phosphorylcytidine), or, CDP-linked bases plus 1,2-diacyl-*sn*-glycerol or 1-alkyl(or 1-alkenyl)-2-acyl-*sn*-glycerol ether. For example, CDP-choline and CDP-ethanolamine are converted to phosphatidylcholine (PC) and phosphatidylethanolamine (PE), respectively, in the presence of 1,2-diacylglycerol. Initially the formation of PC involves the phosphorylation of choline by ATP : choline phosphotransferase (EC 2.7.1.32), a kinase enzyme. Phosphorylcholine then reacts with cytidine triphosphate (CTP) to form CDP-choline which subsequently reacts with a diacylglycerol receptor to yield PC and cytidine monophosphate (CMP). Diacylglycerol is formed by the decarboxylation of phosphatidic acid by enzyme (2) (Fig. 3), phosphatidate phosphatase (EC 3.1.3.27). The enzymes involved in the latter two steps are CTP : phosphorylcholine cytidylyltransferase (EC 2.7.7.15) and CDP-choline : 1,2-diacylglyceride choline phosphotransferase (EC 2.7.8.2), respectively. PE is formed in a similar manner by an ethanolamine phosphate cytidylyltransferase (EC 2.7.7.14) and an ethanolamine phosphotransferase (EC 2.7.8.1).

PE can also form from decarboxylation of phosphatidyl serine (PS), and, PE can be converted to PC by a triple transmethylation involving S-adenosylmethionine as the methyl donor. PS can be derived from PE by an exchange reaction with L-serine. Plasmalogens of PC and PE

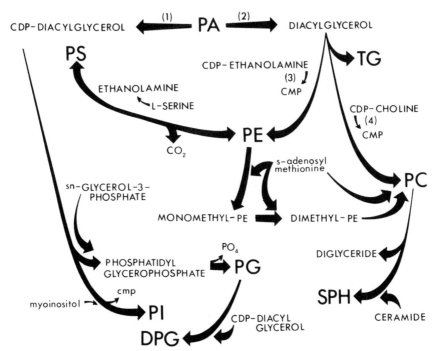

FIG. 3. Phospholipid metabolic pathways. Key enzymes: (1) Microsomal cytidylyl transferase (EC 2.7.7.41), (2) phosphatidate phosphatase (EC 3.1.3.27), (3) CDP-choline:1,2-diacylglyceride choline phosphotransferase (EC 2.7.8.2), (4) CDP-ethanolamine:1,2-diacylglyceride ethanolamine phosphotransferase (EC 2.7.8.1). CDP, Cytidine diphosphate; cmp, cytidine monophosphate; PA, phosphatidic acid; TG, triglyceride; PS, phosphatidylserine; PE, phosphatidylethanolamine; PC, phosphatidylcholine; PG, phosphatidylglycerol; DPG, diphosphatidylglycerol; PI, phosphatidylinositol; SPH, sphingomyelin.

can also be formed in similar biosynthetic pathways. CDP-choline or ethanolamine can react with either an alkyl group or a cis-α,β-unsaturated derivative in the number C-1 position to form glycerol ethers and plasmalogens, respectively. Sphingomyelin can also be formed by group transfer of choline from PC to a ceramide derived from the interaction of sphingosine with acyl-CoA or free fatty acids (Witting, 1975).

Phosphatidylglycerol (PG), diphosphatidylglycerol (DPG), phosphatidylinositol (PI) and higher phosphoinositides are derived from CDP-diacylglycerol. In contrast to the above-mentioned reactions, PA is not dephosphorylated prior to use in this lipid biosynthetic pathway. A microsomal cytidylyltransferase (EC 2.7.7.41) catalyzes the reaction of CTP with PA, releases a pyrophosphate and forms CDP-diacylglycerol. PI is formed by the reaction of CDP-diglyceride with myoinositol catalyzed by a microsomal enzyme. PG is formed by the reaction of CDP-diglyceride and glycerol phosphate (sn-glycerol-3-phosphoric acid) to form an intermediate phosphatidyl-glycerol phosphate, which is subsequently dephosphorylated fo form PG. Condensation of PG with a second molecule of CDP-diacyl-glycerol will form diphosphatidylglycerol or cardiolipin (Gurr and James, 1975; Packter, 1973; Newsholm and Start, 1973; Bishop, 1971).

B. Brain Phospholipids of Mammals Capable of Hibernation

The two major studies of brain phospholipids of hibernating mammals are those of Aloia (1978a) and Goldman (1975), both of which demonstrate significant alterations in the phospholipid class composition commensurate with hibernation. In the brain microsomes of the hamster (*Mesocricetus auratus*), phosphatidylethanolamine was shown to increase from 0.226 ± 0.005 to 0.263 ± 0.003 micromoles lipid/mg protein ($P < 0.001$), and phosphatidylcholine from 0.253 ± 0.005 to 0.285 ± 0.005 micromoles lipid/mg protein ($P < 0.001$) commensurate with hibernation (Goldman, 1975). In the whole brain of the ground squirrel (*Citellus lateralis*), phos-phatidylethanolamine was shown to decrease from 33.71 to 30.08 mole % ($P < 0.001$), whereas phosphatidylcholine increased from 31.48 to 32.80 mole % ($P < 0.01$) during hibernation. These data are illustrated in Table VI. Besides the above-mentioned changes it can be observed that the phosphorus-containing material which remained at the origin during thin layer chromatography is larger by a factor of two in the brain of hibernating squirrels compared to the value found in active animals. This substance probably represents increased polyphosphoinositides which are known to remain at the origin during chromatography with the solvent system employed (Rouser et al., 1967).

Table VII illustrates the comparative brain phospholipid composition of mammalian and nonmammalian species. It can be observed that except for the guinea pig, the quantity of phos-phatidylethanolamine is equal or greater to that of phosphatidyl-choline, the PC/PE ratio being 0.93 for all five mammalian species. This is similar to the relationship of PC to PE in brain tissue of active squirrels, the PC/PE ratio also being 0.93. However, the relationship seems to be reversed in brain tissue from hibernating squirrels, since the level of PC has increased and that of PE has decreased. As can be

TABLE VI
BRAIN PHOSPHOLIPIDS

(n):[b]	Active animals[a]								
	A (4)	B (3)	C (3)	D (5)	E (4)	F (4)	G (5)	H (2)	$\bar{X}$ (28)
PE[c]	33.0	33.0	34.6	34.1	33.9	33.4	33.4	34.3	33.7
PC	31.8	31.0	31.6	30.8	31.6	31.7	31.5	31.9	31.5
SPH	9.6	10.2	9.8	9.3	11.1	10.3	10.6	9.9	10.1
PI	2.1	2.6	3.1	1.9	2.5	2.9	3.0	2.7	2.6
PS	15.4	13.8	14.0	15.1	14.1	13.7	14.6	15.1	14.5
ORIG	2.3	2.1	1.6	2.4	2.1	2.8	1.5	1.7	2.1
Rec[f]	102	100	102	101	100	102	99	100	101 ± 1%

(n):	Hibernating animals[a]									
	A (5)	B (3)	C (4)	D (3)	E (2)	F (2)	G (2)	H (3)	$\bar{X}$ (24)	(P)[e]
PE	31.0	30.2	30.9	28.3	29.0	31.2	31.7	28.4	30.1	(P < 0.001)
PC	33.8	32.2	32.8	32.6	34.1	32.2	33.0	31.6	32.8	(P < 0.01)
SPH	9.9	10.2	10.5	11.0	9.8	9.9	9.9	11.4	10.2	NS
PI	1.6	2.5	2.0	2.3	2.1	2.2	2.4	2.4	2.2	NS
PS	14.3	13.7	13.9	15.2	14.5	14.4	14.3	14.4	14.3	NS
ORIG	4.9	5.2	5.2	5.4	5.5	4.1	2.9	4.9	4.7	(P < 0.001)
Rec[f]	98	99	99	96	99	98	98	101	99 ± 1.5%	

[a] Mole % values of major phospholipid classes modified from Aloia (1978a).
[b] n, number of animals/pool group.
[c] PE, phosphatidylethanolamine; PC, phosphatidylcholine; SPH, sphingomyelin; PI, phosphatidylinositol; PS, phosphatidylserine, ORIG, origin.
[d] Rounded values. Reader is referred to original publication for complete values, standard deviations, and minor lipid values.
[e] Student's t-test of mean values of active and hibernating squirrels. NS, not significant.
[f] Recovery, mole.

observed in the table, the PC/PE ratio for the nonmammalian (frog, goldfish, dogfish, and rattlesnake) brain tissue is 1.23 while that of the hibernating squirrel brain is 1.10.

Several animals from our colony of squirrels were warm-adapted at 35°C for 3 years. Phospholipid analysis of one sample of whole brain tissue revealed a PC/PE ratio which is even lower (0.73) than that found in the active squirrel tissue. Consquently, it appears that the decrease in PC and the increase in PE found in brain tissue from active squirrels compared to that found in hibernating squirrel brain,

TABLE VII

COMPARATIVE ANALYSIS OF MAMMALIAN AND NONMAMMALIAN BRAIN PHOSPHOLIPID
CLASS COMPOSITION

	Adult[a] human	Bovine[a]	Bovine[b]	Rat[a]	Rat[c]	Rabbit[d]	Guinea[e] pig	Active[i] squirrel
PE[h]	35.0%	35.7%	33.2%	37.2%	36.4%	35.2%	30.7%	33.7%
PC	29.2%	29.6%	29.2%	37.5%	36.8%	32.2%	32.6%	31.5%
PS	17.6%	16.7%	16.6%	12.6%	11.8%	15.8%	12.3%	14.5%
PI	2.0%	2.7%	3.2%	3.5%	3.1%	3.0%	2.9%	2.6%
SPH	13.6%	13.2%	12.5%	5.1%	5.7%	12.4%	8.3%	10.1%

	Hiber- nating[i] squirrel	Warm[j] adapted squirrel	Frog[a]	Gold- fish[a]	Gold- fish[f]	Dog- fish[c]	Rattle- snake[g]
PE[h]	30.1%	36.1%	39.0%	33.6%	34.8%	28.1%	36.7%
PC	32.8%	26.4%	43.2%	49.6%	47.4%	33.6%	39.0%
PS	14.3%	15.3%	10.2%	8.8%	8.1%	8.1%	9.1%
PI	2.2%	2.3%	2.4%	3.6%	2.7%	3.0%	3.9%
SPH	10.2%	11.1%	2.7%	1.7%	1.8%	8.2%	10.1%

[a] Data adapted from Rouser et al. (1968).
[b] Data adapted from Dawson et al. (1962).
[c] Data adapted from Wuthier (1966).
[d] Data adapted from Owens (1966).
[e] Data adapted from Eichberg et al. (1964).
[f] Data adapted from Leslie and Buckley (1976).
[g] Data adapted from Cuzner et al. (1965).
[h] See Fig. 2 for abbreviations.
[i] Data modified from Aloia (1978a).
[j] One warm-adapted squirrel brain was analyzed. (Warm adaptation: 3 years at 35°C.)

continues in the same direction when the animals are warm-adapted at an ambient temperature of 35°C, although their body temperature remains at 37°–38°C. Although these results were derived from only one animal, the trend is interesting, especially in view of the same phenomenon occurring in lung tissue (see below).

C. LIVER PHOSPHOLIPIDS OF MAMMALS CAPABLE OF HIBERNATION

The liver of mammals in hibernation is a unique organ in that although protein and lipid biosynthesis are reduced (Burlington, 1972; Denyes and Carter, 1961), gluconeogenesis (Klain and Whitten, 1968), carbohydrate metabolism (Tashima et al., 1970) and succinate oxidation (Chaffee et al., 1966) in liver slices and mitochondrial preparations continue. Liver mitochondria from ground squirrels in their active physiological state, on the other hand, have been shown to undergo a distinct Arrhenius discontinuity at approximately 23°C concomitant with a large increase in activation energy (Lyons and Raison, 1970). If the Arrhenius discontinuity is indicative of a phase shift of membrane lipids from a fluid to a solid state, as indicated above (Lee and Gear, 1974; Raison, 1973; McMurchie et al., 1973; Lyons, 1972; Lenaz et al., 1972), then the liver mitochondrial lipids of squirrels in their hibernating state do not undergo such a shift, since straight-line Arrhenius plots are obtained (Lyons, 1972). In the hibernating bat (*Myotis lucifugus*), liver phospholipids have been shown to be increased by a factor of two in hibernating animals (Esher et al., 1973). However, the liver from both active and hibernating squirrels (*Citellus lateralis*) contains 4.72 and 4.68 mmoles phosphorus per 100 gm wet weight, respectively. This indicates that in this species there is no additional membrane material being synthesized during the hibernating state. However, several significant differences have been demonstrated in liver phospholipid classes of active and hibernating squirrels (Table VIII). Commensurate with the hibernating state, phosphatidylethanolamine appears to increase (16%), while phosphatidylinositol, phosphatidylserine, and diphosphatidylglycerol decrease 9, 23, and 9%, respectively (Aloia, 1978b).

Comparison of the mean and range values of phospholipid classes from the liver of hibernators with those from nonhibernators reveals some interesting relationships. By examining Table IX one observes that the range of values of phosphatidylethanolamine from active animals (23.62–26.81%) falls within that of five vertebrate species, whereas the range of values for the same phospholipid class in hibernating squirrels is clearly higher (29.02–30.71%). The reverse situation is exhibited by phosphatidylinositol, namely, the range of values from the hibernating animals (8.2–8.75%) falls within the range of the other vertebrate species, while that from active squirrels (9.08–9.81%) is higher. Last, the molar % values of diphosphatidylglycerol from both active and hibernating animals fall within the range of the other species. Phospholipid recovery in two-dimensional thin layer chromatography and phosphorus analysis of liver lipids was 99.6 ± 1.5% for active animals and 100.3 ± 1.9% for hibernating animals.

D. KIDNEY PHOSPHOLIPIDS OF MAMMALS CAPABLE OF HIBERNATION

There has only been one investigation of kidney phospholipid classes of hibernating mammals (Aloia, 1978c). The membranes of the kidney are considered to remain functional at temperatures of approximately 5°C since filtration and reabsorption have been demonstrated

TABLE VIII
LIVER PHOSPHOLIPIDS

	Active ground squirrels[a]						
	A	B	C	D	E	F	$\overline{X}$
(n):[b]	(3)	(3)	(3)	(4)	(4)	(3)	(20)
PE[c]	26.5[d]	23.6	25.8	26.0	24.9	26.8	25.6
PC	48.3	51.2	50.8	50.2	49.4	48.8	49.8
DPG	4.3	4.0	4.3	4.3	4.1	4.4	4.2
SPH	3.5	3.5	3.7	2.6	3.5	3.6	3.4
PI	9.0	9.8	9.1	9.2	9.5	9.3	9.3
PS	3.1	3.3	3.5	3.8	3.4	3.7	3.5
Rec[f]	99.6	98.8	98.8	101.6	100.2	100.2	$99.6 \pm 1.5\%$

	Hibernating ground squirrels							
	A	B	C	D	E	F	G	$\overline{X}$
(n):[b]	(3)	(3)	(2)	(3)	(3)	(3)	(4)	(21)	(P)[e]
PE[c]	30.5[d]	29.6	30.7	29.5	29.3	29.2	29.1	29.7	(P < 0.001)
PC	47.9	48.7	48.2	49.1	48.5	48.6	46.8	48.3	NS
DPG	4.0	3.6	3.9	3.4	4.2	3.9	3.8	3.8	(P < 0.01)
SPH	3.7	3.2	3.1	3.3	3.3	3.1	3.5	3.3	NS
PI	8.4	8.6	8.2	8.5	8.5	8.8	8.2	8.5	(P < 0.001)
PS	2.5	2.4	2.8	2.9	2.7	2.7	2.7	2.7	(P < 0.001)
Rec[f]	98.3	102.1	102.4	97.8	101.8	99.0	100.6	$100.3 \pm 1.9\%$	

[a] Mole % values of major phospholipid classes modified from Aloia (1978b).

[b] n, number of animals/pool group.

[c] PE, phosphatidylethanolamine; PC, phosphatidylcholine; SPH, sphingomyelin; PI, phosphatidylinositol; PS, phosphatidylserine; DPG, diphosphatidylglycerol.

[d] Rounded values. Reader is referred to original publication for complete percentages, standard deviations, and minor lipid classes.

[e] Student's t-test of mean values of active and hibernating squirrels. NS, not significant.

[f] Recovery, mole %.

(Tempel and Mussacchia, 1975; Zatzman and South, 1972). The kidney cortex, for example, has been shown to accumulate K^+ during hibernation primarily by sustained $Na^+ + K^+$-ATPase activity and reduced K^+ leakage (Willis and Li, 1969; Willis, 1966). Consequently, the lipid changes which are found in the kidney of the ground squirrel (*Citellus lateralis*) may be such that they contribute to low-temperature membrane tolerance of the cell membranes. These data are presented in Table X.

It can be observed that the steady state levels of phosphatidylethanolamine increase significantly from 33.9 to 35.2 mole %, while the values of sphingomyelin decrease significantly from 13 to 12

TABLE IX
LIVER PHOSPHOLIPIDS

	Rat (9), mouse (7), frog (3), bovine (1), human (1)[a,b]		Active ground squirrel (20) (T_B = 37°C)		Hibernating ground squirrel (21) (T_B = 2–4°C)	
	x̄	Range	x̄	Range	x̄	Range
PE[c]	24.25[d]	22.1–27.9	25.61	23.63–26.81	29.70	29.09–30.71
DPG	3.86	2.0–5.2	4.23	3.99–4.33	3.84	3.44–4.22
PI	8.2	7.3–8.9	9.32	9.02–9.81	8.53	8.20–8.75
PS	3.5	2.8–4.1	3.47	3.12–3.76	2.66	2.41–2.87
Rec[e]	98.4 ± 1.2%		99.6 ± 1.5%		100.3 ± 1.9%	

[a] Data modified from Rouser et al. (1969b).
[b] Number in parentheses indicate number of animals analyzed.
[c] PE, DPG, PI, PS = see explanation of abbreviations in Fig. 2.
[d] Mole % values.
[e] Recovery, mole %.

mole % during the hibernating state. Recovery from experiments involving active and hibernating animals was 99 ± 1.5%. The values for each of the phospholipid classes seem to fall within the range of values for phospholipids from other vertebrate species analyzed by comparable quantitative analytical techniques (Table X).

TABLE X
KIDNEY PHOSPHOLIPIDS[a]

	Rat (7), human (1), bovine (1), mouse (12), frog (9)[a,b]	Active[g] squirrels (31)	Hibernating squirrels (28)
PE[c]	28.1 ± 1.4[d]	28.4 ± 0.5[d]	28.6 ± 0.7
PC	34.2 ± 1.2	33.9 ± 0.8	35.2 ± 0.1[f]
DPG	5.8 ± 1.4	7.3 ± 0.7	6.6 ± 0.6
SPH	11.9 ± 1.2	13.0 ± 0.5	12.0 ± 0.8[f]
PI	6.0 ± 0.7	5.8 ± 0.7	6.0 ± 0.3
PS	6.8 ± 0.6	7.0 ± 0.5	7.4 ± 0.5
Rec[e]		99.4 ± 1.8%	99.2 ± 1.6%

[a] Selected data modified from Rouser et al. (1969b).
[b] Number in parentheses indicate number of animals analyzed.
[c] See Fig. 2 for abbreviations.
[d] Values expressed as mole % ± standard deviation.
[e] Recovery: mole % ± standard deviation.
[f] ($P < 0.05$); Probability: two-tailed Student's t-test comparing active with hibernating animals.
[g] Data modified from Aloia (1978c).

E. Lung Phospholipids of Mammals Capable of Hibernation

To our knowledge no phospholipid studies of lung tissue from hibernation species have been published. Table XI illustrates the mean molar % values of lung phospholipids from 34 squirrels in their active physiological state compared to the mean molar values of 37 squirrels in their hibernating state. Additionally, Table XI also presents the phospholipid class composition of the lung tissue of three squirrels warm-adapted for 3 years at 35°C ambient temperature and sacrificed in the summer of 1976. For comparative purposes, the lung phospholipid compositions of five vertebrate species are also presented. The first point to illustrate is the fact that the molar % quantities of phosphatidylethanolamine from both active and hibernating squirrel lungs are less than that from the other vertebrate species, while the quantities from phosphatidylcholine are greater. However, the molar quantities of phosphatidylcholine and phosphatidylethanolamine from the warm-adapted animals are within the range of values of the five vertebrate species. More important, however, is the apparent trend toward increase levels of phosphatidylcholine and decreased levels of phosphatidylethanolamine at colder acclimation temperatures. There is a slight ($P < 0.1$) increase in the levels of phosphatidylcholine from the lung tissue of the active animal compared to the that of the warm-adapted animal, and a further (significant) increase ($P < 0.01$) in the lung from the hibernating animal. There is a highly significant decrease ($P < 0.001$) in the level of phosphatidylethanolamine in the lungs from active animals compared to warm-adapted animals, which appears to farther decline in the hibernating state, although not significantly. Furthermore, there appears to be a slight decrease in the molar quantities of phosphatidylserine, phosphatidylinosital, and sphingomyelin with decreasing ambient temperatures. Although the results from the warm-adapted animals are derived from analysis of the lungs of only three animals, some of the differences in molar values of phospholipids between these animals and the active and hibernating animals are significant. Furthermore, they appear to be indicative of a trend in steady state levels of phospholipid classes from animals maintained at different ambient temperatures (see Section V,B on Brain).

F. Skeletal Muscle Phospholipids of Mammals Capable of Hibernation

To our knowledge no phospholipid composition of skeletal muscle from a mammal capable of hibernation has been published. Table XII

TABLE XI
COMPARATIVE ANALYSIS OF LUNG PHOSPHOLIPIDS

	Human[a]	Rat[a]	Mouse[a]	Bovine[a]	Frog[a]	Hibernating squirrel[b]	Active squirrel[b]	Warm-adapted squirrel[b,j]
n^c	(1)	(5)	(12)	(1)	(8)	(37)	(34)	(3)
n^d	(8)	(20)	(8)	(8)	(12)	(72)	(64)	(12)
PE[e]	21.2 ± 0.5[f]	22.2 ± 0.4	20.1 ± 0.2	21.2 ± 0.6	21.2 ± 0.5	17.6 ± 2.1	18.5 ± 1.0	21.2 ± 0.9[h]
PC	47.5 ± 1.0	44.9 ± 0.3	43.7 ± 1.2	39.5 ± 0.5	42.6 ± 1.0	54.4 ± 3.6	50.8 ± 2.5[g]	47.9 ± 0.5
PS	7.0 ± 0.3	9.0 ± 0.2	8.1 ± 0.3	9.4 ± 0.2	7.8 ± 0.2	7.2 ± 0.8	8.0 ± 0.6	8.8 ± 0.3
PI	3.2 ± 0.1	3.9 ± 0.1	4.1 ± 0.2	3.3 ± 0.1	3.4 ± 0.2	2.8 ± 0.6	3.0 ± 0.6	3.3 ± 0.4
SPH	11.1 ± 0.3	10.1 ± 0.2	9.3 ± 0.3	16.1 ± 0.7	10.8 ± 0.3	11.4 ± 1.2	13.0 ± 1.0	13.5 ± 0.5
Rec^k	97.0%	96.6%	93.1%	97.6%	93.6%	100.6 ± 1.5%	101.2 ± 1.2%	97.2 ± 1.1%

[a] Data adapted from Baxter et al. (1969).
[b] R. C. Aloia, unpublished observations: (lungs from active and hibernating squirrels were extracted in groups of 2–4 animals and analyzed in quadruplicate).
[c] Number of animal organs analyzed.
[d] Number of 2D-phospholipid analyses.
[e] See Fig. 2 for nomenclature.
[f] All values in this table are expressed as mole %.
[g] ($P < 0.01$); Probability: Student's t-test comparing active with hibernating animals.
[h] ($P < 0.001$); Student's t-test comparing active with warm-adapted animals.
[i] ($P < 0.01$); Student's t-test comparing active with warm-adapted animals.
[j] Warm-adapted 3 years at 35°C.
[k] Recovery: mole % ± standard deviation.

R. C. Aloia and E. T. Pengelley

illustrates the skeletal muscle phospholipid class composition from active and hibernating ground squirrels and similar values from other quantitative studies in the literature. It can be observed that, except for the high level of phosphatidylethanolamine from the muscle of the hibernating squirrel, all values fall within the range of values exhibited by the other mammalian species. The higher phosphatidylethanolamine value from the hibernating squirrel is similar to the higher value in the poikilothermic vertebrate (frog). More importantly, however, the molar quantity of phosphatidylethanolamine increased from 26.8% in the active squirrel to 30.3% in the hibernating squirrel $(P < 0.001)$, and that of diphosphatidylglycerol increased

TABLE XII
COMPARATIVE PHOSPHOLIPID COMPOSITION FROM SKELETAL MUSCLE

$(n)^c$	Human[a] (3)	Human[b] (1)	Guinea[a] pig (7)	Bovine[b] (1)	Rat[b] (5)	Mouse[b] (5)	Frog[b] (10)
PE[e]	21.9	26.4	27.7	26.6	22.2	25.8	29.0
PC	55.7	48.0	50	46.5	51.1	52.4	55.2
DPG	4.7	6.6	6.4	8.9	1.4	5.8	2.3
SPH	5.6	4.0	4.2	4.5	2.7	3.6	3.1
PI	6.0	8.8	7.9	5.6	8.9	6.7	5.5
PS	4.2	3.3	3.6	4.1	3.7	4.0	3.4

	Active squirrels	Hibernating squirrels	Probability[g]
n^c	(16)	(26)	
n^d	(52)	(64)	
PE[e]	26.8 ± 1.3^f	30.3 ± 1.6	$(P < 0.001)$
PC	48.9 ± 0.9	46.6 ± 1.5	$(P < 0.05)$
DPG	8.5 ± 0.6	9.6 ± 1.0	$(P < 0.05)$
SPH	3.5 ± 0.3	3.4 ± 0.3	NS
PI	5.5 ± 0.6	5.2 ± 0.4	NS
PS	3.1 ± 0.3	3.0 ± 0.2	NS
Rec[g]	$100.0 \pm 2.5\%$	$99.5 \pm 2.9\%$	

[a] Data adapted from Hof and Simon (1970).
[b] Data adapted from Simon and Rouser (1969).
[c] Number of animals analyzed
[d] Number of 2D-TLC phospholipid analyses.
[e] See Fig. 2 for abbreviations.
[f] R.C. Aloia, unpublished observations: mole % values.
[g] Probability: Student's two-tailed t-test comparing active with hibernating animals. NS, not significant.

from 8.5 mole % to 9.6 mole % ($P < 0.05$). On the other hand, the molar % of phosphatidylcholine decreased from 48.9 to 46.6% ($P < 0.05$) during hibernation.

VI. Discussion

It appears that various hibernating species as well as different organs within the same animal experience different modifications in lipid composition of their cellular membrane commensurate with hibernation. A primary objective of the research concerning low temperature membrane function in hibernators is to determine the molecular mechanism of such adaptation and the precise role of the membrane lipid changes therein. Many organs of the hibernator have been shown to remain active, although at a much reduced level, in the hibernating state at body temperatures approaching 0°C. In order for an organ to remain functional, by definition, the cellular membranes of that organ must also remain functional. If, however, the activity of a particular organ is severely reduced at low temperatures and/or the organ ceases to function altogether, functional activity of the membranes may not be required. However, membrane structural integrity would appear to be a *sine qua non* for renewed membrane function at higher temperatures during arousal. Electron microscopic studies have demonstrated that membrane structure is maintained during hibernation in retina (Kuwabara, 1975), kidney (Zimney, 1968), thyroid (Nuñez *et al.*, 1974), and cardiac muscle (Aloia and Platt-Aloia, 1976; Hagopian *et al.*, 1974).

A. BRAIN PHOSPHOLIPIDS

Since phospholipid alterations are observed in animals, a question arises concerning their relationship to the functional capacity of the brain at low temperatures. The electrical activity of the brain of nonhibernating mammals is severely reduced below 20°C (Massopust *et al.*, 1965; Kayser and Malan, 1963), while that of poikilotherms and hibernating mammals is maintained (Hensel *et al.*, 1973). Movement of ear pinna and postural adjustments of the head and body in hibernating squirrels has been reported to occur in response to sound stimuli (Strumwasser, 1959). Continued synaptic transmission has been demonstrated in the hibernating hedgehog below 10°C, while synaptic inhibition was observed in the rabbit (Saarikoski, 1969). Additionally, conduction time through the rabbit superior cervical ganglion is markedly increased at reduced temperatures compared to that of the hibernating hedgehog (Saarikoski, 1975). Furthermore, spontaneous arousals occur, not only on a regular periodic basis during the hibernating season, but also when elicited by appropriate mechanical, auditory, and/or thermal stimulation (Willis *et al.*, 1972). It appears therefore, that the nervous system of hibernating animals remains functionally active during the hibernating state.

Both central and peripheral cutaneous receptors are active and have been shown to signal arousal. The dormouse, for example, hibernates in a supine position with paws extending upward (Kayser and Malan, 1963). Peripheral cutaneous receptors on the footpads signal the arousal process when temperature changes are detected, even when the animal's body temperature is 5°C. Electrophysiological studies have corroborated these observations of continued central nervous system activity by demonstrating that continuous cortical electrical recordings can be made in some species at temperatures of approximately 5°C, while in others, although spontaneous activity is not present, activity can be recorded with appropriate stimuli (Kayser and Malan, 1963; Strumwasser, 1959).

Maintenance of brain tissue activity at temperatures less than 5°C implies that cellular ionic gradients are also maintained. In other excitable tissue from nonhibernating mammals, such as the cardiac muscle, a drop in temperature can result in a loss of cellular K^+, a gain of Na^+, and a fall in resting potential which thereby produces a nonexcitable membrane (Marshall and Willis, 1962; Hoffman, 1956). The key enzyme responsible for the maintenance of cellular K^+ and Na^+ electrochemical gradients is the $Na^+ + K^+$-ATPase enzyme (Skou, 1972). The brain of at least one hibernating mammal, the Syrian hamster, has been shown to maintain K^+ gradients and residual activity of $Na^+ + K^+$-ATPase during hibernation (Goldman and Albers, 1975; Goldman and Willis, 1973a,b). Prior to and up through approximately 5 or 6 weeks of cold exposure, $Na^+ + K^+$-ATPase activity is cold sensitive and inhibited below 5°C, similar to that of obligate homeotherms (Bowler and Duncan, 1968, 1969). However, by the time the hamster hibernates, the enzyme has become cold tolerant (Bowler and Duncan, 1969). Analysis of the enzyme activity from the brain tissue of those few hibernators which, for some unknown reason, did not hibernate during the period of cold exposure (several months), has revealed that the ATPase enzyme was cold sensitive and apparently did not convert to the cold tolerant form (Bowler and Duncan, 1969).

Also, recently an increase in the specific activity of brain microsomal $Na^+ + K^+$-ATPase of the ground squirrel (*Spermophilus richardsonii*) during hibernation, has been reported (Charnock, 1978). However, although the specific activity for this enzyme appears to be increased during hibernation compared to the enzyme activity from the active animal, in both the Syrian hamster (Goldman and Albers, 1975) and the ground squirrel (Charnock, 1978), the Arrhenius relationship of the activity of these enzymes appears to be distinctly nonlinear. For example, Charnock has recently recalculated original data from Goldman and Willis (1973b) and Goldman and Albers (1975) and determined that the phase transition temperature (T_c) for $Na^+ + K^+$-ATPase from both active and hibernating hamsters in the former study was 17°C, while T_c for awake and hibernating hamsters from the second study was 16.5 and 13.5°C, respectively (Charnock, 1978). Furthermore, T_c for the ground squirrel brain $Na^+ + K^+$-ATPase was determined to be 20°C (Charnock, 1978). This nonlinearity is distinctly different from the linear Arrhenius relationship reported for the poikilothermic toad heart $Na^+ + K^+$-ATPase (McMurchie et al., 1973), as well as the linear Arrhenius plots for mitochondrial succinate oxidation in hibernating ground squirrel liver (Raison and Lyons, 1971) and various poikilotherms (catfish, rainbow trout; Lyons and Raison, 1970). As mentioned above, there is usually an increase in activation energy below the break temperature and an assumed increased energy requirement for the enzyme reaction to proceed. Consequently, from this particular thermodynamic perspective the hibernator appears no different from the nonhibernating mammal. In some cases, however, the activation energy (E_a) below T_c is actually reduced in the hibernating animal compared to the active animal. For example, $E_{a_{II}}$, the activation energy for $Na^+ + K^+$-ATPase from brain microsomes below the phase transition temperature, was 42 kcal/mole compared to 34 kcal/mole for awake and hibernating hamsters, respectively (recalculated by Charnock, 1978). However, $E_{a_{II}}$ for $Na^+ + K^+$-ATPase from the active ground squirrel brain averages 29 kcal/mole compared to 31 kcal/mole for the hibernating ground squirrel brain (Charnock, 1978). Consequently, the adaptation of the $Na^+ + K^+$-ATPase enzyme system of brain microsomes may not be a simple reversion to a temperature-insensitive, cold-tolerant form which exhibits linear Arrhenius plots similar to poikilotherms (Lyons, 1972); see Chapter 2 by Aloia for discussion of Arrhenius plots).

As mentioned above, $Na^+ + K^+$-ATPase has been shown to be a lipid-requiring enzyme (Fourcans and Jain, 1974; Coleman, 1973). Recent evidence indicates that it requires either the serine or glycerol polar headgroup for maximal activity (Tanaka, 1974; Kimelberg and Papahadjopoulos, 1974). Consequently, a shift in the steady state levels of these two phospholipids, or a change in the degree of fatty acid unsaturation, might be expected if the number of enzyme molecules is altered or an isozymic variant is utilized during adaptation for hibernation. Although the steady state levels of these two phospholipids does not change, the level of phosphatidylcholine is elevated significantly while that of phosphatidylethanolamine is depressed (see Table VI). Also, although the Unsaturation Index is low, there is a 19% increase in unsaturation of the fatty acids from PS (see Table III). Palmitoleic (16:1) and linoleic acid (18:2) increase significantly in PS during hibernation and may be required for continued low temperature function of a specific enzyme such as $Na^+ + K^+$-ATPase.

However, the phospholipid alterations which are observed may be required to enhance the bulk phase lipid fluidity properties of the brain cell membranes and thereby enhance total membrane function at reduced temperatures. It has been demonstrated by differential scanning calorimetry that for a given fatty acid composition, phosphatidylcholine melts approximately 25°C lower than phosphatidylethanolamine. For example, dimyristoyl phosphatidylcholine undergoes a LC–C phase transition at 23°C, whereas dimyristoyl phosphatidylethanolamine undergoes a similar transition at approximately 48°C (Papahadjopoulos, 1977; Kimelberg, 1976; Chapman, 1976; Hayashi et al., 1973). This difference in phase transition temperatures cannot be explained on the basis of charge-repulsion (Gingell, 1976) since both lipids are zwitterions (Slotboom and Bonsen, 1970). However, the hydrated polar region of phosphatidylcholine apparently weakens whatever polar interactions are present, thus permitting greater flexibility of the fatty acyl chains at reduced temperatures (Papahadjopoulos, 1977; Michaelson et al., 1974; Chapman and Urbina, 1974). It seems therefore, that if the bulk phase lipids of the membranes of the brain of active hibernators are in such a physical state that they would solidify at the temperatures encountered during hibernation, then given the fact that the changes in fatty acid composition between active and hibernating states are slight, the trend in phospholipid alterations demonstrated in the brain of the hibernating ground squirrel could be interpreted as a mechanism to counter this tendency. There is also a reduction in palmitic (16:0) and stearic acid (18:0) in phosphatidylethanolamine and phosphatidylcholine during hibernation in ground squirrel brain (see Table III). This trend could also facilitate low temperature membrane function since the saturated hydrocarbon chains have a higher melting point than the unsaturated, e.g., in distearoylphosphatidylcholine (18:0), melting point 54°C (Shimshick and McConnell, 1973), dioleoylphosphatidylcholine, mp $-20°C$ (Steim et al., 1969).

In hamster brain, the observed significant elevation in both phosphatidylcholine and phosphatidylethanolamine (Goldman, 1975) may also enhance the low-temperature tolerant capacities of the brain membranes. Possible negative effects on membrane fluidity properties caused by decreased PC/PE ratios observed in the hibernating state (1.12 in the active state versus 1.08 in the hibernating state) may be more than countered by the significant increases in oleic acid (18: 1; 11.4 to 13.7%) and arachidonic acid (20: 4; 15.4 to 20.3%) in phosphatidylethanolamine and oleic acid (14.4 to 27.2%) in phosphatidylethanolamine (plasmalogen), (Table II, Goldman, 1975). A similar increase in unsaturation has been reported for phosphatidylethanolamine from the intestine of 6°C-adapted fish, although the increments were mostly of polyunsaturates. For example, arachidonic acid and docosahenaenoic acid (22: 6) were shown to increase from 6.91 to 13.7% and 24.2 to 34.5%, respectively, in the 2-position of phosphatidylethanolamine (Miller et al., 1976). Other tissues also demonstrate a higher content of unsaturated fatty acids esterified to phosphatidylethanolamine than to other lipids (Breckenridge et al., 1972). In hamster brain phosphatidylethanolamine is five times more unsaturated than phosphatidylcholine (UI = about 275 versus 50, respectively) and in ground squirrel brain, phosphatidylethanolamine is approximately two times more unsaturated. Consequently any decrease in PC/PE ratio will augment the percent unsaturation within the membrane.

B. LIVER PHOSPHOLIPIDS

If liver membranes of active squirrels were not capable of tolerating the low body temperatures of hibernation, then alterations in phospholipid classes may be expected which would increase membrane disorder and thus sustain a higher degree of fluidity at the reduced temperatures. Since phosphatidylethanolamine melts approximately 25°C higher than phosphatidylcholine (see above), a reduction in the former (relative to the latter) might be expected during hibernation. However, the reverse is actually observed (see Table VIII). Therefore, unless the fatty acid complement of phosphatidylethanolamine is sufficiently unsaturated to compensate for the ordering effect of its polar headgroup, higher ethanolamine levels should be detrimental to low temperature membrane function. However, increased quantities of phosphatidylethanolamine may be required for some other membrane function and the membrane ordering properties of the molecule may be balanced by the observed concomitant reduction of acidic phospholipids (see Table VIII).

Low levels of Ca^{2+} have been shown to interact with negatively charged phospholipids to induce isothermal lateral phase separations and raise the temperature of the LC–C phase transition (Papahadjopoulos, 1977; Sauerheber and Gordon, 1975; Jacobson and Papahadjopoulos, 1975; Verkleij et al., 1974). For example, electron spin resonance studies have shown that interaction of Ca^{2+} with synaptic vesicles decreases the fluidity and increases the Arrhenius break temperature from 15° to 22°C (Viret and Leterrier, 1976). Elevated Ca^{2+} levels have been demonstrated in the heart, skeletal muscle, bone, and cerebrospinal fluid of hibernating mammals (Bito and Roberts, 1974; Ferrin et al., 1971). In heart muscle, for example, Ca^{2+} levels increase from 9.67 to 16.1 mg/100 gm dry weight in hamster (Ferrin et al., 1971) and from 24.51 to 32.33 mg/100 gm dry weight in ground squirrels (R. C. Aloia, unpublished observations). Similar elevations of liver Ca^{2+} have recently been reported in hiberating Arctic ground squirrels, C. undulatus (Behrisch, 1978). Thus, the reduction in acidic phospholipids observed in the liver of the ground squirrel during hibernation may be a compensatory change required to retain a specific degree of membrane fluidity necessary for low temperature survival. In other words, the ordering effect of the combination of high levels of liver Ca^{2+} and acidic membrane lipids may be detrimental to continued membrane function at low temperatures and thus, the levels of these lipids are reduced. Furthermore, there is some evidence that certain membrane lipids have a preference for unstable, nonbilayer configurations, such as hexagonal (H_{11}), cubic or inverted micellar phases (Cullis and deKruijff, 1978). Diphosphatidylglycerol and Ca^{2+} has been shown to be one of these systems (Rand and Sengupta, 1972). Thus the higher levels of tissue calcium may lead to a reduction in levels of diphosphatidylglycerol in order to maintain the required degree of membrane bilayer for proper membrane function during hibernation (see Chapter 2 for further discussion of membrane destabilization). However, the demonstration of phospholipid alterations occurring during hibernation which have opposite effects on membrane fluidity precludes a definitive interpretation. It may be that the observed changes in steady state levels of phospholipid are unrelated to fluidity changes.

If the observed changes in phospholipid class composition are not required to sustain bulk phase lipid fluidity, and thus membrane function at reduced temperatures, they may simply result from nonspecific low temperature-induced alterations in biosynthetic pathways. As shown in Fig. 3, if microsomal CTP-cytidylyltransferase, which forms CDP-diacylglycerol from phosphatidic acid, were slightly temperature sensitive [Fig. 3, enzyme (2)], turnover of CDP-diacylglycerol would be reduced and, all other factors remaining equal, less of the acidic phospholipids, diphosphatidylglycerol, and phosphatidylinositol, would be synthesized. Reduced steady state levels of these two lipids are observed in hibernating squirrel liver (Table VIII).

It has been demonstrated that the enzyme glucose-6-phosphatase, which dephosphorylates glucose 6-phosphate to glucose in gluconeogenesis, is maximally activated by phosphatidylethanolamine (Gurr and James, 1975; Duttera et al., 1968; however, see Nordlie and Jorgenson, 1976). Since gluconeogenesis is elevated by approximately 100% in some hibernating species (Burlington, 1972; Klain and Whitten, 1968), the increased quantities of phosphatidylethanolamine present in liver of hibernating squirrels, may be required for the continued function of glucose-6-phosphatase. This would account for a possible preferential reaction of diacyl glycerol with CDP-ethanolamine as opposed to CDP-choline and the formation of increased quantities of phosphatidylethanolamine (Fig. 3). Additionally, an increased rate of decarboxylation of phosphatidyl serine would contribute to the elevated levels of phosphatidylethanolamine, reduce the level of phosphatidylserine and possibly provide CO_2 for gluconeogenesis (see Fig. 3). Conversely, the exchange reaction of phosphatidyl ethanolamine and serine may be temperature sensitive and the level of the former would increase due to a reduced shunting to form phosphatidylserine (Fig. 3 and Table VIII).

C. KIDNEY PHOSPHOLIPIDS

While enzymes such as $Na^+ + K^+$-ATPase from brain tissue of hibernating species are not cold resistant unless the animal is prepared for hibernation (Goldman and Willis, 1973a,b), the same enzyme from kidney of awake hamsters demonstrates a considerable rate of K^+ transport, which increases twofold upon hibernation (Fang and Willis, 1974). Both the forward reaction, Na^+-dependent phosphorylation ($Na^+ + K^+$-ATPase) and the reverse K^+-dependent dephosphorylation (assayed by p-nitrophenyl phosphatase) exhibited an increase in specific enzyme activity during hibernation when assayed at 37°, 25°, 15°, and 5°C (Fang and Willis, 1974). The increased activity occurs after cold exposure and before the actual period of hibernation when the kidney cells would be exposed to the cold temperatures. Similar acclimation was shown to be necessary for adaptation of brain $Na^+ + K^+$-ATPase (Bowler and Duncan, 1969). Again, as with the

brain enzyme, there was no adaptation of the enzyme in those recalci-
trant hamsters which did not hibernate after 3–4 months of cold exposure (Fang and Willis, 1974).

Since this enzyme ($Na^+ + K^+$-ATPase) is a lipid-activated enzyme
requiring a lipid in a liquid-crystalline state for maximal activity
(Kimelberg, 1976), perhaps lipid alterations which occur in the hibernating physiological state are involved in the enzyme adaptation. It
can be observed that the steady state level of phosphatidylcholine
increases while that of sphingomyelin decreases during hibernation.
As mentioned above, if the membranes of active animals are not sufficiently fluid to remain functional at low temperatures, the augmenta-
tion of phosphatidylcholine (compared to other lipids such as
phosphatidylethanolamine) in the kidney membrane lipids during
hibernation could be interpreted as a trend to enhance membrane
disorder and fluidity. If a greater degree of membrane fluidity is required by the kidney membranes to remain functional during hibernation, the observed reduction in sphingomyelin may also contribute to
enhance membrane function. Although both phosphatidylcholine and
sphingomyelin contain the same choline headgroup, phosphatidylcholine differs from sphingomyelin in that it contains two ester bonds
while sphingomyelin contains a trans-double bond, an amide linkage,
and an hydroxyl group. Furthermore, the fatty acids of sphingomyelin
are primarily high melting point lipids (Rouser *et al.*, 1968). For example, brain sphingomyelin contains 70–80% of the membrane fatty
acids as 18 carbons and longer, only 15–25% of the fatty acids as a
monounsaturated species (nervonic acid; 24:1) and essentially no
polyunsaturated molecules (Shinitsky and Barenholz, 1974; Shipley *et
al.*, 1974). Additionally, and perhaps more importantly, direct physical
measurement by differential scanning calorimetry and fluorescence
polarization substantiates this evidence for a high melting point by
demonstrating that a LC–C phase transition occurs between 30° and
40°C (Shinitsky and Barenhoz, 1973; Shipley *et al.*, 1974). Consequently, a reduction in sphingomyelin levels in the kidney of hibernating mammals may serve to reduce those factors which would induce a phase shift of the membrane lipids to the solid state at reduced
temperatures.

Furthermore, high levels of hydroxysphingomyelin are known to be
present in mammalian kidney (Karlsson *et al.*, 1973). It has been demonstrated that considerable rigidity is imparted to a membrane bilayer
by lateral hydrogen bonding of these sphingomyelin hydroxyl groups
which can act as both hydrogen donors and acceptors (Pascher, 1976).
Perhaps this factor contributes to the viscosity at 25°C of sphin-

gomyelin liposomes being about 6 poises, which is six times greater than the viscosity of PC liposomes (Shinitsky and Barenholz, 1974). Therefore, an increase in the PC/SPH ratio in the kidney of hibernating mammals (PC/SPH = 2.61 in active squirrels and 2.93 in hibernating squirrels) may be required if an increased membrane fluidity is necessary for continued kidney function at low temperatures. It may be that the inverse relationship in steady state levels of phosphatidylcholine and sphingomyelin, also evident in other tissues (Rouser and Solomon, 1969; deGier and Van Deenen, 1961), may serve as a synergistic mechanism of low temperature adaptation. Conversely, this inverse relationship could merely be the result of low temperature inhibition of the choline-group transfer reaction from phosphatidylcholine to ceramide, which would elevate the former and reduce the level of sphingomyelin (see Fig. 3).

D. LUNG AND SKELETAL MUSCLE PHOSPHOLIPIDS

In lung and skeletal muscle from hibernating squirrels steady state levels of phosphatidylcholine are modified in opposite directions, increasing in the lung and decreasing in skeletal muscle. Consequently, the interpretation of the contribution of these changes toward maintenance of the hibernating state is difficult without further information, such as the fatty acid composition of each phospholipid class. Furthermore, the quantity of phosphatidylethanolamine increases significantly in skeletal muscle during hibernation (26.8 mole % in the active state vs. 30.3 mole % in the hibernating state; Table XII). Based solely on an interpretation of the role of the polar head group's contribution toward membrane fluidity, the changes observed to occur in skeletal muscle would counter any change toward increasing membrane fluidity at reduced temperatures. Consequently, it may be that the cell membranes of skeletal muscle are previously adapted to continue functioning at reduced temperatures or the fatty acids of phosphatidylethanolamine become or are more unsaturated to overcome the ordering effect of a decreased PC/PE ratio.

In lung tissue from hibernating squirrels there is a significant increase in phosphatidylcholine commensurate with hibernation (Table XI). Since the level of phosphatidylethanolamine essentially remained unchanged (decreased slightly, but not significantly) during hibernation, the increased PC/PE ratio, in and of itself, with other factors such as fatty acid composition remaining unchanged, would tend to increase the general fluidity of the lung cell membranes. However, there may be other factors responsible for membrane phospholipid alterations besides the requirement to sustain the appropriate degree of membrane fluidity needed to function at reduced temperatures. Some of these are discussed in Chapter 2 by Aloia.

ACKNOWLEDGMENTS

This investigation was supported in part by grants to the senior author from the National Science Foundation (BMS 75-07667), the American Heart Association (75-711), the California Heart Association, Riverside Affiliate, and the Department of Anesthesiology, Loma Linda University School of Medicine. Partial support was also provided by an N. I. H. grant (PHS-HL-13532-06) to E. T. Pengelley. The senior author specifically thanks Dr. George Rouser for advice and consultation and Mr. Frank Awender for valuable technical assistance. The senior author is also indebted to Dr. Bernard Brandstater for providing a conducive research environment, Ms. Debbie Yetter for her dedication to the task of typing this manuscript, and Ms. Dorothy Whitson for her devotion to the research project.

REFERENCES

Ackman, R. G. (1972). *Prog. Chem. Fats Other Lipids* **12**, 165–285.
Aloia, R. C. (1978a). *J. Therm. Biol.* (in press).
Aloia, R. C. (1978b). *Lipids* (submitted for publication).
Aloia, R. C. (1978c). *Comp. Biochem. Physiol.* **60B**, 19–26.
Aloia, R. C., and Platt-Aloia, K. (1976). *J. Cell Biol.* **70**, 355a.
Aloia, R. C., and Rouser, G. (1975). *In* "Modification of Lipid Metabolism" (E. G. Perkins and L. A. Witting, eds.), pp. 225–245. Academic Press, New York.
Bailey, E. D., and Davis, D. E. (1965). *Can. J. Zool.* **43**, 701–707.
Baranska, J., and Wlodawer, P. (1969). *Comp. Biochem. Physiol.* **28**, 553–570.
Baxter, C. F., Rouser, G., and Simon, G. (1969). *Lipids* **4**, 243–244.
Behrisch, H. W. (1978). *In* "Strategies in Cold: Natural Torpidity and Thermogenesis." Academic Press, New York (in press).
Bishop, D. G. (1971). *In* "Biochemistry and Methodology of Lipids" (A. R. Johnson and J. B. Davenport, eds.), pp. 391–424. Wiley (Interscience), New York.
Bito, L. Z., and Roberts, J. C. (1974). *Comp. Biochem. Physiol. A* **47**, 173–193.
Bligh, E. G., and Dyer, W. J. (1959). *Can. J. Biochem. Physiol.* **37**, 911–921.
Bowler, K., and Duncan, C. J. (1968). *Comp. Biochem. Physiol.* **24**, 1043–1054.
Bowler, K., and Duncan, C. J. (1969). *Physiol. Zool.* **42**, 211–219.
Breckenridge, W. C., Gombos, G., and Morgan, I. G. (1972). *Biochim. Biophys. Acta* **266**, 695–702.
Brenner, F. J., and Lyle, P. D. (1974). *Trans. N.Y. Acad. Sci.* [2] **36**, 273–280.
Burlington, R. F. (1972). *In* "Hibernation and Hypothermia, Perspectives and Challenges" (F. E. South *et al.*, eds.), pp. 3–16. Am. Elsevier, New York.
Caldwell, R. S., and Vernberg, F. J. (1970). *Comp. Biochem. Physiol.* **34**, 179–191.
Cannon, B., and Polnaszek, C. F. (1976). *In* "Regulation of Depressed Metabolism and Thermogenesis" (L. Jansky and X. J. Musacchia, eds.), pp. 93–116. Thomas, Springfield, Illinois.
Chaffee, R. R. J., Pengelley, E. T., Allen, J. R., and Smith, R. E. (1966). *Can. J. Physiol. Pharmacol.* **44**, 217–223.
Chapman, D. (1976). *In* "Mammalian Cell Membranes" (G. A. Jamieson and D. M. Robinson, eds.), pp. 97–137. Butterworth, London.
Chapman, D., and Urbina, J. (1974). *J. Biol. Chem.* **249**, 2512–2521.

Charnock, J. S. (1978). *In* "Strategies in Cold: Natural Torpidity and Thermogenesis." Academic Press (in press).

Coleman, R. (1973). *Biochim. Biophys. Acta* **300**, 1–30.

Cotmore, J. M., Nichols, G., Jr., and Werthier, R. E. (1971). *Science* **172**, 1339–1342.

Cullis, P. R., and deKruijff, B. (1978). *Biochim. Biophys. Acta* **507**, 207–218.

Cuzner, M. L., Davison, A. N., and Gregson, N. A. (1965). *J. Neurochem.* **12**, 469–477.

Davis, D. E. (1967). *Ecology* **48**, 683–689.

Dawe, A. R., and Spurrier, W. A. (1975). *In* "Temperature Regulation and Drug Action" (P. Lomax *et al.*, eds.), pp. 209–217. Karger, Basel.

Dawson, R. M. C., Hemington, N., and Davenport, J. B. (1962). *Biochem. J.* **84**, 497–503.

deGier, J., and Van Deenen, L. L. M. (1961). *Biochim. Biophys. Acta* **49**, 286–296.

Denyes, A., and Carter, J. D. (1961). *Am. J. Physiol.* **200**, 1043–1050.

Dill, D. B., and Forbs, W. H. (1941). *Am. J. Physiol.* **132**, 685–698.

Driedzic, W. M., and Roots, B. I. (1975). *J. Therm. Biol.* **1**, 7–11.

Duttera, S. M., Byrne, W. L., and Ganoza, M. C. (1968). *J. Biol. Chem.* **243**, 2216–2228.

Eichberg, J., Whittaker, V. P., and Dawson, R. M. C. (1964). *Biochem. J.* **92**, 91–96.

Eletr, S., and Inesi, G. (1972). *Biochim. Biophys. Acta* **290**, 178–185.

Eletr, S., Zakim, D., and Vessey, D. A. (1973). *J. Mol. Biol.* **78**, 351–362.

Esher, R. J., Fleischman, A. I., and Lenz, P. H. (1973). *Comp. Biochem. Physiol. A* **45**, 993–938.

Fang, L. S. T., and Willis, J. S. (1974). *Comp. Biochem. Physiol. A* **48**, 687–698.

Farkas, T., and Herodek, S. (1964). *J. Lipid Res.* **5**, 369–373.

Fawcett, D. W., and Lyman, C. P. (1954). *J. Physiol. (London)* **126**, 235–249.

Feldman, G., Feldman, L. S., and Rouser, G. (1965). *J. Am. Oil Chem. Soc.* **42**, 742–749.

Ferrin, L. G., South, F. E., and Jacobs, H. K. (1971). *Cryobiology* **8**, 506–510.

Folch, J., Lees, M., and Sloane-Stanley, G. H. (1957). *J. Biol. Chem.* **226**, 497–512.

Fourcans, B., and Jain, M. K. (1974). *Adv. Lipid Res.* **12**, 147–226.

Fox, C. F. (1975). *Biochem., Ser. One* **2**, 279–307. Butterworths, Baltimore.

Fox, C. F., and Tsukugoshi, N. (1972). *Membr. Res., ICN/UCLA Symp. Mol. Biol., Proc., 1st, 1972* pp. 145–154.

Fraenkel, G., and Hof, H. S. (1940). *Biochem. J.* **34**, 1085–1092.

Gates, D. M. (1962). "Energy Exchange in the Biosphere." Harper, New York.

Gingell, D. (1976). *Mamm. Cell Membr.* **1**, 198–223.

Goldman, S. S. (1975). *Am. J. Physiol.* **228**, 834–839.

Goldman, S. S., and Albers, R. W. (1975). *Arch. Biochem. Biophys.* **169**, 540–544.

Goldman, S. S., and Willis, J. S. (1973a). *Cryobiology* **10**, 211–217.

Goldman, S. S., and Willis, J. S. (1973b). *Cryobiology* **10**, 218–224.

Grant, C. W. M., and McConnell, H. M. (1974). *Biochim. Biophys. Acta* **363**, 151–158.

Gruener, N., and Avi-dor, Y. (1966). *Biochem. J.* **100**, 762–767.

Gurr, M. I., and James, A. T. (1975). "Lipid Biochemistry: An Introduction." Chapman & Hall, London.

Hagopian, M., Anversa, P., and Nuñez, E. A. (1974). *Anat. Rec.* **178**, 599–616.

Hayashi, M., Muramatsu, T., and Hara, I. (1973). *Biochim. Biophys. Acta* **291**, 335–343.

Hazel, J. R. (1972). *Comp. Biochem. Physiol. B* **43**, 837–861.

Hazel, J. R., and Prosser, C. L. (1974). *Physiol. Rev.* **54**, 620–677.

Hazel, J. R., and Schuster, V. L. (1976). *J. Exp. Zool.* **195**, 425–438.

Hensel, J., Bruck, K., and Raths, P. (1973). *In* "Temperature and Life" (H. Precht *et al.*, eds.), pp. 505–732. Springer-Verlag, Berlin and New York.

Himms-Hagen, J., Behrens, W., Muirhead, M., and Hbous, A. (1975). *Mol. Cell. Biochem.* **6**, 15–31.

44 R. C. Aloia and E. T. Pengelley

Hirvonen, L. (1956). Acta Physiol. Scand. 36, 38–46.
Hof, H., and Simon, R. G. (1970). Lipids 4, 243–249.
Hoffman, B. F. (1956). N.A.S.-N.R.C., Publ. 451, 302–327.
Horiuchi, S. (1977). Comp. Biochem. Physiol. B 56, 135–138.
Horvath, A. (1876). Arch. Gesamte Physiol., Menschen Tiere 12, 278–282.
Horwitz, A. F. (1972). In "Membrane Molecular Biology" (C. F. Fox and A. D. Keith, eds.), pp. 164–191. Sinauer Assoc., Stamford, Connecticut.
Horwitz, B., and Eaton, M. (1975). Eur. J. Pharmacol. 34, 241–245.
Hudson, J. W. (1967). Mamm. Hibernation 3, Proc. Int. Symp., 1965 pp. 30–46.
Hulbert, A. J., and Hudson, J. W. (1976). Am. J. Physiol. 230, 1211–1216.
Hulbert, A. J., Augee, M. L., and Raison, J. K. (1976). Biochim. Biophys. Acta 455, 597–601.
Illanes, A., and Marshall, J. M. (1964). Naunyn-Schmiedebergs Arch. Exp. Pathol. Pharmakol. 248, 15–26.
Inesi, G., Millman, M., and Eletr, S. (1973). J. Mol. Biol. 81, 433–454.
Jacobson, K., and Papahadjopoulos, D. (1975). Biochemistry 14, 152–161.
Johansson, B. (1967). Mamm. Hibernation 3, Proc. Int. Symp., 1965 pp. 200–219.
Johansson, B. W. (1969). In "Depressed Metabolism" (X. J. Musacchia and J. F. Saunders, eds.), pp. 313–375. Am. Elsevier, New York.
Karlsson, K. A., Samuelsson, B. E., and Steen, G. O. (1973). Biochim. Biophys. Acta 306, 317–328.
Kates, M., and Paradis, M. (1973). Can. J. Biochem. 51, 184–193.
Kayser, C. (1953). Annee Biol. 29, 109–150.
Kayser, K., and Malan, A. (1963). Experientia 19, 441–451.
Kimelberg, H. K. (1976). Mol. Cell Biochem. 10, 171–190.
Kimelberg, H. K., and Papahadjopoulos, D. (1972). Biochim. Biophys. Acta 282, 277–292.
Kimelberg, H. K., and Papahadjopoulos, D. (1974). J. Biol. Chem. 249, 1071–1080.
Klain, G. J., and Whitten, B. K. (1968). Comp. Biochem. Physiol. 25, 363–366.
Knipprath, W. G., and Mead, J. F. (1966). Lipids 1, 113–117.
Kodoma, A. M., and Pace, N. (1964). J. Appl. Physiol. 19, 863–867.
Kramm, K. R. (1973). BioScience 23, 516–521.
Kreps, E. M., Chebatareva, M. A., and Akulin, V. N. (1969). Comp. Biochem. Physiol. 31, 419–430.
Kruger, S., and Mendler, M. (1970). J. Neurochem. 17, 1313.
Krulin, G. S., and Sealander, J. S. (1972). Comp. Biochem. Physiol. A 42, 537–549.
Kumamoto, J., Raison, J. K., and Lyons, J. M. (1971). J. Theor. Biol. 31, 47–51.
Kuwabara, T. (1975). Invest. Ophthalmol. 14, 457–467.
Lacko, L., Wittke, B., and Geck, P. (1973). J. Cell. Physiol. 82, 213–218.
Landau, B. R. (1956). Diss. Abstr. 16, 2195.
Lee, M. P., and Gear, A. R. L. (1974). J. Biol. Chem. 249, 7541–7549.
Lenaz, G., Sechi, A. M., Parenti-Castelli, G., Lindi, L., and Bertoli, E. (1972). Biochem. Biophys. Res. Commun. 49, 536–543.
Lerner, E., Shug, A. L., Elson, C., and Shrago, E. (1972). J. Biol. Chem. 247, 1513–1519.
Leslie, J. M., and Bucklie, J. T. (1976). Comp. Biochem. Physiol. B 53, 335–337.
Levine, Y. K. (1976). In "Mammalian Cell Membranes" (G. A. Jamieson and D. M. Robinson, eds.), pp. 78–96. Butterworth, London.
Lindberg, O., Bieber, L. L., and Houstek, J. (1976). In "Regulation of Depressed Metabolism and Thermogenesis" (L. Jansky and X. J. Musacchia, eds.), pp. 117–136. Thomas, Springfield, Illinois.

Linden, C. D., Wright, K. L., McConnell, H. M., and Fox, C. F. (1973). *Proc. Natl. Acad. Sci. U.S.A.* **70**, 2271–2275.

Lyman, C. P. (1958). *Am. J. Physiol.* **194**, 83–91.

Lyman, C. P., and O'Brien, R. C. (1960). *Bull. Mus. Comp. Zool.* **174**, 353–372.

Lyman, C. P., and O'Brien, R. C. (1963). *J. Physiol. (London)* **168**, 477–499.

Lyons, J. M. (1972). *Cryobiology* **9**, 341–350.

Lyons, J. M., and Raison, J. K. (1970). *Comp. Biochem. Physiol.* **37**, 405–411.

McElhaney, R. N., deGier, J., and vander Neur-kak, E. C. M. (1973). *Biochim. Biophys. Acta* **298**, 500–512.

McMurchie, E. J., Raison, J. K., and Cairncross, K. D. (1973). *Comp. Biochem. Physiol. B* **44**, 1017–1026.

Marshall, J. M., and Willis, J. S. (1962). *J. Physiol. (London)* **164**, 64–76.

Martin, H. N., and Applegarth, E. C. (1890). *Johns Hopkins Univ. Stud. Biol. Lab.* **4**, 275.

Massey, V., Curti, B., and Ganther, H. (1966). *J. Biol. Chem.* **241**, 2347–2357.

Massopust, L. C., Jr., Wolin, L. R., and Meder, J. (1965). *Exp. Neurol.* **12**, 25–32.

Melchior, D. L., and Steim, J. M. (1976). *Annu. Rev. Biophys. Bioeng.* **5**, 205–238.

Melhorn, R. J., and Keith, A. D. (1972). *In* "Membrane Molecular Biology" (C. F. Fox and A. D. Keith, eds.), pp. 192–227. Sinauer Assoc., Stamford, Connecticut.

Menaker, M. (1962). *J. Cell. Comp. Physiol.* **59**, 163–173.

Michaelson, D. M., Horwitz, A. F., and Klein, M. P. (1974). *Biochemistry* **13**, 2605–2612.

Miller, N. G. A., Hill, M. W., and Smith, M. W. (1976). *Biochim. Biophys. Acta* **455**, 644–654.

Morris, R. W., and Schneider, M. J. (1969). *Comp. Biochem. Physiol.* **28**, 1461–1465.

Morrison, W. R., and Smith, L. M. (1964). *J. Lipid Res.* **5**, 600–608.

Mrosovsky, N. (1976). *Am. Zool.* **16**, 685–697.

Newsholm, E. A., and Start, C. (1973). "Regulation in Metabolism." Wiley, New York.

Nordlie, R. C., and Jorgenson, R. A. (1976). *Enzymes Biol. Membr.* **2**, 465–492.

Nuñez, E. A., Wallis, J., and Gershon, M. D. (1974). *Am. J. Anat.* **141**, 179–202.

Owens, K. (1966). *Biochem. J.* **100**, 354–360.

Packter, N. M. (1973). "Biosynthesis of Acetate-derived Compounds." Wiley, New York.

Papahadjopoulos, D. (1977). *J. Colloid Interface Sci.* **58**, 459–470.

Pascher, I. (1976). *Biochim. Biophys. Acta* **455**, 433–451.

Paulsrud, J. R., and Dryer, R. L. (1968). *Lipids* **3**, 340–345.

Pengelley, E. T. (1967). *Mamm. Hibernation 3, Proc. Int. Symp., 1965* pp. 1–30.

Pengelley, E. T., and Asmundson, S. J. (1974). *In* "Circannual Clocks" (E. T. Pengelley, ed.), pp. 95–161. Academic Press, New York.

Plattner, W. S., Patnayak, B. C., and Musacchia, X. J. (1972). *Comp. Biochem. Physiol. A* **42**, 927–938.

Plattner, W. S., Steffen, D. G., Temple, G., and Musacchia, X. J. (1976). *Comp. Biochem. Physiol. A* **53**, 279–283.

Raison, J. K. (1972). *Bioenergetics* **4**, 285–309.

Raison, J. K. (1973). *In* "Rate Control of Biological Processes" (D. D. Davis, ed.), pp. 485–512. Cambridge Univ. Press, London and New York.

Raison, J. K., and Lyons, J. M. (1971). *Proc. Natl. Acad. Sci. U.S.A.* **68**, 2092–2095.

Raison, J. K., and McMurchie, E. J. (1974). *Biochim. Biophys. Acta* **363**, 135–140.

Raison, J. K., Lyons, J. M., Melhorn, R. J., and Keith, A. D. (1971a). *J. Biol. Chem.* **246**, 4036–4040.

Raison, J. K., Lyons, J. M., and Thomson, W. W. (1971b). *Arch. Biochem. Biophys.* **142**, 83–90.

Rand, R. P., and Sengupta, S. (1972). *Biochim. Biophys. Acta* **255**, 484–492.

Reinert, J. C., and Steim, J. M. (1970). *Science* **168**, 1580–1583.

Roots, B. I. (1968). *Comp. Biochem. Physiol.* **25**, 457–466.

Rouser, G., and Solomon, R. D. (1969). *Lipids* **4**, 232–234.

Rouser, G., Kritchevsky, G., Galli, C., Yamamoto, A., and Knudson, A. G. (1967). *In* "Inborn Errors of Sphingolipid Metabolism" (S. M. Aronson and B. M. Volk, eds.), pp. 303–316. Pergamon, Oxford.

Rouser, G., Nelson, G., Fleisher, S., and Simon, G. (1968). *In* "Biological Membranes" (D. Chapman, ed.), pp. 5–69. Academic Press, New York.

Rouser, G., Kritchevsky, G., Yamamoto, A., Simon, G., Calli, C., and Bauman, A. J. (1969a). *In* "Methods in Enzymology" (J. M. Lowenstein, ed.), Vol. 14, pp. 272–317. Academic Press, New York.

Rouser, G., Simon, G., and Kirtchevsky, G. (1969b). *Lipids* **4**, 599–606.

Rouser, G., Kritchevsky, G., Siakotos, A. N., and Yamamoto, A. (1970). *In* "Neuropathology: Methods and Diagnosis" (C. G. Tedeschi, ed.), pp. 691–753. Little, Brown, Boston, Massachusetts.

Rouser, G., Kritchevsky, G., Yamamoto, A., and Baxter, C. F. (1972). *Adv. Lipid Res.* **10**, 261–360.

Saarikoski, J. (1969). *Ann. Acad. Sci. Fenn.* **145**, 3–7.

Saarikoski, J. (1975). *Commun. Biol.* **75**, 113.

Sauerheber, R. D., and Gordon, L. M. (1975). *Proc. Soc. Exp. Biol. Med.* **150**, 28–31.

Shimshick, E. J., and McConnell, H. M. (1973). *Biochemistry* **12**, 2351–2360.

Shinitzky, M., and Barenholz, Y. (1974). *Biochemistry* **12**, 2351–2360.

Shipley, G. G:, Avecilla, L. S., and Small, D. M. (1974). *J. Lipid Res.* **15**, 124–131.

Simon, G., and Rouser, G. (1969). *Lipids* **4**, 607–614.

Sinensky, M. (1974). *Proc. Natl. Acad. Sci. U.S.A.* **71**, 522–525.

Skou, J. C. (1972). *Bioenergetics* **4**, 203–232.

Slotboom, A. J., and Bonsen, P. P. M. (1970). *Chem. Phys. Lipids* **5**, 301–398.

South, F. E., Heath, J. E., Luecke, R. H., Mihilovic, Lj. T., Myers, R. D., Panuska, J. A., Williams, B. A., Hartner, W. C., and Jacobs, H. K. (1972). *In* "Hibernation and Hypothermia, Perspectives and Challenges" (F. E. South *et al.*, eds.), pp. 629–633. Am. Elsevier, New York.

Steffen, D. G., and Plattner, W. S. (1976). *Am. J. Physiol.* **231**, 650–654.

Steim, J. M. (1974). *In* "Methods in Enzymology" (S. Fleischer and L. Packer, eds.), Vol. 32, Part B. pp. 262–272. Academic Press, New York.

Steim, J. M., Tourtellotte, M. E., Reinert, J. C. McElhaney, R. N., and Rader, R. L. (1969). *Proc. Natl. Acad. Sci. U.S.A.* **63**, 104–108.

Strumwasser, F. (1959). *Am. J. Physiol.* **196**, 23–30.

Sturtevant, J. M. (1974). *In* "Quantum Statistical Mechanics in the Natural Sciences" (B. Kursunoglu *et al.*, eds.), pp. 63–86. Plenum, New York.

Suomalainen, P., and Saarikoskic, D. L. (1967). *Experientia* **23**, 457.

Swan, H. (1975). "Thermoregulation and Bioenergetics." Am. Elsevier, New York.

Swanson, P. D. (1966). *J. Neurochem.* **13**, 229–236.

Tait, J. (1922). *Am. J. Physiol.* **59**, 467.

Tanaka, R. (1974). *In* "Reviews of Neurosciences" (S. Ehrenpreis and I. J. Kopin, eds.), pp. 181–230. Raven, New York.

Tanaka, R., and Teruya, A. (1973). *Biochim. Biophys. Acta* **323**, 584–591.

Tashima, L. S., Adelstein, S. J., and Lyman, C. P. (1970). *Am. J. Physiol.* **218**, 303–309.

Tempel, G., and Musacchia, X. J. (1975). *Am. J. Physiol.* **228**, 602–607.

Thorpe, R. F., and Ratledge, C. (1973). *J. Gen. Microbiol.* **78**, 203–206.

Tsukugoshi, N., and Fox, C. F. (1973). *Biochemistry* **12**, 2822–2828.

Van Dijck, P. W. M., deKruyff, B., Van Deenen, L. L. M., deGier, J., and Demel, R. A. (1976). *Biochim. Biophys. Acta* **455**, 576–587.

Verkleij, A. J., deKruyff, B., Ververgaert, P. H. J. Th., Tocanne, J. F., and Van Deenen, L. L. M. (1974). *Biochim. Biophys. Acta* **339**, 432–437.

Vessey, D. A., and Zakim, D. (1974). *Horizons Biochem. Biophys.* **1**, 138–174.

Viret, J., and Leterrier, F. (1976). *Biochim. Biophys. Acta* **436**, 811–824.

Watson, K., Bertoli, E., and Griffiths, D. E. (1975). *Biochem. J.* **146**, 401–407.

Wells, H. J., Makita, M., Wells, W. W., and Drutzach, P. H. (1965). *Biochim. Biophys. Acta* **98**, 269.

Williams, D. D., and Plattner, W. S. (1967). *Am. J. Physiol.* **212**, 167–172.

Willis, J. S. (1966). *J. Gen. Physiol.* **49**, 1221–1239.

Willis, J. S. (1967). *Mamm. Hibernation 3, Proc. Int. Symp.*, *1965* pp. 356–381.

Willis, J. S., and Li, N. M. (1969). *Am. J. Physiol.* **217**, 321–326.

Willis, J. S., Fang, L. S. T., and Foster, R. F. (1972). *In* "Hibernation and Hypothermia, Perspectives and Challenges" (F. E. South *et al.*, eds.), pp. 123–148. Am. Elsevier, New York.

Wisnieski, B., Huang, Y. O., and Fox, C. F. (1974a). *J. Supramol. Struct.* **2**, 593–608.

Wisnieski, B., Parks, J. G., Huang, Y. O., and Fox, C. F. (1974b). *Proc. Natl. Acad. Sci. U.S.A.* **71**, 4381–4385.

Witting, L. A. (1975). *In* "Modification of Lipid Metabolism" (E. G. Perkins and L. A. Witting, eds.), pp. 1–41. Academic Press, New York.

Wood, L., and Beutler, E. (1967). *J. Lab. Clin. Invest.* **70**, 287–294.

Wuthier, R. E. (1966). *J. Lipid Res.* **7**, 544–549.

Yellin, N., and Levin, I. W. (1977). *Biochemistry* **16**, 642–647.

Zakim, D., and Vessey, D. A. (1976). *Biochemical J.* **157**, 667–673.

Zatzman, M. L., and South, F. E. (1972). *Am. J. Physiol.* **22**, 1035–1039.

Zimmer, G., and Schirmer, H. (1974). *Biochim. Biophys. Acta* **345**, 314–320.

Zimney, M. (1968). *Comp. Biochem. Physiol.* **27**, 859–863.

CHAPTER 2

Membrane Structure and Function during Hibernation

R. C. Aloia

I. Introduction 49
II. Lipid Alterations Unrelated to Low Temperature
   Membrane Function 51
III. Nonspecific Temperature Effects and Lipid Metabolism
   during Hibernation 52
IV. Membrane Compartmentation and Relationship to the Lipids of
   Hibernating Mammals 53
   A. Membrane Asymmetry 54
   B. Cholesterol 56
   C. Lipid Domains 60
   D. Boundary Lipid 67
V. Lipid Alterations and Low Temperature Membrane Function 67
VI. Conclusion and Future Research 69
   References 71

I. Introduction

Hibernation in mammals is a unique phenomenon since the animal's normal body temperature is lowered by over 98% in a metabolically controlled manner from 37°C to temperatures approaching 0°C. The animal can remain in a torpid, metabolically depressed state for days to weeks at a time and maintain vital physiological functions, such as heartbeat, respiration, kidney and brain activity. To maintain such a vital functional capacity requires that cellular ionic gradients be sustained. This is apparently accomplished by the presence of cold-tolerant cellular membranes; membranes that function as optimal permeability barriers and ion pumps efficient at low temperatures. Key membrane-bound enzymes, such as $Na^+ + K^+$-ATPase have been shown to become cold tolerant during the animal's preparatory phase for hibernation (Goldman and Abers, 1975). Whether the entire cellular membrane or simply the loci containing the enzymes responsible for ion transport become resistant to low temperatures, the membrane lipids may well be involved since they are responsible in part for membrane permeability barriers and are required by many membrane-bound enzymes for activity (Kimelberg, 1976). However,

49

CHEMICAL ZOOLOGY, VOL. XI
Copyright © 1979 by Academic Press, Inc.
All rights of reproduction in any form reserved.
ISBN 0-12-261040-5

there may be lipid alterations commensurate with hibernation which are not related to the functional capacity of the membrane at reduced temperatures.

There seem to be three obvious possibilities relating membrane lipid alterations found in the tissues of hibernators during hibernation (see Chapter 1 by Aloia and Pengelley) to the low temperature functional capacity of the cell membranes: (1) lipid alterations occur during hibernation which are not related to the capacity of the membrane to function at low temperatures and the membrane lipids undergo a liquid-crystalline to crystalline (LC–C) phase transition and become nonfunctional; (2) lipid alterations occur which are unrelated to low temperature tolerance of the cell membranes because the cell membranes are already adapted to function at reduced temperatures; and (3) lipid alterations occur commensurate with hibernation which specifically confer a low temperature functional capacity to the membrane. Table I presents a summary of the alterations in phospholipid classes which are found in various organs of the hibernating animal. Increments and decrements in steady state levels of each phospholipid class are shown as a comparison to the levels found in the organs of animals in their active physiological state. Each organ examined was found to have an altered phospholipid class composition in

TABLE I

PERCENTAGE CHANGE IN STEADY STATE LEVELS OF PHOSPHOLIPIDS FROM VARIOUS TISSUES OF HETEROTHERMIC MAMMALS IN THEIR HIBERNATING PHYSIOLOGICAL STATE COMPARED TO THEIR ACTIVE PHYSIOLOGICAL STATE[a]

Tissue	PC	PE	DPG	SPH	PI	PS
Brain[b]	13↑	16↑	—	—	—	—
Brain[c]	4↑	11↓	—	—	—	—
Liver[d]	—	16↑	9↓	—	9↓	23↓
Kidney[e]	4↑	—	—	8↓	—	—
Lung[f]	7↑	—	—	—	—	—
Skeletal muscle[g]	5↓	13↑	13↑	—	—	—

[a] PC, phosphatidylcholine; PE, phosphatidylethanolamine; DPG, diphosphatidylglycerol (cardiolipin); SPH, sphingomyelin; PI, phosphatidylinositol; PS, phosphatidylserine. (↑) Increased or (↓) decreased level found in organ during hibernation. All increments or decrements are in comparison to the active physiological state.

[b] Data from Goldman, 1975, *M. auratus;* see Chapter 1.

[c] Data from Aloia, 1978a, *C. lateralis;* see Chapter 1.

[d] Data from Aloia, 1978b, *C. lateralis;* see Chapter 1.

[e] Data from Aloia, 1978c, *C. lateralis;* see Chapter 1.

[f] Data from R. C. Aloia, unpublished, *C. lateralis;* see Chapter 1.

[g] Data from R. C. Aloia, unpublished, *C. lateralis;* see Chapter 1.

the hibernating animal when compared to the active animal at 37°C. As stated in Chapter 1 by Aloia and Pengelley, active or awake animals are those which were sacrificed during the summer months with a body temperature of 37°C, whereas hibernating animals are those sacrificed in the winter months while they were in a hibernating state with a body temperature of 3°–5°C. Throughout the chapter, phrases such as lipid alterations occurring in hibernation, commensurate with hibernation, or during hibernation are not to be taken literally since they simply refer to the lipid composition which was found when the hibernating animal was sacrificed and the lipids analyzed. There is no intent to imply that these changes in steady state levels of lipids found in the hibernating animals, as compared to the active animals, occurred only when the animal hibernated since they may well have been preparatory, preseasonal changes, as has been shown for some membrane-bound enzymes (Goldman and Albers, 1975; Fang and Willis, 1974). The following discussion will therefore consider the above-mentioned possibilities relating lipid alterations to membrane function and focus upon new ideas in membrane structure.

II. Lipid Alterations Unrelated to Low Temperature Membrane Function

Regarding the first two alternatives, it has been demonstrated for cell membranes of *Acholeplasma laidlawii,* that complex membrane transport functions required for growth can continue at a normal rate until approximately 50% of the membrane lipid is in a solid state (McElhaney, 1974). Additionally, the organism can continue to grow at a reduced rate until approximately 90% of the membrane lipids are in a crystalline state. Consequently, a similar degree of membrane function in hibernator tissues also may be possible with a fraction of the lipids in a solid phase. If some of the membrane lipids present in normothermic hibernator tissues are in such a physical state that they would undergo liquid crystalline to crystalline (LC–C) phase transition at low temperatures, that percentage of solid phase lipid resulting from the transition may be tolerated quite well depending upon the degree of residual activity required of the cell membranes during hibernation. For example, the brain, heart, and lungs may require a higher degree of residual membrane activity than other organs due to their ostensible physiological function during hibernation. Conversely, some organs may tolerate a higher degree of inactivity than others and their membrane lipids may enter a predominantly solid state during hibernation.

There is evidence, however, that the cell membranes of some obligate homeotherms do not undergo a LC–C phase transition and remain in a liquid-crystalline or fluid state even at temperatures approaching 0°C. Differential scanning calorimetry (DSC) has detected the LC–C phase transition of the membrane lipids of rat liver mitochondria and microsomes and rabbit sarcoplasmic reticulum at approximately 0°C with a range of ±10°C (Hackenbrock, 1976; Martonosi, 1974; Blayzk and Steim, 1972). For example, whole rat liver mitochondria undergo a phase transition to the solid state beginning at 0°C and ending at −20°C. In fact, the energy-transducing inner mitochondrial membrane does not begin a phase transition to the solid state until −4°C (Hackenbrock, 1976). In other words, it appears that the lipids of these mitochondrial membranes exist in such a physical state of disorder that a solid phase does not appear until the temperature reaches 0°C or less. These temperatures are similar to those reported for the LC–C phase transition of nonmammalian poikilothermic membranes, such as the frog and squid photoreceptors (Mason and Abrahamson, 1974). Consequently, the membranes of tissues from hibernating mammals may be similar to these membranes and may well be sufficiently fluid to accommodate the low temperatures extant during hibernation and not solidify at all or to any appreciable degree.

III. Nonspecific Temperature Effects and Lipid Metabolism during Hibernation

If this is the case, that the hibernator's membrane lipids are cold tolerant during its active physiological state, then the observed changes in lipid composition which accompany hibernation may be the result of a nonspecific low temperature-induced inhibition and/or activation of specific lipid biosynthetic pathways. A slight temperature inhibition of microsomal cytidyltransferase (EC 2.7.7.41) would cause a reduction in CDP-diglyceride and those phospholipids synthesized therefrom, namely phosphatidylinositol, phosphatidylglycerol, and diphosphatidylglycerol, as is observed in the liver (see Fig. 3 and Table VII in Chapter 1 by Aloia and Pengelley). Also, for example in the kidney, there appears to be an inverse relationship between phosphatidylcholine and sphingomyelin, the former increasing and the latter decreasing during hibernation. Until recently it was thought that sphingomyelin was formed from the reaction of CDP-choline with a ceramide (Scribney and Kennedy, 1958). However it has been demonstrated that this reaction was highly active only with the "threo" sphingosine moiety and much less so with the natural "erythro" form. Subsequently, it was shown that sphingomyelin was formed by a group transfer reaction from phosphatidylcholine to erythroceramide (Witting, 1975). Consequently, a slight inhibition of this reaction would result in lower levels of sphingomyelin and higher levels of phosphatidylcholine, as is indeed observed in the kidney (see Table IX in Chapter 1 by Aloia and Pengelley).

On the other hand, various lipid biosynthetic pathways could be enhanced when the end product of that pathway is required for the activation of a specific enzyme within the membrane and is removed from its metabolic site. Phosphatidylcholine, for example, which is elevated in the brain and kidney of hibernating squirrels, has been shown to be required for the activity of β-hydroxybutyrate dehydrogenase (Grover et al., 1974) among others (Fourcans and Jain, 1974; Coleman, 1973). Also, phosphatidylethanolamine has been shown to be the most effective lipid for activating glucose-6-phosphatase (Gurr and James, 1975; Duttera et al., 1968; however, see Nordlie and Jorgenson, 1976). Since gluconeogenesis is elevated in some hibernating species (Burlington, 1972), the enhanced enzyme activity could require a greater quantity of enzyme, and hence, a greater quantity of lipid. Conversely, an isozymic form of the enzyme may be utilized to continue activity at low temperatures (Burlington, 1972; Somero, 1972), and this variant form may require greater quantities of a particular lipid. As was pointed out by Behrisch (1978), alteration of a small portion of a multimeric enzyme could be responsible for a conformational change resulting in a new isozymic form with different temperature characteristics. As with pyruvate kinase and fructose-1,6-diphosphatase, membrane enzymes may undergo alterations on a seasonal basis to prepare the animal for entry into and emergence from the hibernating state (Behrisch, 1978). Since certain phospholipids have been shown to function as positive allosteric effectors for various membrane-bound enzymes (Farias et al., 1975), different conformational variants of the same enzyme may require different lipids or quantities of lipids for allosteric modulation. A higher quantity of a particular phospholipid may be required for the activation of greater numbers of a particular enzyme molecule, an isozymic variant of the enzyme, or for enhanced positive (or negative) modulation needed to sustain the appropriate degree of enzyme activity.

IV. Membrane Compartmentation and Relationship to the Lipids of Hibernating Mammals

In addition to specifically affecting enzyme function, the changes in phospholipid class composition observed during hibernation could be required to modify a segment or compartment within the membrane to sustain or inhibit passive permeability properties or to maintain various functional parameters in a specific region within the membrane. For example, it has been demonstrated that prostaglandins are synthesized from 20-carbon fatty acids eicosatri-, tetra- and pentaenoic acids (Schneider, 1976). One of the most thoroughly studied is arachidonic acid (20 : 4) and its incorporation into the PG_2 series. There appear to be no cellular stores for the prostaglandins and when required for a particular cellular function they are synthesized *de novo* in response to the appropriate hormonal stimulation. It appears that the proper hormone can stimulate phospholipase A_2 which in turn cleaves arachidonic acid from membrane phospholipids. The free fatty acid is then acted on by cyclooxygenase to form the endoperoxides and prostaglandins (Schoene and Iacono, 1976). Normally, phospholipase A_2 cleaves whatever fatty acid is present in the C-2 position of the phospholipid (Bishop, 1971). However, it has recently been demonstrated that in response to cellular trauma, tissue injury, or appropriate hormonal stimulation, phospholipase A_2 acts predominantly on a single phospholipid class (phosphatidylcholine in rabbit heart: Hsueh *et al.*, 1977; phosphatidylinositol in blood platelets: Schoene and Iacano, 1976), and, specifically on the 2-position, cleaving arachidonic acid (Schoene and Iacono, 1976). This is a specific cleavage of only arachidonic acid primarily from one phospholipid class, other lipids and their 2-position fatty acids being unaffected. One interpretation of these data (although not the only interpretation) is that the activated phospholipase A_2 attacks arachidonate-containing glycerophosphatides of a particular lipid class which are segregated to or sequestered within a specific region of the membrane, perhaps close to the cyclooxygenase enzyme system (Hsueh *et al.*, 1977). It can readily be envisioned that a similar sequestration of phospholipids to a specific region within a membrane of hibernators may be required for such a unique function as prostaglandin biosynthesis. The following discussion will consider further evidence to support the concept of membrane compartmentation.

A. MEMBRANE ASYMMETRY

From the pioneering studies of Gorter and Grendel (1925) on erythrocyte lipids, Fricke's (1925) studies on erythrocyte electrical capacitance, and Danielli, Harvey, and Davson's work on surface tension measurements of fish egg oil, the biological membrane has been considered to be a bilayer structure of approximately 40Å in thickness (Danielli and Harvey, 1935). Robertson's (1957) unit membrane model emphasized the universality of the bilayer concept, Singer and Nicolson's (1972) fluid mosaic model emphasized the dynamic highly mobile nature of the membrane components and the hydrophobic aspect of protein–lipid interactions, and recent physical studies have elaborated on the fluidity aspects of membrane function (Melchior and Steim, 1976). Membrane phospholipids and proteins have been shown to be highly mobile and to move freely and rapidly within the plane of the membrane (Jain and White, 1977; Petit and Edidin, 1974). However, recent studies have demonstrated specific regions within the membrane where stable, nonmobile, lipid–lipid and lipid–protein relationships must exist (Aloia and Rouser, 1975b; Wallach et al., 1975). In other words, there is evidence to support the existence of membrane compartments of differing composition. As such, this concept must be considered when evaluating the role of membrane lipids in hibernation.

It has been amply demonstrated that bilayer lipid asymmetry exists in some cell membranes. This has been shown by the application of linker molecules such as FMMP [formylmethionyl(sulfonyl)methyl phosphate] and TNBS (trinitrobenzenesulfonic acid) which bind to primary amine groups on the membrane surface (Bretscher, 1971, 1972), phospholipase C, which cleaves off the polar headgroup of phospholipids (Rothman et al., 1976), and cholesterol exchange between vesicles and cells (Lenard and Rothman, 1976). For example, in human erythrocytes, the total phospholipids seem to be distributed equally between both bilayer halves. However, the predominant quantities of phosphatidylcholine and sphingomyelin (about 75%) have been localized to the outer half of the erythrocyte bilayer while the ethanolamine (about 75%) and serine glycerophosphatides (99%) seem to reside mostly in the inner bilayer half (Kahlenberg et al., 1974; Renooij et al., 1974; Verkleij et al., 1973; Bretscher, 1971, 1972). Recently, the membrane surrounding the influenza virus grown in bovine kidney cells has been shown to have an asymmetric architecture. As the viral nucleocapsid buds from the cell plasma membrane it becomes enveloped by a single modified region of the cell membrane which contains host-cell lipids and viral-coded proteins (Rothman and Lenard, 1977). Although the viral proteins may associate with specific phospholipids and alter their distribution, it nevertheless can be reasonably assumed that the viral membrane reflects that of the host-cell plasma membrane (Rothman and Lenard, 1977). The viral membrane was shown to contain about 70% of the total phospholipid in the inner bilayer half. Additionally, 50% of phosphatidylcholine (PC) and 70–80% of phosphatidylethanolamine (PE), phosphatidylserine (PS), and sphingomyelin (SPH) are situated in the inner bilayer (Rothman and Lenard, 1977). The asymmetry seems to be a stable condition since "flip-flop" rates of phospholipids from one side of the bilayer are on the order of hours or days and the process requires a high energy of activation (Singer, 1977). For example, the $t_{1/2}$ of 30°C for a nitroxide spin-labeled phosphatidylcholine molecule across an egg PC liposome is about 6.5 hours (Kornberg and McConnell, 1971). For radiolabeled phosphatidylcholine in the outer bilayer of a liposome, the flip-flop rate is about 11 days (Rothman and Davidowicz, 1975). Estimated lower limit for cholesterol exchange between inner and outer monolayers is 3 and 6 days for influenza virus membranes and PC–cholesterol vesicles (Rothman and Lenard, 1977).

Not only has asymmetry in the location of membrane phospholipids been demonstrated, but also an asymmetry in the distribution of the lipid biosynthetic enzymes. For example, using enzyme latency to various proteases as a criterion for the distribution of the enzyme on the inner or outer half of the bilayer, the enzymes involved in the biosynthesis of phosphatidylethanolamine and phosphatidylcholine were shown to be localized on the cytoplasmic side of the microsomal membrane (Coleman and Bell, 1978).

If the phospholipid distribution in the erythrocyte can be considered typical for asymmetric arrangement, then some of the phospholipid alterations which are found in hibernating animals may occur predominantly in only one half of the membrane. For example, the increase in PC and decrease in SPH in the kidney found in hibernating squirrels could possibly be occurring in the outer monolayer, thus influencing the physiochemical behavior of this half of the membrane.

B. Cholesterol

Cholesterol (cholest-5-en-3β-ol) is considered to have a dual role in membrane function, creating an intermediate state of membrane fluidity (Demel and deKruyff, 1976). For example, in Langmuir trough experiments expanded monolayer films such as 18 : 0/18 : 1 phosphatidylcholine occupy an area/molecule of about 63 Å^2 at a surface pressure of 30 dynes/cm. However, with cholesterol added to the system at a ratio of 1 : 1, this area is reduced to approximately 50 Å^2/mole. Conversely, long-chain saturated species in a solid state below their phase transition temperature will become more fluid with the addition of cholesterol (Demel and deKruyff, 1976). This duality has also been supported by physical studies such as electron paramagnetic resonance. Using a bilayer system of egg yolk phosphatidylcholine which is fluid at 20°C and dipalmitoylphosphatidylcholine (DPPC), which is solid at 20°C, it was shown that the addition of cholesterol to a level of 1 : 1 caused spectral line shifts such that the order parameter ($2T_{11}$) decreased in DPPC bilayer and increased in egg PC bilayers (Oldfield and Chapman, 1972). The increase in the order parameter (solidification) for egg PC liposomes and the decrease in order parameter for DPPC liposomes (fluidization) resulted in an intermediate state of fluidity with order parameters of similar values. The planar steroid ring structure, the hydrocarbon chain at the C-17 position and 3β-hydroxy group have been found to be essential for an effective modification of the membrane by cholesterol. Steroids which lack these molecular features are either less effective or ineffective (Demel et al., 1972a,b). Although it has not been found to hydrogen bond with phosphorus or polar head group components, cholesterol is thought to be positioned between phospholipid molecules from the polar region of the membrane centrally for approximately 10 carbon atoms (Demel and deKruyff, 1976). The hydrocarbon side chain at C-17 is important in hydrophobic interactions and the planar nucleus appears necessary to maximize acyl chain interaction. It thus appears to form a membrane wedge which can disrupt hydrophobic interactions and van der Waals forces. Consequently, it can reduce cooperativity between hydrocarbon acyl chains and sustain a greater degree of thermal motion at lower temperatures than would be possible without its presence. By this same physical mode of action, the thermal motion of the acyl chains above their phase transition temperature, in the fluid state, can be hindered.

It has been demonstrated that 32 mole % cholesterol essentially eliminates the LC–C phase transition in dipalmitoylphosphatidylcholine bilayers detected by differential scanning calorimetry (Chapman, 1976; Ladbrooke et al., 1968). However, laser–Raman spectroscopy shows that the phase transition is not eliminated but merely extensively broadened (Lippert and Peticolas, 1971). Lesser amounts of cholesterol have been shown to lower the Arrhenius break temperatures for various membrane-bound enzymes (Colbon and Haslam, 1973; deKruyff et al., 1973). For example, incorporation of 9 mole % cholesterol into the membranes of Acholeplasma laidlawii lowered the phase transition of ATPase activity from 18° to 10°C (deKruyff et al., 1973). Erythrocytes and liver microsomal membranes have been shown to exhibit phase transitions in the range of 20°–30°C even though their cholesterol content is ≥32 moles % (Kairns et al., 1974; Gottlieb and Eanes, 1974). These transitions and those of mitochondrial membranes and several others are depicted in Table II (see below). If cholesterol were evenly distributed within the membrane, than one would not expect to observe such phase transitions between 20° and 30°C for membranes with such high quantities of cholesterol

since the buffering effect of cholesterol would be dissipated throughout the entire membrane. For each cholesterol nucleus to be surrounded by phospholipids requires 7 nearest neighbor acyl chains, a
condition extant at a 33 mole % cholesterol level (Engleman and
Rothman, 1972). However, at greater than 33 mole % cholesterol,
cholesterol–cholesterol interactions must occur and hence differential
distribution results. Differential distribution of cholesterol is also considered to result from 33 mole % cholesterol in PC liposomes in which
phases of pure PC and mixed PC/cholesterol (2:1) would develop
(Hinz and Sturtevant, 1972). Others have considered regions of cholesterol and PC to phase-separate when the ratio of lipid/cholesterol is
1:1 (Phillips and Finer, 1974). Last, both transmission and freeze-
fracture electron microscopy have demonstrated nonrandom distribu-
tions of cholesterol in artificial bilayer systems below the phase transition temperature of the bilayer lipids (Hui and Parson, 1975; Verkley
et al., 1974). Additionally, in various *in vitro* bilayer systems, cholesterol has been shown to associate preferentially with the lower melting point lipids (Demel and deKruyff, 1976). Consequently, it appears
that cholesterol is probably not distributed equally throughout the
membrane domain and must be segregated within the membrane (see
Aloia and Rouser, 1975b). Recently it has been demonstrated that
cholesterol is distributed predominantly to the outer monolayer in
Mycoplasma membranes (Bittman and Rottem, 1976) and preferentially associates with phosphatidylcholine (deKruyff *et al.*, 1974).

   Cholesterol has also been shown to be excluded from the region of
the annular lipid surrounding the $Ca^{2+} + Mg^{2+}$-ATPase enzyme from
sarcotubular reticulum membranes (Metcalf *et al.*, 1976). Since
cholesterol is considered to have a dual role in membrane function,
namely fluidizing the membrane below its phase transition temperature and solidifying it above its phase transition temperature, the
distribution of cholesterol appears to be quite selective and its membrane buffering capacity restricted to specific regions. In *in vitro*
bilayer systems, perhaps cholesterol is distributed uniformly and
symmetrically between bilayers and directly modifies bulk phase
fluidity properties by creating an intermediate fluid condition (Demel
and deKruyff, 1976). In natural biological membranes the distribution
of cholesterol is such that it does not abolish phase transitions for many
membrane transport processes (see Table II).

   The level of cholesterol in hamster brain microsomes is reduced
from 624 μmole/mg protein in the active state to 550 μmole/mg protein
in the hibernating state (Goldman, 1975). This reduction in membrane
cholesterol is difficult to interpret in terms of membrane function, not

only because of the possible compartmentation of cholesterol within the membrane, but also because of the lack of information about the state of fluidity of brain microsomes in the awake hibernator at normothermic body temperatures. If, for example, hamster brain microsomal membranes are similar to liver mitochondrial and microsomal membranes from rat and rabbit (Hackenbrock, 1976; Martonosi, 1974; Blazyk and Steim, 1972; also see above), and remain in a liquid-crystalline state from 37°C to approximately 0°C, then a reduction in cholesterol content might enhance, or at least not inhibit, the degree of fluidity resulting at temperatures approaching 0°C. If, on the other hand, the phase transition detectable for brain microsomal $Na^+ + K^+$-ATPase in ground squirrel (Charnock, 1978); (see Chapter 1 by Aloia and Pengelley) at 20°C, represents a phase transition of the membrane to the solid state during hibernation, then a decrease in cholesterol would seem to be counterproductive. In this situation, an increase in cholesterol would have induced a state of greater fluidity below the phase transition temperature of the membrane lipids. However, if this phase transition of ground squirrel brain microsomes at 20°C (and that of hamster brain microsomes at 19°C, cited in Chapter 1) is not a LC–C phase transition, but rather some other type of phase transition, perhaps a Critical Viscosity transition (see below) to a more viscous state, but not a solid state, then a reduction in membrane cholesterol may not counter this transition.

On the other hand, the reduction in cholesterol levels during hibernation in the hamster brain microsomes may be related to the maintenance of membrane bilayer stability and not to the maintenance of the appropriate degree of fluidity required for low temperature function. Recent studies have emphasized the fact that certain membrane lipids have a tendency to form unstable, nonbilayer configurations, such as hexagonal, cubic, or inverted micellar phases. Highly unsaturated species of phosphatidylethanolamine tend to form hexagonal (H_{11}) phases within the central plane of the bilayer and destabilize its planar configuration between 0° and 50°C (Cullis and deKruyff, 1976). This tendency appears to be enhanced by the addition of cholesterol, which also favors an hexagonal phase with polar regions directed inward and acyl chains of both cholesterol and PE directed outward to interact with the surrounding bilayer hydrocarbons (see Fig. 6, Cullis and deKruyff, 1978). Phosphatidylcholine, regardless of its fatty acid species, seems to counter the destabilization by inducing a bilayer or intermediate configuration when added to membranes containing PE and cholesterol (Cullis and deKruyff, 1978). Furthermore, differential scanning calorimetry has shown that PE has a lower affinity for cholesterol than does PC (van Dijck *et al.*, 1976), and, membranes containing high quantities of PE usually contain low levels of cholesterol (Demel *et al.*, 1976).

In hamster brain PE is a highly unsaturated species with approximately five times as much total unsaturation as PC. Although the Unsaturation Index (% occurrence × number of double bonds) does not change appreciably to accommodate the hibernating state (see Table II, Chapter 1 by Aloia and Pengelley), there are significant increases in oleic acid (18 : 1) and arachidonic acid (20 : 4). Since the steady state level of PE was found to be increased in hibernating hamster brain microsomes (Goldman, 1975) the concomitant increase in PC and reduction of cholesterol levels may therefore be necessary alterations required to counter the destabilizing tendency of PE and hence maintain an appropriate degree of bilayer domain for proper membrane function.

C. LIPID DOMAINS

1. Arrhenius Plots

One of the most convenient and most frequently used methods of expressing fluidity data is by an Arrhenius plot: log of an activity versus the reciprocal of the absolute temperature. When membrane enzymes were studied as a function of temperature in the late 1960's and early 1970's, Arrhenius breaks were thought to be induced by phase changes in membrane lipids from an L-β-liquid crystalline state to an L-α-crystalline state as temperature was reduced (Lyons, 1972). Perhaps, however, this concept may have to be elaborated since extensive membrane protein–lipid interactions have been demonstrated. Chapter 1 by Aloia and Pengelley cited several examples of protein-induced phase transitions (e.g., Ca^{2+} + Mg^{2+}-ATPase: Inesi *et al.*, 1973), and similar phase transitions have been shown to result from conformational changes in soluble enzymes such as pyruvate kinase (Behrisch, 1978). Furthermore, many membrane-bound enzymes, such as Ca^{2+} + Mg^{2+}-ATPase and Na^+ + Mg^+-ATPase are now known to exhibit sigmoidal kinetics and are influenced by various allosteric effectors (Farias *et al.*, 1975). Membrane fluidity effects on such allosteric proteins are perhaps better plotted as Hill plots which measure the degree of cooperativity in the enzyme reaction. Changes in interaction energies of about 3 kcal/mol are required to shift the temperature of an Arrhenius discontinuity while only 0.7–0.8 kcal/mol are needed to change cooperative binding activities of enzyme molecules

(Farias *et al.*, 1975). Nevertheless, to date significant results have been obtained in studies of hibernating animals by the use of Arrhenius plots (Charnock, 1978; Goldman and Albers, 1975; Lyons, 1972; Raison, 1972).

The Arrhenius discontinuity is considered to be the point at which two mutually exclusive enzyme reactions, catalyzed by different conformers of the same enzyme, are in equilibrium (Kumamoto *et al.*, 1971). Below the break temperature enzyme activity is usually reduced and the apparent activation energy (E_a) for the system is increased. Such a reaction with an increased E_a is theoretically energically less feasible. However, E_a does not seem to be a reliable guide to the probability that a particular enzyme reaction will ensue since the real energy barrier is the magnitude of the free energy change, $\Delta G\ddagger$. However, $\Delta G\ddagger$ is difficult to ascertain since the number of enzyme molecules involved in the reaction must be known, and this is usually not a feasible task in most biological reactions. E_a should be used as a valid criterion for the ability of an enzyme to reduce the energy barrier to a reaction only if the activation entropy $\Delta S\ddagger$ remains constant (Low *et al.*, 1973; Somero, 1972). Lactate dehydrogenase (LDH), for example, was found to have large differences in E_a for ectothermic and homeothermic enzyme systems when assayed at 5° and 35°C. However, the magnitude of these differences was drastically reduced by differences in $\Delta S\ddagger$. Consequently $\Delta G\ddagger$ for the LDH reaction of the poikilothermic and homeothermic enzymes was similar (Low *et al.*, 1973). Simply because E_a above and below the phase transition of homologous reactions in hibernators and nonhibernating mammals is similar, does not necessarily imply that both reactions are energetically as feasible, since a greater reliance on the enthalpy (ΔH) contributions (in homeothermic enzyme systems) or on entropy ($\Delta S\ddagger$) contributions (as in ectothermic systems) can change the value of the free energy of the reaction (Low *et al.*, 1973). Conversely, a lower E_a found for a homologous enzyme reaction does not necessarily imply that the reaction is energically more probable.

Furthermore, it has recently been shown that changes in the K_m values with temperature for enzymes such as $Na^+ + K^+$-ATPase, can induce artificial Arrhenius breaks, change the slope above and below the break temperature, and reduce the temperature at which the Arrhenius discontinuity occurs (Silvius *et al.*, 1978). For example, the ratio v (at some fixed substrate concentration)/V_{max} was shown to decrease with increasing temperature. This induced an artificially flattened curve above the Arrhenius break and resulted in a lower E_a

value than actually existed (Silvius *et al.*, 1978). Consequently, E_a values may be even less desirable as a criterion for the energetic feasibility of a reaction if K_m changes are not controlled.

In the recent review paper by Charnock (1978), substrate concentrations and K_m fluctuations seem to be controlled adequately. Nevertheless, Arrhenius discontinuities were observed at 20°C for brain microsomal $Na^+ + K^+$-ATPase from awake and hibernating ground squirrels (*Spermophilus richardsonii*), and approximately 17°C for the same enzyme system from hamster brain (Charnock, 1978; and recalculated by Charnock, 1978, respectively; see Discussion in Chapter 1 by Aloia and Pengelly). Furthermore, the activation energies were similar in preparations from awake and hibernating animals, being about 15 kcal/mole above and 30 kcal/mole below the break temperature (Charnock, 1978). However, the E_a values below the break were slightly lower than those for bovine brain, being 32 versus 37 kcal/mole, respectively. The important point to illustrate from these examples is that caution must be exercised in interpreting the energetic feasibility of the reaction of enzyme preparations from hibernating versus active hibernator and hibernating versus nonhibernating mammal. This method of calculating the thermodynamic parameters of reactions, namely, comparing E_a values, is useful but has its limitations. Simply because E_a values above and below an Arrhenius break temperature are similar in hibernator and nonhibernator does not imply that both reactions are energetically as feasible, since it is obvious that enzymes from hibernators, such as brain microsomal $Na^+ + K^+$-ATPase, function well at 5°C and those from awake hibernators are more inhibited (Goldman and Albers, 1975).

2. Phase Transitions

Table II presents the reported phase transitions from several thoroughly studied membrane preparations; mitochondria, microsomes, sarcoplasmic reticulum, and erythrocytes. The majority of these transitions are between 20° and 30°C. In the original publications, they were interpreted as representing LC–C or fluid-to-solid phase transitions of the membrane lipids. However, as pointed out earlier (Aloia and Rouser, 1975b), they may well represent a different form of phase transition to a more viscous state. For example, *Acholeplasma laidlawii* membranes containing approximately 65% oleic acid (McElhaney, 1974) have a LC–C phase transition temperature around $-15°$ to $-20°C$, when measured by differential scanning calorimetry (DSC) or X-ray diffraction (Steim *et al.*, 1969; Engleman, 1971; respec-

TABLE II
ARRHENIUS BREAK TEMPERATURES

System	Temperature (°C) and animal[a,b]	
Mitochondria		
Respiration	23°(rb)[a,19]	22°(r)[23]
Substrate: succinate, β-OH-butyrate, α-keto-	24°(r)[23,29]	23°(gs/r)[1,23]
glutarate	27°(b)[24]	24°, 8°(r)[10]
	29°, 16°(r)[10]	25°(r)[25]
	18.5°(b)[24]	17°(r)[40]
Enzymes		
Succ.-oxidase, dehydrogenase, cytochrome *c*	16–17°(y)[33]	20°(p)[20]
oxidoreductase	20°(b)[24]	28°(r)[29]
NADH-oxidase, dehydrogenase	16–17°(y)[33]	
cytochrome *c* oxidoreductase		
ATPase and other enzymes	18°(b)[24]	25°, 16°(r)[29]
	20°(y)[32,33]	18°(b)[24]
Other		
Ca^{2+} uptake, respiration and ATP stimulated	27°, 13°(r)[29]	27°, 19°(r)[29]
K$^+$ uptake, valinomycin-induced	29°, 12°(r)[29]	
High amplitude swelling	27°, 13°(r)[29]	
Membrane viscosity	31°, 22°(r)[2]	
[^{14}C]-leucine incorporation (liver)	23°(r)[21]	
Spin-labeled mitochondrial membranes	24°(r)[26]	29°, 17°(s)[10]
	22°(r)[19]	24°, 8°(r)[10]
	23°(r)[19]	
Microsomes		
Na$^+$ K$^+$-ATPase	18°(r)[6,8]	19°(b)[7]
	17°(b)[9]	22°(rb)[22]
		17°(rb)[41]
	20°(rb)[38]	20°(s)[40]
	42°, 20°(f)[42]	20°(r)[45]
	18°(b)[46]	22(s)[47]
UDP-glucuronyltransferase	19°, 32°(gp)[5]	
Glucose-6-phosphatase	17°(t)[34]	
Spin-labeled microsomes	17°(t)[34]	20°(s)[40]
	23°(r)[28]	33°, 11°(s)[10]
[^{14}C]-leucine incorporation	21°(r)[28]	
8-ANS (8-anilino-1-naphthalene sulfonate)		
fluorescence	17°(t)[34]	
Cytochrome P_{450} reductase	30°(rb)[36]	
Liver monooxygenase	24°(r)[48]	
Sarcoplasmic reticulum (SR)		
Ca^{2+}-ATPase activity	29°(rb)[11]	25°(rb)[11]
	20°, 35°(rb)[4]	17°(rb)[35]
	11.5°(1)[35]	
Ca^{2+} uptake	19°, 42°(rb)[4]	
Spin-labeled SR membranes	25°(rb)[11]	22°(rb)[19]
	20°(r)[27]	

(Continued)

TABLE II (*Continued*)

System	Temperature (°C) and animal[a,b]
Other membrane systems	
Red blood cells	
Membrane viscosity	19°(h)[2]
Na^{++} K$^+$-ATPase activity	15–20°(h)[30]
Glucose transport	20°(h)[49]
Acetylcholine esterase	24°(h)[37]
Fibroblasts	
α-Aminoisobutyrate transport	37°, 29°, 20°, 16°[13]
Plasma membrane Na^{++} K$^+$-ATPase	37°, 30°, 23°, 15°[13]
Spin-labeled plasma membrane	37°, 30°, 21°, 15°[13]
Spin-labeled endoplasmic reticulum	38°, 32°, 22°, 16°[13]
Spin-labeled New Castle disease virus	38°, 32°, 22°, 14°[32]
Cessation of growth	23°[13]
[^{14}C]-leucine incorporation	20°[43]
Miscellaneous	
Antigen mixing, mouse–human heterokaryon	21°, 15°[12]
Liver plasma membrane, hormone-stimulated adenylate cyclase	32°(r)[16]
Con-A binding by and agglutination by Py 3T3 cells	15°[14]
Synaptic membranes (spin labeled)	
With Ca^{2+}	20°[44]
Without Ca^{2+}	15°[44]

[a] rb, Rabbit; r, rat; gs, ground squirrel; b, bovine, y, yeast; p, porcine, s, sheep; c, canine; t, *Tetrahymena;* gp, guinea pig; 1, lobster; h, human; f, fish.

[b] Key to references:

1. Brinkman and Parker, 1970
2. Zimmer and Schirmer, 1974
3. Lacko *et al.*, 1973
4. Inesi *et al.*, 1973
5. Eletr *et al.*, 1973
6. Charnock *et al.*, 1973
7. Taniguchi and Iida, 1972
8. Charnock *et al.*, 1971
9. Tanaka and Teruya, 1973
10. Raison and McMurchie, 1974
11. Lee *et al.*, 1974
12. Petit and Edidin, 1974
13. Wisnieski *et al.*, 1974a
14. Noonan and Burger, 1973a
15. Noonan and Burger, 1973b
16. Kairns *et al.*, 1974
17. Rittenhouse and Fox, 1974
18. Rittenhouse *et al.*, 1974
19. McMurchie *et al.*, 1973
20. Zeylemaker *et al.*, 1972
21. Towers *et al.*, 1973
22. Raison *et al.*, 1971a
23. Lyons and Raison, 1970
24. Lenaz *et al.*, 1972
25. Raison *et al.*, 1971b
26. Williams *et al.*, 1972
27. Eletr and Inesi, 1972
28. Towers *et al.*, 1972
29. Lee and Gear, 1974
30. Wood and Beutler, 1967
31. Bertoli *et al.*, 1973
32. Wisnieski *et al.*, 1974b
33. Watson *et al.*, 1975
34. Wunderlich *et al.*, 1975
35. Madeira *et al.*, 1975
36. Stier and Sackman, 1973
37. Aloni and Livne, 1974
38. Kimelberg and Papahadjopoulos, 1974
39. Kemp *et al.*, 1969
40. Grisham and Barnett, 1973

tively), yet, spin labels detect a phase transition around +20°C (Rottem *et al.*, 1970). Similarly, *Escherichia coli* membranes containing about 55% oleic acid (Overath and Trauble, 1973) exhibit no DSC-detectable phase transition above 0°C, although dichloroindophenol reductase undergoes a phase transition at 18°C (Heast *et al.*, 1972). Last, when the spin label TEMPO (2,2,6,6-tetramethylpiperidine-1-oxyl) was used to monitor the fluidity of dioleoylphosphatidylcholine (DOPC), an Arrhenius discontinuity was observed at 30°C which is approximately 50°C above the LC–C phase transition temperature for this lipid (Lee *et al.*, 1974). More importantly, however, CA^{2+} + Mg^{2+}-ATPase from rabbit muscle exhibits an Arrhenius break between 25° and 29°C, whether combined with its native lipid or delipidated with detergent and reconstituted with DOPC (Lee *et al.*, 1974). Consequently, these phase transitions between 25° and 30°C must result from some other form besides a transition of the membrane lipids to the solid state. They must result from a phase shift to a slightly more viscous state in this temperature range.

This phase transition was thought to result from the formation of lipid clusters within the membrane (Lee *et al.*, 1974). We have previously coined the term, "critical viscosity" or "critical volume" transitions since there is an increase in viscosity and a decrease in volume at the transition (Aloia and Rouser, 1975a,b). Since these transitions were frequently observed with phospholipids containing one or more olefinic groups we considered them to form by the interaction of double bonds and to cause the lipid molecules to move together is a more coordinated fashion. Since acyl chain motion is not stopped, the transition is not a fluid-to-solid or LC–C phase transition (Aloia and Rouser, 1975b). However, such critical viscosity transitions could also result from coupled, coherent fluctuation of several kink isomers (2 gl kink) from adjacent hydrocarbon chains moving together in a coordinated fashion as a kink block (Trauble, 1972). Normally kink isomers (±120° rotation by adjacent C-C atoms) are thought to undergo rapid displacement along the hydrocarbon chain with a diffusion coefficient of $D = 1 \times 10^{-5}$ cm/sec (Trauble, 1972) and thus create defects and increase the volume in the hydrocarbon phase. The formation of kink blocks between several adjacent hydrocarbon chains would decrease

41. Charnock *et al.*, 1975
42. Kohonen *et al.*, 1973
43. Craig and Fahrman, 1977
44. Viret and Leterrier, 1976
45. Gruener and Avi-Dor, 1966

46. Charnock and Simonson, 1977
47. Charnock and Bashford, 1975
48. Yang *et al.*, 1977
49. Lacko and White, 1977

the effective free volume of the system and increasé the apparent viscosity in a specific microlocus within the membrane. The important point to emphasize, however, is that these critical viscosity-type transitions are apparently localized to specific regions or compartments within the membrane. This appears to be the case since in the above cited membranes, DSC, which detects the transition from the L-β-crystalline phase to the L-α-liquid crystalline phase, does not record a phase transition until well below 0°C, while other phase transitions are detected for specific enzyme systems and spin probe perturbations which are well above 0°C. Furthermore, the extensive list of phase transitions presented in Table II are mostly between 20° and 30°C, yet many membranes do not exhibit a LC–C phase transition until 0°C or below (Hackenbrock, 1976; Martonosi, 1974; Blazyk and Steim, 1972). It has recently been demonstrated by enzyme analysis, freeze-fracture electron microscopy, and electron spin resonance, that *Tetrahymena pyriformis* cultured on unsaturated fatty acids exhibits critical viscosity-type phase transitions between 20° and 30°C. However, proton nuclear magnetic resonance could not detect a LC–C phase transition above 0°C (Wunderlich *et al.*, 1975; Wunderlich and Ronai, 1975).

Further evidence for membrane compartmentation is derived from studies of thermotropic lateral phase separations. Temperature-induced phase separations of membrane lipids into regions of similar lipid composition have been amply demonstrated (Kimelberg, 1976; Chapman, 1976; Grant *et al.*, 1974; Shimshick *et al.*, 1973). As temperature of a binary or more complex system of lipids is reduced, a point is reached at which some of the higher melting point lipids undergo a phase transition to the solid or crystalline state. As the temperature is further reduced, small regions of solid phase lipid act as nucleation sites for the accretion of newly formed solid lipid and a phase separation of solid lipids from lipids in a liquid crystalline phase ensues in a direction lateral to the plane of the membrane. Isothermal lateral phase separations have also been demonstrated by the addition of various ions and/or proteins to membrane bilayers (Papahadjopoulos, 1977; Papahadjopoulos *et al.*, 1975; Verkleij *et al.*, 1974), thus illustrating that isolated segments can form within a functionally active membrane system at a constant temperature. For example, binary mixtures composed of phosphatidylcholine and phosphatidylserine usually exhibit smooth fracture faces in freeze-fracture micrographs. However, with the addition of Ca^{2+}, patches and the ridges characteristic of pure PC are apparently observed, indicating that separate PC domains coexist laterally with PS–Ca^{2+} domains in the same membrane (Papahadjopoulos, 1977). Monovalent cations such as Na^+ seem to have the opposite effect and have been shown to induce order to fluid or crystalline to liquid crystalline (C–LC) phase transitions in acidic bilayers (Trauble and Eibl, 1974).

D. BOUNDARY LIPID

Similar conclusions can be drawn from the numerous studies of annular or boundary lipid which has been shown to surround many membrane-bound enzymes (Steir and Sackman, 1973; Jost *et al.*, 1973). Ca^{2+}-ATPase from sarcoplasmic reticulum, for example, has been shown to require and tightly bind a shell of approximately thirty phospholipids, and its activity is severely reduced when fewer molecules are present (Metcalfe *et al.*, 1976).

Furthermore, the dissociation between the temperature at which a membrane-bound enzyme and its surrounding lipid undergo a phase transition and that at which the bulk phase lipid undergoes such a transition clearly illustrates the existence of two separate phases. For example, delipidated $Na^+ + K^+$-ATPase which is reconstituted with dipalmitoylphosphatidylglycerol (DPPG) exhibits a phase transition in enzyme activity at 30°C, yet DPPG does not undergo a DSC-detectable phase transition until 41°C (Kimelberg and Papahadjopolous, 1974). This is also an example of a critical viscosity transition; a transition observable at a temperature other than that at which the membrane components enter the solid state.

The concept of membrane compartmentation seems to be amply supported. Interpretation of phospholipid and fatty acid alterations which occur in tissues of hibernators as a function of low temperatures, should therefore be interpreted not simply as alterations which affect the total membrane "fluidity" per se. Rather, the specific distribution of the lipids within the membrane must be evaluated in order to determine possible functions of lipid alterations.

V. Lipid Alterations and Low Temperature Membrane Function

Last, it seems that if the cell membranes of hibernators are not capable of continued activity at temperatures approaching 0°C with the complement of membrane lipids present in their normothermic state, and if the lipid changes observed are not required for enzyme activation, or are not simply the result of nonspecific low temperature modification of biosynthetic pathways, then the lipid alterations may be required primarily to offset a phase shift of the bulk phase lipids to the solid or to a more viscous state (alternative 3 above). The lipid modifications observed in hibernating species were discussed in Chapter 1 by Aloia and Pengelley in regard to a bulk phase lipid fluidization and its requirement to obviate a LC–C phase transition. Furthermore, it is entirely possible that membrane lipid alterations observed in the tissues of hibernating mammals may be required to induce or eliminate a critical viscosity phase transition. If a real advantage is achieved by the organism by the presence or absence of critical viscosity or LC–C phase transitions, membrane phospholipid content could be modified in the appropriate direction. At the point of a critical viscosity transition, or in other words, at the point of a decrease in membrane volume due to the sequestration of a small number of phospholipids from the fluid phase domain into a reduced volume, solid phase, or more viscous phase, a reduction in internal membrane pressure results (Trudell, 1977; J. R. Trudell, personal communication). Consequently, the membrane experiences an increase in lateral compressibility which could result in an enhanced rate of protein conformational change. This may allow a significant change in the transport of those molecules dependent upon the conformational changes of a transport protein. An example of this phenomenon is illustrated by the transport of glycoside by $E.$ $coli$ cultured on elaidic acid, $18:1_t$. At the upper phase transition temperature, T_h, at which point some of the membrane lipids undergo a phase transition, transport of the β-glucoside doubles (Fox, 1975). In other words, at the temperature of a phase transition in which some of the membrane lipids become more viscous or solid, a reduction in volume and increased membrane transport occurs. Phase transitions detected in the cells of hibernating animals may therefore be critical viscosity transitions which are required to reduce internal membrane pressure and thus allow greater enzyme activity at low temperatures.

Membrane phospholipids and/or the fatty acids of each phospholipid class could be altered to achieve the appropriate membrane physical state. These shifts in lipid and fatty acid composition could represent synergistic low temperature adaptive mechanisms or they could result from a sequence of mutually dependent metabolic events. In $Tetrahymena$, for example, a temperature shift from 28° to 10°C results in a cessation of growth and a clustering of the intramembraneous particles as observed by freeze-fracture electron microscopy (Wunderlich et $al.$, 1973). After several hours at 10°C, cellular growth was shown to resume and the intramembraneous particles were shown to be redispersed to their normal loci. Analysis of membrane lipids at this time revealed an increase in unsaturated fatty acids (Wunderlich et $al.$, 1973). Recent experiments with $Tetrahymena$ involve culturing the protozoon on lipids of various degrees of saturation and/or at 39° and 15°C prior to a shift-up (to 39°C) or a shift-down (to

15°C) in growth temperature (Martin *et al.*, 1976). These experiments result in cellular membranes whose degree of fluidity is either supraoptimal or suboptimal for the growth temperature to which the organism is shifted. It was shown that the fatty acid desaturase system was only highly active when the degree of membrane fluidity was suboptimal for the new growth temperature. Low temperature per se did not increase the fatty acid desaturase activity since supraoptimal membranes showed a reduction in activity even at low (15°C) temperatures (Kasai *et al.*, 1976).

If this form of metabolic regulation were operative in other organisms, such as hibernating mammals, it would have specific implications regarding levels of unsaturated fatty acids present in tissues from animals in their hibernating physiological state. For example, it may be that the steady state levels of fatty acids are the result of alterations in the physical state of the membrane brought about by different ratios of particular phospholipid classes. If for example, the membrane physical state is made more ordered by an increase in the phosphatidylethanolamine/choline ratio, activation of fatty acid desaturase may result in greater levels of unsaturated fatty acids. Similarly, increased levels of Ca^{2+} which occurs in the heart and liver during hibernation may cause an increase in order in the lipids which may thus activate the desaturase system.

VI. Conclusion and Future Research

In order to truly assess the significance of the phospholipid alterations observed during the hibernating physiological state and their contribution, if any, to continued low temperature membrane function, several avenues of investigation must be pursued. First, membranes and organelles from summer active ($T_b = 37°C$), winter active and winter aroused ($T_b = 37°C$) and winter hibernating ($T_b = 3°–4°C$) animals must be analyzed by various physical techniques such as differential scanning calorimetry, electron paramagnetic resonance, nuclear magnetic resonance, and others to determine the extent of membrane fluidity in each physiological state. These studies should be coupled with precise, accurate, and reproducible assays of molar quantities of membrane lipids and fatty acids so that the membrane chemical composition can be correlated with the physical state of the membrane detected by these techniques. Perhaps a chronological study throughout the year will be required to correlate altered lipid composition with membrane fluidity and low temperature tolerance. It is only with this type of correlation that the role of membrane lipids in low temperature membrane function and adaptation can be determined.

Second, lipids of the membranes must be modified either by drug, diet, or temperature exposure and the effect of these changes assessed in terms of the capacity of the animal to hibernate. A temperature shift has been shown to alter the membrane lipid composition of hamsters. Chronic exposure to 35°C has been shown to result in changes in both phospholipids and fatty acids of erythrocyte membranes (Kuiper et al., 1971). In a preliminary study, five squirrels were maintained for 10 weeks on a beef tallow-supplemented fat-free diet and five siblings (matched for litter mate and sex) were maintained on a soy oil-supplemented fat-free diet. Lung, heart, and erythrocyte phospholipids and erythrocyte fatty acids were analyzed. Phosphatidylethanolamine was found to be elevated by 3% in lung and sphingomyelin by 8% in erythrocytes of the soy oil-fed animals. Linolenic acid (18:2) was found to be three times higher in erythrocytes from the soy oil-fed animals than in those from the tallow-fed animals. Furthermore, the soy oil-fed animals were shown to hibernate 35% more days and to spend 65% more time in long bouts of hibernation lasting 5–10 days than the tallow-fed animals (R. C. Aloia, unpublished observations). Longer exposure to such dietary regimens may alter membrane lipid composition to a greater degree than that observed in this preliminary study and more profoundly affect the capacity of the animal to hibernate. A precise analysis of the membrane lipids of these diet-modified and high temperature-exposed animals and subsequent careful evaluation of the animal's hibernating capacity may yield significant information concerning the effect of membrane lipid composition on membrane low temperature tolerance.

Third, these *in vivo* studies must be coupled with *in vitro* studies of purified enzymes reconstituted into bilayers and liposomes of different lipid composition. These studies should include enzymes from awake and hibernating hibernators, as well as homologous enzymes from closely related species. Furthermore, liposomes composed of synthetic phospholipids, the composition of which can be stoichiometrically controlled, and liposomes composed of lipids extracted from hibernators should be employed. It is only by studies such as these that the effect of varying lipid composition on the temperature dependence of various hibernator enzymes can be determined.

ACKNOWLEDGMENTS

This investigation was supported in part by grants from the National Science Foundation (BMS 75-07667), the American Heart Association (75-711), the California Heart Association, Riverside Affiliate, and the Department of Anesthesiology, Loma Linda

University School of Medicine. I would like to thank Dr. George Rouser for advice and consultation and Dr. Bernard Brandstater for providing a conducive research environment. The author gratefully acknowledges the assistance of Ms. Debbie Yetter for typing this manuscript.

REFERENCES

Aloia, R. C. (1978a). *J. Therm. Biol.* (in press).

Aloia, R. C. (1978b). *Lipids* (submitted for publication).

Aloia, R. C. (1978c). *Comp. Biochem. Physiol.* (in press).

Aloia, R. C., and Rouser, G. (1975a). *Intra-Sci. Symp., 1974.*

Aloia, R. C., and Rouser, G. (1975b). *In* "Modification of Lipid Metabolism" (E. G. Perkins and L. A. Witting, eds.), pp. 225–245. Academic Press, New York.

Aloni, B., and Livne, A. (1974). *Biochim. Biophys. Acta* **339**, 359–366.

Behrisch, H. W. (1978). "Strategies in Cold: Natural Torpidity and Thermogenesis." Academic Press, New York (in press).

Bertoli, E., Watson, K., and Griffiths, D. (1973). *Protides Biol. Fluids, Proc. Colloq.* 263–266.

Bishop, D. G. (1971). *In* "Biochemistry and Methodology of Lipids" (A. R. Johnson and J. B. Davenport, eds.), pp. 391–424. Wiley (Interscience), New York.

Bittman, R., and Rottem, S. (1976). *Biochem. Biophys. Res. Commun.* **71**, 318–324.

Blayzk, J. F., and Steim, J. (1972). *Biochim. Biophys. Acta* **266**, 737–741.

Bretscher, M. (1971). *Nature (London), New Biol.* **236**, 11–13.

Bretscher, M. (1972). *Nature (London)*, New Biol. **231**, 229–232.

Brinkman, K., and Packer, L. (1970). *J. Bioenerg.* **1**, 523–526.

Burlington, R. F. (1972). *In* "Hibernation and Hypothermia, Perspectives and Challenges" (F. E. South *et al.*, eds.), pp. 3–16. Am. Elsevier, New York.

Chapman, D. (1976). *In* "Mammalian Cell Membranes" (G. A. Jamieson and D. M. Robinson, eds.), pp. 97–137. Butterworth, London.

Charnock, J. S. (1978). "Strategies in Cold: Natural Torpidity and Thermogenesis." Academic Press, New York.

Charnock, J. S., and Bashford, C. L. (1975). *Mol. Pharmacol.* **11**, 766–774.

Charnock, J. S., and Simonson, L. P. (1977). *Comp. Biochem. Physiol. B* **58**, 381–387.

Charnock, J. S., Cook, D. A., and Casey, R. (1971). *Arch. Biochem. Biophys.* **147**, 323–330.

Charnock, J. J., Cook, D. A., Almeida, A. F., and To, R. (1973). *Arch. Biochem. Biophys.* **159**, 393–399.

Charnock, J. S., Almeida, A. F., and To, R. (1975). *Arch. Biochem. Biophys.* **167**, 480–487.

Colbon, G. S., and Haslam, J. M. (1973). *Biochem. Biophys. Res. Commun.* **52**, 320–326.

Coleman, R. (1973). *Biochim. Biophys. Acta* **300**, 1–30.

Coleman, R., and Bell, R. M. (1978). *J. Cell Biol.* **76**, 245–253.

Craig, N., and Fahrman, C. (1977). *Biochim. Biophys. Act* **474**, 478–490.

Cullis, P. R., and deKruyff, B. (1976). *Biochim. Biophys. Acta* **436**, 523–540.

Cullis, P. R., and deKruyff, B. (1978). *Biochim. Biophys. Acta* **507**, 207–218.

Danielli, J. F., and Harvey, E. M. (1935). *J. Cell. Physiol.* **5**, 483–494.

deKruyff, B., Van Dijck, P. W. M., Goldback, R. W., Demel, R. A., and Van Deenen, L. L. M. (1973). *Biochim. Biophys. Acta* **330**, 269–282.

deKruyff, B., Van Dijck, P. W. M., Demel, R. A., Schuijff, A., Brants, F., and Van Deenen, L. L. M. (1974). *Biochim. Biophys. Acta* **356**, 1–7.

Demel, R. A., and deKruyff, B. (1976). *Biochim. Biophys. Acta* **457**, 109–132.
Demel, R. A., Bruckdorfer, K. R., and Van Deenen, L. L. M. (1972a). *Biochim. Biophys. Acta* **255**, 311–320.
Demel, R. A., Bruckdorfer, K. R., and Van Deenen, L. L. M. (1972b). *Biochim. Biophys. Acta* **255**, 321–330.
Demel, R. A., Jansen, J. W. C. M., Van Dijck, P. W. M., and Van Deenen, L. L. M. (1976). *Biochim. Biophys. Acta* **465**, 1–10.
Duttera, S. M., Byrne, W. L., and Ganoza, M. C. (1968). *J. Biol. Chem.* **243**, 2216–2228.
Eletr, S., and Inesi, G. (1972). *Biochim. Biophys. Acta* **290**, 178–185.
Eletr, S., Zakem, D., and Vassey, D. A. (1973). *J. Mol. Biol.* **78**, 351–362.
Engleman, D. M. (1971). *J. Mol. Biol.* **58**, 153–165.
Engleman, D. M., and Rothman, J. E. (1972). *J. Biol. Chem.* **247**, 3697–3700.
Fang, L. S. T., and Willis, J. S. (1974). *Comp. Biochem. Physiol. A* **48**, 687–698.
Farias, R. N., Bloj, B., Morero, R. D., Sineriz, F., and Trucco, R. E. (1975). *Biochim. Biophys. Acta* **415**, 231–251.
Fourcans, B., and Jain, M. K. (1974). *Adv. Lipid Res.* **12**, 147–226.
Fox, C. F. (1975). *Biochem., Ser. One* **2**, 279–307.
Fricke, H. (1925). *J. Gen. Physiol.* **9**, 137–152.
Goldman, S. S. (1975). *Am. J. Physiol.* **228**, 834–839.
Goldman, S. S., and Albers, R. W. (1975). *Arch. Biochem. Biophys.* **169**, 540–544.
Gorter, E., and Grendel, F. (1925). *J. Exp. Med.* **41**, 439–443.
Gottlieb, M. H., and Eanes, E. D. (1974). *Biochim. Biophys. Acta* **373**, 519–522.
Grant, C. W. M., and McConnell, H. M. (1974). *Biochim. Biophys. Acta* **363**, 151–158.
Grisham, C. M., and Barnett, R. E. (1973). *Biochemistry* **12**, 2635–2637.
Grover, A. K., Slotboom, A. J., DeHaas, G. H., and Hammes, G. G. (1974). *J. Biol. Chem.* **250**, 31–38.
Gruener, N., and Avi-dor, Y. (1966). *Biochem. J.* **100**, 762–767.
Gurr, M. I., and James, A. T. (1975). "Lipid Biochemistry: An Introduction." Chapman & Hall, London.
Hackenbrock, C. R. (1976). *In* "Structural of Biological Membranes" (S. Abrahamsson and I. Pasher, eds.), pp. 199–235. Plenum, New York.
Heast, C. W. M., Verkleij, A. J., deGier, J., Scheik, R., Ververgaert, P. H. J. Thos and Van Deenen, L. L. M. (1972). *Biochim. Biophys. Acta* **288**, 43–54.
Hinz, H. J., and Sturtevant, J. M. (1972). *J. Biol. Chem.* **247**, 3697–3700.
Hsueh, W., Isakson, P. C., and Needleman, P. (1977). *Prostaglandins* **13**, 1073–1091.
Hui, S. W., and Parson, D. F. (1975). *Science* **190**, 383–384.
Inesi, G., Millman, M., and Eletr, S. (1973). *J. Mol. Biol.* **81**, 433–454.
Jain, M. K., and White, H. B., III (1977). *Adv. Lipid Res.* **15**, 1–60.
Jost, P., Griffith, O. H., Capaldi, R. A., and Vanderkooi, G. (1973). *Biochim. Biophys. Acta* **311**, 141–149.
Kahlenberg, A., Walker, C., and Rohrlick, R. (1974). *Can. J. Biochem.* **52**, 803–806.
Kairns, J. J., Kreinen, P. W., and Bitensky, M. W. (1974). *J. Supramol. Struct.* **2**, 368–379.
Kasai, R., Kitajima, Y., Martin, C. E. Nozawa, Y., Skriver, L., and Thompson, G. A., Jr. (1976). *Biochemistry* **15**, 5228–5233.
Kemp, A., Jr., Grout, G. S. P., and Reitsma, H. J. (1969). *Biochim. Biophys. Acta* **180**, 28–34.
Kimelberg, H. K. (1976). *Mol. Cell. Biochem.* **10**, 171–190.
Kimelberg, H. K., and Papahadjopoulos, D. (1974). *J. Biol. Chem.* **249**, 1071–1080.
Kohonen, J., Lagerspetz, K. Y. H., and Tirri, R. (1973). *Acta Physiol. Scand.* **87**, 19a–20a.
Kornberg, R. D., and McConnell, H. M. (1971). *Biochemistry* **10**, 1111–1120.

Kuiper, P. J. C., Livne, A., and Meyerstein, M. (1971). *Biochim. Biophys. Acta* **248**, 300–307.

Kumamoto, J., Raison, J. K., and Lyons, J. M. (1971). *J. Theor. Biol.* **31**, 47–51.

Lacko, L., and White, B. (1977). *Experientia* **33**, 191–192.

Lacko, L., Wittke, B., and Geck, P. (1973). *J. Cell. Physiol.* **82**, 213–218.

Ladbrooke, B. D., Williams, R. M., and Chapman, D. (1968). *Biochim. Biophys. Acta* **150**, 333–340.

Lee, A. G., Birdsall, N. J. M., Metcalfe, J. C., Toon, P. A., and Warren, G. B. (1974). *Biochemistry* **13**, 3699–3705.

Lee, M. P., and Gear, A. R. L. (1974). *J. Biol. Chem.* **249**, 7541–7549.

Lenard, J., and Rothman, J. E. (1976). *Proc. Natl. Acad. Sci. U.S.A.* **73**, 391–395.

Lenaz, G., Sechi, A. M., Parenti-Castelli, G., Lindi, L., and Bertoli, E. (1972). *Biochem. Biophys. Res. Commun.* **49**, 536–543.

Lippert, J. L., and Peticolas, W. L. (1971). *Proc. Natl. Acad. Sci. U.S.A.* **68**, 1572–1576.

Low, P. S., Bada, J. L., and Somero, G. N. (1973). *Proc. Natl. Acad. Sci. U.S.A.* **70**, 430–432.

Lyons, J. M. (1972). *Cryobiology* **9**, 341–350.

Lyons, J. M., and Raison, J. K. (1970). *Comp. Biochem. Physiol.* **37**, 405–411.

McElhaney, R. N. (1974). *J. Mol. Biol.* **84**, 145–15.

McMurchie, E. J., Raison, J. K., and Cairncross, K. D. (1973). *Comp. Biochem. Physiol. B* **44**, 1017–1026.

Madeira, V. M. C., and Antuno-Madeira, M. C. (1975). *Biochem. Biophys. Res. Commun.* **65**, 997–1003.

Martin, C. E., Hiramitsu, K., Kitajima, Y., Nozawa, Y., Skriver, L., and Thompson, G. A., Jr. (1976). *Biochemistry* **15**, 5218–5227.

Martonosi, M. A. (1974). *FEBS Lett.* **47**, 327–329.

Mason, W. T., and Abrahamson, E. W. (1974). *J. Membr. Biol.* **15**, 383–392.

Melchior, D. L., and Steim, J. M. (1976). *Annu. Rev. Biophys. Bioeng.* **5**, 205–238.

Metcalfe, J. C., Bennett, J. P., Hesketh, T. R., Houslay, M. A., Smith, G. A., and Warren, G. B. (1976). *In* "The Structural Basis of Membrane Function" (Y. Hatefi and L. Djavadi-Ohaniance, eds.), pp. 57–67. Academic Press, New York.

Noonan, K. D., and Burger, M. M. (1973a). *J. Biol. Chem.* **248**, 4286–4292.

Noonan, K. D., and Burger, M. M. (1973b). *J. Cell Biol.* **59**, 134–142.

Nordlie, R. C., and Jorgenson, R. A. (1976). *Enzymes Biol. Membr.* **2**, 465–492.

Oldfield, E., and Chapman, D. (1972). *FEBS Lett.* **23**, 285–297.

Overath, P., and Trauble, H. (1973). *Biochemistry* **12**, 2625–2630.

Papahadjopoulos, D. (1977). *J. Colloid Interface Sci.* **58**, 459–470.

Papahadjopoulos, D., Moscarello, M., Eylar, E. H., and Isac, T. (1975). *Biochim. Biophys. Acta* **401**, 317–335.

Petit, V. A., and Edidin, M. (1974). *Science* **184**, 1183–1185.

Phillips, M. C., and Finer, E. G. (1974). *Biochim. Biophys. Acta* **356**, 199–206.

Raison, J. K. (1972). *Bioenergetics* **4**, 285–309.

Raison, J. K., and McMurchie, E. J. (1974). *Biochim. Biophys. Acta* **363**, 135–140.

Raison, J. K., Lyons, J. M., Melhorn, R. J., and Keith, A. D. (1971a). *J. Biol. Chem.* **246**, 4036–4040.

Raison, J. K., Lyons, J. M., and Thomson, W. W. (1971b). *Arch. Biochem. Biophys.* **142**, 83–90.

Renooij, W., Van Golde, L. M. G., Zwaal, R. F. A., Roelofsen, B., and Van Deenen, L. L. M. (1974). *Biochim. Biophys. Acta* **363**, 287–296.

Rittenhouse, H. G., and Fox, C. F. (1974). *Biochem. Biophys. Res. Commun.* **57**, 323–330.

Rittenhouse, H. G., Williams, R. E., Wisnieski, B., and Fox, C. F. (1974). *Biochem. Biophys. Res. Commun.* **58**, 222–228.

Robertson, J. D. (1957). *J. Biophys. Biochem. Cytol.* **3**, 1043–1047.

Rothman, J. E., and Davidowicz, E. A. (1975). *Biochemistry* **14**, 2809–2816.

Rothman, J. E., and Lenard, J. (1977). *Science* **195**, 743–753.

Rothman, J. E., Tsai, D. K., Davidowicz, E. A., and Lenard, J. (1976). *Biochemistry* **15**, 2361–2367.

Rottem, S., Hubbell, W. L., Hayflick, L., and McConnell, H. M. (1970). *Biochim. Biophys. Acta* **219**, 104–110.

Schneider, W. P. (1976). *In* "Prostaglandins: Chemical and Biochemical Aspects" (S. M. M. Karim, ed.), pp. 1–24. Univ. Park Press, Baltimore, Maryland.

Schoene, N. W., and Iacano, J. M. (1976). *Prostaglandins Thromboxane Res.* **2**, 763–766.

Scribney, M., and Kennedy, E. P. (1958). *J. Biol. Chem.* **233**, 1315–1321.

Shimshick, E. J., Kleeman, W., Hubbell, W. L., and McConnell, H. M. (1973). *J. Supramol. Struct.* **1**, 285–294.

Silvius, J. R., Reed, B. D., and McElhaney, R. N. (1978). *Science* **199**, 902–904.

Singer, S. J., and Nicolson, G. L. (1972). *Science* **175**, 720–731.

Singer, S. J. (1977). *J. Supramolec. Struct.* **6**, 313–323.

Somero, G. N. (1972). *In* "Hibernation and Hypothermia, Perspectives and Challenges" (F. E. South *et al.*, eds), pp. 55–80. Am. Elsevier, New York.

Steim, J. M., Tourtellotte, M. E., Reinert, J. C., McElhaney, R. N., and Rader, R. L. (1969). *Proc. Natl. Acad. Sci. U.S.A.* **63**, 104–108.

Stier, A., and Sackman, E. (1973). *Biochim. Biophys. Acta* **311**, 400–408.

Tanaka, R., and Teruya, A. (1973). *Biochim. Biophys. Acta* **323**, 584–591.

Taniguchi, K., and Iida, S. (1972). *Biochim. Biophys. Acta* **247**, 536–541.

Towers, N. R., Raison, J. K., Kellerman, G. M., and Linnane, A. W. (1972). *Biochim. Biophys. Acta* **287**, 301–311.

Towers, N. R., Kellerman, G. M., Raison, J. K., and Linnane, A. W. (1973). *Biochim. Biophys. Acta* **299**, 153–161.

Trauble, H. (1972). *Biomembranes* 197–229.

Trauble, J., and Eibl, H. (1974). *Proc. Natl. Acad. Sci. U.S.A.* **71**, 214–219.

Trudell, J. R. (1977). *Anesthesiology* **46**, 5–10.

Van Dijck, P. W. M., deKruyff, B., Van Deenen, L. L. M., deGier, J., and Demel, R. A. (1976). *Biochim. Biophys. Acta* **455**, 576–587.

Verkleij, A. J., Zwaal, R. F. A., Roelofsen, B., Comfurius, P., Kastelijn, D., and Van Deenen, L. L. M. (1973). *Biochim. Biophys. Acta* **323**, 178–193.

Verkleij, A. J., deKruyff, B., Ververgaert, P. H. J. Th., Tocanne, J. F., and Van Deenen, L. L. M. (1974). *Biochim. Biophys. Acta* **339**, 432–437.

Verkley, A. J., Ververgaert, P. H. J. Th., deKruyff, B., and Van Deenen, L. L. M. (1974). *Biochim. Biophys. Acta* **373**, 495–501.

Viret, J., and Leterrier, F. (1976). *Biochim. Biophys. Acta* **436**, 811–824.

Wallach, D. F. H., Beri, V., Verma, S. P., and Schmidt-Ullrich, R. (1975). *Ann. N.Y. Acad. Sci.* **264**, 142–160.

Watson, K., Bertoli, E., and Griffiths, D. E. (1975). *Biochem. J.* **146**, 401–407.

Williams, M. A., Stancliff, R. C., Packer, L., and Keith, A. D. (1972). *Biochim. Biophys. Acta* **267**, 444–456.

Wisnieski, B., Huang, Y. O., and Fox, C. F. (1974a). *J. Supramol. Struct.* **2**, 593–608.

Wisnieski, B., Parks, J. G., Huang, Y. O., and Fox, C. F. (1974b). *Proc. Natl. Acad. Sci. U.S.A.* **71**, 4381–4385.

Witting, L. A. (1975). *In* "Modification of Lipid Metabolism" (E. G. Perkins and L. A. Witting, eds.), pp. 1–41. Academic Press, New York.

Wood, L., and Beutler, E. (1967). *J. Lab. Clin. Invest.* **70**, 287–294.

Wunderlich, F., and Ronai, A. (1975). *FEBS Lett.* **55**, 237–241.

Wunderlich, F., Speth, V., Batz, W., and Kleinig, K. (1973). *Biochim. Biophys. Acta* **298**, 39–49.

Wunderlich, F., Ronai, A., Speth, V., Seelig, J., and Blume, F. (1975). *Biochemistry* **14**, 3370–3375.

Yang, C. S., Strickhart, F. S., and Kicha, L. P. (1977). *Biochim. Biophys. Acta* **465**, 362–370.

Zeylemaker, W. P., Jansen, H., Verger, C., and Slater, E. C. (1972). *Biochim. Biophys. Acta* **242**, 14–22.

Zimmer, G., and Schirmer, H. (1974). *Biochim. Biophys. Acta* **345**, 314–320.

CHAPTER 3

Mineral Metabolism and Bone

Melvin L. Moss and Letty Moss-Salentijn

I.	Introduction	77
II.	General Theory of Mineral Metabolism	78
III.	Structural Aspects of Mineral Metabolism	80
	A. Extraskeletal	80
	B. Skeleton	81
	C. Bone Matrix	83
IV.	Calcification Processes of Bone	86
	A. Cell	86
	B. Matrix	87
V.	Bone Resorption	88
VI.	Regulation	89
	A. Vitamin D	89
	B. Parathyroid Hormone (PTH)	90
	C. Calcitonin	91
VII.	Mode of Action of Mineral Homeostatic Agents	92
	A. Vitamin D Metabolites	92
	B. Parathyroid Hormone	93
	C. Calcitonin	94
	D. Interaction between Homeostatic Agents	95
	References	98

I. Introduction

Mammalian mineral metabolism is an extraordinarily complex field for a number of pertinent reasons. Obviously, it is the common focus of many diverse disciplines, ranging from ultrastructural morphology, X-ray diffraction, biological and physical chemistry on one hand, to clinical nutrition, endocrinology, dentistry, and orthopedic surgery on the other. Additional complications arise for other reasons. It is now quite clear that many aspects of mineral metabolism exhibit highly significant species differences; hence, it is difficult to easily compare interspecific data. Further, a significant portion of the work on skeletal tissues has utilized the cartilaginous growth plates of long bones as a test site, while significant differences exist between cartilaginous and osseous tissues in mineral ion metabolism. Partly as a result of this heterogenicity and also because of the often highly sophisticated na-

CHEMICAL ZOOLOGY, VOL. XI
Copyright © 1979 by Academic Press, Inc.
All rights of reproduction in any form reserved.
ISBN 0-12-261040-5

ture of the data base and consequent hypotheses of each discipline, it is difficult, if not impossible, for any single worker to encompass the complete field on a level of personal experience and competence.

The field of mineral metabolism is very active, and the extensive literature continues to present oftentimes seemingly conflicting data, with the result that there is little consensus concerning many critical points. Another result of this proliferation of data, especially during the past decade or so, has been the major revision of many of the more general hypotheses about the regulation of the specific processes involved in mineral metabolism.

In view of this situation this chapter can present only a summary report of the "state of the art" concerning calcium and phosphorus metabolism and this summary is subject to subsequent modification of many of the critical details and hypotheses reported. Finally, it is a chapter written by authors whose research endeavors lie in the areas of bone morphology and growth, with the result that we shall attempt to provide a structural basis for the data of our colleagues.

Comprehensive, often encyclopedic, reviews of many aspects of the field of mammalian mineral metabolism are available, but even here, no single publication provides, or could provide, equal coverage of the entire spectrum of scientific and clinical disciplines encompassed within the topic (Comar and Bronner, 1964; Bourne, 1972; Irving, 1973; Little, 1973; Vaughan, 1975; Nordin, 1976).

II. General Theory of Mineral Metabolism

While there is not yet a comprehensive, consensual understanding of the totality of mammalian mineral metabolism, it is possible to suggest a generalized theoretical framework within which pertinent data could be organized.

1. Mineral homeostasis is characteristic of all living forms. Calcium and phosphate ions are required for a broad range of intracellular functions (Rodan, 1973; Robertson, 1976). These extracellular ions may, theoretically, pass through the cellular unit membrane by either active or passive transport processes. In the former case we expect to find appropriate cellular constituent structures to assist inward flux, while in the latter case, "pumping" systems should provide outward flux.

2. Increasing metazoan morphological complexity, of necessity, introduces additional ingestive, conductive, and excretory complexities, perceived in site-specific organo- and cytodifferentiations. These same structural specializations are more closely interrelated by

increasing numbers of ever more sophisticated regulatory systems and processes, thus permitting diversity in the homeostatic levels of mineral ions in the various somatic compartments.

3. In addition to mineral utilization in both general and specific cellular functions, many plant and animal species, both uni- and multicellular, utilize these salts in the construction of a wide variety of extracellular structural elements. In addition to their biomechanical protective or supportive roles these same constructions frequently serve also as homeostatically significant reservoirs of mineral ions (Moss, 1964a,b, 1968a,b; Halstead, 1974; Aaron, 1976).

With respect to biological mineralizations, we call attention to the distinction between three terms. *Mineralization* is the deposition of various inorganic ions (calcium, strontium, magnesium, iron, and silicate) as well as several types of organic materials (oxalates, bilirubinates, cholesterol, and uric acids) in an organic matrix produced by a vital cell. *Calcification* is the deposition of any calcium salt (generally either a carbonate or a phosphate), while *ossification* specifically implies the deposition of calcium salts in a unique vertebrate tissue histologically identified as bone. Accordingly, most studies of mammalian skeletal metabolism must include the latter two terms, while all three types of processes obviously require consideration in any completely comprehensive study of mammals.

4. All biological mineralizations require the participation of a living cell, generally a scleroblast. In mammals the varieties of this cell type include osteoblasts, chondroblasts, odontoblasts, cementoblasts, and ameloblasts, a diverse group developmentally derived from ectoderm, mesoderm, and ectomesenchyme (neural crest).

5. All biological mineralizations consist of an organic and an inorganic phase. The organic phase is formed by the vital activity of the scleroblast. This same cell, it now appears, plays a significant role in the initiation of the mineral component. This last point was not evident from the data available several years ago (Moss, 1964a,b, 1968a,b).

6. The organic phase generally consists of both fibrous and nonfibrous components. The latter usually are viscous, highly hydrated, polyionic colloids. In bone and cartilage this phase consists of collagen, proteoglycans, glycoproteins, and other noncollagenous proteins. It is noted that the specific calcium phosphate salt usually found in bone (hydroxyapatite) has no causally necessary relationship to either the presence of collagen or to vertebrates, since the same salt is found in vertebrate tooth enamel and in several invertebrates (such as the brachiopod, *Lingula*).

7. The gross and microscopic size and shape of the mineralized tissue is determined by its organic matrix. Further, it appears rather conclusive now that the same matrix significantly regulates the nature of the mineral salt deposited on (or in) it.

Mammalian skeletal mineral metabolism, viewed against this background, is but a specific, albeit more complex, example of a generalized event.

III. Structural Aspects of Mineral Metabolism

A. EXTRASKELETAL

Calcium and phosphate are absorbed from the gut primarily in the duodenum, by linked processes of passive diffusion, facilitated diffusion, and by active transport; and the phosphate probably moves with the calcium as an accompanying anion (Avioli, 1972; Omdahl and DeLuca, 1973; Wilkinson, 1976). While some calcium is excreted into the gut (in gastric juice and bile) and a portion is lost by sweat, most is excreted through the urine. Phosphate is excreted principally by the kidney, after about a 90% proximal tubular reabsorption of the amount filtered through the glomerulus (Carr *et al.*, 1973; Bijvoet and van der Sluys Veer, 1972). An excellent review of phosphate homeostasis is given by Fleisch *et al.* (1976).

Specific details of calcium absorption in sheep are given by Lueker and Lofgreen (1961) and by Young *et al.* (1966). Calcium partition in sheep plasma between bound and free forms are reported by Giese and Comar (1964), and the mobilization of both calcium and phosphate from goat blood by Symonds and Treacher (1968), while an older paper provides comparative data on plasma phosphate levels of a number of mammalian species (Kay, 1928).

All cells regulate and utilize calcium and phosphate.

In addition to the function of calcium in the skeletal system, it is well known that this ion is exceedingly important in muscular contraction, axonal conduction, and in synaptic transmission, including neuromuscular transmission. This latter event appears to bear great similarities to cellular extrusion processes in general (Rodan, 1973; Thorn, 1976). It is also suggested that calcium (and magnesium) ions somehow play a role in the initiation of DNA synthesis, eventually leading to cellular mitosis (Perris and Morgan, 1976).

The mitochondria are among the cell organelles involved in the regulation of calcium concentrations in the cytosol. The cellular morphology associated with such homeostasis has been recently reviewed

(Davis *et al.*, 1976). The movement of mineral ions between the bloodstream and the small intestinal lumen is processed through epithelial cells, whose morphology resembles that of the renal proximal convoluted tubule cells responsible for mineral ion reabsorption from the urine. Both types of epithelial cells are responsible for active as well as passive ion transport. They are characterized by prominent brush borders, consisting of microvilli which enlarge their luminal surfaces for maximum uptake and by numerous mitochondria, indicative of the high energy requirements for ion transport in their cytoplasm.

The surface coat of the microvilli of intestinal cells is rich in a specific calcium-binding protein (CaBP), and it is intimately associated with calcium transport across intestinal epithelial plasma membranes (Wasserman and Taylor, 1966, Taylor and Wasserman, 1970).

B. SKELETON

The postnatal mammalian skeleton consists of both intramembranously and endochondrally derived bones. With the exception of calvarial and facial bones, almost all the rest are structural complexes of osseous, cartilaginous, and investing fibroelastic envelopes. While the gross structure of the cartilaginous and osseous tissues variably reflects imposed functional demands (Moss and Salentijn, 1969), these factors are not related significantly to mineral homeostasis. At the histological and ultrastructural levels, however, several pertinent structures are observed.

Mammalian bone consists of a mineralized organic matrix within which osteocytes are enclosed, lying in lacunae. The several lacunae are seen to interconnect via patent canaliculi through which the cytoplasmic processes of the osteocytes extend. This tissue is not a true syncytium; adjacent processes interconnect by either desmosomal contacts or gap junctions (Holtrop and Weinger, 1972). An uncalcified region always exists between the bone matrix and canalicular and lacunar walls and their enclosed processes and cells. Just as all internal bone surfaces are enclosed within a continuous covering of osteocytic membranes, so are all external bone surfaces covered by a similar membrane of osteoblasts. This concept of a bone membrane has many interesting properties (Neuman and Ramp, 1971; Davis *et al.*, 1975; Scarpace and Neuman, 1976a,b), foremost of which is that it permits us to speak of a bone tissue fluid (Owen and Melick, 1973; Vaughan, 1975), which differs from both plasma and extracellular fluid in its composition. Finally, bone is a vascular tissue. No osteocyte is more than 200 μm from an afferent blood vessel.

Osteoblasts, osteocytes, and osteoclasts form a functionally modulating series of bone cells, whose particular structural (Aaron, 1976) and histochemical (Doty and Schofield, 1976) attributes correlate well with their respective functions of forming, maintaining, and resorbing bone.

Active osteoblasts are cuboidal or columnar in shape with a basally placed single nucleus and a well-developed Golgi apparatus, many mitochondria, and an extensive rough endoplasmic reticulum. Cytoplasmic microspherules contain acid protein–polysaccharide complexes and both calcium and phosphate (Kashiwa, 1970; Matthews *et al.*, 1973). More recently, membrane-bound vesicles were described, enzymatically rich, which may well contain calcium–phosphate salts (Anderson, 1973). Alkaline phosphatase reactions are strongly present. These attributes are generally found in the newly enclosed, young osteocyte, while in the mature osteocyte there are significant reductions in these structural attributes of cytoplasm, mitochondria, and Golgi apparatus.

Bone resorption is associated with three cell types: the large, multinucleated osteoclast, a resorbing osteocyte, and a monocyte (possibly a histiocyte). Any bone surface denuded of its membrane is subject to resorption. The osteoclast contacts the bone with a ruffled brush-border membrane near which are numerous mitochondria. Both the endoplasmic reticulum and Golgi apparatus are reduced. Acid phosphatase reactions are present in an acidophilic cytoplasm (see Doty and Schofield, 1972, for further ultrastructural correlates of function). The ability of an osteocyte to initiate resorption of its surrounding bone is undoubted, as evidenced by the normal maturational alterations of osteocytic lacunar shape. The further ability of a mature osteocyte to function resorptively is termed osteolysis (Bélanger *et al.*, 1963), although the reality of this phenomenon is questioned by some (Cameron, 1972). The possibility of monocytes in bone resorption is not generally accepted. However, available data suggest the reality of their role in bone resorption (Hancox, 1972). A recent histochemical survey of the bone cells is available (Doty and Schofield, 1976).

In postcranial adult long bones, skeletal cartilage normally exists only on articular surfaces and in the epiphyseal ends of immature long bones. While an extraordinarily rich literature reports on the structure and processes of mineralization of the maturing zones of growth plate cartilages, there is little reason to believe that this transient tissue plays any significant role in the totality of mineral metabolism. What such studies do, of course, is to provide a model system which, however, suffers from the defect that it is by no means certain that the chemistry of cartilage (and the bovine nasal septal cartilage is most frequently studied biochemically and biophysically) is homologous to that of bone (Vaughan, 1975). There are good reasons to suggest that growth cartilages are not avascular as usually stated, a fact of some significance in studies of their mineral ion transport (Moss-Salentijn, 1976).

C. BONE MATRIX

Bone matrix is composed of a gel of polymeric compounds to which the presence of collagen gives coherence. The obvious differences in physical properties of bone and hyaline cartilage matrix are related directly to their collagen content (Little, 1973); an excellent review is given by Herring (1972).

The role of bone in mineral metabolism is related directly to the structure and composition of its intercellular, ossified matrix. This organic matrix varies considerably in relation to age, the particular bone, and to the specific part of the bone studied (Herring, 1972). Further, very significant species differences are found. In this regard it is noteworthy that the rat, a most commonly studied mammal, has an exceptionally high degree of osseous mineralization [Moss, 1964a; see also Dickerson (1962) and Rowland et al. (1959) for similar reports in other species].

As a very rough approximation, a well-calcified, dry, fat-free, long bone has an ash content of about 65% and about 90% of the organic matter is collagen. The remaining organic matter is about 0.5% lipid and 9.5% "mucosubstances" (Irving, 1973), perhaps better described as noncollagenous organic matter.

Bone collagen differs from other collagens, including those of cartilage, the former consisting of packed tropocollagen macromolecules (300 nm × 15 nm) made up of a triple helix consisting of two alpha 1 and one alpha 2 polypeptide chains (Grant and Prockop, 1972).

There is considerable controversy concerning the three-dimensional geometry of tropocollagen packing in the collagen microfilaments (Veis and Bhatnager, 1970). The principal point of interest here is whether, as suggested by these several models of packing, a space, or "hole" exists in the microfilament capable of serving as a physical site for the initiation of mineralization. One point should be emphasized in all the discussions of the role of collagen in bone mineral metabolism, and that is that collagen per se is not a biologically necessary condition for such mineralization, since similar processes occur in mammalian tooth enamel in the absence of collagen. Nevertheless, a normal bone collagen is prerequisite for bone calcification. While the complete biosynthetic pathway of collagen is not yet fully comprehended, it is reasonably clear that polypeptide assembly occurs on the ribosomes, with subsequent hydroxylation of the lysine and proline taking place in the cisternae of the rough endoplasmic reticulum and procollagen extrusion occurring through the plasma membrane. It is suggested that the principal difference between the osseous and nonosseous collagens may be related to the extent of lysine hydroxylations (Barnes, 1973).

Native collagen, per se, does not exist in vital bone matrix, but rather is found complexed with a variety of noncollagenous, protein–carbohydrate substances. The specific nomenclature of these complexes is not yet settled, although it is generally agreed to classify them as either proteoglycans or as glycoproteins. The former (a high molecular weight protein–polysaccharide molecule) consist of a central protein chain attached to which are a number of polysaccharides formed by long chains of repeating dissacharide units (the glycosaminoglycans); chondroitin sulfate is a typical example of such a polysaccharide. The glycoproteins, on the other hand, are covalently linked protein–polysaccharide complexes, which usually contain a great many types of sugars, arrayed in short oligosaccharide chains which are attached to a noncollagenous protein core.

Additionally, bone matrix contains about 1% of a specific bone sialoprotein (Herring, 1968), whose relatively numerous sialic acid and carboxyl groups are of interest in that they may be responsible for the high metal ion binding ability of this substance, a significant property shared by the skeletal glycoproteins in general (Taylor, 1972). With maturation and increasing mineralization of bone matrix, the concentrations of noncollagenous constituents steadily decrease, suggesting that they play a role in the initiation of mineralization, but that they are relatively unimportant for the subsequent processes by which the skeleton is utilized in somatic mineral ion homeostasis.

Evidence is increasing that some portion or portions of the osseous organic matrix can, under appropriate experimental circumstances, repeatedly demonstrate the property of inducing osteogenesis, and this material is spoken of as a "bone morphogenetic polypeptide" (Urist *et al.*, 1976).

Mineral

The principal mineral form of calcium–phosphate salt found in mature bone is usually termed hydroxyapatite, being closely related to, but not identical with, the mineralogical material of that name $(Ca_{10}(PO_4)_6 \cdot (OH)_2)$. Calcium carbonates are normally present as well as smaller amounts of magnesium, sodium, and fluoride (Eanes and Posner, 1970).

While the Ca/P molar ratio of the mineralogic apatite is 1.67, the biological apatites range in value from 1.37 to 1.71 (Woodard, 1962). It appears that bone apatite is formed as microcrystals providing the possibility of a very high surface area for the bone mineral. Further, each microcrystal is conceptualized as being surrounded by an externally bound hydration shell (Neuman and Neuman, 1958), so that in terms of mineral ion movement, we may think of three crystalline domains, the hydration shell, the crystalline surface unit cells, and the interior unit cells. Ions incapable of entering the apatite crystal lattice, but associated with bone mineral (such as potassium), are presumed to be adsorbed in the surface hydration layer.

There is good reason to believe that biologically active hydroxyapatites are all "calcium-deficient" in formulation and indeed, much of the sequences of bone mineral maturation may be thought of as involving increases in the calcium ion content of the biological apatite, i.e., as increasing mineralogic perfection. This concept is supported strongly by data derived by A. S. Posner and co-workers, showing that probably the first form of biological mineralization in bone is in the form of an "amorphous" calcium phosphate salt. The term "amorphous" is used only to indicate that this salt gives X-ray diffraction evidence of only short-order repeats of unit cell organization, not the relatively long-order repeat characteristic of the microcrystalline apatite (Harper and Posner, 1966; Termine and Posner, 1967). The Ca/PO_4 molar ratio of these amorphous salts ranges from 1.44 to 1.55, when prepared *in vitro*, and they are rapidly converted to apatite, with a higher calcium content, both *in vivo* and *in vitro*. It is currently believed that, *in vivo*, some substantial portion of the bone mineral may persist in the amorphous form, even in otherwise mature bone. Whatever the actual case may be, the principal point is that the calcium–phosphate salts in the mammalian skeleton possess divergent formulations and physical forms.

It is not yet certain that the amorphous calcium phosphate produced synthetically is identical with that same material as it occurs in bone (Blumenthal *et al.*, 1975).

While there is some evidence to support the suggestion that a calcium carbonate salt exists in bone, its significance is still unclear (Irving, 1973), and a similar statement is possible for citrate (Taylor, 1960).

IV. Calcification Processes of Bone

A. CELL

As in so many aspects of skeletal mineral metabolism, much of the data on calcification is derived from studies of epiphyseal cartilaginous growth plates, and to a lesser extent from bone tissue itself, with the presumption that the processes are homologous.

The correlation between chemical and structural attributes of calcification becomes clearer when we consider the relation of hydroxyapatite to the several fluid compartments. Both serum and extracellular fluids are supersaturated with calcium and phosphate ions in relation to the solution of already formed biological and synthetic apatites, but undersaturated with respect to their initial formation. Accordingly, apatite deposition cannot be a simple precipitation, nor the removal of that salt a simple solution.

Considering the initiation of apatite formation, while the complete description of the processes involved, and their regulation, is not totally agreed upon, it is possible to give a reasonable general summary.

All bone tissue is surrounded by a continuous cellular membrane. The osteoblasts (-cytes) forming this membrane unquestionably are the sites of synthesis of the precursor components forming the bone matrix, and following their extrusion through the cellular membrane subsequent maturation of the bone matrix occurs. As noted, many of the protein–polysaccharide molecules avidly bind mineral ions. Further, collagen itself is thought, by some at least, to possess "pores" or "holes" within which the first sites of calcification might conveniently occur. In an older view, collagen, per se, was thought to play a major role in initiation of calcification. This crystalline protein was thought to bind calcium (and perhaps phosphate also) which was then combined with phosphate. This concept presumed that the nucleation of the first unit cells of hydroxyapatite was aided by the specific physicochemical and electric configuration of the "hole" region in the collagen, which provided an appropriate epitactic environment for nucleation (Glimcher and Krane, 1968; Marino and Becker, 1970; Hohling *et al.*, 1971).

A further aspect of this theory required an explanation of why all collagens did not calcify, and it was then proposed that specific crystallization inhibitors (phosphonates) were present, especially pyrophosphates, whose enzymatic removal must precede apatite nucleation (Fleisch, 1964; Fleisch *et al.*, 1966). That phosphonates markedly affect mineral formation in both *in vitro* and experimental situations is now quite clear (Jung *et al.*, 1973) and the use of such diphosphonates in therapy and in studies of calcium metabolism should prove of value. However, their biological role *in vivo* has not been unequivocally established.

B. MATRIX

More recently, the role of unique noncollagenous macromolecules of the bone matrix has been emphasized (Vitter *et al.*, 1972), and there is electron-microscopic evidence to support the relationship between apatite nucleation and these protein–carbohydrate components (Bernard and Pease, 1969). However, it is not clear whether, or how, such components as sialoprotein act to initiate nucleation.

A possible resolution of this problem may be provided by the activity of the osteoblast in calcium and phosphate ion transport to the bone matrix they previously formed; it is suggested further that these same cellular processes act to sufficiently increase the concentration of these mineral ions locally to permit nucleation to occur.

The osteoblast is presented with previously absorbed calcium and phosphate ions by the bloodstream. Both of these ions are passively transported through the cell membrane of the osteoblasts, which also possess an active (pumping) transport mechanism to lower the intracellular concentration of calcium (Rasmussen, 1972). The cytoplasmic mineral ions are also capable of being actively transported to the mitochondria. It is suggested that these cellular organelles are responsible for the concentration of both calcium and phosphate ions; the Ca/PO_4 ratio here is 1.7 (Matthews *et al.*, 1973; Shapiro and Greenspan, 1969, see Halstead, 1974). It is suggested also that in bone cells there may be two types of mitochondria, those associated with mineral ion concentration, showing electron-dense bodies, and others associated primarily with the established role of these organelles in the regeneration of adenosine triphosphate (Matthews *et al.*, 1973).

While these mitochondrial "micro-packets" have not been visualized as such, abundant data do show that many cells associated with mineralization are functionally associated with mineral-rich matrix vesicles. These unit membrane-bound vesicles presumably are produced by a pinching off or by a budding off of peripheral pseudopodial portions of the actively motile osteoblastic surface, and, in fact, have been found in relation to other cells related to mineralization, odontoblasts and chondroblasts (Anderson, 1973; Bernard and Pease, 1969).

The precise nature of the calcium–phosphate salts in these vesicles is as yet unclear; they may be apatite, or amorphous calcium phosphate, or perhaps some other salt. Whatever the case, it is suggestive that these matrix vesicles, themselves, may act as initial sites of apatite nucleation which may occur first intravesicularly or occur after the breakdown of the vesicle membrane in the bone matrix. In the former case, we are left without a role for the matrix itself in the initiation of calcification, while in the latter case, we may view the vesicles as only transport vehicles. In any event, it is also reasonably certain that vesicles are found only when calcification of the bone matrix is initiated. Subsequent mineralization of latter-formed matrix does not appear to utilize this mechanism; here the role of the matrix as a nucleation site seems clearer, while we must conceptualize that the osteoblasts are now actively pumping bone calcium and phosphate out of their cytoplasms.

In a recent summary, Vaughan (1975) has suggested an interesting summarizing hypothesis capable of unifying much of the available data. If the matrix vesicles released amorphous calcium phosphate, the matrix macromolecules, such as sialoprotein, could act as an "ion-buffer," capable of supplying mineral ions to available nucleation sites after appropriate enzymatic activity which would degrade these macromolecules, a concept supported by conformable histochemical and histological data (Hooff, 1964; Williamson and Vaughan, 1967).

V. Bone Resorption

Mineral metabolism requires ionic flux both into and out of the skeletal system. The cell types associated with the removal of bone tissue were noted above. It only remains to note here that whenever the cellular membrane surrounding bone tissue becomes disrupted, that area of "uncovered" bone is subject to volumetric removal. While resorption, per se, requires both demineralization and degradation of the organic matrix, it is possible for mineral ions to be removed from bone without loss of the affected matrix, a process termed diminution (Marshall, 1969).

The skeletal system, at all ages, is constantly adjusting its internal and external configuration to best meet fluctuating functional demands. It does so primarily by using the procedures of deposition and resorption; accordingly, some aspects of observed bone resorption are related to this structural homeostasis and may be independent of mineral homeostasis. On the other hand, it seems clear that some portions of the bone tissues are constantly being remodeled, at all ages. The effect of this is to constantly provide a small fraction of the bone mass (about 4% in an adult) in a condition of less than complete mineralization; presumably this fraction is more readily available for mineral homeostasis.

VI. Regulation

Vaughan (1975) makes a clear and useful distinction between skeletal (structural) and mineral homeostasis. The following discussion deals with the latter, although it is clear that the same processes, of necessity, are involved at the cellular level in each case; in skeletal homeostasis, however, we deal with other types of epigenetic stimuli and an additional order of response.

As presently used, mineral homeostasis concerns the regulation of the constituent ions of the bone salts, in terms of their absorption and excretion, their concentration in the several body fluid compartments, as well as the mineral ionic environment of the cells. This type of homeostasis is skeletally nonspecific. In so far as the skeletal salts are utilized in any way in mineral homeostasis, they are potentially utilized without regard to any particular site, or osseous histology, in any bone.

Skeletal homeostasis, on the other hand, reflects the regulation of the response of the skeleton to a wide variety of stimuli, hormonal and mechanical. This type of homeostasis is bone and bone site-specific, regulating as it does the quality, quantity, rate, and direction of skeletal changes usually described as the processes of growth, development, repair, or adaptation.

Mineral ion regulation is accomplished by a series of complex physiological interactions between three pharmacologically active substances: vitamin D, parathyroid hormone (PTH), and calcitonin. Skeletal homeostasis involves the additional interaction of growth hormone, thyroid hormones, ovarian and adrenocortical steroids as well as other factors (Vaughan, 1975).

It will be useful to discuss the "state of the art" concerning several of these substances prior to reviewing their specific activities.

A. VITAMIN D

It is clear that the active metabolites of vitamin D play a major role in the close regulation of plasma calcium concentrations, interacting in not completely understood ways with PTH in the gut, skeleton, and kidney. These metabolites regulate calcium absorption and bone resorption, and probably play a similar, though less precise, role for phosphate ions.

Cholecalciferol (vitamin D_3) is obtained either from the diet or by the action of sunlight on 7-dehydrocholesterol. In the liver it is hydroxylated to 25-hydroxycholecalciferol (25-OHD_3), following which further hydroxylation takes place in the proximal convoluted tubules of the kidney to either $1,25$-$(OH)_2D_3$ or to $24,25$-$(OH)_2D_3$. The former product seems produced in hypocalcemic or hypophosphatemic states, while the latter product is produced under either normal or "hyper" states of both ions. The $24,25$-$(OH)_2D_3$ is a less active, and not fully understood metabolite (MacIntyre *et al.*, 1976). It may then be further converted to $1,24,25$-$(OH)_3D_3$; and other metabolites of vitamin D are also known. It is emphasized that this is a rapidly expanding and very active research area, and revisions of these statements are to be expected. In the face of a low calcium diet, $1,25$-$(OH)_2D_3$ is the effective metabolite, while $1,24,25$-$(OH)_3D_3$ is effective when calcium levels are normal or high (Holick *et al.*, 1973).

The active metabolites of vitamin D_3 are structurally like other steroid hormones (Norman and Henry, 1974), and they act in similar concentrations. There is reason to believe that the D_3 metabolites act at the transcriptional level in their target cells, producing their effect by acting upon the synthesis of mineral ion carriers and on changes of the cell membranes (Kodicek, 1973).

B. PARATHYROID HORMONE (PTH)

It has long been recognized that the parathyroid gland plays a critical role in mineral metabolism; first with the discovery of PTH, and more recently with the addition of calcitonin to its hormonal repertoire. Earlier clinical observations demonstrated that total parathyroidectomy was followed by tetany and a marked hypocalcemia, while tumors of that gland were associated with marked bone resorption, so characteristic of von Recklinghausen's disease. The further finding that appropriately prepared parathyroid extracts elevated serum calcium levels to normal values in parathyroidectomized animals and that hypercalcemia inhibited, while hypocalcemia stimulated the parathyroids to hormonal production, led to the dual concepts that PTH played a major role in mineral homeostasis, and that a feedback process including calcium ions existed *in vivo* (McLean, 1957).

The structure of PTH has been well studied, and there are species-specific differences in the amino acid composition of this polypeptide (Talmage and Bélanger, 1968). PTH is produced by the chief cells of the gland, and between cords of acini. It is stored as a prohormone which, after release, is cleaved to produce a group of related smaller peptides. It is probable that this cleavage occurs in a number of other viscera (Parsons and Potts, 1972).

It is generally agreed that PTH has two primary target organs, kidney and bone, in each of which a different mineral ion is regulated. In the kidney, PTH significantly decreases the reabsorption of phosphate ions by the distal tubules. In the skeleton, PTH causes a flux of calcium from the bone to the extracellular fluids, and this in both a rapid and a slower manner. In the former, calcium ions (and necessarily phosphate also) are removed from, presumably incompletely calcified, bone and passed through the skeletal membrane cells into the extracellular fluids, a process that is not necessarily correlated with histological evidence of bone resorption. In the slower process such resorption is seen associated with resorptive cells. In both instances, PTH acts to increase the flux of calcium ions into the cytoplasm of both the osteoblasts of the skeletal membrane, as well as into the bone resorptive cells. The possible role of PTH in the renal regulation of calcium ion excretion, on vitamin D_3 metabolite synthesis, and on intestinal flux of calcium remains a matter of active research, but of little consensus at present.

Indeed, much of the experimental and clinical work on PTH is difficult to fit within a common conceptual framework for a variety of reasons, as noted in detail in recent reviews (Irving, 1973; Vaughan, 1975). For example, data on the role of PTH in urinary calcium ion excretion show conflicting species differences, the hormone increases this ion in rat urine (Bacon et al., 1956), and decreases it in mice (Buchanan et al., 1959). Further, the calcium mobilization response of an individual animal to PTH may show temporal variation (Parsons and Potts, 1972), and this response is marked in young rats, while insignificant in adults (Garel, 1969; see Vaughan, 1975); and as is so often true, nonphysiologic concentrations of PTH produce effects other than those found in long-term studies of lower hormone dosage (Vaughan, 1975).

Of parenthetic interest, but significant for certain studies, is the fact that in the goat the thyroid and parathyroid glands have independent blood supplies (Foster et al., 1964).

C. CALCITONIN

The presence of a hypocalcemic hormone derived from the thyroid was shown recently (Copp et al., 1961; Copp and Chaney, 1962). Derived from the C (parafollicular) cells of the thyroid (and possibly from similar cells in the human thymus and parathyroid, Foster et al., 1972), this hormone was earlier termed thyrocalcitonin. Calcitonin is a straight-chained polypeptide, whose precise composition is species-specific (Niall *et al.*, 1969), and there are also species differences in measured effectiveness (Irving, 1973). It is suggested that calcitonin is present in the circulation both in a protein-bound and a free state (Leggate *et al.*, 1969).

The most commonly agreed upon action of calcitonin is its ability to stop resorption of bone, and specifically its inhibition of osteoclastic cell function (Milhaud *et al.*, 1965; MacIntyre *et al.*, 1967); and it is suggested that this inhibition acts on both the removal of the bone mineral and matrix (Anast and Conaway, 1972). Not only existing osteoclasts appear affected, but also the differentiation of such new cells is prevented (Reynolds and Dingle, 1970).

Although it was earlier believed that calcitonin exerted a positive regulatory effect upon bone deposition, this point remains considerably in doubt (Anast and Conaway, 1972). Indeed, the precise physiological role of calcitonin in calcium homeostasis of an intact and normal animal remains unclear at present.

VII. Mode of Action of Mineral Homeostatic Agents

A. Vitamin D Metabolites

1. Intestinal Cells

The flux of calcium ions from the intestinal lumen across the microvillar membrane of the mucosal cells is a process of facilitated absorption (Robertson, 1976), associated with the presence of CaBP in the brush border, and with a possible ATPase–alkaline phosphatase calcium ion pump (Omdahl and DeLuca, 1973). Intracellularly it is reasonable to presume that the calcium ions are bound to proteins (CaBP, Robertson, 1976) and possibly to mitochondria. At the basal membrane of the intestinal calcium absorbing cell, the free calcium ions actively pass this serosal membrane, in exchange for sodium, by a rate-limiting process. There is evidence that $1,25\text{-}(OH)_2D_3$ (in hypocalcemia) and $1,24,25\text{-}(OH)_3D_3$ (in normal or hypercalcemic states) act at the transcriptional level of protein synthesis, effecting eventually the formation of CaBP and of other calcium ion transport proteins and enzymes (Kodicek, 1973; Tsai *et al.*, 1973; Robertson, 1976; Wilkinson, 1976).

2. Kidney

While vitamin D hydroxylation occurs here, there is some doubt about a direct effect of these metabolites at this site. It has been suggested that some slight increase of tubular reabsorption of both calcium and phosphate ions is brought about, but to a much lesser extent, than that occasioned by PTH (Omdahl and DeLuca, 1973).

3. Bone

The primary effect of vitamin D deficiency is the hypocalcification of the matrix of both cartilage and bone, a condition, in children, of rickets. In bone, the rate of osteoblastic matrix production is decreased, and the time required for its subsequent calcification is increased (Baylink et al., 1970). Additionally and importantly, $1,25-(OH)_2D_3$ is an effective agent in promoting bone resorption, far more so than is $1,24,25-(OH)_3D_3$, and this by apparent stimulation of osteoclastic-mediated resorptive processes. It remains disputed whether these metabolites play any role in calcium ion flux into bone.

The precise processes by which vitamin D metabolites act upon bone cells is not clear, although it is presumed that, as in the case of intestinal cells, it is the transcriptional stage of DNA activity that is affected since osseous responses to this vitamin are blocked by actinomycin D, a specific inhibitor of such transcription.

B. PARATHYROID HORMONE

1. Intestinal Cells

As noted, the role of PTH in the regulation of calcium absorption is highly controversial. Irving (1973) concludes that if this hormone does exert such an effect, it is minor compared to its role on bone and kidney, while Vaughan (1975) believes that PTH has no direct action in the gut.

2. Kidney

The characteristic renal response to PTH of decreased phosphate resorption seems to utilize the adenyl cyclase mechanism (Chase and Aurbach, 1968), while the possible role PTH plays in renal calcium reabsorption apparently does not involve this same mechanism (Frick et al., 1965).

3. Bone

As in the kidney, the adenyl cyclase mechanism is involved in PTH action in bone. It is suggested that PTH acts as a "first messenger" to increase cyclic AMP, and that the intracellular level of this substance then regulates calcium mobilization from the bone matrix, first raising the intracellular concentration of calcium in the cells of the bone envelope, and then in the extracellular fluid compartments. In this concept, PTH activates the membrane bound adenyl cyclase, enhancing the passive calcium ion flux into the bone cells. Presumably the membrane becomes more permeable to calcium, and at the same time, the activated AMP also inhibits the active transport processes by means of which intracellular calcium ions are moved to mitochondria (Rasmussen, 1972).

As noted previously, PTH has both an immediate calcium mobilization action and a slower cellularly modulated resorptive effect. The mechanism postulated above presumably accomplishes the former rapid effect.

The slow effect is perceived histochemically by the increase of many lysosomal enzymes in the increased osteoclasts, which may play some role in matrix disaggregation (Vaes, 1968), and biochemically by increases in extracellular compartments of characteristic collagenous amino acids.

It is possible that PTH diminishes osteoblastic metabolism, while the osteocytic osteolysis of Bélanger *et al.* (1963) may similarly reflect a direct effect of PTH on these cells. In the latter case, it may be postulated, but not proved, that metabolic changes occur similar to those observed in osteoclasts.

However this may be, it would seem reasonable to suggest that PTH involves some alteration in the structure of calcified bone matrix and its complex structural relationships with apatite, which permits the slower phase of calcium mobilization to occur (Richelle and Bronner, 1963).

C. CALCITONIN

In contradistinction to what is known concerning the mode of action of vitamin D and of PTH, very little is understood about the processes by which calcitonin exerts its effects. The failure of actinomycin D to inhibit the action of calcitonin suggests that this agent does not utilize control of RNA synthesis (Tashjian, 1965). Although Rasmussen (1972) has presented a model in which calcitonin activates the centripetal calcium pump of bone cells, and also activates some postulated intracellular messenger presumed to aid in the flux of calcium into mitochondria, the factual basis for this is not completely present. Vaughan (1975) suggests acceptance of the view of Robison *et al.* (1971) that calcitonin opposes the action of PTH by mechanisms not yet understood, and probably not involving cyclic AMP.

1. Intestinal Cells

It is not clear whether calcitonin plays any role in mineral ion flux in the intestine. While it has been suggested that a relation exists between this hormone and gastrin (in man), to prevent hypercalcemia after eating, others believe that the hypocalcemic action of calcitonin does not involve the gut (Foster *et al.*, 1972).

2. *Kidney*

Here, again, it is unclear if calcitonin has any direct renal effect (Foster *et al.*, 1972; Nordin, 1976). Phosphaturia results, however, after the injection of calcitonin.

3. *Bone*

The processes by which calcitonin markedly inhibits osteoclastic resorption likewise are uncertain. Reynolds (1972) suggests that inhibition of a collagenase may be involved. However this may be, calcitonin does block the bone resorption following PTH administration *in vitro*.

The possible action of calcitonin to increase bone deposition is suggested. *In vitro*, an increased synthesis of glycosaminoglycans follows calcitonin addition (Martin *et al.*, 1969). However, while calcitonin does seem to produce clinically demonstrable increases in bone formation in certain diseases (Doyle *et al.*, 1974), the effect of this hormone on osteogenesis remains uncertain to others (Anast and Conaway, 1972).

D. INTERACTION BETWEEN HOMEOSTATIC AGENTS

It is evident that mineral homeostasis is the result of a complicated interaction of at least three principal agents, involving at least three different organ systems, whose details are not yet fully understood. Although any synthesis of this topic is, at once, incomplete and probably incorrect in some of its details, it is still helpful to try to present some unifying schematic representation. Figure 1 is intended to give a graphic summary of mineral homeostasis. The growth in complexity of any such schematic presentation is well illustrated by comparison of Figs. (7A) and (7B) of Norman and Henry (1974).

As defined above, none of the biological agents regulative of skeletal homeostasis probably play any direct role in mineral ion homeostasis. Nevertheless, a few comments on some of these agents should be included here for the sake of completeness. The interactions between the several hormones regulative of skeletal homeostasis are still unclear at this time.

1. *Somatotropin*

Growth hormone is secreted from the anterior pituitary in response to growth hormone-releasing factor produced in the hypothalamus.

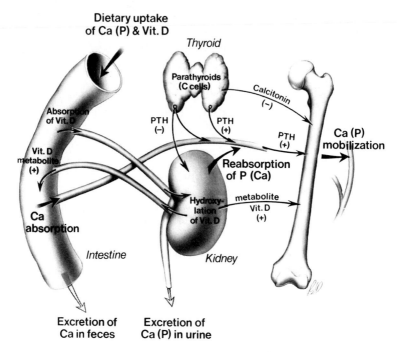

FIG. 1. Diagrammatic representation of mammalian mineral metabolism. The specific details are given in the text; (+) indicates stimulation and (−) indicates inhibition of effect.

The principal effect of growth hormone is on chondrogenesis. To a lesser extent it may affect osteogenesis (Li, 1972; see also Tanner, 1972; Urist, 1972). The hormone acts after the release of another substance, somatomedin (the former sulfation factor), from the liver (and possibly kidney) (Grant, 1972). The site of action of somatomedin is felt to be the protein core of the protein–polysaccharide complexes of the cartilage matrix (Salmon, 1971).

The maturation of all cells of the epiphyseal growth plate are affected by somatomedin, effectively regulating the rate of elongation of long bones (Kember, 1971). In bone it is suggested that the rate of soluble collagen synthesis is stimulated (Aer and Kivirikko, 1969).

It remains unclear whether the skeletal growth-promoting effects of growth hormone result from a direct effect upon cartilage and bone, or whether these responses are but compensatory changes following the primary effect of somatomedin on skeletally related muscles (and viscera).

2. Thyroid Hormones

As growth hormone affects chondrogenesis, so thyroid hormones apparently regulate osteogenesis. Abundant evidence makes it obvious that thyroid hormones are required for normal skeletal growth and development, the hypothyroid dwarf being uniformly retarded, while the somatomedin-deficient dwarf attains relatively mature skeletal proportions. The hyperthyroid child shows acceleration of attainment of skeletal growth and developmental stages. In adults, while hypothyroidism leads to little skeletal change, hyperthyroid states can produce profound changes, with both bone deposition and resorption occurring at increased rates (Adams and Jowsey, 1967). These two authors also suggest that thyroid hormones have a direct, calcium-mobilizing effect on bone cells. Presumably RNA synthesis is the target process.

However, once again, there is evidence that thyroid hormones lack any effect on bone cells *in vitro* (Raisz, 1965), possibly suggesting that the action of these hormones is primarily on the function of related nonskeletal tissues (such as muscles); the effect of these hormones on such tissues is reported by Tata (1966).

Whatever the primary site of action, it is clear that as growth hormone regulates long bone growth plate elongation, thyroid hormones more nearly regulate the maturation of this cartilage, and of its subsequent endochondrally replaced bone. The actions of both hormones are interlocked normally in man so that "the normal proportions of calcified cartilage to primary spongiosa to metaphysis are maintained only when both human growth hormone and thyroid hormones are present" (Vaughan, 1975).

3. Other Hormonal Effects—Osteoporosis

At this place it is appropriate to discuss both ovarian hormones and adrenocortical steroids with specific reference to their ability to somehow be involved in osteoporosis. This condition, in man, is at once a matter of considerable clinical importance and of great confusion, extending to the very definition of the term. The ovarian hormonal skeletal effects are reviewed extensively by Silberberg and Silberberg (1971), and those of the adrenocorticosteroids on bone by Jowsey and Riggs (1970).

It is unclear whether ovarian hormones act directly upon bone cells, and wherever they act, the processes involved at both the molecular and cellular level are uncertain. A similar statement is equally valid, at present, for the corticosteroids (see Jee *et al.*, 1972).

4. Vitamin A

It is long known (Mellanby, 1945) that vitamin A deficiency leads to a severe lack of osseous resorption, while allowing osteogenesis to proceed (Reynolds, 1972), a condition reversed by vitamin A feeding. In cartilage, vitamin A has a papainlike effect of depolymerization of the matrix glycoproteins (Fell and Thomas, 1960). This vitamin apparently acts to release lysosomal enzymes (especially cathepsin D) (Weston *et al.*, 1969), while others believe the active agent is a protease (Woessner, 1973).

In bone tissue, vitamin A, *in vitro*, stimulates osteoclastic proliferation and activity but it is doubtful whether vitamin A has any direct effect upon osteocytes (Reynolds and Dingle, 1970; Reynolds, 1972).

5. Vitamin C

The association of this vitamin with both collagen and glycosaminoglycan synthesis is well known (Kodicek, 1965; Gould, 1970; Reynolds, 1972). The collagen effect of ascorbic acid deficiency is related to the fact that this vitamin is a reducing co-factor in the hydroxylation of both proline and lysine (Barnes, 1973).

REFERENCES

Aaron, J. E. (1976). *In* "Calcium, Phosphate and Magnesium Metabolism" (B. E. C. Nordin, ed.), pp. 298–356. Churchill-Livingstone, London.

Adams, P., and Jowsey, J. (1967). *Endocrinology* **81**, 735–740.

Aer, J., and Kivirikko, K. I. (1969). *Hoppe-Seyler's Z. Physiol. Chem.* **350**, 87–90.

Anast, C. S., and Conaway, H. H. (1972). *Clin. Orthop. Relat. Res.* **84**, 207–262.

Anderson, H. C. (1973). *Hard Tissue Growth, Repair Remineralization, Ciba Found. Symp.* No. 11 (New Ser.), pp. 213–225.

Avioli, L. V. (1972). *Arch. Intern. Med.* **129**, 345–355.

Bacon, J. A., Patrick, H., and Hansard, S. L. (1956). *Proc. Soc. Exp. Biol. Med.* **93**, 349.

Barnes, M. J. (1973). *Hard Tissue Growth, Repair Remineralization, Ciba Found. Symp.* No. 11 (New Ser), pp. 247–258.

Baylink, D., Stauffer, M., Wergedal, J., and Rich, C. (1970). *J. Clin. Invest.* **49**, 1122–1134.

Bélanger, L. F., Robichon, J., Migicovsky, B., Copp, D., and Vincent, J. (1963). *In* "Mechanisms of Hard Tissue Destruction" (R. F. Sognnaes, ed.), Publ. No. 75, pp. 521–576. Am. Assoc. Adv. Sci., Washington, D.C.

Bernard, G. W., and Pease, D. C. (1969). *Am. J. Anat.* **125**, 271–290.

Bijvoet, O. L. M., and van der Sluys Veer, J. (1972). *Clin. Endocrinol. Metab.* **1**, 217–237.

Blumenthal, N. C., Betts, F., and Posner, A. (1975). *Calcif. Tissue Res.* **18**, 81–90.

Bourne, G. H., ed. (1972). "The Biochemistry and Physiology of Bone," 2nd ed. Academic Press, New York.

Buchanan, G. D. *et al.* (1959). *Proc. Soc. Exp. Biol. Med.* **101**, 306–309.

Cameron, D. A. (1972). *Biochem. Physiol. Bone, 2nd Ed.* **1**, 191–236.

Carr, T. E. F., Harrison, G., and Nolan, J. (1973). *Calcif. Tissue Res.* **12**, 217–226.

Chase, L. R., and Aurbach, G. D. (1968). *In* "Parathyroid Hormone and Thyrocalcitonin (Calcitonin)" (R. V. Talmage and L. F. Belanger, eds.), Int. Congr. Ser. No. 159, pp. 247–257. Excerpta Med. Found., Amsterdam.

Comar, C., and Bronner, F., eds. (1964). "Mineral Metabolism," Vol. 2. Academic Press, New York.

Copp, H., and Chaney, B. (1962). *Nature (London)* **193**, 381–382.

Copp, H., Chaney, B., and Davidson, A. (1961). *Proc. Can. Fed. Biol. Sci.* **4**, 17.

Davis, W. L., Matthews, J., Martin, J., and Talmage, R. (1975). *In* "Calcium-regulating Hormones" (R. V. Talmage, M. Owen, and J. A. Parsons, eds.), Int. Congr. Ser. No. 346, pp. 275–284. Excerpta Med. Found., Amsterdam.

Davis, W. L., Matthews, J., Talmage, R., and Martin, J. (1976). *Calcif. Tissue Res.* **21**, Suppl., 59–69.

Dickerson, J. W. T. (1962). *Biochem. J.* **82**, 47–55.

Doty, S. B., and Schofield, B. H. (1972). *Histochem. J.* **4**, 245–258.

Doty, S. B., and Schofield, B. H. (1976). *Prog. Histochem. Cytochem.* **8**, 38.

Doyle, F., Pennock, J., Greenberg, P., Joplin, G., and MacIntyre, I. (1974). *Br. J. Radiol.* **47**, 1–8.

Eanes, E. D., and Posner, A. S. (1970). *In* "Biological Calcification" (H. Schraer, ed.), pp. 1–26. North-Holland Publ., Amsterdam.

Fell, H. B., and Thomas, L. (1960). *J. Exp. Med.* **111**, 719–743.

Fleisch, H. (1964). *Clin. Orthop. Relat. Res.* **32**, 170–180.

Fleisch, H., Schenk, F., Bisaz, R., and All Gower, M. (1966). *Am. J. Physiol.* **211**, 821–825.

Fleisch, H., Bonjour, J. P., and Troehler, V. (1976). *Calcif. Tissue Res.* **21**, Suppl., 327–331.

Foster, G. V., Baghdiantz, A., Kumar, M., Slack, E., Soliman, H., and MacIntyre, I. (1964). *Nature (London)* **202**, 1303–1305.

Foster, G. V. Byfield P., and Gudmundson, T. (1972). *Clin. Endocrinol. Metab.* **1**, 93–124.

Frick, A., Runrich, G., Ullrich, K., and Lassiter, W. (1965). *Pfluegers Arch. Gesamte Physiol. Menschen Tiere* **286**, 109–117.

Garel, J. M. (1969). *C.R. Hebd. Seances Acad. Sci., Ser. D* **268**, 2932–2933.

Giese, W., and Comar, C. L. (1964). *Nature (London)* **202**, 31–33.

Glimcher, M. J., and Krane, S. M. (1968). *In* "Treatise on Collagen" (G. N. Ramachandr, ed.), vol. 2, Part B, pp. 67–251. Academic Press, New York.

Gould, B. S. (1970). *Chem. Mol. Biol. Intercell. Matrix, Adv. Study Inst., 1969* vol. 1, pp. 431–437.

Grant, D. B. (1972). *Clin. Endocrinol. (N.Y.)* **1**, 387–398.

Grant, M. E., and Prockop, D. J. (1972). *N. Engl. J. Med.* **286**, 194–199, 242–249, and 291–300.

Halstead, L. B. (1974). "Vertebrate Hard Tissues." Wykeham Publications (London) Ltd., London.

Hancox, N. M. (1972). "Biology of Bone." Cambridge Univ. Press, London and New York.

Harper, R. A., and Posner, A. S. (1966). *Proc. Soc. Exp. Biol. Med.* **122**, 137–142.

Herring, G. M. (1968). *Clin. Orthop. Relat. Res.* **60**, 261–299.

Herring, G. M. (1972). *Biochem. Physiol. Bone, 2nd Ed.* **1**, 128–190.

Hohling, H. J., Kreilos, R., Neubauer, G., and Boyde, A. (1971). *Z. Zellforsch. Mikrosk. Anat.* **122**, 36–52.

Holick, M. F., Kleiner-Bossaller, A., and Schnoes, H. (1973). *J. Biol. Chem.* **248**, 6691–6696.

Holtrop, M. E., and Weinger, J. M. (1972). *In* "Calcium, Parathyroid Hormone and the Calcitonins" (R. V. Talmage and P. L. Munson, eds.), Int. Congr. Ser. No. 243, pp. 365–374. Excerpta Med. Found., Amsterdam.

Hooff, van den A. (1964). *Acta Anat.* **57**, 16–28.

Irving, J. T. (1973). "Calcium and Phosphorus Metabolism." Academic Press, New York.

Jee, W. S. S., Roberts, W., Park, N., Julian, G., and Kramer, M. (1972). *In* "Calcium, Parathyroid Hormone and the Calcitonins" (R. V. Talmage and P. L. Munson, eds.), Int. Congr. Ser. No. 243, pp. 430–439. Excerpta Med. Found., Amsterdam.

Jowsey, J., and Riggs, B. L. (1970). *Acta Endocrinol. (Copenhagen)* **63**, 21–28.

Jung, A., Bisaz, D. D., and Fleisch, H. (1973). *Calcif. Tissue Res.* **11**, 269–280.

Kashiwa, H. K. (1970). *Clin. Orthop. Relat. Res.* **70**, 200–211.

Kay, H. H. (1928). *J. Physiol. (London)* **65**, 374–380.

Kember, N. F. (1971). *Clin. Orthop. Relat. Res.* **76**, 213–230.

Kodicek, E. (1965). *Struct. Funct. Connect. Skeletal Tissue, Proc. Adv. Study Inst., 1964* pp. 307–319.

Kodicek, E. (1973). *Hard Tissue Growth, Repair Remineralization, Ciba Found. Symp., 1972* No. 11, pp. 359–366. 11(New Ser), pp. 359–366.

Leggate, J., Care, A., and Frazer, S. (1969). *J. Endocrinol.* **43**, 73–81.

Li, C. H. (1972). *Clin. Orthop. Relat. Res.* **89**, 123–128.

Little, K. (1973). "Bone Behavior." Academic Press, New York.

Lueker, C. E., and Lofgreen, G. P. (1961). *J. Nutr.* **74**, 233–238.

MacIntyre, I., Parsons, J., and Robinson, C. (1967). *J. Physiol. (London)* **191**, 393–405.

MacIntyre, I., Colston, K., and Evans, I. (1976). *Calcif. Tissue Res.* **21**, Suppl., 136–141.

McLean, F. C. (1957). *Clin. Orthop.* **9**, 46–60.

Marino, A. A., and Becker, R. O. (1970). *Nature (London)* **226**, 652–653.

Marshall, J. H. (1969). *Miner. Metab.* **3**, 1–122.

Martin, D. L., Melancon, M., and DeLuca, H. (1969). *Biochem. Biophys. Res. Commun.* **35**, 819–823.

Matthews, J. L., Martin, J., Kennedy, J., and Collins, E. (1973). *Hard Tissue Growth, Repair Remineralization, Ciba Found. Symp., 1972* No. 11 (New Ser.), pp. 187–211.

Mellanby, E. (1945). *Proc. R. Soc. London, Ser. B* **132**, 28–46.

Milhaud, G., Moukhtar, M., Bourichow, J., and Perault, A. (1965). *C.R. Hebd. Seances Acad. Sci.* **261**, 4513–4516.

Moss, M. L. (1964a). *Am. J. Phys. Anthropol.* **22**, 155–162.

Moss, M. L. (1964b). *Int. Rev. Gen. Exp. Zool.* **1**, 297–331.

Moss, M. L. (1968a). *In* "Biology of Oral Tissues" (P. Person, ed.), Publ. No. 89, pp. 37–66. Am. Assoc. Adv. Sci., Washington, D.C.

Moss, M. L. (1968b). *Curr. Probl. Lower Vertebrate Phylogeny, Proc. Nobel Symp., 4th, 1968* pp. 359–371.

Moss, M. L., and Salentijn, L. (1969). *Am. J. Orthod.* **56**, 477–490.

Moss-Salentijn, L. (1976). "The Epiphyseal Vascularization of Growth Plates." University of Utrecht. Thesis.

Neuman, W. F., and Neuman, M. W. (1958). "The Chemical Dynamics of Bone Mineral." Univ. of Chicago Press, Chicago, Illinois.

Neuman, W. F., and Ramp, W. K. (1971). *In* "Cellular Mechanisms for Calcium Transfer and Homeostasis" (G. Nichols and R. H. Wasserman, eds.), pp. 197–206. Academic Press, New York.

Niall, H. D., Keutmann, H., Copp, D., and Potts, J. (1969). *Proc. Natl. Acad. Sci. U.S.A.* **64**, 771–778.

Nordin, B. E. C., ed. (1976). "Calcium, Phosphate and Magnesium Metabolism." Churchill-Livingstone, London.

Norman, A. W., and Henry, H. (1974). *Clin. Orthop. Relat. Res.* **98**, 258–287.

Omdahl, J. L., and DeLuca, H. F. (1973). *Phys. Rev.* **53**, 327–372.

Owen, M., and Melick, R. A. (1973). *Hard Tissue Growth, Repair Remineralization, Ciba Found. Symp., 1972* No. 11 (New Ser.), pp. 263–293.

Parsons, J. A., and Potts, J. T., Jr. (1972). *Clin. Endocrinol. Metab.* **1**, 33–78.

Perris, A. D., and Morgan, J. I. (1976). *Calcif. Tissue Res.* **21**, Suppl., 15–20.

Raisz, L. G. (1965). *Proc. Soc. Exp. Biol. Med.* **119**, 614–617.

Rasmussen, H. (1972). *Clin. Endocrinol. Metab.* **1**, 3–20.

Reynolds, J. J. (1972). *Biochem. Physiol. Bone 2nd Ed.* **1**, 69–126.

Reynolds, J. J., and Dingle, J. T. (1970). *Calcif. Tissue Res.* **4**, 339–349.

Richelle, L. J., and Bronner, F. (1963). *Biochem. Pharmacol.* **12**, 647–659.

Robertson, W. G. (1976). *In* "Calcium, Phosphate and Magnesium Metabolism" (B. E. C. Nordin, ed.), pp. 230–256. Churchill-Livingstone, London.

Robison, G. A., Sutherland, E. W., and Butcher, R. W. (1971). "Cyclic AMP," pp. 363–373. Academic Press, New York.

Rodan, G. A. (1973). *In* "Calcium and Phosphorus Metabolism" (J. T. Irving, ed.), pp. 187–206. Academic Press, New York.

Rowland, R. E., Jowsey, J., and Marshall, H. (1959). *Radiol. Res.* **10**, 234–242.

Salmon, W. D. (1971). *In* "Growth Hormone" (A. Pecile and E. E. Müller, eds.), Int. Congr. Ser. No. 236, p. 7. Excerpta Med. Found., Amsterdam.

Scarpace, P. J., and Neuman, W. F. (1976a). *Calcif. Tissue Res.* **20**, 137–149.

Scarpace, P. J., and Neuman, W. F. (1976b). *Calcif. Tissue Res.* **20**, 151–158.

Silberberg, M., and Silberberg, R. (1971). *Biochem. Physiol. Bone, 2nd Ed.* **3**, 401–484.

Symonds, H. W., and Treacher, R. J. (1968). *J. Physiol. (London)* **198**, 193–201.

Talmage, R. V., and Bélanger, L. F., eds. (1968). "Parathyroid Hormone and Thyrocalcitonin (Calcitonin)," Int. Cong. Ser. No. 159. Excerpta Med. Found., Amsterdam.

Tanner, J. M. (1972). *Nature (London)* **237**, 433–439.

Tashjian, A. H. (1965). *Endocrinology* **77**, 375–381.

Tata, J. R. (1966). *Prog. Nucleic Acid Res. Mol. Biol.* **5**, 191–250.

Taylor, A. N., and Wasserman, R. H. (1970). *J. Histochem. Cytochem.* **18**, 107–115.

Taylor, D. M. (1972). *Health Phys.* **22**, 575–581.

Taylor, T. G. (1960). *Biochim. Biophys. Acta* **39**, 148–149.

Termine, J. D., and Posner, A. S. (1967). *Calcif. Tissue Res.* **1**, 8–23.

Thorn, N. A. (1976). *Calcif. Tissue Res.* **21**, Suppl., 11–14.

Tsai, H. C., Midgett, R., and Norman, A. (1973). *Arch. Biochem. Biophys.* **157**, 339–347.

Urist, M. R. (1972). *Biochem. Physiol. Bone, 2nd Ed.* **2**, 155–195.

Urist, M. R., Nogami, H., and Mikulski, A. (1976). *Calcif. Tissue Res.* **21**, Suppl., 81–87.

Vaes, G. (1968). *J. Cell Biol.* **39**, 676–697.

Vaughan, J. M. (1975). "The Physiology of Bone," 2nd ed. Oxford Univ. Press (Clarendon)

Veis, A., and Bhatnagar, R. S. (1970). *Chem. Mol. Biol. Intercell. Matrix, Adv. Study Inst., 1969* Vol. 1, pp. 279–286.

Vitter, F., Pugliarello, M., and Bernard, B. (1972). *Biochem. Biophys. Res. Commun.* **48**, 143–152.

Wasserman, R. H., and Taylor, A. N. (1966). *Science* **152**, 791–793.

Weston, P. D., Barrett, A., and Dingle, J. (1969). *Nature (London)* **222**, 285–287.
Wilkinson, R. (1976). *In* "Calcium, Phosphate and Magnesium Metabolism" (B. E. C. Nordin, ed.), pp. 36–112. Churchill-Livingstone, London.
Williamson, M., and Vaughan, J. (1967). *Nature (London)* **215**, 711–714.
Woessner, J. F. (1973). *Fed. Proc., Fed. Am. Soc. Exp. Biol.* **32**, 1485–1488.
Woodard, H. Q. (1962). *Health Phys.* **8**, 513–517.
Young, V. R., Lofgreen G., and Luick, J. (1966). *Br. J. Nutr.* **20**, 795–805.

CHAPTER 4

Mammalian Endocrines

Bradley T. Scheer

I.	Introduction	103
II.	Pituitary	105
	A. Neurohypophysis	105
	B. Adenohypophysis	111
III.	Adrenal Glands	118
	A. Medulla and Sympathetic Neurons	118
	B. Cortex	120
IV.	Sex Hormones	123
	A. Androgens	123
	B. Ovarian Hormones	125
V.	Thyroid Gland	128
VI.	Parathyroid Glands	138
VII.	Endocrine Pancreas	139
	A. Insulin	139
	B. Glucagon	142
	C. Peripheral Actions	145
VIII.	Pineal Body	148
	A. Vasotocin	148
	B. Serotonin, Melatonin	150
IX.	Kidney	152
	A. Renin, Angiotensin	152
	B. Erythropoietin	154
	References	154

I. Introduction

The most distinctive features of the Mammalia are the details of reproduction and of homeostatic regulation. The patterns of cyclic regulation characteristic of all animals are also present. In all of these, internal secretions play a large part, and mammalian endocrinology has become a major research field in biochemistry and physiology, nearly replacing bacteriology in its implications for physical medicine. In consequence, the recent research literature is too vast for comprehensive review here. Rather, this chapter comprises a critical interpretation of recent work that has come to the attention of the writer, used to illustrate and exemplify the present state of knowledge.

103

The references cited, and the standard textbooks in the field of general and comparative endocrinology (Turner, 1955; Gorbman and Bern, 1962; and more recent editions of these), will provide more complete background information.

Many recent awards of Nobel Prizes in Physiology and Medicine have been made for work in endocrinology, with the usual result of stimulating a burst of research in this field. The 1977 award to Guillemin, Schally, and Yalow recognizes a definite revolution in endocrinology. The classical technique of endocrine research has been the surgical removal of a structure, suspected on grounds of morphology or pathology of having an endocrine function. Inferences concerning possible endocrine function were then tested by replacement of the structure with implants, extracts, or purified components of extracts. Once purified, the effective factor was studied chemically, and structure verified by synthesis. Hormones are active in very low concentrations, and ordinary methods of microanalysis were not suitable for estimating the concentration or even detecting the presence of the hormone in the blood, or demonstrating changes in concentration under physiological conditions. The only recourse was to the time-consuming, expensive, and relatively imprecise methods of biological assay. The technique of radioimmune assay, in the development of which Yalow had a major part, has changed all that. Many of the results reviewed here have been obtained by application of this sensitive and precise technique.

The perfection of techniques for the culture *in vitro* of cells, tissues, and organs has made experiments under controlled conditions possible beyond the range of the classical *in vivo* methods. The present writer has objected (Scheer and Langford, 1976) to the uncritical use of Student's *t* test in the evaluation of experimental results, and has proposed the confidence limit test as being more rigorous. These matters are treated in textbooks of statistical methods. The confidence limit test used in analyzing many of the published results presented in this chapter uses the confidence limit of a mean, which is Student's *t* for a given degrees of freedom and probability, multiplied by the standard error of the mean. The confidence interval for a probability of 1%, used in this chapter, is the range within which 99% of sample means drawn from a population with the mean and standard error of the given sample may be expected to fall. The test applied is that two sample means are considered to represent different populations if their confidence intervals do not overlap. Means with overlapping confidence intervals are considered not to differ significantly.

II. Pituitary

The pituitary body or hypophysis consists of two major regions: the adenohypophysis, comprising the anterior and intermediate lobes of the pituitary, has the characteristics of a gland with no ducts. The neurohypophysis, or posterior lobe of the pituitary, consists essentially of nervous tissue and is continuous through its stalk and the median eminence of the midbrain with the hypothalamus. The hypothalamic nerve centers (nuclei) contain the cell bodies of neurosecretory neurons which send axons through the median eminence into the neurohypophysis. A portal circulatory system exists, connecting capillary beds in the neurohypophysis with capillary beds in the adenohypophysis. The anatomical arrangements of the terminations of the neurosecretory axons in relation to the portal system suggest that the products of neurosecretion are transferred through this system to the adenohypophysis.

A. NEUROHYPOPHYSIS

Some of the axon termini of hypothalamic neurons are located on capillaries which are not part of the portal system, and the hypothesis that the systemic hormones of the posterior lobe, oxytocin and vasopressin in mammals, are discharged into these capillaries is reasonable but has not been tested until recently. Oxytocin has the function of stimulating uterine contractions during parturition. Vasopressin is the antidiuretic hormone of mammals, acting to decrease urine formation by its effect in increasing the permeability of the renal tubules to water. It also, as its name suggests, has the effect of increasing blood pressure. Comparison of the concentration of vasopressin in the portal capillaries of the median eminence with that in the systemic circulation (Zimmerman et al., 1973) shows a ratio of about 300 : 1 in favor of the portal capillaries. This suggests that vasopressin may have a function in the anterior lobe in addition to its systemic antidiuretic and pressor functions.

The posterior lobe hormones have diverse functions. Oxytocin stimulates contractions of the uterus during parturition. Vasopressin has two effects: it increases blood pressure by stimulating tonic contraction of arterial muscles, and it decreases the elimination of urine by its effect on the permeability of renal tubular walls, increasing the reabsorption of water by osmosis from the glomerular filtrate. Vasopressin is thus the antidiuretic hormone (ADH) of mammals. Both hormones are octapeptides.

The arrangements of the portal circulation noted above suggested the possibility of neurosecretory control of the anterior lobe. This suggestion was confirmed by the discovery of "releasing factors" by Guillemin, Schally, and collaborators (Schally *et al.*, 1973, 1977). Several specific factors or hormones have been demonstrated, each of which brings about or inhibits the release of a specific hormone by the anterior lobe (Table I).

1. *Somatostatin (SRIH)*

The inhibitor of release of somatotropin, the growth hormone, was demonstrated by Brazeau *et al.* (1973). The fact that crude extracts of the hypothalamus inhibit secretion of growth hormone had been known since 1968, and these workers in Guillemin's laboratory were able to isolate from the hypothalamus of sheep a cyclic peptide with the amino acid sequence:

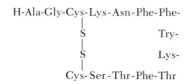

This peptide, designated SRIF, SRIH, or somatostatin, inhibits somatotropin formation by a cell culture from rat pituitaries comparably with crude extract of the hypothalamus (Table II). *In vivo* results were less well-defined because of the high variance, but in one experiment synthetic SRIH action was comparable to that of hypothalamic extract (Table III). The source of SRIH has been established by Pelletier *et al.* (1977) as certain neurons in the periventricular nucleus of the rat hypothalamus.

The term hormone was first used by Bayliss and Starling as a general category into which secretin, the postulated agent formed in the duodenum in response to acid and eliciting secretion of digestive fluid by the pancreas, could be classified. Their pioneering study (Bayliss and Starling, 1902) was made with dogs, but Vega *et al.* (1977) have recently demonstrated that commercial preparations of secretin are effective in stimulating pancreatic secretion in two species of monkeys. Boden *et al.* (1975) review evidence that SRIH inhibits the release of thyroid-stimulating hormone (TSH) from the anterior lobe, of insulin and glucagon from the islet tissue of the pancreas, and of gastrin from the stomach. They find that SRIH inhibits release of secretin, and also inhibits eccrine pancreatic secretion in the absence of secretin.

TABLE I
Neurohypophysial Factors Controlling the Release of Adenohypophysial Hormones[a]

Symbols	Names	Anterior lobe hormone	Target	Product of target
CRF, CRH, ACTHRH	Corticotropin-releasing factor of hormone	ACTH, corticotropin, adrenocorticotropic hormone	Adrenal cortex	Corticosterone and other steroids
GHRF, GHRH, STRF, STRH	Growth hormone-releasing factor or hormone, somatotropin-releasing factor, hormone	Growth hormone, somatotropin	Body cells generally	Growth, anabolism
GHRIF, GHRIH, SRIF, SRIH	Growth hormone release-inhibiting factor or hormone, somatostatin	Growth hormone, somatotropin	Body cells generally	Growth, anabolism
PRF, PRH, LTRF, LTRH	Prolactin- or luteotropin-releasing factor or hormone	Prolactin, luteotropin	Mammary glands, corpus luteum	Development and function, secretion of progesterone
PRIF, PRIH, LTIF, LTIH	Prolactin or luteotropin release-inhibiting factor or hormone	Prolactin, luteotropin	Mammary glands, corpus luteum	Development and function, secretion of progesterone
LHRF, LHRH, FSHRF, FSHRH, LH/FSHRH	Luteinizing hormone-releasing factor or hormone, follicle-stimulating hormone-releasing factor or hormone, or combination	Luteinizing hormone, follicle-stimulating hormone	Ovarian follicles	Development of *corpus luteum*, growth and development of follicles, ovulation
MRH	Melanotropin releasing hormone	Melanotropin from intermediate lobe	Melanophores	Dispersion of pigment
MRIF, MRIH	Melanotropin release-inhibiting factor or hormone	Melanotropin from intermediate lobe	Melanophores	Dispersion of pigment

[a] After Schally *et al.* (1973, 1977).

TABLE II

EFFECT OF NATURAL (OVINE) AND SYNTHETIC SOMATOTROPIN RELEASE-INHIBITING
HORMONE (SRIH) ON PRODUCTION OF SOMATOTROPIN BY CULTURED
PITUITARY CELLS FROM RATS[a]

Medium	Concentration	Somatotropin production (ng/hr)	
		Mean	Confidence interval
Saline control	—	355	245, 465
Extract of hypothalamus	—	51	−111,213
Ovine SRIH	25	53	− 47,100
Synthetic SRIH	5	110	43,178
	25	100	40,161
	330	29	−116,174

[a] Means and 1% confidence intervals, calculated from data of Brazeau *et al.* (1973).
Values are all significantly less than the values for saline controls.

The report noted above, of an effect of SRIH on release of insulin,
was tested by Koerker *et al.* (1974). They administered SRIH to fasted
male baboons and observed a subsequent hypoglycemia. Their curves
suggest that plasma levels of insulin and of glucagon (p. 142) decrease
significantly after administration of SRIH, but numerical data were not
presented. They also tested the effect of SRIH on release of insulin
induced by arginine, but their results (Table IV) show no significant
effect.

The stomach and pancreas of normal rats contain SRIH in concentrations comparable to those in the hypothalamus (Arimura *et al.* 1975).

TABLE III

EFFECTS OF EXTRACTS OF SHEEP BRAIN REGIONS AND OF SYNTHETIC SRIH ON
PLASMA SOMATOTROPIN LEVELS IN RATS[a]

Material applied	Plasma somatotropin	
	Mean	Confidence interval
Experiment a		
Saline control	74.1	56.1, 92.1
Extract of ovine cerebellum	73.8	48.7, 98.9
Extract of ovine hypothalamus	32.1	15.9, 48.3[b]
Experiment b		
Saline control	65.6	38.3, 92.9
Extract of ovine hypothalamus	41.2	24, 58.4[b]
Synthetic SRIH, 0.1	70.8	47.4, 94.2
Synthetic SRIH, 10	32	7.6, 56.4[b]

[a] Means and 1% confidence intervals calculated from data of Brazeau *et al.* (1973).
[b] Less than control value.

TABLE IV

EFFECT OF ARGININE, WITH AND WITHOUT THE SOMATOTROPIN RELEASE-INHIBITING HORMONE (SRIH) ON PLASMA LEVELS OF INSULIN AND GLUCAGON IN BABOONS[a]

Treatment	Insulin (μU/ml)		Glucagon (pg/ml)	
	Mean	Confidence interval	Mean	Confidence interval
Arginine	72	−20, 165	299	−23, 621
Arginine + SRIH	9	− 5, 23	134	−68, 336

[a] Means and 1% confidence intervals calculated from data of Koerker et al. (1974)

This is taken as evidence that this substance or some related one is synthesized in the digestive tract. No further evidence is presented in support of this hypothesis.

The literature showing that "stress" inhibits secretion of growth hormone in rats is reviewed by Terry et al. (1976). They confirmed these results, and also claim that injection of an antiserum against SRIH prevents the effects of stress. Their results, Table V, do not support this claim when evaluated by the confidence interval test.

2. Thyrotropin-Releasing Hormone (TRH)

Immunochemical techniques were used by Winokur and Utiger (1974) and Brownstein et al. (1974) to determine the site of secretion of TRH in the brain of rats. The immune reaction to a specific antiserum was detected in all parts of the brain except the cerebellum. The highest concentration (38.4 ± 48.3 ng/mg, mean ± confidence limits)

TABLE V

CONCENTRATIONS OF SOMATOTROPIN (STH, GROWTH HORMONE, GH) IN THE SERUM OF RATS, IN RELATION TO STRESS (SWIMMING) AND THE EFFECT OF AN ANTISERUM AGAINST SOMATOSTATIN (SRIH)[a]

Treatment	Serum STH (ng/ml), 3.5 hour after treatment	
	Mean	Confidence interval
Unstressed controls	105.6	55.2, 156
Saline-injected controls	62.9	33.9, 91.9
Stressed		
Rabbit serum controls[b]	7.6	2.4, 12.8
SRIH antiserum[b]	25.5	3.3, 47.7

[a] Means and 1% confidence intervals calculated from data of Terry et al. (1976).
[b] Significantly different from unstressed.

was found in the median eminence, with lower (9 ± 19) but not significantly lower concentration in the ventromedial nucleus of the hypothalamus. The concentration in the preoptic nucleus, traditional site of neurosecretion, was 1.09 ± 0.22, not significantly higher than that in the hypothalamus as a whole (1.29 ± 0.02 ng/mg). This pattern suggests to Utiger and associates in these studies that TRH has some function in the ventral midbrain, in addition to its role in stimulating release of thyrotropin from the anterior lobe of the pituitary.

Introduction of TRH into medical practice has revealed a number of side effects, including symptoms in the gastrointestinal tract which the present writer has observed in a case of complete regression of the thyroid treated with thyroid preparations. Smith *et al.* (1977) review the clinical literature, and report experimental studies with rabbits in which they observed increased colonic motility of central origin in response to administration of TRH. It is conceivable to the present writer, as Neary *et al.* (1976) have suggested in a review of the literature, that many of the nervous effects attributed to thyroid hormone may result from changes in blood levels of TRH.

Reports that the blood plasma of adult rats inactivates TRH rapidly have been confirmed by Neary *et al.* (1976). They find that this effect develops only after the fifth day of life. Its adaptive significance is evidently that of confining the effect of TRH primarily to the adenohypophysis in the portal circulation.

The occurrence of a gonadotropin-releasing factor comparable to LH/FSH-RH in the milk of women, cows, and rats has been reported by Baram *et al.* (1977). They suggest that this factor may influence secretion of gonadotropins (LH and FSH) in newborn mammals, but no evidence is offered in support of this suggestion. They also report the presence of a thyrotropin-releasing factor (TSH-RH) in milk.

3. Corticotropin-Releasing Hormone (CRH)

The first releasing factor to be demonstrated was that for corticotropin, the adrenocorticotropic hormone, ACTH, and a recent symposium was devoted to CRH (Sayers, 1977). In rats deprived of the normal source of CRH by lesions in the hypothalamus, Brodish (1977) reports the presence of a substance similar to CRH, presumably produced in other tissues, possibly in response to "stress." The extrahypothalamic CRH has a more prolonged action than does the product of the hypothalamus.

4. Gonadotropin-Releasing Hormone (LH/FSH-RH)

The effects of hormones, especially the polypeptide hormones, on behavior have been reviewed by Strand (1975). The lordosis behavior characteristic of estrus in rodents has been shown by Pfaff (1973) to be potentiated by LH/FSH-RH. More generally, this factor has been shown to induce mating behavior in rats by Moss and McCann (1973), independent of the gonadotropins (LH, FSH) and of TSH, which are released from the anterior pituitary by LH/FSH-RH.

B. ADENOHYPOPHYSIS

The intermediate lobe of the pituitary, a relatively thin layer of secretory tissue between the anterior and posterior lobes, secretes a single hormone, intermedin, the melanophore-stimulating hormone, MSH. This occurs in mammals, as it does in all vertebrates, but its function in mammals is not established. In those vertebrates capable of changes of color (or rather of hue) in the skin, due to the concentration or dispersion of melanin in the pigment cells, chromatophores, MSH acts to disperse the melanin, thus darkening the skin. The melanin pigment of mammalian skin (see Chapter 8 by Goodwin, this volume) is contained in melanophores, but is not capable of concentration or dispersion.

The anterior lobe secretes six hormones, all proteins; for easy reference these are listed in Table VI. In general, the use of two or more names and corresponding acronymic abbreviations for the same hormone reflects the diverse function of the individual hormones. Many of these functions are "tropic," in that the hormone serves to maintain and to stimulate secretion by specific endocrine organs.

1. Growth Hormone (GH) Somatotropin, (STH)

In general, this hormone acts to promote anabolic processes and growth in all cells of the body, acting on specific cells according to their state of development. The demonstration of a double hypothalamic control of secretion of GH through a releasing factor, SRH, and a release-inhibiting factor, SRIH, has stimulated considerable research, results of some of which have been noted above. Any control system of this type may exhibit oscillations under specified conditions (Milsum, 1966). Plasma GH of adult male rats oscillates in concentration with a period of about 1 hour (Martin et al., 1974). This rhythm was suppressed by lesions in the hypothalamus or by administration of somatostatin. These results suggest that the dual control system includes feedback mechanisms. A possible role of the pineal gland, which generally functions in control of circadian (24 hour) rhythms, is suggested by the results of Smythe and Lazarus (1974), was found that oral administration of melatonin, a product of pineal secretion, caused an increase in plasma STH levels of 8 of 9 human subjects.

TABLE VI
HORMONES OF THE ANTERIOR LOBE OF THE PITUITARY IN MAMMALS

Common names	Acronymic symbols	Target organs	Function or effect
Adrenocorticotropic hormone, corticotropin	ACTH	Adrenal cortex	Maintenance, release of steroid hormones
Growth hormone, somatotropin	GH, STH	Body cells generally	Stimulation of anabolism and growth
Thyroid-stimulating hormone, thyrotropin	TSH	Thyroid gland	Maintenance: production and release of iodinated thyronines
Gonadotropins			
Interstitial cell-stimulating hormone, luteinizing hormone	ICSH, LH	Interstitial cells of testes	Stimulation of production and release of male sex hormone (androgen, testosterone)
		Ovarian follicles	Stimulates formation of corpus luteum and secretion of progesterone after ovulation
Luteotropic hormone,	LTH	Corpus luteum of ovary,	Maintenance
Prolactin		Mammary glands	With other hormones, stimulates development and lactogenesis
Follicle-stimulating hormone	FSH	Ovarian follicles	Stimulates development and production of female sex hormone, estrogen
		Spermatic tubules of testes	Stimulates spermatogenesis

The literature showing that many polypeptide hormones occur in multiple forms, all with the same immunological properties, has been reviewed by Gorden *et al.* (1973). Their study of human growth hormone shows two components of different molecular size in the plasma. These two give the same immune reactions, but differ in their affinity for tissue receptors of STH. Studies of the effects of various preparations of purified hormones, by Golde *et al.* (1977), show considerable species specificity. When the preparations were applied to cultures of bone marrow cells from mice, bovine STH was more effective in stimulating growth than were fragments of human STH. In cultures of human bone marrow cells, intact human STH was more effective than fragments, or than bovine STH. These specificities, combined with the danger of sensitization through repeated injections, eliminate the possibility of using STH preparations from food animals in treatment of human deficiency of STH. Sources of human STH are too limited to provide a reliable supply, and, now that the structure is fully known and methods of synthesis of polypeptides are improving, commercial production of synthetic STH becomes a possibility.

The question whether STH has an effect on developmental changes distinct from the effect on growth, has been considered by Croskerry and Smith (1975). They showed that STH prolongs the gestation period of rats, with the result that the offspring are farther advanced toward maturity than usual. The differences are then attributable to the effect of STH on gestation rather than to a developmental effect as such.

2. Corticotropin (ACTH)

The response to the adrenocorticotropic hormone has been studied in cultured cells from the adrenal cortex by Sayers and Beall (1973). When the cells were derived from rats which had been hypophysectomized 14–18 days previously, they required less ACTH to induce synthesis of cyclic 3',5'-adenosine monophosphate (cAMP) than did cells derived from intact rats. They also found that synthesis of the characteristic steroid hormones, elicited by cAMP, was more sensitive to ACTH in cells from hypophysectomized rats. The influence of a variety of agents on rate of synthesis of ACTH by a culture of cells from the anterior pituitary of rats was studied by Vale and Rivier (1977). Their results are partly summarized in Table VII. Increased concentration of K^+ in the medium increased ACTH production significantly. Theophyllin, the alkaloid of tea, inhibits the cyclic nucleotide diphosphoesterase, which hydrolyzes cAMP in cells. This alkaloid, or brominated cAMP, increased the rate of synthesis of ACTH, though the effect of Br^8-cAMP decreased with increasing dose. Cyclic guanosine monophosphate (cGMP) had little effect, but maximal rates of synthesis were obtained in the presence of phorbol myristyl acetate, said to increase synthesis of cGMP in other cells. The time course of appearance of cAMP, cGMP, and the cortical steroids in adrenal cells suggests to Perchellet et al. (1978) that cGMP, rather than cAMP, is the mediator of the effects of ACTH.

3. Prolactin, Luteotropic Hormone (LTH)

The anterior pituitary hormone first named prolactin for its role in the initiation of lactation in mammals, also has the function of main-

TABLE VII

PRODUCTION OF ACTH BY CULTURED RAT PITUITARY CELLS *in Vitro*, AS INFLUENCED
BY VARIOUS AGENTS[a]

Experiment	Agent	Concentration	ACTH production (ng/dish/hr)	
			Mean	Confidence interval
1	None, control		3.9	2.3, 5.5
	K[+]	16× control	7.1	5.5, 8.7
		4× control	7.2	6, 8.4
	Theophyllin[b,c]	10 mM	15.8	13, 18.6
	Br[s]-cAMP	3 mM	17.1	12.7, 21.5
2	Br[s]-cAMP	100 μM	0.9	0.5, 1.3
		10 μM	1.0	0.2, 1.8
		1 nM	1.1	0.3, 1.9
		10 nM	1.2	0.8, 1.6
	None, control		1.2	0.4, 2
	Br[s]-cAMP	100 nM	1.3	0.9, 1.7
		1 mM	1.7	1.3, 2.1
	PMA[b,d]	0.3 nM	2.7	1.9, 3.5
	Br[s]-cAMP	3 mM	3.8	2.6, 5[f]
	Br[s]-cAMP[b]	1 mM	13	11.8, 14.2
	Br[s]-cAMP + Br[s]-cGMP[b]	1 mM ea.	13.2	12, 14.4
	PMA[b,d]	30 nM	22.7	19.9, 25.5
		3 nM	24.4	21.2, 27.6
3	None, control		6.8	3.8, 10.8[g]
	Hypothalamus extract (HE)		7.8	4.5, 11.2
	Aminophyllin, 5 mM, + HE		13.3	10, 16.7
	Aminophyllin, 5 mM		14.5	10.5, 18.5
	Br[s]-cAMP[b] 5 mM		35	29, 42
4	Dexamethasone[e] 20 nM (Dex)		2.2	1, 3.4 ng/dish/hr
	None, control		3.5	1.9, 5.1
	Dex + brain extract		4.2	2.6, 5.8
	Dex + isobutylmethylxanthine		4.3	2.6, 5.8
	Dex + hypothalamus extract		6.1	4.1, 8.1
	Brain extract		8.3	6.3, 10.3[f]
	Hypothalamus extract[b]		12.7	8.7, 16.7
5	Cycloheximide (CH)		1.5	0.3, 2.7
	Dexamethasone (Dex)		1.7	0.5, 2.9
	None, control		2.6	1.4, 3.8
	Dex + HE		3.7	2.1, 5.3
	CH + Dex + HE		6.1	4.1, 8.1[f]
	CH + Dex		8.3	6.3, 10.3
	Hypothalamus extract (HE)		8.5	4.9, 12.1

[a] Means and 1% confidence intervals, calculated from data of Vale and Rivier
(1977), assuming $t = 4$.

[b] Significantly different values.

[c] Inhibitor of cyclic nucleotide phosphodiesterase.

[d] Phorbol myristate acetate, increases level of cGMP in some cells.

[e] Synthetic glucocorticoid.

[f] Greater than control value.

[g] Data in ng/10^5 cells/hr.

taining the corpus luteum in the ovary during pregnancy. The growing importance of mammary carcinoma and the possibility of endocrine therapy have directed attention to the hormone receptors of the cells of the mammary glands. Costlow *et al.* (1974) studied a mammary tumor of rats which has prolactin receptors comparable to those of normal mammary gland. Rat liver also has a receptor for prolactin (Posner *et al.*, 1975). The level of this receptor in females decreases after hypophysectomy and the decrease is reversed by implants of pituitary tissue, parallel to the increase in prolactin in the blood plasma. This suggests that prolactin induces the synthesis of its own receptor. Other evidence of such induction has been reviewed by Moltz and Leidahl (1977). The receptor protein of rabbits has been isolated, and used as an antigen to produce an antiserum. The antiserum blocked the induction, by prolactin, of synthesis of casein in mammary glands of rabbits (Shiu and Friesen, 1976).

The concept of pheromones—chemical factors comparable to hormones in acting specifically in low concentrations, but acting externally, on individuals other than the producer, is applicable to mammals, especially in connection with reproduction. Lactating rats, exhibiting maternal behavior, produce a pheromone which attracts the young. Foster parents, nonlactating females but not males, also emit the pheromone after more than 2 weeks of contact with young. Moltz and Leidahl (1977) have reviewed the literature on this subject, and report that bile from lactating females actively emitting the pheromone induces emission of the substance by males 4 days after it has been injected into them. In females, the evidence is that changes in the liver are induced by prolactin during a period of 16 days. These changes, parallel to the induction of prolactin receptors in the liver, result in appearance of the ability of the bile to induce pheromone production in other individuals, male or female.

The secretion of prolactin and of other reproductive hormones by the anterior pituitary is under feedback control according to Scheme 1.

The possibility that prolactin, release of which is stimulated by a hypothalamic releasing hormone, is part of a regulatory feedback loop of this sort has been tested by Fuxe *et al.* (1977). They demonstated immune reactivity to prolactin in nerve termini in many parts of the hypothalamus. Hypophysectomy, removing the normal source of prolactin, did not alter the distribution or amounts of the reactive material. This suggests that it is produced, independent of the anterior lobe, in the hypothalamus, and stored in the nerve terminals, perhaps pending release. This observation evidently requires confirmation by experimental tests.

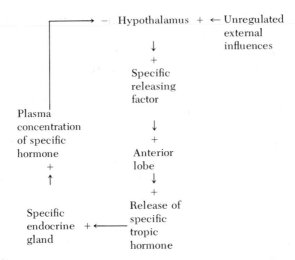

Scheme 1. Secretion of hormone by anterior pituitary. Arrows ending in + indicate a direct relationship, in which a change in one variable is associated with a change in the same direction or sense in the variable to which the arrow points. Arrows ending in — indicate an inverse relationship, in which the two changes occur in opposite senses.

4. Gonadotropins (FSH, ICSH, LH)

Two gonadotropins, FSH, the follicle-stimulating hormone, and ICSH or LH—a single hormone which acts to stimulate the interstitial cells of the testes to produce their hormone and also acts to promote formation of the corpus luteum from the ovarian follicle after ovulation and its secretion of progesterone, are produced in the anterior lobe of both sexes. Their target organs and effects are different as determined by the sex chromosomes of the individual. The evidence for feedback control of the production of gonadotropins is reviewed by Sar and Stumpf (1973). They find that the pituitary cells which secrete the gonadotropins, and the "castration cells" which appear in the pituitary of castrated male rats, concentrate labeled testosterone, the male sex hormone, within one hour after injection.

The rates of secretion of LH and FSH in women with secondary amenorrhea were estimated by Koninckx *et al.* (1976) during infusion of the gonadotropin-releasing hormone (LH/FSH-RH). They observed a biphasic pattern of secretion which suggested that there are two pools of these hormones with different patterns of release. Samples of blood collected continuously during 24 hours from young human subjects were assayed for LH. Pre- and postpuberal young males had a higher concentration of LH/ICSH during the sleep period than during waking hours (daylight). No such pattern of variation was evident in young females (Lee *et al.*, 1976).

The only hormone secreted by the interstitial cells of the testes is an androgen, testosterone. The ovaries, in contrast, secrete two female hormones and small and variable amounts of testosterone. An antiserum specific for testosterone was prepared by Gay and Tomacari (1974) and injected into female rats. There was a consequent decrease in the concentration of testosterone in the serum, and the increase in concentration of FSH which normally occurs every morning during estrus was prevented. This treatment did not alter the increase in FSH and LH which normally precedes estrus, nor did it prevent ovulation. The authors conclude that the cyclic increase in FSH in female rats during estrus is stimulated by testosterone.

The reproductive cycles of female mammals are governed by feedback mechanisms in which the concentrations of the sex hormones in the plasma affect the secretion of the hypothalamic releasing hormone and thus the secretion of the gonadotropins. Five girls who had not yet commenced menstruation, and six just postmenarche, were studied by Hansen *et al.* (1975). They found coincident surges of LH and FSH, with periods of 20–40 days, in approximately half of their subjects. They suggest that the cyclic interactions of hypothalamus, pituitary, and ovaries begin before menarche. Presumably the amplitude of the cyclic variations and their regularity increase when menstrual cycles are established.

The human placenta produces gonadotropins in addition to those produced by the anterior lobe of the pituitary. The chorionic gonadotropin of the human placenta resembles the luteinizing hormone, LH, in general, and its appearance in the urine is the basis for one test for pregnancy. The literature on the structure of this glycoprotein, now fully known, is reviewed by Bahl (1977). A receptor structure, with the properties of a membrane protein specific for gonadotropin, has now been purified.

A "radioreceptor" assay for human LH and chorionic gonadotropin has been developed by Saxena *et al.* (1974), using the plasma membranes of bovine corpora lutea as source of the receptor. This assay will detect pregnancy by the end of the first week. Haour and Saxena (1974) have used it to detect gonadotropin in the rabbit blastocyst before its implantation in the uterine wall. Dickmann *et al.* (1977) cite literature demonstrating the presence of estrogen in the rabbit blastocyst, and that the permeability of uterine capillaries is increased by estrogen. These observations have been confirmed by George and

Wilson (1978), who found that estrogen first appears in the blastocyst immediately before the time of implantation. This suggests that the estrogen in the blastocyst is essential to implantation, through its effects on uterine capillaries. The significance of the gonadotropin in the blastocyst is uncertain.

After the mammary glands, the cervix uteri is the most common site of cancer in human females. A cultured cell line from such a carcinoma has been studied by Patillo *et al.* (1977), and they find that it produces a subunit of human chorionic gonadotropin. This subunit was also present in the blood of the patient from whom the cells were derived, and has been found in 17–33% of cancers of other organs.

III. Adrenal Glands

A. Medulla and Sympathetic Neurons

The medullary or chromaffin tissue of the mammalian adrenal glands is homologous with similar tissue in the interrenal bodies of lower vertebrates, and is structurally and chemically related to the sympathetic division of the autonomic nervous system. Adrenaline or epinephrine was one of the first hormones to be isolated and synthesized, independently by Aldrich, Takamine, and Stolz in the early years of the present century. Epinephrine was identified as the transmitter at the neuroeffector junctions of the sympathetic neurons, and the effects of stimulation of these neurons, since known as adrenergic, were not always the same as those of stimulation of the adrenal medulla. The neural transmitter was identified as noradrenaline or norepinephrine by von Euler (1948), and epinephrine as the product solely of the adrenal medulla. Sympathetic neurons have since been strictly referred to as noradrenergic.

Recent studies to be discussed here have emphasized the importance of noradrenergic neurons in the brain in control of feeding and obesity and in activation of neurosecretion in the hypothalamus. Effects on the pineal body are also considered on p. 150. The control of feeding in rats by a discrete tract of noradrenergic neurons in the brain, and the effect of lesions in the hypothalamus, have been studied by Ahlskog (Ahlskog and Hoebel, 1973; Ahlskog *et al.*, 1975) Food consumption increased 25–40% and body weight 40–60% after damage to this tract or the injection of 6-hydroxydopamine, a precursor of epinephrine. Hypothalamic lesions which did not affect this tract or alter secretion of norepinephrine also caused overeating. These results suggest that two different mechanisms participate in control of feeding. In contradiction, Gold (1973) claims that the only hypothalamic lesions which cause overeating are those which damage the noradrenergic tract.

Evidence that norepinephrine can influence neuro-secretion was obtained by Shimamoto and Miyahara (1976). They infused norepinephrine into normal human subjects, and obtained a diuresis associated with decreased plasma concentration of vasopressin (ADH, antidiuretic hormone). Evidently, norepinephrine can suppress release of ADH from the hypothalamo–neurohypophysial system. This may explain the diuresis associated with certain emotional states, and possibly that associated with exposure to low ambient temperatures. As we shall note below, acclimation of mammals to cold is accompanied by release of norepinephrine.

Epinephrine, which among other effects stimulates glycogenolysis and the release of glucose from the liver into the blood, is an antagonist to insulin (p. 145) in this respect. The results of the study of Stricker et al. (1977), who examined the interaction of these two hormones in rats, will be considered in detail in connection with the endocrine pancreas. These authors found that neither mannose nor β-hydroxybutyrate could replace glucose in maintaining blood sugar levels in rats infused with insulin. The infusion of glucose and insulin together resulted in an increase in levels of catecholamines (epinephrine and norepinephrine) in the blood, and earlier work showing a similar effect in dogs is cited.

An important function of the catecholamines (epinephrine and norepinephrine) in mammals is the stimulation of "nonshivering thermogenesis," the production of heat by oxidative metabolism in brown fat and other tissues. The literature on this topic was reviewed by Berberich et al. (1977), who cite references showing that excretion of catecholamines increases in rats acclimated to low temperatures. They studied two species of Arctic lemmings, Lemmus trimucronatus and Dicrostonyx groenlandicus, in which mechanisms for acclimation to cold may be expected to be highly developed. They found a significant increase in secretion of both epinephrine and norepinephrine in both species after 5 days of acclimation to cold. These increases were considerably greater than those previously reported for rats.

At another ecological and zoological extreme, the effect of catecholamines has been studied in the potoroo, Potorous tridactylus (Kerr), by Nicol (1978). This dwarf kangaroo (body weight 1.2 kg) had earlier been shown to regulate body temperature in a hot environment. In this report, ability to regulate in cool environments is demonstrated, and the effects of epinephrine and norepinephrine on

TABLE VIII

Oxygen Consumption and Core Temperature of Dwarf Kangaroos [Potoroo, *Potorous tridactylus* (Kerr)] in Relation to Ambient Temperature and the Effect of Injection of Norepinephrine (9 μM/kg)[a]

	Ambient temperature			
	17°C		29°C	
Parameter	Mean	Confidence interval	Mean	Confidence interval
Initial oxygen consumption (ml/gm/hr)	0.64	0.53, 0.75	0.60	0.51, 0.69
Maximum oxygen consumption after epinephrine	0.73	0.62, 0.84	0.84	0.66, 1.02
Change in O_2 consumption[b]	0.096	0.024, 0.168	0.296	0.213, 0.379
Initial core temperature (°C)	36.9	35.8, 38	36.7	19.2, 54
Change in core temperature[b]	−0.45	−0.86, −0.04	0.70[c]	0.27, 1.13

[a] Means and 1% confidence intervals calculated from data of Nicol (1978).

[b] Mean of individual differences.

[c] Significantly different from 17°C by confidence interval test or *t* test for paired differences.

the regulatory process are examined. The animals responded to either hormone with an increase in oxygen consumption. However, a small increase in body temperature was noted after injection of norepinephrine when the animals were in warm air (29°C), while a small decrease followed this injection in cool air (17°C). The results of this study are partly presented in Table VIII.

B. Cortex

The cortex of the adrenal glands is clearly differentiated in structure and function from the medulla. The hormones secreted by the cortex are steroids, and it has become customary to categorize them as (a) glucocorticoids, which increase gluconeogenesis, the synthesis of glycogen from noncarbohydrate precursors, especially amino acid residues, and (b) mineralocorticoids, which cause retention of sodium ion by decreasing its urinary excretion and have other effects on salt and water balance related to the increase in rate of active transport of Na^+. These categories are too limited, as are most simplistic categories in biology. There is some overlap in function, and steroids in both categories have functions or effects not included in the definitions above.

The importance of vitamin C, which is especially concentrated in the cells of the adrenal cortex, has long been recognized in the synthesis of cortical steroids. Gruber *et al.* (1976) have reviewed the literature on this point, and raised the question of the possible importance of vitamin A. In nutritional experiments with rats, they find that the latter substance has no effect on synthesis of steroids as such, but is essential for the synthesis of vitamin C by rats. This relationship does not apply to those species (cavy, anthropoids, man) which are unable to synthesize vitamin C under any circumstances. Part of the human requirement for vitamin C is therefore in connection with its role in steroidogenesis in the adrenal cortex.

The steroid hormones are often secreted as sulfate esters. The sulfated molecules are also intermediates in the further metabolism of the hormones. Einarsson *et al.* (1976) used preparations of microsomes from human livers, and showed that the sulfated steroids are hydroxylated in a different pattern from that of the same steroids in unesterified form. Only the hydrophobic, i.e., not sulfated, parts of the molecule are hydroxylated.

The receptor hypothesis of tissue specificity for hormones has been tested in detail for the steroids, and this topic is reviewed by O'Malley and Schrader (1976). It appears that steroids penetrate the plasma membrane, and then are bound in the cytoplasm of target cells by specific receptor proteins. They then enter the nucleus in conjugated form, and there may activate specific genes.

Aldosterone is the principal salt-retaining hormone of terrestrial vertebrates, and has been shown to act by specific activation of the synthesis of the $Na^+ + K^+$-dependent ATPase which is the mechanism of active transport of Na^+ and K^+ in cells. This mineralocorticoid is secreted by a distinct layer of cortical tissue in mammals, and there is considerable evidence suggesting that this tissue differs from other cortical tissue in not being subject to the influence of ACTH. The hypothesis that angiotensin, the pressor substance formed in plasma by the action of renin secreted by the kidney (p. 152), stimulates release of aldosterone has been tested by Campbell *et al.* (1974). They found that, in rats, anesthesia causes release of aldosterone, and note that this accounts for divergent results reported earlier. They injected angiotensin into unanesthetized ambulatory rats, and observed increases in plasma aldosterone in a clear dose–response pattern.

Secretion of aldosterone is increased in hypoglycemia of human beings. The adaptive significance of this is not clear, since aldosterone has little glucocorticoid activity. The time course of concentrations of glucose and aldosterone in the blood of patients injected with insulin was followed by Hata *et al.* (1976). Plasma aldosterone increased parallel with the increase in blood sugar, and both returned to normal in about 3 hours. Plasma cortisol (17-hydroxycorticosterone), a glucocorticoid with some mineralocorticoid activity, followed the same pattern. Administration of dexamethasone, a synthetic glucocorticoid, suppressed the increase in cortisol. Activity of renin (p. 152) in plasma increased after injection of insulin, and was not influenced by dexamethasone. Propanolol, a suppressor of the renin–angiotensin system, suppressed the increase in renin and delayed the return of aldosterone to normal levels. Injection of ACTH, with or without dexamethasone, increased levels of cortisol and aldosterone. The authors conclude that ACTH causes the secretion of aldosterone in hypoglycemia. From the other effects noted, and the fact that secretion of renin and of aldosterone are related to the concentration of Na^+ in plasma, the present writer concludes that insulin influences sodium metabolism.

The clinical condition known as diabetes insipidus, unrelated to diabetes mellitus or sugar diabetes, is characterized by severe diuresis with dilute urine. Experimentally, it can be duplicated in animals by lesions of the hypothalamus, and is caused by a deficiency in secretion of ADH. Schalekamp *et al.* (1976) studied a case of diabetes insipidus with adipsia, cessation of drinking, and resultant dehydration. In this case, they were able to show that the response of the adrenal cortex in secreting aldosterone on stimulus by angiotensin (p. 154) depends on plasma $[Na^+]$. Changes in plasma renin were not correlated with those of aldosterone.

Many body activities and plasma components exhibit rhythmic variation with a period of about 24 hours (diurnal, diel, or circadian). One such rhythm was noted by Sulzman *et al.* (1978) in squirrel monkeys, *Saimiri sciureus*. These authors earlier observed a spontaneous rhythm of plasma cortisol concentration, coordinated with a rhythm in urinary elimination of K^+. In the experiment reported here, they produced an artificial rhythm by periodic administration of cortisol to adrenalectomized monkeys kept in constant light. They found that the spontaneous rhythms of feeding and sleep were not disturbed in these conditions, but the rhythm of K^+ elimination was coordinated with the new rhythm of cortisol. The central control of body rhythms will be considered further under the pineal gland (p. 148).

The classical clinical effect of the glucocorticoids is antiphlogistic—antiinflammatory. The synthetic glucocorticoid fluoro β-methasone 17,21-diacetate was used by Hammarström *et al.* (1977) to study the relief of psoriasis, an inflammatory proliferative skin disease. This disorder had previously been shown to be associated with the presence of polyunsaturated fatty acids in the lesions. Topical application of the steroid resulted in a decrease in concentration of these fatty acids by 50%, as compared with application of a placebo.

The diversity of known effects of the corticosteroids continues to increase. Evidence that cortisol stimulates release of the hormone of the parathyroid gland (p. 138) was reviewed by Au (1976). Using cultured cells from the parathyroid glands of rats, this author showed by radioimmune assay that addition of cortisol to the culture medium increases the rate of synthesis of parathormone. This effect was evident either in cultures with low concentrations of Ca^{2+} in the medium, in which case secretion rate was low, or with high $[Ca^{2+}]$ and high secretion rates. The increase in parathormone was correlated with an incorporation of labeled amino acids into protein.

Studies of the initiation of parturition in mammals have shown that, in sheep at least, the fetus has a role in this initiation, through an increased secretion of cortical steroids. This evidence is reviewed by Sumar *et al.* (1978), who took advantage of the peculiar conditions of birth in the Andean alpaca, *Lama pacos*. These animals deliver their young at high altitudes only in the morning hours, between 06:30 and 12:30. These authors developed a technique for catheterization of a fetal vein, and found that this operation results in premature delivery. They found that the concentration of cortisol in fetal plasma increases just before the onset of parturition, and the level in newborn, spontaneously delivered animals is higher than that in the fetus. The authors conclude that production of cortisol by the fetus causes initiation of parturition.

IV. Sex Hormones

A. ANDROGENS

The steroid hormones androsterone and testosterone, tautomeric forms of the same steroid, were isolated from extracts of testes and synthesized independently by several groups of workers, in 1934–1935. Testosterone is the more potent form, and is probably the usual endocrine product of the interstitial cells of mammalian testes. This and related androgens elicit the development of male secondary sexual characteristics, probably including certain patterns of behavior. The embryonic differentiation of the common gonadal *Anlage* into ovaries or testes is determined by the sex chromosomes, and has recently been traced to a single gene (H-Y) on the Y chromosome. This gene is recognizable in a specific antigen which occurs on the surface of the lymphoid cells of males. This antigen has been shown, by Silvers and Kiang (1973), to be homologous in mice and rats at least. Wachtel (1977) has demonstrated the presence of this antigen in males of many species of mammals, and Silvers and Wachtel (1977) have reviewed the extensive literature on this subject. Koo *et al.* (1977) have examined the X chromosome-linked testicular feminization syndrome, in which genotypic (XY) males develop female phenotypes. The effect of the gene for this syndrome is a deficiency of the receptor for androgens. These authors examined six cases of the syndrome, and found that the leukocytes carry the H-Y antigen, despite the female phenotype. The androgen receptor of a cell-line derived from a tumor of the ductus deferens of a hamster was examined by Norris and Kohler (1976). They find evidence that the receptor molecule is a dimer of two polypeptide monomers.

Among the secondary characteristics of rats are included large differences in the metabolism of steroids in the liver (Gustafsson and Stenberg, 1976). These differences in males are determined by "imprinting" with androgens. Testosterone controls sexual differentiation of the hypothalamus, suppressing the cyclical characteristics of females. Rubin *et al.* (1976) review a large literature showing abnormalities in sexual and endocrine function in human male alcoholics. They administered alcohol to rats for 24 days and showed that the activity of testosterone 5α-reductase in the microsomal fraction of liver was increased. In a group of human volunteers given a daily intake of 3 gm/kg body weight of alcohol, 43% of total caloric intake, they report an increase in testosterone reductase activity of liver. Due to the high variance of the data, the 1% confidence limits of the mean value for the alcohol group (29: -14,71) overlaps those for the same group before alcohol (8: 2,14) A t test based on paired comparisons would have been the appropriate statistical method here.

Carefully defined innate behavior of males is released by testosterone. Copulatory behavior of rats is released in castrated males by daily injections of testosterone, and this is not modified by simultaneous injection of estrogens. The estrogen estradiol, alone or in a mixture with testosterone and dihydrotestosterone, decreases the proportion of males ejaculating in copulation (Baum and Vreeburg, 1973).

Exposure of bulls, men, or male rabbits or hamsters, to females of their own species is followed by an increase in secretion of testosterone. When house mice (*Mus musculus*) were grouped with females for a week, plasma testosterone was not different from that of grouped males. Introduction of strange females to grouped or isolated

TABLE IX

EFFECTS OF CANNABINOIDS ON SECRETION OF TESTOSTERONE BY DECAPSULATED
TESTES OF MICE in Vitro[a]

Age of donors	Agent and concentration (μg/ml)	Control		Treated	
		Mean	Confidence interval	Mean	Confidence interval
Immature	THC 2.5	253	152,254	118	69,167
Adult	THC 0.25	517	322,712	386	295,387
	2.5	517	322,712[b]	426	302,550[b]
	12.5	225	150,300[c]	159	128,190
	25	517	322,712	71	17,125[d]
	C 25	368	280,456	32.5	66,132[d]
	250	368	280,456	17	1, 35[d]

[a] Means and 1% confidence intervals, calculated from data of Dalterio et al. (1977).
THC, tetrahydro cannabinol; C, cannabinol.
[b] Greater than immature.
[c] Less than other controls.
[d] Less than control value.

males for 30–50 minutes resulted in increased testosterone in plasma of the males (Macrides et al., 1975).

As a scientific contribution to the large body of untested lore concerning the effects of marijuana, Dalterio et al. (1977) examined the effects of cannabinol and tetrahydrocannabinol, the active principles of this plant, on secretion of testosterone in vitro by mouse testes. The results are summarized in Table IX. Both principles decrease synthesis of testosterone at the higher dose levels. On the basis of a review of the literature, these authors conclude that the effects of cannabinoids in vivo are exerted through the pituitary.

B. OVARIAN HORMONES

The female hormones are of two types, estrogen and progesterone, produced in the ovarian follicles (estrogen) or in the corpora lutea formed from the follicles after ovulation (progesterone). These two categories determine the development and maintenance of the secondary sexual characteristics of females, including the sexual cycles and some aspects of behavior, under control by the gonadotropins of the anterior pituitary.

Asian musk shrews (Suncus murinus) are among the most primitive of living placental mammals. After castration of males, the accessory structures of the reproductive tract regress to some extent (Table X),

TABLE X

EFFECTS OF GONADECTOMY AND REPLACEMENT WITH SEX HORMONES ON WEIGHTS
OF ACCESSORY STRUCTURES OF THE REPRODUCTIVE TRACTS OF
ASIAN MUSK SHREWS, *Suncus murinus*[a]

Sex	Treatment	Weights (mg)			
		Uterus		Vagina	
		Mean	Confidence interval	Mean	Confidence interval
Female	None: control	8.6	0.6, 16.6	38	22, 54
	Gonadectomy	8.3	1.9, 14.7	40	20, 60
	Gonadectomy + estrogen	10.1	7.3, 12.8	41	21, 61
		Prostate		Ampullae		Epididymides	
		Mean	Confidence interval	Mean	Confidence interval	Mean	Confidence interval
Male	None: control	44	12, 76	47	19, 73	23	15, 3[b]
	Gonadectomy	11	7, 15	8	4, 12[b]	7	3, 11[b]
	Gonadectomy + testosterone	42	14, 70	33	5, 61	19	3, 35

[a] Means and 1% confidence intervals calculated from data of Dryden and Anderson (1977).
[b] Less than controls.

and may be maintained by testosterone, which is present in normal plasma (Dryden and Anderson, 1977). Estrogen cannot be detected in the blood of females, and neither ovariectomy or treatment with estrogen has any effect on the female reproductive tract. There is no estrous cycle and ovulation occurs only in response to copulation.

Maternal behavior is generally considered to be dependent on progesterone. The latent period for onset of maternal behavior in virgin female or male rats exposed to newborn young is decreased by removal of ovaries or testes (Leon *et al.*, 1973). These authors conclude that estrogen or testosterone suppress or delay the onset of maternal behavior, but no separate experimental evidence is supplied in support of this hypothesis. The mammary tumor of rats, not dependent on estrogen or prolactin for growth, studied by Costlow *et al.* (1974), had very little receptor activity for estrogen, though it did have prolactin receptors.

The estrogen–receptor interaction was reviewed by Jensen and De

Sombre (1973). It follows the pattern characteristic of steroids in general. Monomeric protein molecules in the cytosol are bound by the steroid to form a dimeric complex, which then enters the nucleus. The relationship of the response of human mammary neoplasms to endocrine therapy and the presence of estrogen receptors was studied by Horwitz *et al.* (1975). The absence of receptors indicates resistance to endocrine therapy, but their presence is no assurance that it will be effective. Of the tumors studied, 56% had receptors for estrogen and progesterone, and three out of five cases responded favorably to endocrine therapy. Six cases, of nine studied, lacked progesterone receptors and failed to respond to therapy. A cell line was cultured by Zara *et al.* (1977) from a human mammary carcinoma, and estrogen receptors in the nucleus and receptors for testosterone and progesterone in the cytosol were found. Estrogen was not required for growth, and estrogen antagonists inhibited growth in the absence of estrogen. These authors conclude that the specific anomaly of neoplastic cells is the stimulation of growth by the estrogen receptor in the absence of estrogen.

Endocrine factors act on development of mammals only after birth in general, and prenatal exposure to hormones may result in abnormalities. Mice were exposed *in utero* by McLachlan *et al.* (1975) to the synthetic estrogen diethylstilbestrol, and 60% of the males so treated were sterile. The changes associated with this sterility included intraabdominal (undescended) testes and nodular testicular masses which were possibly proneoplastic. Noller (1976) reviews the literature on the effect of this synthetic steroid in pregnant women where it is used, with variable efficacy, to prevent spontaneous abortion. There is evidence of neoplastic changes in both males and females exposed to this steroid *in utero*.

Diethylstilbestrol has also been administered to cattle, where it promotes greater gain in weight with smaller food consumption. Jukes (1976) points out that the increase in production of protein for food thus obtained must (somehow) be balanced against the risk of possible carcinogenic effect on consumers of the meat. This latter risk is very difficult to assess, and conservatives are generally opposed to any procedure which carries any risk, even hypothetical, of carcinogenesis.

The existence of a specific endocrine agent, termed relaxin, which dilates the cervix uteri and relaxes the pelvic joints of women during parturition, has been known for some time, and a peptide with this effect has been isolated. The literature on this material is reviewed by Weiss *et al.* (1976), who have demonstrated the existence of an immunoreactive relaxin in the blood of women at caesarean section, with a high concentration in ovarian veins. The authors suggest that this material is produced in the corpus luteum.

V. Thyroid Gland

As noted above, the secretion and maintenance of the thyroid gland are dependent on thyrotropin, the thyroid-stimulating hormone TSH. This in turn is released from the anterior pituitary by the thyrotropin-releasing factor, TRH, of the neurohypophysis. The commonly used analgesic and antipyretic aspirin is reported by Ramey *et al.* (1976) to inhibit the response of the thyroid to TRH, presumably through an inhibition of release of TSH.

After a long history extending to Hippocrates who treated the hypertrophy of the thyroid gland known as goiter with burnt sponge or algae, the substance thyroxine was isolated in 1914, and its structure as tetraiodothyronine was established by synthesis in 1927. More recently, triiodothyronine has been identified as a product of the gland and a constituent of normal plasma. Both hormones, designated for convenience by T_4 and T_3, have the physiological effects associated with the gland which, in mammals, include a general stimulus of oxidative metabolism. The results of Ramey *et al.* (1976) are summarized in Table XI. They compared aspirin with the recently introduced analgesic indomethacin. Aspirin decreased both T_4 and T_3 concentration in the plasma during a control period, and during treatment with TRH. Indomethacin had no effect in either period, except to increase the level of T_3 during treatment with TRH.

A clearcut circadian rhythm of plasma TSH was found by Weeke, and Weeke and Laurberg (1976) found no such rhythm in severely hypothyroid patients unless they were treated with thyroxine—dose regime unspecified. Then the rhythm was again evident.

The receptor for TSH in the thyroid gland has not been isolated, but Mullin *et al.* (1978) have isolated a ganglioside from thyroid tissue that inhibits binding of TSH to thyroid cell membranes. They suggest that this material may react with a glycoprotein receptor for TSH.

The question of a second messenger between TSH reception and the response of thyroid cells has been investigated by Fallon *et al.* (1974). Using an immunofluorescence technique, they showed that cyclic 3′,5′-adenosine monophosphate (cAMP) is distributed through the cytoplasm of the cells of the thyroid follicles, and increases when the cells are stimulated by TSH. Cyclic guanosine monophosphate (cGMP) is also present in the cells, but showed no response to TSH.

TABLE XI

PLASMA CONCENTRATIONS OF THYROID HORMONES AS INFLUENCED BY
TREATMENT WITH ASPIRIN, INDOMETHACIN, AND TRH[a]

	No TRH				TRH			
	Aspirin		Indomethacin		Aspirin		Indomethacin	
Plasma hormone	Mean	Confidence interval	Mean	Confidence interval	Mean	Confidence interval	Mean	Confidence interval
T_4 Before treatment	6.9	6.5, 7.3	7	6.7, 7.3	7.2	6.8, 7.6	7.5	7.2, 7.8
After treatment	5.5	5.2, 5.8[b]	7	6.7, 7.3	5.6	5.2, 6.0[b]	7.4	7.1, 7.7
T_3 Before treatment	69	63, 75	65	59, 71	96	89, 103[c]	87	79, 95[c]
After treatment	42	38, 46[b]	77	69, 83	72	65, 79[b,c]	106	100, 112[c,d]

[a] Means and 1% confidence intervals calculated from data of Ramey et al. (1976). T_4, thyroxine, tetraiodothyronine; T_3, triiodothyronine.

[b] Less than before treatment.

[c] Greater than without TRH.

[d] Greater than before treatment.

An elaborate experiment with thyroid cells in culture was carried out by Burke et al. (1973) to investigate the relation of prostaglandins and cyclic nucleotides to the response of the cells to TSH. Their results are summarized in Table XII. Treatment of the cells in vitro with TSH caused an increased in cAMP in proportion to log dose, further showing that cAMP is part of the response to TSH. Quazodine, an inhibitor of the phosphodiesterase which hydrolyzes cAMP, did not alone increase levels of the nucleotide in cells, but increased levels above those with TSH alone when used in conjunction with TSH. Theophyllin, the alkaloid of tea, which also inhibits the diesterase, had effects similar to those of quazodine.

These experiments also included measurements of the production of several prostaglandins, the ubiquitous hydrocarbon derivatives which have attracted much attention in recent years. Only those examples which showed significant change are included in Table XII. These are prostaglandin A_1 (PGA$_1$) and E_1 (PGE$_1$). Increases in prostaglandin A_1 were less consistently observed, and PGF$_{2\alpha}$, which was also studied, showed little significant change.

TABLE XII

EFFECTS OF TSH, CYCLIC ADENOSINE MONOPHOSPHATE (cAMP), AND INHIBITORS OF cAMP PHOSPHODIESTERASE (QUAZODINE, THEOPHYLLIN) ON cAMP AND PROSTAGLANDIN (PGA$_1$, PGE$_1$) CONCENTRATIONS IN A CULTURE OF THYROID CELLS[a]

	cAMP (μM/ml)		PGA$_1$ (ng/ml)		PGE$_1$ (ng/ml)	
Treatment	Mean	Confidence interval	Mean	Confidence interval	Mean	Confidence interval
1. Basal (control)	14	13, 15	25.5	12, 39	50	35, 65
2. TSH 5 mU	14	13, 15	25.5	19, 32	49.5	36, 63
3. Quazodine 5×10^{-5}	14	13, 15	26	14, 38	50	38, 62
4. TSH 10 mU	21	19, 23[b]	41	30, 52	74	59, 89
5. Quazodine 10^{-4}	23	20, 26	37.5	26, 49	88	68, 108[b]
6. Dibutyryl cAMP 0.5 mM	—	— —	39	25, 53	80.5	65, 96
7. (2) + (3)	26	21, 31	42	26.5, 57.5	91.5	79, 104
8. cAMP 3 mM + (5)	—	— —	49.5	39, 65[c]	108.5	93, 124
9. Dibutyryl cAMP 1.5 mM	—	— —	54	37, 51	107	90, 124
10. (5) + (6)	—	— —	52.5	39, 66[c]	120	101, 139[d]
11. Theophyllin 10^{-3}	24	20, 28	56	41, 71[e]	114	94, 134
12. (4) + (5)	33	28, 38[f]	53	41, 65[e]	109.5	93, 124
13. (4) + (12)	34	28, 40[f]	74	59, 89[e,f]	130	112, 148[d,g]
14. cAMP 3 mM + (12)	—	— —	68	53, 83[e,f]	126	112, 140[d,g]
15. Dibutyryl cAMP 3 mM	—	— —	65	45, 85[c]	131	107, 155[d]
16. Quazodine 5×10^{-4}	39	35, 43[f]	56	44, 68[e]	149	122, 176[d,g]
17. TSH 50 mU	34	31, 37[f]	66	50, 82[e]	148	128, 168[d,g]
18. Quazodine 10^{-3}	46	42, 50[b,f]	80	59, 101[b,e,f]	176	146, 206[d,g]
19. (4) + (18)	56	48, 64[h]	131	112, 150[h]	255	224, 286[b,h]

[a] Means and 1% confidence intervals calculated from data of Burke *et al.* (1973).
[b] Significantly different values.
[c] Greater than (2), (3).
[d] Greater than (6).
[e] Greater than (1), (2), (3).
[f] Greater than (4).
[g] Greater than (5).
[h] Maximum value.

The application of the confidence interval test to the results of Table XII justifies the authors' conclusion that prostaglandins mediate the effects of the release of cAMP which is the first step in the response to TSH, according to Scheme 2.

The proposed role of prostaglandins has been confirmed with intact human subjects by Shenkman *et al.* (1974). They injected prostaglan-

TSH thyrotropin → Synthesis of cAMP
↓
Synthesis of prostaglandins
↓
Secretion of thyroid hormones

SCHEME 2

din $F_{2\alpha}$ into pregnant women, intravaginally or intravenously, and observed an increase in circulating thyroid hormone as a result, with no change in TSH.

Many studies *in vivo* have demonstrated that injected thyroxine (T_4) is converted in the body to triiodothyronine (T_3). The literature on this question was reviewed by Sterling *et al.* (1973), who showed that cultured cells from human liver or kidney convert T_4 to T_3 *in vitro*. This conversion was not inhibited by added T_3. The literature on *in vivo* conversion is also reviewed by Maeda *et al.* (1976). They found that in cases of severe hypothyroidism, serum TSH concentrations decrease with increasing dose levels of T_4 or T_3 (Table XIII).

The target cells of thyroid hormones, probably all the cells of the body, presumably bind the effective hormone by means of a specific receptor molecule. Isolated nuclei and extracts of the soluble material of the nuclei, of cells of the liver and kidney of rats or of a culture of pituitary cells, were shown by Samuels *et al.* (1973, 1974) to bind T_3.

TABLE XIII
SERUM TSH OF SEVERELY HYPOTHYROID
PATIENTS DURING TREATMENT WITH THYROXINE[a]

Thyroxine dose (μg/day)	Serum TSH (μU/ml) duration of treatment (months)			
	1		2	
	Mean	Confidence interval	Mean	Confidence interval
50	82	60, 104	44	28, 60
75	34.5[b]	24, 45	13.5[b,c]	9, 18
100	24.8	18, 32	21	15, 27
125	7.6[b]	5, 10	5.7[b]	4, 7
150	4	1.5, 6.5	8.7	6, 11

[a] Means and 1% confidence intervals calculated from data of Maeda *et al.* (1976).
[b] Significant differences.
[c] Less than one month value.

The binding agent has the properties of a protein which is not a histone.

The binding of T_3 labeled with ^{125}I in the nuclei of liver cells of rats *in vivo* was studied by Schussler and Orlando (1978). They found that the ratio of label in the nuclei to that in the serum decreases after a 5-day period of fasting, and suggest that this may be an adaptation conserving energy during deprivation of food. A similar adaptation in human beings is probably an obstacle to persons attempting to reduce body weight by fasting.

The cellular mechanism of action of thyroid hormones remains obscure, despite a large amount of effort expended in investigation. The classical effect of thyroid hormone in mammals is that of increasing metabolic rate, and measurement of BMR (basal metabolic rate) was long the primary clinical test for thyroid function. Development of microchemical methods for estimating the circulating thyroid hormone has largely eliminated the cumbersome and imprecise technique for measurement of whole body metabolism. The effect of thyroid hormones in accelerating the metamorphosis of amphibians has not been associated with any developmental effect in mammals, nor with the stimulation of metabolism.

The evidence cited above that the thyroid hormone receptors are in the nucleus suggests that the site of cellular action may be in the transcription process, rather than in direct effects on mitochondrial oxidation as has long been postulated. The use of the fact that the antibiotic puromycin reverses the increase in BMR produced by thyroxine as evidence concerning the nature of the calorigenic effect of the hormone *in vivo* has been criticized by Bilder and Denckla (1974). They compared the effect of puromycin on rats at an ambient temperature of 22°C with its effect at the temperature of thermoneutrality, the highest ambient temperature at which the rectal temperature of an anesthetized rat is maintained within 1°–2°C of 38°C. Their results are presented in Table XIV. At 22°C, puromycin decreases metabolic rate in the presence or absence of thyroxine, but this can be accounted for by the decrease in body temperature caused by puromycin. At a thermoneutral ambient temperature, puromycin has no effect on body temperature or on metabolic rate.

The emphasis in hypotheses and observations concerning thyroid action has shifted from cellular oxidation and energetics to molecular and developmental. The activity of lactase is high in the intestines of newborn rats, and decreases during the first 3 weeks of life (Yeh and Moog, 1974). These authors removed the hypophysis at 6 days of age and found that the decrease is arrested. Thyroidectomy has the same

TABLE XIV

EFFECTS OF PUROMYCIN AND THYROXINE ON BODY TEMPERATURE AND RATE
OF OXYGEN CONSUMPTION (ML/100 GM BODY WEIGHT/MIN) IN ANESTHETIZED
RATS IN RELATION TO AMBIENT TEMPERATURE[a]

	Ambient temperature						
	22°C				Thermoneutrality		
	Oxygen consumption		Rectal temperature (°C)		Oxygen consumption		Rectal temperature (°C)
	Before puromycin						
Thyroxine dose (mg)	Mean	Confidence interval	Mean	Confidence interval	Mean	Confidence interval	Mean	Confidence interval
0	4.88	3.85, 5.91	37.8	37.5, 38.1	3.34	3.12, 3.56	37.8	37.5, 38.1
1	6.44	5.41, 7.47[b]	38	37.6, 38.4[b]	5.54	5.04, 6.04	37.9	37.8, 38
	After puromycin						
0	2.84	2.40, 3.28[b]	33.5	33.2, 33.8[b]	3.27[c]	3.01, 3.53	37.9[c]	37.8, 38
1	4.76[c]	4.22, 5.30	36	35.6, 36.4	6.09[c]	5.37, 6.81	38.1[c]	38, 38.2

[a] Means and 1% confidence intervals calculated from data of Bilder and Denckla (1974).
[b] Significantly different values, before/after.
[c] Not significantly different from value before puromycin.

effect, and the normal decrease can be restored in either case by administration of thyroxine. Cortisol has no effect.

Evidence of binding of T_3 to cultured rat myocardial cells has been obtained by Tsai and Chen (1976). They found that incubation of the cells in a hypothyroid medium, using serum from hypothyroid animals, decreases the rate of increase in glucose utilization in response to thyroxine. The T_3 is bound in the nuclei of the cells, and cells derived from adult rats bind less T_3 than do cells from newborn rats.

Clinical and personal problems of obesity have directed attention to the thyroid in relation to a possible role in metabolic obesity. The genetically obese strain, *ob/ob*, of mice has been used by York *et al.* (1978) in a study of this question. The authors review earlier reports suggesting that the mice of this strain are hypothyroid; the circulating levels of TSH are comparable to those of other strains, and the adrenal cortex has also been implicated. These authors estimated the levels of circulating thyroid hormones in three strains of mice at different ages (Table XV). The concentration of T_4 was 15 to 20 times that of T_3 in all

TABLE XV

CONCENTRATIONS OF THYROID HORMONES IN THE SERUM OF
THREE STRAINS OF MICE IN RELATION TO AGE[a]

| | | Age in months | | | | | |
| | | 2 | | 3 | | 6 | |
Hormone	Strain	Mean	Confidence interval	Mean	Confidence interval	Mean	Confidence interval
T_3	lean	0.95	0.42, 1.48	0.89	0.45, 1.33	1.21	0.74, 1.68
	ob/ob	1.02	0.32, 1.72	0.88	0.63, 1.13	0.56	0.15, 0.97
	GTG			1.32	0.91, 1.73		
T_4	lean	28.4	0.6, 56.2	20.9	9.8, 32	16.7	3.3, 30.1
	ob/ob	18.8	5.4, 32.2	16.1	10.8, 21.4	16.4	−1.1, 33.9
	GTG			18.8	10.6, 27		

[a] Means and 1% confidence intervals calculated from data of York *et al.* (1978).

strains, but there was no significant difference in concentration with age or strain. Evidently the obesity of the *ob* strain is not related to a difference in the level of circulating thyroid hormone.

The results of studies of thyroid function in relation to adrenalectomy are presented in Table XVI. In female mice, the uptake of labeled hormone was less in the obese strain than in a lean strain; adrenalectomy eliminated the strain difference in that the confidence intervals for the two strains overlap after adrenalectomy. There was no significant difference between the strains in the biological half-life of the labeled hormone. There was likewise no effect of adrenalectomy on the concentration of thyroid hormones in the serum.

Metabolic effects of administration of T_3 were also compared. The results of study of the release of glycerol, presumably from lipolysis, in response to injections of T_3 and of epinephrine are shown in Table XVII. Release of glycerol is stimulated by T_3 in lean mice, but there was no dose/response relationship over the range of doses studied. Injection of epinephrine did not change the response to T_3, except at the lowest dose. The obese mice were definitely resistant to T_3, in that a very large dose was required to stimulate glycerol release at or above the levels observed in lean mice. Moreover, the release was in proportion to the dose level used. Again, no consistent change in response resulted from injection of epinephrine, with or without T_3. In both strains, some differences were observed between groups treated with epinephrine and those not treated, but these differences were not dis-

TABLE XVI

Thyroid Function Tests in Female Mice in Relation to Strain and Adrenalectomy[a]

Treatment strain	^{131}I uptake (% of dose)		Half-life of ^{131}I in serum (hr)		Serum T$_3$ (µg-liter)		Serum T$_4$ (µg/liter)	
	Mean	Confidence interval	Mean	Confidence interval	Mean	Confidence interval	Mean	Confidence interval
Control								
lean	7.98	4.48, 11.52[b]	101.6	74.3, 128.9	0.8	0.4, 1.2	21	9, 33
ob/ob	3.56	2.87, 4.25[b]	139.8	100.3, 149.3	0.9	0.4, 1.4	17	10, 24
Adrenalectomized								
lean	8.17[c]	5.19, 11.15	103.4[c]	68.3, 138.5	0.8[c]	0.3, 1.3	24[c]	14, 34
ob/ob	3.47[c]	1.6, 5.34	101.3[c]	28.3, 174.3	0.9[c]	0.5, 1.3	15[c]	10, 20

[a] Means and 1% confidence intervals calculated from data of York et al. (1978).

[b] Significant differences.

[c] Not significantly different from control values or values for the lean strain.

TABLE XVII

Glycerol Release *in Vivo*, Percentage of Control Values, in Relation to Strain of Mice and to Administration of T_3 and Epinephrine[a]

Thyroxine dose (μg/kg body wt./day)	Strain and epinephrine administration							
	lean				ob/ob			
	0		+		0		+	
	Mean	Confidence interval	Mean	Confidence interval	Mean	Confidence interval	Mean	Confidence interval
0	0.44	0.18, 0.7	1.4	0.7, 2.1	3[b]	2, 4	3.7[b]	2.4, 5
62.5	162	112, 212	33[c]	7, 59	4.1[d]	0.7, 3.5	−2[d]	−10, 6
125	153	116, 190	109	44, 174	0.1[d]	−0.2, 0.4	−0.1[d]	−0.7, 5
250	154	115, 193	138[e]	77, 199	10.8[d,e]	2.4, 19.2	27[d,e,f]	21, 33
500	—	—	—	—	242[e]	144, 342	82.4[c,e]	80.8, 84

[a] Means and 1% confidence intervals calculated from data of York *et al.* (1978).
[b] Greater than lean value.
[c] Less than value without epinephrine.
[d] Less than lean value.
[e] Greater than value for thyroxine dose 62.5.
[f] Greater than value without epinephrine.

tributed according to any consistent pattern with regard to strain or to dose of T_3.

The thyroid gland secretes an additional hormone, calcitonin, independently of the iodinated thyronines. This substance stimulates deposition of Ca^{2+} in the bones, accounting for the fact that deficiency in formation of bones and teeth is sometimes associated with hypothyroidism. The relationship of calcitonin secretion to hibernation of the 13-lined ground squirrel, *Citellus tridecimlineatus*, has recently been studied by Kenny and Musacchia (1977). Their results are presented in Table XVIII, but in terms of the confidence interval test, the only significant differences established are the well-known differences in body temperature between active and hibernating animals, and a difference in the concentration of Mg^{2+} in the plasma correlated with this. No variation either in calcitonin secretion or in concentration of Ca^{2+} is demonstrated (see Chapter 3, this volume).

Autoimmune diseases, in which deterioration of a specific organ or carcinogenesis can be traced to autosensitization of an individual to antigens produced by that organ, have attracted considerable attention recently. The thyroid has been observed to be a source of autosensitization, and Silverman and Rose (1974) report that 13% of the rats of the

TABLE XVIII

BODY TEMPERATURE, CALCITONIN, AND DIVALENT CATIONS OF PLASMA IN RELATION TO HIBERNATION IN THE 13-LINED GROUND SQUIRREL, *Citellus tridecimlineatus*[a]

Month	Condition of animals	Rectal temperature (°C) Mean:Confidence interval	Thyroid plasma concentrations of		
			Calcitonin (mU/mg) Mean:Confidence interval	Ca²⁺ (mg/dl) Mean:Confidence interval	Mg²⁺ (mg/dl) Mean:Confidence interval
September	Active	35.3 : 32.7, 37.9[b]	201 : −7, 409	9.4 : 8.1, 10.7	2.5 : 2, 3
	Hibernating	8.2 : 7.9, 8.5[b]	212 : −8, 432	10.2 : 9.2, 11.2	4 : 3.4, 4.6[b]
November	Active	37.1 : 36.4, 37.8[b]	40 : 32, 48	11.4 : 8.4, 14.4	2.6 : 2, 3.2
	Hibernating	8.4 : 7.1, 9.7[b]	45 : −6, 96	11 : 6.6, 15.4	4.5 : 3, 6
December	Active	34.5 : 29.4, 39.6[b]	27 : 2, 52	11 : 9.2, 13.2	2.8 : 2.2, 3.4
	Hibernating	6.5 : 5, 8[b]	42 : 13, 71	11.8 : 9.9, 13.7	4.9 : 3.9, 5.9[b]
	Acclimated to cold	34.5 : 28.4, 40.6	57 : −18, 132	11.2 : 9.9, 12.5	2.9 : 1.9, 3.9
February	Active	37 : 36.9, 37.1	92 : −68, 252	10.4 : 9.9, 10.9	2.5 : 1.6, 3.1
	Hibernating	6.8 : 5.9, 7.7[b]	137 : 7, 267	10.5 : 9.7, 11.3	4 : 3.3, 4.7[b]
	Acclimated to cold	34.6 : 31.6, 37.6	—	10.6 : 9.7, 11.5	2.2 : 1.7, 2.7

[a] Means and 1% confidence intervals calculated from data of Kenny and Musacchia (1977).
[b] Significant differences, active/hibernating.

Buffalo strain show evidence of autoimmune thyroiditis. Feeding of the carcinogen methylcholanthrene increases the incidence to 42%. If the thymus gland is removed at birth, the incidence even without methylcholanthrene is 87%. The authors conclude that the thymus, known to have an essential role in immune processes, also suppresses the autoimmune responses to some degree.

VI. Parathyroid Glands

These structures, as their name indicates, are closely associated structurally and functionally with the thyroid. Between 1890 and 1900, they were established as histologically and physiologically distinct structures, embedded in or closely applied to the thyroid. The most striking effect of their removal, which sometimes occurred incidental to thyroid surgery when they were not known, is the spasmodic twitching of skeletal muscles known as tetany. This is the result of hyperexcitability of the nervous system, caused by a marked decrease in the concentrations of Ca^{2+} and Mg^{2+} in the blood plasma. An increase in phosphate concentration of the blood is associated with the decrease in bivalent cations, and the urinary elimination of phosphate is decreased.

Preparation of active extracts of the parathyroids, and partial purification of their hormone, a protein termed parathormone, were accomplished around 1925. Administration of the hormone has effects opposite to that of removal of the glands: plasma Ca^{2+} increases and phosphate decreases, and urinary elimination of phosphate is increased. Hypotheses of the mechanism of action of the hormone have placed the primary site in the bones, stimulating decalcification, or in the kidneys, stimulating excretion of phosphate, and most of the known effects can be deduced from either hypothesis. Probably, as in many such cases in the past, the answer will be found in a combination of the two hypotheses. The parathyroid hormone has been purified and identified as a protein, but Kemper *et al.* (1974) present evidence that bovine parathyroid tissue *in vitro* secretes a second protein in addition to the known hormone (see Chapter 3, this volume).

The secretion of parathormone, and of the protein described by Kemper *et al.* (1974), is directly responsive to the concentration of Ca^{2+} in the medium, increasing when this concentration falls, and decreasing when it rises. As noted above, Au (1976) has shown that parathyroid cells *in vitro* are stimulated to secrete by cortisol. Some of his results are summarized in Table XIX.

TABLE XIX

EFFECTS OF CORTISOL ON PARATHYROID HORMONE SECRETION BY RAT
PARATHYROID CELLS *in Vitro*[a]

Cortisol concentration	Duration of treatment period (hr)	
	24	48
	Mean : Confidence interval	Mean : Confidence interval
0	1.46 : 0.95, 1.97	1.29 : 0.74, 1.84
10^{-8}	2.69 : −2.88, 6.24	2.20 : −2.53, 6.93
10^{-7}	4.83 : 2.64, 7.02[b]	4.68 : 1.66, 7.7
10^{-6}	5.8 : 2.68, 8.92[b]	8.05 : 3.04, 13.06[b]

[a] Values are ratios of secretion during the treatment period to that during a preceding control period. Means and 1% confidence intervals calculated from data of Au (1976)

[b] Greater than zero concentration value.

VII. Endocrine Pancreas

A. INSULIN

The endocrine function of the pancreas holds a central place in the history of physiology and medicine. Claude Bernard's discovery of glycogen, and his demonstration that the concentration of glucose in the blood is regulated by release of glucose from the liver led him to the concept of the internal environment and to the concept of internal secretion. The discovery, essentially accidental, by von Mering and Minkowski that removal of the pancreas from dogs produced a condition comparable to human diabetes mellitus provided the classical "model" for the experimental study of this disease which had 100% mortality even when treated by the control of diet. The description of the islet tissue of the pancreas by Paul Langerhans provided the histological evidence that this gland has both exocrine (or eccrine) and endocrine functions, and provided the basis for the demonstration by Banting, with Best and MacLeod, that extracts of the pancreas may be prepared which cause a decrease in the elevated blood sugar levels which characterize diabetes. This was the major step towards the conquest of the disease. Then Sanger completed the list of major advances by determining the sequence of amino acids in the insulin molecule, the first such sequence to be determined, and opened the way to the eventual synthesis of the protein hormones or other natural products. I shall follow here the pattern used previously, of citing and discussing critically only a few recent reports.

With increasing knowledge of the amino acid sequences of proteins, it has become evident that insulin shares important features with the gastrointestinal hormones secretin and gastrin. The developmental history of the β cells of the pancreatic islets, known as the source of insulin, has been traced by Pictet *et al.* (1976), showing that, in cultures of neural crest cells of rat embryos, cells with the structural characteristics of β cells differentiate, and insulin is produced. The neural crest tissue is also the source of the gastrointestinal cells which produce hormones.

New techniques of electron microscopy have been applied to the study of insulin release from the β cells. Using the freeze-etching technique, Orci *et al.* (1973a) showed that the phenomenon of emio- or exocytosis is often evident in the cells of islet tissue. Insulin is known to be stored in membrane-bound granules in these cells, and these authors present evidence of coalescence of the granule membrane with the plasma membrane and consequent release of granule contents. In another report (Orci *et al.* 1973b), members of this research group show that incubation of isolated islet tissue with the proteolytic enzyme pronase results in increased release of insulin from the tissue. The release of insulin from islet cells of rats has also been correlated by Pipeleers *et al.* (1976) with polymerization of the cellular protein tubulin, and formation of microtubules. They conclude that microtubules participate in the transport of granules and the release of insulin.

In the present state of endocrinological thought, a study of the role of cyclic adenosine monophosphate (cAMP) in the release of insulin is almost inevitable, and has been carried out by Charles *et al.* (1973). They found that the level of cAMP in perfused islet tissue from rats is correlated with release of insulin stimulated by glucose. Theophyllin, which inhibits the diesterase hydrolyzing cAMP, also stimulates a release of insulin, though this was not correlated with increased levels of cAMP. The authors suggest that cAMP modulates the release of insulin, rather than serving as a "second messenger" between the stimulus of glucose and the release of hormone. The polymerization of tubulin in β cells may be related to calcium ion. Early studies showing that glucose induces a release of Ca^{2+} from binding in these cells, and that isotopic Ca^{2+} is taken up by the cells and bound to the secretory granules, are cited by Hellman *et al.* (1976). These authors then present evidence for the existence of two pools of Ca in the cells, from only one of which Ca is readily displaced by La. They consider that only the displaceable pool participates in coupling insulin release to glucose. This latter pool is assigned to the plasma membrane, and the stable pool to the secretory granules.

The question whether there is any difference between the α and β anomers of D-glucose in stimulating release of insulin was considered by Niki *et al.* (1974). They report that the α anomer is the more effective during 5 min of incubation *in vitro*. Mutarotation occurred in the medium used, and was complete in 5 min. The equilibrium mixture was slightly more effective than the β anomer alone, and slightly less effective than the α anomer alone. The conclusion that α-D-glucose is the stimulus for insulin release was also reached by Grodsky *et al.* (1974) using perfused rat pancreas. The purine alloxan, which causes necrosis of the islet tissue specifically, has been a valuable tool in the production of experimental diabetes. Rossini *et al.* (1974) have used this tool in a further consideration of the problem of anomeric specificity. The results of their experiments are presented in Table XX. Administration of alloxan was followed, as usual, by a marked increase in concentration of glucose in the blood. Administration of α-D-glucose with the alloxan resulted in lower concentrations than that after alloxan alone at all dose levels of glucose used; the amount of protection is proportional to the dose, and the maximum dose used protects completely. Administration of β-D-glucose also protected against alloxan, and at the highest dose level used, the blood glucose concentration was less than that of untreated controls. These results suggest to the

TABLE XX

CONCENTRATIONS OF GLUCOSE IN BLOOD OF RATS 24 HOURS AFTER ADMINISTRATION OF ALLOXAN AND ANOMERS OF D-GLUCOSE[a]

Alloxan	Anomer dose (mg/kg)	Anomer	
		α, D-glucose	β, D-glucose
		Mean : Confidence interval	
0	0	164 : 158, 170	
+	0	533 : 476, 590	
		Mean : Confidence interval	Mean : Confidence interval
+	250	266 : 170, 362	409 : 313, 505[b]
+	500	240 : 199, 281	341 : 287, 315
+	750	163 : 152, 174[c]	23 : −3, 49[d]

[a] Means and 1% confidence intervals calculated from data of Rossini *et al.* (1974).

[b] No significant difference from alloxan controls.

[c] No significant difference from alloxan-free controls.

[d] Less than alloxan-free controls.

present writer that, whereas α-D-glucose reacts specifically with the receptors of the β cells, stimulating release of insulin, β-D-glucose may react with receptors of the α cells to inhibit secretion of glucagon. This suggestion evidently is subject to further test.

The fact that glucose acts directly to stimulate release of insulin from the β cells of the islet tissue provides a simple feedback mechanism, since the primary effect of insulin is to stimulate utilization of glucose by body tissues, especially muscular and adipose tissue. (See Scheme 3.)

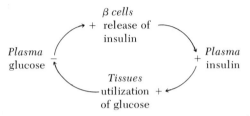

SCHEME 3. Arrows ending in + signify a direct relationship, and those ending in − signify an inverse relationship.

There is no central control of insulin secretion. The diabetogenic effect of pituitary extracts is exerted through growth hormone and ACTH, in inhibition of glucose utilization and stimulating release of glucagon.

B. GLUCAGON

This hormone was first discovered as a component of insulin preparations with an opposing effect. Insulin causes a decrease in blood glucose concentration, and glucagon causes an increase. The α cells of the islet tissue have been shown to be the source of glucagon. A considerable literature, reviewed by Goodner *et al.* (1977), reports oscillations in blood glucose concentration. These authors studied rhesus monkeys (*Macaca mulatta*) and found that, after an overnight fast, their blood sugar exhibited regular oscillations with a period of 9 min and amplitude of 4%. A cycle of plasma insulin with an amplitude of 51% was in phase with the glucose cycle. There was also a cycle of plasma glucagon, amplitude 20%, out of phase with the other two. The complete control system for blood glucose, which can be represented as in Scheme 4, is of a type which, under certain conditions, might develop oscillatory characteristics (Milsum, 1966). The numerical values in this scheme are taken from the data of Goodner *et al.* (1977). Human values for blood sugar level are higher. The operation of the system in the oscillatory fashion reported can be accounted for in the

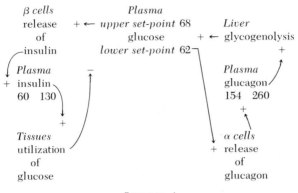

β cells
release   + ←   *Plasma*
of   *upper set-point* 68   *Liver*
−insulin   glucose   + ← glycogenolysis
        lower set-point 62   +

Plasma   −   *Plasma*
+ insulin   glucagon
60   130   154   260

       +   +

Tissues   *α cells*
utilization   + release
of   of
glucose   glucagon

SCHEME 4

following terms: When blood sugar levels reach the lower set-point, as after an overnight fast, glucagon secretion would be activated, and within about 4 min could increase plasma glucose to the level of the upper set-point, at which insulin release is stimulated and release of glucagon inhibited. If a somewhat longer time, about 6 min, is required for the release of insulin and decrease of blood sugar to the lower set-point, the oscillatory phenomena reported by Goodner *et al.* (1977) could be accounted for. The oscillation results from a difference in the time constants of the two processes which were found to be out of phase.

In an elaborate experiment with human subjects, Gerich *et al.* (1973) considered the question whether secretion of glucagon is altered in diabetes mellitus. They compared a group of normal subjects with a group of diabetics in whom the disease had developed early in life. After injection of insulin, blood sugar, significantly higher in the diabetics than in the normals, fell to about the same level in both groups. Normal subjects soon returned to levels near those before injection, but the blood sugar of the diabetics remained low. These results are summarized in Table XXI. Observed values for plasma glucagon clearly account for the difference between the two groups in the recovery after insulin injection. Concentrations in normal plasma were maximal at 45 min, when glucose levels are minimal. In the diabetics, there was no significant change in glucagon level during the entire 2-hour period. Plasma cortisol did not change in either group. Growth hormone levels increased in the normal group during the first 45 min, and remained high thereafter. The mean values for concentration of growth hormone in the diabetic group were similar to those in normals, but the high variance of the means for diabetics renders the

TABLE XXI

EFFECTS OF INSULIN ON CONCENTRATIONS OF GLUCOSE AND HORMONES IN PLASMA OF
NORMAL SUBJECTS (N) COMPARED WITH SUBJECTS WITH JUVENILE DIABETES (D)[a]

Time (min)	Subjects	Glucose (mg%)	Glucagon (pg/ml)	Cortisol (µg%)	Growth hormone (ng/ml)
		Mean: Confidence interval	Mean: Confidence interval	Mean: Confidence interval	Mean: Confidence interval
−10	N	85 : 79, 91	144 : 23, 165	16 : 1, 31	2.5 : 1, 4
	D	163[b] : 95, 231	154 : 77, 231	20 : 8, 32	2.6 : −1, 6.2
0	N	85 : 79, 91	142 : 121, 163	15 : 9, 21	2.7 : 1.2, 4.2
	D	165[b] : 97, 233	148 : 88, 208	17 : 5, 29	2.4 : −1.4, 6.2
30	N	26 : 20, 32	391[b] : 328, 454	—	—
	D	80 : 32, 128	130[b] : −15, 275	—	—
45	N	36 : 27, 45	472[b] : 427, 517	26 : 20, 32	18 : 2.1, 23.9
	D	54 : 22, 86	108[b] : 56, 160	—	—
60	N	40 : 14, 66	415[b] : 332, 498	29 : 23, 35	35 : 20, 50
	D	39 : 16, 67	120[b] : 44, 196	24 : 4, 44	21 : −3, 45
75	N	45 : 19, 71	346[b] : 269, 423	28 : 19, 37	33 : 21, 45
	D	28 : 12, 44	124[b] : 27, 221	26 : 2, 50	28 : 4, 52
90	N	53 : 44, 62	304[b] : 238, 369	28 : 2, 34	31 : 22, 40
	D	25 : 13, 37	119[b] : 59, 179	27 : −5, 59	30 : 2, 58
105	N	61 : 52, 70	—	—	—
	D	22[b] : 18, 26	119 : 67, 171	31 : −1, 63	36 : 4, 68
120	N	74[b] : 65, 83	188 : 164, 212	24 : 15, 33	22 : 4, 40
	D	26[b] : 18, 34	105 : 32, 178	34 : 10, 58	29 : −3, 61

[a] Insulin was administered at time 0. Means and 1% confidence intervals, calculated from data of Gerich *et al.* (1973).

[b] Significantly different values, N/D.

significance of any change questionable. Glucagon may be secreted in direct response to a decrease in blood sugar, or, as has been suggested, may be under indirect control by growth hormone, which is secreted in response to low blood sugar levels. The fact that, in normals, the concentration of glucagon decreased as blood sugar increased between 45 and 120 min, and that levels of growth hormone remained elevated during this period, suggests a direct response of the α cells to glucose concentration. The defect in the α cells of the diabetics may not be inability to secrete glucagon, which was present in the same concentrations in normals and diabetics before beginning of insulin infusion. However, the secretion of glucagon in diabetics did not respond to a change in glucose concentration, and secretion of growth hormone was highly variable.

A considerable literature indicating that there are sources of glucagon other than the α cells is reviewed by Lawrence *et al.* (1977). These authors have found a substance which reacts with an antigluca- gon immune serum in the salivary glands of mice, rats, rabbits, and human beings, in the sublingual gland of dogs and men, and in the parotid gland of rats. This substance was released from the glands *in vitro* in response to addition of arginine. Its significance as an effective hormone remains to be determined, but it may account for the gluca- gon in the blood of diabetics.

The participation of Ca^{2+} in the release of insulin in response to glucose was noted above. During brief periods of deprivation of Ca, Leclercq-Meyer *et al.* (1975) report that the production of glucagon by the perfused pancreas of rats increases in response to increased con- centration of glucose. Under the same conditions, secretion of insulin decreases. The basis for this reversal of normal responses is not clear.

C. PERIPHERAL ACTIONS

The tissue receptor for insulin has been identified as a glycoprotein (Hollenberg and Cuatrecasas, 1975). The overall effect of insulin has long been recognized as that of permitting the utilization of glucose, especially by muscular and adipose tissue, which are the largest users. The present writer has suggested that, in the lower vertebrates, where blood sugar is not closely regulated and the effects of variation in concentration of blood glucose are minor, the regulation of cellular utilization is the primary function of insulin (Scheer, 1963).

The close regulation of concentration of glucose in the plasma in mammals has reduced this primary function to a secondary one, in that glucose utilization becomes essentially a means of removing glucose from the blood, complementing the independent action of the liver in glycogenesis. As this is written, the precise cellular mechanism by means of which insulin exerts its action is still under debate, as is the means by which the reaction of glucose with the receptor is transmit- ted to the cellular mechanisms of response (Hollenberg and Cua- trocasas, 1975).

The resistance to insulin has been shown to increase with age in human subjects, and Rosenbloom *et al.* (1976) report a positive corre- lation between binding of insulin to human fibroblasts in culture and the age of the donor. Six patients with normal insulin secretion and a high resistance to insulin, such that very high doses were required for an effect, were studied by Flier *et al.* (1975). Serum from these persons was incubated with cultured lymphocytes, and found to decrease their affinity for insulin. The monocytes of the patients bound only 5–30% of the normal amounts of insulin. The suggestion was offered that the impaired binding of insulin, and resistance to its action, are consequences of autoantibodies directed against the insulin receptor. A form of diabetes which is a form of autoaggression, in that lymphocytes infiltrate the islets and destroy the β cells was described by Huang and Maclaren (1976). The possibility of a relationship between binding of insulin and obesity was considered by Amatruda *et al.* (1975) on the basis of a review of the literature. They then studied the relatively large adipocytes from obese persons in comparison with the smaller cells from persons of normal status, and found no difference in the number of receptor sites or their affinities for insulin between the two types of cell.

There is a considerable literature concerning the relationships between electrolyte balance of cells and insulin, but this is largely neglected at present. The phenomenon of insulin shock is distinct from the diabetic coma of hypoglycemia, and is the result of a direct action of insulin on nervous tissue. Reports that insulin may set off episodes of paralysis have been considered by Kao and Gordon (1975). They suggested that this is a consequence of a decrease in the K^+ conductance of neurons by the action of insulin. They tested this hypothesis in nuscles from rats depleted of K^+, and showed that insulin decreases the K^+ conductance of such fibers, rendering them inexcitable. The plant product, concanavalin A, one of a class of substances known as lectins, blocks the insulin receptors. This substance, applied to the rat muscles, also induces a paralysis due to decreased K^+ conductance. These authors suggest that any changes in cellular Na^+ caused by insulin are attributable to this effect on K^+ conductance. We (Scheer and Langford, 1976) have demonstrated that insulin injected into eels increases the efflux of isotopic Na^+ from the whole animal, and that this increase is entirely in the Na^+-dependent component of the efflux, the so-called exchange diffusion. The importance of this phenomenon in mammals must be entirely at the cellular level, and the effect of insulin on it has not been considered worth investigation by the experts.

The endocrine pancreas also seems to influence metabolism of calcium. The literature reporting osteoporosis (softening of the bone) as a sympton of human diabetes has been reviewed by Schneider *et al.* (1977). They also report that experimental diabetes, induced in rats by alloxan or streptozotocin, is accompanied by a decrease in absorption of Ca^{2+} from the intestine. Associated with this effect is a decrease of concentration of 1,25-dihydroxy vitamin D, and of a protein which binds Ca^{2+} in the presence of this vitamin. Insulin restores all these to normal, and the authors infer that the defect in diabetes is in the hydroxylation of vitamin D.

There is evidence to the effect that changes in concentration of glucose in the plasma influence appetite, and Stricker *et al.* (1977) have studied the relation of infusion of sugars and insulin to blood glucose and food consumption of unanesthetized rats. When insulin was infused with a sugar, either glucose, fructose, or mannose, food consumption was depressed below that in the controls, infused only with 0.15 *M* NaCl. At the molar concentration used for the sugars, neither NaCl nor β-hydroxy butyrate decreased food intake below that of the controls, when infusion of hydroxybutyrate was accompanied by that of insulin. The effect on food intake was not parallel to the effect on blood glucose concentration, however. Insulin and glucose together maintained blood glucose within the confidence interval for the controls. Neither fructose nor mannose nor hydroxybutyrate could do this. Possibly the depression of food intake by the sugars is due to their sweet taste rather than to any metabolic effect (Table XXII).

In addition to its many other effects, insulin modifies metabolism of nitrogenous compounds. Secretion of insulin by the islets occurs in response to arginine as well as to glucose, and Stein *et al.* (1976) have reviewed evidence that loss of nitrogen in the urine increases in human diabetes, and that insulin decreases the concentration of branched-chain amino acids in the plasma, with no effect on those having straight chains, such as alanine. Insulin is also reported to inhibit proteolysis in perfused rat liver, with no effect on protein syn-

TABLE XXII

FOOD INTAKE AND BLOOD GLUCOSE CONCENTRATION OF RATS (UNANESTHETIZED) INFUSED WITH INSULIN AND METABOLITIES[a]

Infusion	Concentration (*M*)	Food intake (gm) Mean : Confidence interval	Blood glucose (mg%) Mean : Confidence interval
Control			
NaCl	0.15	2.15 : 1.26, 3.04	108 : 97, 119
NaCl	1.2	1.19 : −0.17, 2.55	— — —
Insulin			
Glucose	1.2	0.27[b] : −0.34, 0.91	96 : 53, 139
Fructose	1.2	0.38[b] : −0.07, 0.83	53[b] : 48, 58
Mannose	1.2	0.20[b] : −0.21, 0.61	34[b,c] : 17, 52
β-hydroxybutyrate	1.2	0.73 : −0.16, 1.62	33[b,d] : 29, 37

[a] Means and 1% confidence intervals calculated from data of Stricker *et al.* (1977).
[b] Less than control.
[c] Less than glucose.
[d] Less than fructose.

thesis; this results in a decrease in urea synthesis and elimination. In their experimental studies with rats, these authors injected insulin into animals infused continually with ^{15}N-labeled glycine. The effect was a temporary (1 hour) decrease in the labeling of other amino acids and of ammonia. Similarly, during continuous infusion of ^{15}N-labeled aspartate, the rate of labeling of NH_3, urea, and amino acids was decreased by injection of insulin. Infusion of $^{15}NH_4Cl$ resulted in increased urinary elimination of ^{15}N, and this was interrupted by insulin injection. These results are interpreted in terms of an effect of insulin on synthesis of glutamine or on its distribution in the plasma.

VIII. Pineal Body

The pineal body in the roof of the midbrain is homologous with the "third eye" which functions in certain lower vertebrates as a photoreceptor, though not as an eye. Its endocrine function in mammals was long suspected—after Descartes' opinion that it is the seat of the soul was abandoned. This suspicion has been confirmed since 1960 when Farrell reported an effect on the secretion of steroids by the adrenal cortex, and Lerner and Case (1960) reported the secretion of melatonin (N-acetyl-5-methoxytryptamine) by the ependymal cells of the pineal. Since that time there has been a large body of experimental work devoted to this organ, and again this presentation will be confined to recent papers.

A. Vasotocin

A culture of ependymal cells was prepared by Pavel *et al.* (1973) from male fetuses obtained in therapeutic abortions. After 8–38 days, the culture exhibited the pharmacological properties of vasotocin, the water-balance principle of the neurohypophysis of amphibians, which occurs in the neurohypophysis of other lower vertebrates. These authors review the evidence suggesting that this peptide is secreted by the pineal body of mammals.

Evidence has accumulated that the pineal controls photoperiodic aspects of reproductive activity of rodents, and Vaughan *et al.* (1974) have examined the effect of vasotocin on the development of the reproductive organs of mice. Their results are presented in Table XXIII. Vasotocin evidently decreases the weight of the gonads of both sexes, and also decreases the weight of the male accessory organs, during a period of general body growth.

The hypothalamic factor, MRIH, the MSH-release–inhibiting hormone, has been shown to increase the pharmacological properties as-

TABLE XXIII

WEIGHTS OF REPRODUCTIVE STRUCTURES OF MICE 15–25 DAYS OF AGE, AS INFLUENCED BY INJECTION OF ARGININE VASOTOCIN FOR 3–4 DAYS[a]

	Female				Male			
	Control		Vasotocin		Control		Vasotocin	
	Mean	Confidence interval	Mean	Confidence interval	Mean	Confidence interval	Mean	Confidence interval
Body weight (gm)	20.3	19.3, 21.3	18.5	16.5, 20.5	24.3[b]	22.7, 25.8	21.7[b]	23.3, 24.9
Gonad weight (mg)	4.93	4.28, 5.58	3.44[c]	3.44, 4.19	166.51[b]	135.1, 177.9	107.21[b,c]	120.5, 133.8
Uterus or ventral prostate (mg)	17.9	11.1, 24.7	12.6	15.2, 17.8	4.15	3.86, 4.44	1.72[c]	2.34, 2.96
Male accessory organs (mg)					22.4	18.8, 26	11.75[c]	15, 18.25

[a] Means and 1% confidence intervals calculated from data of Vaughan et al. (1974).
[b] Greater than female values.
[c] Less than control values.

sociated with vasotocin—hydroosmotic activity and antidiuretic activity—in the cerebrospinal fluid (CSF) of rats by Pavel *et al.* (1977). Synthetic MRIH was injected into the carotid artery, and CSF and the pineal body were assayed for these activities, with the results shown in Table XXIV. The antidiuretic activity of the pineal increased 30–60 min after injection, but the hydroosmotic activity remained unchanged. Both activities appeared in the CSF only 5–15 min after injection, but declined thereafter.

B. SEROTONIN, MELATONIN

The amine known as serotonin, 5-hydroxytryptamine or 5HT, was first discovered as a neural transmitter substance in mollusks, but it also occurs in the pineal of vertebrates. The literature reporting a circadian rhythm of serotonin, of the *N*-acetyltransferase which catalyzes its conversion to melatonin, and of melatonin, has been reviewed by Brownstein and Axelrod (1973). The rhythms of serotonin and of the acetylating enzyme are endogenous, in that they persist in blindness or in darkness. The pineal is innervated by a sympathetic nerve tract from the superior cervical ganglion, and denervation abolishes the pineal rhythms. Stimulation of the nerve tract increases enzyme activity in the daytime when it is normally low, and blocking of noradrenergic transmission prevents the usual decrease in enzyme activity at night. These authors demonstrated a persistent circadian rhythm of turnover of norepinephrine in the sympathetic tract innervating the pineal. They and Romero and Axelrod (1974) propose that the circadian rhythms of the pineal originate in the suprachiasmatic nucleus of the hypothalamus, and are transmitted through the sympathetic tract innervating the pineal. The innervation of hypothalamic control of feeding by a sympathetic tract has been noted earlier (p. 118). A cell culture of ependymal cells from the pineal has been shown by Klein *et al.* (1978) to respond to norepinephrine with an increase in acetyltransferase activity. This was reversed by propanolol and the reversal is associated with a decrease in cAMP. The authors suggest that a sudden decrease in cAMP is the signal for a decrease in enzyme activity.

Early studies of the pineal suggested that this structure influences reproductive activity. Short photoperiods (light : dark, 1 : 23 hour) were shown by Reiter *et al.* (1974) to result in involution of the male reproductive tract of hamsters (*Mesocricetus auratus*). When effects of removal of the pineal or of implants of melatonin were compared, the only significant effect demonstrated was that implants of melatonin increased plasma prolactin levels over those of the controls. In a re-

TABLE XXIV

HYDROOSMOTIC AND ANTIDIURETIC ACTIVITY OF CEREBROSPINAL FLUID (CSF) AND PINEAL BODY OF RATS AFTER INTRAARTERIAL INJECTION OF MELANOPHORE-STIMULATING HORMONE RELEASE-INHIBITING HORMONE (MRIH)[a]

Tissue or fluid	Time after injection (min)	Activity assayed							
		Hydroosmotic				Antidiuretic			
		Bovine MRIH		Synthetic MRIH		Bovine MRIH		Synthetic MRIH	
		Mean	Confidence interval	Mean	Confidence interval	Mean	Confidence interval	Mean	Confidence interval
Pineal	−5			84	53, 116			55	40, 70
	5	43	21, 64	45.5	22, 69	28	19, 37	30	15, 45
	30	54	34, 75	52	26, 79	100[b]	50, 150	95[b]	51, 139
	60	64	37, 91	67	43, 91	177[b]	116, 238	181[b]	114, 248
CSF	−5	0		0		0		0	
	5	52	25, 79	42.5	23, 64	32	17, 47	27	18, 36
	15	40.5	16, 65	47	28, 66	30	18, 42	25	16, 34
	30	17[c]	13, 21	13[c,d]	6, 20	11[c,d]	5, 17	9[c,d]	3, 15

[a] Means and 1% confidence limits calculated from data of Pavel et al. (1977).
[b] Greater than 5 min value.
[c] Less than 5 min value.
[d] Less than 15 min value.

view of this and other work, Turek *et al.* (1975) conclude that the effect of melatonin on the reproductive systems of mammals depends on dose and on photoperiod, and the nature of the effect, if any, remains indeterminate. Some instances of an effect of injection of melatonin on testicular weight of hamsters have been demonstrated by Tamarkin *et al.* (1977). These depend irregularly on dose and on time of injection, and there was one instance in which pinealectomy resulted in a lower value of testis weight in animals injected with melatonin than that for unoperated animals receiving the same dose at the same time (Table XXV). The meaning of this single value is questionable, even though the difference is statistically significant.

Cells from the anterior pituitaries of juvenile (5-day-old) female rats were cultured by Martin and Klein (1976), and shown to respond, by release of LH, to LHRH administration. This response was clearly inhibited, in proportion to dose, by melatonin. In human subjects, Smythe and Lazarus (1974) have shown that melatonin causes an increase in growth hormone (GH, STH) levels. Comparison with other studies which they review suggests that this effect is not direct, but results from stimulation of the nervous system by melatonin. This may account for many inconsistent or inconclusive results, since serotonin is certainly known to be concerned in many functions of the nervous system (Fuller, 1977).

A circadian rhythm in concentration of melatonin in the blood of calves is reported by Hedlund *et al.* (1977), who review an extensive literature concerning such cycles in birds and mammals. The melatonin concentration of human pineal glands has been determined by Greiner and Chan (1978).

IX. Kidney

A. RENIN, ANGIOTENSIN

The kidney is traditionally regarded as an organ of external secretion. The syndrome, recognized by Bright in 1827, of chronic hypertension associated with kidney disease, was duplicated experimentally in 1934 in animals by Goldblatt and associates by producing ischemia with clamps on the renal arteries. Tigerstedt and Bergman had already, in 1898, used the term renin for the component of extracts of kidney which produces hypertension (Goldblatt, 1947). Renin is now known as a tetradecapeptide (Skeggs *et al.*, 1977) produced in the juxtaglomerular cells, closely applied to the glomerular capillaries of the kidney. This substance acts upon a protein component of blood

TABLE XXV

Effect of Melatonin on Reproductive Structures and Gonadotropins in Male Hamsters, in Relation to Time of Administration and Pinealectomy (OP)[a]

Melatonin dose (μg per injection)	Time of injection	Pineal status	Weight of testes (mg)		Serum LH (ng/ml)		Serum FSH (ng/ml)	
			Mean	Confidence interval	Mean	Confidence interval	Mean	Confidence interval
Oil control	Night	Op	1.8	1.6, 2.0	88	−87, 263	167	70, 264
25	Night	Op	0.7[b]	−0.1, 1.5	35	24, 36	106	65, 147
75	20:00	Op	1.4	1.1, 1.7	99	25, 173	245	119, 371
Oil control	Night	N	1.7	1.4, 2.0	—	—	—	—
25	Night		0.5[b]	0.1, 0.9	—	—	—	—
Oil control	Day	Op	1.6	1.3, 1.9	139	20, 258	213	129, 297
25	Day	Op	0.3[b]	0.1, 0.5	110	23, 197	115	3, 227
75	10:00	Op	0.6[b]	−0.6, 1.2	49	−10, 108	86[b]	49, 123
Oil control	Day	N	1.4	1.1, 1.7	98	−5, 205	181	14, 348
25	Day	N	1.3[c]	1.1, 1.5	95	−16, 206	69	36, 102

[a] Night injections were given at 20:00, 23:00, and 02:00. Day injections were given at 10:00, 13:00, and 16:00. Blood samples were taken only at 15:00. Means and 1% confidence intervals calculated from data of Tamarkin et al. (1977).
[b] Less than control.
[c] Greater than value for operated.

plasma, angiotensinogen, to produce a decapeptide, angiotensin I, which is then converted by enzymes in plasma and in the lungs to angiotensin II, which has the specific effect of increasing blood pressure, and is the most powerful pressor substance known. Renin is a hormone, in terms of the classical definition, in that it is a specific substance, produced in cells remote from its site of action (target), and is transported in the blood to that site which is the plasma itself.

The secretion of renin and the action of angiotensin bear complex relationships to the concentration of Na^+ in plasma and to blood volume, and these will not be considered in detail here. The matter has been reviewed by Gavras et al. (1973, 1975) and by many others.

Angiotensin has been implicated as a possible stimulus for the release of aldosterone from the adrenal cortex. This question has been reexamined by Campbell et al. (1974), using unanesthetized rats, on the grounds that anesthesia may cause release of aldosterone. They infused angiotensin II (AT II) and angiotensin III (deaspartyl AT III), and showed that either increases plasma aldosterone of unanesthetized rats, in proportion to log dose. This action was blocked by an antagonist to AT II. AT II is bound to particulate receptors from bovine adrenal cortex (Glossmann et al., 1974). The binding is enhanced by Na^+ and K^+, and sodium ion elicits the appearance of new sites with high affinity for AT II. These properties help to explain the dependence of aldosterone secretion on plasma $[Na^+]$, since the renin–angiotensin system is also dependent on Na^+.

B. ERYTHROPOIETIN

The kidney also produces an erythropoietin, a glycoprotein which is concentrated in the lysosomes of kidney cells and which, on release into the plasma, reacts with a component of plasma to produce a product which stimulates the production of erythrocytes in bone marrow (Libbin et al., 1974; Peschle and Condorelli, 1975). A similar product is also formed in regenerating liver in response to hypoxia (Naughton et al., 1977).

REFERENCES

Ahlskog, J. E., and Hoebel, B. G. (1973). *Science* **182**, 166–169.
Ahlskog, J. E., Randall, P. K., and Hoebel, B. G. (1975). *Science* **190**, 399–401.
Amatruda, J. M., Livingston, J. N., and Lockwood, D. H. (1975). *Science* **188**, 264–266.
Arimura, A., Sato, H., Dupont, A., Nishi, N., and Schally, A. V. (1975). *Science* **189**, 1007–1009.
Au, W. Y. W. (1976). *Science* **193**, 1015–1017.
Bahl, O. P. (1977). *Fed. Proc., Fed. Am. Soc. Exp. Biol.* **36**, 2119–2127.
Baram, T., Koch, Y., Hazum, E., and Fridkin, M. (1977). *Science* **198**, 300–301.

Baum, M. J., and Vreeburg, J. T. M. (1973). *Science* **182**, 283–285.

Bayliss, W. M., and Starling, E. H. (1902). *J. Physiol. (London)* **28**, 325–353.

Berberich, J. J., Andrews, R. V., and Folk, G. E., Jr. (1977). *Comp. Biochem. Physiol. C.* **58**, 133–135.

Bilder, G. E., and Denckla, W. D. (1974). *Science* **185**, 1060–1061.

Boden, G., Sivitz, M. C., Owen, O. E., Essa-Koumar, N., and Landor, J. H. (1975). *Science* **190**, 163–165.

Brazeau, P., Vale, W., Burgos, R., Ling, N., Butcher, M., Rivier, J., and Guillemin, R. (1973). *Science* **179**, 77–79.

Brodish, A. (1977). *Fed. Proc., Fed. Am. Soc. Exp. Biol.* **36**, 2088–2093.

Brownstein, M., and Axelrod, J. (1973). *Science* **184**, 163–165.

Brownstein, M. J., Palkovits, M., Saavedra, J. M., Bassiri, R. M., and Utiger, R. D. (1974). *Science* **185**, 267–269.

Burke, G., Chang, L. C., and Szabo, M. (1973). *Science* **180**, 872–875.

Campbell, W. B., Brooks, S. M., and Pettinger, W. A. (1974). *Science* **184**, 994–996.

Charles, M. A., Fanska, R., Schmid, F. G., Forsham, P. H., and Grodsky, G. M. (1973). *Science* **179**, 569–571.

Costlow, M. E., Buschow, R. A., and McGuire. W. L. (1974). *Science* **184**, 85–86.

Croskerry, R. G., and Smith, G. K. (1975). *Science* **189**, 648–650.

Dalterio, S., Bartke, A., and Burstein, S. (1977). *Science* **196**, 1472–1473.

Dickmann, Z., Gupta, J. S., and Dey, B. K. (1977). *Science* **195**, 687–688.

Dryden, G. L., and Anderson, J. N. (1977). *Science* **197**, 782–784.

Einarsson, K., Gustafsson, J. A., Hive, T., and Ingelman-Sundberg, M. (1976). *J. Clin. Endocrinol. Metab.* **43**, 56–63.

Fallon, E. F., Agrawal, R., Furth, E., Steiner, A. L., and Cowden, R. (1974). *Science* **184**, 1089–1091.

Farrell, G. (1960). *Fed. Proc., Fed. Am. Soc. Exp. Biol.* **19**, 601.

Flier, J. S., Kahn, C. R., Roth, J., and Bar, R. S. (1975). *Science* **190**, 63–65.

Fuller, R. W. (1977). *Fed. Proc., Fed. Am. Soc. Exp. Biol.* **36**, 2133–2134.

Fuxe, K., Hokfelt, T., Enereth, P., Gustafsson, J. A., and Skett, P. (1977). *Science* **196**, 899–900.

Gavras, H., Brunner, H. R., Vaughan, E. J., Jr., and Laragh, J. H. (1973). *Science* **180**, 1369–1372.

Gavras, H., Brunner, H. R., Thurston, H., and Laragh, J. H. (1975). *Science* **188**, 1316–1317.

Gay, J. L., and Tomacari, R. L. (1974). *Science* **184**, 75–77.

George, F. W., and Wilson, J. D. (1978). *Science* **199**, 200–201.

Gerich, J. E., Langlois, M., Noacco, O., Karum, J. H., and Forsham, P. H. (1973). *Science* **162**, 171–173.

Glossmann, H., Baukal, A., and Catt, K. J. (1974). *Science* **185**, 281–283.

Gold, R. M. (1973). *Science* **182**, 488–489.

Goldblatt, H. (1947). *Physiol. Rev.* **27**, 120–165.

Golde, D. W., Bersch, N., and Li, C. H. (1977). *Science* **196**, 1112–1113.

Goodner, C. J., Walike, B. C., Koerker, D. J., Ensinck, J. W., Brown, A. C., Chideckel, E. W., Palmer, J., and Kalnasy, L. (1977). *Science* **195**, 177–179.

Gorbman, A., and Bern, H. A. (1962). "Textbook of Comparative Endocrinology." Wiley, New York.

Gorden, P., Lesniak, M. A., Hendricks, C. M., and Roth, T. (1973). *Science* **182**, 829–831.

Greiner, A. C., and Chan, S. C. (1978). *Science* **199**, 83–84.

Grodsky, G. M., Fanska, R., West, L., and Manning, M. (1974). *Science* **186**, 536–538.

Gruber, K. A., O'Brien, L. V., and Gerstner, R. (1976). *Science* **191**, 472–475.

Gustafsson, J. A., and Stenberg, A. (1976). *Science* **191**, 203–204.

Hammarström, S., Hamberg, M., Duell, E. A., Stawiski, M. A., Anderson, T. F., and Voorhees, J. J. (1977). *Science* **197**, 994–996.

Hansen, J. W., Hoffman, H. J., and Ross, J. T. (1975). *Science* **190**, 161–163.

Haour, F., and Saxena, B. B. (1974). *Science* **185**, 444–445.

Hata, S., Kunita, H., and Okamoto, M. (1976). *J. Clin. Endocrinol. Metab.* **43**, 173–177.

Hedlund, L., Lischko, M. M., Rollag, M. D., and Niswender, G. D. (1977). *Science* **195**, 686–687.

Hellman, B., Stehlin, J., and Täljedal, I. B. (1976). *Science* **194**, 1421–1423.

Hollenberg, M. D., and Cuatrecasas P. (1975). *Fed. Proc., Fed. Am. Soc. Exp. Biol.* **34**, 1556–1563.

Horwitz, K. B., McGuire, W. L., Pearson, O. H., and Segaloff, A. (1975). *Science* **189**, 726–727.

Huang, S. W., and Maclaren, N. K. (1976). *Science* **192**, 64–66.

Jensen, E. V., and De Sombre, E. R. (1973). *Science* **182**, 126–134.

Jukes, T. H. (1976). *BioScience* **26**, 544–547.

Kao, I., and Gordon, A. M. (1975). *Science* **188**, 740–741.

Kemper, B., Habener, J. F., Rich, A., and Potts, J. T., Jr. (1974). *Science* **184**, 167–169.

Kenny, A. D., and Musacchia, X. J. (1977). *Comp. Biochem. Physiol. A* **57**, 485–489.

Klein, D. C., Buda, M. J., Kapoor, C. L., and Krishna, G. (1978). *Science* **199**, 309–311.

Koerker, D. J., Ruch, W., Chideckel, E. W., Palmer, J., Goodner, C. J., Ensinck, J., and Gale, C. C. (1974). *Science* **184**, 482–484.

Koninckx, P., de Hertogh, R., Heyns, W., Meulepas, E., Brosens, L., and de Moor, P. (1976). *J. Clin. Endocrinol. Metab.* **43**, 159–167.

Koo, G. C., Wachtel, S. S., Saenger, P., New, M. I., Dosik, K., Amerose, A. P., Dorus, E., and Ventruto, V (1977). *Science* **196**, 655–656.

Lawrence, A. M., Tan, S., Hojvat, S., and Kirstens, L. (1977). *Science* **195**, 70–72.

Leclercq-Meyer, V., Rebolledo, O., Marchand, J., and Malaisse, W. J. (1975). *Science* **189**, 897–899.

Lee, P. A., Plotnik, L. P., Steele, R. E., Thompson, K. G., and Blizzard, R. M. (1976). *J. Clin. Endocrinol. Metab.* **43**, 168–172.

Leon, M., Numan, M., and Moltz, H. (1973). *Science* **179**, 1018–1019.

Lerner, A. B., and Case, J. D. (1960). *Fed. Proc., Fed. Am. Soc. Exp. Biol.* **19**, 590.

Libbin, R. M., Person, P., and Gordon, H. S. (1974). *Science* **185**, 1174–1176.

McLachlan, J. A., Newbold, R. R., and Bullock, B. (1975). *Science* **190**, 991–992.

Macrides, F., Bartke, A., and Dalterio, S. (1975). *Science* **189**, 1104–1106.

Maeda, M., Kuzuya, N., Masuyama, Y., Imai, Y., Ikeda, H., Uchimura, H., Matsuzaki, F., Kumagai, C. F., and Nagataki, S. (1976). *J. Clin. Endocrinol. Metab.* **43**, 10–17.

Martin, J. B., Renaud, L. P. and Brazeau, P., Jr. (1974). *Science* **186**, 538–540.

Martin, J. E., and Klein, D. C. (1976). *Science* **191**, 301–302.

Milsum, J. H. (1966). "Biological Control Systems Analysis." McGraw-Hill, New York.

Moltz, H., and Leidahl, L. C. (1977). *Science* **196**, 81–83.

Moss, R. L., and McCann, S. M. (1973). *Science* **181**, 177–179.

Mullin, B. R., Pacuszko, T., Lee, G., Kohn, L. D., Brady, R. O., and Fishman, P. H. (1978). *Science* **199**, 77–79.

Naughton, R. A., Kaplan, S. M., Roy, M., Burdowski, A. J., Gordon, A. S., and Piliero, S. J. (1977). *Science* **196**, 301–302.

Neary, J. T., Kieffer, J. D., Federico, P., Mover, H., Maloof, F., and Soodak, M. (1976). *Science* **193**, 403–405.

Nicol, S. C. (1978). *Comp. Biochem. Physiol. C* **59**, 33–37.
Niki, A., Niki, H., Miwa, I., and Okuda, J. (1974). *Science* **186**, 150–151.
Noller, K. L. (1976). *BioScience* **26**, 541–543.
Norris, J. S., and Kohler, P. O. (1976). *Science* **192**, 898–900.
O'Malley, B. W., and Schrader, W. T. (1976). *Sci. Am.* **234**, 32–43.
Orci, L., Amherdt, M., Malaisse-Lague, F., and Renold, A. E. (1973a). *Science* **179**, 82–84.
Orci, L., Amherdt, M., Henquin, J. C., Unger, R. H., and Renold, A. E. (1973b). *Science* **180**, 647–649.
Patillo, R. A., Hussa, R. O., Story, M. T., Ruckert, A. C. F., Shalaby, M. R., and Mattingly, R. F. (1977). *Science* **196**, 1465–1458.
Pavel, S., Dorescu, M., Petrescu-Holban, R., and Ghinea, A. (1973). *Science* **181**, 1252–1253.
Pavel, S., Goldstein, R., Gheorghiu, C., and Calb, M. (1977). *Science* **197**, 179–180.
Pelletier, G., Dudé, D., and Pardani, R. (1977). *Science* **196**, 1469–1470.
Perchellet, J. P., Shanker, G., and Sharma, R. K. (1978). *Science* **199**, 311–312.
Peschle, C., and Condorelli, M. (1975). *Science* **190**, 910–912.
Pfaff, D. W. (1973). *Science* **182**, 1148–1149.
Pictet, R. L., Rall, L. B., Phelps, P., and Rutter, W. J. (1976). *Science* **191**, 191–192.
Pipeleers, D. G., Pipeleers-Marchal, M. A., and Kipnis, D. M. (1976). *Science* **191**, 88–90.
Posner, B. I., Kelly, P. A., and Friesen, H. G. (1975). *Science* **188**, 57–59.
Ramey, J. N., Burrow, G. N., Spaulding, S. W., Donabedian, R. K., Speroff, L., and Frantz, A. G. (1976). *J. Clin. Endocrinol. Metab.* **43**, 107–114.
Reiter, R. J., Vaughan, M. K., Blask, D. E., and Johnson, L. Y. (1974). *Science* **185**, 1169–1171.
Romero, J., and Axelrod, J. (1974). *Science* **184**, 1091–1092.
Rosenbloom, A. L., Goldstein, S., and Yip, C. L. (1976). *Science* **192**, 64–66.
Rossini, A. A., Berger, M., Shadden, J., and Cahill, G. F., Jr. (1974). *Science* **183**, 424–425.
Rubin, E., Lieber, C. S., Altman, K., Gordon, G. G., and Southren, A. L. (1976). *Science* **191**, 563–564.
Samuels, H. H., Tsai, J. S., and Cintron, R. (1973). *Science* **181**, 1253–1256.
Samuels, H. H., Tsai, J. S., and Casanova, J. (1974). *Science* **184**, 1188–1191.
Sar, M., and Stumpf, W. E. (1973). *Science* **179**, 389–391.
Saxena, B. B., Hasan, S. H., Haour, F., and Schmidt-Gollwitzer, M. (1974). *Science* **184**, 793–795.
Sayers, G. (1977). *Fed. Proc., Fed. Am. Soc. Exp. Biol.* **36**, 2087–2088.
Sayers, G., and Beall, R. J. (1973). *Science* **179**, 1330–1331.
Schalekamp, M. A. D. H., Donker, S. C. B., Jansen-Goemans, A., Fawzi, T. D., and Muller, A. (1976). *J. Clin. Endocrinol. Metab.* **43**, 287–294.
Schally, A. V., Arimura, A., and Kastin, A. T. (1973). *Science* **179**, 341–350.
Schally, A. V., Kastin, A. J., and Arimura, A. (1977). *Am. Sci.* **65**, 712–719.
Scheer, B. T. (1963). "Animal Physiology." Wiley, New York.
Scheer, B. T., and Langford, R. W. (1976). *Gen. Comp. Endocrinol.* **30**, 313–326.
Schneider, L. E., Schedl, H. P., McCain, T., and Haussler, M. R. (1977). *Science* **196**, 1452–1454.
Schussler, G. C., and Orlando, J. (1978). *Science* **199**, 686–687.
Shenkman, L., Imai, Y., Kataoka, K., Hollander, C. S., Wan, L., Tang, S. C., and Avruskin, T. (1974). *Science* **184**, 81–82.

Shimamoto, K., and Miyahara, M. (1976). *J. Clin. Endocrinol. Metab.* **43**, 201–207.
Shiu, R. P. C., and Friesen, H. G. (1976). *Science* **192**, 259–261.
Silverman, D. A., and Rose, N. R. (1974). *Science* **184**, 162–163.
Silvers, W. K., and Kiang, S.-L. (1973). *Science* **181**, 570–572.
Silvers, W. K., and Wachtel, S. S. (1977). *Science* **195**, 956–960.
Skeggs, L. T., Levine, M., Lentz, K. E., Kahn, J. R., and Dorer, F. E. (1977). *Fed. Proc., Fed. Am. Soc. Exp. Biol.* **36**, 1755–1759.
Smith, J. R., Latham, T. R., Chesnut, R. M., Carino, M. A., and Morita, A. (1977). *Science* **196**, 660–662.
Smythe, G. A., and Lazarus, L. (1974). *Science* **184**, 1373–1374.
Stein, T. P., Leskiw, M. J., Wallace, H. W., and Blakemore, W. S. (1976). *J. Clin. Endocrinol. Metab.* **43**, 18–25.
Sterling, K., Brenner, M. A., and Suldanha, V. F. (1973). *Science* **179**, 1000–1001.
Strand, F. L. (1975). *BioScience* **25**, 568–577.
Stricker, E. M., Rowland, N., Saller, C. F., and Friedman, M. I. (1977). *Science* **196**, 79–81.
Sulzman, F. M., Fuller, C. A., and Moore-Ede, M. C. (1978). *Comp. Biochem. Physiol. A* **59**, 279–283.
Sumar, J., Smith, G. W., Mayhua, E., and Nathanielsz, P. W. (1978). *Comp. Biochem. Physiol. A.* **59**, 79–84.
Tamarkin, L., Hollister, C. W., Lefebvre, N. G., and Goldman, B. D. (1977). *Science* **198**, 953–955.
Terry, L. C., Willoughby, J. O., Brazeau, P., Martin, J. B., and Patel, Y. (1976) *Science* **192**, 565–567.
Tsai, J. S., and Chen, A. (1976). *Science* **194**, 202–204.
Turek, F. W., Desjardins, C., and Menaker, M. (1975). *Science* **190**, 280–282.
Turner, C. D. (1955). "General Endocrinology," 2nd ed. Saunders, Philadelphia, Pennsylvania.
Vale, W., and Rivier, C. (1977). *Fed. Proc., Fed. Am. Soc. Exp. Biol.* **36**, 2094–2099.
Vaughan, M. K., Vaughan, G. M., and Klein, D. C. (1974). *Science* **186**, 938–939.
Vega, D. F., Martinez-Victoria, E., Esteller, A., and Murillo, A. (1977). *Comp. Biochem. Physiol. A* **58**, 259–264.
von Euler, U. S. (1948). *Acta Physiol. Scand.* **16**, 63.
Wachtel, S. S. (1977). *Science* **198**, 797–799.
Weeke, J., and Laurberg, P. (1976). *J. Clin. Endocrinol. Metab.* **43**, 32–37.
Weiss, G., O'Byrne, E. M., and Steinetz, B. G. (1976). *Science* **194**, 948–949.
Winokur, A., and Utiger, R. D. (1974). *Science* **185**, 265–267.
Yeh, K.-Y., and Moog, F. (1974). *Science* **183**, 77–79.
York, D. A., Otto, W., and Taylor, T. G. (1978). *Comp. Biochem. Physiol. B* **59**, 59–65.
Zara, D. T., Chamness, G. C., Horwitz, K. B., and McGuire, W. L. (1977). *Science* **196**, 663–664.
Zimmerman, E. A., Carmer, P. W., Husain, M. K., Ferin, M., Tannenbaum, M., Frantz, A. G., and Robinson, A. G. (1973). *Science* **182**, 952–957.

CHAPTER 5

Biochemical Aspects of Viviparity*†

Sander Shapiro and Dwain D. Hagerman

I.	Introduction	159
II.	Placental Peptide Hormones	161
	A. Gonadotropins	162
	B. Lactogens	166
	C. Thyrotropin	170
	D. Adrenocorticotropin	171
III.	Placental Steroidogenesis	173
IV.	Placental Enzymes	177
V.	The Uterine Environment of the Fetus	185
VI.	Summary	188
	References	189

I. Introduction

Viviparity, the bearing of living young, exists today in several classes of the animal kingdom: elasmobranchs, teleosts, reptiles, and mammals (Amoroso, 1952). Only in mammals does this form of embryologic development occupy a central place in the identity of the class. With the exception of the thinly populated order Monotremata, all mammals are viviparous. This unanimity in reproductive physiology suggests an advantage in Darwinian terms that is especially marked when one considers the paucity of true viviparas among reptiles and their absence within the class Amphibia (Van Tienhoven, 1968).

The evolutionary advantages of viviparity that are commonly considered include protection of the developing embryo from predators and deleterious alterations in the environment, concentration of effort in small numbers of eggs with an inherent conservation of metabolic investment, decreased risk to the mother by eliminating the need for nesting with its resultant vulnerability, and an increased ratio of fertilized to unfertilized ova by the concomitant necessity of internal fertilization. However, the infrequency of this reproductive technique

* The literature review for this article was completed in April, 1977.

† Research in the authors' laboratories was supported by grants from NIH, NICHD, Grant Numbers HD 11730 and HD 11726.

CHEMICAL ZOOLOGY, VOL. XI

in animals of the other major vertebrate classes may make these assumptions of an evolutionary improvement somewhat tenuous. Since the capacity to sustain and mature an embryo within the body exists in elasmobranchs and fishes one might expect that if it were an advantageous practice, those species possessing it would have flourished. Hence, this method of reproduction is either of no real advantage to aquatic animals or it becomes a significant advantage only when present in conjunction with other evolutionary alterations. It may be of interest in this context to note that there are no true viviparas among the amphibians.

There are a number of reptiles, including some snakes, that are viviparous, but the origin of the first mammalian viviparas remains clouded. Whether the class Mammalia is monophyletic or polyphyletic continues to be a controversial issue among paleozoologists (Morriss, 1975; Hopson and Cromton, 1969). What is more certain is that the mammalian placenta, the major anatomic adaptation of viviparity, has developed in many diverse ways. Accompanying these changes in placental structure and function a great many variations in genital tract structure and function have developed. As a result many facets of mammalian reproduction may be greatly different from one species to another.

The topics encompassed within the general area of viviparous reproduction are varied and the material voluminous. Genital and embryologic anatomy (Mossman, 1953; Arey, 1965; Rudolph and Ivy, 1931; Witschi, 1959) have been extensively explored. Placental anatomy has also excited interest for many years (Mossman, 1937; Wimsatt, 1962; Amoroso, 1959; Ramsey, 1975). The physiology and molecular biology of the ovary has been studied extensively (Mossman and Duke, 1973; Perry, 1971; James *et al.*, 1976), especially in relation to the neurohumoral control of ovulation and the maintenance of pregnancy (Gorski, 1968; Priedkalns, 1975). Placental transport has developed into a major subspecialty of the general field of physiology (Dancis and Schneider, 1975; Boreus, 1973; Meschia, 1974). It includes subjects such as the transfer of oxygen across the placenta, the pharmacodynamics of placental perfusion, and the transport of metabolites. Genital tract motility (Pauerstein, 1974; Croxatto, 1974) and the factors involved in the maintenance of pregnancy (Heap *et al.*, 1973) also interest the reproductive physiologist. Significant physiologic and metabolic alterations that occur within the host (maternal) organism have begun to be identified (Hytten and Thomson, 1968; Hytten, 1976; Bergstein, 1973; Biezenski, 1975). The intracellular actions of steroid hormones (Chan and O'Malley, 1976; Gorski and

Gannon, 1976) and the immunologic tolerance of foreign protein (Lanman, 1975; Beer and Billingham, 1974; Borland, 1975) have been studied extensively in systems involving the reproductive tract. In all of these areas the lines between the classic disciplines of anatomy, physiology, and biochemistry have become blurred. Topics such as the etiology of the onset of labor have been studied from the perspective of each of these disciplines.

The accumulation of anatomic and physiologic data concerning the reproductive process in mammals had led to an appreciation for the diversity and complexity of this process. What at first appeared to be essentially the same event among various mammalian species has been recognized as an integrated series of anatomic, physiologic, and endocrine events. The techniques of molecular biology and biochemistry have emphasized the many minute alterations that have evolved in the parallel development of the reproductive processes of various present-day mammals. We have selected, for discussion, several topics in the biochemistry of viviparity in which significant progress has been made in recent years using the techniques of the molecular biologist and biochemist. Space limitations prevent any attempt at an encyclopedic approach to the literature. The reader who wishes to pursue these topics further will find that the bibliographic references to specific subjects include review articles whenever possible.

II. Placental Peptide Hormones

During the second quarter of this century, after the concept of a cyclic alteration in the morphology of the mammalian reproductive tract had become established, it became evident that the fetus or its membranes were capable of influencing these cyclic changes. Not only did these structures have the capacity to produce steroid hormones like those produced by the ovary but they could also produce factors capable of directing ovarian function. These latter factors had biological actions similar to certain secretions of the pituitary. The question thus arose whether these placental factors were a new set of hormones or whether the placenta had captured the manufacture of existing pituitary products. Ovarian function was carefully scrutinized to determine any evolutionary advantage of such a scheme, and the human species became the most frequently studied in this regard. The placental factors that cause alterations in ovarian function have been isolated and identified as polypeptides or proteins. In addition, other placental peptide hormones have been identified that have no effect on ovarian function. The discovery of these hormones has led to the realization that the placenta has captured or developed potent control mechanisms for regulating the metabolic status of the pregnant hostess.

Detailed understanding of these hormones has only become possible as advances were made in the techniques for isolating, measuring, and analyzing the biologic characteristics of peptides and proteins in general. Because these hormones are relatively new in evolution, interest in their structures has developed in a wide circle beyond that of the reproductive biologist.

A. GONADOTROPINS

Human chorionic gonadotropin (hCG) is a glycoprotein hormone produced by the placenta and excreted in the urine throughout pregnancy (Diczfalusy, 1953; Varma et al., 1971). It is detectable in maternal blood before implantation (Saxena et al., 1974). The concentration of hCG is ten times higher in maternal serum than in fetal serum. Synthesis of hCG occurs in the syncytiotrophoblast of the chorionic villi (Jones et al., 1943). Maximum serum and urinary levels are reached during the tenth week of pregnancy. The gonadotropin is generally believed to be responsible for maintaining the human corpus luteum beyond its nonpregnant life span of 14 days. There is ample experimental evidence that hCG rescues the corpus luteum from involution (Brown et al., 1938; Vande Wiele et al., 1970) but its role in human pregnancy beyond that point remains obscure. It has also been suggested that the tropin stimulates the fetal testis in a male embryo to secrete testosterone which plays an important role in prenatal sexual differentiation.

In recent years the recognition of structural similarities between hCG and the pituitary gonadotropins, combined with the ready availability of the placental hormone in large quantities, has fostered detailed study of its structure. Numerous methods of purification, using urine as the source and involving gel filtration, ion exchange chromotography, and isoelectric focusing have been described (Bahl, 1969a; Van Hell et al., 1968; Qazi et al., 1974; Bell et al., 1969). Preparations assaying up to 18,500 IU/mg have been obtained (Grasslin et al., 1972). These preparations are homogeneous by immunodiffusion, polyacrylamide gel electrophoresis, and gel filtration; nevertheless biological heterogeneity has been reported (Qazi et al., 1974). The molecular weight of the hormone as ascertained by gel filtration has been calculated as 59,000 (Bahl, 1969a) or 65,000 (Qazi et al., 1974). Ultracentrifugation studies have given weights from 37,700 (Mori, 1970) to 62,000 (Grasslin et al., 1972). The most recent molecular weights reported (from sedimentation equilibrium) are 46,000 (Grasslin *et al.*, 1972) or 47,000 (Bahl, 1969a). Extinction measurements have given a weight of 39,000 (Morgan *et al.*, 1974). This wide dispersion of results may reflect the limitation of these physicochemical techniques when applied to a glycoprotein which contains large segments of carbohydrate within the molecule.

Human chorionic gonadotropin has been separated into two subunits of different size and amino acid sequence (Swaminathan and Bahl, 1970; Morgan and Canfield, 1971). The dissociation is accomplished in 10 M urea which eliminates disulfide bridging as a possible mode of subunit binding. A more specific determination of the forces and sites of binding has not been made. The smaller subunit (designated α) is remarkably similar, if not identical, to the α subunit of human LH and TSH (Morgan and Canfield 1971, Sairam and Li, 1973; Sairam *et al.*, 1972) which, in turn, are very similar to the corresponding subunits of the cow, pig, and sheep (Liu *et al.*, 1972; Pierce *et al.*, 1971; Maghuin-Rogister *et al.*, 1972). The α subunit contains 92 amino acids with polysaccharide side chains at residues 52 and 78 (Morgan *et al.*, 1974; Bellisario *et al.*, 1973). The two carbohydrate side chains have molecular weights of about 3000 and structures for them have been tentatively proposed (Kennedy and Chaplin, 1976). Cell-free synthesis of α subunits has been accomplished, but the intracellular site and time at which the carbohydrate moieties are attached to the subunit has not been established (Landefeld *et al.*, 1976). The β subunit is reported to have either 145 (Morgan *et al.*, 1975) or 147 (Carlsen *et al.*, 1973) amino acids and five or six oligosaccharide chains. There are no free sulfhydryl groups in the peptide and the sites of the disulfide bonds are not established. Human chorionic gonadotropin carbohydrate chains are both asparagine- and serine-linked as distinguished from other β subunit carbohydrate chains which are linked solely to asparagine (Bahl, 1969b). The average sequence in a single branch of the asparagine-linked units is NeuAc–(Fuc)–Gal–GlcNac–Man– and that in the serine-linked units is NeuAc–Gal–GalNac–(Bahl, 1969b). The sialic acid content of hCG is much higher than that of its pituitary homologue (Bahl, 1969a). Removal of a relatively small fraction (0.3%) of this sugar from the nonreducing end of the carbohydrate chains will cause a reduction in biological potency by about one-third (Van Hall *et al.*, 1971; Van Hell *et al.*, 1968). Extensive desialylation lowers biological activity still further, but does not eliminate it and the desialylated hormone remains immunologically active (Van Hall *et al.*, 1971; Gershey and Kaplan, 1974). Removal of a portion of the carbohydrate component of the hCG molecule has also been shown to decrease the molecule's capacity to induce cyclic AMP production in Leydig cells without altering binding of the hormone to the cells (Moyle, 1975). This peculiarity has not yet been adequately explained.

Significant conformational changes in the molecule have been demonstrated upon dissociation into subunits by the technique of circular dichroism spectral analysis (Holladay and Puett, 1975). The β subunit of hCG has minimal biological activity. Such activity as is present can be removed by incubation with antibody to the whole hormone but not by antibody to the β subunit, suggesting that the residual activity is the result of a contaminant (Vaitukaitis *et al.*, 1972). Recombination of the subunits restores biological activity as it should; combining the α subunit of hLH with the β subunit of hCG also restores activity (Morgan and Canfield, 1971; Morgan *et al.*, 1974; Aloj *et al.*, 1973). Since the β subunits of hLH and hCG are very similar, except for an additional 30 amino acids at the N terminal end of the hCG molecule, it may be assumed that interunit binding forces in this region of the molecule are not significant (Carlsen *et al.*, 1973).

It should be emphasized that almost all of the work on the structure of hCG has been done on material derived from the pooled urine of pregnant women (Ashitaka *et al.*, 1970). Comparative studies using placental extracts or urinary material from women at various specific times during gestation have not been reported (Donini *et al.*, 1975). The longer half-life of hCG as compared to hLH (the half times for disappearance are 11 hours and 21 minutes, respectively) may reflect a protective action of sialic acid. Removal of the sugar results in an increased rate of liver metabolism (Morell *et al.*, 1971). The relative rates of subunit uptake and catabolism in the liver have not yet been studied.

The presence of FSH-like activity in urinary hCG has been suggested (Albert, 1969). Evidence for heterogeneity in purified preparations has directed attention to the possibility that several forms of the hormone exist and that they may possess different biological activities (Van Hell *et al.*, 1968). Two purified hCG fractions from first trimester urine examined for LH and FSH activity differed in potency. These two fractions are dissimilar in amino acid content and the differences tend to resemble those between the pituitary gonadotropins (Ashitaka *et al.*, 1970). This dual activity within purified hCG has been shown to vary with the length of the gestation, with LH-like activity becoming more dominant late in pregnancy (Ashitaka *et al.*, 1970).

In subhuman primates [gorilla (Turner and Gray, 1968), chimpanzee (Hobson, 1975; Nixon *et al.*, 1972), rhesus monkey (Hodgen *et al.*,

1975b; Tullner and Hertz, 1966), stump tail monkey (Tullner, 1969), and marmoset (Hobson and Wide, 1972)] a gonadotropin is produced by the placenta which is detectable in the urine. The period of pregnancy during which the hormone is detectable varies with the species. Rhesus monkey urine contains mCG up to the fortieth day of gestation; gorilla urine contains the hormone throughout pregnancy (Hodgen *et al.*, 1975b). To date, no chemical analysis data on mCG have been published. However, immunologic cross reactivity with antibodies to hCG and hFSH suggests that there are common antigenic determinants on the three glycoproteins (Hobson and Wide, 1972; Wide and Newton, 1971; Tullner *et al.*, 1969), as might have been anticipated.

Pregnant mare serum gonadotropin (PMSG) has been used extensively for the induction of ovulation in laboratory animals. This use and the commercial availability of concentrated preparations have stimulated investigation of its chemical composition. The hormone is produced in the endometrial cups of the horse uterus. These structures were originally considered to be of maternal origin but more recent evidence strongly suggests that the hormone producing cells in these cups are of fetal origin (Allen and Moor, 1972; Allen, 1969). Secretion of PMSG is detectable on the fortieth day of gestation which is just 4 days after the endometrial cups begin to develop. Urinary levels of the hormone are less than 1% of that in serum. Commercial PMSG has been further purified using gel filtration and Sephadex chromatography to an activity of 8000–1300 IU/mg (Gospodarowicz, 1972; Schams and Papkoff, 1972) and such preparations have been shown to be electrophoretically homogeneous. A molecular weight of 53,000 was determined by sodium dodecyl sulfate gel electrophoresis. After reduction and alkylation the molecular weight was found to be 23,000 (Gospodarowicz, 1972). Subunits of PMSG have been separated after incubation in 10 M urea and will recombine with the β subunit of hCG, hLH or hFSH with regeneration of biological activity. This finding would suggest a similarity of structure in the α subunits of these hormones. The carbohydrate and sialic acid content of PMSG have also been determined and were found to be quantitatively similar to that of the human hormone (Gospodarowicz, 1972). As yet, the amino acid and carbohydrate sequences of the hormone have not been established. Nor has the catabolic pathway that leads to the near-absence of the biologically active hormone from the urine been established (Cole *et al.*, 1967). A gonadotropin has also been identified in the donkey but in much lower concentration than in the horse. Other perissodactyls (giraffe and elephant) may also produce a placental gonadotropin (Rowlands, 1964). Cross breeding (i.e., donkey, horse, and zebra) has been used to demonstrate that the quantities of gonadotropin produced are the result of genetic programming within the fetus (Allen, 1969). Here again, studies of the biological activity and chemical variation within the genus have not been made.

Placental gonadotropin has not been identified in assayed tissues of the cow, sheep, sow, rabbit, or guinea pig (Catchpole, 1969). A more extensive search for placental gonadotropins among the various orders of eutherian mammals has not been carried out.

B. LACTOGENS

A substance with somatotropic and prolactinlike activity is produced by the human placenta and circulates primarily in the maternal blood (Ito and Higashi, 1961; Fukushima, 1961; Sciarra *et al.*, 1963). It has been designated as human placental lactogen (hPL) (Josimovich and MacLaren, 1962) or human chorionic somatomammotropin (hCS) (Li *et al.*, 1968). Synthesis of the hormone by placental fragments has been demonstrated by means of standard tissue culture techniques (Grumbach and Kaplan, 1964; Gusdon and Yen, 1967). The placenta contains 50–125 mg of the hormone which suggests a rather low capacity for storage (Cohen *et al.*, 1964). This hormone is produced by the syncitiotrophoblast during the whole of human pregnancy (Beck, 1970; Sciarra *et al.*, 1963). Its turnover rate has been estimated to be between 300 mg and 10 gm per day (Kaplan *et al.*, 1968; Kaplan and Grumbach, 1965) and its half-life in the maternal bloodstream is about 30 minutes (Samaan *et al.*, 1966). The amounts produced increase throughout pregnancy as a result of both increased placental mass and increased messenger RNA activity (Boime *et al.*, 1976). Production of the peptide has been shown to occur on membrane-bound rather than on free polysomes (Ver Eecke *et al.*, 1974). The major site of degradation is not known. However, it is presumed to be the liver, as has been shown to be the case for hGH (Taylor *et al.*, 1972). In addition to the major actions of hPL, luteotropic, erythropoietic, and antidiuretic effects have also been documented. While the primary biologic action of this hormone of pregnancy has not been established with certainty, most investigators would probably agree that its role in preparation of the maternal organism for lactation is very important.

Highly purified hPL from placental tissue has been obtained by exclusion chromatography on Sephadex (Li *et al.*, 1971) and precipitation in the B_1 fraction during immunoglobulin fractionation (Kasperska-Dworak, 1975; Josimovich and MacLaren, 1962; Friesen, 1965). These methods give a product with 2–4 IU/mg by bioassay. The molecular weight of the hormone is about 22,000 (Li *et al.*, 1971;

Florini *et al.*, 1966; Andrews, 1969). Analysis of these purified hPL preparations has established the hormone as a single chain polypeptide of 191 amino acids (Niall *et al.*, 1973; Li *et al.*, 1971; Sherwood *et al.*, 1971). There are two disulfide bridges within the molecule, both of which are at the same relative position as those in human growth hormone (at positions 53–165 and 182–189) (Niall *et al.*, 1971). The similarity of disulfide bridge position is also present in the human prolactin (hPr) molecule. A portion of the molecule forms an α helix (Li *et al.*, 1971). Sequence studies have established further the marked similarity between these peptides (Li *et al.*, 1973; Niall, 1971). Only 32 of the residues in hPL and hGH differ and most of these differences are conservative replacements (Niall *et al.*, 1973). Moreover, there are repetitive sequences within both of these molecules and also within human prolactin that suggest a common evolutionary origin from a smaller peptide (Niall *et al.*, 1971; Niall, 1971).

The accumulation of evidence for heterogeneity in the pituitary peptide hormones has prompted a search for similar molecular variations among the placental peptide hormones. The presence of a large molecule that is indistinguishable from hPL by radioimmunoassay, that is not a polymer or aggregate, and that is found in both maternal serum and placenta, has been documented (Schneider *et al.*, 1975). Its biologic activity and physiologic role have not been established. "Big" hPL, as it has been called, may be a nonfunctional precursor of the active hormone. A similar molecule has also been produced in a cell-free generating system using messenger RNA extracted from placenta (Boime *et al.*, 1976) but a detailed comparison between "Big" hPL and the *in vitro* product has not been made. There is also evidence for two forms of the hPL molecule that are separable on Sephadex G-50. These moieties are interconvertible and differ only in the number of their thiol groups. The form with two thiol groups has significantly greater biologic activity than that with only one thiol group. The tertiary structures of the two groups is also different (Belleville *et al.*, 1975). Such complexities as these emphasize the need for further structural studies of the active form of this hormone and the active sites on the molecule.

Searches for a human placental lactogenlike hormone in other species has resulted in the identification of similar substances in a number of eutherians. The high frequency with which lactogens are found in placental mammals is remarkable when contrasted with the limited distribution of placental gonadotropins. All of the primates thus far studied have been found to produce a placental lactogen (mPL). Like their human counterparts, and in contrast to mCG, the lactogenic hormone is produced throughout pregnancy. Extraction procedures have yielded about one-tenth the amount of lactogen found in the human organ on a weight basis (Friesen, 1973). Extracts of monkey placenta contained two immunoreactive components when examined by exclusion gel chromatography. Since the estimated molecular weight of one of these proteins is twice that of the second the possibility of dimerization was suggested (Grant *et al.*, 1970; Vinik *et al.*, 1973). Polyacrylamide gel analysis of purified mPL also revealed two distinct components (Shome and Friesen, 1971). However, these have approximately the same molecular weight (22,000), and their amino acid compositions are similar with the relative contents of histidine, tyrosine, and tryptophan contributing the major differences between them (Grant *et al.*, 1970). Placental lactogen from rhesus monkeys and baboons is acidic, migrating ahead of both hCG and hPL on gel electrophoresis (Vinik *et al.*, 1973; Josimovich *et al.*, 1973). Monkey placental lactogen also differs from hPL in its greater somatotropic activity (Friesen *et al.*, 1969). As yet the question as to whether there are in fact two active placental lactogens secreted by the simian placenta has not been clearly answered; nor has the relationship between these products and the two intracellular lactogens found upon incubations of human placentas been elucidated (Suwa and Friesen, 1969).

Hybrid radioimmunoassays using antibodies to hGH and hPL have pointed to a high degree of antigenic similarity between mPL and these human hormones (Grant *et al.*, 1970; Vinik *et al.*, 1973; Shome and Friesen, 1971). Such studies suggest that mPL shares immunologic determinants with hGH to a greater extent than it does with hPL. In this regard it is interesting to note that ovine prolactin (oPr), which is quite similar to the human hormones in amino acid sequence, does not cross-react significantly with mPL in either an oPr or mPL assay system (Vinik *et al.*, 1973).

A sheep placental lactogen (oPL) has also been identified by both bioassay and radioreceptor assay (Handwerger *et al.*, 1974). The hormone is detectable in maternal blood shortly after implantation and throughout pregnancy (Kelly *et al.*, 1974). It has been purified to apparent homogeneity on disc gel electrophoresis (Chan *et al.*, 1976; Martal and Djiane, 1975). While oPL binds to growth hormone receptors of liver tissue and prolactin receptors of mammary tissue as does hPL, it fails to cross-react immunologically with hPL or hGH (Handwerger *et al.*, 1974; Chan *et al.*, 1976). These results suggest that although there is a similarity of those portions of the molecule that effect biologic activity, hPl and oPL have significant differences in tertiary structure that prevents immunologic cross-reactivity. OPL is equipotent with hGH in the radioreceptor assay suggesting that these hormones, while not antigenically similar, do have a partial likeness that results in a functional cross-reactivity (Chan *et al.*, 1976).

Purified oPL, which migrates as a single electrophoretic band, co-migrates with oGH but upon electrophoresis moves more slowly than oPR and hPL. Its molecular weight has been estimated at about 21,000 (Martal and Djiane, 1975). However, no work on the amino acid content or sequences of this hormone has been reported.

Bovine placental lactogen (bPL) has been purified nearly to homogeneity (Bolander and Fellows, 1976). The best preparation gives two closely spaced bands on gel electrophoresis and two peaks on cellulose chromatography. In contradistinction to the two components of hPL, these molecular forms of the bovine hormone have almost identical amino acid contents. Bovine PL migrates between bGH and bPr on gel electrophoresis and is very similar to these hormones in amino acid composition (it is higher in serine and glycine and lower in leucine). Its molecular weight has been estimated as 22,150 on Sephadex (Bolander and Fellows, 1976). On radioreceptor assay bPL shows low lactogenic activity which, together with the other data given above, has led investigators to postulate that it is a structural intermediate between bGH and bPr.

Evidence for production of a placental lactogen by the rabbit is not conclusive. Animals immunized against hPL are capable of pregnancy but they show a high rate of fetal absorption after implantation (El Tomi *et al.*, 1971). Placental extracts from term rabbits have shown significant reactivity in a hemagglutination inhibition system using rabbit anti-hPL serum (Gusdon *et al.*, 1970). However, coculture experiments with rabbit placental tissue have failed to elicit the characteristic histological alterations of a lactogen in mouse mammary tissue.

The rat placenta contains a substance that has mammotrophic, luteotropic, and luteolytic activity but does not evoke a response in the tibia test for growth hormone activity (Matthies, 1967; Matthies and Lyons, 1971; Lyons, 1958). This placental lactogen (rPL or chorionic mammotropin) is found in maternal serum from day 9 of pregnancy through term. Two peaks of concentration on days 12 and 18 were found using the radioreceptor assay (Kelly *et al.*, 1975). A similar result was found in the mouse placenta (Talamantes, 1975). This radioreceptor positive material can also be extracted from the placenta throughout the latter half of pregnancy but the concentration dip present in serum was not demonstrated. The discrepancy between serum and placenta concentrations suggests that the assay method may be measuring two different compounds. Furthermore, the biologic luteotropic activity which parallels the rise in radioreceptor activity in day 12 of pregnancy does not parallel the day 18 increase. A more extensive analysis of these placental and serum components is now possible, as a method of purification of rPL has been worked out using the radioreceptor assay to identify the active agent (Robertson and Friesen, 1975). Purification of 1300-fold was achieved but did not produce homogeneity, for purified substance from placenta gave four bands on gel electrophoresis, all active by radioreceptor assay. Similar examples of heterogeneity have been reported for monkey placental lactogen (Shome and Friesen, 1971; Chrambach *et al.*, 1973). Gel electrophoresis in sodium dodecyl sulfate resulted in an estimated molecular weight of 22,000 for the purified product (Robertson and Friesen, 1975).

Other eutherian mammals investigated for the production of a placental lactogen include the horse, pig, chinchilla, hamster, mouse, guinea pig, goat, fallow deer, and domestic dog. The methods used have included hemagglutination inhibition tests (Kelly *et al.*, 1976), coculture with mammary tissue (Forsyth, 1974; Talamantes, 1975) and radioreceptor assay (Talamantes, 1975). All the species listed above have given positive results in at least one of these systems. However, pig and dog placentas were negative for lactogenic activity when analyzed by the coculture method (Talamantes, 1975).

The seeming universality of placental lactogen in eutherians is remarkable. Compare the striking contrast between the very limited number of mammals which produce a placental gonadotropin and the thus far unanimous finding of a placental lactogen. Likewise, contrast the seeming variety of ontogenetic routes taken to establish a functional placenta and the capacity of these variously formed placentae to produce what appears to be a physiologically and biochemically similar protein hormone. The questions of pre- and prohormones, similarity of immunogenic sites, structure of biologically active sites, and the relationship between the hormones in various species will be clarified by the establishment of the amino acid sequences for these hormones. The availability of such information should shed much new light on the problems of taxonomy and phylogeny within the mammalian class.

C. Thyrotropin

The appearance of thyroid gland enlargement during human pregnancy and the recognition of hyperthyroidism in a relatively large percentage of women with hydatidiform moles has stimulated searches for chorionic thyrotropin in the placenta and the plasma of pregnant women (Josimovich and MacLaren, 1962; Cohen and Utiger, 1970; Tojo *et al.*, 1974). The data collected so far suggest that there are two agents produced by the human placenta that are capable of evoking a thyroid stimulatorlike response upon bioassay. One of these is a glycoprotein with properties similar to those of human pituitary thyroid stimulating hormone (hTSH) and the other is most probably hCG. In patients with high serum levels of hCG secondary to hydatidiform mole or choriocarcinoma there is a good correlation between bioassayable thyroid-stimulating hormone and radioimmunoassayable hCG (Higgens et al., 1975). Purification of material from molar tissue results in the maintenance of the ratio between these two activities (Kenimer et al., 1975). Material obtained from disc gels and electrofocusing shows that the two activities migrate together. Moreover, regeneration of hCG from subunits has been found to restore both gonadotropin and thyroid-stimulatory activity (Nisula et al., 1974). The thyroid-stimulating activity of hCG has been calculated as 1/400th that of TSH on a molar basis (Kenimer et al., 1975).

The other human placental thyroid-stimulator, human chorionic thyrotropin (hCT), has been extracted from normal human placentas (Hershman and Starnes, 1971; Hennen, 1965) and from pregnancy serum (Hennen et al., 1969). Placental hCT has been purified 3500-fold but remains heterogeneous on gel electrophoresis with a bioassayable activity of 0.35 IU/mg (Hennen and Freychet, 1974). Antisera to the α subunit of hCG will neutralize hTSH but not hCT (Nisula et al., 1973). Antibody to intact hTSH will not neutralize hCT. However, antibody to the bovine pituitary product (bTSH) will neutralize hCT if the antibody–antigen ratio is twice that used for hTSH (Hennen and Freychet, 1974). These results are consistent with the assumption that hCT and hTSH have similar but not identical structures. Gel filtration experiments have given a molecular weight for hCT that is close to that for hTSH (Hershman and Starnes, 1971). More extensive purification of hCT will be necessary before the amino acid and sugar sequences of this glycoprotein are known. Questions concerning the existence of prohormones and the mechanisms for control of production for this hormone may be answered when purification to homogeneity is achieved (Hershman et al., 1973). Left unanswered is perhaps the most fascinating question concerning this hormone; what, if any, is its physiologic role in pregnancy?

Searches for a placental thyrotropin in other primates apparently have not been carried out. Bioassayable thyrotropin activity has been found in guinea pig placentas but not in those of rats or pigs. Studies have not been reported for other eutherian mammals (Friesen, 1973).

D. ADRENOCORTICOTROPIN

Perhaps the least well-studied of the pituitarylike peptide hormones produced by the placenta is the corticotropin (Little et al., 1958). Evidence for such a hormone was first presented over 25 years ago (Jailer and Knowlton, 1950), but good data to substantiate the claim for its existence had to await the development of radioimmunoassay methods and modern techniques for protein separation. Human plasma from pregnant women and placental extracts contain a substance that cross-reacts with human adrenocorticotropic hormone in a nonparallel fashion during radioimmunoassay (Genazzani *et al.*, 1975; Rees *et al.*, 1975). Organ culture experiments were used to demonstrate that this human chorionic corticotropin (hCC) is synthesized by placental cells (Rees *et al.*, 1975). Placental tissue has also been shown to contain bioassayable corticotropic material. Purification has not, however, been successfully carried out.

In the rhesus monkey, studies on hypophysectomized, fetectomized animals have supported the existence of a placental corticotropin (Hodgen *et al.*, 1975a). There is also some evidence that the rat placenta produces an agent capable of partially sustaining the adrenal glands after hypophysectomy (Knobil and Briggs, 1955). More definite studies directed toward establishing the existence of a corticotropin in the placenta of rodents have not been published, nor have such studies been carried out in other eutherian mammals.

In addition to the peptide hormones already described, several other secretory proteins with hormonal characteristics may be produced by the placenta. Preliminary evidence for both a melanocyte-stimulating hormone and a relaxin of placental origin have been advanced (Zarrow *et al.*, 1955; Karkren and Sen, 1963). Extrapolation of findings from humans to other species, especially nonprimates, is not justified. The seemingly highly specialized human placenta may not be representative of chorioallantoic placentas in general, and the specialized adaptive mechanisms involved in protein hormone synthesis by the human placenta may be a very singular evolutionary event.

The majority of investigations of placental peptide hormones have been done with the aim of identification and purification. In several instances the amino acid sequence of the complete peptide chains or portions thereof has been established. In these instances rapid progress toward a better understanding of the relationship of tertiary structure to function should be forthcoming. As sequences become known for the various species, a further appreciation of the evolutionary and functional relationships among structure and biological actions will also develop. The study of sequences in chorionic gonadotropins of various primate species and especially of the widespread placental lactogens in all mammals has already given greater insight into the phylogenetic relationships between species. A much broader survey of species in the major orders of eutherian mammals would certainly extend this knowledge.

The development of the placental capacity to produce polypeptide hormones similar to those of the pituitary gland initially leads one to consider this an evolutionary step toward self-sufficiency. Control over maternal and possibly fetal metabolism may offer certain selective advantages. However, knowledge concerning the primary physiologic actions of these agents and dependency of various species upon them is not available. It is presently not possible to state whether the primary sites of action for these proteins is in the fetus, the placenta, or the maternal host (the seemingly unidirectional secretion of these proteins into the host may merely represent an effective way of removing them from the placenta). Moreover, the control mechanisms for the placental peptide hormones are not presently known (Gibbons *et al.*, 1975). As functional groups on the peptide chains are recognized and manipulated, a more concise and accurate picture of the place these hormones hold in the mammalian reproductive process may become known. It seems likely that the placental peptide hormones assume a triple role: regulation of certain aspects of fetal development; initiation and regulation of special maternal adaptations to the pregnant state; and coordination of the complex biochemical and physiological activities of two virtually independent, albeit symbiotic, organisms.

III. Placental Steroidogenesis

Sex steroids are secreted by the ovary of species of every vertebrate class and are necessary for implantation in all viviparous mammals studied to date with the exception of the guinea pig and armadillo (Ryan and Ainsworth, 1966). A continuing source of progesterone and in some cases estrogen, also appears to be necessary for the maintenance of mammalian pregnancy (Van Tienhoven, 1968). In many animals the ovary remains the source of these steroid hormones throughout gestation while in others the placenta assumes this role to a variable extent. The capacity of the placenta of various species to secrete estrogens has been related to the length of gestation: placental production is usually present where gestation is longer than pseudopregnancy (Ryan, 1973; Ainsworth and Ryan, 1966). By timed surgical removal of the ovary followed by steroid replacement, the extent to which pregnancy is dependent on sex steroids supplied by the ovary has been established for a number of species (Heap, 1972). In those species that do not abort after oophorectomy, either estrogen, progesterone or both have been demonstrated to arise from the placenta (Ryan and Ainsworth, 1966). The results of physiologic studies using extirpative techniques have been clarified and extended in recent years by intensive investigation of the steroid enzymology of the placenta (Ferguson and Christie, 1967). These studies have supplemented and reinforced the concepts derived from earlier work.

As is the case with the placental peptide hormones, the most thoroughly investigated species with regard to placental steroidogenesis is man. Human placenta is steroidogenically the most complete of those studied to date. It secretes large amounts of progesterone and the three classic estrogens (estrone, estradiol-17β, and estriol) from early gestation to term. Nevertheless it does not have the capacity to duplicate all of the synthetic reactions of the ovary and adrenal cortex. Absence of the ability to hydroxylate the C-11, C-17, and C-21 carbons of pregnenolone prevent the formation of glucocorticoids and mineralocorticoids in this tissue (Villee, 1969). There is evidence that the enzymatic capability to convert acetate to cholesterol is present within the placenta (Zelewski and Villee, 1966), but perfusion studies have failed to demonstrate significant production of cholesterol for secretion by this organ (Van Leusden *et al.*, 1973). Presumably the cholesterol synthesized by placental cells is used for construction of membranes required for continual existence of the cell in which synthesis takes place. The prodigious amounts of progesterone secreted by the placenta are derived from maternal blood cholesterol via pregnenolone (Ryan *et al.*, 1966; Edwards *et al.*, 1976). The maternal cholesterol is derived from the diet and biosynthesis in various organs, especially the liver. Enzymes needed to carry out the oxidation of ring A of 5-ene-steroid molecules (3β-hydydroxy-Δ^5-steroid dehydrogenase isomerase complex) have been localized to the particulate fraction of the tissue (Koide and Torres, 1965). Direct conversion of C-21 to C-19 steroids within the human placenta and that of all other species studied is prevented by the lack of a C-17,20 lyase (Villee, 1969; Sobrevilla *et al.*, 1964). Therefore the extensive steroid-aromatizing capacity of the organ cannot be utilized without an external supply of C-19 steroids. This enzymatic restriction dictates that precursor material be obtained from the maternal and fetal circulations. Dehydroepiandrosterone sulfate, as well as androstenedione and testosterone, from both sources is converted to estrone and estradiol-17β by the placental microsomes (Morand *et al.*, 1975). Estriol, the most abundant of the human pregnancy estrogens, cannot be synthesized by this pathway because the placenta is also deficient in 16α-hydroxylase (Ryan and Engel, 1953). In this instance the estrogen precursor is 16α-hydroxy-dehydroepiandrosterone sulfate (Diczfalusy and Mancuso, 1969). Its source is the liver and adrenal glands of the conceptus by way of the fetal circulation. An ample capacity for desulfation of 3β-hydroxy-5-ene steroids at the three position is a concomitant of the aromatization process in the human placenta.

There are other quantitatively less significant pathways that contribute to the plasma pools of progesterone and estrogen during human pregnancy (Diczfalusy and Mancuso, 1969; Villee, 1969). Relative contributions to the various steroid pools by each of these pathways changes during the course of gestation. Enzymatic capabilities that go unused for want of precursor also exist in the placenta (Starka et al., 1966). The products of steroid synthesis may be metabolized directly (Kitchin et al., 1967) or secreted into either the fetal or the maternal circulations. Thus placental progesterone becomes a major source of glucocorticoid precursor for the fetal adrenal, and estrogens are preferentially secreted into the maternal circulation. The interdependence of the fetal and placental steroid-synthesizing machinery, together with their dependence on maternal precursor substrate, has been recognized in the formulation of the human feto–placental unit (Diczfalusy and Mancuso, 1964). This concept has been investigated to some extent in other primates but only minimal attention has been given to the interactions in steroid production in other mammalian orders (Ryan and Hopper, 1974).

The placentas of nonhuman primates produce progesterone and estrogens. Their capabilities vary both qualitatively and quantitatively from species to species (Ryan and Hopper, 1974) and also with the sex of the fetus (Hagemenas et al., 1976). In general, progesterone levels in serum and placental tissue of nonhuman primates are lower than those found in the human. The differences observed have been attributed to both a lower production rate and an increased capacity to metabolize progesterone within the organ. Estrogen levels in the urine of nonhuman primates have also generally been lower than those of humans. All three classic estrogens have been observed in serum or urine of these animals but with varying concentrations and ratios among the species. Estriol is generally found in quite small amounts. The relative importance of the three classic estrogens to the maintenance of pregnancy has not been established for any species.

In the horse and other equine species there is evidence for placental production of both progesterone and estrogens. The placental concentrations of progesterone are higher than those found in other domestic animals (Ainsworth and Ryan, 1970), but the levels in plasma are lower (Schomberg et al., 1967). The plasma levels of this hormone fall in mid-pregnancy and then rise again late in gestation, probably as a reflection of a change in source from ovary to placenta. In addition to estrone and estradiol-17β the horse produces two B-ring unsaturated estrogens, equilin and equilinen. Labeled precursor studies have shown that the pathway to these compounds differs from that of estrone and estradiol-17β (Starka et al., 1966; Ainsworth and Ryan,

1966). The absence of a 17,20-desmolase (Ainsworth and Ryan, 1966) makes it necessary for the sheep placenta to procure C-19 compounds from an extraplacental source. As with the primates, the sheep fetus appears to be the supplier of these androgens (Findlay and Seamark, 1971). The placenta also is capable of conversion of C-21 precursors to progesterone. Little of this steroid is seen in the plasma due to rapid metabolism within the organ (Schomberg *et al.*, 1967).

The rat maintains its corpora lutea until gestation and oophorectomy at any time results in failure of the pregnancy. Neither estrogens nor progesterone have been isolated from the rat placenta. *In vitro* conversion of pregnenolone to progesterone has been demonstrated (Chan and Leathem, 1975), suggesting that small amounts of progesterone may be produced by this placenta but the amount is insufficient to maintain pregnancy. Like many other enzyme systems, those for steroid metabolism have been shown to be age dependent in the rat placenta (Sanyal and Villee, 1976). The absence of any aromatizing capability has been demonstrated in the rat placenta (Sybulski, 1969).

A number of other mammalian placentas have been evaluated for their steroid-synthesizing capabilities (Ryan and Ainsworth, 1966). Those of the seal, guinea pig, rhinoceros, and giraffe were found to contain progesterone and those of the cow, goat, pig, dog, and rabbit did not (Ryan and Ainsworth, 1966). Evidence for the existence of progesterone, together with the presence of a 3β-hydroxy-Δ^5-steroid dehydrogenase in the preimplantation blastocyst of several "non-progesterone" producing species has been presented (Dickmann *et al.*, 1976). Estrogen synthesis in microsomal preparations prepared from the placentas of the cow, pig, and goat have also been demonstrated (Ainsworth and Ryan, 1966). The guinea pig, rabbit, and mouse, like the rat, seem to be unable to synthesize estrogens within their placentas (Ainsworth and Ryan, 1966).

The capacity to synthesize steroid hormones within the placenta is associated with a longer gestational period. In those species that have gestational periods longer than the nonpregnant luteal phase either the ovary, the placenta or both have developed into producers of the required hormones. The feedback and control mechanisms for ovarian steroid synthesis are partially understood (Heap *et al.*, 1973) but those for the placenta are yet to be established. In several species there is a suggestion that local progesterone metabolites may be involved in feedback control of progesterone synthesis (Wiener, 1976). There is also suggestive evidence that hCG plays a stimulatory role in placental steroid synthesis (Tabei and Troen, 1975; Cédard *et al.*, 1970; Macome *et al.*, 1972). Another area that needs further investigation is that of the importance for the maintenance of pregnancy of the various natural estrogens that seem unique to placentas of certain orders (estriol in primates, B-ring unsaturated estrogens in equines and estradiol-17α in ruminants). The possibility that sex steroids, whether of ovarian or placental origin, effect the placenta directly also needs investigation (McCormack and Glasser, 1976). Incubation studies have not always been consistent with inferences that were made on the basis of oophorectomy experiments (Chan and Leathem, 1975) concerning the presence or absence of certain steroid metabolizing enzymes. Therefore, it is not unreasonable to expect evidence of additional placental enzyme capability will be developed in the future. While these studies may reveal capacities that are only marginal, they can contribute data that will eventually be of use in plotting the paths covered during evolution of the mammalian placenta and its steroid-synthesizing systems.

IV. Placental Enzymes

The enzyme content of the placenta has been studied in great detail and most of the voluminous literature on the subject deals with human tissue. In fact the placenta may be enzymologically the most extensively explored human tissue. Interest in the innate character of the human placenta is only partially responsible for this large body of information. Because of the requirements for large quantities of normal tissue and the limits placed on *in vivo* experimentation there has been substantial impetus to the use of the placenta in enzymatic studies of human tissue. Many specific human enzymes required for various kinds of biochemical or clinical investigations can be obtained quite easily from placentas which are readily available in large numbers and which are conveniently processed by standard techniques for enzyme isolation.

Both the "housekeeping" (i.e., common to all cells) and special function enzymes of the placenta have been investigated extensively. The enzymes for glycolysis, fatty acid synthesis, pentose shunt metabolism, hexosamine synthesis, and cholesterol synthesis have been identified. Many of the enzymes of steroid metabolism and of drug catabolism are established as present in the mammalian placenta. As might be expected, cytoplasmic, nuclear, and particulate sites have been found for the various enzymes. In certain instances either identical or very similar enzymes have been found in more than one of these sites (Ferre *et al.*, 1975). Of the nearly 1800 enzymes from all sources catalogued by the IUPAC/IUB Commission on Biochemical Nomenclature, over 130 have been identified in placental tissue by direct assay (see Table I). A number of these have been highly purified, some even to homogeneity. Nevertheless there are relatively few instances where the work of identification and analysis has been accomplished using nonhuman placental tissue. There is also only a very limited literature concerning the comparative aspects of placental enzymology.

Histological studies of the placenta have shown it to be a continuously evolving, maturing organ. Recent biochemical studies of its enzyme content have bolstered this concept (Hempel and Geyer, 1969). Isoenzymes of human alkaline phosphatase (Fishman *et al.*, 1976), hexokinase (Dean and Gusseck, 1976) and lactate dehydrogenase (Edlow *et al.*, 1971) have been shown to undergo changes in electrophoretic pattern that are temporally related. A similar relationship has been shown to exist in the development of tRNA-aminoacylating enzymes (Gusseck, 1973). In the latter case a dynamic alteration in gene expression is suggested. In the former example the change in isoenzyme content of the tissue could represent either a translationally imposed restriction on enzyme synthesis or a direct effect of altered gene expression. The rapidly changing metabolism (i.e., aging) of the placenta is a conspicuous characteristic of the organ that has only been minimally exploited.

With the availability of purified placental enzymes the analysis of their physicochemical characteristics has become possible (Fjellstedt and Robinson, 1975b). The amino acid content of several purified enzymes has been examined; however, amino acids sequencing has not been reported (Badger and Sussman, 1976). Nor have comparisons been made between the amino acid composition of the enzymes in various species. Such comparisons as well as the evaluation of isoenzyme patterns from placentas of animals from different orders, might give us greater insight into the evolutionary pattern of placental development.

In one instance the comparison of enzymes with the same function from the placenta and other tissues has cast doubt on the identity of the control mechanisms for their synthesis. In the rat both the ovary and placenta contain 3β-hydroxysteroid dehydrogenase (Marcal *et al.*, 1975). The activity of the enzyme from these two sources has been measured and found to be temporally discordant. The level of activity in the two organs is not reciprocal, nor does the placental enzyme show any alteration when gonadotropin levels are varied. These data may be interpreted to indicate that the control mechanisms for the activity of the enzyme in placenta and ovary are different. Furthermore, the dissimilar behavior of the enzyme from these two sources when inhibited

TABLE I
PLACENTAL ENZYMES

Enzyme number[a]	Name	Species	Reference
1.1.1.1	Alcohol dehydrogenase	Human	Hagerman, 1969
1.1.1.14	Sorbitol dehydrogenase	Human, sheep, cow	Rama et al., 1973; Hastein and Velle, 1969; Britton et al., 1967
1.1.1.21	Aldose reductase	Sheep, human	Hastein and Velle, 1969; Clements and Winegrad, 1972
1.1.1.22	UDPglucose dehydrogenase	Human	Grygiel and Kratzsch, 1974
1.1.1.27	Lactate dehydrogenase	Human	Edlow et al., 1971; Mino et al., 1968; Hagerman, 1969
1.1.1.40	Malate dehydrogenase (decarboxylating) (NADP⁺)	Human, rat	Diamant et al., 1975; Diamant and Shafrir, 1972; Hagerman, 1969
1.1.1.42	Isocitrate dehydrogenase (NADP⁺)	Human	Hagerman, 1969
1.1.1.43	Phosphogluconate dehydrogenase	Human, rat, guinea pig	Diamant et al., 1975; Edlow et al., 1971, 1975; Diamant and Shafrir, 1972
1.1.1.47	Glucose dehydrogenase	Human	Edlow et al., 1971
1.1.1.49	Glucose-6-phosphate dehydrogenase	Human, rat, guinea pig	Diamant and Shafrir, 1972; Menzel et al., 1970; Edlow et al., 1975; Hagerman, 1969
1.1.1.51	β-Hydroxysteroid dehydrogenase	Human, pig	Thomas and Veerkamp, 1976
1.1.1.62	Estradiol 17β-dehydrogenase	Human, mare, ewe, cat, dog, ferret, rat, pig, rabbit, guinea pig, sheep, cow	Jarabak and Street, 1971
1.1.1.139	Polyol dehydrogenase (NADP⁺)	Bovine	Hagerman, 1969
1.1.1.141	15-Hydroxyprostaglandin dehydrogenase	Human	Schlegel and Greep, 1975
1.1.1.145	3β-Hydroxy-Δ⁵-steroid dehydrogenase	Human, pig, rat, cow, goat, bat, monkey, mouse, cat, guinea pig, armadillo	Diamant et al., 1975; Marcal et al., 1975; Wiener, 1976; Hagerman, 1969

(Continued)

TABLE I (Continued)

Enzyme number[a]	Name	Species	Reference
1.1.1.146	11β-Hydroxysteroid dehydrogenase	Human	Hall and Giroud, 1971; Hagerman, 1969
1.1.1.149	20α-Hydroxysteroid dehydrogenase	Human	Billiar and Little, 1971; Hagerman, 1969
1.1.1.151	21-Hydroxysteroid dehydrogenase (NADP[+])	Human	Monder and Martinson, 1968
1.1.99.5	Glycerol-3-phosphate dehydrogenase	Human	Swierczynski et al., 1976
1.2.1.14	IMP dehydrogenase	Human	Holmes et al., 1974
1.3.99.1	Succinate dehydrogenase	Human	Hagerman, 1969
1.3.99.5	3-Oxo-5α-steroid-Δ[4]-dehydrogenase	Human	Hagerman, 1969
1.4.1.3	Glutamate dehydrogenase (NAD(P)[+])	Human	Hagerman, 1969
1.4.3.4	Amine oxidase (flavin-containing)	Human	Castren and Saarikoski, 1974; Hagerman, 1969
1.4.3.6	Amine oxidase (pyridoxal-containing)	Human	Crabbe et al., 1976
1.5.1.3	Tetrahydrofolate dehydrogenase	Human	Jarabak and Bachur, 1971
1.5.1.9	Saccharopine dehydrogenase (NAD[+], L-glutamate-forming)	Human	Fjellstedt and Robinson, 1975a
1.6.1.1	NAD(P)[+] transhydrogenase	Human, rat	Hagerman, 1969
1.6.2.4	NADPH-cytochrome reductase	Human	Hagerman, 1969
—	NADH and NADPH diaphorases	Human	Hagerman, 1969
—	Glutathione-oxytocin transhydrogenase	Human	Small and Watkins, 1974
1.9.3.1	Cytochrome oxidase	Human, rat	Schultz and Jacques, 1971; Hagerman, 1969
1.11.1.6	Catalase	Human	Matkovics et al., 1975
1.11.1.7	Peroxidase	Human	Matkovics et al., 1975
1.14.14.1	Aryl 4-monooxygenase	Human, rat	Gough et al., 1975; Bogdan and Juchau, 1970
1.14.99.7	Squalene monooxygenase (2,3-epoxidizing)	Human	Tabacik et al., 1973
—	Steroid 16α-monooxygenase	Human	Hagerman, 1969
1.14.99.9	Steroid 17α-monooxygenase	Human	Hagerman, 1969
1.15.1.1	Superoxide dismutase	Human	Matkovics et al., 1975

2.1.1.6	Catechol methyltransferase	Human, rat	Chen et al., 1974
—	Imidazole methyltransferase	Human	Hagerman, 1969
2.2.1.1	Transketolase	Human, rat	Hagerman, 1969
2.2.1.2	Transaldolase	Human, rat	Hagerman, 1969
2.3.1.6	Choline acetyltransferase	Human, monkey, lemur, cow, mongoose, horse, sheep, goat, pig, hamster, rat, rabbit, guinea pig	Rama Sastry and Henderson, 1972; Roskoski et al., 1975; Hagerman, 1969
2.3.1.21	Carnitine palmitoyltransferase	Human	Karp et al., 1971
2.3.1.23	Lysolecithin acyltransferase	Human	Robertson and Sprecher, 1966
2.4.1.1	Phosphorylase	Human, rat	Ross and Welsh, 1972; Patillo et al., 1970
2.4.1.11	Glycogen synthase	Rat	Grillo, 1972
2.4.1.17	UDPglucuronosyltransferase	Human, rabbit, rat, guinea pig	Aitio, 1974; Berte et al., 1969
2.4.2.14	Amidophosphoribosyltransferase	Human	Holmes et al., 1973
2.5.1.6	Methionine adenosyltransferase	Human	Hagerman, 1969
2.6.1.1	Aspartate aminotransferase	Human, rat	Diamant et al., 1975; Jaroszewicz et al., 1971; Diamant and Shafrir, 1972
2.6.1.2	Alanine aminotransferase	Human, rat	Jaroszewicz, et al., 1971, Diamant and Shafrir, 1972
2.6.1.4	Glycine aminotransferase	Human	Jaroszewicz, et al., 1971
2.6.1.5	Tyrosine aminotransferase	Rat	Wade and Gusseck, 1976
2.6.1.6	Leucine aminotransferase	Human	Jaroszewicz et al., 1971
2.7.1.1	Hexokinase	Human, rabbit, monkey, rat, guinea pig	Dean and Gusseck, 1976; Hagerman, 1969
2.7.1.4	Fructokinase	Human	Hagerman, 1969
2.7.1.6	Galactokinase	Human	Srivastava et al., 1972
2.7.1.11	6-Phosphofructokinase	Human	Hagerman, 1969
2.7.1.23	NAD$^+$ kinase	Human	Hagerman, 1969

181

(Continued)

TABLE I (Continued)

Enzyme number[a]	Name	Species	Reference
2.7.1.35	Pyridoxal kinase	Human	Contractor and Shane, 1969
2.7.1.40	Pyruvate kinase	Human, rat	Diamant et al., 1975; Diamant and Shafrir, 1972; Spellman and Fottrell, 1973
2.7.2.9	Carbamoyl-phosphate synthase (glutamine)	Rat	Shambaugh et al., 1971
2.7.3.2	Creatinine kinase	Human	Hagerman, 1969
2.7.4.3	Adenylate kinase	Human	Hagerman, 1969
2.7.5.1	Phosphoglucomutase	Human	Gustke and Kowalewski, 1975; Edlow et al., 1975
2.7.7.6	RNA nucleotidyltransferase	Human	Mertelsmann, 1969
2.7.7.7	DNA nucleotidyltransferase	Rat, mouse	Sherman and Rang, 1973; Velasco and Brasel, 1975
—	Reverse transcriptase (RNA-dependent DNA polymerase)	Rhesus monkey, baboon, mouse	Mayer et al., 1974; Sherman and Rang, 1973
2.8.2.1	Aryl sulfotransferase	Human, guinea pig, cow	Hagerman, 1969
2.8.2.4	Estrone sulfotransferase	Cow	Adams and Low, 1974
3.1.1.1	Carboxylesterase	Human	Hagerman, 1969
3.1.1.3	Triacylglycerol lipase	Human	Hagerman, 1969
3.1.1.4	Phospholipase	Rat	East et al., 1975
3.1.1.6	Acetylesterase	Human	Hagerman, 1969
3.1.1.8	Cholinesterase	Human, guinea pig, mouse	Kosharkji et al., 1974
3.1.1.13	Cholesterol esterase	Human	Chen and Morin, 1971
3.1.1.34	Diacylglycerol lipase	Human, rat	Hagerman, 1969
3.1.3.1	Alkaline phosphatase	Human, rat, guinea pig	Hempel and Geyer, 1969; Badger and Sussman, 1976; Doellgast and Fishman, 1974; Fishman et al., 1976; Messer et al., 1975
3.1.3.2	Acid phosphatase	Human, rat	Swallow and Harris, 1972; Schultz and Jacques, 1971

182

3.1.3.5	5′-Nucleotidase	Human	Fox and Marchant, 1975
3.1.3.9	Glucose-6-phosphatase	Human	Hagerman, 1969
3.1.4.5	Deoxyribonuclease I	Human	Sarkar and Mukherjee, 1976
3.1.4.6	Deoxyribonuclease II	Rat, mouse	Schultz and Jacques, 1971; Bertini et al., 1974
3.1.4.17	3′,5′-Cyclic-AMP phosphodiesterase	Human	Breuiller and Cédard, 1975
3.1.4.22	Ribonuclease I	Human, mouse	Sarkar and Mukherjee, 1976; Bertini et al., 1974
—	Correxonuclease	Human	Doniger and Grossman, 1976
3.1.6.1	Arylsulfatase	Human, rat	Lewicki and Trzeciak, 1972
3.1.6.2	Sterol-sulfatase	Human, rat	Townsley, 1973; Hagerman, 1969
3.2.1.17	Lysozyme	Human	Izaka et al., 1971
3.2.1.20	α-Glucosidase	Human	Thanavala et al., 1974
3.2.1.21	β-Glucosidase	Human	Pentchev et al., 1973
3.2.1.22	α-Galactosidase	Human	Beutler et al., 1975
3.2.1.23	β-Galactosidase	Human, rat	Beutler et al., 1975
3.2.1.24	α-Mannosidase	Human	Beutler et al., 1975
3.2.1.30	β-N-Acetylglucosaminidase	Human, rat	Edlow et al., 1971; Tallman et al., 1974; Vladutiu et al., 1975
3.2.1.31	β-Glucuronidase	Human, rat, mouse	Contractor and Shane, 1972; Bertini et al., 1974
3.2.1.51	α-L-Fucosidase	Human	Alhadeff et al., 1974
3.4.11.1	Aminopeptidase	Human, rat	Oya et al., 1975; Hagerman, 1969
3.4.11.3	Cystyl-aminopeptidase	Human	Oya et al., 1975
3.4.12.1	Carboxypeptidase C	Human	Hagerman, 1969
3.4.13.1	Glycyl-glycine dipeptidase	Human	Hagerman, 1969
3.4.22.1	Cathepsin B	Human	Swanson et al., 1974
3.4.23.5	Cathepsin D	Rat, mouse	Schultz and Jacques, 1971; Bertini et al., 1974
3.4.99.19	Renin	Rabbit	Hagerman, 1969
3.5.1.2	Glutaminase	Human	Hagerman, 1969
3.5.3.1	Arginase	Human	Hagerman, 1969
3.5.4.3	Guanine deaminase	Human	Hagerman, 1969
3.5.4.4	Adenosine deaminase	Human, rat, guinea pig, rabbit, cow	Sim and Maguire, 1970; Hagerman, 1969
3.5.99.2	Thiaminase	Human	Hagerman, 1969

(Continued)

TABLE I (Continued)

Enzyme number[a]	Name	Species	Reference
3.6.1.7	Acylphosphatase	Human	Hagerman, 1969
3.6.1.8	ATP pyrophosphatase	Human, guinea pig	Shami and Radde, 1971; Miller and Berndt, 1973
3.6.1.9	Nucleotide pyrophosphatase	Human	Hagerman, 1969
4.1.1.17	Ornithine decarboxylase	Mouse	Jones et al., 1972
4.1.1.28	Aromatic-L-amino-acid decarboxylase	Human	Hagerman, 1969
4.1.1.32	Phosphoenolpyruvate carboxykinase (GTP)	Human, rat	Diamant et al., 1975; Diamant and Shafrir, 1972
4.1.3.8	ATP citrate-lyase	Human	Diamant et al., 1975; Diamant and Shafrir, 1972
4.2.1.2	Fumarate hydratase	Human	Hagerman, 1969
4.2.1.3	Aconitate hydrase	Human	Hagerman, 1969
4.6.1.1	Adenylate cyclase	Human	Satoh and Ryan, 1971
5.1.3.1	Ribulosephosphate 3-epimerase	Human	Hagerman, 1969
5.3.1.6	Ribosephosphate isomerase	Human	Hagerman, 1969
5.3.1.9	Glucosephosphate isomerase	Mouse	Chapman et al., 1972
6.1.1.2	Tryptophanyl-tRNA synthetase	Human	Penneys and Muench, 1974
6.3.4.4	Adenylosuccinate synthetase	Human	Van Der Weyden and Kelly, 1974
6.4.1.2	Acetyl-CoA carboxylase	Human, rat	Diamant et al., 1975; Diamant and Shafrir, 1972
—	L-Lysine-2-oxoglutarate reductase	Human	Fjellstedt and Robinson, 1975b
—	Steroid 20-22 lyase	Human	Mason and Boyd, 1971; Hagerman, 1969
—	Steroid 17-20 lyase	Human, sheep	John and Pierrepoint, 1975; Hagerman, 1969
—	Steroid aromatase	Human, sheep, cow, horse, sow	Bellino and Osawa, 1974; Siiteri and Thompson, 1975; Mann et al., 1975; Hagerman, 1969

[a] Enzyme identification numbers are those recommended (1972) by the International Union of Pure and Applied Chemistry and the International Union of Biochemistry.

by cyanoketone would suggest that there is a structural difference as well. If further study substantiates this difference, a search for the transcriptional or translational event that causes it will be in order.

Some steroid hormones are known to be inducers of liver enzymes. It has also been suggested that gonadotropins from either the pituitary or the placenta play an important role in the modulation of placental steroid production. Comparative studies using species chosen for their particular placental hormone composition might go a long way to determining the effects of placental steroids on placental enzyme synthesis and the effects of chorionic peptide hormones on placental enzyme synthesis. However, to date such studies have not been reported.

The large number of enzymes that have been studied in mammalian placental tissue prevents a comprehensive discussion of each enzyme here. Instead, those enzymes that have been demonstrated to be present are listed in tabular form (Table I). This list has been prepared in accordance with the system of the Commission on Biochemical Nomenclature, IUPAC–IUB. The list was prepared on the basis of a previous tabulation (Hagerman, 1969) and indirect citation by way of this previous tabulation has been used in order to conserve space. The references cited are meant as an introduction to the literature and are not intended to imply any definitive report.

V. The Uterine Environment of the Fetus

Alterations in the secretions of the uterus and oviduct have long been recognized as important in sustaining the free-floating early embryo, while changes in uterine histology have signified preparation of an acceptable site for implantation (Murray et al., 1971; Aitken, 1974; Pincus and Kirsch, 1956). This process of differentiation has been described for many mammalian species and appears to be essential since the uterus is otherwise hostile to conception and implantation. In addition to providing nutrition for the conceptus, there is in many mammals a uterine mechanism by which a period of dormancy (diapause) in the embryo is induced and then terminated (Enders, 1963; Daniel and Krishnan, 1969; Wimsatt, 1975). The genital tract changes that affect sperm activity, ova and embryo metabolism and eventually implantation are primarily under sex steroid control. In recent years investigation of uterine and oviducal fluid have resulted in the recognition of several proteins that appear to be specific to the genital tract. Their secretion patterns, composition and possible roles in reproduction have come under intensive study. In addition, specific intracellular biochemical changes in the uterus have begun to be identified and quantitated.

Several proteins found within the uterus of the rabbit are specific to that structure. However, attention has been focused on one particular protein (uteroglobin) because it makes up a significant fraction of the postovulatory secretions. Uteroglobin was first recognized in the secretory fluids of the 5-day pregnant rabbit uterus. Subsequently it was found to be present during pseudopregnancy as well and in smaller amounts during estrus. There is also evidence that it is produced by the oviduct (Arthur *et al.*, 1972). Castrated animals will secrete the protein after progesterone administration (Beier and Beier-Hellwig, 1973). Immunohistologic evidence as well as incubation studies have established that it is synthesized within the endometrial epithelium (Joshi and Eberts, 1976). The fur seal secretes a protein of identical chemical character (Daniel, 1972a). Evidence that this protein promotes blastulation resulted in the application of the name blastokinin (El-Banna and Daniel, 1972). It has a molecular weight of about 15,000 by gel filtration and equilibrium ultracentrifugation (Beato, 1976). Its migration pattern on sodium dodecyl sulfate polyacrylamide gels suggests a subunit structure (Murray *et al.*, 1972). Uteroglobin has a high specific-binding-affinity for progesterone (Arthur *et al.*, 1972). It is found within the blastocyst in high concentration which has suggested to some investigators that its primary function is to facilitate progesterone transport. The protein contains some carbohydrate (5.7%) but is devoid of sialic acid (Gulyas and Kristman, 1971). Messenger RNA for uteroglobin has been extracted from the rabbit uterus and translated in both oocyte and cell-free systems (Beato, 1975; Levey and Daniel, 1976; Bullock *et al.*, 1976). The cell-free product appears to be larger than the secreted one suggesting that translation produces a preprotein (Bullock *et al.*, 1976). The site of its cleavage and the fate of the short chain are not established.

The endometrial secretory proteins of the pig have also been studied in some detail (Bazer, 1975). Seven separate bands on polyacrylamide gels develop during late estrus (Squire *et al.*, 1972). One of these has a natural lavender color, makes up 15% of the total, and migrates to the cathode of standard gels. It can be induced by progesterone administration (Knight *et al.*, 1973). The protein has a molecular weight of about 32,000 and stains periodic-acid-Schiff positive. Allantoic fluid of the pig contains an immunologically identical protein (Chen *et al.*, 1973) and this material has been purified to homogeneity by serial chromatography. A large fraction of the purified product is made up of glutamine and asparagine. Its carbohydrate content is estimated at 7.5% by weight (Chen *et al.*, 1973) and each molecule contains a single iron atom (Schlosnagle *et al.*, 1974). Phosphatase and pyrophosphatase activity are present, but in such low levels that assignment of phosphatase ester hydrolysis as a primary action must be questioned (Roberts and Bazer, 1976). The exact role of this luminal protein and of those not yet extensively investigated in the reproductive process is no more clear than that of uteroglobin. In both cases an effect on embryonic development has been postulated. For the pig, this suggestion has been supported by immunization studies that resulted in passive immunity to the protein and development of infertility (Daniel, 1972b). Proteins that are unique to uterine fluid have also been demonstrated in the rat (Pincus and Kirsch, 1956), human (Wolf and Mastroianni, 1975), and wallaby (Renfree, 1973). In each case it appears that progesterone is the agent responsible for changes in specific protein synthesis. However in none of these species does a single protein increase in concentration or as a fraction of the total luminal protein to the extent of those mentioned for the rabbit and pig. Studies so far have not established the functional significance of these luminal proteins. No investigation of such proteins in animals having an obligatory diapause has been reported. Such studies might be informative.

In addition to changes in the luminal fluids of the uterus and oviduct during the estrous cycle and pregnancy there are marked alterations in the endometrium itself. The deciduoma reaction (those changes that occur in steroid-primed endometrium after trauma) and its relationship to implantation has been studied extensively in the rat. The knowledge gained in these studies may well be only partially applicable to other species since the manner of implantation in terms of site, timing, endocrine status and histology vary immensely among mammals (Wimsatt, 1975; Boyd and Hamilton, 1952). Common to all however, is the characteristic of developmental synchrony (Beato, 1976). There is only a small period of time during which the embryo and its maternal endometrium will interact. The biochemical events that precede this time frame in both tissues are only partially understood for the rat and comparative studies in this area are limited (Psychos, 1973).

In the rat, deciduoma formation differs during pseudopregnancy and pregnancy (Glasser and Clark, 1975; Reid and Heald, 1971) so that those qualitative and quantitative alterations that have been studied in the ovariectomized, steroid-treated animal cannot be assumed to necessarily take place during pregnancy. Deciduoma formation in the rat is marked by an initial 100% increase in weight as well as extensive mitotic activity (Yochim, 1975). There are increases in DNA and RNA content of the tissue (O'Grady *et al.*, 1974; Shelesnyak and Tic, 1963) as well as new protein synthesis (Denari *et al.*, 1976). Intrauterine oxygen tension, endometrial glucose metabolism, and pentose shunt activity are also altered at this time (Yochim, 1975). All of these changes are facilitated by the presence of estrogens and progesterone. However, the steroid concentration relationships and their induced alterations that result in the multiplicity of events making up the deciduoma reaction are far from clear (Glasser and Clark, 1975). Whereas the commonly studied model systems for estrogen (ovariectomized rat uterus) and progesterone (chick oviduct) actions isolate the effects of the two steroids, the implantation reaction requires an interplay of actions and effects that has not yet been worked out on a molecular level.

VI. Summary

Oviparity and viviparity are the main methods of reproduction in the animal kingdom. In most species the young are relatively mature when hatched or born so that the parents have little or no further responsibility to the offspring. In the primates dependency of the young becomes greater as man is approached along the path of evolution and this development may be associated with the development of a more complex scheme of viviparity. No survival advantage is evident in the increasingly complex event of viviparous reproduction as one proceeds from the "lower" to the "higher" mammals. The increasing immaturity at birth of the "higher" forms may be associated with the development of more complex central nervous systems which require a longer time to mature and can use a prolonged period of learning which is ensured by the dependency of the young offspring. Presumably, the resulting increase in "learned adaptability" provides an evolutionary advantage.

Considering the whole range of events in viviparity as briefly reviewed above, it is quite clear that the progression has been toward more and more complex patterns. This may be of subtle survival value; if so, it is so subtle as to be presently invisible. Alternatively, it may be simply a reflection of the random-walk character of evolution which is permitted by two fundamental principles of operation of the whole process. One, anything that works biologically is permissible in evolution. Here, works is equivalent to survival. Two, almost any mode of reproduction is possible if a sufficient number of offspring are developed to permit the survival of some members of the species so that they can reproduce again.

The central problem in viviparous reproduction lies in the maintenance of biologic individuality while utilizing a common, integrated system for metabolism. Marsupials (and viviparous fish) avoid the problem by reducing to a minimum the contact between parent and offspring; the Eutheria have developed an interposed organ which separates that which must be separated while permitting free access to all else.

REFERENCES

Adams, J. B., and Low, J. (1974). *Biochim. Biophys. Acta* **370**, 189–196.

Ainsworth, L., and Ryan, K. J. (1966). *Endocrinology* **79**, 875–883.

Ainsworth, L., and Ryan, K. J. (1970). *Steroids* **16**, 553–559.

Aitio, A. (1974). *Biochem. Pharmacol.* **23**, 2203–2205.

Aitken, R. J. (1974). *J. Reprod. Fertil.* **39**, 225–233.

Albert, A. (1969). *J. Clin. Endocrinol. Metab.* **29**, 1504–1509.

Alhadeff, J. A., Miller, A. L., and O'Brien, J. S. (1974). *Anal. Biochem.* **60**, 424–430.

Allen, W. R. (1969). *Nature (London)* **223**, 64–66.

Allen, W. R., and Moor, R. M. (1972). *J. Reprod. Fertil.* **29**, 313–316.

Aloj, S. M., Edelhoch, H., and Ingham, K. C. (1973). *Arch. Biochem. Biophys.* **159**, 497–504.

Amoroso, E. C. (1952). *In* "Marshall's Physiology of Reproduction" (A. S. Parkes, ed.), Vol. 2, pp. 127–311. Longmans, Green, New York.

Amoroso, E. C. (1959). *Ann. N.Y. Acad. Sci.* **75**, 855–872.

Andrews, P. (1969). *Biochem. J.* **111**, 799–800.

Arey, L. B. (1965). "Developmental Anatomy." Saunders, Philadelphia, Pennsylvania.

Arthur, A. T., Cowan, B. D., and Daniel, J. C. (1972). *Fertil. Steril.* **23**, 69–77.

Ashitaka, Y., Tokura, Y., Tane, M., Mochizuki, M., and Tojo, S. (1970). *Endocrinology* **87**, 233–244.

Badger, K. S., and Sussman, H. H. (1976). *Proc. Natl. Acad. Sci. U.S.A.* **73**, 2201–2205.

Bahl, O. P. (1969a). *J. Biol. Chem.* **244**, 567–574.

Bahl, O. P. (1969b). *J. Biol. Chem.* **244**, 575–583.

Bazer, F. W. (1975). *J. Anim. Sci.* **41**, 1376–1382.

Beato, M. (1975). *FEBS Lett.* **59**, 305–309.

Beato, M. (1976). *J. Steroid Biochem.* **7**, 327–334.

Beck, J. S. (1970). *N. Engl. J. Med.* **283**, 189–190.

Beer, A. E., and Billingham, R. E. (1974). *In* "The Placenta; Biological and Clinical Aspects" (K. S. Moghissi and E. S. E. Hafez, eds.), pp. 346–367. Thomas, Springfield, Illinois.

Beier, H. M., and Beier-Hellwig, K. (1973). *Acta Endocrinol. (Copenhagen), Suppl.* **180**, 404–425.

Bell, J., Canfield, R. E., and Sciarra, J. J. (1969). *Endocrinology* **84**, 298–307.

Belleville, F., Peltier, A., Paysant, P., and Nabet, P. (1975). *Eur. J. Biochem.* **51**, 429–435.

Bellino, F. L., and Osawa, Y. (1974). *Biochemistry* **13**, 1925–1931.

Bellisario, R., Carlsen, R. B., and Bahl, O. P. (1973). *J. Biol. Chem.* **248**, 6796–6809.

Bergstein, N. A. M. (1973). "Liver and Pregnancy," pp. 159–166. Excerpta Med. Found., Amsterdam.

Berte, F., Manzo, L., De Bernardi, M., and Benzi, G. (1969). *Arch. Int. Pharmacodyn. Ther.* **182**, 182–185.

Bertini, F., Sacerdote, F. L., and Del Campo, H. (1974). *J. Reprod. Fertil.* **37**, 7–15.

Beutler, E., Guinto, E., and Kuhl, W. (1975). *J. Lab. Clin. Med.* **85**, 672–677.

Biezenski, J. J. (1975). *In* "Obstetrics and Gynecology Annual 1975" (R. Wynn, ed.), pp. 39–70. Appleton, New York.

Billiar, R. B., and Little, B. (1971). *Endocrinology* **8**, 263–267.

Bogdan, D. P., and Juchau, M. R. (1970). *Eur. J. Pharmacol.* **10**, 119–126.

Boime, I., McWilliams, D., Szczesna, E., and Camel, M. (1976). *J. Biol. Chem.* **251**, 820–825.

Bolander, F. F., Jr., and Fellows, R. E. (1976). *J. Biol. Chem.* **251**, 2703–2708.

Boreus, L. (1973). *In* "Fetal Pharmacology" (L. Boreus, ed.), pp. 111–126. Raven, New York.

Borland, R. (1975). *In* "Comparative Placentation" (D. H. Steven, ed.), pp. 268–281. Academic Press, New York.

Boyd, J. D., and Hamilton, W. S. (1952). *In* "Marshall's Physiology of Reproduction" (A. S. Parkes, ed.), Vol. 2, pp. 1–126. Longman's, Green, New York.

Breuiller, F. F., and Cédard, L. (1975). *FEBS Lett.* **52**, 295–299.

Britton, H. G., Huggett, A. St. G., and Nixon, D. A. (1967). *Biochim. Biophys. Acta* **136**, 427–440.

Brown, W. R., Ades, H., and Mettler, F. (1938). *Am. J. Physiol.* **123**, 26.

Bullock, D. W., Woo, S. L. C., and O'Malley, B. W. (1976). *Biol. Reprod.* **15**, 435–443.

Carlsen, R. B., Bahl, O., and Swaminathan, N. (1973). *J. Biol. Chem.* **248**, 6810–6827.

Castren, O., and Saarikoski, S. (1974). *Acta Obstet. Gynecol. Scand.* **53**, 41–47.

Catchpole, H. R. (1969). *In* "Reproduction in Domestic Animals" (H. H. Cole and P. T. Cupps, eds.), 2nd ed., pp. 430–341. Academic Press, New York.

Cédard, L., Alsat, E., Urtasum, M., and Varangot, J. (1970). *Steroids* **16**, 361–375.

Chan, J. S. D., Robertson, H. A., and Friesen, H. G. (1976). *Endocrinology* **98**, 65–76.

Chan, L., and O'Malley, B. W. (1976). *N. Engl. J. Med.* **294**, 1322–1328, 1372–1381, and 1430–1437.

Chan, S. W. C., and Leathem, J. H. (1975). *Endocrinology* **96**, 298–303.

Chapman, V. M., Ansell, J. D., and McLaren, A. (1972). *Dev. Biol.* **29**, 48–54.

Chen, C. H., Klein, D. C., and Robinson, J. C. (1974). *J. Reprod. Fertil.* **39**, 407–410.

Chen, L., and Morin, R. (1971). *Biochim. Biophys. Acta* **231**, 194–197.

Chen, T. T., Bazer, F. W., Cetorelli, J., Pollard, W. E., and Roberts, R. M. (1973). *J. Biol. Chem.* **248**, 8560–8566.

Chrambach, A., Yadley, R. A., Ben-David, M., and Rodbard, D. (1973). *Endocrinology* **93**, 848–857.

Clements, R. S., and Winegrad, A. I. (1972). *Biochem. Biophys. Res. Commun.* **47**, 1473–1479.

Cohen, H., Grumbach, M. M., and Kaplan, S. L. (1964). *Proc. Soc. Exp. Biol. Med.* **117**, 438–441.

Cohen, J. D., and Utiger, R. D. (1970). *J. Clin. Endocrinol. Metab.* **30**, 423–429.

Cole, H. H., Bigelow, M., Finkel, J., and Rupp, G. R. (1967). *Endocrinology* **81**, 927–930.

Contractor, S. F., and Shane, B. (1969). *Clin. Chim. Acta* **25**, 465–474.

Contractor, S. F., and Shane, B. (1972). *Biochem. J.* **128**, 11–18.

Crabbe, M. J. C., Waight, R. D., Bardsley, W. G., Baker, R. W., Kelly, I. D., and Knowles, P. F. (1976). *Biochem. J.* **155**, 679–687.

Croxatto, H. B. (1974). *In* "Physiology and Genetics of Reproduction" (E. M. Coutinho and F. Fuchs, eds.), pp. 159–166. Plenum, New York.

Dancis, J., and Schneider, J. (1975). *In* "The Placenta and Its Maternal Supply Line" (P. Gruenwald, ed.), pp. 98–124. Univ. Park Press, Baltimore, Maryland.

Daniel, J. C. (1972a). *Fertil. Steril.* **23**, 78–80.

Daniel, J. C. (1972b). *Experientia* **28**, 700–701.

Daniel, J. C., and Krishnan, R. S. (1969). *J. Exp. Zool.* **172**, 278–281.

Dean, J. D., and Gusseck, D. J. (1976). *Arch. Biochem. Biophys.* **172**, 130–134.

Denari, J. H., Germino, N. I., and Rosner, J. M. (1976). *Biol. Reprod.* **15**, 1–8.

Diamant, Y. Z., and Shafrir, E. (1972). *Biochim. Biophys. Acta* **279**, 424–430.

Diamant, Y. Z., Mayorek, N., Neuman, S., and Shafrir, E. (1975). *Am. J. Obstet. Gynecol.* **121**, 58–61.

Dickmann, Z., Dey, S. K., and Gupta, J. S. (1976). *Vitam. Horm.* (*N.Y.*) **34**, 215–242.

Diczfalusy, E. (1953). *Acta Endocrinol.* (*Copenhagen*) **12**, Suppl., 7–81.

Diczfalusy, E., and Mancuso, S. (1964). *Fed. Proc., Fed. Am. Soc. Exp. Biol.* **23**, 791–798.

Diczfalusy, E., and Mancuso, S. (1969). *In* "Foetus and Placenta" (A. Klopper and E. Diczfalusy, eds.), pp. 191–248. Blackwell, Oxford.

Doellgast, G. J., and Fishman, W. H. (1974). *Biochem. J.* **141**, 103–112.

Doniger, J., and Grossman, L. (1976). *J. Biol. Chem.* **251**, 4579–4587.

Donini, S., D'Alessio, I., and Donini, P. (1975). *Acta Endocrinol.* (*Copenhagen*) **79**, 749–766.

East, J. M., Chepenik, K. P., and Waite, M. B. (1975). *Biochim. Biophys. Acta* **388**, 106–112.

Edlow, J. B., Huddleston, J. F., Lee, G., Peterson, W. F., and Robinson, J. C. (1971). *Am. J. Obstet. Gynecol.* **111**, 360–364.

Edlow, J. B., Ota, T., Relacion, J. R., Kohler, P. O., and Robinson, J. C. (1975). *Am. J. Obstet. Gynecol.* **121**, 674–681.

Edwards, D. P., O'Connor, J. L., Bransome, E. D., and Braselton, W. C. (1976). *J. Biol. Chem.* **251**, 1632–1638.

El-Banna, A. A., and Daniel, J. C. (1972). *Fertil. Steril.* **23**, 105–114.

El-Tomi, A. E. F., Crystle, C. D., and Stevens, V. C. (1971). *Am. J. Obstet. Gynecol.* **109**, 74–77.

Enders, A. C. (1963). "Delayed Implantation." Univ. of Chicago Press, Chicago, Illinois.

Ferguson, M. M., and Christie, G. A. (1967). *J. Endocrinol.* **38**, 291–306.

Ferre, F., Breuiller, M., Cedard, L., Duchesne, M., Saintot, M., Descomps, B., and Crastes de Paulet, A. (1975). *Steroids* **26**, 551–570.

Findlay, J. K., and Seamark, R. F. (1971). *J. Reprod. Fertil.* **24**, 141–142.

Fishman, L., Miyayama, H., Driscoll, S., and Fishman, W. H. (1976). *Cancer Res.* **36**, 2268–2273.

Fjellstedt, T. A., and Robinson, J. C. (1975a). *Arch. Biochem. Biophys.* **168**, 536–548.

Fjellstedt, T, A., and Robinson, J. C. (1975b). *Arch. Biochem. Biophys.* **171**, 191–196.

Florini, J. R., Tonelli, G., Breuer, C. B., Coppola, J., Ringler, I., and Bell, R. H. (1966). *Endocrinology* **79**, 692–708.

Forsyth, I. A. (1974). *In* "Lactogenic Hormones, Fetal Nutrition, and Lactation" (J. B. Josimovich, M. Reynolds, and E. Cobo, eds.), pp. 49–67. Wiley, New York.

Fox, I. H., and Marchant, P. J. (1975). *Can. J. Biochem.* **54**, 462–469.

Friesen, H. (1965). *Endocrinology* **76**, 369–381.

Friesen, H. (1973). *Handb. Physiol. Sect. 7: Endocrinol.* **2**, Part 2, 295–309.

Friesen, H., Suwa, S., and Pare, P. (1969). *Recent Prof. Horm. Res.* **25**, 161–187.

Fukushima, M. (1961). *Tohoku J. Exp. Med.* **74**, 161–170.

Genazzani, A. R., Fraioli, F., Hurlimann, J., Fioretti, P., and Felber, J. P. (1975). *Clin. Endocrinol.* (Oxford) **4**, 1–14.

Gershey, E. L., and Kaplan, I. (1974). *Biochim. Biophys. Acta* **342**, 322–332.

Gibbons, J. M., Mitnick, M., and Chieffo, V. (1975). *Am. J. Obstet. Gynecol.* **121**, 127–131.

Glasser, S. R., and Clark, J. H. (1975). *Symp. Soc. Dev. Biol.* **33**, 311–345.

Gorski, J., and Gannon, F. (1976). *Annu. Rev. Physiol.* **38**, 425–450.

Gorski, R. A. (1968). *In* "Biology of Gestation" (N. S. Assali, ed.), Vol. I, pp. 2–66. Academic Press, New York.

Gospodarowicz, D. (1972). *Endocrinology* **91**, 101–106.

Gough, E. D., Lowe, M. C., and Juchau, M. R. (1975). *J. Natl. Cancer Inst.* **54**, 819–822.

Grant, D. B., Kaplan, S. L., and Grumbach, M. M. (1970). *Acta Endocrinol. (Copenhagen)* **63**, 736–746.

Grasslin, D., Gygan, P. J., and Weise, H. (1972). *In* "Structure Activity Relationship of Protein and Polypeptide Hormones" (M. Margoulies and F. C. Greenwood, eds.), pp. 366–368. Excerpta Med. Found., Amsterdam.

Grillo, T. A. I. (1972). *Histochemie* **30**, 13–23.

Grumbach, M. M., and Kaplan, S. L. (1964). *Trans. N.Y. Acad. Sci* [2] **27**, 167–188.

Grygiel, I. H., and Kratzsch, E. (1974). *Arch. Gynaekol.* **217**, 173–188.

Gulyas, B. J., and Kristman, R. S. (1971). *In* "The Biology of the Blastocyst" (R. J. Blandau, ed.), pp. 261–276. Univ. of Chicago Press, Chicago, Illinois.

Gusdon, J. P., and Yen, S. S. C. (1967). *Obstet. Gynecol.* **30**, 635–638.

Gusdon, J. P., Leake, N. H., Van Dyke, A. H., and Atkins, W. (1970). *Am. J. Obstet. Gynecol.* **107**, 441–444.

Gusseck, D. J. (1973). *20th Annu. Meet. Soc. Gynecol. Invest.* p. 12.

Gustke, H., and Kowalewski, S. (1975). *Enzyme* **19**, 154–164.

Hagemenas, F. C., Baughman, W. L., and Kittinger, G. W. (1976). *Endocrinology* **96**, 1059–1061.

Hagerman, D. D. (1969). *In* "Foetus and Placenta" (A. Klopper and E. Diczfalusy, eds.), pp. 413–470. Blackwell, Oxford.

Hall, C. St.-G., and Giroud, C. J. P. (1971). *Can. J. Biochem.* **49**, 1384–1387.

Handwerger, S., Maurer, W., Barrett, J., Hurley, T., and Fellows, R. E. (1974). *Endocr. Res. Commun.* **1**, 403–413.

Hastein, T., and Velle, W. (1969). *Biochim. Biophys. Acta* **178**, 1–10.

Heap, R. B. (1972). *In* "Reproduction in Mammals," Book 3 (C. R. Austin and R. V. Short, eds.), pp. 73–105. Cambridge Univ. Press, London and New York.

Heap, R. B., Perry, J. S., and Challis, J. R. G. (1973). *Hanb. Physiol. Sect. 7: Endocrinol.* **2**, Part 2, 217–260.

Hempel, V. E., and Geyer, V. (1969). *Acta Histochem.* **34**, 138–147.

Hennen, G. P. (1965). *Arch. Int. Physiol. Biochim.* **73**, 689–695.

Hennen, G. P., and Freychet, P. (1974). *Isr. J. Med. Sci.* **10**, 1332–1347.

Hennen, G. P., Pierce, J. G., and Freychet, P. (1969). *J. Clin. Endocrinol. Metab.* **29**, 581–594.

Hershman, J. M., and Starnes, W. R. (1971). *J. Clin. Endocrinol. Metab.* **32**, 52–58.

Hershman, J. M., Kojima, A., and Friesen, H. G. (1973). *J. Clin. Endocrinol. Metab.* **36**, 497–501.

Higgens, H. P., Hershman, J. M., Kenimer, J. G., Patillo, R. A., Bayley, A., and Walfish, P. (1975). *Ann. Intern. Med.* **83**, 307–311.

Hobson, B. M., and Wide, L. (1972). *J. Endocrinol.* **55**, 363–368.

Hobson, B. M. (1975). *Folia Primatol.* **23**, 135–139.

Hodgen, G. D., Gulyas, B. J., and Tullner, W. W. (1975a). *Steroids* **26**, 233–240.

Hodgen, G. D., Niemann, W. H., and Tullner, W. W. (1975b). *Endocrinology* **96**, 789–791.

Holladay, L. A., and Puett, D. (1975). *Arch. Biochem. Biophys.* **171**, 708–720.

Holmes, E. W., McDonald, J. A., McCord, J. M., Wyngaarden, J. B., and Kelley, W. N. (1973). *J. Biol. Chem.* **248**, 144–150.

Holmes, E. W., Pehlke, D. M., and Kelly, W. N. (1974). *Biochim. Biophys. Acta* **364**, 209–217.

Hopson, J. A., and Cromton, A. W. (1969). *Evol. Biol.* **3**, 15–68.

Hytten, F. E. (1976). *In* "The Kidney in Pregnancy" (R. R. de Alvarez, ed.), pp. 23–44. Wiley, New York.

Hytten, F. E., and Thomson, A. M. (1968). *In* "Biology of Gestation" (N. S. Assali, ed.), Vol. 1, pp. 450–480. Academic Press, New York.

Ito, Y., and Higashi, K. (1961). *Endocrinology Japonica* **8**, 279–287.

Izaka, K., Shirakawa, H., Yamada, M., and Suyama, T. (1971). *Anal. Biochem.* **42**, 299–309.

Jailer, J. W., and Knowlton, A. I. (1950). *J. Clin. Invest.* **29**, 1430–1436.

James, V. H. T., Seris, M., and Giusti, G., eds. (1976). "The Endocrine Function of the Human Ovary." Academic Press, New York.

Jarabak, J., and Bachur, N. R. (1971). *Arch. Biochim. Biophys.* **142**, 417–425.

Jarabak, J., and Street, M. A. (1971). *Biochemistry* **10**, 3831–3835.

Jaroszewicz, L., Jozwik, M., and Jaroszewicz, K. (1971). *Biochem. Med.* **5**, 436–439.

John, B. M., and Pierrepoint, C. G. (1975). *J. Reprod. Fertil.* **43**, 559–562.

Jones, D., Hampton, J. K., and Preslock, J. P. (1972). *Anal. Biochem.* **49**, 147–154.

Jones, G. E., Gey, G. O., and Gey, M. R. (1943). *Johns Hopkins Med. J.* **72**, 26–38.

Joshi, S. G., and Eberts, K. M. (1976). *Fertil. Steril.* **27**, 730–739.

Josimovich, J. B., and MacLaren, J. A. (1962). *Endocrinology* **71**, 209–220.

Josimovich, J. B., Levitt, M. J., and Stevens, V. C. (1973). *Endocrinology* **93**, 242–244.

Kaplan, S. L., and Grumbach, M. M. (1965). *J. Clin. Endocrinol. Metab.* **25**, 1370–1374.

Kaplan, S. L., Gurpide, E., Sciarra, J., and Grumbach, M. M. (1968). *J. Clin. Endocrinol. Metab.* **28**, 1450–1460.

Karkren, J. N., and Sen, D. P. (1963). *Ann. Biochem. Exp. Med.* **23**, 81–84.

Karp, W., Sprecher, H., and Robertson, H. (1971). *Biol. Neonate* **18**, 341–347.

Kasperska-Dworak, A. (1975). *Acta Med. Pol.* **16**, 171–184.

Kelly, P. A., Robertson, H. A., and Friesen, H. G. (1974). *Nature (London)* **248**, 435–438.

Kelly, P. A., Shiu, R. P. C., Robertson, M. C., and Friesen, H. G. (1975). *Endocrinology* **96**, 1187–1195.

Kelly, P. A., Tsuchima, T., Shiu, R. P. C., and Friesen, H. G. (1976). *Endocrinology* **99**, 765–774.

Kenimer, J. G., Hershman, J. M., and Higgens, H. P. (1975). *J. Clin. Endocrinol. Metab.* **40**, 482–491.

Kennedy, J. F., and Chaplin, M. F. (1976). *Biochem. J.* **155**, 303–315.

Kitchin, J. D., Pion, R. J., and Conrad, S. H. (1967). *Steroids* **9**, 263–274.

Knight, J. W., Bazer, F. W., and Wallace, H. D. (1973). *J. Anim. Sci.* **36**, 546–553.

Knobil, E., and Briggs, F. N. (1955). *Endocrinology* **57**, 147–152.

Koide, S. S., and Torres, M. T. (1965). *Biochim. Biophys. Acta* **105**, 115–120.

Kosharkji, R. P., Sastry, R. V. B., and Harbison, R. D. (1974). *Res. Commun. Chem. Pathol. Pharmacol.* **9**, 181–184.

Landefeld, T., Bogieslawski, S., Corash, L., and Boime, I. (1976). *Endocrinology* **98**, 1220–1227.

Lanman, J. T. (1975). *In* "The Placenta and Its Maternal Supply Line" (P. Grueniwald, ed.), pp. 145–157. Univ. Park Press, Baltimore, Maryland.

Levey, I. L., and Daniel, J. C. (1976). *Biol. Reprod.* **14**, 163–174.

Lewicki, J., and Trzeciak, W. H. (1972). *Am. J. Obstet. Gynecol.* **112**, 881–885.

Li, C. H., Grumbach, M. M., Kaplan, S. L., Josimovich, J. B., Friesen, H., and Catt, K. J. (1968). *Experientia* **24**, 1288.

Li, C. H., Dixon, J. S., and Chung, D. (1971). *Science* **173**, 56–58.
Li, C. H., Dixon, J. S., and Chung, D. (1973). *Arch. Biochem. Biophys.* **155**, 95–110.
Little, B., Smith, O. W., Jessiman, A. G., Selenkow, H. A., Van't Hoff, W., Eglin, J. M., and Moore, F. D. (1958). *J. Clin. Endocrinol. Metab.* **18**, 425–443.
Liu, W. K., Nahm, H. S., Sweeney, C. M., Lankin, W. M., Baker, H. N., and Ward, D. N. (1972). *J. Biol. Chem.* **247**, 4351–4364.
Lyons, W. R. (1958). *Proc. R. Soc. London, Ser. B* **149**, 303–325.
McCormack, S. A., and Glasser, S. R. (1976). *Endocrinology* **99**, 701–712.
Macome, J. C., Bischoff, K., UmaBai, R., and Diczfalusy, E. (1972). *Steroids* **20**, 469–485.
Maghuin-Rogister, G., Combarnous, Y., and Hennen, G. (1972). *FEBS Lett.* **25**, 57–59.
Mann, M. R., Curet, L. B., and Colas, A. E. (1975). *J. Endocrinol.* **65**, 117–125.
Marcal, J. M., Chew, N. J., Solomon, D. S., and Sherman, M. I. (1975). *Endocrinology* **96**, 1270–1279.
Martal, J., and Djiane, J. (1975). *Biochem. Biophys. Res. Commun.* **65**, 770–778.
Mason, J. I., and Boyd, G. S. (1971). *Eur. J. Biochem.* **21**, 308–321.
Matkovics, B., Fodor, S., and Kovacs, K. (1975). *Enzyme* **19**, 285–293.
Matthies, D. L. (1967). *Anat. Rec.* **159**, 55–67.
Matthies, D. L., and Lyons, W. R. (1971). *Proc. Soc. Exp. Biol. Med.* **136**, 520–523.
Mayer, R. J., Smith, R. G., and Gallo, C. R. (1974). *Science* **185**, 864–867.
Menzel, P., Gobbert, M., and Oertel, G. W. (1970). *Horm. Metab. Res.* **2**, 225–227.
Mertelsmann, R. (1969). *Eur. J. Biochem.* **9**, 311–318.
Meschia, G. (1974). *In* "The Placenta; Biological and Clinical Aspects" (K. S. Moghissi and E. S. E. Hafez, eds.), pp. 81–88. Thomas, Springfield, Illinois.
Messer, H. H., Shami, Y., and Copp, D. H. (1975). *Biochim. Biophys. Acta* **391**, 61–66.
Miller, R. K., and Berndt, W. O. (1973). *Proc. Soc. Exp. Biol. Med.* **143**, 118–122.
Mino M., Takai, T., and Yamaguchi, T. (1968). *Acta Paediatr. Jpn.* **10**, 1–5.
Monder, C., and Martinson, F. A. E. (1968). *Biochim. Biophys. Acta* **171**, 217–228.
Morand, P., Williamson, D. G., Layne, D. D., Lompa-Krzymien, L., and Salvador, J. (1975). *Biochemistry* **14**, 635–638.
Morell, A. G., Grezoriadis, G., Scheinberg, H. I., Hickman, J., and Ashwell, G. (1971). *J. Biol. Chem.* **246**, 1461–1467.
Morgan, F. J., and Canfield, R. E. (1971). *Endocrinology* **88**, 1045–1053.
Morgan, F. J., Canfield, R. E., Vaitukaitis, J. L., and Ross, C. T. (1974). *Endocrinology* **94**, 1601–1605.
Morgan, F. J., Birken, S., and Canfield, R. E. (1975). *J. Biol. Chem.* **250**, 5247–5258.
Mori, K. F. (1970). *Endocrinology* **86**, 97–106.
Morriss, G. (1975). *In* "Comparative Placentation" (D. H. Steven, ed.), pp. 87–107. Academic Press, New York.
Mossman, H. W. (1937). *Contrib. Embryol. Carnegie Inst.* **26**, 129–246.
Mossman, H. W. (1953). *J. Mammal.* **34**, 289–298.
Mossman, H. W., and Duke, K. L. (1973). "Comparative Morphology of the Mammalian Ovary." Univ. of Wisconsin Press, Madison.
Moyle, W. R. (1975). *J. Biol. Chem.* **250**, 9163–9169.
Murray, F. A., Bazer, F. W., Rundell, J. W., Vincent, C. K., Wallace, H. D., and Warnick, A. C. (1971). *J. Reprod. Fertil.* **24**, 445–448.
Murray, F. A., McGaughey, R. W., and Yarus, M. J. (1972). *Fertil. Steril.* **23**, 69–77.
Niall, H. D. (1971). *Nature (London), New Biol.* **230**, 90–91.
Niall, H. D., Hogan, M. L., Sauer, R., Rosenblum, I. Y., and Greenwood, F. C. (1971). *Proc. Natl. Acad. Sci. U.S.A.* **68**, 866–869.

Niall, H. D., Hogan, M. L., Tregéar, G. W., Segré, G. V., Hwang, P., and Friesen, H. (1973). *Recent Prog. Horm. Res.* **29**, 387–404.

Nisula, B. C., Kohler, P. O., Vaitukaitis, J. L., Hershman, J. M., and Ross, G. T. (1973). *J. Clin. Endocrinol. Metab.* **37**, 664–669.

Nisula, B. C., Morgan, F. J., and Canfield, R. E. (1974). *Biochem. Biophys. Res. Commun.* **59**, 86–91.

Nixon, W. E., Hodgen, G. D., Niemann, W. H., Ross, G. T., and Tullner, W. W. (1972). *Endocrinology* **90**, 1105–1109.

O'Grady, J. E., Moffat, G. F., and Heald, P. J. (1974). *J. Endocrinol.* **61**, I–II.

Oya, M., Yoshino, M. O., and Mizutani, S. (1975). *Experientia* **31**, 1019–1020.

Patillo, R. A., Hussa, R. O., Delfs, E., Garancis, J., Bernstein, R., Ruckert, A. C. F., Huang, W. Y., Gey, G. O., and Mattingly, R. F. (1970). *In Vitro* **6**, 205–214.

Pauerstein, C. J. (1974). "The Fallopian Tube: A Reappraisal." Lea & Febiger, Philadelphia, Pennsylvania.

Penneys, N. S., and Muench, K. H. (1974). *Biochemistry* **13**, 560–565.

Pentchev, P. G., Brady, R. O., Hibbert, S. R., Gal, A. E., and Shapiro, D. (1973). *J. Biol. Chem.* **248**, 5256–5261.

Perry, J. S. (1971). "The Ovarian Cycle of Mammals." Oliver & Boyd, Edinburgh.

Pierce, J. G., Liao, T., Carlsen, R. B., and Reimo, T. (1971). *J. Biol. Chem.* **246**, 866–872.

Pincus, G., and Kirsch, R. E. (1956). *Am. J. Physiol.* **115**, 219–228.

Priedkalns, J. (1975). *In* "Comparative Placentation" (D. H. Steven, ed.), pp. 189–213. Academic Press, New York.

Psychos, A. (1973). *Handb. Physiol. Sect. 7: Endocrinol.* **2**, Part 2, 187–216.

Qazi, M. H., Mukherjee, G., Javidi, K., Pala, A., and Diczfalusy, E. (1974). *Eur. J. Biochem.* **47**, 219–223.

Rama, F., Castellano, M. A., Germino, N. I., Miducci, M., and Ohanian, C. (1973). *J. Anat.* **114**, 109–113.

Rama Sastry, B. V., and Henderson, G. I. (1972). *Biochem. Pharmacol.* **21**, 787–802.

Ramsey, E. M. (1975). "The Placenta of Laboratory Animals and Man." Holt, New York.

Rees, L. H., Burke, C. W., Chard, T., Evans, S. W., and Letchworth, A. J. (1975). *Nature (London)* **254**, 620–622.

Reid, R. J., and Heald, P. J. (1971). *J. Reprod. Fertil.* **27**, 73–82.

Renfree, M. B. (1973). *Dev. Biol.* **32**, 41–49.

Roberts, R. M., and Bazer, F. W. (1976). *Biochem. Biophys. Res. Commun.* **68**, 450–455.

Robertson, A., and Sprecher, H. (1966). *Pediatr. Res.* **38**, 1028–1038.

Robertson, M. C., and Friesen, H. G. (1975). *Endocrinology* **97**, 621–627.

Roskoski, R., Lum, C., and Roskoski, L. M. (1975). *Biochemistry* **14**, 5105–5110.

Ross, E., and Welsh, D. (1972). *Biochim. Biophys. Acta* **264**, 490–496.

Rowlands, I. W. (1964). *In* "Gonadotropins, Their Chemical and Biological Properties and Secretory Control" (H. H. Cole, ed.), pp. 94–95.

Rudolph, L., and Ivy, A. C. (1931). *Am. J. Obstet. Gynecol.* **21**, 65–82.

Ryan, K. (1973). *Handb. Physiol. Sect. 7: Endocrinol.* **2**, Part 2, 285–294.

Ryan, K. J., and Ainsworth, L. (1966). *In* "Comparative Aspects of Reproductive Failure" (K. Benirschke, ed.), pp. 154–169. Springer-Verlag, Berlin and New York.

Ryan, K. J., and Engel, L. L. (1953). *Endocrinology* **52**, 287–291.

Ryan, K. J., and Hopper, B. R. (1974). *Contrib. Primatol.* **3**, 258–283.

Ryan, K. J., Meigs, R., and Petro, Z. (1966). *Am. J. Obstet. Gynecol.* **96**, 676–686.

Sairam, M. R., and Li, C. H. (1973). *Biochem. Biophys. Res. Commun.* **51**, 336–342.

Sairam, M. R., Papkoff, H., and Li, C. H. (1972). *Biochem. Biophys. Res. Commun.* **48**, 530–537.

Samaan, N., Yen, S. C. C., Friesen, H., and Pearson, O. H. (1966). *J. Clin. Endocrinol. Metab.* **26**, 1303–1308.

Sanyal, M. K., and Villee, C. A. (1976). *Endocrinology* **99**, 249–259.

Sarkar, G. C., and Mukherjee, G. (1976). *Indian J. Med. Res.* **64**, 792–795.

Satoh, K., and Ryan, K. J. (1971). *Biochim. Biophys. Acta* **244**, 618–624.

Saxena, B. B., Hasan, S. H., Haour, F., and Schmidt-Gollwitzer, M. (1974). *Science* **184**, 793–795.

Schams, D., and Papkoff, H. (1972). *Biochim. Biophys. Acta* **263**, 139–149.

Schlegel, W., and Greep, R. O. (1975). *Eur. J. Biochem.* **56**, 245–252.

Schlosnagle, D. C., Bazer, F. W., Tsibris, J. C. M., and Roberts, R. M. (1974). *J. Biol. Chem.* **249**, 7574–7579.

Schneider, A. B., Kowalski, K., and Sherwood, L. M. (1975). *Endocrinology* **97**, 1364–1372.

Schomberg, D. W., Coudert, S. P., and Short, R. V. (1967). *J. Reprod. Fertil.* **14**, 277–285.

Schultz, R. L., and Jacques, P. J. (1971). *Arch. Biochem. Biophys.* **144**, 292–303.

Sciarra, J. J., Kaplan, S. L., and Grumbach, M. M. (1963). *Nature (London)* **199**, 1005–1006.

Shambaugh, G. E., Metzger, B. E., and Freinkel, N. (1971). *Biochem. Biophys. Res. Commun.* **42**, 155–158.

Shami, Y., and Radde, I. C. (1971). *Biochim. Biophys. Acta* **249**, 345–352.

Shelesnyak, M. C., and Tic, L. (1963). *Acta Endocrinol. (Copenhagen)* **43**, 462–468.

Sherman, M. I., and Rang, H. S. (1973). *Dev. Biol.* **34**, 200–210.

Sherwood, L. M., Handwerger, S., McLaurin, W. D., and Lanner, M. (1971). *Nature (London) New Biol.* **233**, 59–61.

Shome, B., and Friesen, H. G. (1971). *Endocrinology* **89**, 631–641.

Siiteri, P. K., and Thompson, E. A. (1975). *J. Steroid Biochem.* **6**, 317–322.

Sim, M. K., and Maguire, M. H. (1970). *Biol. Reprod.* **2**, 291–298.

Small, C. W., and Watkins, W. B. (1974). *Nature (London)* **251**, 237–239.

Sobrevilla, L., Hagerman, D. D., and Villee, C. A. (1964). *Biochim. Biophys. Acta* **93**, 665–667.

Spellman, C., and Fottrell, P. F. (1973). *FEBS Lett.* **37**, 281–284.

Squire, G. D., Bazer, F. W., and Murray, F. A. (1972). *Biol. Reprod.* **7**, 321–325.

Srivastava, S. K., Blume, K., Van Loon, C., and Beutler, E. (1972). *Arch. Biochem. Biophys.* **150**, 191–198.

Starka, L., Breuer, H., and Cédard, L. (1966). *J. Endocrinol.* **34**, 447–456.

Suwa, S., and Friesen, H. (1969). *Endocrinology* **85**, 1037–1045.

Swallow, D. M., and Harris, H. (1972). *Ann. Hum. Genet.* **36**, 141–152.

Swaminathan, N., and Bahl, O. P. (1970). *Biochem. Biophys. Res. Commun.* **40**, 422–427.

Swanson, A. A., Martin, B. J., and Spicer, S. S. (1974). *Biochem. J.* **137**, 223–228.

Swiercyznski, J., Scislowski, P., and Aleksandrowicz, Z. (1976). *FEBS Lett.* **64**, 303–306.

Sybulski, S. (1969). *Steroids* **14**, 427–440.

Tabacik, C., Descomps, C. T., and Crastes de Paulet, A. C. (1973). *FEBS Lett.* **34**, 238–242.

Tabei, T., and Troen, P. (1975). *J. Clin. Endocrinol. Metab.* **40**, 697–704.

Talamantes, F., Jr. (1975). *Gen. Comp. Endocrinol.* **27**, 115–121.

Tallman, J. F., Brady, R. O., Quirk, J. M., Villalba, M., and Gal, A. E. (1974). *J. Biol. Chem.* **249**, 3489–3499.

Taylor, A. L., Lipman, R. L., Salam, A., and Mintz, D. H. (1972). *J. Clin. Endocrinol. Metab.* **34**, 395–399.

Thanavala, Y. M., Sheth, A. R., Thakur, A. N., Rao, S. S., and Purandare, M. (1974). *Am. J. Obstet. Gynecol.* **120**, 285–289.

Thomas, C. M. G., and Veerkamp. J. H. (1976). *Acta Endocrinol. (Copenhagen)* **82**, 150–163.

Tojo, S., Mochizuki, M., and Kanazawa, S. (1974). *Acta Obstet. Gynecol. Scand.* **53**, 369–373.

Townsley, J. D. (1973). *Endocrinology* **93**, 172–181.

Tullner, W. W. (1969). *In* "Fifth Rochester Trophoblast Conference" (C. Lund and J. W. Choate, eds.), pp. 363–377. Rochester Univ. Press, Rochester.

Tullner, W. W., and Hertz, R. (1966). *Endocrinology* **78**, 204–207.

Tullner, W. W., Rayford, P. L., and Ross, G. T. (1969). *Endocrinology* **84**, 908–911.

Turner, W. W., and Gray, C. W. (1968). *Proc. Soc. Exp. Biol. Med.* **128**, 954–955.

Vaitukaitis, J. L., Braunstein, G. D., and Ross, G. T. (1972). *Am. J. Obstet. Gynecol.* **113**, 741–757.

Van Der Weyden, M. B., and Kelly, W. N. (1974). *J. Biol. Chem.* **249**, 7282–7289.

Vande Wiele, R. L., Bogumil, J., Dyrenfurth, I., Ferin, M., Jewelewiz, R., Warrn, M., Rizkallah, T., and Mikhail, G. (1970). *Recent Prog. Horm. Res.* **26**, 63–95.

Van Hall, E. V., Vaitukaitis, J. L., and Ross, G. T. (1971). *Endocrinology* **89**, 11–15.

Van Hell, H., Matthijsen, R., and Homan, J. D. H. (1968). *Acta Endocrinol. (Copenhagen)* **59**, 89–104.

Van Leusden, H. A., Telegdy, T. G., and Diczfalusy, E. (1973). *J. Steroid Biochem.* **4**, 349–354.

Van Tienhoven, A. (1968). *In* "Reproductive Physiology of Vertebrates," pp. 311–354. Saunders, Philadelphia, Pennsylvania.

Varma, K., Larraga, L., and Selenkow, H. A. (1971). *Obstet. Gynecol.* **37**, 10–17.

Velasco, E. G., and Brasel, J. A. (1975). *J. Pediatr.* **86**, 274–279.

Ver Eecke, T., Vanduffel, L., Peeters, B., and Rombauts, W. (1974). *Arch. Int. Physiol. Biochim.* **82**, 1024–1025.

Villee, D. B. (1969). *N. Engl. J. Med.* **281**, 473–483.

Vinik, A. I., Kaplan, S. L., and Grumbach, M. M. (1973). *Endocrinology* **92**, 1051–1064.

Vladutiu, G. D., Carmody, P. J., and Rattazi, M. C. (1975). *Prep. Biochem.* **5**, 147–159.

Wade, R. S., and Gusseck, D. J. (1976). *Biochem. J.* **154**, 245–247.

Wide, L., and Newton, J. (1971). *J. Reprod. Fertil.* **27**, 103–106.

Wiener, M. (1976). *Biol. Reprod.* **14**, 306–313.

Wimsatt, W. A. (1962). *Am. J. Obstet. Gynecol.* **84**, 1568–1594.

Wimsatt, W. A. (1975). *Biol. Reprod.* **12**, 1–40.

Witschi, E. (1959). *Ann. N.Y. Acad. Sci.* **75**, 412–435.

Wolf, D. P., and Mastroianni, L. (1975). *Fertil. Steril.* **26**, 240–247.

Yochim, J. M. (1975). *Biol. Reprod.* **12**, 106–133.

Zarrow, M. X., Holmström, E. G., and Salhanick, H. A. (1955). *J. Clin. Endocrinol. Metab.* **15**, 22–24.

Zelewski, L., and Villee, C. A. (1966). *Biochemistry* **5**, 1805–1814.

CHAPTER 6

Lactation

F. J. Reithel

I.	Introduction	199
II.	Mammary Gland Structure and Its Changes	200
III.	Milk	202
IV.	Hormonal Control of Growth and Function	204
	A. Mammary Gland Growth	205
	B. Lactation	206
	C. Studies in Model Systems	207
	D. The Prolactin Receptor	208
	E. Regulation of Lactose Synthetase Activity	210
V.	Regulation of Genetic Expression	211
	Control of Prolactin Release	216
VI.	Secretion of Milk	216
VII.	The Process of Milk Ejection	219
VIII.	Selected Chemical Features Involved in Synthesis	219
	A. Lactose	219
	B. Fat	220
	C. Milk Proteins	222
IX.	Regression and Involution	223
X.	Epilogue	224
	References	225

I. Introduction

With the formulation of biochemistry as a subdivision of physiology, there arose an enthusiasm for demonstrations of unity in biochemistry. For example, in every tissue investigated, the process of glycolysis was discerned; at some stage all cells were found to contain mitochondria, and, more recently, it was found that the processes of protein synthesis are very similar in even the most diverse organisms. The number of examples that could be cited is large. Comparative biochemistry has been somewhat less sedulously pursued but has provided convincing evidence of specialization in agreement with the observations of biologists. Latterly it has become evident that much of what appears to be specialization is due to differences in regulatory mechanisms rather than qualitative differences in enzyme systems. It will be seen in what follows that much of the singular biochemistry of mammary gland

CHEMICAL ZOOLOGY, VOL. XI
Copyright © 1979 by Academic Press, Inc.
All rights of reproduction in any form reserved.
ISBN 0-12-261040-5

involves enzyme systems encountered in other cells. Specialization of function has resulted from emphasis on certain regulatory features. These in turn result from, or accompany, multihormonal controls emanating from several organs and possibly coordinated by the hypothalamus.

The gland characteristic of mammals is outstanding in its responses to a variety of control systems. Not only is its function finely attuned to the needs of the newborn, indeed its entire structure literally rises to the occasion when pregnancy occurs and then it regresses after weaning. Hence both macro- and microstructure, as well as secretory activity, are a function of two living creatures rather than one individual. Since the relation is not symbiotic, perhaps the phenomenon of lactation should be viewed as a prototype of social biochemistry or physiology.

II. Mammary Gland Structure and Its Changes

Extensive studies of the secretory tissue structure (Hollmann, 1974) have revealed a unity of a kind often encountered in cellular biology. The secretory cells, epithelial in origin, when actively secreting, are found to contain the subcellular structures providing the synthetic sites (Golgi apparatus, ergastoplasm*), those for supplying the energy required for synthesis (mitochondria) and for transporting fat and protein aggregates out of the cell (transport vacuoles, lysosomes). A large nucleus as well as a complex cell membrane are presumably involved in the mechanisms of hormonal control. Synthesized fat coalesced into droplets of substantial size (highly variable, up to 7000 nm) and protein aggregates may also be seen (30–300 nm, depending on species).

Excepting the subclass Prototheria (monotremes), the mammary gland, when lactating, can be described as follows. It is a compound gland composed of epithelial cells and of connective tissue whose ratios may vary so that considerable changes in size (of the whole gland) may occur. In the human, size of the breast correlates very poorly with milk production (Hytten, 1976). In some species, adipose tissue may replace connective tissue in part (Mayer and Klein, 1961). The connective tissue partitions off the secretory portions of the gland into lobules, the number of which varies considerably with the species. Within each lobule may be found a number of alveoli† con-

---

* Ergastoplasm is the name used by some workers for the membrane-ribosome complex that is the site of protein synthesis. Fatty acid esterification also occurs here.

† Alveolus, dim. of *alveus*, a cavity, like a honeycomb cell.

taining clusters of secretory cells (Nemanic and Pitelka, 1971). Within each lobule are also acini* which form intercalated ducts. These in turn fuse into intralobular ducts which, in turn, empty into collecting ducts terminating in a teat. (See Fig. 1.1 in Cowie and Tindal, 1971.) The parenchyma, the tissue characteristic of the gland, consists of clusters of secretory cells associated with collecting tubules, together with myoepithelial cells (Richardson, 1949) that serve as the contractile element of the gland. This complex, plus the associated circulatory system, may be seen in Fig. 82 of Turner (1952). The gross appearance of the mammary gland, the number of teats and the position on the torso are all species-variable. For discussion and description of the mammary glands of the Ungulata, see Turner (1952). For Prototheria, Metatheria, and other Eutheria, see Cowie and Tindal (1971).

Since the secretion of milk in quantity usually involves two individuals, it is to be expected that the teat and the surrounding areola would have sensory endings. There are also receptors sensitive to changes in intramammary pressure. Hence afferent discharges arise from a combination of stimuli. Efferent innervation has been shown in a number of species to involve sympathetic fibers of the autonomic system (Grosvenor and Mena, 1974).

With sexual maturity, either the adipose tissue or the connective tissue component of the mammary gland begins to proliferate. This process is species-variable. Estrogen stimulates duct growth; progesterone stimulates alveolar development. After implantation, further enlargement ensues at a rate that is a function of the gestation period. (See Figs. 1 and 2 in Smith and Abraham, 1975.) In the human, substantial alveolar development does not occur until the second trimester (Hytten, 1976). Not only is there an increase in the number of cells but, in the later weeks of pregnancy, there are changes in the subcellular complement. The Golgi (Whaley, 1975) and endoplasmic reticulum membranes become more abundant as do the mitochondria. Some protein granules and abundant fat droplets are formed but remain in the cytoplasm. Secretion occurs abruptly shortly before delivery and forms the colostrum which precedes lactation proper, induced by suckling or milking.

Lactation may continue for a period of days as in guinea pigs, months as in dairy cattle, or years as in some humans. Eventually regression occurs with a reversal of those phenomena that attended development during pregnancy. Another cycle is initiated by an ensuing pregnancy.

---

* Acinus, a berry, esp. a grape, a small saclike dilation.

III. Milk

The compound which characterizes milk as a secretion, and is common to nearly all species, is lactose [β-D-galactosyl-(1 → 4)-D-glucose], one of the very few naturally occurring disaccharides. Equally characteristic of milk, but species-variable in composition, are the caseins. They represent a group of closely related proteins that form micellar aggregates. Third, there is characteristically present "fat," a lipid mixture of great complexity and variability in composition. This is present in globules ranging in size from less than 1 μm to about 10 μm and surrounded by a membrane of phospholipid–protein complexes. A large number of other compounds with a wide spread of molecular weights may be found. These include most of the vitamins, organic phosphates, nonprotein nitrogen compounds, a long list of ions, a number of enzymes, immunoglobins and noncasein proteins. Of the latter, α-lactalbumin is noteworthy for its role in lactose biosynthesis (see Section IV,E). β-Lactoglobulin, in contrast, has no known function, although a well-known milk protein, but present only in the milk of ruminants.

Those interested in comparing the composition of milks from all the various mammalia may begin by consulting the compilation of Jenness (1974) and, for other pertinent comments on milk, Linzell and Peaker (1971). Data are being added yearly. Due to difficulties of sampling, some data are unreliable. In other cases analytical methods lacking in specificity have been used. The analytical difficulties are by no means superficial and may render doubtful the older values of even those low molecular weight compounds occurring in quantity, such as lactose. There is adequate justification for continuing effort in this analytical enterprise to support a well-founded comparative biochemistry. Figure 1 is intended to convey at a glance the range of values likely to be encountered in milk samples and a convenient way of comparing milk of one species with that of another.

One may inquire why lactose should have arisen as a particularly advantageous carbohydrate for the infant. It is seldom encountered except in milk (Venkataraman and Reithel, 1958) although β-galactosides of other kinds are ubiquitous. In seeking to rationalize the function of lactose, one is tempted to emphasize the lack of HCl secretion by the infant stomach. In contrast, low pH keeps the adult stomach relatively inhospitable toward bacterial growth. Since many bacteria do not grow well, if at all, on lactose, the presence of this sugar at pH values near neutrality may well enhance the establishment of the characteristic flora of *Lactobacillus*, presumably a favorable, if

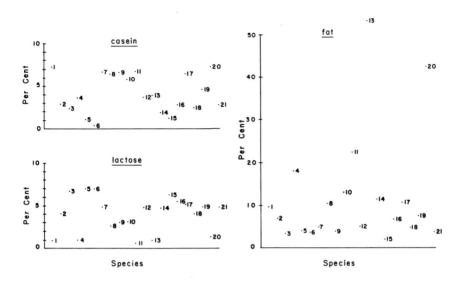

No.	Common name	Genus and species	Casein	Lactose	Fat
1	Echidna	*Tachyglossus aculeatus*	7.3	0.9	9.6
2	Opossum	*Didelphis marsupialis*	2.8	4.1	7.0
3	Kangaroo	*Mageleia rufa*	2.3	6.7	3.4
4	Cottontail	*Sylvilagus floridanus*	3.6	1.0	17.9
5	Monkey	*Macaca mulatta*	1.1	7.0	4.0
6	Woman	*Homo sapiens*	0.4	7.0	3.8
7	Hamster	*Mesocricetus auratus*	6.7	4.9	4.9
8	Rat	*Rattus norvegicus*	6.4	2.6	10.3
9	Guinea pig	*Cavia porcellus*	6.6	3.0	3.9
10	Dog	*Canis familiaris*	5.8	3.1	12.9
11	Grizzly bear	*Ursus arctos horribilis*	6.7	0.6	22.3
12	Cat	*Felius catus*	3.7	4.8	4.8
13	Harp seal	*Pagophilus groenlandicus*	3.8	0.9	52.5
14	Elephant	*Elephas maximus*	1.9	4.7	11.6
15	Horse	*Equus caballus*	1.3	6.2	1.9
16	Pig	*Sus scrofa*	2.8	5.5	6.8
17	Deer	*Odocoileus hemionus*	6.4	5.4	10.9
18	Goat	*Capra hircus*	2.5	4.1	4.5
19	Sheep	*Ovis aries*	4.6	4.8	7.4
20	Whale	*Balaenoptera musculus*	7.2	1.3	42.3
21	Cow	*Bos taurus*	2.8	4.8	3.7

FIG. 1. Comparison of milk samples from various species. From Jenness (1974, pp. 56–76).

temporary, symbiosis. It has been shown that the meconium* of the infant contains a growth factor for *Lactobacillus bifidus*. (See also the suggestions of Lindsay in Linzell and Peaker, 1971, p. 590.) Whether this is of importance in all mammals is a moot point.

While most human infants thrive on milk, many adults do not tolerate it. Lactose intolerance, almost invariably observed in persons of African or Asiatic origin above the age of 10, is due to lack of the enzyme lactase in the intestine (Paige *et al.*, 1975). Since lactose as such cannot be absorbed it will behave like magnesium salts, also nonabsorbable, and thus acts as a laxative. Although lactose is not a remarkably soluble sugar, even solutions of moderate concentration are osmotically powerful because of the low molecular weight of the solute. Problems of lactose intolerance are infrequent among those of European origin since the synthesis of lactase in the upper intestine persists through childhood into adulthood.

It has been found that lactose is very low in the milk of the echidna (*Tachyglossus aculeatus*) and the platypus (*Ornithorhynchus anatinus*) (Messer and Kerry, 1973). Instead, the characteristic carbohydrate found in echidna milk is a tetrasaccharide, difucosyllactose. Lactose is absent, as are other carbohydrates, from the Otariidae. It has been established that, in *Zalophus californianus* (sea lion) (Johnson *et al.*, 1972), this absence of lactose is matched by an absence of α-lactalbumin without which biosynthesis of lactose cannot occur.

IV. Hormonal Control of Growth and Function

As noted previously, lactation is a physiological process involving stimuli by two or more individuals. Hormonal controls are accordingly complex and the specificity exhibited by various species is marked. The physiological and biochemical adjustments necessitated by suckling variable numbers of young in the most variable environmental stresses must elicit a symphony of controls. The secretory glands must develop at an appropriate rate, begin secretion on demand, and maintain a level of secretion (galactopoiesis) consonant with the demand. Thus controls emanate from the endocrines of the mother, from the placenta, and from the stimuli of suckling by the young. Not only are circulating hormones involved but neural controls as well.

It is now widely recognized that humoral regulation involves receptor proteins in the cell membrane of the target cell. In some cases, (steroids in particular) the hormone is carried into the nucleus so that a

* Contents of the fetal intestinal tract.

direct effect on the genetic material is attained, in others a secondary messenger such as cyclic AMP is involved. Some hormones elicit a very specific effect, others can be considered metabolic modifiers. It may be concluded from the following discussion that the mammary parenchyma has a rich and varied complement of hormonal receptors. Finally, it should be borne in mind that hormone controlled systems are poised systems with activators and inhibitors opposing each other in their actions.

A. MAMMARY GLAND GROWTH

The development of the gland following the events of puberty involve several of the endocrine glands. The anterior pituitary polypeptides, prolactin and growth hormone, appear to be primarily mammogenic. In addition there are synergistic effects involving the adrenal cortex (ACTH)*, the thyroid (TSH), and the ovaries (FSH and LH). Variation can be observed particularly with respect to the type of estrus: cyclic (rat, guinea pig, sheep, pig, horse, cow, human) or continuous (rabbit, cat).

After implantation and development of a placenta, another polypeptide hormone comes into play, namely placental lactogen. For a comparative review, see Forsyth (1974) and Chapter 5, this volume. Since an increase of secretory tissue during pregnancy implies or suggests cell proliferation, it was suggested by Davidson and Leslie (1950) that DNA levels be used to assess parenchymal growth. Studies of DNA levels involving a variety of species have shown that the mammary parenchyma increases throughout pregnancy and in some cases continues into the period of lactation. The anterior pituitary hormones, together with progesterone, are primarily involved in this growth process.

In the rat and the mouse, the fetus is not aborted if the pituitary is removed during pregnancy. In such animals, the placenta alone furnishes the mammogenic hormone and nearly normal mammary development occurs. It also seems likely that human placental lactogen (hPL) is a major stimulus to mammary development.

Many studies have been undertaken to determine the action of the ovarian steroids on mammary development. At present, the evidence suggests that they have an indirect effect via the pituitary but differences of opinion exist concerning this. For example insulin effects on mouse mammary explants are strongly modified by 17β-estradiol (Turk-

* ACTH, adrenocorticotropic hormone; TSH, thyroid-stimulating hormone; FSH, follicle-stimulating hormone; LH, luteinizing hormone.

ington, 1972a). There is evidence, in the mouse, that ductal cell proliferation requires estrogen and prolactin while alveolar formation requires progesterone in addition (Topper & Oka, 1974; Freeman and
Topper, 1977). Somewhat different results were obtained with rat
mammary explants by Jones and McCarty (1977).

Both *in vivo* studies (Freeman and Topper, 1977) as well as mammary tissue culture experience (Ichinose and Nandi, 1966) indicate
that insulin is required for normal mammary growth and differentiation, but the data for thyroid hormone are equivocal. Again Ichinose
and Nandi (1966) found aldosterone to influence mammary differentiation *in vitro*. McGrath (1971) demonstrated that the replication and
release of mammary tumor virus was induced by both insulin and
hydrocortisone. The hormonal effects of each were separable but each
involved a modification of cell–cell interaction in the cultures of mammary adenocarcinoma from BALB/cfC3H mice. Prolactin had no effect. This calls attention to the possibility that hormones somehow
effect the assembly of a three-dimensional array of cells.

B. LACTATION

In rats, Lyons *et al.* (1958) showed that prolactin and the adrenal
glucocorticoids were required for the initiation of lactation (lactogenesis) in animals from which pituitaries, adrenals, and ovaries
were removed. However, Rivera (1974) has warned against ignoring
ovarian steroid hormone effects. Gardner and Wittliff (1973) have
demonstrated specific cytoplasmic estrogen receptors in the lactating
gland of the rat. In other species, growth hormone and thyroxine are
needed or, alternatively, prolactin alone is sufficient. The latter may be
true for humans but has been demonstrated for the rabbit. In the cow it
has been shown that parturition is accompanied by a large but transitory increase in prolactin, growth hormone (Ingalls *et al.*, 1973), and
adrenal corticoids in the serum, and a marked decrease in progesterone. It has recently been reported by Cowie (1976) that intense
lactation in the cow is more readily correlated with levels of growth
hormone than prolactin. The latter may be involved with initiation of
lactation, the former with maintenance. The relative importance of the
adrenal corticoids, despite extensive experimentation, is not yet
agreed upon. Moreover, in the mouse, the polyamine, spermidine,
mimics glucocorticoid action (Kano and Oka, 1976). Shyämalä (1973)
found specific cytoplasmic glucocorticoid receptors in the lactating
gland of the rat, and Chomczynski and Zwierzchowski (1976) demonstrated that the activity of the receptor changes during lactogenesis.
Payne *et al.* (1976) have found corticosteroid-binding proteins both in

human colostrum and milk and in rat milk. In contrast, progesterone has been clearly demonstrated to inhibit lactation. Specifically, α-lactalbumin synthesis (Turkington and Hill, 1969) and lactose synthesis (Kuhn, 1969) have been shown to be inhibited.

C. Studies in Model Systems

The refinement of this general picture requires a judgment concerning unequivocal evidence. For pertinent comments see Rivera (1974). Several experimental systems have been used:

1. Whole animal experiments
    (a) extirpation of endocrines (deletion)
    (b) administration of hormones (addition)
    (c) comparison of genetic variants
2. Transplantation of glands into a host animal (deletion of neural stimuli)
3. Dispersed cell culture studies (short term but well defined) (Kraehenbuhl, 1977)
4. Organ culture studies, (fragments of tissue floated on nutrient media) (Elias, 1957)
5. Mathematical or dynamic simulation (Plucinski and Baldwin, 1977)
6. Preparations of purified acinar epithelial cells for correlation studies (Scalise *et al.*, 1977).

A relatively large body of data has been obtained with the organ culture technique and has led to some concept of the mechanism by which prolactin can modify gene expression.

Diagram 1 (Turkington, 1972b) indicates the point at which prolactin and other factors are thought to act on the *in vitro* system.

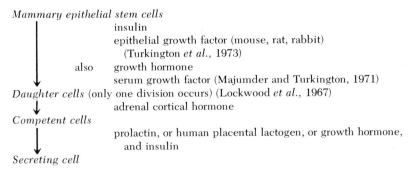

Mammary epithelial stem cells
               insulin
               epithelial growth factor (mouse, rat, rabbit)
                  (Turkington *et al.*, 1973)
       also   growth hormone
               serum growth factor (Majumder and Turkington, 1971)
Daughter cells (only one division occurs) (Lockwood *et al.*, 1967)
               adrenal cortical hormone
Competent cells
               prolactin, or human placental lactogen, or growth hormone,
                 and insulin
Secreting cell

DIAGRAM 1

A specific binding of prolactin to mammary plasma membrane has been demonstrated (Turkington, 1972b). Mouse membranes were incubated with [^{125}I]prolactin.* After washing, the prolactin bound was estimated by measuring the radioactivity released by specific displacement when unlabeled hormone was added. When labeled prolactin was covalently bonded to Sepharose beads (much larger than the epithelial cells) the hormone still elicited a cell response. Such an experiment suggests that the action of the hormone is restricted to the cell membrane and that the response of the cell is elicited by a secondary effector. The putative secondary effector does not appear to be cAMP. At parturition, cAMP concentration decreases and cGMP increases (Rillema, 1976). Moreover, recent evidence (Grosvenor and Whitworth, 1977; Nolin and Witorsch, 1976) indicates that prolactin enters the alveolar cells and milk-containing ducts in rats.

Shortly after the addition of prolactin to competent cells, an increase of nuclear RNA (nRNA) can be demonstrated, as well as an increase in RNA polymerase (DNA-dependent). After about 8 hours, the biosynthesis of α-lactalbumin and casein is stimulated. This, together with other supporting evidence, indicated that prolactin stimulated transcription of RNA from the chromosomal DNA. It was shown that the induced ribosomal RNA was to be found in polysomes, to be expected if a new genetic expression of protein synthesis were to occur. Before these polysomes can participate in the synthesis of new milk proteins, it has been proposed that they must be bound to the endoplasmic reticulum (Gaye and Denamur, 1970). It was presumed, but not proved, that a new messenger RNA was also transcribed. Studies on the appearance of new transfer RNA (4 S) suggests that insulin and adrenal corticoids as well as prolactin are required. Moreover, synchronous increases in transfer RNA and transfer RNA-methylating enzymes were observed.

D. THE PROLACTIN RECEPTOR

With the demonstration of a receptor site for prolactin in the mammary cell membrane, there arose the challenge to isolate a membrane fragment, or possibly a protein, with a capacity for binding prolactin at high dilution and with great specificity. Both have been proved possible (Shiu and Friesen, 1974a). A membrane preparation obtained by differential centrifugation of rabbit mammary homogenate was found to be solubilized by Triton X-100 (a non-ionic detergent) and to yield a

---

* Proteins can be labeled with ^{125}I without affecting the biological properties.

partially purified protein with a molecular weight of about 200,000. At the concentration of Triton X-100 that was necessary, $[^{125}I]$prolactin cannot be used, but ^{125}I-labeled human growth hormone proved to be a satisfactory surrogate. Thus the formation of a polypeptide hormone–receptor complex was demonstrated and could be used as an assay for the receptor since it was found in the void volume of a Sephadex G-100 column, or alternatively, could be precipitated by polyethylene glycol (Shiu and Friesen, 1976).

It is currently assumed that the interaction (binding) of a polypeptide hormone with its protein receptor is an equilibrium process. Hence it is just as important to demonstrate that hormone dissociated from the complex retains full biological activity. This has been shown for the membrane–hormone complexes (Shiu and Friesen, 1974b).

Presumably, the isolation of this model hormone–receptor system will allow further inquiry into the network of interactions (mechanism) involved. As Cuatrecasas and Hollenberg (1976) have called to our attention: "The extraordinary specificity, selectivity, potency at exquisitely low concentrations, and susceptibility to rapid on–off control, so characteristic of the physiological action of hormones, pose special apparent paradoxes that must be reconciled in order to understand the molecular basis of hormone action." At least some of the answers to these questions will come from protein chemistry but as far as prolactin is concerned, there are, in addition, many other problems of a more biological nature. As noted by Nicoll and Bern (1972), prolactin has been reported to produce scores of different effects in a very wide range of species. The name prolactin is semantically misleading. For example, teleost prolactin (Farmer et al., 1977) does not stimulate mammary tissue although it may be related, on an evolutionary basis, to mammalian prolactin. Moreover, Shiu and Friesen (1974b) found cell membranes from organs other than mammary gland to show substantial specific binding of ovine prolactin. Adrenal showed a specific binding capacity twice that of mammary gland. Ovary, kidney, and liver membranes all showed substantial binding. The presence of prolactin receptor activity in liver has led to the working hypothesis that the receptor is induced by prolactin itself (Posner et al., 1975).

During the recent past a variety of protein phosphorylation reactions have been shown to be involved with changes induced by polypeptide hormones and catecholamines (Rubin and Rosen, 1975). A number of cyclic nucleotide-dependent protein kinases are now recognized. Although phosphorylation looks like a rather general regulatory mechanism there is as yet no evidence that prolactin is involved in such a process.

E. REGULATION OF LACTOSE SYNTHETASE ACTIVITY

Brodbeck and Ebner (1966) demonstrated that the biosynthesis of lactose was due to "lactose synthetase," an enzyme system in which a particulate galactosyltransferase protein is modified in its action by α-lactalbumin. The transferase occurs in many other tissues, but the α-lactalbumin modifier is restricted in its synthesis to mammary gland. Galactosyltransferase and α-lactalbumin form a 1:1 protein complex in the presence of either glucose or N-acetylglucosamine (Ivatt and Rosemeyer, 1976). The K_m values are such that the galactosyl moiety of UDP galactose is transferred only to glucose in mammary gland.

The findings of Coffey and Reithel (1968) indicated that lactose synthetase activity was associated with the Golgi membranes in bovine mammary gland and this was confirmed by Keenan *et al.* (1970a) for rat tissue. As discussed by Hill and Brew (1975) the evidence to date suggests that the galactosyltransferase forms part of the Golgi membrane and provides one example of membrane control (Whaley *et al.*, 1975). The transferase is a glycoprotein (Ebner and Mager, 1975) and it is generally accepted that the major fraction of the glycosylation of polypeptide chains in glycoprotein synthesis occurs in the Golgi apparatus. In the case of this galactosyltransferase, evidently the completed glycoprotein remains associated with the membrane. The modifying component of the synthetase, α-lactalbumin, is a small, soluble protein which is synthesized in the rough endoplasmic reticulum (ergastoplasm) and penetrates the cisternae of the Golgi membranes to form a protein complex with the transferase. α-Lactalbumin has been designated as a peripheral membrane protein (Singer, 1974). Thus the initiation of lactose synthesis is at the molecular level and its continuation may depend on an adequate steady state concentration of α-lactalbumin, not only in the cell, but in a specific subcellular component.

A variety of studies in mice, in rats, and in rabbits, both *in vivo* and *in vitro*, show that galactosyltransferase is synthesized during pregnancy but that α-lactalbumin levels are low until parturition. It is believed that progesterone inhibits α-lactalbumin synthesis. Studies on mouse mammary gland explants (Turkington *et al.*, 1968) indicated that insulin, hydrocortisone, and prolactin were all necessary for the biosynthesis of galactosyltransferase and of α-lactalbumin and that synthesis was asynchronous. Prolactin was replaceable with placental lactogen. Both synchronous and asynchronous appearance of synthetase have been observed, depending on conditions. Moreover, all three hormones seem to be necessary not only for induction of the synthesis of lactose synthetase but also its continuance.

V. Regulation of Genetic Expression

It may be profitable to differentiate two types of regulatory processes operative in all tissues. The continual adjustments of rate that must occur in a metabolic network are those adaptable to a branched system. In such systems, many of the reactions are reversible, and fine control is a consequence of rapidly altering fluxes and rates. In contrast, the much slower regulation of genetic expression is that of a linear, nonreversible and nonbranching, system. This may involve some types of feedback loop. A "protein mutation" such as that postulated recently by Lewin (1976) would provide such a feedback effect. This involves changes in regulatory or structural proteins of the gene at the posttranslational stage so that the genetic expression would in turn be modulated. There must be safeguards to prevent realization of all the possible modes of interaction which are so often realized in the small molecular transmutations of metabolism. In a linear system, one might expect the analog of an on–off switch rather than the analog of a voltage divider. Accordingly we note the importance of suppressors in genetic systems and the puzzling off–on phenomena of differentiation. The chemical nature of nucleic acids reduces the likelihood of side reactions since these polymers exhibit a relatively small number of interactions. In contrast, proteins exhibit and respond to every change in the microenvironment.

It may then be expected that regulations of genetic expression must have developed in accord with a unidirectional, nonbranched system consisting of a large number of highly specified reactions. Each reaction must be highly directional and, since it involves polymers, specificity is easy to visualize. What is not so evident are the safeguards against side reactions, particularly when proteins are involved.

In the foregoing section has been described a rather unique example of control, the description of which has afforded much satisfaction to those concerned with mammary gland function. However, the initiation and control of lactose synthesis depends finally on control of protein biosynthesis in the secretory epithelial cells. Some of the proteins characteristic of milk are synthesized in the mammary gland during late pregnancy and the formation of colostrum* testifies to this. At parturition, there is an abrupt change in the rate of protein biosynthesis, as if a repressor had suddenly been removed. Milk production soon begins with a very nice regulation of composition and rate. Several metabolic factors must enter into the complex steady state of milk

---

* A mammary secretion similar to, but differing from, milk. Colostrum may appear without accompanying pregnancy. It contains milk components, desquamated epithelium, transudate, and a characteristic cytology.

production, but this steady state can only be achieved if the enzymes assured by the genetic potential are present. Metabolic control in general can be attributed to a limiting enzyme level at one or more steps. Eventually the control modifiers of enzyme biosyntheses must be discovered.

There is no reason to believe that the fundamental chemistry of protein synthesis in mammary gland differs from that in other types of eukaryotic cells. There is a distinct possibility, however, that the regulation of protein synthesis may differ in this tissue and from one species to another.

Cells from lactating mammary tissue, when maintained in dispersed cell culture, lose their characteristic differentiation (Larson and Jorgensen, 1974). Some synthesis of β-lactoglobulin persists for weeks, but that of β-casein and of α-lactalbumin dwindles more rapidly and at differing rates. The addition of hormones tends to offset these losses but the implication of the results reported is that the coordination of control points is lost in such cultures.

Because small amounts of blood proteins are usually found in milk, it was thought formerly that all the major milk proteins were not made in the mammary gland. A substantial body of data now supports the view that the major portion of the milk proteins are indeed synthesized in the mammary secretory tissue.

For protein synthesis to occur, there must exist a system for introducing the amino acids in the proper conformation and state of reactivity. This energy-demanding process involves uptake of amino acids from the blood and, with the expenditure of ATP, results in a supply of aminoacyl-transfer RNA complexes. For each polypeptide sequence, there must be an agent that provides genetic information for the unique translation into a special amino acid sequence. This agent is messenger RNA.

In eukaryotes, newly formed nuclear RNA is heterogeneous (Darnell *et al.*, 1973). A fraction of this is lengthened in the presence of poly(A) synthetase to polyadenylic messenger RNA which appears to be associated with protein both in the nucleus and later in the cytoplasm (mRNP) (Barrieux *et al.*, 1975).

Regardless of concentration, transcription of the "message" of mRNP cannot occur in the absence of ribosomes. In the cytoplasm mRNP (or mRNA?) associates with the 60 S subunit of the ribosome (synthesized in the nucleolus). This association involves several protein factors and ATP (Schreier and Staehelin, 1973). Initiation of polypeptide formation is potentiated by a [Met-tRNA$_f$ · 40 S ribosome subunit] complex, the formation of which requires GTP and two other proteins (Barrieux and Rosenfeld, 1977). In the presence of yet another protein factor, the functional [mRNA-Met-tRNA$_f$ · 80 S] ribosome complex forms.

At present, it is tempting to postulate that the poly(A) segment of mRNA provides the site for interaction with the membrane of the endoplasmic reticulum. With the ribosome or polysome positioned on the ER, electron micrographs reveal rough endoplasmic reticulum (RER). At this point in the organization of the transcription process, it is not entirely clear what the transcript is. In the case of insulin, a preproinsulin is formed with a very transitory amino terminal end that, within seconds, becomes proinsulin. This, then, is converted to insulin at a much slower rate. A variety of proteins have been shown to be synthesized as precursors of the final form, including prolactin (Maurer et al., 1976). Palade (1975) has suggested that secretory proteins are discharged from the ribosome into the cisternae of the endoplasmic reticulum. Evidence of a precursor of milk protein is as yet preliminary. Gaye and Houdebine (1975) have noted that the casein mRNA isolated from ewe mammary gland is larger than one might expect for the polypeptide sequences of the caseins. A similar observation was made by Rosen (1976) with respect to purified rat casein mRNA. More directly, Craig et al. (1976) have reported the formation of an α-lactalbumin precursor in an in vitro system from guinea pig.

Gaye et al. (1973) have isolated an mRNA from bound polyribosomes of mammary gland (ewe) which they have reported to be active in the absence of initiation factors. Rosen et al. (1975) isolated and partially purified casein mRNA (rat), only 40% of which appeared to possess a poly(A) sequence. It is unclear as yet whether this is due to a variation in synthesis, an aging effect, or an artifact. It has been shown that the phenol extraction process as such is not responsible (Houdebine, 1976b).

Beitz et al. (1969) have demonstrated cell-free synthesis of milk proteins in a preparation from bovine mammary gland. In this study, no hormonal effects were observed, indicating that hormonal control is at the transcriptional level and requires cell organization. Moreover, it has been pointed out (Keenan et al., 1970b) that in bovine mammary gland all of the cells in an alveolus are either active or inactive. This suggests that hormonal effects are sensed by all of the cells in a morphological unit to produce an all-or-none effect.

In all likelihood, there exist two or more sets of protein synthesis controls. The biosynthesis of enzymes normally involved in energy metabolism, much of carbohydrate metabolism, and fat metabolism, must be correlated if anything like normal metabolism is to be attained. As already described, the unique carbohydrate of mammary gland is initiated by a unique specific protein that is continually synthesized and secreted in milk. In contrast the limiting enzyme for fatty acid synthesis (and hence fat synthesis) is thought to be acetyl-CoA carboxylase, an enzyme normally present in many cells and not lost in milk secretion. Presumably, the flux of protein in each case is very different. Thus it is possible that one set of synthesis controls serves the normal complement of enzymes having a low or moderate turnover rate. For those proteins being synthesized in quantity and secreted into the milk, we have the analogy of a very high turnover rate, and we might expect a different modulation for each individual protein. Viewed in this way, it is to be expected that galactosyltransferase, normally present in many cells, would be governed by one set of controls and α-lactalbumin, a unique protein continually leaving the cell, might be governed by another. In fact, the data of Turkington *et al.* (1968) support this view. Linzell and Peaker (1971) have emphasized that wide variations in the composition of milk have been observed and that this suggests a certain biochemical autonomy for lactose, for fats, and for proteins. The work of Gaye and Denamur (1970) suggested that secreted proteins such as α-lactalbumin are synthesized on bound polyribosomes. This provides a type of control in itself. A similar situation exists in liver where serum albumin is primarily synthesized on bound polyribosomes (Redman, 1969). It would be reassuring if there were unambiguous evidence that the total amount of bound polyribosomes increases with the secretory phase of lactation, although the results of Turkington and Riddle (1970) show the coupling of hormone-dependent polysome formation and casein synthesis to increases in transcriptional activity, and Kano and Oka (1977) have evidence that prolactin promotes the formation of membrane-bound polysomes. Moreover, Houdebine and Gaye (1975a) have demonstrated the absence of casein mRNA in free polysomes from lactating ewe mammary gland.

It is tempting to assume that the control of milk protein synthesis by prolactin is due to the formation of a burst of new messenger RNA. There is, however, an alternative mechanism that might prevail. The burst of new RNA noted by Turkington (1972b) might have resulted largely in an increase in the opportunity for the translation of the level of mRNA that was present before prolactin treatment. This would be a species of "post transcriptional regulation" discussed by Darnell *et al.* (1973). The results of Rosen *et al.* (1975) support this view. These investigators found that cells from early and mid-pregnancy rat mammary gland contained relatively high levels of casein mRNA. This mRNA was not active *in situ* but was active in protein synthesizing systems. Rosen and Barker (1976) have concluded that regulation of casein synthesis does not involve chemical activation of previously inactive mRNA. Kano and Oka (1977) have also proposed that prolactin is effective at the posttranscriptional level.

Recent developments may make possible a fruitful and intensive investigation of specific transcripts. Wood and Lingrel (1977) have succeeded in devising a procedure for purifying specified mRNA from crude extracts. This involved isolating a labeled poly(A)-containing mRNA which was used with an RNA-dependent DNA polymerase and an oligo(dT)-cellulose to produce a cDNA-cellulose. This solid state complementary DNA was then used to purify crude mRNA by annealing. By this procedure, a biologically active globin mRNA was isolated from nucleated erythroid cells. A similar procedure has been devised by Houdebine (1976b) for milk protein mRNA and by Rosen and Barker (1976) for purified rat casein mRNA. With such procedures at hand for identifying specific transcription, there is also a bright prospect for identifying the regulatory action of prolactin.

Houdebine and co-workers have conducted several studies on the rabbit and the ewe to determine the effect of prolactin on casein synthesis. Using translation in a reticulocyte lysate as well as hybridization to cDNA to assay for mRNA, Shuster *et al.* (1976) found casein mRNA even in the mammary gland of virgin rabbits. At day 5 of pregnancy, casein mRNA was already found associated with polysomes. The casein mRNA level increased 80 times between coitus and mid-pregnancy and 900 times between mid-pregnancy and lactation. (Similar data for the rat have been obtained by Rosen and Barker, 1976.) As the capacity for casein synthesis increased, there was a corresponding shift of ribosomes from monomeric to polymeric forms. The authors concluded that lactogenesis involved an increase in mRNA concentration rather than activation of a previously inactive gene. Thus it is proposed that lactogenesis is not an all-or-none phenomenon but involves a gradual increase in a transcription process whose rate is determined by a hormonal balance. Using similar methods, Houdebine (1976a) showed that prolactin injected into pseudopregnant rabbits induced lactogenesis and that progesterone, injected simultaneously, reduced the effect. Again, casein mRNA was present before stimulation and there was evidence of inefficient translation of newly formed mRNA. Some maturation process could be involved. One probability is that association of mRNA with polysomes is necessary for efficient translation (Houdebine and Gaye, 1975b). It has also been found that the mRNA for α-lactalbumin in lactating rat mammary gland is far more abundant than anticipated from the rate of α-lactalbumin synthesis (Qasba et al., 1977).

CONTROL OF PROLACTIN RELEASE

The current view is that prolactin continually enters the blood from the anterior pituitary but only at a rate permitted by a prolactin-inhibiting factor (PIF) of unknown structure, originating in the hypothalamus (Horrobin, 1973; Hytten, 1976). An increase in prolactin flow requires a diminution of, or antagonism of, the inhibitor. In rats, injection of dopamine is followed by an increase in PIF in the circulation. An increase in LH and FSH has also been noted. This suggests that a poised system exists in which the ratio of prolactin to PIF is a function of dopamine levels. There are indications, but as yet no convincing proofs, of a prolactin-releasing factor in the hypothalamus (Clemens and Meites, 1974). However, a body of evidence exists supporting the thesis that the hypothalamus coordinates many of the regulatory mechanisms of lactation (Sulman, 1970).

Unfortunately, the subject of prolactin regulation is still conjectural (Reichlin et al., 1976). It is undisputed, however, that prolactin levels rise just before parturition and suckling or milking causes a sharp rise in prolactin secretion. The neural pathways involved in the rat and the rabbit have been investigated (Tindal and Knaggs, 1977) and there is reason to believe that species differences may exist. It is also possible that prolactin secretion may be stimulated in more than one fashion. As noted recently (Antelman and Caggiula, 1977) catecholamine systems may mediate many interactions between an organism and the environment. In the suckling reflex norepinephrine–dopamine interactions may be involved. A specific suppressor of prolactin secretion, 2-Br-α-ergocryptine (bromocriptine), has been found in ergot alkaloids. This compound inhibits lactation in the woman, rabbit, dog, pig, rat, and mouse but not in cows or goats (Cowie, 1976).

VI. Secretion of Milk

Before full-scale lactation is achieved, there is secretion of colostrum, a modified "milk," modified in that the concentration of lactose is low; the fat and protein concentrations are high. More important is the high content of immunoglobins. It has been shown that these immune proteins can be absorbed by the intestine of the newborn so that passive immunity is conferred (Brambell, 1971). In man and in the rabbit, IgA is the main component; in the cow, pig, and sheep, IgM and IgG are more prominent. The placenta is not permeable to these proteins.

When full lactation has been achieved, a secretion rate of 1–2 ml of milk per gm tissue per day is observed (Linzell and Peaker, 1971). Milk is isosmotic with plasma but differs in ionic composition. It has a higher concentration of K^+ and a lower concentration of Na^+. Milks may vary from 50 to 92% water and since lactose is responsible for most of the osmotic pressure, it seems quite possible that the secretion of water accompanies that of lactose.

There has been much dispute concerning the details of milk secretion, how the main components of milk are "packaged" for extrusion into the alveolar lumen, what happens to the cell membrane during secretion, how much of the cell contents are lost during secretion, how many cells are destroyed.

It has been proposed by Brew (1969) that lactose is formed in Golgi vesicles which presumably begin to swell with water as the chemical potential of the sugars rise. Kuhn and White (1975) have presented evidence that lactose is concentrated in the Golgi lumen. In some fashion these vesicles might move toward the apical membrane and fuse with it whereby the lactose solution is extruded, together with ions and some proteins. Microtubules have also been implicated recently (Guérin and Loizzi, 1977). A somewhat similar process is envisioned for transfer of proteins from the endoplasmic reticulum to the apical membrane (Kurosumi et al., 1968) via the Golgi vesicles. Linzell et al. (1976) have proposed that citrate also is secreted into milk via the Golgi apparatus, in which citrate appears to be concentrated. The extrusion of fat globules is a process not fully characterized. Kurosumi et al. (1968) interpreted their electron micrographs to show that lipid droplets were formed in the smooth endoplasmic reticulum of rat mammary gland. In contrast, Stein and Stein (1967) found that the rough endoplasmic reticulum of mouse mammary gland was involved and Powell et al. (1977) have found Golgi marker enzymes in fat globule membranes. In any case, droplets of various sizes appear in the alveolar lumen, and scanning electron micrographs (Nemanic and Pitelka, 1971) show the evidence of their extrusion. These fat droplets are surrounded by a portion of the plasma membrane and occasionally have associated with them some cytoplasm and subcellular components.

The role of membranes in milk secretion has been extensively discussed by Keenan et al. (1974). There is adequate biochemical evidence to support the view that the fat globules of milk are surrounded by plasma membrane (Mather and Keenan, 1975). This represents a substantial loss of membrane from the cell. At present it is proposed that the secretory vesicles of Golgi membranes replace the membrane material extruded into the alveolar lumen. The Golgi membranes have been found to be intermediate in composition between that of the plasma membrane and that of the endoplasmic reticulum in liver. Hence it is proposed that there is a flow of membrane components from endoplasmic reticulum to the apical cell surface (Morré *et al.*, 1971). In this latter reference much evidence is presented for the view that the endomembrane system consisting of the membranes of the nucleus, endoplasmic reticulum, Golgi and secretory vesicles are a continuum and that they supply material for the plasma membrane. The possible relationship between the mechanism of synthesis and the insertion of polypeptides into the cisternae with subsequent transport to the plasma membrane has been discussed by Rothman and Lenard (1977). Unfortunately, purified membrane preparations have been difficult to obtain from mammary gland. A procedure has been reported recently that provided highly purified plasma membranes from the lactating mammary gland of the rat (Huggins and Carraway, 1976).

It is possible to imagine a gradient of chemical potential reaching from nucleus to periphery, which would imply material movement in that direction. Thus one might view milk secretion in the following manner. First, that syntheses of lactose, fat, and proteins are very active and that a substantial gradient forms from the inner portions of the cell out to the lumen. Unidirectional movement of products may be associated with the gradient of chemical potential of solutes alone but in addition, there is a membrane flux. The evidence is that a unidirectional gradient of membrane composition also obtains so that the membrane system should be considered not so much a structural feature, but a "visible" steady state. The cellular relations in mammary gland thus allow a steady flow of carbon and energy sources from the blood into the cell to be transformed into the products secreted through the apical membrane into the alveolar lumen. The final product which we call milk contains water and ions from blood, products of synthesis of mammary secretory cells, membrane in substantial amount, and small amounts of blood and cytoplasmic components as leakage or entrainment substances. Uninhibited synthesis guarantees milk flow after parturition in the presence of the proper hormones. A rather detailed discussion of secretory processes in general (but based on data from pancreatic cells) has been provided by Palade (1975).

VII. The Process of Milk Ejection

It is common observation that lactation is not maintained in the absence of suckling or milking. Two kinds of signals flow from such stimuli: afferent impulses from the mammary glands to the hypothalamic–hypophysial system and exteroceptive (external, non-tactile) stimuli.

In some mammals, such as the cow and the goat, the small ducts empty into larger ones that terminate in a cistern from which milk can be drawn passively. Insertion of a cannula through the teat can remove a substantial percentage of the milk. Most species do not possess such cisterns and the importance of the ejection reflex is thus heightened. Suckling, or milking, neural impulses from the teat and areola effect a release of oxytocin from the neurohypophysis. This humoral agent effects contraction of the alveolar myoepithelium in less than a minute and the alveolar milk is expelled under pressure.

This is a simplistic model of a process that can be modified either positively or negatively by an impressive array of stimuli amounting to a societal reaction. For observations by an investigator esteemed by his contemporaries in lactation research, see Folley (1969). Milk ejection can be inhibited by central nervous system responses to unpleasant environmental events. Alternatively, the smell, sight, and sound of suckling young amplifies milk ejection as do a variety of tactile responses and conditioned reflexes (Grosvenor and Mena, 1974). Continued suckling or milking is required for the maintenance of lactation and in this sense lactation is a truly social phenomenon.

VIII. Selected Chemical Features Involved in Synthesis

A. LACTOSE

In most cells the following compounds are maintained at steady state concentrations during aerobic glycolysis.

$$\begin{array}{l}\text{Glucose} \quad \rightarrow \text{Glucose-6-P} \rightleftarrows \text{Glucose-1-P} \\ \text{(from blood)} \qquad\qquad\qquad \updownarrow \\ \qquad\qquad\qquad\qquad\quad \text{UDP glucose} \rightarrow \text{Glycogen} \\ \qquad\qquad\qquad\qquad\qquad \updownarrow \\ \text{Galactosyl groups} \leftarrow \text{UDPgalactose} \end{array}$$

In most cells, UDPgalactose is the source of the galactosyl moiety for a variety of compounds that become incorporated into glycoproteins in or on the Golgi membranes or into structures forming connective tissue, as well as glycolipids.

In the mammary gland, there is a diminishingly small glycogen content. Evidently the flux of lactose synthesis reduces the concentration of UDPgalactose (and accordingly UDPglucose) to a very low level since, in the presence of α-lactalbumin, glucose is the prime acceptor. However, the carbohydrate side chains of glycoproteins are fashioned in mammary gland as elsewhere and there must exist a fine regulation of the amount of UDPgalactose available to the various receptors. There is one suspected case of "leakage." Milk contains many lactose-containing oligosaccharides, at least twenty in the human. Several have been utilized in studies on the chemical bases of blood type (Ginsburg, 1972). The A, B, H, and Lewis antigens are oligosaccharides found in various secretions. The glycosyltransferases that catalyze the formation of these antigens occur in milk where, it has been suggested (Kobata et al., 1970), lactose in very high concentration is a surrogate acceptor. The milk oligosaccharides are thus viewed as rare products due to the action of ubiquitous enzymes in a singular chemical environment. Recently Hallgren et al. (1977) have reported extensive studies on the excretion of milk oligosaccharides in the urine of pregnant and lactating women. Glöckner et al. (1976) have shown that the oligosaccharides from the glycoproteins of erythrocyte membranes having antigenic activity differ from those of the milk fat globule membranes.

B. FAT

The intermediary reactions involved in the synthesis of the characteristic fatty acids and triglycerides of milk are similar to those in other tissues. The same may be said of phospholipid synthesis (Infante and Kinsella, 1976). As noted in Fig. 2, the aliphatic acids are synthesized by a repetitive mechanism primed by acyl-CoA, adding two-carbon fragments from malonyl-CoA during each cycle until a chain length is attained that is destined to be liberated. The electron donor necessary for the reductions is NADPH (Smith and Abraham, 1970; Bartley, 1976).

The enzyme involved has been termed a *multifunctional* enzyme (Kirschner and Bisswanger, 1976) since it contains only two polypeptide chains but catalyzes at least eight reactions. This differentiates it from a *multienzyme complex* of several types of polypeptide chains, each of which has a single catalytic function. A detailed study of the enzyme present in liver, which appears to be very similar, has been presented by Lornitzo et al. (1975). A homogeneous synthase has been obtained by Knudsen (1972) from lactating bovine mammary gland.

All purified preparations of mammalian fatty acid synthase, when

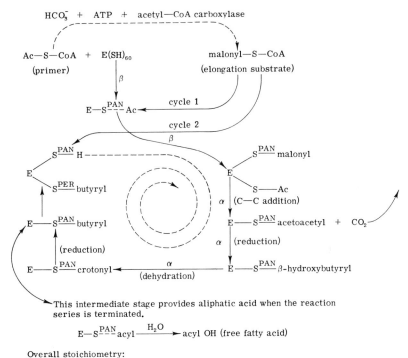

HCO$_3^-$ + ATP + acetyl—CoA carboxylase

Ac—S—CoA + E(SH)$_{60}$ malonyl—S—CoA
(primer)   (elongation substrate)

β

E—S$\overset{PAN}{-}$Ac ← cycle 1

cycle 2

β

S$\overset{PAN}{-}$H   S$\overset{PAN}{-}$malonyl

E   E

S$\overset{PER}{-}$butyryl   S——Ac
                                   α (C—C addition)

E——S$\overset{PAN}{-}$butyryl   E—S$\overset{PAN}{-}$acetoacetyl + CO$_2$

(reduction) α (reduction)

E——S$\overset{PAN}{-}$crotonyl ← α ——— E—S$\overset{PAN}{-}$$\beta$-hydroxybutyryl
               (dehydration)

This intermediate stage provides aliphatic acid when the reaction
series is terminated.

E—S$\overset{PAN}{-}$acyl $\xrightarrow{\text{H}_2\text{O}}$ acyl OH (free fatty acid)

Overall stoichiometry:

AcCoA + n malonyl—CoA + $2n$ NADPH + $2n$ H$^+$ $\Longrightarrow$ Et (CH$_2$—CH$_2$)$_{n-1}$ CH$_2$COOH

+

n CO$_2$ + $2n$ NADP$^+$ + n H$_2$O

FIG. 2. Fatty acid synthesis catalyzed by fatty acid synthase. The enzyme protein is
dimeric, consisting of two similar but distinct polypeptide sequences. One of these,
denoted by α in the figure, contains the cofactor 4′-P-panthetheine (acyl intermediate
binding site), two NADPH binding sites, and the reductase activity. The other, β, contains no pantetheine but rather cysteine loading or binding sites. It also shows acetyl-
CoA and malonyl-CoA : pantetheine transacylase activity. The holoenzyme has about 60
sulfhydryl groups. One sulfhydryl group is symbolized as S^{PAN} to designate the pantetheine sulfhydryl, another is symbolized as S^{PER}, referring to a peripheral sulfhydryl.
The dashed spiral line indicates that n cycles of the reaction set occur once the enzyme
is primed. Butyryl-CoA may also function as a primer. The properties of the other
multifunctional protein involved, acetyl-CoA carboxylase, have been described by
Tanabe *et al.* (1975).

combined with the proper substrates, yield principally palmitic acid
(Carey, 1977) and yet milk fat contains notable quantities of short and
medium length fatty acid triglycerides. Recent evidence (Carey, 1977;
Smith and Abraham, 1975) reveals that a chain length modifying protein is present in mammary gland that interacts with the fatty acid

synthase. The detailed mechanism by which a spectrum of aliphatic chain lengths results is not yet known.

In man, the synthesis and distribution of fat in various tissues involves a variety of dietary factors and hormonal controls as well as neural and genetic influences. It has been reported that the range of fat content in human milk exceeds that of other mammals (Hytten, 1976).

The mammary gland, like most tissues, receives both glucose and fatty acids from the blood. In nonruminants, fats found in milk derive from both sources. Long chain fatty acids largely derive from blood chylomicrons and are made available by the action of lipoprotein lipase. Evidence from both normal and lactating rats indicates that the activity of this enzyme increases during lactation, is low in hypophysectomized rats and can be restored by injections of prolactin (Scow, 1977). The characteristic short chain fatty acids of milk are synthetic products of the mammary gland. Lin *et al.* (1976) have shown that the specificities of the mammary gland acyltransferases are responsible for the nonrandom arrangement of fatty acids in milk triglycerides. The preferred substrates are 1,2-diglycerides substituted with long chain fatty acids. To these substances are added short or medium chain length fatty acids to form triglycerides. Since all of the latter arise in mammary gland much of the former (long chain) likely derives from the blood. In ruminants, glucose is not a source of carbon for fatty acid synthesis, but rather acetate and β-hydroxybutyrate from the rumen. [For the ruminant mammary gland, gluconeogenesis (liver and kidney) is the major source of glucose.]

There is also a difference between ruminants and nonruminants in the source of NADPH which, for fatty acid synthesis, must originate outside the mitochondrion. In nonruminants, the oxidation of glucose-6-P and 6-P-gluconate (catalyzed by the respective dehydrogenases) has proven to be an important source if the levels of the enzymes involved are related. In ruminants, NADP-isocitrate dehydrogenase seems more likely to provide the reducing power needed (Bauman *et al.*, 1970).

There is as yet no agreement on the site of triglyceride synthesis. Both the endoplasmic reticulum and the mitochondria have been proposed. The glycerol involved derives from blood glucose via glycolysis.

C. MILK PROTEINS

The caseins in particular require comment because there are several, because they are found in milk as complex micelles, and because they are modified phosphoproteins. In general, caseins have a high proline* content, hence little helix structure, and are poorly soluble at pH 4–5. The α-caseins†, of which there are many genetic variants, are restricted to ruminants. The β-caseins are of lower electrophoretic mobility. The κ-caseins are glycosylated. The γ-caseins may be derived from β-caseins by proteolysis. All contain ester-bound phosphate (Ser-P) groups, variable in number. Phosphorylation is catalyzed by a cyclic nucleotide-independent phosphoprotein kinase (Rubin and Rosen, 1975; Bingham et al. 1977).

The addition of phosphate to the caseins and of the carbohydrate moiety to κ-casein, is thought to occur in the Golgi region (Bingham et al., 1972). This also appears to be the region of the cell where the casein micelles form. These micelles (which in bovine milk are about 140 nm in diameter) contain thousands of the various casein monomers as well as calcium phosphate. The detailed structure of these micelles is still a subject for investigation (Taborsky, 1974). The many negatively charged residues, phosphate and carboxyl, are restricted to one portion of the primary structure. It is those groups that interact with Ca ion. Since the polypeptide chains are relatively hydrophobic in character there are many residue interactions leading to aggregation.

IX. Regression and Involution

With normal or premature weaning of the young, changes occur in the mammary gland. The several effects are termed regression. In the absence of suckling or milking, the milk accumulates in the gland and there follows both an inhibition of secretion and a diminution in the number of secretory cells. The rate at which changes occur varies with the species and depends on a variety of factors. Undoubtedly hormonal regulation is involved since regression is slowed by injections of prolactin, cortisol, and oxytocin. However, regardless of hormone presence, tissue explants show changes in function similar to those in regression (Rivera, 1974). It also has been noted that removal of the pituitary from lactating rats is accompanied by the same enzyme changes that occur during regression (Cowie and Tindal, 1971).

* The observation that lactating mammary gland had a very high arginase activity (Folley and Greenbaum, 1947) posed a long standing puzzle. Recently Yip and Knox (1972) have submitted evidence that this enzyme is involved in the conversion of arginine into proline. Actually two forms of the enzyme have been found in the mouse, one of which may be involved in the synthesis of spermidine (Oka and Perry, 1977).

† The extensive nomenclature of the caseins has been revised by McKenzie (1967). Because of polymorphism there are several α-caseins. The subscript-modified designation $α_s$ signifies calcium-sensitive.

Early in regression there are changes in the number and character of the subcellular components in the secretory cells (Hollmann, 1974). Particularly noticeable is the increase in lysosome* content. There ensues a reduction of the parenchyma and it is believed that lysosomal autophagocytosis is marked. Some macrophage activity is also noted. Obviously, whatever is the mechanism of the autolysis or cell destruction, most of the products must be removed by the circulatory blood or the lymph.

In discussions of regression there is an emphasis on the degradative processes that effect the removal of much of the parenchyma. Nevertheless, as Hollmann (1974) has emphasized, there is a considerable synthesis of connective tissue and of fat cells. In the human there is an analogous process that has a much longer time constant. After age 35 the mammary lobules shrink and are gradually replaced by fat cells. The total organ size may even increase but becomes less and less functional. This aspect of mammary change is often denoted as involution. Since such change occurs with the diminution of reproductive vigor, involution is likely delayed by ovarian hormones.

Thus, in regression, there is a return of the structure of the gland to the state immediately before pregnancy. This involves both degradative and synthetic processes and relatively rapid alterations in balance. Involution represents an altered ordering of tissue components and an elimination of secretory cells. The regulation of cell types within a tissue is a topic consisting of many inchoate queries and the mammary gland furnishes a number of examples. One such situation occurs during regression. Alveoli may be seen in which some epithelial cells remain metabolically vigorous while others are necrotic. Despite the vigor of the processes eliminating the secretory cells, there is little effect on the myoepithelial cells that aid in preserving the alveolar pattern. Changes that now dispose to the formation of new connective tissue or adipose tissue must involve an alteration in metabolism and new cell-contact relations. This presents many opportunities for scientific inquiry and may involve some form of surface modulation as proposed by Edelman (1976).

X. Epilogue

Despite its complexities, indeed because of them, lactation is a process eminently worthy of study. If one wishes to determine the regulatory mechanisms in a steady state system it is necessary to perturb the system to emphasize those mechanisms. This kind of perturbation is

* Occasionally called cytosomes or cytosegresomes.

automatically provided during lactation. Again, if one wishes to delineate the changes during cell development a study of mammary gland development provides a system that is uniquely reversible.

The hormonal control of lactation has received unusual attention, particularly in a descriptive sense. The recent investigations documented here have shown how hormonal control is integrated into the intracellular regulatory mechanisms. The initiation of lactose synthesis is undoubtedly effected by α-lactalbumin. How that synthesis, once initiated, is regulated is not yet so clear. Again, the synthesis of the characteristic aliphatic acid length distribution seems to be dictated by another modulating protein unique to mammary gland. Finally, both of these, and other, modulations at the molecular level are dependent on the synthesis of unique proteins. All proteins synthesized are the result of genetic expression. It is clear that in mammary gland genetic expression is modulated by other tissues and even other individuals. The current density of papers on lactation is adequate testimony to the experimental opportunities afforded by this mammalian phenomenon.

REFERENCES

Antelman, S. M., and Caggiula, A. R. (1977). *Science* **195**, 646–653.

Barrieux, A., and Rosenfeld, M. G. (1977). *J. Biol. Chem.* **252**, 392–398.

Barrieux, A., Ingraham, H. A., David, D. N., and Rosenfeld, M. G. (1975). *Biochemistry* **14**, 1815–1821.

Bartley, J. C. (1976). *Lipids* **11**, 774–777.

Bauman, D. E., Brown, R. E., and Davis, C. L. (1970). *Arch. Biochem. Biophys.* **140**, 237–244.

Beitz, D. C., Mohrenweiser, H. W., Thomas, J. W., and Wood, W. A. (1969). *Arch. Biochem. Biophys.* **132**, 210–222.

Bingham, E. W., Farrel, H. M., Jr., and Basch, J. J. (1972). *J. Biol. Chem.* **247**, 8193–8194.

Bingham, E. W., Groves, M. L., and Szymanski, E. S. (1977). *Fed. Proc., Fed. Am. Soc. Exp. Biol.* **36**, 634.

Brambell, F. W. R. (1971). "The Transmission of Passive Immunity from Mother to Young." North-Holland Publ., Amsterdam.

Brew, K. (1969). *Nature (London)* **222**, 671–672.

Brodbeck, U., and Ebner, K. E. (1966). *J. Biol. Chem.* **241**, 5526–5532.

Carey, E. M. (1977). *Biochim. Biophys. Acta* **486**, 91–102.

Chomczynski, P., and Zwierzchowski, L. (1976). *Biochem. J.* **158**, 481–483.

Clemens, J. A., and Meites, J. (1974). *In* "Lactogenic Hormones, Fetal Nutrition and Lactation" (J. B. Josimovich, M. Reynolds, and E. Cobo, eds.), pp. 111–140. Wiley, New York.

Coffey, R. G., and Reithel, F. J. (1968). *Biochem. J.* **109**, 169–176 and 177–183.

Cowie, A. T. (1976). *Arch. Int. Physiol. Biochim.* **84**, 877–908.

Cowie, A. T., and Tindal, J. S. (1971). "The Physiology of Lactation." Arnold, London.

Craig, R. K., Brown, P. A., Harrison, O. S., McIlreavy, D., and Campbell, P. N. (1976). *Biochem. J.* **160**, 57–74.

Cuatrecasas, P., and Hollenberg, M. D. (1976). *Adv. Protein Chem.* **30**, 251–451.

Darnell, J. E., Jelinek, W. R., and Molloy, G. R. (1973). *Science* **181**, 1251–1221.

Davidson, J. N., and Leslie, L. (1950). *Nature (London)* **165**, 49–53.

Ebner, K. E., and Mager, S. C. (1975). *In* "Subunit Enzymes" (K. E. Ebner, ed.), Vol. 2, pp. 137–179. Dekker, New York.

Edelman, G. M. (1976). *Science* **192**, 218–226.

Elias, J. J. (1957). *Science* **126**, 842–843.

Farmer, S. W., Papkoff, H., Bewley, T. A., Hayashida, T., Nishioka, R. S., Bern, H. A., and Li, C. H. (1977). *Gen. Comp. Endocrinol.* **31**, 60–71.

Folley, S. J. (1969). *J. Endocrinol.* **44**, ix–xx.

Folley, S. J., and Greenbaum, A. L. (1947). *Biochem. J.* **41**, 261–269.

Forsyth, I. A. (1974). *In* "Lactogenic Hormones, Fetal Nutrition, and Lactation" (J. B. Josimovich, M. Reynolds, and E. Cobo, eds.), pp. 49–67. Wiley, New York.

Freeman, C. S., and Topper, Y. J. (1977). *Fed. Proc., Fed. Am. Soc. Exp. Biol.* **36**, 295.

Gardner, D. G., and Wittliff, J. L. (1973). *Biochemistry* **12**, 3090–3096.

Gaye, P., and Denamur, R. (1970). *Biochem. Biophys. Res. Commun.* **41**, 266–272.

Gaye, P., and Houdebine, L. M. (1975). *Nucleic Acids Res.* **2**, 707–722.

Gaye, P., Houdebine, L., and Denamur, R. (1973). *Biochem. Biophys. Res. Commun.* **51**, 637–644.

Ginsburg, V. (1972). *Adv. Enzymol.* **36**, 131–149.

Glöckner, W. M., Newman, R. A., Dahr, W., and Uhlenbruck, G. (1976). *Biochim. Biophys. Acta* **443**, 402–413.

Grosvenor, C. E., and Mena, F. (1974). *In* "Lactation" (B. L. Larson and V. R. Smith, eds.), Vol. 1, pp. 227–276. Academic Press, New York.

Grosvenor, C. E., and Whitworth, N. S. (1977). *Fed. Proc., Fed. Am. Soc. Exp. Biol.* **36**, 367.

Guérin, M. A., and Loizzi, R. G. (1977). *Fed. Proc., Fed. Am. Soc. Exp. Biol.* **36**, 343.

Hallgren, P., Lindberg, B. S., and Lundblad, A. (1977). *J. Biol. Chem.* **252**, 1034–1040.

Hill, R. L., and Brew, K. (1975). *Adv. Enzymol.* **43**, 411–490.

Hollmann, K. H. (1974). *In* "Lactation" (B. L. Larson and V. R. Smith, eds.), Vol. 1, pp. 3–95. Academic Press, New York.

Horrobin, D. F. (1973). "Prolactin: Physiology and Clinical Significance." Med. Tech. Publ. Co. Ltd., Lancaster, England.

Houdebine, L. M. (1976a). *Eur. J. Biochem.* **68**, 219–225.

Houdebine, L. M. (1976b). *Nucleic Acids Res.* **3**, 615–630.

Houdebine, L. M., and Gaye, P. (1975a). *Nucleic Acids Res.* **2**, 165–177.

Houdebine, L. M., and Gaye, P. (1975b). *Mol. Cell. Endocrinol.* **3**, 37–55.

Huggins, J. W., and Carraway, K. L. (1976). *J. Supramol. Struct.* **5**, 59–63.

Hytten, F. E. (1976). *J. Hum. Nutr.* **30**, 225–232.

Ichinose, R. R., and Nandi, S. (1966). *J. Endocrinol.* **35**, 331–340.

Infante, J. P., and Kinsella, J. E. (1976). *Lipids* **11**, 727–735.

Ingalls, W. G., Convey, E. M., and Hafs, H. D. (1973). *Proc. Soc. Exp. Biol. Med.* **143**, 161–164.

Ivatt, R. J., and Rosemeyer, M. A. (1976). *Eur. J. Biochem.* **64**, 233–242.

Jenness, R. (1974). *In* "Lactation" (B. L. Larson and V. R. Smith, eds.), Vol. 3, pp. 3–107. Academic Press, New York.

Johnson, J. D., Christiansen, R. O., and Kretchmer, N. (1972). *Biochem. Biophys. Res. Commun.* **47**, 393–397.

Jones, R. F., and McCarty, K. S. (1977). *Fed. Proc., Fed. Am. Soc. Exp. Biol.* **36**, 913.

Kano, K., and Oka, T. (1976). *J. Biol. Chem.* **251**, 2795–2800.

Kano, K., and Oka, T. (1977). *Fed. Proc., Fed. Am. Soc. Exp. Biol.* **36**, 648.

Keenan, T. W., Morré, D. J., and Cheetham, R. D. (1970a). *Nature (London)* **228**, 1105–1106.

Keenan, T. W., Saacke, R. G., and Patton, S. (1970b). *J. Dairy Sci.* **53**, 1349–1352.

Keenan, T. W., Morré, D. J., and Huang, C. M. (1974). *In* "Lactation" (B. L. Larson and V. R. Smith, eds.), Vol. 2, pp. 191–233. Academic Press, New York.

Kirschner, K., and Bisswanger, H. (1976). *Annu. Rev. Biochem.* **45**, 143–166.

Knudsen, J. (1972). *Biochim. Biophys. Acta* **280**, 408–414.

Kobata, A., Grollman, E. F., Torain, B. F., and Ginsburg, V. (1970). *In* "Blood and Tissue Antigens" (D. Aminoff, ed.), pp. 497–506. Academic Press, New York.

Kraehenbuhl, J. P. (1977). *J. Cell Biol.* **72**, 390–405.

Kuhn, N. J. (1969). *J. Endocrinol.* **45**, 615–616.

Kuhn, N. J., and White, A. (1975). *Biochem. J.* **148**, 77–84.

Kurosumi, K., Kobayashi, Y., and Baba, N. (1968). *Exp. Cell Res.* **50**, 177–192.

Larson, B. L., and Jorgensen, G. N. (1974). *In* "Lactation" (B. L. Larson and V. R. Smith, eds.), Vol. 1, pp. 115–146. Academic Press, New York.

Lewin, S. (1976). *Biochem. Soc. Trans.* **4**, 68–69.

Lin, C. Y., Smith, S., and Abraham, S. (1976). *J. Lipid Res.* **17**, 647–656.

Linzell, J. L., and Peaker, M. (1971). *Physiol. Rev.* **51**, 564–597.

Linzell, J. L., Mepham, T. B., and Peaker, M. (1976). *J. Physiol. (London)* **260**, 739–750.

Lockwood, D. H., Stockdale, F. E., and Topper, Y. J. (1967). *Science* **156**, 945–947.

Lornitzo, F. A., Qureshi, A. A., and Porter, J. W. (1975). *J. Biol. Chem.* **250**, 4520–4529.

Lyons, W. R., Li, C. H., and Johnson, R. E. (1958). *Recent Progr. in Hormone Research* **14**, 219–248.

McGrath, C. M. (1971). *J. Natl. Cancer Inst.* **47**, 455–467.

McKenzie, H. A. (1967). *Adv. Protein Chem.* **22**, 55–234.

Majumder, G. C., and Turkington, R. W. (1971). *Endocrinology* **88**, 1506–1510.

Mather, I. H., and Keenan, T. W. (1975). *J. Membr. Biol.* **21**, 65–85.

Maurer, R. A., Stone, R., and Gorski, J. (1976). *J. Biol. Chem.* **251**, 2801–2807.

Mayer, G., and Klein, M. (1961). *In* "Milk: The Mammary Gland and Its Secretion" (S. K. Kon and A. T. Cowie, eds.), Vol. 1, pp. 47–126. Academic Press, New York.

Messer, M., and Kerry, K. R. (1973). *Science* **180**, 201–203.

Morré, D. J., Mollenhauer, H. H., and Bracker, C. E. (1971). *Results Probl. Cell Differ.* **2**, 82–126.

Nemanic, M. K., and Pitelka, D. R. (1971). *J. Cell Biol.* **48**, 410–413.

Nicoll, C. A., and Bern, H. A. (1972). *Lactogenic Horm., Ciba Found. Symp., 1971* pp. 299–317.

Nolin, J. M., and Witorsch, R. J. (1976). *Fed. Proc., Fed. Am. Soc. Exp. Biol.* **35**, 1365.

Oka, T., and Perry, J. W. (1977). *Fed. Proc., Fed. Am. Soc. Exp. Biol.* **36**, 738.

Paige, D. M., Bayless, T. M., Huang, S.-S., and Wexler, R. (1975). *In* "Physiological Effects of Food Carbohydrates" (A. Jeanes and J. Hodge, eds.), pp. 191–206. Am. Chem. Soc., Washington, D.C.

Palade, G. (1975). *Science* **189**, 347–358.

Payne, D. W., Peng, L.-H., Pearlman, W. H., and Talbert, L. M. (1976). *J. Biol. Chem.* **251**, 5272–5279.

Plucinski, T. M., and Baldwin, R. B. (1977). *Fed. Proc., Fed. Am. Soc. Exp. Biol.* **36**, 1159.

Posner, B. I., Kelly, P. A., and Friesen, H. G. (1975). *Science* **188**, 57–59.

Powell, J. T., Jarlfors, U., and Brew, K. (1977). *J. Cell Biol.* **72**, 617–627.

Qasba, P. K., Chakrabartty, P. K., and Adler, R. G. (1977). *Fed. Proc., Fed. Am. Soc. Exp. Biol.* **36**, 930.

Redman, C. M. (1969). *J. Biol. Chem.* **244**, 4308–4315.

Reichlin, S., Saperstein, R., Jackson, I. M. D., Boyd, A. E., III, and Patel, Y. (1976). *Annu. Rev. Physiol.* **38**, 389–424.

Richardson, K. C. (1949). *Proc. R. Soc. London, Ser. B* **136**, 30–45.

Rillema, J. A. (1976). *Proc. Soc. Exp. Biol. Med.* **151**, 748–751.

Rivera, E. M. (1974). In "Lactogenic Hormones, Fetal Nutrition, and Lactation" (J. B. Josimovich, M. Reynolds, and E. Cobo, eds.), pp. 279–295. Wiley, New York.

Rosen, J. M. (1976). *Biochemistry* **15**, 5263–5271.

Rosen, J. M., and Barker, S. W. (1976). *Biochemistry* **15**, 5272–5280.

Rosen, J. M., Woo, S. L. C., and Comstock, J. P. (1975). *Biochemistry* **14**, 2895–2903.

Rothman, J. E., and Lenard, J. (1977). *Science* **195**, 743–753.

Rubin, C. S., and Rosen, O. M. (1975). *Annu. Rev. Biochem.* **44**, 831–887.

Scalise, M. M., Thorp, R. B., Jr., Pretlow, T. G., and Murad, T. M. (1977). *Am. J. Pathol.* **86**, 123–132.

Schreier, M. H., and Staehelin, T. (1973). In "Regulation of Transcription and Translation in Eukaryotes" (E. K. F. Bautz, P. Karlson, and H. Kersten, eds.), pp. 335–49. Springer-Verlag, Berlin and New York.

Scow, R. O. (1977). *Fed. Proc., Fed. Am. Soc. Exp. Biol.* **36**, 182–185.

Shiu, R. P. C., and Friesen, H. G. (1974a). *J. Biol. Chem.* **249**, 7902–7911.

Shiu, R. P. C., and Friesen, H. G. (1974b). *Biochem. J.* **140**, 301–311.

Shiu, R. P. C., and Friesen, H. G. (1976). In "Cell Membrane Receptors for Viruses, Antigens and Antibodies, Polypeptide Hormones, and Small Molecules" (R. F. Beers, Jr. and E. G. Bassett, eds.), pp. 105–115. Raven, New York.

Shuster, R. C., Houdebine, L. M., and Gayle, P. (1976). *Eur. J. Biochem.* **71**, 193–199.

Shyämalä, G. (1973). *Biochemistry* **12**, 3085–3089.

Singer, S. J. (1974). *Annu. Rev. Biochem.* **43**, 805–833.

Smith, S., and Abraham, S. (1970). *J. Biol. Chem.* **245**, 3209–3217.

Smith, S., and Abraham, S. (1975). *Adv. Lipid Res.* **13**, 195–239.

Stein, O., and Stein, Y. (1967). *J. Cell Biol.* **34**, 251–263.

Sulman, F. G. (1970). "Hypothalamic Control of Lactation." Springer-Verlag, Berlin and New York.

Taborsky, G. (1974). *Adv. Protein Chem.* **28**, 1–210.

Tanabe, T., Wada, K., Okazaki, T., and Numa, S. (1975). *Eur. J. Biochem.* **57**, 15–24.

Tindal, J. S., and Knaggs, G. D. (1977). *Brain Res.* **119**, 211–221.

Topper, Y. J., and Oka, T. (1974). In "Lactation" (B. L. Larson and V. R. Smith, eds.), Vol. 1, pp. 327–348. Academic Press, New York.

Turkington, R. W. (1972a). *Biochem. Actions Horm.* **2**, 55–80.

Turkington, R. W. (1972b). *Lactogenic Horm., Ciba Found. Symp., 1971* pp. 111–135.

Turkington, R. W., and Hill, R. L. (1969). *Science* **163**, 1458–1460.

Turkington, R. W., and Riddle, M. (1970). *J. Biol. Chem.* **245**, 5145–5152.

Turkington, R. W., Brew, K., Vanaman, T. C., and Hill, R. L. (1968). *J. Biol. Chem.* **243**, 3382–3387.

Turkington, R. W., Majumder, G. C., Kadohama, N., MacIndoe, J. H., and Frantz, W. L. (1973). *Recent Prog. Horm. Res.* **29**, 417–455.

Turner, C. W. (1952). "The Mammary Gland," Vol. I. Lucas Bros., Columbia, Missouri.

Venkataraman, R., and Reithel, F. J. (1958). *Arch. Biochem. Biophys.* **75**, 443–452.

Whaley, W. G. (1975). "The Golgi Apparatus." Springer-Verlag, Berlin and New York.

Whaley, W. G., Dauwalder, M., and Leffingwell, T. P. (1975). *Curr. Top. Dev. Biol.* **10**, 161–186.

Wood, T. G., and Lingrel, J. B. (1977). *J. Biol. Chem.* **252**, 457–463.

Yip, M. C. M., and Knox, W. E. (1972). *Biochem. J.* **127**, 893–899.

CHAPTER 7

Helminth Parasites and the Host Inflammatory System

R. Wesley Leid, Jr. and Jeffrey F. Williams

I. Introduction 229
II. Pathways to Inflammation 230
  A. Hageman Factor Dependent and Independent Pathways 231
  B. The Complement Cascade (Classical and
    Alternative Pathways) 234
  C. Mast Cells/Basophils and the Chemical Mediators of
    Immediate Hypersensitivity 236
III. Tissue Reactions to Helminth Parasites 243
  A. Nematodes 243
  B. Cestodes 245
  C. Trematodes 246
IV. Parasites and the Pathways to Inflammation 247
  A. The Coagulation System 247
  B. The Kinin System 252
  C. The Complement System 254
  D. Mast Cells/Basophils and the Chemical Mediators of
    Immediate Hypersensitivity 259
V. Conclusion 265
  References 266

I. Introduction

From the perspective of the biologist interested in host–parasite relationships, one of the most exciting advances in immunology and biochemistry in recent years has been the emergence of a comprehensive account of the pathways which lead to the inflammatory response in mammals. The isolation and characterization of the proteins of the coagulation and complement cascades, the discovery of the mechanisms of kinin production and fibrinolysis, and the identification of the major chemical mediators of the immediate hypersensitivity reaction, provide us with an opportunity to view the interrelationships of these systems and their contributions to host responses to injury and infectious organisms. This is a particularly important development for parasitologists, who now have a firm base from which to explore experimentally the unique features of tissue and cellular reactions during the intimate association between metazoan helminth parasites and their

229

CHEMICAL ZOOLOGY, VOL. XI
Copyright © 1979 by Academic Press, Inc.
All rights of reproduction in any form reserved.
ISBN 0-12-261040-5

hosts. The insights gained thereby into the evolution of parasitism, and the ways in which the delicate balance between hosts and their worms can be tipped in favor of either party, should have a significant impact on the course of both fundamental and applied work in this area in the future.

Our purpose in this chapter is to focus on each of the major components of the pathways to acute inflammation and to examine the evidence presently available concerning interactions between helminth parasites and these host response mechanisms. In many cases, of course, this evidence is indirect because the parasitologic observations predate the critical technological procedures which have been so valuable in the experimental analysis of inflammatory pathways. However, the literature on host–parasite relationships serves as a rich source of detailed and stimulating descriptions of both clinical and experimental aspects of tissue responses to helminths. We propose to interpret some of these past observations in the context of contemporary understanding of inflammation, and, in this way, to point out some of the most promising directions for future research.

In the first part of the chapter we present an account of the characteristics and biological importance of the coagulation cascade, the complement and kinin-generating systems, and the principal cellular elements of the acute inflammatory response in mammals. Selected examples are then drawn from studies on three classes of parasitic helminths, the Nematoda, Cestoda, and Trematoda, in order to develop a broad view of the known and potential interactions between these organisms and inflammatory pathways, and to highlight some of the pecularities of each group. The organisms chosen for discussion are, in almost all instances, either responsible for important disease problems in human or veterinary medicine, or occur in the laboratory animal models commonly used to study the major parasitoses of man and domesticated animals.

II. Pathways To Inflammation

Many homeostatic physiological processes depend on the enzymic activation of plasma proteins. Often these proteins are synthesized and circulate as inactive precursors, or zymogens, which are converted to their active forms by limited proteolytic cleavage. In some instances important biologic products result from the activation of single zymogens. In other cases, sequential multiple steps or "cascades" are triggered by a single stimulus, creating an amplification effect. The inflammatory response in mammals involves two such cascades; the coagulation and complement systems.

Inflammation represents the sum of the host responses to injury (Ryan and Majno, 1977). The cardinal signs of increased vascular flow to the affected area and the development of swelling and pain result largely from the biologic activities of the proteins and their cleavage products which make up these two cascades. These activities lead to changes in the permeability of blood vessels, the tone of smooth muscle in blood vessel walls, the movement and secretory functions of cells which are recruited to participate from local or circulating pools, and the stimulation of pain receptors. In Section II,A we shall describe the consequences of initiation of coagulation and, in Section II,B, the effects of complement fixation. It should be remembered, however, that these two systems do not operate independently. They function in the same milieu and are therefore interconnected in many ways which influence and regulate the inflammatory response.

A. HAGEMAN FACTOR DEPENDENT AND INDEPENDENT PATHWAYS

The coagulation, fibrinolytic, and kinin generating systems are included under a single heading because of the central role of Hageman Factor activation in all three pathways (Fig. 1; Davie and Fujikawa,

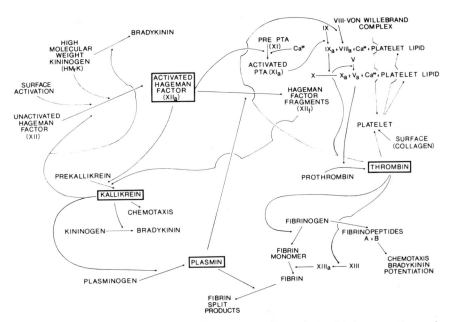

FIG. 1. The Hageman Factor-dependent systems of coagulation, kinin generation, and fibrinolysis. The four enzymes within boxes are Hageman Factor, thrombin, kallikrein, and plasmin and represent the most important activated proteins in these pathways. All serve to amplify portions of these inflammatory cascades.

1975; Kaplan and Austen, 1975). These pathways are initiated by the conversion of native Hageman Factor (HF) or coagulation Factor XII to its biologically active form HFa (XIIa). Hageman Factor circulates in mammalian plasma and is activated by contact with negatively charged surfaces. The events which follow form a part of the HF-dependent or "intrinsic" coagulation pathway. High molecular weight kininogen (HMrK) has recently been shown to be necessary for this activation process (Griffen and Cochrane, 1976). *In vivo* the initial stimulus may be provided by contact between HF and collagen or vascular basement membrane (Cochrane *et al.*, 1972; Wilner *et al.*, 1968), platelets (Walsh, 1972), endotoxin, pyrophosphate or uric acid crystals (Morrison and Cochrane, 1974; Kellermeyer and Brecken-ridge, 1965). Enzymic activation of Factor XII may also occur, and trypsin, plasmin, kallikrein, and plasma thromboplastin antecedent (PTA, Factor XIa) have all been shown capable of doing this (Cochrane and Wuepper, 1971; Cochrane *et al.*, 1973; Weiss *et al.*, 1974). Once the HF zymogen is activated, all three pathways leading to coagula-tion, kinin generation, and fibrinolysis may be triggered.

Activated Hageman Factor converts another protein, prekallikrein, to kallikrein, an extremely important enzyme with many amplifying effects (Fig. 1). Kallikrein can stimulate directed cell movement of neutrophilic or mononuclear leukocytes (Kaplan *et al.*, 1972; Gallin and Kaplan, 1974). It also cleaves the nonapeptide, bradykinin, from the parent molecule, kininogen, and further feeds back to enzymically activate more HF. Bradykinin increases vascular permeability, con-tracts smooth muscle, causes a profound hypertension and is a potent vasodilator as well as an elicitor of pain (Rocha e Silva, 1973). Bradykinin can also release histamine from mast cells (Johnson and Erdös, 1973) and this may amplify its ability to increase vascular per-meability. These activities indicate that it probably has a role in the development of inflammation.

Kallikrein enzymically cleaves a precursor, plasminogen, to its bio-logically active form, plasmin. Plasmin also plays an important part in this scheme because it affects all three pathways to acute inflammation (Fig. 1), and furthermore provides an entrance into the complement cascade through its effects on the complement components C3 and C1 (Ratnoff and Naff, 1967; Bokisch *et al.*, 1969). Activated HF is con-verted by plasmin to Hageman Factor fragments (HFf) which in turn can cleave additional prekallikrein to kallikrein, or feed back to pro-duce more HFa. Plasmin functions in fibrinolysis by enzymic diges-tion of fibrin to small molecules, the so-called fibrin split-products.

Activated HF converts the zymogen pre-PTA to its active form (PTA,

XIa). PTA activates Factor IX enzymically and generates IXa. Factor IXa, in the presence of platelet phospholipids or tissue phospholipids and Factor VIIa, will enzymically cleave Factor X to its active form Xa. Factor Xa in the presence of Ca^{2+}, phospholipid, and activated Factor V (Va) cleaves prothrombin to thrombin. Thrombin is another key enzyme in the coagulation cascade since it not only feeds back to aggregate platelets, and thus releases further phospholipids (Mustard and Packham, 1970), but also converts more Factor VIII to VIIIa and activates additional Factor V to Va (Fig. 1). All these steps serve to amplify the coagulation reaction, so that a few initially activated molecules have the potential to rapidly escalate the host response. Thrombin carries the coagulation pathway to completion by converting fibrinogen to fibrin monomers. These monomers become the cross-linked fibrin clot after interaction with Factor XIIIa, a protein which is simultaneously activated by thrombin. The thrombin cleavage products of fibrinogen left after fibrin monomer generation are termed the fibrinopeptides A and B. One of these, fibrinopeptide B, has chemotactic activity for neutrophilic leukocytes and potentiates the action of bradykinin (Gladner et al., 1963; Kay et al., 1973).

Many of the active enzymes associated with coagulation, fibrinolysis, and kinin generation are susceptible to control by circulating plasma protein inhibitors such as α_2-macroglobulin, α_1-antitrypsin, C1 inhibitor and antithrombin III. The rate-limiting control points in all three pathways are HF activation and the conversion of prothrombin, prekallikrein, and plasminogen to their active forms. Were any of these enzymes to be regulated by an invading organism, very marked effects on coagulation, kinin generation, and fibrinolysis would occur.

Damage to tissues can lead to initiation of clot formation via the "extrinsic" pathway, which bypasses the sequential activation of the HF dependent intrinsic route. In this process an unidentified tissue factor interacts with Factor VII to convert the precursor Factor X to its active form Xa. After that point the coagulation cascade proceeds as outlined above for the intrinsic pathway. Whether Factor VII is activated by tissue factor, or instead exists in an activated form, is not clear at present. It is obvious, however, that mechanical damage to tissues by helminth parasites, as well as the release of parasite products capable of tissue destruction, could give rise to fibrin deposition through this mechanism. The generation of other biologically active host components could also be initiated by these interactions. Additional evidence suggests that kallikrein may activate Factor VII and lead to fibrin formation through a slight modification of the extrinsic pathway (Gjonnaess, 1972a,b).

B. The Complement Cascade (Classical and Alternative Pathways)

The complement cascade can be activated by either the classical or alternative pathways, both of which have a common terminal portion, designated the "attack sequence" (Fig. 2) (Müller-Eberhard, 1975; Fearon et al., 1976; Fearon and Austen, 1978). The nomenclature which has evolved to describe these two pathways is bewildering, but some recent efforts to standardize the terminology have clarified the situation. The classical pathway proteins are now conventionally identified by the capital letter C, followed by the numbers 1–9. Subunits are indicated by lower-case letters, and the active form of each component is shown by a bar over the letter or number. Plasma proteins

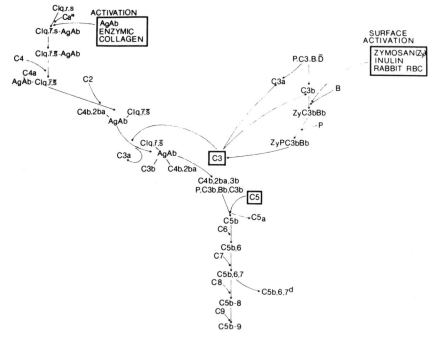

Fig. 2. The mechanisms of activation for the complement cascade, including both the classical and the alternative pathways. Activated components are indicated by a bar over each (e.g., $\bar{B}$), and the cleavage fragments are noted as lower case letters (e.g., Bb). The classical pathway proteins are designated by a capital C followed by the numbers 1–9 (e.g., C2), while the proteins unique to the alternative pathway are D, P and Factor B. C3 and C5 are placed within boxes to emphasize the enormous biological potential of these molecules in inflammation. Also enclosed with a box are the activators or initiators of both the classical and the alternative pathways.

unique to the alternative pathway are identified by the capital letters, P, B, and D, and the active forms are again indicated by use of a bar. Fragments of these components are signified by lower-case letters (e.g., Bb, represents a fragment remaining after cleavage of Factor B).

Activation of the classical pathway is initiated by the binding of C1q to a unique region, the Fc portion, of either the IgM or IgG antibody molecules in immune complexes. This binding results in the activation of C1r, which in turn converts C1s to its active form, C1s̄. The activation of C1s to C1s̄ may also proceed by proteolysis using the enzymes trypsin or plasmin (Ratnoff and Naff, 1967). C1s̄ alone or in the form of the trimolecular complex C1q̄r̄s̄, enzymically cleaves two more proteins, C4 and C2 to form a bimolecular complex, C4̄2̄. This enzyme cleaves a plasma protein C3 into a major fragment C3b and a smaller peptide, C3a. C3b alters the specificity of the C4̄2̄ enzyme so that it is now capable of cleaving C5 plasma protein into two fragments, C5b the major cleavage product and a smaller peptide C5a. The C5b then initiates the nonenzymatic, noncovalent assembly of the multimolecular complex C5b6789. This combination of the terminal complement components is the membrane attack sequence.

C3a and C5a are the so-called "anaphylatoxins." They have many potent biological activities which probably contribute to the development of the inflammatory response. C3a and C5a increase vascular permeability, contract smooth muscle, and release histamine from mast cells (Dias de Silva et al., 1967; Valotta and Müller-Eberhard, 1973; Johnson et al., 1975). Both C3a and C5a are chemotactic for polymorphonuclear leukocytes as well as mononuclear cells (Ward, 1974). Additionally, the C5̄6̄7̄ fluid phase complex is chemotactic for polymorphonuclear leukocytes (Ward and Becker, 1967). C3b, the major cleavage fragment of C3 promotes the adherence of complexes to cells bearing the immune adherence (IA) receptor, and can induce enhanced phagocytosis of the complex in cells bearing such receptors. C3a and C3b-like molecules can also be generated through enzymic activation of the native proteins with proteolytic enzymes such as trypsin or plasmin (Ratnoff and Naff, 1967). C3a and C5a generation by plasmin permits complement activation through the Hageman Factor dependent systems. The anaphylatoxins generated can have many if not all of the striking biologic properties described for the peptides produced by classical or alternative pathway C3-converting enzymes (Ward, 1974).

The alternative pathway is activated after interaction of complex polysaccharides (e.g., zymosan, inulin), IgA cryoproteins, and rabbit red blood cells with the serum proteins, C3, factor B, activated D (D̄), and properdin (P) (Fearon *et al.*, 1976; Fearon and Austen, 1978). This initial C3-converting enzyme is capable of cleaving native C3 to C3b and C3a and is constantly being generated and destroyed. The C3b formed from native C3 interacts with B and $\bar{D}$ to form the amplification converting enzyme. This convertase is extremely short-lived due to its intrinsic decay properties but is stabilized both by properdin and C3NeF, an abnormal protein which circulates in certain glomerulonephritides. The amplification convertase is expressed normally in the fluid phase but can be protected from decay by formation on certain surfaces such as yeast cell walls. In this case, zymosan serves to protect the amplification convertase from control by either endogenous decay or by the protein $\beta 1H$. $\beta 1H$ serves as an alternative pathway control protein and is capable of reversing the stabilization induced by P but not that caused by C3NeF. Another alternative pathway control protein in plasma is the C3b inactivator (C3bINA), an enzyme which cleaves C3b to its inactive components C3c and C3d.

Stabilization of the amplification convertase and the abrogation of normal control by decay or inactivation allows the alternative pathway to feed on itself and generate additional substrate on which the terminal attack sequence can assemble. The alternative pathway activation system is therefore quite different from that of the classical pathway, in which zymogens are converted to their active state in a sequential manner. Alternative pathway initiation comes about through circumvention of host control. The system is normally in a constant state of activation and decay, and only when an "alternative pathway activator" is present to provide a protected surface for the amplification converting enzyme to be stabilized does this pathway outstrip inherent control by proteins C3bINA and $\beta 1H$. It is clear that agents which activate C3 and C5 would have major biologic importance, and interruption of control of alternative pathway activation by $\beta 1H$ or C3bINA could have serious consequences to the host or parasite. In addition, interaction between parasites and C1 inhibitor (C1Inh), the major control protein of the classical pathway, would affect the expression of this sequence, and thereby affect the myriad of biological effector molecules derived from the complement cascade.

C. Mast Cells/Basophils and the Chemical Mediators of Immediate Hypersensitivity

The cascades of proteolytic enzymes of the plasma which generate cleavage products with chemotactic and vascular permeability activity, are supplemented in the acute inflammatory process by cells recruited from either the extravascular tissue pool or the blood. Tissue histiocytes (macrophages) and mast cells are the participants of extravascular origin, and both probably arise from the differention of mesenchymal cells. Monocytic, neutrophilic, eosinophilic, and basophilic leukocytes form part of the population of intravascular cells, and are derived from precursors in the bone marrow. Circulating non-nucleated thrombocytes, or platelets, make up the spectrum of cellular elements which become involved in localized acute inflammation.

White blood cells in the lymphocytic series may be of bone marrow or thymic origin, and are responsible for the specific recognition and effector mechanisms of the immunologic system. While immunologic (antigen–antibody) reactions often serve as primary stimuli for acute inflammatory events, the influx of mononuclear lymphocytic cells is generally regarded as a distinguishing characteristic of chronic inflammation and delayed hypersensitivity. We realize that this distinction is increasingly difficult to defend (Askenase, 1977a), but, for the purposes of this review we propose to limit our discussion to the granulocytes in general, with particular reference to the mast cell, which stands at the crossroads between chemical and cellular pathways to acute inflammation in helminth parasite infections.

A great many of the advances in research on acute inflammation have come about through the study of antigen- or antibody-activated tissue mast cells and their circulating counterparts, the basophils. As a result it has become clear that these two "target" cell types can be induced to synthesize and/or release a diverse array of products which can influence the chemical and cellular composition of the microenvironment. Collectively these products have come to be known as "the chemical mediators of immediate hypersensitivity." Although the experimental manipulations which have led to their characterization have depended largely on immunologic stimuli, the profound effects which these mediators exert on the cellular elements of the inflammatory response and, in turn, the modulating effects which inflammatory cells and the coagulation and complement systems exert on mast cells, provide a basis for understanding the molecular processes which occur in all acute inflammatory reactions. The nature of target cell activation and the biologic activities of the chemical mediators are therefore discussed in detail in this section.

Mast cell immunologic activation can occur by two different mechanisms involving antibodies in two distinct classes of immunoglobulins. Antibodies in the IgE class are commonly associated with parasitic helminth infection (Jarrett, 1973). IgE antibodies persist after fixation to the mast cell or basophil for extended periods of time, up to several months in some cases (Ishizaka and Ishizaka, 1971). The target cell receptors are specific for the Fc region of the IgE molecules, and have high binding constants. The resultant low level of IgE dissociation from the cellular surface contributes to persistence of antibody on sensitized cells. Release of the chemical mediators from mast cells occurs after antigen binding and bridging of two adjacent IgE molecules on the cell membrane. This triggers a sequence of biochemical reactions which culminate in secretion of vasoactive amines and other inflammatory substances. For those interested in a detailed discussion of the biochemical events subsequent to antigen bridging, several reviews have been published recently (Kaliner and Austen, 1974; Austen and Orange, 1975; Soter and Austen, 1976; Austen et al., 1976).

A second antibody of the IgG class capable of mast cell activation has been recognized in several animals (Stechschulte et al., 1971), although this biologic activity resides in different IgG subclasses, depending on the species in question. Antibodies in this class have also been shown to develop in helminth infections but they have been less well-studied (Catty, 1969; Leid and Williams, 1974; Musoke and Williams, 1975a,b). All the subclasses, however, appear to activate the target cell in a similar manner, functioning as an immune complex with specific antigen. The target cells do not appear to have an Fc receptor with high affinity for IgG subclasses, and as a result antibodies bind only very weakly to the cell surface. This property of limited interaction with the target cell has given rise to the designation of these antibodies as "short term" skin-sensitizing immunoglobulins. The IgG–antigen complex functions as a target cell activator only in the presence of complement and neutrophils in the animal species so far examined (Stechschulte et al., 1971; Henson and Cochrane, 1969).

As we have mentioned in our discussion of coagulation and complement, mast cells and basophils can be activated by factors generated via these two cascades, as well as by immunologic reactions. In either case, a broad spectrum of chemical mediators is released after mast cells are activated. These are depicted in Fig. 3, which is based on the model developed by Austen and co-workers (Austen et al., 1976). It is most convenient to separate mediators into two groups; those which are preformed and the total available at any one time is stored within the cell, and those that must be generated de novo and released after immunologic activation. The most universally known of the stored mediators is histamine. Histamine can enhance random motility of eosinophils, as well as enhance the directed migration of these cells in vitro after stimulation with a chemotactic factor (Clark et

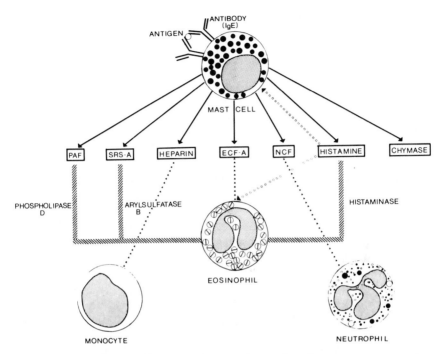

FIG. 3. The potential chemical mediators of immediate hypersensitivity reactions released as a result of antigen–antibody triggering of mast cells/basophils. The regulatory role of the eosinophil is shown by the hatched lines, which indicate enzymic inactivation of SRS-A, PAF, and histamine. Chemotactic attraction of monocytes, neutrophils and eosinophils by mast cell products is shown by the solid arrows.

al., 1975). Histamine contracts smooth muscle, increases vascular permeability and is a potent vasodilator. It is enzymically inactivated by deamination or methylation.

Serotonin (5-hydroxytryptamine) is one of the principal chemical mediators in mast cells of rodents, but it is not associated with these cells in other species (Austen, 1971). It probably plays an important role in systemic anaphylaxis of mice. Serotonin has many of the biologic activities of histamine and is also inactivated by deamination.

The next most commonly present mediators are the tetrapeptides which comprise the eosinophilic chemotactic factor of anaphylaxis (ECF-A). These molecules appear to be present in many animal species but have only been purified and sequenced in humans and rats (Goetzl and Austen, 1975; Boswell *et al.*, 1978). ECF-A causes the directed migration of eosinophils *in vitro* as well as *in vivo* (Goetzl and Austen, 1976; Austen *et al.*, 1976). ECF-A has no other presently recognizable biologic function and is enzymically inactivated by car-boxypeptidases and certain aminopeptidases (Goetzl and Austen, 1975).

A neutrophil chemotactic factor (NCF) has been recently appreciated as a mediator released from mast cells and basophils (Austen *et al.*, 1976; Wasserman *et al.*, 1977). NCF has no known biologic functions other than its attractiveness for neutrophils, but the characterization of this activity has not been pursued as extensively as for ECF-A.

Macromolecular heparin or other highly charged polysaccharides are the intracellular glue holding the mast cell or basophil granules together. Heparin, a glycosaminoglycan, can be released in macromolecular form after immunologic activation (Yurt *et al.*, 1977). It functions as a cofactor with antithrombin III by accelerating inhibition of thrombin by antithrombin III, and this effectively abrogates clot formation. Heparin appears to have chemotactic activity *in vitro* for mononuclear leukocytes and could provide a chemotactic stimulus for infiltration by these cells (R. W. Yurt, unpublished). A variety of other biologic activities are recognized (Jacques, 1975) and the parent molecule can be degraded by heparinases and disaccharidases (Ögren and Lindahl, 1975; Silva and Dietrich, 1975). Other animal cells may have different glycosaminoglycans for the intragranular matrix, but these have not been extensively characterized.

A granule-associated chymotrypsinlike enzyme, termed chymase, has been detected in highly purified rat mast cells. Chymase is released after immunologic activation and may serve to affect several inflammatory pathways. Chymase seems to have all the substrate specificity of the native pancreatic chymotrypsins. Pancreatic chymotrypsin is a potent mast cell activator (Lagunoff *et al.*, 1975), and the release of chymase could feed back to intensify and localize an inflammatory reaction. So far chymase has been characterized only for rat mast cells (Lagunoff and Benditt, 1963; Kawiak *et al.*, 1971; Yurt and Austen, 1978), and the limited information available suggests that different species, as well as basophils and mast cells from the same species may contain either trypsin or chymase. In the dog, tissue mast cells contain both enzymes within the same differentiated cell. Inactivation of released enzymes might depend on further proteolysis or on inhibition by the major serum protein inhibitors previously mentioned in the discussion of complement.

Of the chemical mediators which are unstored and generated *de novo,* slow reacting substance of anaphylaxis (SRS-A) is the most important. SRS-A probably plays a decisive role in asthma and asthma-like syndromes in man and other animals. SRS-A can increase vascular permeability in guinea pigs, rats, and monkeys (Brocklehurst, 1968; Orange *et al.,* 1969); produce a profound and prolonged contraction of smooth muscle (Brocklehurst, 1962) and through this effect it may initiate pronounced bronchoconstriction. It is an acidic lipid which contains a sulfur moiety that appears to be critical for the expression of biologic activity. SRS-A is susceptible to enzymic inactivation by arylsulfatases from a variety of tissues (Orange *et al.,* 1974; Wasserman *et al.,* 1975; Wasserman and Austen, 1976).

Platelet activating factor (PAF) is representative of a family of lipids which have recently been recognized as unstored mediators. PAF causes the release of intracellular contents of platelets without aggregating them. These intracellular contents include the phospholipids necessary for coagulation, collectively called platelet factor 3; a protein capable of neutralizing heparin, termed platelet factor 4, the cathepsin and acid hydrolase enzymes, calcium, serotonin, ATP, ADP, and, probably most importantly, the prostaglandins, endoperoxides, and thromboxanes (Weiss, 1975; Hamberg *et al.,* 1975; Samuelsson *et al.,* 1976). PAF has no known smooth-muscle contractile properties, and is inactivated enzymically by a phospholipase D enzyme which is present in eosinophils (Kater *et al.,* 1976).

Another group of unstored lipid mediators are the lipid chemotactic factors (Valone *et al.,* 1977). One of these lipids serves to attract the eosinophils as well as the neutrophils *in vitro,* and another increases the random motility of these leukocytes *in vitro.* Chemotactic lipids, ECF-A tetrapeptides, NCF and heparin could obviously be major determinants of the character of local cellular infiltration around the triggered mast cell. Whether lipid chemotactic factors have other biologic properties has not yet been determined. Finally, the prostaglandins and the thromboxanes, in conjunction with their various intermediary compounds, can produce dramatic biologic effects in the host. These substances have been the subject of recent reviews (Samuelsson *et al.,* 1975; Hamberg *et al.,* 1976), and the reader is encouraged to seek them out as we will not attempt to cover developments in this area.

A unifying hypothesis has been developed recently by Austen and his collaborators (Austen and Orange, 1975; Austen *et al.,* 1976) which serves to place target cells, phagocytes, and chemical mediators into a meaningful picture of the acute inflammatory process. Most of their

observations are based on the situation in man and the common laboratory rodents, and a direct extrapolation to other mammals may not be
appropriate. However, target cells and mediators in other species
have not been well-characterized and for the moment the scheme devised by Austen can be viewed as a very useful model for discussion.
After activation of tissue mast cells by either IgE- or IgG-dependent
mechanisms or by nonimmunologic means, they propose that there is
an initial period of increased vascular permeability in the affected
area. This period will only last for minutes and is termed the
"humoral" phase. The "cellular" phase begins within minutes and
lasts for hours and is the result of a cellular influx into the area around
the site of released chemotactic factors. Platelets will be activated to
release their intracellular constituents by contact with PAF. After cellular infiltration has begun, control mechanisms come into play which
limit the response and keep it within the range required for response
and repair. Eosinophils now seem likely to be very important as control cells for immediate hypersensitivity reactions. Recently, they have
been shown to contain histaminase (Zeiger *et al.*, 1976) which is capable of deaminating histamine; arylsulfatase B which cleaves SRS-A
(Wasserman *et al.*, 1975); and phospholipase B which inactivates PAF
(Kater *et al.*, 1976). Whether enzymes are present in the eosinophil to
control other chemical mediators remains to be determined. Neutrophilic and mononuclear cell infiltration probably is vital for the
phagocytic control of invading organisms and damaged cell debris.
Should any of these regulatory mechanisms become deranged, the
inflammatory reactions may proceed into aggravated acute inflammatory diseases or progress to a chronic stage.

Askenase and his collaborators have recently emphasized the central role of the mast cell (Askenase, 1977a). In their studies mast cells
were shown to be necessary for the expression of the immune response
(Gershon *et al.*, 1975). Indeed, Askenase points out that mast cells or
basophils may be essential for effective resistance to a variety of invasive organisms. An illustration of how these interrelationships might
be envisioned is depicted in Fig. 4. These ideas have been expressed
by Askenase at a recent symposium on immunoparasitology (Askenase, 1977b). The interaction of T and B lymphocytes and the possibility of products from such a reaction having a direct action on mast
cells, as well as the proposed effect of macrophage products on mast
cells should provide a fruitful area of investigation in the coming
years. Recent reviews cover the field of lymphocyte products very well
and should be examined by those with a further interest in this area
(David and David, 1972; Bloom *et al.*, 1974).

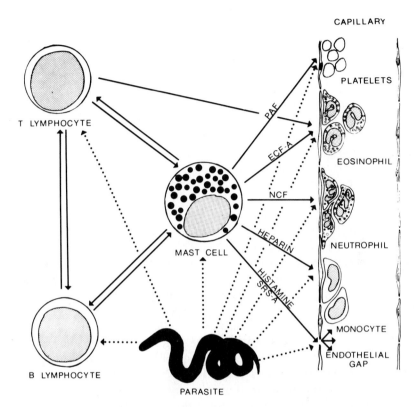

FIG. 4. The possible relationships between parasites and the inflammatory system, including lymphocytes, mast cells/basophils, neutrophils, eosinophils, platelets, and monocytes. Hypothesized direct effects of parasite products on each inflammatory component are indicated by dotted lines, while established interactions between cellular constituents are shown by a solid line.

III. Tissue Reactions to Helminth Parasites

A. NEMATODES

Helminths in the class Nematoda are probably the most successful parasites and collectively the most important causes of clinical helminthiasis in both man and domesticated animals. They parasitize a remarkably wide range of tissues and organ systems, often following complex migratory pathways *en route* to their specific predilection sites, so that opportunities for interaction with host inflammatory mechanisms are plentiful. In some instances these interactions undoubtedly are responsible for profound pathologic changes, particularly in sensitized hosts, but in other cases recognition and rejection systems of the host are evidently not called into play. This is especially true where prolonged survival of either immature or mature nematodes occurs in the host as an essential feature of the life cycle. The phenomenon permits the eventual passage of larval forms to subsequent hosts via carnivorism, or vertical transmission to offspring pre- or post-natally. It also provides for delayed maturation of larvae into egg-producing adults at times which are more propitious for survival of offspring in the environment. In the case of adult worms, it permits shedding of offspring through some portal of exit from the host for extended periods of time.

The traditional hallmarks of host reaction to nematodes have long been recognized as eosinophilia and acute allergic inflammation, though the reasons for the nature of this host response are not clear. The impressive feeding and digestive systems with which many nematodes are equipped obviously lead to some physically destructive effects on cells and tissues, which have important pathologic and clinical consequences. However, recent work suggests that more subtle and complex processes are also involved, and that these may underlie the pecularities of the host response. The contemporary concepts of pathways to the inflammatory response outlined in the Section II of this review form a framework not only for the interpretation of some of the observations on host reactions in nematodiasis, but also for the development of hypotheses on the prolonged survival of worms in tissues. The fact that many intense inflammatory responses to tissue-dwelling parasites are provoked only in the event of death of the worms suggests that these pathways are actively held in check by the living organisms, and some of the means whereby this may be achieved are explored below.

In histopathologic studies of host reactions to migrating nematode larvae, the cellular composition of the area adjacent to the parasites is typically characterized by monocytic and polymorphonuclear neutrophilic inflammatory cells, whereas at the tail end of the worms, eosinophilic leukocytes predominate (Archer, 1963; Poynter, 1966). In the tracks left behind as nematodes move through tissues, eosinophils, and plasma cells become most prominent. In some instances, where adult worms grow in the tissues, eosinophilic infiltration is characteristic of the migratory phase, but once the adult settles at the site of preference, the inflammatory response subsides. Abrupt recurrence of a new inflammatory response at the site is usually associated with death of the worms. This sequence of events is typical of dracunculosis in man, for example (Faust et al., 1970).

Nematodes which live in the gut often cause the local accumulations of mast cells and polymorphonuclear granulocytes, including large numbers of eosinophils (Wells, 1962; Poynter, 1966; Miller, 1971a,b). At times of immunologic rejection of gastrointestinal nematodes there is a very marked alteration in vascular permeability in the mucosa (Stewart, 1953; Murray *et al.*, 1971). The events which lead to expulsion of intestinal worms are still not understood, but recent experimental evidence suggests that the terminal effector mechanisms are not necessarily immunologically specific (Dineen *et al.*, 1977). The reason for the local accumulation of mast cells is also not clear, although so little is known of the origin, life span and the fate of mast cells at present that this situation is not surprising. In fact, experimental work on parasitic infections may provide a means of examining these characteristics, since infections with nematodes and other helminths in man and animals have been shown to result in highly significant increases in mast cell numbers in tissues all over the body (Fernex, 1962; Fernex and Bèzes, 1962; Fernex and Sarasin, 1962). Possibly some unique features of helminth products lead to an effect both locally and systemically on the differentiation of mast cells from precursor cells and/or on their replication.

B. CESTODES

A great deal of the literature on host reactions to cestode helminths concerns the tissue phases of larvae of tapeworms in the family Taeniidae. Here we are dealing with organisms which enter the tissues of the intermediate host by penetrating the intestinal epithelial surface, and which then travel hematogenously to their sites of predilection (Smyth and Heath, 1970). At those sites they become sessile, undergoing rapid morphogenesis into fluid-filled bladders. In some cases a moderate degree of migration occurs before the cystic organisms eventually settle in the serous, peritoneal cavity (Smyth and Heath, 1970). By assuming a sessile cystic form, they present an inviting target for both nonspecific and specifically triggered inflammatory defense mechanisms.

The balance of this relationship seems all the more precarious for the cestodes in view of the fact that they possess no tough outer cuticle, as do the nematodes. Instead they expose to their hosts a delicate plasma membrane, or "tegument" (Lumsden, 1975). Across this tegument all nutrients, secretions, and excretions must pass if the parasites are to survive and grow. The tegumentary surface is covered with microvilli, rather like those of intestinal epithelial cells (Beguin, 1966; Smyth, 1969), and represents the plasma membrane of the cytoplasm of syncytial cells which make up the bulk of the parasite bladder wall.

During the migration and tissue-establishment phases of metaces-
tode infections, eosinophils accumulate at the local site and increase
enormously in numbers in the blood (Arvy, 1950; Weinstein et al.,
1954; Freeman, 1962, 1964; Soule et al., 1971; Ansari and Williams,
1976). Chronic larval cestode infections in man and domesticated
animals are often associated with mild tissue reactions, although on
death of the worms acute inflammatory responses quickly follow
(MacArthur, 1934; Dixon and Hargreaves, 1944; Silverman and Hulland, 1961; Sweatman and Henshall, 1962). Peripheral blood
eosinophilia and local eosinophilic infiltration in the intestinal mucosa
also occur in cases of adult tapeworm infections (Lapierre, 1953;
Adonajlo and Bonczak, 1961; for review, see Pawlowski and Schultz,
1972). As with the nematodes, mast cell numbers are much increased
in the tissues of animals infected with cestodes (Fernex and Fernex,
1962; Varute, 1971; Cook and Williams, 1978), and some of the host
responses in cestode infections, which are manifested as pruritus, urticaria, and/or asthma (Blamoutier, 1952), could be produced by release of inflammatory mediators from these cells.

C. TREMATODES

Host responses to trematodes share many of the pecularities of those
which occur in nematode and cestode infections. Eosinophils participate conspicuously in local tissue infiltrates and generally circulate in
increased numbers at some point in the infection (Fernex and Fernex,
1962; Sinclair, 1962), though mononuclear and polymorphonuclear
neutrophils predominate in the immediate area of migrating organisms. Mast cells also increase in numbers, and not only at the sites
where parasites are lodged (Fernex and Fernex, 1962).

Flukes living in mammalian hosts are generally in either the
metacercarial or adult form. In each case they tend to migrate extensively from their point of entry through the bowel or skin to a tissue of
preference. There they settle down to a lengthy period of quiescence,
in the case of metacercaria, or egg production in the instance of adults,
such as in the medically important liver, lung, and blood flukes. As for
the cestodes, their interface with host tissue occurs at a "tegument"
(Erasmus, 1977) and they have no acellular protective cuticle, comparable to the nematodes. They do, however, feed via an oral apparatus and a blind cecal gut system. Their physically destructive
feeding habits, and their requirement for migration obviously lead to
inflammation, though the kinds of mechanisms which shape the
character of the host reaction have not been well studied in this group.
It is clear, however, that host tissue changes are not wholly attributable to local physical damage, and that cell infiltration and proliferation

may occur at some distance from the flukes themselves (Dawes, 1963; Dawes and Hughes, 1964; Anonymous, 1974). The ability of both larval and adult forms to survive in some tissues for extended periods suggests that host responses may be held in abeyance at times during infection.

IV. Parasites and the Pathways to Inflammation

A. THE COAGULATION SYSTEM

1. Nematodes

In view of the fact that many nematodes are dependent upon the ingestion of host blood for survival and that many others live for extended periods of time in blood, remarkably little research has been done on the effects of these parasites on the coagulation system. This is particularily striking since some of the earliest experimental observations on the pathology of intestinal nematodes involved the influence of parasite extracts or products on blood clotting.

Loeb and Smith (1904) found that extracts of the canine hookworm, *Ancylostoma caninum*, inhibited the coagulation of dog blood for periods up to 24 hours. The factor responsible was resistant to boiling and was present only in the anterior half of the worms. Later Weinberg (1907) showed that a similar agent was present in *Strongylus* spp., the so-called "intestinal bloodworms" of the horse, and Leroy (1910) and Weil and Boyé (1910) found that the blood of animals which had been inoculated with perienteric fluid of *Parascaris equorum* clotted much more slowly than normal. Flury (1912) reported that this fluid would delay coagulation of both human and dog blood "*in vitro*," and later Schwartz (1921) was able to obtain similar results with extracts of *Bunostomum phlebotomum*, the hookworm of cattle, and *Haemonchus contortus*, the abomasal worm of cattle and sheep. Schwartz also confirmed the observations of earlier workers on the anticoagulant effects of strongyles and ascarids.

In a number of instances of infection with the nematodes studied by these early workers, blood loss from intestinal lesions is a very prominent clinical sign. However, with the exception of *A. caninum*, little effort has been made in the intervening years to characterize either the agents responsible for inhibition of clot formation or the site or sites at which the pathways to coagulation are disrupted. *Haemonchus contortus*, for example, is probably the most important parasite of domesticated ruminants throughout the world, but nothing is known of the role of any interaction between this parasite and the host clotting systems in terms of the pathogenesis of clinical haemonchosis. That some defect does occur *in vivo* is suggested by the observation of Boughton and Hardy (1935) that after detachment of the worms from the abomasal wall, the site continued to bleed for 7 minutes. Some comparable interaction in *B. phlebotomum* infections in cattle is indicated by the finding that blood from heavily infected animals failed to clot normally (Sprent, 1946).

As in the case of *A. caninum*, the anticoagulant in *B. phlebotomum* is apparently present only in the anterior portion of the parasites (Hoeppli and Feng, 1933). In further work on the canine hookworms, the activity has been shown to be localized in the cephalic glands (Thorson, 1956; Eiff, 1966). Extracts of the cephalic glands were found to prolong the terminal portion of the coagulation cascade (prothrombin time) but had no effect on the intrinsic pathway (partial thromboplastin time) (Eiff, 1966). The activity in these extracts was stable at 100°C and was not precipitable by either 10% trichloroacetic acid or saturated ammonium sulfate. Spellman and Nossel (1971) found slightly different results using extracts of whole worms. In their experiments prolongation of both prothrombin and partial thromboplastin times were observed, but thrombin times were unaltered. The anticoagulant inhibited activated Factor X, and inhibition was more striking when activation was produced by thromboplastin and Factor VII. The extract also inhibited platelet aggregation. The authors raised the possibility that clotting system factors other than Factor X might also be affected by *Ancylostoma*. Apparently this has not been pursued to date, nor is it known if migrating larvae of these parasites are capable of affecting the coagulation cascade.

In some cases of intestinal nematodiasis, blood and protein losses into the gut are so great that plasma protein levels fall. Under these circumstances, it is difficult to know whether clotting deficiencies are due to active inhibition by parasite products or inadequate concentrations of coagulation factors in the plasma. However, the severe coagulation defect described in dogs infected with *Angiostrongylus vasorum* by Dodd (1973) was not associated with blood-loss anemia, and it seems likely that some agent released by the adult worms living in the heart and the pulmonary artery actively inhibits coagulation. Both prothrombin and partial thromboplastin times were prolonged in blood from infected dogs, and freshly drawn blood failed to clot in 24 hours while in contact with glass. Dodd concluded that many clotting factors including Factor XII were affected by the worm, but it is not clear how this determination was made. After successful treatment of the infection with an anthelminthic, clotting parameters returned to normal. Dodd suggested that although the mechanisms of clot inhibition were unknown, the phenomenon had "obvious benefits for the parasite, minimizing any tendencies for blood clots to form around them." It is surprising that comparable studies have not been done on other nematodes, since a similar advantage would accrue to many parasites, especially those which live or travel in blood vessels.

The significance of clotting defects in blood from animals given intravenous inoculations of nematode extracts is difficult to assess because, again, investigations on the mechanisms involved have not been conducted. Normal dogs given extracts of *Ascaris suum* have been shown to go into shock, and elevated levels of blood histamine occur within several minutes after injection (Rocha e Silva and Graña, 1946). Blood drawn in the severe shock phase does not clot for at least 24 hours *in vitro*. Addition of full strength thrombin does not enhance coagulation, and the circulating anticoagulant persists for several hours. Similar changes occur in normal guinea pigs, and the effect seems likely to be due at least in part to some primary pharmacologic effect of parasite-derived substances. It is possible that mast-cell degranulation *in vivo* initiated either nonimmunologically or immunologically, in sensitized animals (Yurt *et al.*, 1977), leads to massive heparin release and interference with clotting. The fact that addition of perienteric fluid to blood *in vitro* will inhibit coagulation (Schwartz, 1921) suggests that some more direct effect on the clotting system might be responsible, but no detailed analysis of this seems to have been undertaken.

The central position occupied by Hageman Factor in the initiation of the coagulation and kinin-generation pathways to inflammation suggests that any inhibiting effects of parasite substances on this component would have broad significance in the host–parasite relationship. In addition to the advantages ensuing from an uninterrupted flow of blood into the digestive tract of hematophagous nematodes, the prevention of Hageman Factor-dependent activation of biologically active molecules would exert a significant modulatory effect on the local tissue response. The influx of inflammatory cells might be impaired and the local accumulation of humoral effector molecules such as antibody and complement might be avoided. This area of research clearly needs attention in the future.

2. Cestodes

The effects of tissue-dwelling stages of cestode parasites on the coagulation cascade of host blood have not been investigated to any significant extent. Many years ago Rocha e Silva and Graña (1944) noticed that intravenous administration of fluid from *Echinococcus granulosus* in dogs produced a profound shock, during which the blood showed a greatly impaired ability to clot on contact with glass. *In vivo* studies were not carried out and, as with the nematode extracts, it seems possible that the *in vivo* effect results from massive immunologic release of heparin from IgE-sensitized mast cells (Yurt *et al.*, 1977). However, a direct anticoagulant effect of cestode-derived factors cannot be ruled out.

In recent studies, Hammerberg *et al.* (1978) have shown that substances released by metacestodes of *T. taeniaeformis in vitro* can cause a significant prolongation of partial thromboplastin times in normal human or dog plasma, but no effects on prothrombin time were observed. Prothrombin times were also unaffected in the presence of *E. granulosus* cyst fluid. The site at which the intrinsic coagulation system is affected by these cestode factors has not been clarified, but it seems possible that release of such agents *in vivo* could influence intrinsic pathway-mediated inflammatory processes. This could be a particularly important consequence for a sessile organism which might be susceptible to intense accumulation of humoral and cellular components of the host defense mechanism. Identification of the active principle and quantitation of its release *in situ* will be required before an assessment of its role in the host–parasite relationship can be achieved. Recently, Leid (1977) has found that culture products collected from *in vitro* cultivation of *T. taeniaeformis* seem to have an effect on Factor XIII, preventing the stabilization of fibrin clots. The precise role of this product or products in the host–parasite relationship also remains to be clarified.

3. *Trematodes*

Until recently very little information has been available on the interaction between trematodes and the coagulation cascade, despite the fact that liver and blood fluke infections in animals and man are among the most serious and widely distributed parasitic diseases. Blood loss and hypoproteinemia are cardinal signs of *Fasciola hepatica* infections (Dawes and Hughes, 1964) but there are no reports of investigations on clotting defects in infected hosts. In experimental studies on *Schistosoma mansoni* in mice, significant prolongations in prothrombin times were observed in chronic heavy infections, but this was attributed to hepatic dysfunction and deficiencies in plasma protein synthesis rather than active inhibition of the coagulation mechanisms of the blood (DeWitt and Warren, 1959). However, Tsang *et al.* (1977) and Tsang and Damian (1977) have reinvestigated the relationship between schistosomes and clot formation and suggest that these parasites may indeed inhibit the coagulation system, through an effect on activated Hageman Factor.

Schistosomes, of course, are constantly bathed in all the elements of the coagulation cascade by virtue of their intravascular location in the host as adult worms. There is no evidence of thrombin formation around living parasites *in vivo*, although death of the worms rapidly leads to the development of thromboembolic lesions wherever the dead parasites lodge. Tsang and Damian (1977) prepared extracts of adult flukes and showed that these contained an anticoagulant principle which caused a highly significant prolongation of partial thromboplastin time, indicating inhibition of the intrinsic coagulation pathway. They were able to show that the anticoagulant did not interfere with activation of Factor XII but inhibited activation of Factor XI by XIIa. The agent was stable at 100°C, and sedimented after centrifugation at 27,000 g, but further physicochemical characterizations were not obtained. These preliminary data are consistent with the presence of a heparin-like molecule. The authors do not indicate whether the blood from which the adult flukes were collected by perfusion was anticoagulated with heparin. If this was the case, the possibility of heparin contaimination of their extracts needs to be examined since this agent binds very avidly with charged surfaces (Ehrich and Stivala, 1973).

In discussing the significance of their findings, Tsang and Damian point to the advantage which might be derived from inhibition of coagulation at an early stage, rather than at the terminal portion of the cascade, because of the positive amplification mechanisms which enhance the steps following initial Hageman Factor activation. It seems possible that the factor described by Tsang and Damian may have some role to play in the prevention of clot formation around the parasites *in vivo*, although this remains to be seen. The promising results obtained thus far certainly suggest that similar studies may be worthwhile in other fluke infections, particularly since the migrating juveniles of trematodes, which ultimately reside in tissues such as the lung, liver, or pancreas, must be exposed to plasma coagulation factors during the course of their travels. Localized thrombus development around these organisms would be expected to severely hamper their migratory efforts, especially if other components of the inflammatory systems were called forth.

The possibility has recently been raised that inhibition of coagulation may be responsible for the bleeding tendencies shown by patients with chronic schistosomiasis (Tawfik et al., 1977); these workers found evidence of high levels of fibrinogen degradation products in the blood of affected patients, suggestive of uncontrolled fibrinolysis. Further work will be necessary to determine whether this change is directly attributable to some product of the parasite, or due to a derangement of hemostatic mechanisms resulting from liver damage during the infection.

B. THE KININ SYSTEM

1. Nematodes

Studies on the interaction of parasites with the kinin forming system have been confined primarily to protozoan organisms, with little experimental work conducted on helminths (Boreham and Wright, 1976). The only report on helminths is that of Fal (1974) in which depressed plasma kininogen levels were demonstrated in rats infected with *Trichinella spiralis*. However, the potential for initiation of kinin generation by direct parasite involvement is clearly evident. It is known that proteolytic enzymes of parasite origin are secreted into the environment surrounding nematodes (Lee, 1965). These secreted enzymes could generate bradykinin via Hageman Factor activation and initiation of the sequential enzymic cleavage process. Enzymic generation of kallikrein from prekallikrein followed by kallikrein cleavage of kininogens, as well as the direct proteolytic cleavage of kininogen, may be produced by parasite secretions, or by proteases released from physically damaged host cells.

It seems likely that in many nematode infections the focus of investigative work on the adult worms might not be as instructive, in terms of kinin generation, as work on the invasive stages. Skin penetration by invasive larval stages of nematodes such as *A. caninum* or *Strongyloides* occurs very rapidly (Lewert and Lee, 1954; Stirewalt, 1963, 1966), and the pathologic picture of the host response to the invading organisms certainly indicates that permeability factors are activated (Lewert and Lee, 1954). Kinin formation could be responsible for local increases in a vascular permeability. In fact, penetration of other tissues by migrating nematodes may involve the generation of permeability increasing factors, whether by means of bradykininlike molecules, complement-derived peptides, or mast cell products. The notion that enzymic dissolution of tissue ground substances is the sole means of penetration seems inadequate to explain the rapid time course of tissue invasion. Enzymic production of kinins or other vascular permeability factors by parasite secretions may be extremely important in facilitating this process because of the amplification effect of inflammatory mediators. Parasite penetration and migration in tissues has not been approached experimentally by many workers and could be one of the most fruitful areas in terms of understanding the mechanisms of infection by parasitic nematodes, and even other helminths. The potential effect of such permeability factors as bradykinin in allowing host protein to accumulate around tissue parasites for nutritive purposes also needs to be examined.

2. Cestodes

There is no experimental evidence to date on kinin generation in cestode infections. However, as has been pointed out for nematodes, the generation of vascular permeability factors, such as bradykinin, could provide a mechanism whereby more nutrients could be made available for development of worms in the tissues or even in the intestinal lumen. Furthermore, tissue penetration by invading oncospheres of most cestodes takes only a few minutes (Silverman, 1954; Banerjee and Singh, 1969; Smyth and Heath, 1970) and, again, amplification of enzymic effects by kinins could be a facilitating factor in this process. We have indirect evidence implicating bradykinin in intestinal penetration in rats exposed to *T. taeniaeformis*. Rat intestines were exposed by laparotomy and 2 cm sections of the duodenum were ligated before intraluminal injection of synthetic bradykinin. This was followed by challenge with *T. taeniaeformis* eggs. Animals given bradykinin had a greater number of cysts developing in the liver than cestodes given intraluminal inoculations of buffer solution (A. J. Musoke, R. W. Leid, and J. F. Williams, unpublished observations). These results suggest that increases in local vascular permeability allowed more activated oncospheres to migrate from the intestinal epithelium to the liver. Further characterization of such systems may reveal the importance of amplification reactions in the host–parasite relationship.

3. Trematodes

No experimental results are available which incriminate the activation of kininogen and kinin generation in trematode infections, although certain pathologic changes are suggestive of this. Once again rapid invasion of gut or skin by juvenile flukes occurs and this may depend on the types of mechanisms proposed for other helminths. The marked facial edema seen in patients heavily infected with *Fasciolopsis buski* (Faust et al., 1970) could be the result of vascular permeability factors being generated systemically in the host, rather than being due to directly toxic agents absorbed from the worm.

C. The Complement System

Most of the research on complement in helminth infections has concerned exploitation of the occurrence of complement-fixing antibodies as a means of detecting parasitism serologically (for reviews, see Kagan, 1974; Sadun, 1976). Little work has been done on the contribution of this enormously potent amplification system to nonspecific or specific defense mechanisms in helminth infections, or to the development of the pathological changes associated with tissue phases of parasitic worms. The biological activities of molecules which are generated by complement fixation are quite capable of influencing the cellular composition of the immediate tissue microenvironment and, by virtue of their effects on vascular permeability, may cause the extravasation of many other plasma components. These might be either beneficial or detrimental to the parasite depending on the nature of further interactions with the organism or its products.

1. Nematodes

At this time there are no detailed accounts available of changes in complement components in response to the presence of nematodes. An attempt was made by Jones and Ogilvie (1971) to implicate complement in the effector mechanism of resistance to *Nippostrongylus brasiliensis*, by measuring the response to challenge infection in complement-depleted immune rats. Cobra venom treatment, which can be used to lower circulating plasma complement levels to less than 5% of normal values (Maillard and Zarco, 1968), did not affect the ability of these rats to resist challenge. Whether complement is required in the attack system in animals or humans immune to other nematodes is presently not known.

The fact that complement-fixing antibodies commonly develop in nematodiasis, certainly indicates that some complement-mediated processes are likely to be occurring *in vivo*, especially when parasites persist in tissue. Migrating nematodes are known to contain and release a variety of proteolytic enzymes (Lee, 1965, 1972) some of which might be expected to initiate the complement cascade nonspecifically. Other enzymes might inactivate biologically active peptides as soon as they are formed. Some recent results obtained by Leventhal *et al.* (1977) point to a role for complement in the inflammatory response to *Ascaris* larvae, although they shed no light on the possible mechanisms involved. They found that the eosinophilic response to migrating ascarids in complement-depleted guinea pigs was far greater than in normocomplementary animals. This suggests some regulatory role for complement in influencing cellular infiltration, but further work will be necessary to clarify this phenomenon, which at present seems at odds with current understanding of inflammatory pathways.

A more conventional view of the amplification potential of the complement cascade, outlined in the first part of this review, can be used to develop a rationale for understanding the sequence of events which characterize nematode-associated inflammation. Kallikrein, fibrinopeptides, C5a and C3a anaphylatoxins, the C5b67 complex, and NCF from mast cells could all be involved in determining the degree of attraction of neutrophilic leukocytes. Generation of these chemotactic factors could come about via direct enzymic cleavage of precursor molecules such as native C5 and C3, as has been described for bacteria and viruses (Ward, 1974), or direct activation of kallikrein and fibrinogen to generate chemotactically active principles.

The infiltration of mononuclear cells into areas around nematodes could be produced by the effects of kallikrein, C5a, C3a, or possibly by macromolecular heparin (R. W. Yurt, unpublished observations) on the directed movement of this cell type. Alternative pathway fixation of complement by nematode surface collagen-like molecules (Michaeli et al., 1972) interacting with C3b, or fluid phase stabilization of the C3 amplification convertase by parasite products could also lead to anaphylatoxin production. In addition, chemotactic factor generation could result from immunologic or nonimmunologic activation of the entire complement, coagulation, and kinin generation sequences.

Presently, the most important chemotactic factors for eosinophils appear to be the ECF-A tetrapeptides (Goetzl and Austen, 1975). Although C5a, C3a, and kallikrein can also be chemotactic for eosinophils, the selective activity of the ECF-A tetrapeptides in attracting eosinophils from mixed cell populations would provide for local accumulation of these cells. Another possible mechanism is via direct chemotactic activity of parasite components. An allergen present in A. suum has recently been shown to cause the directed cell movement of eosinophils (Torisu et al., 1975). The isolation and characterization of parasite products and examination of their ability to stimulate chemotaxis has received little attention in the past. Another prominent feature of nematode and other helminthic infections is peripheral eosinophilia. Parasite products may have eosinophilopoietic as well as eosinophilotactic effects and this is another area open to experimental manipulation.

Clearly, parasite-dependent modulation of the effects of host regulatory proteins or complement consumption could serve to either amplify or attenuate the biological potential of this enzymic cascade. The end result would depend on the stage of the organism, its migratory intentions, or the presence of antibody, provided the host has been immunologically sensitized. The seemingly contradictory positive or negative influences which we can postulate for the complement system, in terms of its effect on the outcome of host–parasite interactions in tissues, reflect both the complexities of this defense mechanism, and present uncertainties regarding the specific types of chemical reactions which might be initiated by the worms. What is certain, however, is that complement is a component of the microenvironment of tissue nematodes, and that its role as a determinant of the reactions which have been observed and described can no longer be ignored in experimental work.

This potential for complement involvement need not necessarily be confined to microcirculatory or cellular infiltration changes. Grossly visible effects of nematode parasitism, such as diarrhea, may be the result of local complement-mediated efflux of intravascular fluid into the extravascular space, and thence into the gut. Leakage of plasma proteins in this way might again serve in either a protective way or as a source of nutrients for the parasite in the lumen. Other signs of nematode parasitism such as the labial and facial edema which occurs in hemonchosis (Levine, 1968) could also be derived from an effect on the complement system, similar to that which results in facial edema in hereditary angioneurotic edema in man (Valentine *et al.*, 1971). Facial edema in gastrointestinal parasitism is usually attributed to hypoproteinemia after loss of plasma protein into the gut. The peculiar location of this edema might be the result of causal mechanisms which have not yet been explored. The brisk reversal of peripheral edema sometimes within hours after anthelminthic treatment to kill the worms in cases of severe haemonchosis (J. F. Williams, personal observations), certainly suggests that some circulating permeability factor, dependent on the presence of living worms in the host, may be contributing to the vascular disturbance.

2. Cestodes

Many of the ways in which complement might affect host responses to nematodes can be proposed equally for the tissue-dwelling taeniid metacestodes.

Chemotactic factors produced by interactions between metacestodes and the complement system, as well as via kinin-generating mechanisms, may well be involved in the pattern of host response which we have described. The pathways whereby these events may come about have been adequately explained in the discussion of the host reaction to nematodes. Likewise, the effects of local vascular permeability changes and the roles which these may play in enhancing the local accumulation of plasma proteins, or facilitating the migratory efforts of the early oncospheral stages have also been considered above.

In our studies on the tissue response of *T. taeniaeformis* in rats, eosinophilic leukocytes appeared around blood vessels in the liver within a few days of primary infection (Cook and Williams, 1978). In this instance, it seems likely that chemotactic attractants, generated via metacestodes in the liver sinusoids, pass to the portal triad area in the bile and cause local margination and diapedesis of the eosinophils. Over the succeeding weeks as the infection progressed, eosinophilic leukocytes accumulated in massive numbers around the parasites, but the severity of the infiltration then subsided and became mild, although eosinophils remained in somewhat increased numbers in the local area (Ansari and Williams, 1976). Whether complement is involved in the attraction of these cells is not clear, but there is now good evidence that complement plays an important role in specific resistance to *T. taeniaeformis* in rats (Musoke and Williams, 1975a). In our experiments complement-depleted rats were not resistant to challenge infections even though they had received adequate doses of protective immune serum. This fact led us to propose that taeniid larvae which survive in tissues have evolved a means of interfering with complement-mediated attack on their membranes. We found that *T. taeniaeformis* and many other tissue cestodes contain substances which are able to fix complement nonimmunologically (Hammerberg *et al.*, 1976, 1977; Hammerberg and Williams, 1978a,b). It seems possible that such factors could be released into the local microenvironment around the sessile parasites and consume complement in the fluid phase so that attack sequences would not be assembled on the parasite membrane. This consumption, which occurs in large part via alternative pathway activation in the case of *T. taeniaeformis*, might also generate anaphylatoxins, but here again the biological effects of these mediators could be modified by parasite-derived factors such as anaphylatoxin inactivators or carboxypeptidase B-like enzymes.

In vitro, cestodes seem to be able to interact with complement nonimmunologically so that membrane fixation does occur and lytic processes follow (Herd, 1976; Kassis and Tanner, 1976, 1977; Rickard *et al.*, 1977). Similar results occur with *T. taeniaeformis* (Picone and Williams, 1976), but the effect can be blocked by the addition of parasite products to the serum prior to interaction with the living metacestodes. It appears likely, therefore, that the rate of release of complement-fixing factors is crucial in determining whether this interaction has detrimental effects on the tissue tapeworms. Hammerberg and Williams (1978b) have characterized a complement-consuming principle released from the larvae of T. taeniaeformis, and suggested that it is a highly sulfated polyanionic glycosaminoglycan. Substances of this nature are present at the surface of mammalian cells (Chiarugi et al., 1974) and probably play a key role in determining their susceptibility to complement-mediated damage. Gewurz and his colleagues have proposed that interactions between polyanions and polycations may provide one of the regulatory mechanisms which affect the involvement of complement in the inflammatory process (Siegel et al., 1975; Rent et al., 1975, 1976; Feidel et al., 1976; Claus et al., 1977). The amount of these factors at the surface of the worms, and their rates of synthesis and release at different stages of development will need to be determined before the full implication of these reactions can be appreciated in cestodiasis.

The occurrence of blood eosinophilia (Lapierre, 1953) and systemic symptoms such as urticaria and itching (Blamoutier, 1952), attest to the fact that host responses do develop in adult cestode infections of the gut lumen. Traditionally, absorbed substances which stimulate host responses have been referred to as "toxins" though whether they act in this manner directly or through inflammatory pathways, such as complement, is not yet known. Shock-like syndromes, consistent with wholesale mediator release, are seen in dogs given extracts of adult cestodes intravenously, and animals which are actively infected are said to be resistant to these reactions. There is now evidence that complement can be deposited on the tegument of adult cestodes (Befus, 1974, 1975), but the significance of this is not yet clear.

3. Trematodes

In recent years an enormous amount of research has been done on schistosomiasis in an effort to define host defense mechanisms which might be susceptible to artificial stimulation by a vaccine. A number of in vitro models for immunologically mediated attack have implicated complement (Dean et al., 1974; Perez et al., 1974; McLaren et al., 1975), but little is known of the role of complement in vivo or of nonimmunologic interactions with this system. There is good evidence that the cercariae (free-living) forms of schistosomes can initiate complement fixation directly through the alternative pathway (Gazzinelli et al., 1969; Machado et al., 1975), but the significance of this phenomenon is difficult to assess because as soon as the cercariae penetrate the skin and become parasitic schistosomuli, they lose their susceptibility to complement (Kusel, 1970). This may indicate that complement fixation no longer occurs on the parasite surface, but it does not rule out the possibility that alternative pathway activation could proceed in the fluid phase surrounding the organism. By this means, anaphylatoxin generation might alter the local vasculature in such a way as to enhance movement of the worms into the circulation. There is a suggestion in a report on the pulmonary phase of schistosomiasis (von Lichtenberg *et al.*, 1977), that this may indeed be the case, although no direct evidence is cited in support. In *in vitro* experiments mast cells can be shown to accumulate or "rosette" at the schistosome surface and this reaction is complement-dependent (Sher, 1976). How this might relate to the release of complement-consuming factors by the parasite is not at all clear at the moment.

The "uncontrolled" inflammatory phase of host reactions to migrating juveniles of *Fasciola hepatica* (Dawes, 1963) far exceeds that which one might expect solely from the destruction of hepatic cells. It seems likely that some derangement of the regulatory systems which operate in inflammation occurs at this time, though whether wholesale activation of the complement system is a contributing factor is not known. It would be interesting to determine the effect of prolonged complement depletion on the characteristics of the host reaction at this critical phase of the infection.

D. MAST CELLS/BASOPHILS AND THE CHEMICAL MEDIATORS OF IMMEDIATE HYPERSENSITIVITY

It is becoming increasingly clear that the tissue mast cell and its circulating counterpart, the basophil, occupy a crucial position in the matrix of interconnecting cellular and chemical components which make up the inflammatory system. Their responsiveness to both nonspecific and immunologic stimuli and, conversely, their sensitivity to modulating influences of other effector cells have been incorporated into the scheme devised by Askenase (1977b) to illustrate the inflammatory process. Figure 4 is based on his model and has been modified to include the points at which helminth parasites or their products might participate. It is not our intent to imply that all of these hypothetical modes of interaction do occur, or that these consequences are necessarily favorable to the parasite in each case. Rather, we feel that the model provides a context in which to place and interpret observations on host–parasite relationships, and a basis for experimental analysis of this fascinating area in the future.

In considering this scheme, it would be wrong to imagine that the effects of the organism on mast cell modulation would necessarily be the same at each stage of infection. There are clearly times when vascular changes which accompany inflammation, such as increased permeability and local hyperemia, might facilitate movement of the parasite into blood vessels. Equally well, there are occasions, such as those where tissue larval phases become sequestered for prolonged periods, when a dampening effect on the inflammatory response might be advantageous. The important point here is that these types of interactions can be visualized, and furthermore that they are susceptible to experimental testing.

In the past, much of the work on the mast cell in helminthic infections has focused on the host IgE-immunoglobulin response to parasites and the roles that these reaginic mast cell/basophil-sensitizing antibodies might have in host resistance. Preoccupation with the IgE-dependent pathway to activation has overshadowed the other unique characteristics of basophils and tissue-derived mast cells. These cells may be activated in a variety of ways, all of which could have some bearing on the survival of the parasite and the reaction of the host to helminths. Target cells, for example, may be triggered to release histamine, SRS-A, or other inflammatory mediators by IgG immune complex reactions; direct mast cell degranulators (Uvnäs and Wold, 1967; Thompson, 1972); the phospholipase enzyme present in eosinophils (Archer and Jackas, 1965); the cationic proteins of the neutrophil, which could be released during lysomal secretion (Ranadive and Muir, 1972) or by bradykinin (Johnson and Erdös, 1973). Few of these alternative activation pathways have been cited in the interpretation of the variety of host responses and pathologic events observed in helminthiasis.

1. Nematodes

A great deal of the experimental work on mast cells in parasitic infections has been carried out on intestinal nematodes. *Trichostrongylus colubriformis* in the guinea pig has been explored extensively by Rothwell and Dineen and their colleagues (Dineen and Wagland, 1966; Rothwell *et al.*, 1971, 1974; Rothwell and Dineen, 1972; Kelly and Dineen, 1976). Host reactions in this infection are characterized, as we have come to expect, by the very marked involvement of both eosinophilic and basophilic polymorphonuclear leukocytes. There is a peripheral and bone marrow eosinophilia and basophilia (Rothwell and Dineen, 1972), and an accumulation of these cells at the intestinal sites of parasitism by the adult worms. Degranulation of both types of cells has been described in the tissue reaction (Rothwell, 1975; Huxtable and Rothwell, 1975; Rothwell and Huxtable, 1976). The degranulation of eosinophils and basophils could be attributed to either IgE- or IgG-dependent reactions, as well as to a sequence which might involve parasite effects on eosinophils, followed by the discharge of granules from the mast cells. The degranulation process itself has been proposed as an important step in the rejection of intestinal trichostrongyles. Whether a mast cell degranulator is released by the *T. colubriformis* has not been investigated although this possibility has been raised by Rothwell and Love (1975). It may be that such factors are more prevalent among helminths than heretofore demonstrated, although difficulties in detection may be a result of their lability. Isolation of the degranulator from *A. suum* (Thompson, 1972), for example, requires the presence of sulfhydryl protective reagents, and comparable precautions may need to be taken in other systems.

The sequence of events in *T. colubriformis*-infected guinea pigs is ideally suited for studying the acute rejection phenomenon in laboratory animals, but research on this aspect addresses only a part of the host–parasite relationship under field conditions. Very often in the field, animals are continually exposed to infective larvae, with adult worms being eliminated either periodically or continually and histotropic larvae migrating out of the tissues to replace them (Michel, 1974). This situation is not mimicked in *T. colubriformis*-infected guinea pigs where the acute rejection of adult worms is the norm, and if histotropic larvae are present they certainly have received little attention. The inflammatory processes which accompany this phase of the life cycle in nematodes deserve more analysis in the future, especially since the emergence of sequestered larvae can be the major pathologic event in infection (Armour, 1974).

The apparent incongruity between mast cell activation by a degranulator produced by the worms and the postulated vasoactive amine component of the rejection process in *Trichostrongylus* infections requires some comment. Since comparable acute crises may not be very important events in the overall host–parasite relationship in the sheep, mast cell triggering by worm degranulators in abnormal hosts cannot really be interpreted in terms of its selective value. In the sheep, degranulation could serve as a means of facilitating migration in the tissues by the histotropic nematode larvae. Extravasation of plasma proteins as a result of the release of histamine from mast cells might also serve to augment the nutrient supply around the larval or adult stages.

With respect to *N. brasiliensis* infections in the rat, the role of the mast cell in this host–parasite model system has emerged as a controversial issue in recent years (Wells, 1962; Keller, 1971; Murray *et al.*, 1971; Dineen *et al.*, 1973; Ogilvie and Jones, 1973; Ogilvie and Love, 1974). *N. brasiliensis* infections in the intestine are characterized by eosinophilic and mast cell infiltrates with significant numbers of an additional cell, the globular leukocyte, also present. The nature and origin of the latter have also been the subject of debate, although Murray *et al.* (1968) have suggested that they are derived from tissue mast cells. Whatever their origin, globular leukocytes must be regarded as a potential amine source (Miller, 1971a,b; Miller and Walshaw, 1972), though how amine-containing cells contribute to rejection is not known. Hogarth-Scott and Bingley (1971) have shown that by inoculating an antiserum directed against peritoneal cells, including mast cells, into infected rats they could prolong the retention of adult worms in the intestine. Keller (1971) showed that histamine-containing cells are decreased in number during the host rejection phase, and concluded on the basis of this and other evidence that mast cells are not necessarily a requirement for parasite expulsion. Mast cell activation may be occurring in a manner similar to that described above for *T. colubriformis*. The evidence for the presence of a mast cell degranulator in *N. brasiliensis* has not been reported, although the existence of such a factor has been inferred from the sequence of pathologic changes which occur during infection (Murray *et al.*, 1971; Miller, 1971a,b).

Ascaris spp. infections are potent IgE–antibody inducers and several of the allergens responsible have been characterized (Hussain *et al.*, 1973; Ambler *et al.*, 1972, 1974). There is good evidence in the case of *Ascaris* that a mast cell degranulator is present. Tolone *et al.* (1972, 1974) have shown that *Parascaris equorum* contains, in addition to a mast cell degranulator, a parasite product which is capable of inducing rat eosinophils to release a factor which subsequently degranulates mast cells. The peripheral eosinophilia observed clinically with *Ascaris* spp. infections may be a result of potent mast cell activation and ECF-A tetrapeptide release. The isolation and characterization of parasite products capable of granulopoietic effects has not been investigated in nematode infections or other helminthiases, although nematode parasitoses, in general, and ascarid infections, in particular, have pronounced effects on eosinophilic leukocyte production and mobilization.

2. Cestodes

Although mast cells have been shown to be present around adult and larval cestodes (Fernex and Fernex, 1962; Varute, 1971; Byram, 1974; Cook and Williams, 1978), little is known of the mechanism which leads to this infiltration or what function these cells serve in the area. In intestinal infections, such as *Diphyllobothrium latum*, parasite metabolites are generally considered capable of producing allergic signs (Faust *et al.*, 1970). In infected patients, the presence of the worms may be evidenced by edema of the face, and sometimes even the abdominal wall and extremities. These clinical signs could be a result of mast cell activation, either immunologically or nonimmunologically, with release of vascular permeability factors such as histamine or SRS-A, both of which can initiate edema formation.

Human sparaganosis may also result in edema of the facial area, primarily confined to the eyelids (Weinstein *et al.*, 1954). This again could be due to mast cell activation. A pronounced allergic reaction with intensive eosinophil accumulation occurs upon the death of the worms in tissues. Eosinophils have been observed degranulating, and there is a localized deposition of Charcot–Leyden crystals in the area around the dead parasites. This eosinophil infiltrate may reflect ECF-A release from target cells, although more experimental work is needed to confirm the nature of the chemoattractant.

In *T. taeniaeformis* infections Varute (1971) and Cook and Williams (1978) have reported mast cell infiltration around the developing cysticercus. Whether this concentration is a result of specific leukoattractants or is the product of differentiation of mast cells from local precursors remains to be clarified. Mast cells not only proliferated in the local host reaction to *T. taeniaeformis*, but their numbers were elevated in the gut of infected rats (Cook and Williams, 1978). When infected and uninfected rats were surgically united so as to form parabiotic pairs, the mast cell proliferation occurred in the intestinal tract of the uninfected partner as well as in the infected rat. This suggests that some parasite product which circulates may be able to direct the differentiation of this very important inflammatory cell at a site distant from the worms. The nature of the stimulus for such proliferation is at present undefined but under active investigation. Another aspect of interaction between cestodes and mast cells concerns the possibility that chemical mediators may be inactivated and/or inhibited by parasite products. Leid (1977) has recently presented evidence that in *T. taeniaeformis* such inactivators are present within the developing cysticerci. Cysticerci contained both an arylsulfatase B and a phospholipase D-like enzyme, which can enzymically inactivate SRS-A and PAF, respectively. Whether histamine-degrading enzymes or inactivators of other chemical mediators are present in larvae of *T. taeniaeformis* is unknown. The capacity of a parasite to inactivate or inhibit mast cell products has far-reaching implications in the interpretation of the mechanisms of prolonged survival for tissue stages of cestodes or other helminth parasites. It is our feeling that regulators of secretory system functions of other white blood cells, such as eosinophils, may be a common feature of the local response to helminth parasites in tissues, and this possibility should certainly be investigated in the coming years.

Eosinophils which migrate into tissues surrounding the cysticerci of *T. taeniaeformis* may differ from normal eosinophils in terms of both the content and activity of intracellular enzymes such as arylsulfatase B, and these changes may reflect the effects of parasite products in growing or dying organisms. Rubin *et al.* (1975) have reported that in a filarial infection of humans, parasite-dependent modulation of host leukocyte enzymes does occur systemically. The general significance of this observation in other helminthic infections was not discussed. However, we have preliminary evidence (Leid, 1977) that rat eosinophils obtained during a primary infection with *T. taeniaeformis* show a similar alteration in arylsulfatase activity. The recruitment of specialized cells, including eosinophils, and the modulation of both the activity and content of the intracellular enzymes of leukocytes, might provide an additional mechanism for controlling the local host inflammatory environment to the benefit of the parasite.

3. Trematodes

As with other helminths, there is evidence of systemic increase in mast cell numbers in animals and man infected with trematodes (Fernex and Bèzes, 1962). Mast cell/basophilic leukocytes are present in granulomas of schistosome-infected mice (Moore *et al.*, 1976). Capron and his collegues have provided evidence in *S. mansoni* infections of rats that IgE antibody–antigen complexes can bind to the cell surface of macrophages and arm these cells for the *in vitro* killing of schistosomula (Capron *et al.*, 1975, 1977b). In a further extension of this work Capron *et al.* (1977a) have shown that such IgE complexes are capable of binding to rat eosinophils, a view that had been proposed earlier by Hubscher (1975) for human eosinophils and human IgE. The *in vivo* relevance of these findings to the host–parasite relationship is not yet clear, but the experiments have shown very elegantly that IgE antibody can activate other cells than the mast/cell basophil leukocyte. Another prominent cell in tissues as well as in the circulation of infected mice is the eosinophil. Recent evidence demonstrates that this polymorphonuclear leukocyte in humans is essential for killing schistomsomula, *in vitro* and *in vivo* (Butterworth *et al.*, 1975; Mahmoud *et al.*, 1975), and miracidia within eggs trapped in tissue (James and Colley, 1976). The attraction of an eosinophil to the schistosome or its eggs may indicate depletion of ECF-A from mast cell stores by immunologic or nonimmunologic activation. An eosinophilopoietic substance has been associated with *S. mansoni* infections in mice (Miller *et al.*, 1976), and Colley *et al.* (1977) have recently shown that crude schistosomal cercarial antigen may possess leukotactic activity.

We have remarked previously that *F. buski* is capable of provoking "allergic" clinical signs in affected individuals. In this case vascular permeability changes secondary to mast cell activation could account for the host reactions observed. The possible existence of a degranulator as a means by which activation could be achieved in trematode infections should be addressed in future experimental work.

Recently, it has been shown that the schistosomula of *S. mansoni* have the capacity to initiate rosette formation around mast cells, and that this rosetting is complement-dependent (Sher, 1976). The complement receptor on mast cells has been characterized (Sher and McIntyre, 1977) and it appears to have C3b-binding properties. The functional significance of this binding in relation to mast cell activation and facilitation of invasion by cercariae remains to be defined. However, the possibility that C3b activation of mast cells could provide yet another pathway to release of chemical inflammatory mediators has not escaped our attention.

V. Conclusion

Clearly, very little experimental work has been undertaken so far on the interaction between inflammatory pathways and helminths, yet the observations which have surfaced on the nature of the host responses in these infections make it a ripe field for investigation. We certainly hope that in future work on natural host–parasite relationships, scientists will take advantage of all these recent advances in the areas of mediators of immediate hypersensitivity, complement and coagulation cascades, the fibrinolytic and kinin-generating systems, and delayed hypersensitivity. In return, we feel confident that the unique characteristics of host responses to helminth parasites will offer important insights into the molecular and cellular mechanisms of the inflammatory system in general.

ACKNOWLEDGMENTS

Portions of the work reported from our laboratories were supported by NIH grant AI 10842, by the Michigan Agricultural Experiment Station, and by a NIH postdoctoral fellowship (5-F22-AI-01097) to the senior author in the laboratory of Dr. K. F. Austen in the Department of Medicine, Harvard Medical School, Boston, Mass. We are particularly indebted to our many colleagues at Michigan State University and at Harvard University who have contributed in invaluable ways to the development of our ideas on host–parasite relationships and the inflammatory response.

REFERENCES

Adonajlo, A., and Bonczak, J. (1961). *Przegl. Epidemiol.* **15**, 425–427.

Ambler, J., Doe, J. E., Gemmell, D. K., Roberts, J. A., and Orr, T. S. C. (1972). *J. Immunol. Methods* **1**, 317–327.

Ambler, J., Miller, J. N., and Orr, T. S. C. (1974). *Int. Arch. Allergy Appl. Immunol.* **46**, 427–437.

Anonymous. (1974). *Vet. Rec.* **95**, 572–573.

Ansari, A., and Williams, J. F. (1976). *J. Parasitol.* **62**, 728–736.

Archer, G. T., and Jackas, M. (1965). *Nature (London)* **205**, 599–600.

Archer, R. K. (1963). "The Eosinophil Leukocytes." Blackwell, Oxford.

Armour, J. (1974). *Vet. Rec.* **95**, 391–395.

Arvy, L. (1950). *Rev. Can. Biol.* **9**, 368–381.

Askenase, P. W. (1977a). *Prog. Allergy*, Vol. 23, pp. 199–320. S. Karger, Basel, Switzerland.

Askenase, P. W. (1977b). *Am. J. Trop. Med. Hyg.* **26**, 96–103.

Austen, K. F. (1971). *In* "Immunological Diseases" (M. Samter, ed.), 2nd ed., Vol. 1, pp. 332–355. Little, Brown, Boston, Massachusetts.

Austen, K. F., and Orange, R. P. (1975). *Am. Rev. Respir. Dis.* **112**, 423–436.

Austen, K. F., Wasserman, S. I., and Goetzl, E. J. (1976). *In* "Molecular and Biological Aspects of the Acute Allergic Reaction" (S. G. O. Johansson, K. Strandberg, and B. Uvnäs, eds.), pp. 293–320. Plenum, New York.

Banerjee, D., and Singh, S. K. (1969). *Indian J. Anim. Sci.* **39**, 149–154.

Befus, A. D. (1974). *Trans. R. Soc. Trop. Med. Hyg.* **68**, 273.

Befus, A. D. (1975). Ph.D. Thesis, University of Glasgow.

Beguín, F. (1966). *Z. Zellforsch. Mikrosk. Anat.* **72**, 30–46.

Blamoutier, P. (1952). *Sem. Hop.* **28**, 3278–3279.

Bloom, B. R., Stoner, G., Fischetti, V., Nowakowski, M., Muschel, R., and Rubenstein, A. (1974). *Prog. Immunol.* **3**, 133–144.

Bokisch, V. A., Müller-Eberhard, H. J., and Cochrane, C. G. (1969). *J. Exp. Med.* **129**, 1109–1130.

Boreham, P. F. L., and Wright, I. G. (1976). *Prog. Med. Chem.* **13**, 159–204.

Boswell, R. N., Austen, K. F., and Goetzl, E. J. (1978). *J. Immunol.* **120**, 15–20.

Boughton, I. B., and Hardy, W. T. (1935). *Tex., Agric. Exp. Stn., 48th Annu. Rep.* pp. 236–239.

Brocklehurst, W. E. (1962). *Prog. Allergy* **6**, 539–558.

Brocklehurst, W. E. (1968). *In* "Clinical Aspects of Immunology" (P. F. H. Gell and R. R. A. Coombs, eds.), 2nd ed., pp. 611–632. Blackwell, Oxford.

Butterworth, A. E., Sturrock, R. F., Houba, V., Mahmoud, A. A. F., Sher, A., and Rees, P. H. (1975). *Nature (London)* **256**, 727–729.

Byram, J. E. (1974). *Program Abstr., 49th Annu. Meet. Am. Soc. Parasitol.* p. 23.

Capron, A., Dessaint, J. P., Capron, M., and Bazin, H. (1975). *Nature (London)* **253**, 474–476.

Capron, A., Dessaint, J. P., Joseph, M., Torpier, G., Capron, M., Rousseaux, R., Santoro, F. and Bazin, H. (1977a). *Am. J. Trop. Med. Hyg.* **26**, 39–47.

Capron, A., Dessaint, J. P., Joseph, M., Rousseaux, R., Capron, M., and Bazin, H. (1977b). *Eur. J. Immunol.* **7**, 315–322.

Catty, D. (1969). *Monog. Allergy* **5**, 1–134.

Chiarugi, V. P., Vannucchi, S., and Urbano, P. (1974). *Biochim. Biophys. Acta* **345**, 283–293.

Clark, R. A. F., Gallin, J. I., and Kaplan, A. P. (1975). *J. Exp. Med.* **142**, 1462–1476.

Claus, D. R., Siegel, J., Petras, K., Skor, D., Osmand, A. P., and Gewurz, H. (1977). *J. Immunol.* **118**, 83–87.

Cochrane, C. G., and Wuepper, K. D. (1971). *J. Exp. Med.* **134**, 986–1004.

Cochrane, C. G., Revak, S. D., Aikin, B. S., and Wuepper, K. D. (1972). *In* "Inflammation: Mechanisms and Control" (I. H. Lepow and P. A. Ward, eds.), pp. 119–138. Academic Press, New York.

Cochrane, C. G., Revak, S. D., and Wuepper, K. D. (1973). *J. Exp. Med.* **138**, 1564–1583.

Colley, D. G., Savage, A. M., and Lewis, F. A. (1977). *Am. J. Trop. Med. Hyg.* **26**, 88–95.

Cook, R. W., and Williams, J. F. (1978). *J. Comp. Pathol.* (submitted for publication).

David, J. R., and David, R. R. (1972). *Prog. Allergy* **16**, 300–449.

Davie, E. W., and Fujikawa, K. (1975). *Annu. Rev. Biochem.* **44**, 799–829.

Dawes, B. (1963). *Parasitology* **53**, 135–143.

Dawes, B., and Hughes, D. L. (1964). *Adv. Parasitol.* **2**, 97–168.

Dean, D. A., Wistar, R., and Murrell, K. D. (1974). *Am. J. Trop. Med. Hyg.* **23**, 420–428.

DeWitt, W. D., and Warren, K. S. (1959). *Am. J. Trop. Med. Hyg.* **8**, 440–446.

Dias de Silva, W., Eisele, J. W., and Lepow, I. H. (1967). *J. Exp. Med.* **126**, 1027–1048.

Dineen, J. K., and Wagland, B. M. (1966). *Immunology* **11**, 47–57.

Dineen, J. K., Ogilvie, B. M., and Kelly, J. D. (1973). *Immunology* **24**, 467–475.

Dineen, J. K., Gregg, P., Windon, R. G., Donald, A. D., and Kelly, J. D. (1977). *Int. J. Parasitol.* **7**, 211–215.

Dixon, H. B. F., and Hargreaves, W. H. (1944). *Q. J. Med.* **13**, 107–121.

Dodd, K. (1973). *Vet. Rec.* **92**, 195–197.

Ehrich, J., and Stivala, S. S. (1973). *J. Pharm. Sci.* **62**, 517–544.

Eiff, J. A. (1966). *J. Parasitol.* **52**, 833–843.

Erasmus, D. A. (1977). *Adv. Parasitol.* **15**, 201–242.

Fal, W. (1974). *Wiad. Parazytol.* **20**, 159–166.

Faust, E. C., Russell, P. F., and Jung, R. C. (1970). *In* "Craig and Faust's Clinical Parasitology" (E. C. Faust, P. F. Russell, and R. C. Jung, eds.), p. 641. Lea & Feabiger, Philadelphia, Pennsylvania.

Fearon, D. T., and Austen, K. F. (1978). *In* "Arthritis and Allied Conditions" (J. L. Hollander, ed.) 9th ed., Lea & Feabiger, Philadelphia, Pennsylvania, (in press).

Fearon, D. T., Daha, M. R., and Austen, K. F. (1976). *Proc. Int. Congr. Nephrol., 6th 1975*, pp. 406–418.

Feidel, B. A., Rent, R., Myhrman, R., and Gewurz, H. (1976). *Immunology* **30**, 161–169.

Fernex, M. (1962). *Bull. Soc. Pathol. Exot.* **55**, 508–528.

Fernex, M., and Bèzes, H. (1962). *Acta Trop.* **19**, 252–257.

Fernex, M., and Fernex, P. (1962). *Acta Trop.* **19**, 248–251.

Fernex, M., and Sarasin, R. (1962). *Acta Trop.* **19**, 258–260.

Flury, F. (1912). *Naunyn-Schmiedebergs Arch. Exp. Pathol. Parmakol.* **67**, 275–292.

Freeman, R. S. (1962). *Can. J. Zool.* **40**, 969–990.

Freeman, R. S. (1964). *Can. J. Zool.* **42**, 367–385.

Gallin, J. I., and Kaplan, A. P. (1974). *J. Immunol.* **113**, 1928–1934.

Gazzinelli, G., Ramalho-Pinto, F. J., and Dias de Silva, W. (1969). *Exp. Parasitol.* **26**, 86–91.

Gershon, R. K., Askenase, P. W., and Gershon, M. D. (1975). *J. Exp. Med.* **142**, 732–747.

Gjonnaess, H. (1972a). *Thromb. Diath. Haemorrh.* **28**, 182–193.

Gjonnaess, H. (1972b). *Thromb. Diath. Haemorrh.* **28**, 194–205.

Gladner, J. A., Murtaugh, P. A., Folk, J. E., and Laki, K. (1963). *Ann. N. Y. Acad. Sci.* **104**, 47–52.

Goetzl, E. J., and Austen, K. F. (1975). *Proc. N. Acad. Sci.* **72**, 4123–4127.

Goetzl, E. J., and Austen, K. F. (1976). *In* "Molecular and Biological Aspects of the Acute Allergic Reaction" (S. G. O. Johansson, K. Strandberg, and B. Uvnäs, eds.), pp. 417–435. Plenum, New York.

Griffen, J. H., and Cochrane, C. G. (1976). *Proc. N. Acad. Sci.* **73**, 2554–2558.

Hamberg, M., Svensson, J., and Samuelsson, B. (1975). *Proc. N. Acad. Sci.* **72**, 2994–2998.

Hamberg, M., Svensson, J., and Samuelsson, B. (1976). *Adv. Prostaglandin Thromboxane Res.* **1**, 19–28.

Hammerberg, B., and Williams, J. F. (1978a). *J. Immunol.* **120**, 1033–1038.

Hammerberg, B., and Williams, J. F. (1978b). *J. Immunol.* **120**, 1039–1045.

Hammerberg, B., Musoke, A. J., Hustead, S. T., and Williams, J. F. (1976). *In* "Pathophysiology of Parasitic Infections" (E. J. L. Soulsby, ed.), pp. 233–240. Academic Press, New York.

Hammerberg, B., Musoke, A. J., and Williams, J. F. (1977). *J. Parasitol.* **63**, 327–331.

Hammerberg, B., Dangler, C., and Williams, J. F. (1978). In preparation.

Henson, P. M., and Cochrane, C. G. (1969). *J. Exp. Med.* **129**, 153–165.

Herd, R. D. (1976). *Parasitology* **72**, 325–334.

Hoeppli, R., and Feng, L. C. (1933). *Arch. Schiffs Trop. Hyg.* **37**, 176–182.

Hogarth-Scott, R. S., and Bingley, J. B. (1971). *Immunology* **21**, 87–99.

Hubscher, T. (1975). *J. Immunol.* **114**, 1379–1388.

Hussain, R., Bradbury, S. M., and Strejan, G. (1973). *J. Immunol.* **111**, 260–268.

Huxtable, C. R., and Rothwell, T. L. W. (1975). *Aust. J. Exp. Biol. Sci.* **53**, 437–445.

Ishizaka, K., and Ishizaka, T. (1971). *Clin. Allergy* **1**, 9–24

Jacques, L. B. (1975). *Gen. Pharmacol.* **6**, 235–245.

James, S. L., and Colley, D. G. (1976). *J. Reticuloendothel. Soc.* **20**, 359–374.

Jarrett, E. E. E. (1973). *Vet. Rec.*, 480–483.

Johnson, A. R., and Erdös, E. G. (1973). *Proc. Soc. Exp. Biol. Med.* **142**, 1252–1256.

Johnson, A. R., Hugli, T. E., and Müller-Eberhard, H. J. (1975). *Immunology* **28**, 1067–1080.

Jones, V. E., and Ogilvie, B. M. (1971). *Immunology* **20**, 549–561.

Kagan, I. G. (1974). *Z. Parasitenkd.* **45**, 163–195.

Kaliner, M., and Austen, K. F. (1974). *Biochem. Pharmacol.* **23**, 763–771.

Kaplan, A. P., and Austen, K. F. (1975). *J. Allergy Clin. Immunol.* **56**, 491–506.

Kaplan, A. P., Kay, A. B., and Austen, K. F. (1972). *J. Exp. Med.* **135**, 81–97.

Kassis, A. I., and Tanner, C. E. (1976). *Int. J. Parasitol.* **6**, 25–36.

Kassis, A. I., and Tanner, C. E. (1977). *Immunology* **33**, 1–10.

Kater, L. A., Goetzl, E. J., and Austen, K. F. (1976). *J. Clin. Invest.* **57**, 1173–1180.

Kawiak, J., Vensel, W. H., Komender, J., and Barnard, E. A. (1971). *Biochim. Biophys. Acta* **235**, 172–187.

Kay, A. B., Pepper, O. S., and Ewart, M. R. (1973). *Nature (London), New Biol.* **243**, 56–57.

Keller, R. (1971). *Parasitology* **63**, 473–481.

Kellermeyer, R. W., and Breckenridge, R. T. (1965). *J. Lab. Clin. Med.* **65**, 307–315.
Kelly, J. D., and Dineen, J. K. (1976). *Aust. Vet. J.* **52**, 391–397.
Kusel, J. R. (1970). *Parasitology* **60**, 89–96.
Lagunoff, D., and Benditt, E. P. (1963). *Ann. N. Y. Acad. Sci.* **103**, 185–198.
Lagunoff, D., Chi, E. Y., and Wan, H. (1975). *Biochem. Pharmacol.* **24**, 1573–1578.
Lapierre, J. (1953). *Ann. Parasitol. Hum. Comp.* **28**, 126.
Lee, D. L. (1965). "Physiology of Nematodes". Oliver & Boyd, Edinburgh.
Lee, D. L. (1972). *Parasitology* **65**, 499–505.
Leid, R. W. (1977). *Am. J. Trop. Med. Hyg.* **26**, 54–60.
Leid, R. W., and Williams, J. F. (1974). *Immunology* **27**, 195–208.
Leroy, A. (1910). *Arch. Int. Physiol.* **9**, 276–282.
Leventhal, R., Bonner, H., Soulsby, E. J. L., and Schreiber, A. D. (1977). *Program Abstr., 52nd Annu. Meet., Am. Soc. Parasitol.*, pp. 41–42.
Levine, N. D. (1968). "Nematode Parasites of Domestic Animals and Man." Burgess, Minneapolis, Minnesota.
Lewert, R. M., and Lee, C. (1954). *J. Infect. Dis.* **95**, 13–51.
Loeb, L., and Smith, A. J. (1904). *Proc. Pathol. Soc. Philadelphia* [N.S.] **7**, 173–178.
Lumsden, R. D. (1975). *Exp. Parasitol.* **37**, 267–339.
MacArthur, W. P. (1934). *Trans. R. Soc. Trop. Med. Hyg.* **27**, 343–363.
Machado, A. J., Gazzinelli, G., Pellegrino, J., Dias de Silva, W. (1975). *Exp. Parasitol.* **38**, 20–29.
McLaren, D. J., Clegg, J. A., and Smithers, S. R. (1975). *Parasitology* **70**, 67–75.
Mahmoud, A. A. F., Warren, K. S., and Peters, P. A. (1975). *J. Exp. Med.* **142**, 805–813.
Maillard, J. L., and Zarco, R. M. (1968). *Ann. Inst. Pasteur, Paris* **114**, 756–774.
Michaeli, D., Senyk, G., Maoz, A., and Fuchs, S. (1972). *J. Immunol.* **109**, 103–109.
Michel, J. F. (1974). *Adv. Parasitol.* **12**, 280–343.
Miller, A. M., Colley, D. G., and McGarry, M. P. (1976). *Nature (London)* **262**, 586–587.
Miller, H. R. P. (1971a). *Lab. Invest.* **24**, 339–347.
Miller, H. R. P. (1971b). *Lab. Invest.* **24**, 348–354.
Miller, H. R. P., and Walshaw, R. (1972). *Am. J. Pathol.* **69**, 195–206.
Moore, D. L., Grove, D. I., and Warren, K. S. (1976). *J. Pathol.* **121**, 41–50.
Morrison, D. C., and Cochrane, C. G. (1974). *J. Exp. Med.* **140**, 797–811.
Müller-Eberhard, H. J. (1975). *Annu. Rev. Biochem.* **44**, 697–724.
Murray, M., Miller, H. R. P., and Jarrett, W. F. H. (1968). *Lab. Invest.* **19**, 222–233.
Murray, M., Jarrett, W. F. H., and Jennings, F. W. (1971). *Immunology* **21**, 17–31.
Musoke, A. J., and Williams, J. F. (1975a). *Immunology* **28**, 97–101.
Musoke, A. J., and Williams, J. F. (1975b). *Immunology* **29**, 855–866.
Mustard, J. F., and Packham, M. A. (1970). *Pharmacol. Rev.* **22**, 97–187.
Ogilvie, B. M., and Jones, V. E. (1973). *Prog. Allergy* **17**, 93–144.
Ogilvie, B. M., and Love, R. J. (1974). *Transplant. Rev.* **19**, 147–168.
Ögren, S., and Lindahl, U. (1975). *J. Biol. Chem.* **250**, 2690–2697.
Orange, R. P., Stechschulte, D. J., and Austen, K. F. (1969). *Fed. Proc. Fed. Am. Soc. Exp. Biol.* **28**, 1710–1715.
Orange, R. P., Murphy, R. C., and Austen, K. F. (1974). *J. Immunol.* **113**, 316–322.
Pawlowski, Z., and Schultz, M. G. (1972). *Adv. Parasitol.* **10**, 269–343.
Perez, H., Clegg, J. A., and Smithers, S. R. (1974). *Parasitology* **69**, 349–359.
Picone, J. P., and Williams, J. F. (1976). *Program Abstr., 51st Annu. Meet., Am. Soc. Parasitol.* p. 43.
Poynter, D. (1966). *Adv. Parasitol.* **4**, 321–383.
Ranadive, N. S., and Muir, J. D. (1972). *Int. Arch. Allergy Appl. Immunol.* **42**, 236–249.

Ratnoff, O. D., and Naff, G. B. (1967). *J. Exp. Med.* **125**, 337–358.

Rent, R., Ertel, N., Eisenstein, R., and Gewurz, H. (1975). *J. Immunol.* **114**, 120–124.

Rent, R., Myhrman, R., Fiedel, B. A., and Gewurz, H. (1976). *Clin. Exp. Immunol.* **23**, 264–271.

Rickard, M. D., Mackinlay, L. M., Kane, G. J., Matossian, R. M., and Smyth, J. D. (1977). *J. Helminthol.* **51**, 221–228.

Rocha e Silva, M. (1973). *Proc. Int. Congr. Pharmacol., 5th, 1972* Vol. 5, pp. 250–266.

Rocha e Silva, M., and Graña, A. (1944). *Am. J. Phys.* **143**, 306–313.

Rocha e Silva, M., and Graña, R. (1946). *Arch. Surg. (Chicago)* **52**, 523–537.

Rothwell, T. L. W. (1975). *J. Pathol.* **116**, 51–60.

Rothwell, T. L. W., and Dineen, J. K. (1972). *Immunology* **22**, 733–745.

Rothwell, T. L. W., and Huxtable, C. R. (1976). *Aust. J. Exp. Biol. Med. Sci.* **54**, 329–335.

Rothwell, T. L. W., and Love, R. J. (1975). *J. Pathol.* **116**, 183–194.

Rothwell, T. L. W., Dineen, J. K., and Love, R. J. (1971). *Immunology* **21**, 925–938.

Rothwell, T. L. W., Prichard, R. K., and Love, R. J. (1974). *Int. Arch. Allergy Appl. Immunol.* **46**, 1–13.

Rubin, R. H., Austen, K. F., and Goetzl, E. J. (1975). *J. Infect. Dis.* **131**, Suppl., S98–S103.

Ryan, G. B., and Majno, G. (1977). *Am. J. Pathol.* **86**, 185–276.

Sadun, E. H. (1976). *In* "Immunology of Parasitic Infections" (S. Cohen and E. Sadun, eds.), Chapter 4, pp. 47–57. Blackwell, Oxford.

Samuelsson, B., Granström, E., Green, K., Hamberg, M., and Hammarström, S. (1975). *Annu. Rev. Biochem.* **44**, 669–695.

Samuelsson, B., Hamberg, M., Malmsten, C., and Svensson, J. (1976). *Adv. Prostaglandin Thromboxane Res.* **2**, 737–746.

Schwartz, B. (1921). *J. Parasitol.* **7**, 144–150.

Sher, A. (1976). *Nature (London)* **263**, 334–335.

Sher, A., and McIntyre, S. L. (1977). *J. Immunol.* **119**, 722–725.

Siegel, J., Osmond, J. P., Wilson, M., and Gewurz, H. (1975). *J. Exp. Med.* **142**, 709–721.

Silva, M. E., and Dietrich, C. P. (1975). *J. Biol. Chem.* **250**, 6841–6846.

Silverman, P. H. (1954). *Ann. Trop. Med. Parasitol.* **48**, 207–215.

Silverman, P. H., and Hulland, T. H. (1961). *Res. Vet. Sci.* **2**, 248–252.

Sinclair, K. B. (1962). *Br. Vet. J.* **118**, 37–53.

Smyth, J. D. (1969). "The Physiology of Cestodes." Oliver & Boyd, Edinburgh.

Smyth, J. D., and Heath, D. D. (1970). *Helminthol. Abstr., Ser. A* **39**, 1–23.

Soter, N. A., and Austen, K. F. (1976). *J. Invest. Dermatol.* **67**, 313–319.

Soule, P. C., Calamel, M., Chevrier, L., and Pantaleon, J. (1971). *Red. Med. Vet.* **147**, 1247–1257.

Spellman, G. G., and Nossel, H. L. (1971). *Am. J. Physiol.* **200**, 922–927.

Sprent, J. F. A. (1946). *J. Comp. Pathol.* **56**, 149–159.

Stechschulte, D. J., Orange, R. P., and Austen, K. F. (1971). *New Concepts Allergy Clin. Immunol., Proc. Int. Congr. Allergol., 7th, 1970* Int. Congr. Ser. No. 232, pp. 245–254.

Stewart, D. F. (1953). *Aust. J. Agric. Res.* **4**, 100–117.

Stirewalt, M. A. (1963). *Ann. N. Y. Acad. Sci.* **113**, 36–53.

Stirewalt, M. A. (1966). *In* "Biology of Parasites" (E. J. L. Soulsby, ed.), pp. 41–59. Academic Press, New York.

Sweatman, G. K., and Henshall, T. C. (1962). *Can. J. Zool.* **40**, 1287–1311.

Tawfik, S., El-Sawy, M. A., Abou-Zeina, A., and Azzam, Z. A. (1977). *Trans. R. Soc. Trop. Med. Hyg.* **71**, 359–360.

Thompson, A. R. (1972). *Biochim. Biophys. Acta* **261**, 245–257.

Thorson, R. E. (1956). *J. Parasitol.* **42**, 26–30.

Tolone, G., Brai, M., Bonasera, L., Bellavia, A., and Pontieri, G. M. (1972). *Pathol. Microbiol.* **38**, 192–199.

Tolone, G., Bonasera, L., Brai, M., Ferina, F., and Pontieri, G. M. (1974). Pathol. Microbiol. **41**, 41–50.

Torisu, M., Fukawa, M., Harasaki, H., Kitamura, K., Kai, S., Nagano, M., Ikeda, S., Hisatsugu, T., Nishimura, M., Dohy, H., Baba, T., and Sonozaki, H. (1975). *Clin. Immunol. Immunopathol.* **4**, 467–477.

Tsang, V. C. W., and Damian, R. T. (1977). *Blood* **49**, 619–633.

Tsang, V. C. W., Hubbard, W. J., and Damian, R. T. (1977). *Am. J. Trop. Med. Hyg.* **26**, 243–247.

Uvnäs, B., and Wold, J. K. (1967). *Acta Physiol. Scand.* **70**, 269–276.

Valentine, M. D., Sheffer, A. L., and Austen, K. F. (1971). *In* "Immunological Diseases" (M. Samter, ed.), 2nd ed., Vol. 2, pp. 906–919. Little, Brown, Boston, Massachusetts.

Valone, F. H., Leid, R. W., and Goetzl, E. J. (1977). *Fed. Proc. Fed. Am. Soc. Exp. Biol.* **36**, 1328.

Valotta, E. H., and Müller-Eberhard, H. J. (1973). *J. Exp. Med.* **137**, 1109–1123.

Varute, A. T. (1971). *Indian J. Exp. Biol.* **9**, 200–203.

von Lichtenberg, F., Sher, A., and McIntyre, S. (1977). *Am. J. Pathol.* **87**, 105–124.

Walsh, P. (1972). *Br. J. Haematol.* **22**, 237–254.

Ward, P. A. (1974). *Am. J. Pathol.* **77**, 520–538.

Ward, P. A., and Becker, E. L. (1967). *J. Exp. Med.* **125**, 1001–1020.

Wasserman, S. I., and Austen, K. F. (1976). *J. Clin. Invest.* **57**, 738–744.

Wasserman, S. I., Goetzl, E. J., and Austen, K. F. (1975). *J. Immunol.* **114**, 645–649.

Wasserman, S. I., Soter, N. A., Center, D. M., and Austen, K. F. (1977). *J. Clin. Invest.* **60**, 189–196.

Weil, P. E., and Boyé, G. (1910). *C. R. Seances Soc. Biol. Ses Fil.* **69**, 284–285.

Weinberg, M. (1907). *C. R. Seances Soc. Biol. Ses Fil.* **63**, 13–15.

Weinstein, P. P., Krawczyk, H. J., and Peers, J. H. (1954). *Am. J. Trop. Med. Hyg.* **3**, 112–129.

Weiss, A. S., Gallin, J. I., and Kaplan, A. P. (1974). *J. Clin. Invest.* **53**, 622–633.

Weiss, H. J. (1975). *N. Engl. J. Med.* **293**, 531–588.

Wells, P. D. (1962). *Exp. Parasitol.* **12**, 82–101.

Wilner, G. D., Nossel, H. L., and LeRoy, E. C. (1968). *J. Clin. Invest.* **47**, 2608–2615.

Yurt, R. W., and Austen, K. F. (1978). *In* "Proteolysis, Demineralization and other Degradative Processes in Human Biology and Disease," (I. H. Lepow & R. Berlin, eds.), Academic Press, New York (in press).

Yurt, R. W., Leid, R. W., Spragg, J., and Austen, K. F. (1977). *J. Immunol.* **118**, 1201–1207.

Zeiger, R. S., Twarog, F. J., and Colton, H. R. (1976). *J. Exp. Med.* **144**, 1049–1061.

CHAPTER 8

Pigments of Mammals

T. W. Goodwin

I. Introduction 273
II. Structural Colors 273
   A. Tyndall Colors 273
   B. Iridescence 274
III. Red Colors 274
IV. Melanins 274
   A. Nature and Formation 274
   B. Properties 278
   C. Skin Color 278
   D. Control of Melanization 279
V. Miscellaneous Yellow to Black Pigments: Fuscins 280
VI. Pigments That Play Little or No Part in
   External Coloration 280
   Carotenoids 281
   References 284

I. Introduction

The colors of mammals are much less bright and bizarre than those of other animals, (D. L. Fox, 1976; H. M. Fox and Vevers, 1960; Needham, 1974). Pigments which contribute to these bright colors in other animals include pteridines, carotenoids both free and conjugated to proteins, naphthoquinones, anthraquinones, porphyrins (including heme derivatives) and melanins. It is only the last two which make any major contribution to mammalian pigmentation. *Structural colors* also make a significant contribution and although they are not pigments, their manifestation frequently depends on the presence of melanins, and thus they will be considered briefly in Section II.

II. Structural Colors

A. TYNDALL COLORS

The characteristic blue colors of the buttocks and scrotal regions of monkeys and baboons are structural colors caused by Tyndall scatter-

CHEMICAL ZOOLOGY, VOL. XI
Copyright © 1979 by Academic Press, Inc.
All rights of reproduction in any form reserved.
ISBN 0-12-261040-5

ing in skin areas containing high concentrations of melanin in the dermis. Light of long wavelength (red) penetrates to the dermis where it is absorbed by the melanin. Light of shorter wavelength (blue) is scattered back from particles in the dermis or epidermis and thus gives the appearance of blue to the observer (see Fox, 1976, for details). It is the same phenomenon which makes the sky appear blue. Blue eyes are also due to Tyndall scattering in the iris with the longer wavelength light being absorbed by the dense black melanin deposition in the choroid layer. Brown eyes contain yellow and brown pigments (presumably melanins) on the outer surface of the iris together with some melanin deposits in the interior of the iris. The irides of green eyes contain only the yellow pigment and this, combined with the blue Tyndall color, results in the typical green appearance. The white of the corneal regions of the eye is also the result of Tyndall scattering (Mason, 1924).

B. IRIDESCENCE

Iridescent colors and metallic sheens are rare in mammals. Iridescence is seen mainly in finger nails, hair, and eyes. The effect is caused by reflection from laminated structures in these parts of the body; in this case melanin is not involved. The glittering colors in the eye are caused by light being reflected from complex striated surfaces lying deep within the eye; the varying emergent colors depend on the dimensions of the reflecting layer (Fox, 1976).

III. Red Colors

Bright red colors in mammals are rare and again are confined to monkeys and baboons. They are due to the red pigment hemoglobin in the blood flowing through highly vascularized areas of skin which are devoid of other pigments, particularly melanin (Fox, 1976).

IV. Melanins

A. NATURE AND FORMATION

The term melanin covers a wide range of polymers which vary in color from yellow to black. The black pigments normally found in animals are called eumelanins and are nitrogen-containing indole polymers which, in the simplest terms, have a repeating unit with the general structure (I).

(I)

The basic building block is the amino acid tyrosine and the major polymerizing unit is indole-5,6-quinone formed as indicated in Fig. 1. It is now becoming clear that many of the pigments consist of mixed polymers in that many of the intermediates in the pathway given in Fig. 1 can contribute to the polymer and that indole-5,6-quinone is by no means the only precursor (Blois, 1965). For example, in one naturally occurring melanin 16–24% of the residues was dopachrome (see Fig. 1) and this increased to 50% in melanin from a melanoma (Hempel, 1966). However, these results may to some extent be due to the isolated melanins not being radioactively pure (Swan, 1974). In any case it is clear that melanins from different mammalian sources can vary considerably (Nicolaus, 1968; Lillie, 1969; Swan, 1974).

The intermediate "blond" melanins, which vary in color from yellow to brown are named pheomelanins and contain sulfur as well as nitrogen. The polymers are formed from small units formed by condensation of dopaquinone with the sulfur-containing amino acid cysteine (see Thomson, 1974) as outlined in Fig. 2. The red melanins from hair, erythromelanins, probably contain more cysteine residues than indole residues and a [cysteinyl-DOPA]$_2$-Fe^{3+}-protein complex is considered a key intermediate in the biosynthesis of red melanins (Flesch, 1970). This fits in with the view that all melanins are bonded to proteins (see Needham, 1974).

Black phenolic polymers (allomelanins) occur in plants and are thought to exist in animals although the evidence is not yet very compelling.

The major manifestation of melanin in mammals is in skin and hair but an important deposit is in the choroid of the eye and in some neurons, e.g., in the substantia nigra and substantia ferruginea of primates and other mammals (Marsden, 1969).

The key enzyme in melanin formation is tyrosinase [o-diphenol:O$_2$ oxidoreductase], a copper-containing protein which converts tyrosine first into DOPA by hydroxylation and then by oxidation into dopaquinone. A nonenzymatic reaction converts dopaquinone into the

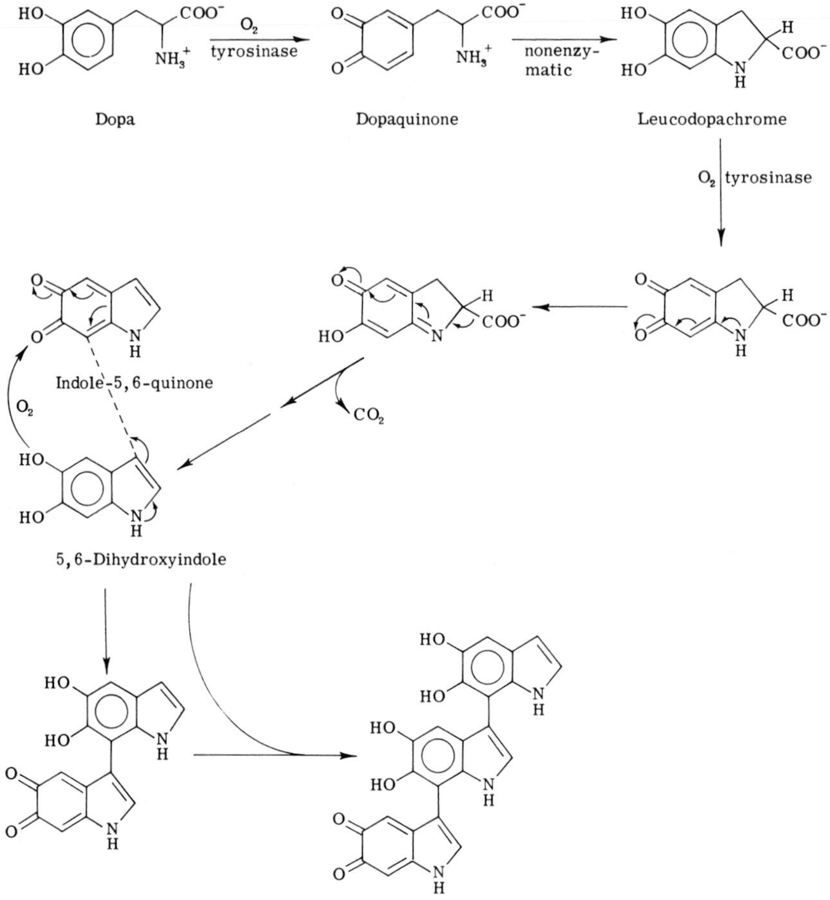

FIG. 1. Formation of melanin from tyrosine with indole 5,6-quinone as the major polymerizing unit.

indole derivative leucodopaquinone which is then oxidized by tyrosine to a quinone which is decarboxylated to yield 5,6-dihydroxyindole which is further oxidized to indole-5,6-quinone. The condensation of one molecule of 5,6-dihydroxyindole with one molecule of indole-5,6-quinone yields a dimer which can grow to a polymer by addition of further 5,6-dihydroxyindole molecules. As indicated previously, molecules other than 5,6-dihydroxyindole can be substituted for this molecule during the polymerization steps.

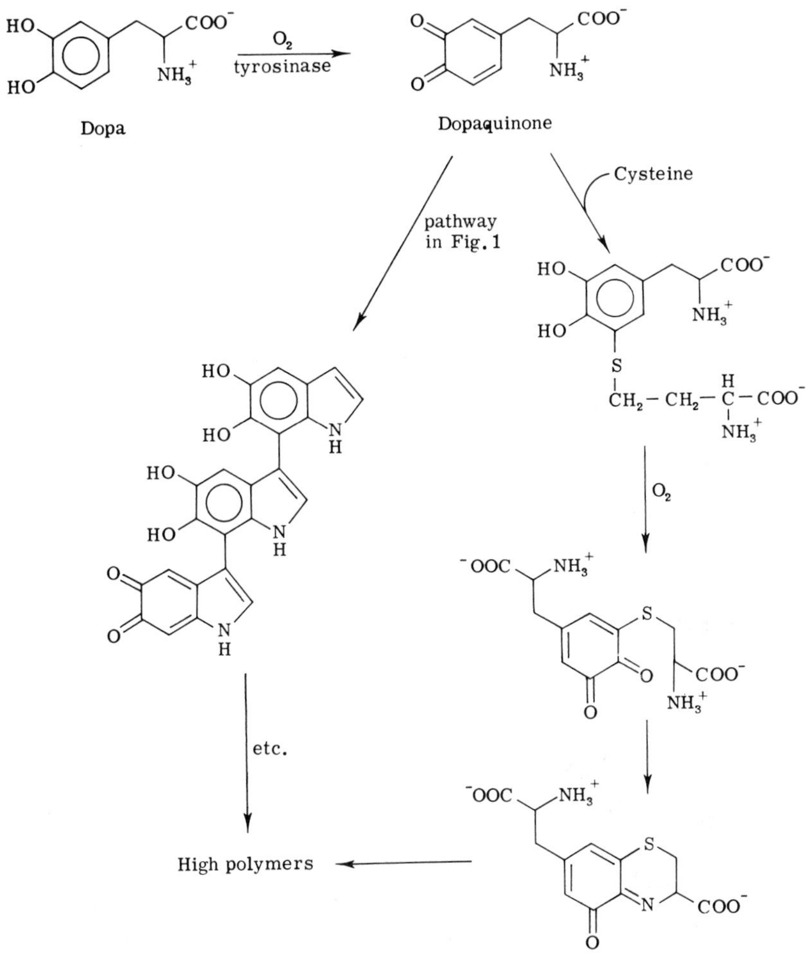

FIG. 2. Formation of blond melanins by condensation of dopaquinone with cysteine.

Melanins accumulate in specific cells named melanocytes. Within the melanocyte is a organelle consisting of a protein matrix, the pre-melanosome, which is the site of deposition of melanin. As the melanin is deposited the premelanosome is transformed into a melanosome (Moyer, 1963; Breathnach, 1969). The structural evolution of the melanosome varies with the tissue under examination (see Needham, 1974). Tyrosinase, which contains four copper atoms per molecule and two active sites (see Brooks and Dawson, 1966), is concentrated within the melanocytes.

B. PROPERTIES

1. Solubility

Eumelanin is soluble in hot alkali and hot concentrated H_2SO_4 but insoluble in water and dilute acids. It is soluble in acetic and formic acids and neutral organic solvents such as diethylamine and ethylene chlorohydrin (Lea, 1945). It can be precipitated from alkaline solution by adjusting the pH to the isoelectric point.

2. Electronic Absorption Spectra

Eumelanins exhibit strong general absorption in the visible region of the spectrum with the intensity increasing toward lower wavelengths. However, erythromelanin from hair exhibits an absorption maximum at 530–560 nm (Nickerson, 1946).

3. Reactions

Melanins are extremely refractory molecules and highly resistant to oxidation and reduction but prolonged treatment with hydrogen peroxide or simultaneous exposure to oxygen and sunlight will bleach eumelanins.

C. SKIN COLOR

The melanocytes in the skin of normal humans are in the epidermis but in certain black races they exist also in the dermis (see Fox, 1976). Variation in the number of melanocytes and the number of melanosomes (melanin granules) within the melanocytes appears to be the main factor controlling skin coloration in humans (see Strong, 1922).

Suntan which occurs on exposure to ultraviolet light is the result of increased melanin production (Lerner and Fitzpatrick, 1950) caused by an increase in tyrosinase activity. Thiol groups in skin which react with copper inhibit tyrosinase activity. Sunlight oxidizes the thiol groups and liberates copper which then activates tyrosinase resulting in excess melanin production and suntan (Flesch, 1949). Tyrosinase activity is also stimulated by the melanophore-stimulating factor (MSH) which can be isolated from the vertebrate pituitary gland (see Needham, 1974). Whether such a factor is activated by light remains to be decided. Early work hinted that the light effect is under hormonal control because castrates burn rather than tan (Cowdry, 1944).

D. CONTROL OF MELANIZATION

1. Biochemical

Excessive melanization observed under pathological conditions is most frequently due to neoplasia of melanocytes but the overall bronzing observed with Addison's disease is due to excessive secretion of MSH from the pituitary. Brown birthmarks (nevi) are due to non-malignant accumulation of melanin while purple birthmarks are due to excess melanin associated with hemoglobin in a highly vascularized tissue.

Absence of melanin from the cells of brainstem nuclei is observed in Parkinson's disease and is particularly associated with the characteristic paralysis agitans of the disease (Lerner, 1959; Issidorides, 1971). Whether this is the cause of the effect is still not certain but impaired synthesis of catecholamines appears to be involved.

Achromotrichia, lack of hair pigmentation, can be brought about experimentally in rats by keeping them on a copper-deficient diet (Hundley, 1950). This clearly correlates with the more recent discovery that copper is an integral component of tyrosinase.

2. Genetics

A great deal of work on the genetics of melanization has been reported (see Fox, 1976; Needham, 1974) which can only be considered in outline here. Hair color and skin color are multigenetically controlled; for example, ten loci control human skin color and at least forty control coat color in the mouse (see Penrose, 1959). The presence of the dominant white gene results in Arctic white races in which the activity of tyrosinase in the skin melanocytes is repressed. The existence of normal eye pigmentation in these races indicates that melanin is effectively synthesized in essential tissues. Presumably in animals which change color with the season (e.g., stoat) tyrosinase is activated and repressed by some factor which is released in response to change in day length.

Mammals carrying the recessive *albino* mutant manufacture no pigment even in essential organs such as the eye, and such animals are overall much more sensitive to the vagaries of normal living. The absence of the enzyme tyrosinase is the lesion in albino animals (Fitzpatrick *et al.*, 1958) and it is due to the failure to synthesize the mRNA required for tyrosinase synthesis, for injection of a nucleoprotein extract from a normal mouse will cause melanin synthesis in albino mice (Ottolenghi-Nightingale, 1969).

V. Miscellaneous Yellow to Black Pigments: Fuscins

A widely distributed group of yellow–brown–black pigments are recognized by histochemists as fuscins. Generally speaking they are chemically ill-defined (Verne, 1926; Needham, 1974) but that of the "brown fat" of mammals is a heme derivative (Prusiner *et al.*, 1970) and some lipid-soluble fuscins may be oxidation products of unsaturated fatty acids in glycerides (Porta and Hartcroft, 1969). Addition of vitamin E to the diet of rats prevents accumulation of a fuscin in rat adrenals and nerve tissue (Weglicki *et al.*, 1968).

VI. Pigments That Play Little or No Part in External Coloration

Colored molecules essential in relatively small amounts for the normal metabolism of mammals but which have no effect on the external appearance of the animals are rather numerous and are listed in Table I. The only group which will be considered in any detail here are the carotenoid pigments.

TABLE I

MAMMALIAN PIGMENTS THAT GENERALLY HAVE NO EFFECT ON EXTERNAL COLORATION

Pigments	Comments
Iron poryphyrins	
Hemoglobin	Respiratory pigment in blood[a]
Myoglobin	Red pigment in muscle
Cytochromes	Pigments in cellular electron transport
Porphyrins	
Bile pigments	Excreted in feces
Iron proteins	
Hemosiderin	Iron storage protein in reticuloendothelial system
Ferritin	Iron transporter
Various enzymes	In respiratory chain
Copper proteins	
Ceruloplasmin	In blood plasma
Cupreins	In erythrocytes, liver, brain, milk
Enzymes	Tyrosinase
Cobalt corrins	
Enzymes	Vitamin B_{12} coenzymes
Flavoproteins	
Enzymes	Various
Urochrome	Urinary excretory product
Carotenoids, see Section VI	

[a] Occasionally obvious as external red pigments; see Section III.

CAROTENOIDS

1. *Nature and Distribution*

Carotenoids are C_{40} terpenoids which are synthesized *de novo* only by plants and some protista. Up to the present time no such synthesis by animals has been substantiated, although colored oxidation products abound, particularly in invertebrates.

Carotenoids are divided into two main groups, carotenes, which are hydrocarbons, and xanthophylls, which contain oxygen in various forms (Goodwin, 1979a). Typical examples are β-carotene (II) and lutein (III). The main, if not the only function of carotenoids in mammals is as a pro-vitamin A. Indeed they are the only source of vitamin A (IV)

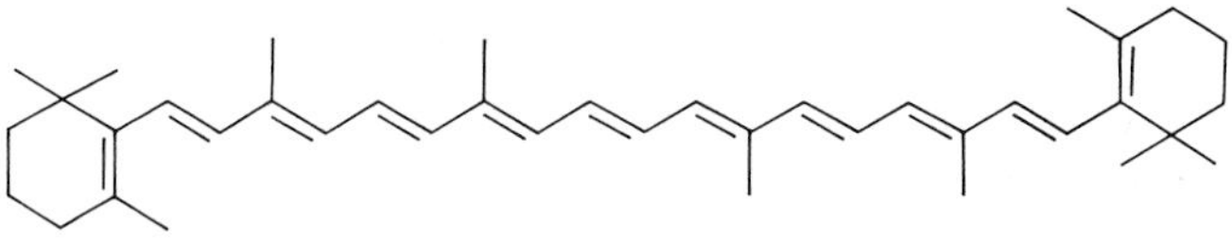

(II)

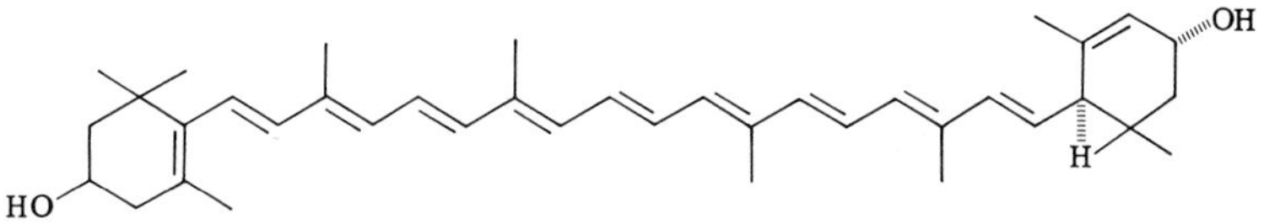

(III)

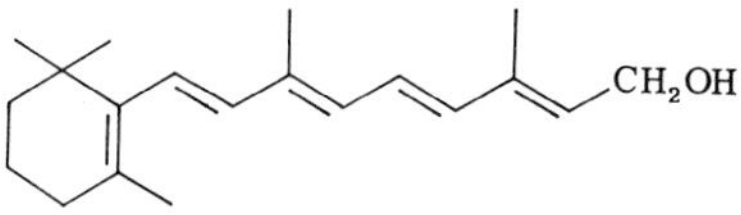

(IV)

in herbivorous animals, for higher plants do not synthesize vitamin A. Structurally a vitamin precursor must contain one-half the structure of β-carotene (II), that is the unsubstituted β-ionone ring and the conjugated double bond system attached to it. Thus it will be clear that lutein (III) has no vitamin A activity. The conversion of vitamin A-active carotenoids into vitamin A takes place mainly in the intestinal mucosa (see Pitt, 1971). The enzyme concerned, β-carotene 15,15′-dioxygenase, catalyzes the addition of oxygen across C-15 and C-15′ to form a peroxide which is rapidly cleaved at the C-15,15′ bond to yield two molecules of retinal (Fidge *et al.*, 1969; Lakshamanan *et al.*, 1972; Singh and Cama, 1974), which are reduced by a relative nonspecific

NADH- or NADPH-dependent reductase also present in the intestinal mucosa (Fidge *et al.*, 1968).

Mammals can be divided into three groups according to whether they accumulate (i) only carotenes in their tissues: (ii) both carotenes and xanthophylls in significant amounts or (iii) do not accumulate any carotenoids in significant amounts. The members of the groups are given in Table II; full details are to be found in Goodwin (1952, 1979). It would seem that in Group I animals xanthophylls are either not absorbed from the gut or rapidly oxidized in the intestinal mucosa, while the carotene is inefficiently converted into vitamin A in the mucosa and the unconverted pigment spills over into the circulating blood. Group II presumably cannot convert carotene into vitamin A or oxidize absorbed xanthophylls effectively and both groups of substances find their way into the bloodstream. Finally Group III must efficiently metabolize pigments of both groups which they absorb. The pathway of oxidative degradation of carotenoids is unknown.

As might be expected, the distribution of carotenoids in mammalian tissues generally reflects the intake of carotenoids and the lipid content of the tissues (Goodwin, 1952, 1979b). This holds for blood plasma, milk, adrenals, seminal vesicles, placenta, heart, liver, pancreas, nerve, and bone marrow in humans; in cattle mainly β-carotene is found in blood plasma, milk, ovaries, testes, adrenals, corpus luteum, corpus rubrum, thymus, liver, retina, iris, pituitary, bile, muscle, kidney, and placenta (Goodwin, 1952). In humans there are no carotenoids in sweat, spermatozoa or cerebrospinal fluid but they are

TABLE II
CAROTENOID DISTRIBUTION IN MAMMALS[a]

Group I (only carotenes)	Group II (carotenes and xanthophylls)	Group III (no significant carotenoids)
Cattle	Man	Goat[b]
Horse	Fitchet	Swine
Hedgehog	Badger	Rat
(?) Ass	Roe deer	Rabbit
		Hare
		Guinea pig
		Sheep
		Dog
		Elephant
		Fox
		Buffalo

[a] From Goodwin (1952, 1970).

[b] Traces of α-carotene, β-carotene, lycopene, and xanthophylls reported (Tricone and Lombardi, 1969).

found in traces in skin (Lee *et al.*, 1975). In cows none is found in the spleen.

In both species there is an unexpectedly high level in the adrenal lipid and in cows in the corpus luteum and corpus rubrum; in fact the last named is one of most concentrated mammalian sources of β-carotene in nature.

In blood, carotenoids are concentrated in the plasma and are probably carried on a component of the low density lipoprotein in human plasma (Chen and Kane, 1974). Fetal blood and that from newborn calves contain less carotenoids than the parent blood (Goodwin, 1952; Emel'yanov, 1973; Surynek *et al.*, 1976) and there is a well-established drop in carotenoid levels in cattle blood at parturition (Goodwin, 1952).

The carotenoid accumulators gradually increase the levels of their stored pigments so that tissues from old animals can have very high levels (Goodwin, 1952; Karnaukhov and Tataryunas, 1972). This suggests that mobilization and oxidation of the stored carotenoids is not very rapid. No obvious excretory degradation products have been reported but their accumulation in the yellow patches of olfactory tissue in the upper region of the nasal cavity and in the ear wax of cows (Goodwin, 1952; Kurihara, 1967) suggests that this may be a possible secretory route.

The carotenoid levels in the plasma and body lipids of cows appear to be much higher than those in bulls. It is not known with certainty whether this reflects a different dietary intake or whether it is due to fundamental sex differences in carotenoid metabolism (see Goodwin, 1952).

2. Genetics

The varying amount of carotenes which accumulate in the blood and milk of different breeds of cows is well known (Goodwin, 1952; Kapoor and Ranjhan, 1974). For example the milk of Guernsey cows contain 5.8 μg/gm milk fat while the corresponding figure for Holsteins is 3.4 (Wise *et al.*, 1947).

Rabbits have been found carrying a recessive gene which results in the accumulation of xanthophylls but not carotenes in the subcutaneous fat (Pease, 1928; Willimott, 1928). Presumably the xanthophyll-oxidizing enzymes are repressed.

3. Pathology

Excessive carotenoid intake in humans can lead to intense yellow pigmentation in the skin, owing to deposition of the pigments in the hypodermis. This pseudoicterus or xanthemia is accompanied by carotenemia (Lawrie *et al.*, 1941). The condition is without ill effect and disappears when carotenoids are removed from the diet (Hoch, 1943). There is one report of a case in which the carotenemia was apparently due to the failure of the patient to convert carotene into vitamin A so that the normal carotene : xanthophyll ratio in the blood was reversed (Cohen, 1949).

Pigmented skin in diabetes "xanthosis diabetica" is also due to sub-cutaneous accumulation of carotenoids with an accompanying carotenemia. Early views that this was due to a failure of conversion of carotenes into vitamin A (see Goodwin, 1952) are now not widely held (Ramachandran, 1973) and it would appear to be merely a reflection of a high carotenoid intake by diabetics [see Hoerer *et al.*, (1975) for a review].

Altered plasma levels have been reported in many other pathologi-cal conditions (see Goodwin, 1952) but it remains to be proved that, as in diabetes, they are not merely a reflection of dietary intake of carotenoids.

REFERENCES

Blois, M. S. (1965). *In* "The Origins of Prebiological Systems" (S. W. Fox, ed.), Aca-demic Press, New York.

Breathnach, A. S. (1969). *In* "Pigments in Pathology" (M. Wolman, ed.), Academic Press, New York.

Brooks, D. W., and Dawson, C. R. (1966). *In* "Biochemistry of Copper" (J. Peisach, P. Aisen, and W. E. Blumberg, eds.), Academic Press, New York.

Chen, G. C., and Kane, J. P. (1974). *Biochemistry* 13, 3330.

Cohen, H. (1949). *Q. J. Med.* 18, 397.

Cowdry, E. V. (1944). "Textbook of Histology." Lea & Febiger, Philadelphia, Pennsylvania.

Emel'yanov, L. V. (1973). *Chem. Abstr.*, 1568.

Fidge, N. H., Shiratori, T., Ganguly, J., and Goodman, de W. S. (1968). *J. Lipid Res.* 9, 103.

Fidge, N. H., Smith, F. R., and Goodman, de W. S. (1969). *Biochem. J.* 114, 689.

Fitzpatrick, T. B., Brunet, P. C. J., and Kukita, A. (1958). *In* "The Biology of Hair Growth" (W. Montagna and R. A. Ellis, eds.), Academic Press, New York.

Flesch, P. (1949). *Proc. Soc. Exp. Biol. Med.* 70, 136.

Flesch, P. (1970). *J. Soc. Cosmet. Chem.* 21, 77.

Fox, D. L. (1976). "Animal Biochromes and Structural Colours," 2nd ed. Univ. of Cali-fornia Press, Berkeley.

Fox, H. M., and Vevers, H. G. (1960). "The Nature of Animal Colours." Sidgwick & Jackson, London.

Goodwin, T. W. (1952). "Comparative Biochemistry of Carotenoids," 1st ed. Chapman & Hall, London.

Goodwin, T. W. (1979a). "Comparative Biochemistry of Carotenoids," 2nd ed., Vol. 1. Chapman & Hall, London.

Goodwin, T. W. (1979b). "Comparative Biochemistry of Carotenoids," 2nd ed., Vol. 2. Chapman & Hall, London.

Hempel, K. (1966). *In* "Structure and Control of the Melanocyte" (G. della Porter and O. Mühlbock, eds.), Springer-Verlag, Berlin and New York.

Hoch, H. (1943). *Biochem. J.* **37**, 430.

Hoerer, *et al.* (1975). *Acta Diabetol. Lat.* **12**, 202.

Hundley, J. M. (1950). *Proc. Soc. Exp. Biol. Med.* **74**, 531.

Issidorides, M. R. (1971). *Brain Res.* **25**, 289.

Kapoor, U. R., and Ranjhan, S. K. (1974). *Indian J. Anim. Sci.* **44**, 22.

Karnaukhov, V. N., and Tataryunas, T. (1972). *Dokl. Akad. Nauk SSSR* **203**, 1197.

Kurihara, K. (1967). *Biochim. Biophys. Acta* **148**, 328.

Lakshmanan, M. R., Chansang, H., and Olson, J. A. (1972). *J. Lipid Res.* **13**, 477.

Lawrie, W. R., Moore, T., and Rajagopol, N. B. (1941). *Biochem. J.* **35**, 825.

Lea, A. J. (1945). *Nature (London)* **156**, 478.

Lee, R., Mathews-Roth, M. M., Pathak, M. A., and Parrish, J. A. (1975). *J. Invest. Dermatol.* **64**, 175.

Lerner, A. B. (1959). *J. Invest. Dermatol.* **32**, 285.

Lerner, A. B., and Fitzpatrick, T. B. (1950). *Physiol. Rev.* **30**, 91.

Lillie, R. D. (1969). *In* "Pigments in Pathology" (M. Wolman, ed.), Academic Press, New York.

Marsden, C. D. (1969). *In* "Pigments in Pathology" (M. Wolman, ed.), Academic Press, New York.

Mason, C. W. (1924). *J. Phys. Chem.* **28**, 498.

Moyer, F. H. (1963). *Ann. N.Y. Acad. Sci.* **100**, 584.

Needham, A. E. (1974). "The Significance of Zoochromes." Springer-Verlag, Berlin and New York.

Nickerson, M. (1946). *Physiol. Zool.* **19**, 66.

Nicolaus, R. A. (1968). "Melanins." Hermann, Paris.

Ottolenghi-Nightingale, E. (1969). *Proc. Natl. Acad. Sci. U.S.A.* **64**, 184.

Pease, M. (1928). *Z. Indukt. Abstamm.-Vererbungsl., Suppl.* **2**, 1153.

Penrose, L. S. (1959). "Outline of Human Genetics." Heinemann, London.

Pitt, G. A. J. (1971). *In* "Carotenoids" (O. Isler, ed.), p. 717. Birkhaeuser, Basel.

Porta, E. A., and Hartcroft, W. S. (1969). *In* "Pigments in Pathology" (M. Wolman, ed.), Academic Press, New York.

Prusiner, S., Cannon, B., and Lindberg, O. (1970). *In* "Brown Adipose Tissue" (O. Lindberg, ed.), Am. Elsevier, New York.

Ramachandram, K. (1973). *Indian J. Med. Res.* **61**, 1831.

Singh, H., and Cama, H. R. (1974). *Biochim. Biophys. Acta* **370**, 49.

Strong, R. M. (1922). *Anat. Rec.* **23**, 40.

Surynek, J., Kucera, A., and Brandejs, P. (1976). *Vet. Med. (Prague)* **21**, 557.

Swan, G. A. (1974). *Fortschr. Chem. Org. Naturst.* **31**, 521.

Thomson, R. H. (1974). *Angew. Chem., Int. Engl. Ed.* **13**, 305.

Tricone, A., and Lombardi, A. (1969). *Ig. Sanita Pubblica* **25**, 497.

Verne, J. (1926). "Les pigments dans l'organisme animale." Doin, Paris.

Weglicki, W. B., Reichel, W., and Nair, P. P. (1968). *J. Gerontol.* **23**, 469.

Willimott, S. G. (1928). *Biochem. J.* **22**, 159.

Wise, G. H., Atkeson, F. W., Caldwell, M. J., Parrish, D. B., and Hughes, J. G. (1947). *J. Dairy Sci.* **30**, 279.

CHAPTER 9

Bile Salts in Mammalia

A. R. Tammar

I. Introduction 287
II. Mammalian Bile Salts 290
    A. Monotremata 290
    B. Marsupialia 290
    C. Insectivora 290
    D. Primates 290
    E. Edentata 291
    F. Pholidota 291
    G. Lagomorpha 291
    H. Rodentia 292
    I. Cetacea 295
    J. Carnivora 295
    K. Tubulidentata 297
    L. Proboscidea 297
    M. Perissodactyla 298
    N. Artiodactyla 298
III. Possible Significance of Mammalian Bile Salt Differences 299
    References 301

I. Introduction

Bile salts are synthesized in vertebrate livers from cholesterol. They are secreted along with phospholipids, cholesterol, and bile pigments in bile. They are surface-active agents and, as such, appear to participate in two distinct physiological phenomena. First, they help to maintain in aqueous solution the cholesterol and phospholipid content of bile, especially gallbladder bile. Second, after secretion into the small intestine they promote the absorption of the products of fat digestion.

For convenience the following trivial names for bile acids are used throughout the chapter: Lithocholic, 3α-hydroxy-5β-cholan-24-oic; chenodeoxycholic, $3\alpha,7\alpha$-dihydroxy-5β-cholan-24-oic; deoxycholic, $3\alpha,12\alpha$-dihydroxy-5β-cholan-24-oic; cholic, $3\alpha,7\alpha,12\alpha$-trihydroxy-5β-cholan-24-oic. All these acids are 5β-steroid structures. The corresponding 5α-structures are designated by the prefix allo, e.g., allocholic acid

287

is $3\alpha,7\alpha,12\alpha$-trihydroxy-5α-cholan-24-oic acid. Other trivial names will be used but will be defined in context.

In mammals the hepatic pathways for cholesterol (C_{27}) breakdown end with the formation of C_{24} bile acids due to the loss of the terminal isopropyl group of the cholesterol side chain. Strictly *bile salts* are formed by conjugation (peptide linkage) of the C-24 carboxyl group of bile acids with the amino group of either glycine or taurine. There is also one report (see Section II,N,2) that ciliatine, the phosphorus analog of taurine, can act as a conjugating amino acid. In addition the hydroxyl groups of the bile acids may be esterified with sulfuric acid to give the bile acid or bile salt sulfates. However, the term bile salt is frequently used loosely to refer to the anionic forms of bile acids encountered at physiological pH.

While in the intestine bile salts are subjected to the action of intestinal microorganisms resulting in partial deconjugation and some modifications to the hydroxylation pattern of the steroid nucleus. The bile acids, original or modified, are mainly reabsorbed in the ileum and returned to the liver by the enterohepatic circulation. Thus a distinction is drawn between primary bile acids which are synthesized in the liver directly from cholesterol and secondary bile acids which have experienced structural alterations in the intestinal tract and which may be further modified by the liver prior to resecretion.

The majority of analyses reported are of gallbladder bile or of duct bile in those species which lack a gallbladder. For this reason it is impossible to say, in most cases, which bile acids found are primary and which are secondary. There are two ways of overcoming this problem. The first is to equip the animal with a bile fistula and to analyze the bile acid content of the bile after allowing sufficient time to elapse for all the bile acids existing in the enterohepatic circulation to be secreted. The objection to this procedure is that, under these nonphysiological conditions, the rate of bile acid synthesis is increased and the relative proportions of primary bile acids found may not reflect those in the intact animal. However from a qualitative point of view the method has provided valuable information. The second approach is to analyze the bile from a germ-free animal. Here, the absence of intestinal microorganisms precludes artifacts of the enterohepatic circulation but does not enable a distinction to be drawn between primary bile acids per se, and primary bile acids modified by the liver prior to resecretion. As a result of applying such techniques it is generally accepted that in Mammalia cholic and chenodeoxycholic acids are primary and that almost everything else, especially deoxycholic acid and any acid with a keto group or groups, is secondary.

TABLE I
SURVEY OF RANGE OF MAMMALIAN BILES ANALYZED

Mammalian order	Number of living genera[a]	Number of genera examined
Monotremata	3	2
Marsupialia	57	3
Insectivora	71	1
Dermoptera	1	—
Chiroptera	118	—
Primates	59	11
Edentata	19	2
Pholidota	1	1
Lagomorpha	10	2
Rodentia	344	16
Cetacea	35	1
Carnivora	114	28
Tubulidentata	1	1
Proboscidea	2	1
Hyracoidea	3	—
Sirenia	2	—
Perissodactyla	6	1
Artiodactyla	86	17

[a] From Simpson, 1945

Paper chromatographic procedures were first applied to bile by Haslewood and Sjövall (1954). Determination of the nature of conjugation of the bile acids prior to this was often based on the sulfur content of a crude bile salt mixture. A low sulfur content implied the presence of glycine conjugates. There is thus considerable doubt concerning the type of conjugation reported in the earlier literature.

The subject of species differences in bile salts has been extensively reviewed (Haslewood, 1967a,b; Tammar, 1974a,b,c).

The mammals will be dealt with in the order in which they appear in Simpson's (1945) classification which lists eighteen mammalian orders. Four eutherian orders Chiroptera, Dermoptera, Hyracoidea, and Sirenia, are distinguished by the fact that no biles of species in them have ever been analyzed. A number of other orders are represented by only a very few species and thus there are large gaps in our knowledge (see Table I). This is less serious than may at first appear since it is apparent that mammalian bile salts, qualitatively at least, exhibit a much greater overall similarity than is found in lower vertebrate classes. Hence the few species differences that do occur are conspicuous.

II. Mammalian Bile Salts

A. Monotremata

The biles of two monotremes, *Ornithorhynchus anatinus* and *Tachyglossus* (= *Echidna*) *aculeatus* have been examined (Bridgwater *et al.*, 1962). Both contained taurocholate, taurochenodeoxycholate, and probably taurodeoxycholate. No trace was found of glycine conjugation or of anything other than C_{24} acids. No detectable amount of allocholic acid was found in hydrolyzed (chemically deconjugated) *T. aculeatus* bile.

B. Marsupialia

Studies on marsupial biles have shown that, despite early reports to the contrary, only taurine-conjugated bile acids are present (Bridgwater *et al.*, 1962). Cholic, chenodeoxycholic, and deoxycholic acids are the bile acids found although no cholic acid has been detected in *Phascolarctos cinereus* (koala) bile. The main bile acid in *P. cinereus*, 3α-hydroxy-7-oxo-5β-cholan-24-oic acid, is almost certainly an artifact of the enterohepatic circulation resulting from oxidation of the chenodeoxycholic acid present in trace amounts in the bile and subsequent failure of the liver to reduce it.

C. Insectivora

The only species of insectivore to be studied to date is *Erinaceus europaeus* (hedgehog) the hydrolyzed bile of which was shown to contain cholic acid in a preliminary examination by Haslewood and Wootton (1950). Conjugation with taurine only was implied.

D. Primates

Interest in primate and particularly human bile has naturally been strong since biles first began to be analyzed. The primary bile acids in man are cholic and chenodeoxycholic acids which are conjugated with both taurine and glycine. Many other acids have been isolated from human bile including lithocholic, deoxycholic, 3α,7β-dihydroxy-5β-cholan-24-oic (ursodeoxycholic) acids as well as acids with one or more hydroxyl groups oxidized to keto groups. Of greater interest than these secondary bile acids are two C_{27} acids, 3α,7α,12α-trihydroxy-5β-cholestan-26-oic acid (Carey and Haslewood, 1963) and 3α,7α-dihydroxy-5β-cholestan-26-oic acid (Hanson and Williams, 1971). These acids, quantitatively of minor importance, are of interest because they are immediate biosynthetic precursors of cholic and chenodeoxycholic acids, respectively (Danielsson and Sjövall, 1975). Their presence in human bile is thus an indication that man shares with other mammals a common metabolic pathway from cholesterol to the primary bile acids. A small quantity of allocholic acid is also to be found in human bile (Tammar, 1966). Other primate species studied have qualitatively the same bile acids conjugated with both glycine and taurine as man. No C_{27} acids however have been detected in nonhuman primate bile.

E. Edentata

Only two species of this order have been investigated (Tammar, 1970). *Myrmecophaga tridactyla* bile contained taurine-conjugated cholic, chenodeoxycholic, deoxycholic, and allocholic acids. In contrast the sample of *Choloepus hoffmani* bile examined contained mainly free cholic acid but there was chromatographic evidence for conjugation with both glycine and taurine.

F. Pholidota

The biles of two species of *Manis,* the only genus of this order extant, have been analyzed (Tammar, 1970). Both *Manis tricuspis* and *Manis pentadactyla* biles contained taurine-conjugated cholic, chenodeoxycholic, and deoxycholic acids. There was no evidence for the presence in these samples of glycine conjugates or 5α-bile salts.

G. Lagomorpha

The bile and bile acid metabolism of *Oryctolagus cuniculus* (domestic rabbit) have been extensively studied. The primary bile acids of this species are (mean values) cholic (93.8%), chenodeoxycholic (1.5%), and allocholic (4.7%) acids as shown by analysis of the gallbladder bile of two germ-free rabbits (Hofmann *et al.,* 1969). These acids appear to be conjugated exclusively with glycine. However the predominant bile acid in conventional rabbits (mean of twelve males) is deoxycholic acid (82.1%) which is accompanied by much smaller quantities of cholic (10.0%) and allodeoxycholic (7.9%) acids (Taylor, 1977). This gross disparity between conventional and germ-free rabbit bile composition is attributed to a particularly active intestinal microflora which cause almost quantitative 7α-dehydroxylation of cholic acid. This affords deoxycholic acid, the arrival of which in the liver appears to suppress the synthesis of chenodeoxycholic acid. The allodeoxycholic acid may arise either by 7α-dehydroxylation of allocholic acid or directly from deoxycholic acid by oxidation to a 3-oxo-4-ene steroid structure and subsequent reduction to a 3α-hydroxy-5α-steroid (Danielsson *et al.,* 1963). Taylor's very thorough analysis (1977) revealed that a small amount (0.5% of total) of the bile salts were sulfate esters in addition to being glycine conjugates. Such lithocholic acid as is present in rabbit bile is in this fraction. The finding of exclusively glycine-conjugated bile acids is consistent with Bremer's (1956) observations on the conjugating ability of rabbit liver microsomes even in the presence of taurine. Lack of taurine conjugation is however not a characteristic of Lagomorpha since *Lepus californicus* bile contains taurine and glycine conjugates of cholic, deoxycholic, and chenodeoxycholic acids (Tammar, 1970).

H. RODENTIA

The primary bile acids of the laboratory rat (*Rattus*) are (mean of fourteen germ-free animals) cholic (51.1%), $3\alpha,6\beta,7\beta$-trihydroxy-5β-cholan-24-oic (β-muricholic; 47.9%) and chenodeoxycholic (0.8%) acids (Madsen *et al.*, 1976). The primary bile acid composition of the laboratory mouse (*Mus*) is more complex: (mean of four determinations on pools from two germ-free animals) cholic (23.6%), β-muricholic (61.7%), chenodeoxycholic (1.5%), allocholic (2.3%), and $3\alpha,6\beta,7\alpha$-trihydroxy-5β-cholan-24-oic (α-muricholic; 8.6%) acids (Eyssen *et al.*, 1976). The 6β-hydroxylation system can use chenodeoxycholate or lithocholate as substrate. The action of this system on chenodeoxy-cholic acid affords α-muricholic acid whereas on lithocholic it results in $3\alpha,6\beta$-dihydroxy-5β-cholan-24-oic acid which is 7β-hydroxylated to give β-muricholic acid. This pathway was originally proposed by Mitropoulos and Myant (1967) who used a rat liver mitochondrial suspension to which soluble supernatant fraction was added. The bile of conventional rats and mice, although more complex in composition, also has cholic and β-muricholic acids as major components as well as deoxycholic and lithocholic acids. The presence of 6α-hydroxylated acids ($3\alpha,6\alpha$-dihydroxy-5β-cholan-24-oic, in the rat only, and $3\alpha,6\alpha,7\beta$-trihydroxy-5β-cholan-24-oic) in conventional rat and mouse bile is clearly a result of exposure to intestinal microorganisms. The laboratory rat is a particularly well investigated species since it has been used in most of the experiments to elucidate mammalian pathways of bile acid biosynthesis. For a review of this the reader is referred to Danielsson and Sjövall (1975). Rat bile acids are conjugated with both taurine and glycine. Hamster, *Cricetus auratus*, bile salts are taurine- and glycine-conjugated cholic, chenodeoxycholic and deoxycholic acids (Prange *et al.*, 1962). However oral administration of 24-^{14}C-chenodeoxycholic acid to a bile fistula hamster gave rise to labeled β-muricholic acid (Tateyama and Katayama, 1976). The 6β-hydroxyla-tion system may therefore be more widely distributed than has hitherto been supposed but it has not been observed in any other rodent species.

Meriones unguiculatus (Mongolian gerbil) bile, for example, after hydrolysis affords cholic, chenodeoxycholic and deoxycholic acids (Noll *et al.*, 1972) as do the biles of six of the seven desert rodents (see Table II) examined by Yousef *et al.* (1973). The seventh, *Neotoma lepida*, lacked chenodeoxycholic acid. There is no information on the nature of conjugation in these species. The predominant bile acid in the guinea pig (*Cavia*) is chenodeoxycholic acid (Danielsson and Einarsson, 1964) but a small amount of cholic acid is present (Schoenfield and Sjövall, 1966). These acids are conjugated with both taurine and glycine.

TABLE II
MAMMALIAN BILE SALTS[a]

Order Species	Acids found	Conju- gation[b]	Reference
Marsupialia			
Phascolarctos cinereus	Chenodeoxycholic (trace)		Tammar (1970)[c]
Macropus rufogriseus	Cholic; chenodeoxycholic; deoxycholic		Tammar (1970)
Primates			
Saimiri sciureus	Cholic; chenodeoxycholic		Tanaka *et al.* (1976)
Macaca irus	Cholic; allocholic; chenodeoxy- cholic; deoxycholic	T	Tammar (1966, 1970)
Macaca maurus	Cholic; chenodeoxycholic; deoxycholic	T,G	Tammar (1970)
Papio ursinus	Cholic; chenodeoxycholic; deoxy- cholic; lithocholic	T,G	Kritchevsky *et al.* (1974)
Simia satyrus	Cholic; chenodeoxycholic; deoxycholic	T,G	Tammar (1970)
Homo sapiens	3α,7α-dihydroxy-5β-cholestan-26-oic		Hanson and Williams (1971)[c,d]
Edentata			
Myrmecophaga tridactyla	Cholic; allocholic; chenodeoxy- cholic; deoxycholic	T	Tammar (1966, 1970)
Choloepus hoffmani	Cholic	T,G	Tammar (1970)
Pholidota			
Manis tricuspis and *M. pentadactyla*	Cholic; chenodeoxycholic; deoxycholic	T	Tammar (1970)
Lagomorpha			
Oryctolagus cuniculus	Allocholic; allolithocholic; others	G,S	Taylor (1977)[c]
Lepus timidus	Cholic; deoxycholic	T,G	Ripatti and Sidorov (1973)[c]
L. californicus	Cholic; deoxycholic	T,G	Tammar (1970)
Rodentia			
Cynomys ludovicianus	Cholic; chenodeoxycholic; deoxy- cholic; lithocholic		Brennerman *et al.* (1972)
Ammospermophilus leucurus	Cholic; chenodeoxycholic; deoxycholic		Yousef *et al.* (1973)
Spermophilus tereticaudus	Cholic; chenodeoxycholic; deoxycholic		Yousef *et al.* (1973)
Perognathus formosus	Cholic; chenodeoxycholic; deoxycholic		Yousef *et al.* (1973)
Dipodomys microps	Cholic; chenodeoxycholic; deoxycholic		Yousef *et al.* (1973)

(*Continued*)

TABLE II (*Continued*)

Order Species	Acids found	Conjugation[b]	Reference
Dipodomys merriami	Cholic; chenodeoxycholic; deoxycholic		Yousef *et al.* (1973)
D. deserti	Cholic; chenodeoxycholic; deoxycholic		Yousef *et al.* (1973)
Castor canadensis	Cholic (trace); chenodeoxycholic; deoxycholic	G,T (trace)	Tammar (1970)
Neotoma lepida	Cholic; deoxycholic		Yousef *et al.* (1973)
Meriones unguiculatus	Cholic; chenodeoxycholic; deoxycholic		Noll *et al.* (1972)
Cetacea			
Balaenoptera physalus		T	Tammar (1970)[c]
Carnivora			
Canis lupus	Cholic; chenodeoxycholic; deoxycholic	T,G (trace)	Ripatti and Sidorov (1973)[c]
Chrysocyon brachyurus	Cholic; chenodeoxycholic; deoxycholic	T,G (trace)	Tammar (1970)
Selenarctos thibetanus	Cholic (trace); chenodeoxycholic; deoxycholic	T	Tammar (1970)[c]
Ursus americanus	Cholic; chenodeoxycholic	T	Tammar (1970)[c]
Thalarctos maritimus	Deoxycholic		Tammar (1970)[c]
Helarctos malayanus	Chenodeoxycholic, deoxycholic	T	Tammar (1970)[c]
Melursus ursinus	Cholic; chenodeoxycholic; deoxycholic		Tammar (1970)
Procyon lotor	Chenodeoxycholic	T	Tammar (1970)[c]
Mustela vison	Cholic, chenodeoxycholic; deoxycholic	T,G (trace)	Ripatti and Sidorov (1973)
Martes martes	Cholic; chenodeoxycholic; deoxycholic	T,G (trace)	Ripatti and Sidorov (1973)
Amblonyx cinerea	Cholic	T	Tammar (1970)
Herpestes edwardsi	Cholic; allocholic (trace); chenodeoxycholic; deoxycholic	T	Tammar (1966, 1970)
Felis concolor	Cholic; allocholic (trace); chenodeoxycholic; deoxycholic	T	Tammar (1966, 1970)
F. leo	Chenodeoxycholic; deoxycholic		Tammar (1970)[c]
F. pardus	Chenodeoxycholic; deoxycholic		Tammar (1970)[c]
F. tigris	Cholic; chenodeoxycholic; deoxycholic	T	Tammar (1970)
F. felis	Cholic; allocholic; chenodeoxycholic; deoxycholic; allodeoxycholic; lithocholic; others	T,S	Taylor (1977)[c]
Acinonyx jubatus	Chenodeoxycholic; deoxycholic	T	Tammar (1970)[c]
Zalophus californianus	Cholic; allocholic; chenodeoxycholic; deoxycholic; other acids	T	Tammar (1966, 1970)[c]
Phoca vitulina	Cholic; chenodeoxycholic; deoxycholic; other acids	T	Tammar (1970)
Tubulidentata			
Orycteropus afer	Cholic; allocholic; chenodeoxycholic; deoxycholic	T	Tammar (1966, 1970)[c]
Perissodactyla			
Equus caballus	Cholic; chenodeoxycholic	T,G	Anwer *et al.* (1975)
Artiodactyla			
Sus sus leucomystax	Hyocholic; hyodeoxycholic; chenodeoxycholic; 3α-hydroxy-6-oxo-5β-cholan-24-oic; lithocholic	G (mostly) T	Morita *et al.* (1970)

TABLE II (*Continued*)

Order Species	Acids found	Conju- gation[b]	Reference
Phacochoerus aethiopicus	Hyodeoxycholic; deoxycholic	T,G	Tammar (1970)[c]
Hippopotamus amphibius	Deoxycholic	G (mostly) T	Tammar (1970)
Strepsiceros kudu	Cholic; chenodeoxycholic; deoxycholic	T (mostly) G	Tammar (1970)
S. imberbis	Cholic; chenodeoxycholic; deoxycholic	T	Tammar (1970)
Bubalus caffer	Cholic; chenodeoxycholic; deoxycholic	T,G	Tammar (1970)
Bos sp.	Allocholic	Ciliatine	Yamasaki *et al.* (1972)[c] Tamari *et al.* (1976)
Oryx beisa	Cholic; allocholic; chenodeoxy- cholic; deoxycholic		Tammar (1966, 1970)
Damaliscus albifrons	Cholic; allocholic; chenodeoxy- cholic; deoxycholic	T,G	Tammar (1966, 1970)
Ourebia ourebi	Cholic; allocholic; chenodeoxy- cholic; deoxycholic	T,G	Tammar (1966, 1970)
Litocranius walleri	Cholic; allocholic; chenodeoxy- cholic; deoxycholic	T, G (trace)	Tammar (1966, 1970)
Gazella bennetti	Cholic; chenodeoxycholic; deoxycholic	T,G	Tammar, (1970)
G. rufifrons	Cholic; chenodeoxycholic; deoxycholic	T, G (trace)	Tammar (1970)
Hemitragus jemlahicus	Cholic; chenodeoxycholic; deoxycholic	T,G	Tammar (1970)
Capra sp. (Grecian wild goat)	Cholic; allocholic (trace); deoxycholic		Tammar (1970)
Ovis musimon	Cholic; allocholic (trace); cheno- deoxycholic; deoxycholic	T,G	Tammar (1966, 1970)

[a] This table summarizes the work published on mammalian bile salts additional to the appendix in "Bile Salts" (Haslewood, 1967a). The intention is both to supplement and complement the table published in that book.

[b] T, Taurine; G, glycine; S, sulfate.

[c] For references to earlier work see Haslewood (1962).

[d] For references to earlier work see Haslewood (1967a).

I. CETACEA

Haslewood and Wootton (1950) extracted cholic acid from *Balaenoptera sibbaldi* bile and cholic and deoxycholic acids from *B. physalus* bile. The acids in *B. physalus* bile are conjugated with taurine only (Tammar, 1970).

J. CARNIVORA

The biles of about 40 species of this order have been examined. It is convenient for this purpose to subdivide them into the family Ursidae, fissipeds other than Ursidae, and pinnipeds.

1. Fissipeda Excluding Ursidae

Many of the observations in the literature on this group of animals date from the prechromatographic era, hence the nature of the conjugation reported is unreliable. However, the overriding impression is of taurine-conjugated cholic, chenodeoxycholic, and deoxycholic acids with cholic acid predominating. The only species in which a trace of glycine conjugation has been found is *Chrysocyon brachyurus*. Allocholic acid, in small amounts, has been found in some fissipeds (Tammar, 1966). A recent very thorough analysis of domestic cat bile (Taylor, 1977) is entirely in accord with earlier work and, in addition, demonstrates that about 2% of cat bile acids are sulfated. The uniformity of the bile salts in this group has given rise to speculation (Haslewood, 1967a) that a high proportion of taurocholate is appropriate to the fat absorption requirements of a carnivorous diet.

2. Ursidae

The reason for treating bears separately is that ursodeoxycholic acid, later shown to be $3\alpha,7\beta$-dihydroxy-5β-cholan-24-oic acid (Iwasaki, 1936), was isolated from *Selenarctos thibetanus japonicus* by Shoda (1927) and possibly by Hammarsten (1902) from *Thalarctos maritimus* (polar bear). The question arises as to whether this acid, known to be a secondary bile acid in other mammalians from which it has since been isolated, is in any way characteristic of bear biles or whether its trivial name is merely misleading. Biles of *Ursus americanus*, *Helarctos malayanus* and *Melursus ursinus* all contained cholic, chenodeoxycholic, and deoxycholic acids conjugated in the first two cases with taurine only (Tammar, 1970). The *M. ursinus* sample was devoid of conjugated material possibly as a result of microbiological deconjugation post mortem. *T. maritimus* bile examined by Hara (1947) and Takuma (1949) contained cholic and chenodeoxycholic acids but no ursodeoxycholic acid. The bile acids are all taurine conjugated (Tammar, 1970). *S. t. japonicus* is the only bear bile from which ursodeoxycholic acid has unequivocally been isolated not only by Shoda but also by Miyaji (1938) and Takuma (1949). However the bile of a member of this species which had died in London Zoo contained taurine-conjugated cholic, chenodeoxycholic and deoxycholic acids (Tammar, 1970). The most recent work on this subject is by Kurozumi *et al.* (1973) who analyzed five bear bile samples. Unfortunately it was not possible to cite the species from which they were derived but they were undoubtedly of Asiatic origin. Both taurine and glycine conjugates were found in the three samples where this analysis was performed and the five samples divided themselves into two groups. One group contained predominantly cholic and deoxycholic acids and the second group (two bears) contained predominantly chenodeoxycholic and ursodeoxycholic acids. For completeness a choleic acid (a complex of deoxycholic acid and a long chain fatty acid) was isolated with glycine, from the hydrolyzed bile of *Ursus arctos isabellinus* (Zumbusch, 1902). Clearly ursodeoxycholic acid is no more characteristic of bear bile than chenodeoxycholic acid is of goose bile.

3. Pinnipedia

The chief distinguishing characteristic of the eleven pinniped biles which have been analyzed is that they have, in addition to acids already met, acids hydroxylated at C-23. Two such acids are known. They are $3\alpha,7\alpha,23\xi$-trihydroxy-5β-cholan-24-oic acid (phocaecholic acid) and $3\alpha,7\alpha,12\alpha,23\xi$-tetrahydroxy-5$\beta$-cholan-24-oic acid ($\alpha$-phocaecholic acid in the older literature). Other acids found in seal biles, though not all have been found in all species, are cholic, chenodeoxycholic, deoxycholic, and small quantities of allocholic acids. With the exception of one early report of a small proportion of glycine conjugation these acids appear to be conjugated exclusively with taurine. There have been no attempts to determine the primary acids of any pinniped bile so the 23-hydroxylation peculiar to these biles may be due either to intestinal microorganisms or to a liver hydroxylation system specific, in mammals, to pinnipeds. It seems improbable that a strain of intestinal microorganisms capable of 23-hydroxylation should be found only in pinnipeds and thus the more likely explanation is that this hydroxylation is hepatic in origin. Certainly in snakes of the genus *Bitis*, where 23-hydroxylated bile acids are also found, liver microsomal preparations are capable of hydroxylating cholic and deoxycholic acids at C-23 (Ikawa and Tammar, 1976).

K. TUBULIDENTATA

The only species of this order extant is the aardvark, *Orycteropus afer*. Its bile contains taurine conjugates of cholic, chenodeoxycholic, deoxycholic, and allocholic acids (Tammar, 1970).

L. PROBOSCIDEA

The hydrolyzed bile of an elephant, presumably Asiatic, afforded cholic and deoxycholic acids (Kuroda and Teroaka, 1953).

M. Perissodactyla

Analysis of the bile from chronic fistulas established in ponies (*Equus caballus*) demonstrated that the primary bile acids were cholic and chenodeoxycholic. Bile salts were present as follows: taurochenodeoxycholate (71%), taurocholate (15%), and glycochenodeoxycholate (3–4%) (Anwer *et al.*, 1975).

N. Artiodactyla

The biles of over twenty species of this large group of animals have been analyzed. It is appropriate to consider separately the family Suidae.

1. Suidae

The chief distinguishing characteristic of biles in this family is the possession of bile acids hydroxylated at C-6α. Two such bile acids, hyocholic (3α,6α,7α-trihydroxy-5β-cholan-24-oic) and hyodeoxycholic (3α,6α-dihydroxy-5β-cholan-24-oic) have been found in the biles of all species of the genus *Sus* so far examined. In these biles glycine conjugates predominate although some taurine conjugation is found and glycohyodeoxycholate is the main bile salt. Very little, if any, cholic acid is present in these biles and it would appear that the ability to form 12α-hydroxylated bile acids is almost totally absent from these species. Although the bile from conventional pigs contains 3α-hydroxy-6-oxo-5β-cholan-24-oic, 3α,6β-dihydroxy-5β-cholan-24-oic, 3β,6α-dihydroxy-5β-cholan-24-oic, and lithocholic acids these acids are absent from the bile of germ-free pigs (Haslewood, 1971). There is a discrepancy between the acids found in pig fistula bile which were reported as chenodeoxycholic and hyocholic acids only (Bergström *et al.*, 1959) and the contents of bile from germ-free pigs which contained hyodeoxycholic acid in addition (Haslewood, 1971). Since chenodeoxycholic acid injected into the bile fistula pig was partially converted to hyocholic acid (Bergström *et al.*, 1959) whereas injected hyodeoxycholic acid was unchanged and was in any case absent from fistula bile normally, it had been assumed that chenodeoxycholic and hyocholic acids were the primary acids in pig bile. Hyodeoxycholic acid in the pig would thus be the equivalent of deoxycholic acid in other species, namely the result of 7α-dehydroxylation of a trihydroxy primary bile acid by intestinal microorganisms. However Haslewood's (1971) finding of an, admittedly small (4% by weight), amount of hyodeoxycholic acid in germ-free pig bile suggests that this acid too is primary, and that its normally low hepatic synthesis rate is augmented by its production from hyocholic acid by intestinal microorganisms. Germ-free pig bile contains a much higher proportion of taurine-conjugated bile salts than conventional pig bile.

The hydrolyzed bile of *Phacochoerus aethiopicus*, warthog, was much more like that of other mammals in that it consisted chiefly of glycine and taurine conjugated cholic, and deoxycholic acids. However although no hyocholic acid was found there was a small proportion, about 3% by weight, of hyodeoxycholic acid (Tammar, 1970).

Species from two of the five extant genera of the family Suidae have been examined and both have bile acids hydroxylated at C-6α. However only those species from the genus *Sus* appear to be almost devoid of 12α-hydroxylating ability.

2. Other Artiodactyla

All the biles in this group which have been examined are very similar and follow the familiar pattern of cholic, chenodeoxycholic, and deoxycholic acids conjugated mainly with taurine but usually accompanied by glycine conjugates. In some cases small amounts of allocholic acid have been found as well (Tammar, 1966; Yamasaki *et al.*, 1972). Conventional ox bile, always available in large quantities, has been exhaustively analyzed and contains a number of keto acids and lithocholic acid. Also a small amount of ciliatocholic acid has recently been isolated from this source (Tamari *et al.*, 1976). Ciliatine has the structure $H_3^+NCH_2P(OH)O_2^-$ and is thus the phosphorus analog of taurine $H_3^+NCH_2CH_2SO_3^-$.

III. Possible Significance of Mammalian Bile Salt Differences

In lower vertebrates it is possible to make some tentative correlations between the chemical structure of the bile salts found in a particular species and the position of the species in the "evolutionary tree" as determined by morphologists. If the assumption is made that the mammalian pathway from cholesterol to cholic acid represents the axis of bile salt development then the bile salts characteristic of a particular lower vertebrate fall into one of two categories. Either (i) they are intermediates in the cholesterol to cholic acid pathway which are not processed further because the species concerned lacks the necessary enzymes or (ii) they are simple modifications of intermediates in the cholesterol to cholic acid pathway (Tammar, 1974a,b,c).

The close similarity in composition of most mammalian biles makes it difficult to pursue such arguments in this Class. Three aspects however are worthy of attention. These are the presence of small quantities of 5α-bile salts, the emergence of glycine conjugation, and unusual hydroxylation systems.

5α-Bile salts, although common in lower vertebrates, appear largely to have been replaced during evolution by 5β-bile salts. The relevant step in the cholic acid pathway is the stereospecific reduction of 7α,12α-dihydroxy-4-cholesten-3-one. Presumably a 5α-reductase evolved first affording 5α-bile salts. Modern mammalians and most lower vertebrates also have a 5β-reductase. To account for the dominance of 5β-bile salts in mammals it is necessary to postulate either that the 5β-reductase is a more efficient enzyme or that its synthesis is less repressed.

The vestigial ability to produce small quantities of 5α-bile salts is widespread among mammals and is therefore unlikely to be of taxonomic significance. It would be interesting in this context to know the relative efficiencies of 5α- and 5β-bile salts as fat solubilizers.

Conjugation with glycine appears to be restricted to eutherian mammals among which its distribution is patchy. Its presence in primate bile suggests, inevitably, that evolutionarily it is an advance. However, not only is taurocholate an efficient surface active agent but the pK of its parent acid is low (ca. 1.4) which ensures that it is in its surface active, anionic form at any likely intestinal pH. Glycocholic acid, on the other hand, is relatively weak (pK ca. 4.5) which means that in the sheep, for example, where the duodenal pH can be as low as 2.5 (Leat and Harrison, 1969), it could not function as a surface active agent. Although the selection pressure on bile salts is small any survival advantage conferred by glycine-conjugated bile salts is unlikely to be concerned with either of the solubility phenomena referred to in the Introduction. It has been suggested that they may help to regulate some intestinal parasites (Smyth and Haslewood, 1963), but no other ideas have been forthcoming. It would be interesting to know if glycine conjugation evolved as a single event or several times. If the former were the case then as Haslewood (1967b) has pointed out the distribution of glycine conjugates would be of interest to students of eutherian mammalian radiation.

The unusual hydroxylation systems met in mammals are 6β-hydroxylation in some rodents, 23-hydroxylation in pinnipeds, and 6α-hydroxylation in suids. The interesting unresolved question concerning these is whether they are relics of hydroxylation patterns once widely distributed or whether they represent relatively recent modifications of bile acid biosynthesis confined to these small, morphologically distinct, groups of animals. The suids are probably the most useful area for further investigation as the biles of three genera,

Hylochoerus, Babirussa and *Tayassu* have yet to be analyzed. In the two genera already investigated 6α-hydroxylation is weakly established in *Phacochoerus* but firmly established, to the almost total exclusion of 12α-hydroxylation in *Sus*.

REFERENCES

Anwer, M. S., Gronwall, R. R., Engelking, L. R., and Klentz, R. D. (1975). *Am. J. Physiol.* **229**, 592–597.

Bergström, S., Danielsson, H., and Göransson, A. (1959). *Acta Chem. Scand.* **13**, 1761–1766.

Bremer, J. (1956). *Biochem. J.* **63**, 507–513.

Brennerman, D. E., Connor, W. E., Forker, E. L., and DenBesten, L. (1972). *J. Clin. Invest.* **51**, 1495–1503.

Bridgwater, R. J., Haslewood, G. A. D., and Tammar, A. R. (1962). *Biochem. J.* **85**, 413–416.

Carey, J. B., and Haslewood, G. A. D. (1963). *J. Biol. Chem.* **238**, 855PC–856PC.

Danielsson, H., and Einarsson, K. (1964). *Acta Chem. Scand.* **18**, 732–738.

Danielsson, H., and Sjövall, J. (1975). *Annu. Rev. Biochem.* **44**, 233–253.

Danielsson, H., Kallner, A., and Sjövall, J. (1963). *J. Biol. Chem.* **238**, 3846–3852.

Eyssen, H. J., Parmentier, G. G., and Mertens, J. A. (1976). *Eur. J. Biochem.* **66**, 507–514.

Hammarsten, O. (1902). *Hoppe-Seyler's Z. Physiol. Chem.* **36**, 525–555.

Hanson, R. F., and Williams, G. (1971). *Biochem. J.* **121**, 863–864.

Hara, K. (1947). *J. Jpn. Biochem. Soc.* **19**, 42; *Chem. Abstr.* **44**, 10860i (1950).

Haslewood, G. A. D. (1962). *Comp. Biochem.* **3A**, 205–229.

Haslewood, G. A. D. (1967a). "Bile Salts." Methuen, London.

Haslewood, G. A. D. (1967b). *J. Lipid Res.* **8**, 535–550.

Haslewood, G. A. D. (1971). *Biochem. J.* **123**, 15–18.

Haslewood, G. A. D., and Sjövall, J. (1954). *Biochem. J.* **57**, 126–130.

Haslewood, G. A. D., and Wootton, V. M. (1950). *Biochem. J.* **47**, 584–597.

Hofmann, A. F., Mosbach, E. H., and Sweeley, C. C. (1969). *Biochim. Biophys. Acta* **176**, 204–207.

Ikawa, S., and Tammar, A. R. (1976). *Biochem. J.* **153**, 343–350.

Iwasaki, T. (1936). *Hoppe-Seyler's Z. Physiol. Chem.* **244**, 181–193.

Kritchevsky, D., Davidson, L. M., Shapiro, I. L., Kim, H. K., Kitagawa, M., Malhotra, S., Nair, P. P., Clarkson, T. B., Bersohn, I., and Winter, P. A. D. (1974). *Am. J. Clin. Nutr.* **27**, 29–50.

Kuroda, M., and Teroaka, H. (1953). *J. Jpn. Biochem. Soc.* **25**, 375.

Kurozumi, K., Harano, T., Yamasaki, K., and Ayaki, Y. (1973). *J. Biochem. (Tokyo)* **74**, 489–495.

Leat, F. A., and Harrison, W. M. F. (1969). *Q. J. Exp. Physiol. Cogn. Med. Sci.* **54**, 187–201.

Madsen, D., Beaver, M., Chang, L., Bruckner-Kardoss, E., and Wostmann, B. (1976). *J. Lipid Res.* **17**, 107–111.

Mitropoulos, K. A., and Myant, N. B. (1967). *Biochem. J.* **103**, 472–479.

Miyaji, S. (1938). *Hoppe-Seyler's Z. Physiol. Chem.* **250**, 34–36.

Morita, S., Yamasaki, K., and Ogura, M. (1970). *Yonago Acta Med.* **14**, 82–86. *Chem Abstr.* **73**, 12849 (1970).

Noll, B. W., Walsh, L. B., Doisy, E. A., Jr., and Elliott, W. H. (1972). *J. Lipid Res.* **13,** 71–77.

Prange, I., Christiansen, F., and Dam, H. (1962). *Z. Ernaehrungswiss.* **3,** 59–78.

Ripatti, P. O., and Sidorov, V. S. (1973). *Dokl. Akad. Nauk SSSR* **212,** 770–773.

Schoenfield, L. J., and Sjövall, J. (1966). *Acta Chem. Scand.* **20,** 1297–1303.

Shoda, M. (1927). *J. Biochem. (Tokyo)* **7,** 505; *Chem. Abstr.* **22,** 981 (1928).

Simpson, G. G. (1945). *Bull. Am. Mus. Nat. Hist.* **85.**

Smyth, J. D., and Haslewood, G. A. D. (1963). *Ann. N.Y. Acad. Sci.* **113,** 234–260.

Takuma, T. (1949). *J. Jpn. Biochem. Soc.* **21,** 74; *Chem. Abstr.* **43,** 6297i (1949).

Tamari, M., Ogawa, M., and Kametaka, M. (1976). *J. Biochem. (Tokyo)* **80,** 371–377.

Tammar, A. R. (1966). *Biochem. J.* **98,** 25P.

Tammar, A. R. (1970). Ph.D. Thesis, University of London.

Tammar, A. R. (1974a). *Chem. Zool.* **8,** 595–612.

Tammar, A. R. (1974b). *Chem. Zool.* **9,** 67–76.

Tammar, A. R. (1974c). *Chem. Zool.* **9,** 337–351.

Tanaka, N., Portman, O. W., and Osuga, T. (1976). *J. Nutr.* **106,** 1123–1134.

Tateyama, T., and Katayama, K. (1976). *Lipids* **11,** 845–847.

Taylor, W. (1977). *J. Steroid Biochem.* **8,** 1077–1084.

Yamasaki, K., Ikawa, S., Ayaki, Y., and Yamamoto, Y. (1972). *J. Biochem. (Tokyo)* **72,** 769–772.

Yousef, I. M., Yousef, M. K., and Bradley, W. G. (1973). *Proc. Soc. Exp. Biol. Med.* **143,** 596–601.

Zumbusch, L. (1902). *Hoppe-Seyler's Z. Physiol. Chem.* **35,** 426–431.

Author Index

Numbers in italics refer to the pages on which the complete references are listed.

A

Aaron, J. E., 79, 82, *98*
Abou-Zeina, A., 251, *270*
Abraham, S., 201, 220, 221, 222, *227, 228*
Abrahamson, E. W., 52, *73*
Ackman, R. G., 13, *42*
Adams, J. B., 182, *189*
Adams, P., 97, *98*
Adelstein, S. J., 26, *46*
Ades, H., 162, *190*
Adler, R. G., 216, *227*
Adonajlo, A., 246, *266*
Aer, J., 96, *98*
Agrawal, R., 128, *155*
Ahlskog, J. E., 118, *154*
Aikin, B. S., 232, *267*
Ainsworth, L., 173, 175, 176, *189, 195*
Aitio, A., 181, *189*
Aitken, R. J., 185, *189*
Akulin, V. N., 10, 17, *44*
Albers, R. W., 9, 17, 34, 35, *43*, 49, 51, 61, 62, *72*
Albert, A., 164, *189*
Aleksandrowicz, Z., 180, *196*
Alhadeff, J. A., 183, *189*
Allen, J. R., 26, *42*
Allen, W. R., 165, 166, *189*
All Gower, M., 86, *99*
Almeida, A. F., 64, 65, *71*
Aloia, R. C., 5, 10, 11, 13, 15, 17, 18, 19, 24, 25, 26, 27, 28, 29, 33, *42*, 50, 55, 58, 62, 65, *71*
Aloj, S. M., 164, *189*
Aloni, B., 64, *71*
Alsat, E., 176, *190*
Altman, K., 124, *157*
Amatruda, J. M., 146, *154*
Ambler, J., 262, *266*
Amerose, A. P., 124, *156*
Amherdt, M., 140, *157*
Amoroso, E. C., 159, 160, *189*
Anast, C. S., 92, 95, *98*
Anderson, H. C., 82, 87, *98*

Anderson, J. N., 126, *155*
Anderson, T. F., 122, *156*
Andrews, P., 167, *189*
Andrews, R. V., 119, *155*
Ansair, A., 246, 257, *266*
Ansell, J. D., 184, *190*
Antelman, S. M., 216, *225*
Antuno-Madeira, M. C., 64, *73*
Anversa, P., 33, *43*
Anwer, M. S., 294, 298, *301*
Applegarth, E. C., 4, *45*
Archer, G. T., 260, *266*
Archer, R. K., 244, *266*
Arey, L. B., 160, *189*
Arimura, A., 106, 107, 108, *154, 157*
Armour, J., 261, *266*
Arthur, A. T., 186, *189*
Arvy, L., 246, *266*
Ashitaka, Y., 164, *189*
Ashwell, G., 164, *194*
Askenase, P. W., 237, 242, 259, *266, 268*
Asmundson, S. J., 2, *45*
Atkeson, F. W., 283, *285*
Atkins, W., 169, *192*
Au, W. Y. W., 123, 138, 139, *154*
Augee, M. L., 7, 19, *44*
Aurbach, G. D., 93, *99*
Austen, K. F., 232, 233, 236, 238, 239, 240, 241, 242, 249, 250, 255, 256, 264, *266, 268, 269, 270, 271*
Avecilla, L. S., 40, *46*
Avi-dor, Y., 5, *43*, 65, *72*
Avioli, L. V., 80, *98*
Avruskin, T., 130, *158*
Axelrod, J., 150, *155, 157*
Ayaki, Y., 295, 296, 299, *301, 302*
Azzam, Z. A., 251, *270*

B

Baba, N., 217, *227*
Baba, T., 255, *271*

Bachur, N. R., 180, *193*
Bacon, J. A., 91, *98*
Bada, J. L., 61, *73*
Badger, K. S., 178, 182, *189*
Baghdiantz, A., 91, *99*
Bahl, O. P., 117, *154*, 162, 163, 164, *189,*
    190, 196
Bailey, E. D., 3, *42*
Baker, H. N., 163, *194*
Baker, R. W., 180, *190*
Baldwin, R. B., 207, *227*
Banerjee, D., 252, *266*
Bar, R. S., 145, *155*
Baram, T., 110, *154*
Baranska, J., 10, 17, *42*
Bardsley, W. G., 180, *190*
Barenholz, Y., 40, 41, *46*
Barker, S. W., 215, *228*
Barnard, E. A., 240, *268*
Barnes, M. J., 84, 98, *98*
Barnett, R. E., 65, *72*
Barrett, A., 98, *102*
Barrett, J., 168, *192*
Barrieux, A., 212, 213, *225*
Bartke, A., 125, *155, 156*
Bartley, J. C., 220, *225*
Basch, J. J., 223, *225*
Bashford, C. L., 65, *71*
Bassiri, R. M., 109, *155*
Batz, W., 68, *75*
Baughman, W. L., 175, *192*
Baukal, A., 154, *155*
Baum, M. J., 124, *155*
Bauman, A. J., *46*
Bauman, D. E., 222, *225*
Baxter, C. F., 11, 31, *42, 46*
Bayless, T. M., 204, *227*
Bayley, A., 171, *192*
Baylink, D., 93, *98*
Bayliss, W. M., *155*
Bazer, F. W., 185, 186, 187, *189, 190, 193,*
    194, 195, 196
Bazin, H., 264, *266, 267*
Beall, R. J., 113, *157*
Beato, M., 186, 187, *189*
Beaver, M., 292, *301*
Beck, J. S., 166, *189*
Becker, E. L., 233, *271*
Becker, R. O., 86, *100*
Beer, A. E., 161, *189*

Befus, A. D., 258, *266*
Beguin, F., 245, *266*
Behrens, W., 4, *43*
Behrisch, H. W., 38, *42*, 53, 60, *71*
Beier, H. M., 186, *189*
Beier-Hellwig, K., 186, *189*
Beitz, D. C., 213, *225*
Bélanger, L. F., 82, 90, 94, *98, 101*
Bell, J., 162, *189*
Bell, R. H., 167, *191*
Bell, R. M., 56, *71*
Bellavia, A., 262, *271*
Belleville, F., 167, *189*
Bellino, F. L., 184, *189*
Bellisario, R., 163, *189*
Ben-David, M., 170, *190*
Benditt, E. P., 241, *269*
Bennett, J. P., 58, 67, *73*
Benzi, G., 181, *189*
Berberich, J. J., 119, *155*
Berger, M., 141, *157*
Bergstein, N. A. M., 160, *189*
Bergström, S., 298, *301*
Beri, V., 55, *74*
Bern, H. A., 104, *155*, 209, *226, 227*
Bernard, B., 87, *101*
Bernard, G. W., 87, *98*
Berndt, W. O., 184, *194*
Bernstein, R., 181, *195*
Bersch, N., 112, *155*
Bersohn, I., 293, *301*
Berte, F., 181, *189*
Bertini, F., 183, *189*
Bertoli, E., 7, 27, *44, 47*, 64, *71, 73, 74*
Betts, F., 85, *98*
Beutler, E., 7, *47*, 64, *74*, 183, *190*
Bewley, T. A., 209, *226*
Bèzes, H., 245, 264, *267*
Bhatnagar, R. S., 83, *101*
Bieber, L. L., 4, *44*
Biezenski, J. J., 160, *190*
Bigelow, M., 165, *190*
Bijvoet, O. L. M., 80, *98*
Bilder, G. E., 132, 133, *155*
Billiar, R. B., 180, *190*
Billingham, R. E., 161, *189*
Bingham, E. W., 223, *225*
Bingley, J. B., 262, *268*
Birdsall, N. J. M., 64, 65, *73*
Birken, S., 163, *194*

Bisaz, D. D., 86, *100*
Bisaz, R., 86, *99*
Bischoff, K., 176, *194*
Bishop, D. G., 24, *42*, 54, *71*
Bisswanger, H., 220, *227*
Bitensky, M. W., 57, 64, *72*
Bito, L. Z., 38, *42*
Bittman, R., 58, *71*
Blakemore, W. S., 147, *158*
Blamoutier, P., 246, 258, *266*
Blask, D. E., 150, *157*
Blayzk, J. F., 52, 59, 66, *71*
Bligh, E. G., 11, *42*
Blizzard, R. M., 117, *156*
Blois, M. S., 275, *284*
Bloj, B., 53, 60, 61, *72*
Bloom, B. R., 242, *266*
Blume, F., 66, *75*
Blume, K., 181, *196*
Blumenthal, N. C., 85, *98*
Boden, G., 106, *155*
Bogdan, D. P., 180, *190*
Bogieslawski, S., 163, *193*
Bogumil, J., 162, *197*
Boime, I., 163, 166, 167, *190, 193*
Bokisch, V. A., 232, *266*
Bolander, F. F., Jr., 169, *190*
Bonasera, L., 262, *271*
Bonczak, J., 246, *266*
Bonjour, J. P., 80, *99*
Bonner, H., 254, *269*
Bonsen, P. P. M., 36, *46*
Boreham, P. F. L., 252, *266*
Boreus, L., 160, *190*
Borland, R., 161, *190*
Boswell, R. N., 239, *266*
Boughton, I. B., 248, *266*
Bourichow, J., 92, *100*
Bourne, G. H., 78, *98*
Bowler, K., 5, 9, 17, 34, 39, *42*
Boyd, A. E. III., 216, *228*
Boyd, G. S., 184, *194*
Boyd, J. D., 187, *190*
Boyde, A., 86, *99*
Boyé, G., 247, *271*
Bracker, C. E., 218, *227*
Bradbury, S. M., 262, *268*
Bradley, W. G., 293, 294, *302*
Brady, R. O., 128, *156*, 183, *195, 196*
Brai, M., 262, *271*

Brambell, F. W. R., 216, *225*
Brandejs, P., 283, *285*
Bransome, E. D., 174, *191*
Brants, F., 58, *71*
Brasel, J. A., 182, *197*
Braselton, W. C., 174, *191*
Braunstein, G. D., 164, *197*
Brazeau, P. Jr., 106, 108, 109, 111, *155, 156, 158*
Breathnach, A. S., 277, *284*
Breckenridge, R. T., 232, *269*
Breckenridge, W. C., 37, *42*
Bremer, J., 292, *301*
Brenner, F. J., 2, *42*
Brenner, M. A., 131, *158*
Brennerman, D. E., 293, *301*
Bretscher, M., 55, *71*
Breuer, C. B., 167, *191*
Breuer, H., 175, *196*
Breuille, M., 177, *191*
Breuiller, F. F., 183, *190*
Brew, K., 210, 214, 217, *225, 226, 227, 228*
Bridgwater, R. J., 290, *301*
Briggs, F. N., 172, *193*
Brinkman, K., 64, *71*
Britton, H. G., 179, *190*
Brocklehurst, W. E., 241, *266*
Brodbeck, U., 210, *225*
Brodish, A., 110, *155*
Bronner, F., 78, 94, *99, 101*
Brooks, D. W., 278, *284*
Brooks, S. M., 121, 154, *155*
Brosens, L., 116, *156*
Brown, A. C., 142, 143, *155*
Brown, P. A., 213, *225*
Brown, R. E., 222, *225*
Brown, W. R., 162, *190*
Brownstein, M. J., 109, 150, *155*
Bruck, K., 1, 4, 33, *43*
Bruckdorfer, K. R., 57, *72*
Bruckner-Kardoss, E., 292, *301*
Brunet, P. C. J., 279, *284*
Brunner, H. R., 154, *155*
Buchanan, G. D., 91, *98*
Bucklie, J. T., 10, 17, 26, *44*
Buda, M. J., 150, *156*
Buetler, E., 181, *196*
Bullock, B., 127, *156*
Bullock, D. W., 186, *190*
Burdowski, A. J., 154, *156*

Burger, M. M., 64, 73
Burgos, R., 106, 108, 155
Burke, C. W., 172, 195
Burke, G., 129, 130, 155
Burlington, R. F., 26, 39, 42, 53, 71
Burrow, G. N., 128, 129, 157
Burstein, S., 125, 155
Buschow, R. A., 115, 126, 155
Butcher, M., 106, 108, 155
Butcher, R. W., 94, 101
Butterworth, A. E., 264, 266
Byfield, P., 91, 95, 99
Byram, J. E., 262, 266
Byrne, W. L., 39, 43, 53, 72

C

Caggiula, A. R., 216, 225
Cahill, G. F., Jr., 141, 157
Caimcross, K. D., 7, 27, 35, 45, 64, 73
Calamel, M., 246, 270
Calb, M., 150, 151, 157
Caldwell, M. J., 283, 285
Caldwell, R. S., 10, 17, 42
Cama, H. R., 281, 285
Camel, M., 166, 167, 190
Cameron, D. A., 82, 98
Cameron, D. P., 157
Campbell, P. N., 213, 225
Campbell, W. B., 121, 154, 155
Canfield, R. E., 162, 163, 164, 171, 189, 194, 195
Cannon, B., 5, 42, 280, 285
Capaldi, R. A., 67, 72
Capron, A., 262, 264, 266, 267
Capron, M., 264, 266, 267
Care, A., 92, 100
Carey, E. M., 221, 225
Carey, J. B., 290, 301
Carino, M. A., 110, 158
Carlsen, R. B., 163, 164, 189, 191, 195
Carmer, P. W., 105, 158
Carmody, P. J., 183, 197
Carr, T. E. F., 80, 99
Carraway, K. L., 218, 226
Carter, J. D., 26, 43
Casanova, J., 131, 157
Case, J. D., 148, 156
Casey, R., 64, 71
Castellano, M. A., 179, 195

Castren, O., 180, 190
Catchpole, H. R., 166, 190
Catt, K. J., 154, 155, 166, 193
Catty, D., 238, 267
Cédard, L., 175, 176, 177, 183, 190, 191, 196
Center, D. M., 240, 271
Cetorelli, J., 186, 187, 190
Chaffee, R. R. J., 26, 42
Chakrabartty, P. K., 216, 227
Challis, J. R. G., 160, 176, 192
Chamness, G. C., 127, 158
Chan, J. S. D., 168, 190
Chan, L., 160, 168, 169, 190
Chan, S. C., 152, 155
Chan, S. W. C., 176, 177, 190
Chaney, B., 91, 99
Chang, L., 129, 130, 155, 292, 301
Chansang, H., 281, 285
Chaplin, M. F., 163, 193
Chapman, D., 7, 8, 36, 42, 57, 66, 71, 73
Chapman, V. M., 184, 190
Chard, T., 172, 195
Charles, M. A., 140, 155
Charnock, J. J., 64, 71
Charnock, J. S., 34, 35, 43, 59, 61, 62, 64, 65, 71
Chase, L. R., 93, 99
Chebatareva, M. A., 10, 17, 44
Cheetham, R. D., 210, 227
Chen, A., 133, 158
Chen, C. H., 181, 190
Chen, G. C., 283, 284
Chen, L., 182, 190
Chen, T. T., 186, 187, 190
Chepenik, K. P., 182, 191
Chesnut, R. M., 110, 158
Chevrier, L., 246, 270
Chew, N. J., 178, 179, 194
Chi, E. Y., 240, 269
Chiarugi, V. P., 258, 267
Chideckel, E. W., 108, 109, 142, 143, 155, 156
Chieffo, V., 173, 191
Chrambach, A., 170, 190
Chomczynski, P., 206, 225
Christiansen, F., 292, 302
Christiansen, R. O., 204, 226
Christie, G. A., 174, 191
Chung, D., 166, 167, 194

Cintron, R., 131, *157*
Clark, J. H., 187, 188, *191*
Clark, R. A. F., 238, 239, *267*
Clarkson, T. B., 293, *301*
Claus, D. R., 258, *267*
Clegg, J. A., 258, *269*
Clemens, J. A., 216, *225*
Clements, R. S., 179, *190*
Cochrane, C. G., 232, 238, *266, 267, 268,*
    269
Coffey, R. G., 210, *225*
Cohen, H., 166, *190*, 284, *284*
Cohen, J. D., 170, *190*
Colas, A. E., 184, *194*
Colbon, G. S., 57, *71*
Cole, H. H., 165, *190*
Coleman, R., 35, *43*, 53, 56, *71*
Colley, D. G., 264, 265, *267, 268, 269*
Collins, E., 82, 87, *100*
Colston, K., 90, *100*
Colton, H. R., 242, *271*
Comar, C. L., 78, 80, *99*
Combarnous, Y., 163, *194*
Comfurius, P., 55, *74*
Comstock, J. P., 213, 214, *228*
Conaway, H. H., 92, 95, *98*
Condorelli, M., 154, *157*
Connor, W. E., 293, *301*
Conrad, S. H., 175, *193*
Contractor, S. F., 182, 183, *190*
Convey, E. M., 206, *226*
Cook, D. A., 64, *71*
Cook, R. W., 246, 257, 262, 263, *267*
Copp, D. H., 82, 91, 92, 94, *98, 99, 101,*
    182, *194*
Coppola, J., 167, *191*
Corash, L., 163, *193*
Costlow, M. E., 115, 126, *155*
Cotmore, J. M., 11, *43*
Coudert, S. P., 175, 176, *196*
Cowan, B. D., 186, *189*
Cowden, R., 128, *155*
Cowdry, E. V., 278, *284*
Cowie, A. T., 201, 206, 216, 223, *225*
Crabbe, M. J. C., 180, *190*
Craig, N., 65, *71*
Craig, R. K., 213, *225*
Crastes de Paulet, A. C., 177, 180, *191,*
    196
Cromton, A. W., 159, *193*

Croskerry, R. G., 113, *155*
Croxatto, H. B., 160, *190*
Crystle, C. D., 169, *191*
Cuatrecasas, P., 145, *156*, 209, *226*
Cullis, P. R., 38, *43*, 59, *71*
Curet, L. B., 184, *194*
Curti, B., 6, *45*
Cuzner, M. L., 26, *43*

D

Daha, M. R., 233, 236, *267*
Dahr, W., 220, *226*
D'Alessio, I., 164, *191*
Dalterio, S., 125, *155, 156*
Dam, Y., 292, *302*
Damian, R. T., 250, 251, *271*
Dancis, J., 160, *190*
Dangler, B., 250, *268*
Daniel, J. C., 185, 186, 187, *189, 190, 191,*
    193
Danielli, J. F., 54, *71*
Danielsson, H., 291, 292, 293, 298, *301*
Darnell, J. E., 212, 214, *226*
Dauwalder, M., 201, 210, *228*
Davenport, J. B., 26, *43*
David, D. N., 212, *225*
David, J. R., 242, *267*
David, R. R., 242, *267*
Davidowicz, E. A., 55, 56, *74*
Davidson, A., 91, *99*
Davidson, J. N., 205, *226*
Davidson, L. M., 293, *301*
Davie, E. W., 231, 232, *267*
Davis, C. L., 222, *225*
Davis, D. E., 2, 3, *42, 43*
Davis, W. L., 81, *99*
Davison, A. N., 26, *43*
Dawe, A. R., 4, *43*
Dawes, B., 247, 250, 259, *267*
Dawson, C. R., 278, *284*
Dawson, R. M. C., 26, *43*
Dean, D. A., 258, *267*
Dean, J. D., 178, 181, *191*
DeBernardi, M., 181, *189*
deGier, J., 9, 15, 41, *43, 45, 47*, 60, 65, *72,*
    74
De Haas, G. H., 53, *72*
deKruyff, B., 9, 38, *43, 47*, 56, 57, 58, 59,
    60, 66, *71, 72, 74*

DelCampo, H., 183, *189*
Delfs, E., 181, *195*
DeLuca, H. F., 80, 92, 93, 95, *100, 101*
Demel, R. A., 9, *47*, 56, 57, 58, 60, *71, 72, 74*
deMoor, P., 116, *156*
Denamur, R., 208, 213, 214, *226*
Denari, J. H., 188, *191*
DenBesten, L., 293, *301*
Denckla, W. D., 132, 133, *155*
Denyes, A., 26, *43*
Descomps, B., 177, *191*
Descomps, C. T., 180, *196*
Desjardins, C., 152, *158*
DeSombre, E. R., 126, 127, *156*
Dessaint, J. P., 264, *266, 267*
DeWitt, W. D., 250, *267*
Dey, B. K., 117, *155*
Dey, S. K., 176, *191*
Diamant, Y. Z., 179, 181, 182, 184, *191*
Dias de Silva, W., 235, 258, *267, 268, 269*
Dickerson, J. W. T., 83, *99*
Dickmann, Z., 117, *155*, 176, *191*
Diczfalusy, E., 162, 174, 175, 176, *191, 194, 195, 197*
Dietrich, C. P., 240, *270*
Dill, D. B., 4, *43*
Dineen, J. K., 245, 260, 262, *267, 269, 270*
Dingle, J. T., 92, 98, *101, 102*
Dixon, H. B. F., 246, *267*
Dixon, J. S., 166, 167, *194*
Djiane, J., 168, 169, *194*
Dodd, K., 248, *267*
Doe, J. E., 262, *266*
Doellgast, G. J., 182, *191*
Dohy, H., 255, *271*
Doisy, E. A., Jr., 293, 294, *301*
Donabedian, R. K., 128, 129, *157*
Donald, A. D., 245, *267*
Doniger, J., 183, *191*
Donini, P., 164, *191*
Donini, S., 164, *191*
Donker, S. C. B., 122, *157*
Dorer, F. E., 152, *158*
Dorescu, M., 148, *157*
Dorus, E., 124, *156*
Dosik, K., 124, *156*
Doty, S. B., 82, *99*
Doyle, F., 95, *99*
Driedzic, W. M., 10. 17, *43*

Driscoll, S., 178, 182, *191*
Drutzach, P. H., 5, *47*
Dryden, G. L., 126, *155*
Dryer, R. L., 5, *45*
Duchesne, M., 177, *191*
Dudé, D., 106, *157*
Duell, E. A., 122, *156*
Duke, K. L., 160, *194*
Duncan, C. J., 5, 9, 17, 34, 39, *42*
Dupont, A., 108, *154*
Duttera, S. M., 39, *43*, 53, *72*
Dyer, W. J., 11, *42*
Dyrenfurth, I., 162, *197*

E

Eanes, E. D., 57, 85, *72, 99*
East, J. M., 182, *191*
Eaton, M., 4, *44*
Eberts, K. M., 186, *193*
Ebner, K. E., 210, *225, 226*
Edelhoch, H., 164, *189*
Edelman, G. M., 224, *226*
Edidin, M., 55, 64, *73*
Edlow, J. B., 178, 179, 182, 183, *191*
Edwards, D. P., 174, *191*
Eglin, J. M., 171, *194*
Ehrich, J., 250, *267*
Eibl, H., 67, *74*
Eichberg, J., 26, *43*
Eiff, J. A., 248, *267*
Einarsson, K., 121, *155*, 293, *301*
Eisele, J. W., 235, *267*
Eisenstein, R., 258, *270*
El-Banna, A. A., 186, *191*
Eletr, S., 8, *43, 44*, 60, 64, *72*
Elias, J. J., 207, *226*
Elliott, W. H., 293, 294, *301*
El-Sawy, M. A., 251, *270*
Elson, C., 19, *44*
El-Tomi, A. E. F., 169, *191*
Emel'yanov, L. V., 283, *284*
Enders, A. C., 185, *191*
Enereth, P., 115, *155*
Engel, L. L., 174, *195*
Engelking, L. R., 294, 298, *301*
Engleman, D. M., 58, 62, *72*
Ensinch, J. W., 108, 109, 142, 143, *155, 156*
Erasmus, D. A., 246, *267*

Erdös, E. G., 232, 260, 268
Ertel, N., 258, 270
Esher, R. J., 5, 27, 43
Essa-Koumar, N., 106, 155
Esteller, A., 106, 158
Evans, I., 90, 100
Evans, S. W., 172, 195
Ewart, M. R., 233, 268
Eylar, E. H., 66, 73
Eyssen, H. J., 292, 301

F

Fahrman, C., 65, 71
Fal, W., 252, 267
Fallon, E. F., 128, 155
Fang, L. S. T., 5, 10, 34, 39, 40, 43, 47, 51, 72
Fanska, R., 140, 141, 155
Farias, R. N., 53, 60, 61, 72
Farkas, T., 10, 17, 43
Farmer, S. W., 209, 226
Farrel, H. M., Jr., 223, 225
Farrell, G., 155
Faust, E. C., 242, 253, 263, 267
Fawcett, D. W., 4, 43
Fawzi, T. D., 122, 157
Fearon, D. T., 233, 236, 267
Federico, P., 110, 156
Feidel, B. A., 258, 267
Felber, J. P., 172, 191
Feldman, G., 12, 43
Feldman, L. S., 12, 43
Fell, H. B., 98, 99
Fellows, R. E., 168, 169, 190, 192
Feng, L. C., 248, 268
Ferguson, M. M., 174, 191
Ferin, M., 105, 158, 162, 197
Ferina, F., 262, 271
Fernex, M., 245, 262, 264, 267
Fernex, P., 246, 262, 267
Ferre, F., 177, 191
Ferrin, L. G., 38, 43
Fidge, N. H., 281, 282, 284
Fiedel, B. A., 258, 270
Findlay, J. K., 176, 191
Finer, E. G., 58, 73
Finkel, J., 165, 190
Fioretti, P., 172, 191
Fischetti, V., 242, 266

Fishman, L., 178, 182, 191
Fishman, P. H., 128, 156
Fishman, W. H., 178, 182, 191
Fitzpatrick, T. B., 278, 279, 284, 285
Fjellstedt, T. A., 178, 180, 184, 191
Fleisch, H., 80, 86, 99, 100
Fleischman, A. I., 5, 27, 43
Fleisher, S., 5, 14, 26, 40, 46
Flesch, P., 275, 278, 284
Flier, J. S., 145, 155
Florini, J. R., 167, 191
Flury, F., 247, 267
Fodor, S., 180, 194
Folch, J., 11, 43
Folk, G. E., Jr., 119, 155
Folk, J. E., 233, 268
Folley, S. J., 219, 223, 226
Forbs, W. H., 4, 43
Forker, E. L., 293, 301
Forsham, P. H., 140, 143, 144, 155
Forsyth, I. A., 170, 191, 205, 226
Foster, G. V., 91, 95, 99
Foster, R. F., 5, 34, 47
Fottrell, P. F., 182, 196
Fourcans, B., 35, 43, 53, 72
Fox, C. F., 6, 7, 8, 9, 15, 43, 45, 46, 47, 64, 68, 72, 74
Fox, D. L., 273, 274, 279, 284
Fox, H. M., 273, 284
Fox, I. H., 183, 191
Fraenkel, G., 5, 43
Fraioli, F., 172, 191
Frantz, A. G., 105, 128, 129, 157, 158
Frantz, W. L., 207, 228
Frazer, S., 92, 100
Freeman, C. S., 206, 226
Freeman, R. S., 246, 267
Freinkel, N., 182, 196
Freychet, P., 171, 192
Frick, A., 93, 99
Fricke, H., 54, 72
Fridkin, M., 110, 154
Friedman, M. I., 119, 147, 158
Friesen, H. G., 115, 157, 158, 166, 167, 168, 169, 170, 171, 190, 191, 192, 193, 195, 196, 208, 209, 227, 228
Fuchs, S., 255, 269
Fujikawa, K., 231, 232, 267
Fukawa, M., 255, 271
Fukushima, M., 166, 191

Fuller, C. A., 122, *158*
Fuller, R. W., 152, *155*
Furth, E., 128, *155*
Fuxe, K., 115, *155*

G

Gal, A. E., 183, *195, 196*
Gale, C. C., 108, 109, *156*
Galli, C., 24, *46*
Gallin, J. I., 232, 238, 239, *267, 271*
Gallo, C. R., 182, *194*
Ganguly, J., 282, *284*
Gannon, F., 160, 161, *192*
Ganoza, M. C., 39, *43*, 53, *72*
Ganther, H., 6, *45*
Garancis, J., 181, *195*
Gärdner, D. G., 206, *226*
Garel, J. M., 91, *99*
Gates, D. M., 2, *43*
Gavras, H., 154, *155*
Gay, J. L., 117, *155*
Gaye, P., 208, 213, 214, 215, *226*
Gazzinelli, G., 258, *268, 269*
Gear, A. R. L., 7, 27, *44*, 64, *73*
Geck, P., 7, *44*, 64, *73*
Gemmell, D. K., 262, *266*
Genazzani, A. R., 172, *191*
George, F. W., 117, 118, *155*
Gerich, J. E., 143, 144, *155*
Germino, N. I., 179, 188, *191, 195*
Gershey, E. L., 163, *191*
Gershon, M. D., 33, *45*, 242, *268*
Gershon, R. K., 242, *268*
Gerstner, R., 121, *156*
Gewurz, H., 258, *267, 270*
Gey, G. O., 162, 181, *193, 195*
Gey, M. R., 162, *193*
Geyer, V., 178, 182, *192*
Gheorghiu, C., 150, 151, *157*
Ghinea, A., 148, *157*
Gibbons, J. M., 173, *191*
Giese, W., 80, *99*
Gingell, D., 36, *43*
Ginsburg, V., 220, *226, 227*
Giroud, C. J. P., 180, *192*
Giusti, G., 160, *193*
Gjonnaess, H., 233, *268*
Gladner, J. A., 233, *268*
Glasser, S. R., 178, 187, 188, *191, 194*

Glimcher, M. J., 86, *99*
Glöckner, W. M., 220, *226*
Glossmann, H., 154, *155*
Gobbert, M., 179, *194*
Göransson, A., 298, *301*
Goetzl, E. J., 238, 239, 240, 241, 242, 255, 264, *266, 268, 270, 271*
Gold, R. M., 119, *155*
Goldback, R. W., 57, *71*
Goldblatt, H., 152, *155*
Golde, D. W., 112, *155*
Goldman, B. D., 152, 153, *158*
Goldman, S. S., 9, 15, 16, 17, 19, 24, 34, 35, 37, 39, *43*, 49, 50, 51, 58, 60, 61, 62, *72*
Goldstein, R., 150, 151, *157*
Goldstein, S., 145, *157*
Gombos, G., 37, *42*
Goodman, de W. S., 281, 282, *284*
Goodner, C. J., 108, 109, 142, 143, *155, 156*
Goodwin, T. W., 281, 282, 283, 284, *284*
Gorbman, A., 104, *155*
Gorden, P., 112, *155*
Gordon, A. M., 146, *156*
Gordon, A. S., 154, *156*
Gordon, G. G., 124, *157*
Gordon, H. S., 154, *156*
Gordon, L. M., 38, *46*
Gorski, J., 160, 161, *192*, 213, *227*
Gorski, R. A., 160, *192*
Gorter, E., 54, *72*
Gospodarowicz, D., 165, *192*
Gottlieb, M. H., 57, *72*
Gough, E. D., 180, *192*
Gould, B. S., 98, *99*
Graña, A., 249, *270*
Graña, R., 249, *270*
Granström, E., 241, *270*
Grant, C. W. M., 9, *43*, 66, *72*
Grant, D. B., 96, *99*, 168, *192*
Grant, M. E., 83, *99*
Grasslin, D., 162, 163, *192*
Gray, C. W., 164, *197*
Green, K., 241, *270*
Greenbaum, A. L., 223, *226*
Greenberg, P., 95, *99*
Greenwood, F. C., 167, *195*
Greep, R. O., 179, *196*
Gregg, P., 245, *267*
Gregson, N. A., 26, *43*

Greiner, A. C., 152, *155*
Grendel, F., 54, *72*
Grezoriadis, G., 164, *194*
Griffen, J. H., 232, *268*
Griffith, O. H., 67, *72*
Griffiths, D. E., 7, *47*, 64, *71*, *74*
Grillo, T. A. I., 181, *192*
Grisham, C. M., 65, *72*
Grodsky, G. M., 140, 141, *155*
Grollman, E. F., 220, *227*
Gronwall, R. R., 294, 298, *301*
Grossman, L., 183, *191*
Grosvenor, C. E., 201, 208, 219, *226*
Grout, G. S. P., 64, *72*
Grove, D. I., 264, *269*
Grover, A. K., 53, *72*
Groves, M. L., 223, *225*
Gruber, K. A., 121, *156*
Gruener, N., 5, *43*, 65, *72*
Grumbach, M. M., 166, 168, *190*, *192*, *193*, *196*, *197*
Grygiel, I. H., 179, *192*
Gudmundson, T., 91, 95, *99*
Guérin, M. A., 217, *226*
Guillemin, R., 106, 108, *155*
Guinto, E., 183, *190*
Gulyas, B. J., 164, 165, 172, 186, *192*
Gupta, J. S., 117, *155*, 176, *191*
Gurpide, E., 166, *193*
Gurr, M. I., 24, 39, *43*, 53, *72*
Gusdon, J. P., 166, 169, *192*
Gusseck, D. J., 178, 181, *191*, *192*, *197*
Gustafsson, J. A., 115, 121, 124, *155*, *156*
Gustke, H., 182, *192*
Gygan, P. J., 162, 163, *192*

H

Habener, J., 138, *156*
Hackenbrock, C. R., 52, 59, 66, *72*
Hafs, H. D., 206, *226*
Hagemenas, F. C., 175, *192*
Hagerman, D. D., 174, 179, 180, 181, 182, 183, 184, 185, *192*, *196*
Hagopian, M., 33, *43*
Hall, C., St.-G., 180, *192*
Hallgren, P., 220, *226*
Halstead, L. B., 79, 87, *99*
Hamberg, M., 122, *156*, 241, 257, *268*, *270*

Hamilton, W. S., 187, *190*
Hammarsten, O., 296, *301*
Hammarström, S., 122, *156*, 241, *270*
Hammerberg, B., 250, 257, 258, *268*
Hammes, G. G., 53, *72*
Hampton, J. K., 183, *193*
Hancox, N. M., 82, *99*
Handwerger, S., 167, 168, *192*, *196*
Hansard, S. L., 91, *98*
Hansen, J. W., 117, *156*
Hanson, R. F., 290, 293, *301*
Haour, F., 117, *156*, *157*, 162, *196*
Hara, I., 36, *43*
Hara, K., 296, *301*
Harano, T., 296, *301*
Harasaki, H., 255, *271*
Harbison, R. D., 182, *193*
Hardy, W. T., 248, *266*
Hargreaves, W. H., 246, *267*
Harper, R. A., 85, *99*
Harris, H., 182, *196*
Harrison, G., 80, *99*
Harrison, O. S., 213, *225*
Harrison, W. M. F., 300, *301*
Hartcroft, W. S., 280, *285*
Hartner, W. C., 3, 4, *46*
Harvey, E. M., 54, *71*
Hasan, S. H., 117, *157*, 162, *196*
Haslam, J. M., 57, *71*
Haslewood, G. A. D., 289, 290, 295, 296, 298, 300, *301*, *302*
Hastein, T., 179, *192*
Hata, S., 121, *156*
Haussler, M. R., 146, *157*
Hayashi, M., 36, *43*
Hayashida, T., 209, *226*
Hayflick, L., 62, 65, *74*
Hazel, J. R., 7, 10, 15, *43*
Hazum, E., 110, *154*
Hbous, A., 4, *43*
Heald, P. J., 187, 188, *195*
Heap, R. B., 160, 173, 176, *192*
Heast, C. W. M., 65, *72*
Heath, D. D., 245, 253, *270*
Heath, J. E., 3, 4, *46*
Hedlund, L., 152, *156*
Hellman, B., 140, *156*
Hemington, N., 26, *43*
Hempel, K., 275, *285*
Hempel, V. E., 178, 182, *192*

Henderson, G. I., 181, *195*
Hendricks, C. M., 112, *155*
Hennen, G. P., 163, 171, *192, 194*
Henquin, J. C., 140, *157*
Henry, H., 90, 95, *101*
Hensel, J., 1, 4, 33, *43*
Henshall, T. C., 246, *270*
Henson, P. M., 238, *268*
Herd, R. D., 257, *268*
Herodek, S., 10, 17, *43*
Herring, G. M., 83, 84, *99*
Hershman, J. M., 171, 176, *192, 193, 195*
Hertogh, R., 116, *156*
Hertz, R., 165, *197*
Hesketh, T. R., 58, 67, *73*
Heyns, W., 116, *156*
Hibbert, S. R., 183, *195*
Hickman, J., 164, *194*
Higashi, K., 166, *193*
Higgens, H. P., 171, *192, 193*
Hill, M. W., 37, *45*
Hill, R. L., 207, 210, 214, *226, 228*
Himms-Hagen, J., 4, *43*
Hinz, H. J., 58, *72*
Hiramitsu, K., 69, *73*
Hirvonen, L., 4, *44*
Hisatsugu, T., 255, *271*
Hive, T., 121, *155*
Hobson, B. M., 164, 165, *192*
Hoch, H., 284, *285*
Hodgen, G. D., 164, 165, 172, *192, 195*
Hoebel, B. G., 118, *154*
Hoeppli, R., 248, *268*
Hoerer, 284, *285*
Hof, H., 5, 32, *43, 44*
Hoffman, B. F., ·34, *44*
Hoffman, H. J., 117, *156*
Hofmann, A. F., 291, *301*
Hogan, M. L., 167, *194, 195*
Hogarth-Scott, R. S., 262, *268*
Hohling, H. J., 86, *99*
Hojvat, S., 145, *156*
Hokfelt, T., 115, *155*
Holick, M. F., 90, *100*
Holladay, L. A., 164, *192*
Hollander, C. S., 130, *158*
Hollenberg, M. D., 145, *156*, 209, *226*
Hollister, C. W., 152, 153, *158*
Hollmann, K. H., 200, 224, *226*
Holmes, E. W., 180, 181, *192, 193*

Holmström, E. G., 172, *197*
Holtrop, M. E., 81, *100*
Homan, J. D. H., 162, 163, 164, *197*
Hooff, van den A., 88, *100*
Hopper, B. R., 175, *195*
Hopson, J. A., 159, *193*
Horiuchi, S., 17, *44*
Horrobin, D. F., 216, *226*
Horvath, A., 4, *44*
Horwitz, A. F., 7, 36, *44, 45*
Horwitz, B., 4, *44*
Horwitz, K. B., 127, *156, 158*
Houba, V., 264, *266*
Houdebine, L. M., 213, 214, 215, *226, 228*
Houslay, M. A., 58, 67, *73*
Houstek, J., 4, *44*
Hsueh, W., 54, *72*
Huang, C. M., 217, *227*
Huang, S.-S., 204, *227*
Huang, S. W., 146, *156*
Huang, W. Y., 181, *195*
Huang, Y. O., 6, 9, *47*, 64, *74*
Hubbard, W. J., 250, *271*
Hubbell, W. L., 62, 65, 66, *74*
Hubscher, T., 264, *268*
Huddleston, J. F., 178, 179, 183, *191*
Hudson, J. W., 2, 19, *44*
Huggett, A. St. G., 179, *190*
Huggins, J. W., 218, *226*
Hughes, D. L., 247, 250, *267*
Hughes, J. G., 283, *285*
Hugli, T. E., 233, *268*
Hui, S. W., 58, *72*
Hulbert, A. J., 7, 19, *44*
Hulland, T. H., 246, *270*
Hundley, J. M., 279, *285*
Hurley, T., 168, *192*
Hurlimann, J., 172, *191*
Husain, M. K., 105, *158*
Hussa, R. O., 118, *157*, 181, *195*
Hussain, R., 262, *268*
Hustead, S. T., *268*
Huxtable, C. R., 261, *268, 270*
Hwang, P., 167, *195*
Hytten, F. E., 160, *193*, 200, 201, 216, 222, *226*

I

Iacano, J. M., 54, *74*
Ichinose, R. R., 206, *226*

Iida, S., 64, *74*
Ikawa, S., 295, 297, 299, *301, 302*
Ikeda, H., 131, *156*
Ikeda, S., 255, *271*
Illanes, A., 4, *44*
Imai, Y., 130, 131, *156, 158*
Inesi, G., 8, *43, 44*, 60, 64, *72*
Infante, J. P., 220, *226*
Ingalls, W. G., 206, *226*
Ingelman-Sundberg, M., 121, *155*
Ingham, K. C., 164, *189*
Ingraham, H. A., 212, *225*
Irving, J. T., 78, 83, 85, 91, 92, *100*
Isac, T., 66, *73*
Isakson, P. C., 54, *72*
Ishizaka, K., 238, *268*
Ishizaka, T., 238, *268*
Issidorides, M. R., 279, *285*
Ito, Y., 166, *193*
Ivatt, R. J., 210, *226*
Ivy, A. C., 160, *195*
Iwasaki, T., 296, *301*
Izaka, K., 183, *193*

J

Jackas, M., 260, *266*
Jackson, I. M. D., 216, *228*
Jacobs, H. K., 3, 4, 38, *43, 46*
Jacobson, K., 38, *44*
Jacques, L. B., 240, *268*
Jacques, P. J., 180, 182, 183, *196*
Jailer, J. W., 171, 172, *193*
Jain, M. K., 35, *43, 53, 55, 72*
James, A. T., 24, 39, *43, 53, 72*
James, S. L., 264, *268*
James, V. H. T., 160, *193*
Jansen-Goemans, A., 122, *157*
Jansen, H., 64, *75*
Jansen, J. W. C. M., 60, *72*
Jarabak, J., 179, 180, *193*
Jarlfors, U., 217, *227*
Jaroszewicz, K., 181, *193*
Jaroszewicz, L., 181, *193*
Jarrett, E. E. E., 237, *268*
Jarrett, W. F. H., 245, 262, *269*
Javidi, K., 162, *195*
Jee, W. S. S., 97, *100*
Jelinek, W. R., 212, 214, *226*
Jenness, R., 202, 203, *226*

Jennings, F. W., 245, 262, *269*
Jensen, E. V., 126, 127, *156*
Jessiman, A. G., 171, *194*
Jewelewiz, R., 162, *197*
Johansson, B. W., 2, 4, *44*
John, B. M., 184, *193*
Johnson, A. R., 232, 233, 260, *268*
Johnson, J. D., 204, *226*
Johnson, L. Y., 150, *157*
Johnson, R. E., 206, *227*
Jones, D., 183, *193*
Jones, G. E., 162, *193*
Jones, R. F., 206, *226*
Jones, V. E., 254, 262, *268, 269*
Joplin, G., 95, *99*
Jorgensen, G. N., 212, *227*
Jorgenson, R. A., 39, *45, 53, 73*
Joseph, M., 264, *266, 267*
Joshi, S. G., 186, *193*
Josimovich, J. B., 166, 168, *193*
Jost, P., 67, *72*
Jowsey, J., 83, 97, *98, 100, 101*
Jozwik, M., 181, *193*
Juchau, M. R., 180, *190, 192*
Jukes, T. H., 127, *156*
Julian, G., 97, *100*
Jung, A., 86, *100*
Jung, R. C., 242, 253, 263, *267*

K

Kadohama, N., 207, *228*
Kagan, I. G., 254, *268*
Kahlenberg, A., 55, *72*
Kahn, C. R., 145, *155*
Kahn, J. R., 152, *158*
Kai, S., 255, *271*
Kairns, J. J., 57, 64, *72*
Kaliner, M., 238, *368*
Kallner, A., 291, *301*
Kalnasy, L., 142, 143, *155*
Kametaka, M., 295, 299, *302*
Kanazawa, S., 170, *197*
Kane, G. J., 257, *270*
Kane, J. P., 283, *284*
Kano, K., 206, 214, 215, *226*
Kao, I., 146, *156*
Kaplan, A. P., 232, 238, 239, *267, 268, 271*
Kaplan, I., 163, *191*
Kaplan, S. L., 166, 168, *190, 192, 193, 196, 197*

Kaplan, S. M., 154, *156*
Kapoor, C. L., 150, *156*
Kapoor, U. R., 283, *285*
Karkren, J. N., 172, *193*
Karlsson, K. A., 40, *44*
Karnaukhov, V. N., 283, *285*
Karp, W., 181, *193*
Karum, J. H., 143, 144, *155*
Kasai, R., 69, *72*
Kashiwa, H. K., 82, *100*
Kasperska-Dworak, A., 166, *193*
Kassis, A. I., 257, *268*
Kastelijn, D., 55, *74*
Kastin, A. T., 106, 107, *157*
Kataoka, K., 130, *158*
Katayama, K., 292, *302*
Kater, L. A., 241, 242, *268*
Kates, M., 17, *44*
Kawiak, J., 240, *268*
Kay, A. B., 232, 233, *268*
Kay, H. H., 80, *100*
Kayser, C., 3, *44*
Kayser, K., 33, 34, *44*
Keenan, T. W., 210, 213, 217, 218, *227*
Keith, A. D., 7, 9, *45*, 64, *73*, *74*
Keller, R., 262, *268*
Kellerman, G. M., 64, *74*
Kellermeyer, R. W., 232, *269*
Kelly, J. D., 180, *190*, 245, 260, 262, 267, *269*
Kelly, P. A., 115, *157*, 168, 169, 170, *193*, 209, *227*
Kelly, W. N., 180, 181, 184, *192*, *193*, *197*
Kember, N. F., 96, *100*
Kemp, A., Jr., 64, *72*
Kemper, B., 138, *156*
Kenimer, J. G., 171, *192*, *193*
Kennedy, E. P., 52, *74*
Kennedy, J. F., 82, 87, *100*, 163, *193*
Kenny, A. D., 136, 137, *156*
Kerry, K. R., 204, *227*
Keutmann, H., 92, *101*
Kiang, S.-L., 124, *158*
Kicha, L. P., 65, *75*
Kieffer, J. D., 110, *156*
Kim, H. K., 293, *301*
Kimelberg, H. K., 14, 35, 36, 40, *44*, 49, 64, 66, 67, *72*
Kinsella, J. E., 220, *226*
Kipnis, D. M., 140, *157*

Kirsch, R. E., 185, 187, *195*
Kirschner, K., 220, *227*
Kirstens, L., 145, *156*
Kitagawa, M., 293, *301*
Kitajima, Y., 69, *72*, *73*
Kitamura, K., 255, *271*
Kitchin, J. D., 175, *193*
Kittinger, G. W., 175, *192*
Kivirikko, K. I., 96, *98*
Klain, G. J., 26, 39, *44*
Kleeman, W., 66, *74*
Klein, D. C., 148, 150, 152, *156*, *158*, 181, *190*
Klein, M., 200, *227*
Klein, M. P., 36, *45*
Kleiner-Bossaller, A., 90, *100*
Kleinig, K., 68, *75*
Klentz, R. D., 294, 298, *301*
Knaggs, G. D., 216, *228*
Knight, J. W., 186, *193*
Knipprath, W. G., 10, 17, *44*
Knobil, E., 172, *193*
Knowles, P. F., 180, *190*
Knowlton, A. I., 171, 172, *193*
Knox, W. E., 223, *228*
Knudsen, J., 220, *227*
Knudson, A. G., 24, *46*
Kobata, A., 220, *227*
Kobayashi, Y., 217, *227*
Koch, Y., 110, *154*
Kodicek, E., 90, 92, 98, *100*
Kodoma, A. M., 5, *44*
Koerker, D. J., 108, 109, 142, 143, *155*, *156*
Kohler, P. O., 124, *157*, 176, 179, 182, *191*, *195*
Kohn, L. D., 128, *156*
Kohonen, J., 65, *72*
Koide, S. S., 174, *193*
Kojima, A., 171, *192*
Komender, J., 240, *268*
Koninckx, P., 116, *156*
Koo, G. C., 124, *156*
Kornberg, R. D., 56, *72*
Kosharkji, R. P., 182, *193*
Kovacs, K., 180, *194*
Kowalewski, S., 182, *192*
Kowalski, K., 167, *196*
Krachenbuhl, J. P., 207, *227*
Kramer, M., 97, *100*
Kramm, K. R., 2, *44*

Krane, S. M., 86, 99
Kratzch, E., 179, 192
Krawczyk, H. J., 246, 263, 271
Kreilos, R., 86, 99
Kreinen, P. W., 57, 64, 72
Kreps, E. M., 10, 17, 44
Kretchmer, N., 204, 226
Krishna, G., 150, 156
Krishnan, R. S., 185, 191
Kristman, R. S., 186, 192
Kritchevsky, D., 293, 301
Kritchevsky, G., 10, 11, 12, 15, 17, 24, 46
Kruger, S., 11, 44
Krulin, G. S., 5, 44
Kucera, A., 283, 285
Kuhl, W., 183, 190
Kuhn, N. J., 207, 217, 227
Kuiper, P. J. C., 70, 73
Kukita, A., 279, 284
Kumagai, C. F., 131, 156
Kumamoto, J., 6, 44, 61, 73
Kumar, M., 91, 99
Kunita, H., 121, 156
Kurihara, K., 283, 285
Kuroda, M., 297, 301
Kurozumi, K., 217, 227, 296, 301
Kusel, J. R., 259, 269
Kuwabara, T., 33, 44
Kuzuya, N., 131, 156

L

Lacko, L., 7, 44, 64, 65, 73
Ladbrooke, B. D., 57, 73
Lagerspetz, K. Y. H., 65, 72
Lagunoff, D., 240, 241, 269
Laki, K., 233, 268
Lakshmanan, M. R., 281, 285
Landau, B. R., 3, 44
Landefeld, T., 163, 193
Landor, J. H., 106, 155
Langford, R. W., 104, 146, 157
Langlois, M., 143, 144, 155
Lankin, W. M., 163, 194
Lanman, J. T., 161, 193
Lanner, M., 167, 196
Lapierre, J., 246, 258, 269
Laragh, J. H., 154, 155
Larraga, L., 162, 197

Larson, B. L., 212, 227
Lassiter, W., 93, 99
Latham, T. R., 110, 158
Laurberg, P., 128, 158
Lawrence, A. M., 145, 156
Lawrie, W. R., 284, 285
Layne, D. D., 174, 194
Lazarus, L., 111, 152, 158
Lea, A. J., 278, 285
Leake, N. H., 169, 192
Leat, F. A., 300, 301
Leathem, J. H., 176, 177, 190
Leclercq-Meyer, V., 145, 156
Lee, A. G., 64, 65, 73
Lee, C., 252, 269
Lee, D. L., 252, 254, 269
Lee, G., 128, 156, 178, 179, 183, 191
Lee, M. P., 7, 27, 44, 64, 73
Lee, P. A., 117, 156
Lee, R., 283, 285
Lees, M., 11, 43
Lefebvre, N. G., 152, 153, 158
Leffingwell, T. P., 201, 210, 228
Leggate, J., 92, 100
Leid, R. W., 238, 240, 241, 249, 250, 263, 264, 269, 271
Leidahl, L. C., 115, 156
Lenard, J., 55, 56, 73, 74, 218, 228
Lenaz, G., 7, 27, 44, 64, 73
Lentz, K. E., 152, 158
Lenz, P. H., 5, 27, 43
Leon, M., 126, 156
Lepow, I. H., 235, 267
Lerner, A. B., 148, 156, 278, 279, 285
Lerner, E., 19, 44
Leroy, A., 247, 269
LeRoy, E. C., 232, 271
Leskiw, M. J., 147, 158
Leslie, J. M., 10, 17, 26, 44
Leslie, L., 205, 226
Lesniak, M. A., 112, 155
Letchworth, A. J., 172, 195
Leterrier, F., 38, 47, 65, 74
Leventhal, R., 254, 269
Levey, I. L., 186, 193
Levin, I. W., 7, 47
Levine, M., 152, 158
Levine, N. D., 256, 269
Levine, Y. K., 7, 44
Levitt, M. J., 168, 193

Lewert, R. M., 252, *269*
Lewicki, J., 183, *193*
Lewin, S., 211, *227*
Lewis, F. A., 265, *267*
Li, C. H., 96, *100*, 112, *155*, 163, 166, 167, *193*, *194*, *195*, 206, 209, *226*, *227*
Li, N. M., 9, 28, *47*
Liao, T., 163, *195*
Libbin, R. M., 154, *156*
Lieber, C. S., 124, *157*
Lillie, R. D., 275, *285*
Lin, C. Y., 222, *227*
Lindahl, U., 240, *269*
Lindberg, B. S., 220, *226*
Lindberg, O., 4, *44*, 280, *285*
Linden, C. D., 8, 15, *45*
Lindi, L., 7, 27, *44*, 64, *73*
Ling, N., 106, 108, *155*
Lingrel, J. B., 215, *228*
Linnane, A. W., 64, *74*
Linzell, J. L., 202, 204, 214, 217, *227*
Lipman, R. L., 166, *196*
Lippert, J. L., 57, *73*
Lischko, M. M., 152, *156*
Little, B., 171, 180, *190*, *194*
Little, K., 78, 83, *100*
Liu, W. K., 163, *194*
Livingston, J. N., 146, *154*
Livne, A., 64, 70, 71, *73*
Lockwood, D. H., 146, *154*, 207, *227*
Loeb, L., 247, *269*
Lofgreen, G. P., 80, *100*, *102*
Loizzi, R. G., 217, *226*
Lombardi, A., 282, *285*
Lompa-Krzymien, L., 174, *194*
Lornitzo, F. A., 220, *227*
Love, R. J., 260, 261, 262, *269*, *270*
Low, J., 182, *189*
Low, P. S., 61, *73*
Lowe, M. C., 180, *192*
Luecke, R. H., 3, 4, *46*
Lueker, C. E., 80, *100*
Luick, J., 80, *102*
Lum, C., 181, *195*
Lumsden, R. D., 245, *269*
Lundblad, A., 220, *226*
Lyle, P. D., 2, *42*
Lyman, C. P., 3, 4, 26, *43*, *45*, *46*
Lyons, J. M., 6, 8, 9, 27, 35, *44*, *45*, 60, 61, 64, *73*
Lyons, W. R., 169, *194*, 206, *227*

M

MacArthur, W. P., 246, *269*
McCain, T., 146, *157*
McCann, S. M., *156*
McCarty, K. S., 206, *226*
McConnell, H. M., 8, 9, 15, 36, *43*, *45*, *46*, 56, 62, 65, 66, 72, *74*
McCord, J. M., 181, *192*
McCormack, S. A., 178, *194*
McDonald, J. A., 181, *192*
McElhaney, R. N., 15, *45*, 51, 61, 62, 73, *74*
McGarry, M. P., 265, *269*
McGaughey, R. W., 186, *194*
McGrath, C. M., 206, *227*
McGuire, W. L., 115, 126, 127, *155*, *156*, *158*
Machado, A. J., 258, *269*
McIlreavy, D., 213, *225*
MacIndoe, J. H., 207, *228*
MacIntyre, I., 90, 91, 92, 95, *99*, *100*
McIntyre, S. L., 259, 265, *270*, *271*
McKenzie, H. A., 223, *227*
Mackinlay, L. M., 257, *270*
McLachlan, J. A., 127, *156*
McLaren, A., 184, *190*
McLaren, D. J., 258, *269*
MacLaren, J. A., 166, 170, *193*
Maclaren, N. K., 146, *156*
McLaurin, W. D., 167, *196*
McLean, F., 90, *100*
McMurchie, E. J., 7, 9, 27, 35, *45*, 64, *73*
Macome, J. C., 176, *194*
Macrides, F., 125, *156*
McWilliams, D., 166, 167, *190*
Madeira, V. M. C., 64, *73*
Madsen, D., 292, *301*
Maeda, M., 131, *156*
Mager, S. C., 210, *226*
Maghuin-Rogister, G., 163, *194*
Maguire, M. H., 183, *196*
Mahmoud, A. A. F., 264, *266*, *269*
Maillard, J. L., 254, *269*
Majno, G., 231, *270*
Majumder, G. C., 207, *227*, *228*
Makita, M., 5, *47*
Malaisse-Lague, F., 140, *157*
Malaisse, W. J., 145, *156*
Malan, A., 33, 34, *44*
Malhotra, S., 293, *301*

Malmsten, C., 241, *270*
Maloof, F., 110, *156*
Mancuso, S., 174, 175, *191*
Mann, M. R., 184, *194*
Manning, M., 141, *155*
Manzo, L., 181, *189*
Maoz, A., 255, *269*
Marcal, J. M., 178, 179, *194*
Marchand, J., 145, *156*
Marchant, P. J., 183, *191*
Marino, A. A., 86, *100*
Marsden, C. D., 275, *285*
Marshall, H., 83, *101*
Marshall, J. H., 88, *100*
Marshall, J. M., 4, 34, *44*, *45*
Martal, J., 168, 169, *194*
Martin, B. J., 183, *196*
Martin, C. E., 69, *72*, *73*
Martin, D. L., 95, *100*
Martin, H. N., 4, *45*
Martin, J., 81, 82, 87, *99*, *100*
Martin, J. B., 109, 111, *156*, *158*
Martin, J. E., 152, *156*
Martinez-Victoria, E., 106, *158*
Martinson, F. A. E., 180, *194*
Martonosi, M. A., 52, 59, 66, *73*
Mason, C. W., 274, *285*
Mason, J. I., 184, *194*
Mason, W. T., 52, *73*
Massey, V., 6, *45*
Massopust, L. C. Jr., 33, *45*
Mastroianni, L., 187, *197*
Masuyama, Y., 131, *156*
Mather, I. H., 218, *227*
Mathews-Roth, M. M., 283, *285*
Matkovics, B., 180, *194*
Matossian, R. M., 257, *270*
Matsuzaki, F., 131, *156*
Matthews, J. L., 81, 82, 87, *99*, *100*
Matthies, D. L., 169, *194*
Matthijsen, R., 162, 163, 164, *197*
Mattingly, R. F., 118, *157*, 181, *195*
Maurer, R. A., 213, *227*
Maurer, W., 168, *192*
Mayer, G., 200, *227*
Mayer, R. J., 182, *194*
Mayhua, E., 123, *158*
Mayorek, N., 179, 181, 182, 184, *191*
Mead, J. F., 10, 17, *44*
Meder, J., 33, *45*
Meigs, R., 174, *195*

Meites, J., 216, *225*
Melancon, M., 95, *100*
Melchior, D. L., 7, *45*, 55, *73*
Melhorn, R. J., 7, 9, *45*, 64, *73*
Melick, R. A., 81, *101*
Mellanby, E., 98, *100*
Mena, F., 201, 219, *226*
Menaker, M., 152, *158*
Mendler, M., 2, 11, *44*, *45*
Menzel, P., 179, *194*
Mepham, T. B., 217, *227*
Mertelsmann, R., 182, *194*
Mertens, J. S., 292, *301*
Meschia, G., 160, *194*
Messer, H. H., 182, *194*
Messer, M., 204, *227*
Metcalfe, J. C., 58, 64, 65, 67, *73*
Mettler, F., 162, *190*
Metzger, B. E., 182, *196*
Meulepas, E., 116, *156*
Meyersteim, M., 70, *73*
Michaeli, D., 255, *269*
Michaelson, D. M., 36, *45*
Michel, J. F., 261, *269*
Midgett, R., 92, *101*
Miducci, M., 179, *195*
Migicovsky, B., 82, 94, 98
Mihilovic, Lj. T., 3, 4, *46*
Mikhail, G., 162, *197*
Mikulski, A., 84, *101*
Milhaud, G., 92, *100*
Miller, A. L., 183, *189*
Miller, A. M., 265, *269*
Miller, H. R. P., 245, 262, *269*
Miller, J. N., 262, *266*
Miller, N. G. A., 37, *45*
Miller, R. K., 184, *194*
Millman, M., 8, *44*, 60, 64, *72*
Milsum, J. H., 111, 142, *156*
Mino, M., 179, *194*
Mintz, D. H., 166, *196*
Mitnick, M., 173, *191*
Mitropoulos, K. A., 296, *301*
Miwa, I., 141, *157*
Miyahara, M., 119, *158*
Miyaji, S., 296, *301*
Miyayama, H., 178, 182, *191*
Mizutani, S., 183, *195*
Mochizuki, M., 164, 170, *189*, *197*
Moffat, G. F., 188, *195*
Mohrenweiser, H. W., 213, *225*

Mollenhauer, H. H., 218, *227*
Molloy, G. R., 212, 214, *226*
Moltz, H., 115, 126, *156*
Monder, C., 180, *194*
Moog, F., 132, *158*
Moor, R. M., 165, *189*
Moore, D. L., 264, *269*
Moore-Ede, M. C., 122, *158*
Moore, T., 284, *285*
Morand, P., 174, *194*
Morell, A. G., 164, *194*
Moreno, R. D., 53, 60, 61, *72*
Morgan, F. J., 163, 164, 171, *194, 195*
Morgan, I. G., 37, *42*
Morgan, J. I., 80, *101*
Mori, K. F., 162, *194*
Morin, R., 182, *190*
Morita, A., 110, *158*
Morita, S., 294, *301*
Morré, D. J., 210, 217, 218, *227*
Morris, G., 160, *194*
Morris, R. W., 10, 17, *45*
Morrison, D. C., 232, *269*
Morrison, W. R., 12, *45*
Mosbach, E. H., 291, *301*
Moscarello, M., 66, *73*
Moss, M. L., 79, 81, 83, *100*
Moss, R. L., *156*
Mossman, H. W., 160, *194*
Moss-Salentijn, L., 83, *100*
Moukhtar, M., 92, *100*
Mover, H., 110, *156*
Moyer, F. H., 277, *285*
Moyle, W. R., 164, *194*
Mrosovsky, N., 2, *45*
Müller-Eberhard, H. J., 232, 233, *266,*
268, 269, 271
Muench, K. H., 184, *195*
Muir, J. D., 260, *269*
Muirhead, M., 4, *43*
Mukherjee, G., 162, 183, *195, 196*
Muller, A., 122, *157*
Mullin, B. R., 128, *156*
Murad, T. M., 207, *228*
Muramatsu, T., 36, *43*
Murell, K. D., 258, *267*
Murillo, A., 106, *158*
Murphy, R. C., 241, *269*
Murray, F. A., 185, 186, *194, 196*
Murray, M., 245, 262, *269*

Murtaugh, P. A., 233, *268*
Musacchia, X. J., 5, 19, 20, 28, *45, 46,* 136,
137, *156*
Muschel, R., 242, *266*
Musoke, A. J., 238, 257, *268, 269*
Mustard, J. F., 233, *269*
Myant, N. B., 292, *301*
Myers, R. D., 3, 4, *46*
Myhrman, R., 258, 267, *270*

N

Nabet, P., 167, *189*
Naff, G. B., 232, 233, *270*
Nagano, M., 255, *271*
Nagataki, S., 131, *156*
Nahm, H. S., 163, *194*
Nair, P. P., 280, *285*, 293, *301*
Nandi, S., 206, *226*
Nathanielsz, P. W., 123, *158*
Naughton, R. A., 154, *156*
Neary, J. T., 110, *156*
Needham, A. E., 273, 275, 277, 278, 279,
280, *285*
Needleman, P., 54, *72*
Nelson, G., 5, 14, 26, 40, *46*
Nemanic, M. K., 201, 217, *227*
Neubauer, G., 86, *99*
Neuman, M. W., 85, *100*
Neuman, S., 179, 181, 182, 184, *191*
Neuman, W. F., 81, 85, *100, 101*
New, M. I., 124, *156*
Newbold, R. R., 127, *156*
Newman, R. A., 220, *226*
Newsholm, E. A., 24, *45*
Newton, J., 165, *197*
Niall, H. D., 92, *101*, 167, *194, 195*
Nichols, G., Jr., 11, *43*
Nickerson, M., 278, *285*
Nicol, S. C., 119, 120, *157*
Nicolaus, R. A., 275, *285*
Nicoll, C. A., 209, *227*
Nicolson, G. L., 54, *74*
Niemann, W. H., 164, 165, *192, 195*
Niki, A., 141, *157*
Niki, H., 141, *157*
Nishi, N., 108, *154*
Nishimura, M., 255, *271*
Nishioka, R. S., 209, *226*
Nisula, B. C., 171, 176, *195*

Niswender, G. D., 152, *156*
Nixon, D. A., 179, *190*
Nixon, W. E., 164, *195*
Noacco, O., 143, 144, *155*
Nogami, H., 84, *101*
Nolan, J., 80, *99*
Nolin, J. M., 208, *227*
Noll, B. W., 293, 294, *301*
Noller, K. L., 127, *157*
Noonan, K. D., 64, *73*
Nordin, B. E. C., 78, 95, *101*
Nordlie, R. C., 39, *45*, 53, *73*
Norman, A. W., 90, 92, 95, *101*
Norris, J. S., 124, *157*
Nossel, H. L., 232, 248, *270*, *271*
Nowakowski, M., 242, *266*
Nozawa, Y., 69, 72, *73*
Numa, S., 221, *228*
Numan, M., 126, *156*
Nuñez, E. A., 33, *43*

O

O'Brien, J. S., 183, *189*
O'Brien, L. V., 121, *156*
O'Brien, R. C., 3, 4, *45*
O'Byrne, E. M., 127, *158*
O'Connor, J. L., 174, *191*
Ögren, S., 240, *269*
Oertel, G. W., 179, *194*
Ogawa, M., 295, 299, *302*
Ogilvie, B. M., 254, 262, *267*, *268*, *269*
O'Grady, J. E., 188, *195*
Ogura, M., 294, *301*
Ohanian, C., 179, *195*
Oka, T., 206, 214, 215, 223, *226*, *227*, *228*
Okamoto, M., 121, *156*
Okazaki, T., 221, *228*
Okuda, J., 141, *157*
Oldfield, E., 57, *73*
Olson, J. A., 281, *285*
O'Malley, B. W., 121, *157*, 160, 168, 169, 186, *190*
Omdahl, J. L., 80, 92, 93, *101*
Orange, R. P., 238, 241, *266*, *269*, *270*
Orci, L., 140, *157*
Orlando, J., 132, *158*
Orr, T. S. C., 262, *266*
Osawa, Y., 184, *189*
Osmand, A. P., 258, *267*

Osmond, J. P., 258, *270*
Osuga, T., 293, 295, *302*
Ota, T., 179, 182, *191*
Otto, W., 133, 134, 135, 136, *158*
Ottolenghi-Nightingale, E., 279, *285*
Overath, P., 65, *73*
Owen, M., 81, *101*
Owen, O. E., 106, *155*
Owens, K., 26, *45*
Oya, M., 183, *195*

P

Pace, N., 5, *44*
Packer, L., 64, *71*, *74*
Packham, M. A., 233, *269*
Packter, N. M., 5, 24, *45*
Pacuszko, T., 128, *156*
Paige, D. M., 204, *227*
Palade, G., 213, 218, *227*
Palkovits, M., 109, *155*
Palmer, J., 108, 109, 142, 143, *155*, *156*
Pantaleon, J., 246, *270*
Panuska, J. A., 3, 4, *46*
Papahadjopoulos, D., 14, 35, 36, 38, *44*, *45*, 64, 66, 67, 72, *73*
Papkoff, H., 163, 165, *195*, *196*, 209, *226*
Paradis, M., 17, *44*
Pardani, R., 106, *157*
Pare, P., 168, *191*
Parenti-Castelli, G., 7, 27, *44*, 64, *73*
Park, N., 97, *100*
Parks, J. G., 6, 9, *47*, 64, *74*
Parmentier, G. G., 292, *301*
Parrish, D. B., 283, *285*
Parrish, J. A., 283, *285*
Parson, D. F., 58, *72*
Parsons, J. A., 91, 92, *100*, *101*
Pascher, I., 40, *45*
Patel, Y., 109, *157*, *158*, 216, *228*
Pathak, M. A., 283, *285*
Patillo, R. A., 118, *157*, 171, 181, *192*, *195*
Patnayak, B. C., 5, *45*
Patrick, H., 91, *98*
Patton, S., 213, *227*
Pauerstein, C. J., 160, *195*
Paulsrud, J. R., 5, *45*
Pavel, S., 148, 150, 151, *157*
Pawlowski, Z., 246, *269*

Payne, D. W., 206, 227
Paysant, P., 167, 189
Peaker, M., 202, 204, 214, 217, 227
Pearlman, W. H., 206, 227
Pearson, O. H., 127, 156, 166, 196
Pease, D. C., 87, 98
Pease, M., 282, 285
Peers, J. H., 246, 263, 271
Peeters, B., 166, 197
Pehlke, D. M., 180, 193
Pellegrino, J., 258, 269
Pelletier, G., 106, 157
Peltier, A., 167, 189
Peng, L.-H., 206, 227
Pengelley, E. T., 2, 3, 26, 42, 45
Penneys, N. S., 184, 195
Pennock, J., 95, 99
Penrose, L. S., 279, 285
Pentchev, P. G., 183, 195
Pepper, O. S., 233, 268
Perault, A., 92, 100
Perchellet, J. P., 113, 157
Perez, H., 258, 269
Perris, A. D., 80, 101
Perry, J. S., 160, 176, 192, 195
Perry, J. W., 223, 227
Person, P., 154, 156
Peschle, C., 154, 157
Peters, P. A., 264, 269
Peterson, W. F., 178, 179, 183, 191
Peticolas, W. L., 57, 73
Petit, V. A., 55, 64, 73
Petras, K., 258, 267
Petrescu-Holban, R., 148, 157
Petro, Z., 174, 195
Pettinger, W. A., 121, 154, 155
Pfaff, D. W., 111, 157
Phelps, P., 140, 157
Phillips, M. C., 58, 73
Picone, J. P., 257, 269
Pictet, R. L., 140, 157
Pierce, J. G., 163, 171, 192, 195
Pierrepoint, C. G., 184, 193
Piliero, S. J., 154, 156
Pincus, G., 185, 187, 195
Pion, R. J., 175, 193
Pipeleers, D. G., 140, 157
Pipeleers-Marchal, M. A., 140, 157
Pitelka, D. R., 201, 217, 227
Pitt, G. A. J., 281, 285

Platt-Aloia, K., 33, 42
Plattner, W. S., 5, 19, 20, 45, 46, 47
Plotnik, L. P., 117, 156
Plucinski, T. M., 207, 227
Pollard, W. E., 186, 187, 190
Polnaszek, C. F., 5, 42
Pontieri, G. M., 262, 271
Porta, E. A., 280, 285
Porter, J. W., 220, 227
Portman, O. W., 293, 295, 302
Posner, A. S., 85, 98, 99, 101
Posner, B. I., 115, 157, 209, 227
Potts, J. T. Jr., 91, 92, 101, 138, 156
Powell, J. T., 217, 227
Poynter, D., 244, 245, 269
Prange, I., 292, 302
Preslock, J. P., 183, 193
Pretlow, T. G., 207, 228
Prichard, R. K., 260, 270
Priedkalns, J., 160, 195
Prockop, D. J., 83, 99
Prosser, C. L., 10, 43
Prusiner, S., 280, 285
Psychos, A., 187, 195
Puett, D., 164, 192
Pugliarello, M., 87, 101
Purandare, M., 183, 197

Q

Qasba, P. K., 216, 227
Qazi, M. H., 162, 195
Quirk, J. M., 183, 196
Qureshi, A. A., 220, 227

R

Raddle, I. C., 183, 196
Raison, J. K., 6, 7, 9, 19, 27, 35, 44, 45, 61, 64, 73, 74
Raisz, L. G., 97, 101
Rajagopol, N. B., 284, 285
Rall, L. B., 140, 157
Rama, F., 179, 195
Ramachandram, K., 284, 285
Ramalho-Pinto, F. J., 258, 268
RamaSastry, B. V., 181, 195
Ramey, J. N., 128, 129, 157
Ramp, W. K., 81, 100
Ramsey, E. M., 160, 195

Ranadive, N. S., 260, 269
Rand, R. P., 38, 46
Randall, P. K., 118, 154
Rang, H. S., 182, 196
Ranjhan, S. K., 283, 285
Rao, S. S., 183, 197
Rasmussen, H., 87, 94, 101
Raths, P., 1, 4, 33, 43
Ratledge, G., 10, 46
Ratnoff, O. D., 232, 233, 270
Rattazi, M. C., 183, 197
Rayford, P. L., 165, 197
Rebolledo, O., 145, 156
Redman, C. M., 214, 227
Reed, B. D., 61, 62, 74
Rees, L. H., 172, 195
Rees, P. H., 264, 266
Reichel, W., 280, 285
Reichlin, S., 216, 228
Reid, R. J., 187, 195
Reimo, T., 163, 195
Reinert, J. C., 7, 46
Reiter, R. J., 150, 157
Reithel, F. J., 202, 210, 225, 228
Reitsma, H. J., 64, 72
Relacion, J. R., 179, 182, 191
Renaud, L., 111, 156
Renfree, M. B., 187, 195
Renold, A. E., 140, 157
Renooij, W., 55, 73
Rent, R., 258, 267, 270
Revak, S. D., 232, 267
Reynolds, J. J., 92, 95, 98, 101
Rich, A., 138, 156
Rich, C., 93, 98
Rickard, M. D., 257, 270
Richardson, K. C., 201, 228
Richelle, L. J., 94, 101
Riddle, M., 214, 228
Riggs, B. L., 97, 100
Rillema, J. A., 208, 228
Ringler, I., 167, 191
Ripatti, P. O., 293, 294, 302
Rittenhouse, H. G., 64, 73
Rivera, E. M., 206, 207, 223, 228
Rivier, C., 113, 114, 158
Rivier, J., 106, 108, 155
Rizkallah, T., 162, 197
Roberts, J. A., 262, 266
Roberts, J. C., 38, 42

Roberts, R. M., 186, 187, 190, 195, 196
Roberts, W., 97, 100
Robertson, A., 181, 195
Robertson, H. A., 168, 181, 190, 193
Robertson, J. D., 54, 74
Robertson, M. C., 169, 170, 193, 195
Robertson, W. G., 78, 92, 101
Robichon, J., 82, 94, 98
Robinson, A. G., 105, 158
Robinson, C., 92, 100
Robinson, J. C., 178, 179, 180, 181, 182, 183, 184, 190, 191
Robison, G. A., 94, 101
Rocha E Silva, M., 232, 249, 270
Rodan, G. A., 78, 80, 101
Rodbard, D., 170, 190
Roelofsen, B., 55, 73, 74
Rohrlick, R., 55, 72
Rollag, M. D., 152, 156
Rombauts, W., 166, 197
Romero, J., 150, 157
Ronai, A., 64, 66, 75
Roots, B. I., 10, 17, 43, 46
Rose, N. R., 136, 158
Rosemeyer, M. A., 210, 226
Rosen, J. M., 213, 214, 215, 228
Rosen, O. M., 209, 223, 228
Rosenbloom, A. L., 145, 157
Rosenblum, I. Y., 167, 194
Rosenfeld, M. G., 212, 213, 225
Roskoski, L. M.., 181, 195
Roskoski, R., 181, 195
Rosner, J. M., 188, 191
Ross, C. T., 163, 164, 194
Ross, E., 181, 195
Ross, G. T., 163, 164, 165, 176, 195, 197
Ross, J. T., 117, 156
Rossini, A. A., 141, 157
Roth, J., 145, 155
Roth, T., 112, 155
Rothman, J. E., 55, 56, 58, 72, 73, 74, 218, 228
Rothwell, T. L. W., 260, 261, 268, 270
Rottem, S., 58, 62, 65, 71, 74
Rouser, G., 5, 10, 11, 12, 14, 15, 17, 24, 26, 31, 32, 40, 41, 42, 43, 46, 55, 58, 62, 65, 71
Rousseaux, R., 264, 266, 267
Rowland, N., 119, 147, 158
Rowland, R. E., 83, 101

Rowlands, I. W., 165, *195*
Roy, M., 154, *156*
Rubenstein, A., 242, *266*
Rubin, C. S., 209, 223, *228*
Rubin, E., 124, *157*
Rubin, R. H., 264, *270*
Ruch, W., 108, 109, *156*
Ruckert, A. C. F., 118, *157*, 181, *195*
Rudolph, L., 160, *195*
Rundell, J. W., 185, *194*
Runrich, G., 93, *99*
Rupp, G. R., 165, *190*
Russell, P. F., 242, 253, 263, *267*
Rutter, W. J., 140, *157*
Ryan, G. B., 231, *270*
Ryan, K. J., 173, 174, 175, 176, 184, *189,*
  195, 196

S

Saacke, R. G., 213, *227*
Saarikoski, J., 33, *46*
Saarikoski, S., 180, *190*
Saarikoskic, D. L., 5, *46*
Saavedra, J. M., 109, *155*
Sacerdote, F. L., 183, *189*
Sadun, E. H., 254, *270*
Saenger, P., 124, *156*
Saintot, M., 177, *191*
Sairam, M. R., 163, *195*
Salam, A., 166, *196*
Salentijn, L., 81, *100*
Salhanick, H. A., 172, *197*
Saller, C. F., 119, 147, *158*
Salmon, W. D., 96, *101*
Salvador, J., 174, *194*
Samaan, N., 166, *196*
Samuels, H. H., 131, *157*
Samuelson, B., 241, 257, 268, *270*
Samuelsson, B. E., 40, *44*
Santoro, F., *266*
Sanyal, M. K., 176, *196*
Saperstein, R., 216, *228*
Sar, M., 116, *157*
Sarasin, R., 245, *267*
Sarkar, G. C., 183, *196*
Sastry, R. V. B., 182, *193*
Sato, H., 108, *154*
Satoh, K., 184, *196*
Sauer, R., 167, *194*

Sauerheber, R. D., 38, *46*
Savage, A. M., 265, *267*
Saxena, B. B., 117, *156, 157*, 162, *196*
Sayers, G., 110, 113, *157*
Scalise, M. M., 207, *228*
Scarpace, P. J., 81, *101*
Schalekamp, M. A. D. H., 112, *157*
Schally, A. V., 106, 107, 108, *154, 157*
Schams, D., 165, *196*
Schedl, H. P., 146, *157*
Scheer, B. T., 104, 145, 146, *157*
Scheik, R., 65, *72*
Scheinberg, H. I., 164, *194*
Schenk, F., 86, *99*
Schirmer, H., 7, 47, 64, *75*
Schlegel, W., 179, *196*
Schlosnagle, D. C., 187, *196*
Schmid, F. G., 140, *155*
Schmidt-Gollwitzer, M., 117, *157*, 162,
  196
Schmidt-Ullrich, R., 55, *74*
Schneider, A. B., 167, *196*
Schneider, J., 160, *190*
Schneider, L. E., 146, *157*
Schneider, M. J., 10, 17, *45*
Schneider, W. P., 54, *74*
Schnoes, H., 90, *100*
Schoener, N. W., 54, *74*
Schoenfield, L. J., 293, *302*
Schofield, B. H., 82, *99*
Schomberg, D. W., 175, 176, *196*
Schrader, W. T., 121, *157*
Schreiber, A. D., 254, *269*
Schreier, M. H., 212, *228*
Schuijff, A., 58, *71*
Schultz, M. G., 246, *269*
Schultz, R. L., 180, 182, 183, *196*
Schussler, G. C., 132, *158*
Schuster, V. L., 7, *43*
Schwartz, B., 247, 249, *270*
Sciarra, J. J., 162, 166, *189, 193, 196*
Scislowski, P., 180, *196*
Scow, R. O., 222, *228*
Scribney, M., 52, *74*
Sealander, J. S., 5, *44*
Seamark, R. F., 176, *191*
Sechi, A. M., 7, 27, *44*, 64, *73*
Seelig, J., 66, *75*
Segaloff, A., 127, *156*
Segré, G. V., 167, *195*

Selenkow, H. A., 162, 171, *194, 197*
Sen, D. P., 172, *193*
Sengupta, S., 38, *46*
Senyk, G., 255, *269*
Seris, M., 160, *193*
Shadden, J., 141, *157*
Shafrir, E., 179, 181, 182, 184, *191*
Shalaby, M. R., 118, *157*
Shambaugh, G. E., 182, *196*
Shami, Y., 182, 183, *194, 196*
Shane, B., 182, 183, *190*
Shanker, G., 113, *157*
Shapiro, D., 183, *195*
Shapiro, I. L., 293, *301*
Sharma, R. K., 113, *157*
Sheffer, A. L., 256, *271*
Shelesnyak, M. C., 188, *196*
Shenkman, L., 130, *158*
Sher, A., 259, 264, 265, *266, 270, 271*
Sherman, M. I., 178, 179, 182, *194, 196*
Sherwood, L. M., 167, *196*
Sheth, A. R., 183, *197*
Shimamoto, K., 119, *158*
Shimshick, E. J., 8, 9, 36, *46,* 66, *74*
Shinitzky, M., 40, 41, *46*
Shipley, G. G., 40, *46*
Shirakawa, H., 183, *193*
Shiratori, T., 282, *284*
Shiu, R. P. C., 115, *158,* 169, 170, *193,* 208, 209, *228*
Shoda, M., 296, *302*
Shome, B., 168, 170, *196*
Short, R. V., 175, 176, *196*
Shrago, E., 19, *44*
Shug, A. L., 19, *44*
Shusten, R. C., 215, *228*
Shyämalä, G., 206, *228*
Siakotos, A. N., 12, 15, 17, *46*
Sidorov, V. S., 293, 294, *302*
Siegel, J., 258, *267, 270*
Siiteri, P. K., 184, *196*
Silberberg, M., 97, *101*
Silberberg, R., 97, *101*
Silva, M. E., 240, *270*
Silverman, D. A., 136, *158*
Silverman, P. H., 246, 253, *270*
Silvers, W. K., 124, *158*
Silvius, J. R., 61, 62, *74*
Sim, M. K., 183, *196*

Simon, G., 5, 10, 11, 14, 26, 31, 32, 40, *42, 46*
Simon, R. G., 32, *44*
Simonson, L. P., 65, *71*
Simpson, G. G., 289, *302*
Sinclair, K. B., 246, *270*
Sinensky, M., 17, *46*
Sineriz, F., 53, 60, 61, *72*
Singer, S. J., 54, *74,* 210, *228*
Singh, H., 281, *285*
Singh, S. K., 252, *266*
Sivitz, M. C., 106, *155*
Sjövall, J., 289, 291, 292, 293, *301, 302*
Skeggs, L. T., 152, *158*
Skett, P., 115, *155*
Skor, D., 258, *267*
Skou, J. C., 34, *46*
Skriver, L., 69, *72, 73*
Slack, E., 91, *99*
Slater, E. C., 64, *75*
Sloane-Stanley, G. H., 11, *43*
Slotboom, A. J., 36, *46,* 53, *72*
Small, C. W., 180, *196*
Small, D. M., 40, *46*
Smith, A. J., 247, *269*
Smith, F. R., 281, *284*
Smith, G. A., 58, 67, *73*
Smith, G. K., 113, *155*
Smith, G. W., 123, *158*
Smith, J. R., 110, *158*
Smith, L. M., 12, *45*
Smith, M. W., 37, *45*
Smith, O. W., 171, *194*
Smith, R. E., 26, *42*
Smith, R. G., 182, *194*
Smith, S., 201, 220, 221, 222, *227, 228*
Smithers, S. R., 258, *269*
Smyth, J. D., 245, 253, 257, *270,* 300, *302*
Smythe, G. A., 111, 152, *158*
Sobrevilla, L., 174, *196*
Soliman, H., 91, *99*
Solomon, D. S., 178, 179, *194*
Solomon, R. D., 41, *46*
Somero, G. N., 53, 61, *73, 74*
Sonozaki, H., 255, *271*
Soodak, M., 110, *156*
Soter, N. A., 238, 240, *270, 271*
Soule, P. C., 246, *270*
Soulsby, E. J. L., 254, *269*
South, F. E., 3, 4, 28, 38, *43, 46, 47*

Southren, A. L., 124, *157*
Spaulding, S. W., 128, 129, *157*
Spellman, C., 182, *196*
Spellman, G. G., 248, *270*
Speroff, L., 128, 129, *157*
Speth, V., 66, 68, *75*
Spicer, S. S., 183, *196*
Spragg, J., 240, 249, 250, *271*
Sprecher, H., 181, *193*, *195*
Sprent, J. F. A., 248, *270*
Spurrier, W. A., 4, *43*
Squire, G. D., 186, *196*
Srivastava, S. K., 187, *196*
Staehelin, 212, *228*
Stancliff, R. C., 64, *74*
Starka, L., 175, *196*
Starling, E. H., *155*
Starnes, W. R., 171, *192*
Start, C., 24, *45*
Stauffer, M., 93, *98*
Stawiski, M. A., 122, *156*
Stechschulte, D. J., 238, 241, *269*, *270*
Steele, R. E., 117, *156*
Steen, G. O., 40, *44*
Steffen, D. G., 19, 20, *45*, *46*
Stehlin, J., 140, *156*
Steim, J. M., 7, *45*, *46*, 52, 55, 59, 66, *71*, *73*
Stein, O., 217, *228*
Stein, T. P., 147, *158*
Stein, Y., 217, *228*
Steiner, A. L., 128, *155*
Steinetz, B. G., 127, *158*
Stenberg, A., 124, *156*
Sterling, K., 131, *158*
Stevens, V. C., 168, 169, *191*, *193*
Stewart, D. F., 245, *270*
Stirewalt, M. A., 252, *270*
Stivala, S. S., 250, *267*
Stockdale, F. E., 207, *227*
Stone, R., 213, *227*
Stoner, G., 242, *266*
Story, M. T., 118, *157*
Strand, F. L., 110, *158*
Street, M. A., 179, *193*
Strejan, G., 262, *268*
Stricker, E. M., 119, 147, *158*
Strickhart, F. S., 65, *75*
Strong, R. M., 278, *285*
Strumwasser, F., 33, 34, *46*

Stumpf, W. E., 116, *157*
Sturrock, R. F., 264, *266*
Sturtevant, J. M., 7, *46*, 58, *72*
Suldanha, V. F., 131, *158*
Sulman, F. G., 216, *228*
Sulzman, F. M., 122, *158*
Sumar, J., 123, *158*
Suomalainer, P., 5, *46*
Surynek, J., 283, *285*
Sussman, H. H., 178, 182, *189*
Sutherland, E. W., 94, *101*
Suwa, S., 168, *191*, *196*
Suyama, T., 183, *193*
Svensson, J., 241, 257, *268*, *270*
Swallow, D. M., 182, *196*
Swaminathan, N., 163, 164, *190*, *196*
Swan, G. A., 275, *285*
Swan, H., 1, 2, 5, *46*
Swanson, A. A., 183, *196*
Swanson, P. D., 5, *46*
Sweatman, G. K., 246, *270*
Sweeley, C. C., 291, *301*
Sweeney, C. M., 163, *194*
Swiercyznski, J., 180, *196*
Sybulski, S., 176, *196*
Symonds, H. W., 80, *101*
Szabó, M., 129, 130, *155*
Szczesna, E., 166, 167, *190*
Szymanski, E. S., 223, *225*

T

Tabacik, C., 180, *196*
Tabei, T., 176, *196*
Taborsky, G., 223, *228*
Tait, J., 4, *46*
Takai, T., 179, *194*
Takuma, T., 296, *302*
Talamantes, F. Jr., 169, 170, *196*
Talbert, L. M., 206, *227*
Taljedal, I. B., 140, *156*
Tallman, J. F., 183, *196*
Talmage, R. V., 81, 90, 99, *101*
Tamari, M., 295, 299, *302*
Tamarkin, L., 152, 153, *158*
Tammar, A. R., 289, 290, 291, 292, 293, 294, 295, 296, 297, 299, *301*, *302*
Tanabe, T., 221, *228*
Tanaka, N., 293, 295, *302*
Tanaka, R. 14, 15, 35, *46*, 64, *74*
Tane, M., 164, *189*

Tang, S. C., 130, 145, *156, 158*
Taniguchi, K., 64, *74*
Tannenbaum, M., 105, *158*
Tanner, C. E., 257, *268*
Tanner, J. M., 96, *101*
Tashima, L. S., 26, *46*
Tashjian, A. H., 94, *101*
Tata, J. R., 97, *101*
Tataryunas, T., 283, *285*
Tateyama, T., 292, *302*
Tawfik, S., 251, *270*
Taylor, A. L., 166, *196*
Taylor, A. N., 81, *101*
Taylor, D. M., 84, *101*
Taylor, T. G., 85, *101*, 133, 134, 135, 136, *158*
Taylor, W., 291, 292, 293, 294, 296, *302*
Telegdy, T. G., 174, *197*
Tempel, G., 19, 20, 28, *45, 46*
Termine, J. D., 85, *101*
Teroaka, H., 297, *301*
Terry, L. C., 109, *158*
Teruya, A., 14, 15, *46*, 64, *74*
Thakur, A. N., 183, *197*
Thanavala, Y. M., 183, *197*
Thomas, C. M. G., 179, *197*
Thomas, J. W., 213, *225*
Thomas, L., 98, *99*
Thompson, A. R., 260, 261, *271*
Thompson, E. A., 184, *196*
Thompson, G. A. Jr., 69, *72, 73*
Thompson, K. G., 117, *156*
Thomson, A. M., 160, *193*
Thomson, R. H., 275, *285*
Thomson, W. W., *45*, 64, *73*
Thorn, N. A., 80, *101*
Thorp, R. B. Jr., 207, *228*
Thorpe, R. F., 10, *46*
Thorson, R. E., 248, *271*
Thurston, H., 154, *155*
Tic, L., 188, *196*
Tindal, J. S., 201, 216, 223, *225, 228*
Tirri, R., 65, *72*
To, R., 64, 65, *71*
Tocanne, J. F., 38, *47*, 66, *74*
Tojo, S., 164, 170, *189, 197*
Tokura, Y., 164, *189*
Tolone, G., 262, *271*
Tomacari, R. L., 117, *155*
Tonelli, G., 167, *191*

Toon, P. A., 64, 65, *73*
Topper, Y. J., 206, 207, *226, 227, 228*
Torain, B. F., 220, *227*
Torisu, M., 255, *271*
Torpier, G., *266*
Torres, M. T., 174, *193*
Towers, N. R., 64, *74*
Townsley, J. D., 183, *197*
Trauble, H., 65, *73, 74*
Trauble, J., 67, *74*
Treacher, R. J., 80, *101*
Tregéar, G. W., 167, *195*
Tricone, A., 282, *285*
Troehler, V., 80, *99*
Troen, P., 176, *196*
Trucco, R. E., 53, 60, 61, *72*
Trzeciak, W. H., 183, *193*
Tsai, D. K., 55, *74*
Tsai, H. C., 92, *101*
Tsai, J. S., 131, 133, *157, 158*
Tsang, V. C. W., 250, 251, *271*
Tsibris, J. C. M., 187, *196*
Tsuchima, T., 170, *193*
Tsukugoshi, N., 8, 9, 15, *43, 46*
Tullner, W. W., 164, 164, 172, *192, 195, 197*
Turek, F. W., 152, *158*
Turkington, R. W., 205, 206, 207, 208, 210, 214, *227, 228*
Turner, C. D., 104, *158*
Turner, C. W., 201, *228*
Twarog, F. J., 242, *271*

U

Uchimura, H., 131, *156*
Uhlenbruck, G., 220, *226*
Ullrich, K., 93, *99*
UmaBai, R., 176, *194*
Unger, R. H., 140, *157*
Urbano, P., 258, *267*
Urbina, J., 8, 36, *42*
Urist, M. R., 84, 96, *101*
Urtasum, M., 176, *190*
Utiger, R. D., 109, *155, 158*, 170, *190*
Uvnäs, B. 260, *271*

V

Vaes, G., 94, *101*
Vaitukaitis, J. L., 163, 164, 176, *194, 195, 197*

Vale, W., 106, 108, 113, 114, *155*, *158*
Valentine, M. D., 256, *271*
Valone, F. H., 241, *271*
Valotta, E. H., 233, *271*
Vanaman, T. C., 210, 214, *228*
Van Deenen, L. L. M., 9, 38, 41, *43*, *47*, 55, 57, 58, 60, 65, 66, *71*, *72*, *73*, *74*
Vanderkooi, G., 67, *72*
vander Neur-kak, E. C. M., 15, *45*
van der Sluys Veer, J., 80, *98*
Van Der Weyden, M. B., 184, *197*
Van de Wiele, R. L., 162, *197*
Van Dijck, P. W. M., 9, *47*, 57, 58, 60, *71*, *72*, *74*
Vanduffel, L., 166, *197*
Van Dyke, A. H., 169, *192*
Van Golde, L. M. G., 55, *73*
Van Hall, E. V., 163, *197*
Van Hell, H., 162, 163, 164, *197*
Van Leusden, H. A., 174, *197*
Van Loon, C., 181, *196*
Vannucchi, S., 258, *267*
Van't Hoff, W., 171, *194*
Van Tienhoven, A., 159, 173, *197*
Varangot, J., 176, *190*
Varma, K., 162, *197*
Varute, A. T., 246, 262, 263, *271*
Vassey, D. A., 64, *72*
Vaughan, E. J. Jr., 154, *155*
Vaughan, G. M., 148, 149, *158*
Vaughan, J. M., 78, 81, 83, 88, 89, 91, 93, 94, 97, *101*, *102*
Vaughan, M. K., 148, 149, 150, *157*, *158*
Veerkamp, J. H., 179, *197*
Vega, D. F., 106, *158*
Veis, A., 83, *101*
Velasco, E. G., 182, *197*
Velle, W., 179, *192*
Venkataraman, R., 202, *228*
Vensel, W. H., 240, *268*
Ventruto, V., 124, *156*
Ver Eecke, T., 166, *197*
Verger, C., 64, *75*
Verkleij, A. J., 38, *47*, 55, 66, *74*
Verkleiu, A. J., 65, *72*
Verkley, A. J., 58, *74*
Verma, S. P., 55, *74*
Vernberg, F. J., 10, 17, *42*
Verne, J., 280, *285*
Ververgaert, P. H. J. Th., 38, *47*, 58, 65, 66, *72*, *74*

Vessey, D. A., 8, *43*, *47*
Vevers, H. G., 273, *284*
Villalba, M., 183, *196*
Villee, C. A., 174, 176, *196*, *197*
Villee, D. B., 174, 175, *197*
Vincent, C. K., 185, *194*
Vincent, J., 82, 94, *98*
Vinik, A. I., 168, *197*
Viret, J., 38, *47*, 65, *74*
Vitter, F., 87, *101*
Vladutiu, G. D., 183, *197*
von Euler, U. S., 118, *158*
von Lichtenberg, F., 259, *271*
Voorhees, J. J., 122, *156*
Vreeburg, J. T. M., 124, *155*

W

Wachtel, S. S., 124, *156*, *158*
Wada, K., 221, *228*
Wade, R. S., 181, *197*
Wagland, B. M., 260, *267*
Waight, R. D., 180, *190*
Waite, M. B., 182, *191*
Walfish, P., 171, *192*
Walike, B. C., 142, 143, *155*
Walker, C., 55, *72*
Wallace, H. D., 185, 186, *193*, *194*
Wallach, D. F. H., 55, *74*
Wallace, H. W., 147, *158*
Wallis, J., 33, *45*
Walsh, L. B., 293, 294, *301*
Walsh, P., 232, *271*
Walshaw, R., 262, *269*
Wan, H., 240, *269*
Wan, L., 130, *158*
Ward, D. N., 163, *194*
Ward, P. A., 233, 255, *271*
Warm, M., 162, *197*
Warnick, A. C., 185, *194*
Warren, G. B., 58, 64, 65, 67, *73*
Warren, K. S., 250, 264, *267*, *269*
Wasserman, R. H., 81, *101*
Wasserman, S. I., 238, 239, 240, 241, 242, 266, *271*
Watkins, W. B., 180, *196*
Watson, K., 7, *47*, 64, *71*, *74*
Weeke, J., 128, *158*
Weglicki, W. B., 280, *285*
Weil, P. E., 247, *271*
Weinberg, M., 247, *271*

Weinger, J. M., 81, *100*
Weinstein, P. P., 246, 263, *271*
Weise, H., 162, 163, *192*
Weiss, A. S., 232, *271*
Weiss, G., 127, *158*
Weiss, H. J., 241, *271*
Wells, H. J., 5, *47*
Wells, P. D., 245, 262, *271*
Wells, W. W., 5, *47*
Welsh, D., 181, *195*
Wergedal, J., 93, *98*
Werthier, R. E., 11, *43*
West, L., 141, *155*
Weston, P. D., 98, *102*
Wexler, R., 204, *227*
Whaley, W. G., 201, 210, *228*
White, A., 217, *227*
White, B., 65, *73*
White, H. B., III., 55, *72*
Whittaker, V. P., 26, *43*
Whitten, B. K., 26, 39, *44*
Whitworth, N. S., 208, *226*
Wide, L., 165, *192*, *197*
Wiener, M., 176, 179, *197*
Wilkinson, R., 80, 92, *102*
Williams, B. A., 3, 4, *46*
Williams, D. D., 5, *47*
Williams, G., 290, 293, *301*
Williams, J. F., 238, 246, 250, 257, 258, 262, 263, *266*, *267*, *268*, *269*
Williams, M. A., 64, *74*
Williams, R. E., 64, *74*
Williams, R. M., 57, *73*
Williamson, D. G., 174, *194*
Williamson, M., 88, *102*
Willimott, S. G., 283, *285*
Willis, J. S., 4, 5, 9, 10, 28, 34, 35, 39, 40, *43*, 51, *72*
Willoughby, J. O., 109, *158*
Wilner, G. D., 232, *271*
Wilson, J. D., 117, 118, *155*
Wilson, M., 258, *270*
Wimsatt, W. A., 160, 185, 187, *197*
Windon, R. G., 245, *267*
Winegrad, A. I., 179, *190*
Winokur, A., 109, *158*
Winter, P. A. D., 293, *301*
Wise, G. H., 283, *285*
Wisnieski, B., 6, 9, *47*, 64, *74*
Wistar, R., 258, *267*
Witorsch, R. J., 208, *227*

Witschi, E., 160, *197*
Witting, L. A., 23, *47*, 53, *74*
Wittke, B., 7, *44*, 64, *73*
Wittliff, J. L., 206, *226*
Wlodawer, P., 10, 17, *42*
Woessner, J. F., 98, *102*
Wold, J. K., 260, *271*
Wolf, D. P., 187, *197*
Wolin, L. R., 33, *45*
Woo, S. L., 186, *190*, 213, 214, *228*
Wood, L., 7, *47*, 64, *74*
Wood, T. G., 215, *228*
Wood, W. A., 213, *225*
Woodard, H. Q., 85, *102*
Wootton, V. M., 290, 295, *301*
Wostmann, B., 292, *301*
Wright, I. G., 252, *266*
Wright, K. L., 8, 15, *45*
Wuepper, K. D., 232, *267*
Wunderlich, F., 64, 66, 68, *75*
Wuthier, R. E., 26, *47*
Wyngaarden, J. B., 181, *192*

Y

Yadley, R. A., 170, *190*
Yamada, M., 183, *193*
Yamaguchi, T., 179, *194*
Yamamoto, A., 11, 12, 15, 17, 24, *46*
Yamamoto, Y., 295, 299, *302*
Yamasaki, K., 294, 295, 296, 299, *301*, *302*
Yang, C. S., 65, *75*
Yarus, M. J., 186, *194*
Yeh, K.-Y., 132, *158*
Yellin, N., 7, *47*
Yen, S. C. C., 166, *196*
Yen, S. S. C., 166, *192*
Yip, C. L., 145, *157*
Yip, M. C. M., 223, *228*
Yochim, J. M., 188, *197*
York, D. A., 133, 134, 135, 136, *158*
Yoshino, M. O., 183, *195*
Young, V. R., 80, *102*
Yousef, I. M., 293, 294, *302*
Yousef, M. K., 293, 294, *302*
Ystefan, F., *157*
Yurt, R. W., 240, 249, 250, *271*

Z

Zakim, D., 8, *43*, *47*, 64, *72*
Zara, D. T., 127, *158*

Zarco, R. M., 254, *269*
Zarrow, M. X., 172, *197*
Zatzman, M. L., 28, *47*
Zeiger, R. S., 242, *271*
Zelewski, L., 174, *197*
Zeylemaker, W. P., 64, *75*

Zimmer, G., 7, *47*, 64, *75*
Zimmerman, E. A., 105, *158*
Zimney, M., 33, *47*
Zumbusch, L., 297, *302*
Zwaal, R. F. A., 55, *73*, *74*
Zwierzchowski, L., 206, *225*

Subject Index

A

Acetylcholine esterase, 64
Acetyl-CoA carboxylase, 184, 214
Acetylesterase, 182
N-Acetylglucosamine, 210
β-N-Acetylglucosaminidase, 183
Acholeplasma laidlawii, 50, 57, 62
Achromotrichia, 279
Acid phosphatase, 182
Acinonyx jubatus, 294
Acinus, 201
Aconitate hydrase, 184
Actinomycin D, 93, 94
Acylphosphatase, 184
Addison's disease, 279
Adenocarcinoma, mammary, 206
Adenosine deaminase, 183
3',5'-Adenosine monophosphate, cyclic,
    113, 128–130, 140
Adenylate cyclase, 64, 93, 184
    kinase, 182
Adenylosuccinate synthetase, 184
Adipose tissue, 200, 201
    brown, 2, 4
    white, 2
Adrenal cortex, 107, 112, 120–123, 205
    cortical hormone, 207
    glands, 118–123
    lipid, 283
    medula, 118–120
Adrenalectomy, 134, 135
Adrenaline, 118
Adrenocortical steroids, 97
Adrenocorticotropic hormone, 107, 112,
    113, 114, 122, 205
Adrenocorticotropin, 171–173
Adenohypophysis, 105, 111–118
Alanine aminotransferase, 181
Albino, 279
Alcohol, 124
    dehydrogenase, 179
Aldose reductase, 179
Aldosterone, 121–122, 154, 206
Alkaline phosphatase, 182
Allocholic acid, 287, 290, 291, 293–299

Allodeoxycholic acid, 291
Allolithocholic acid, 293
Allomelanin, 275
Alloxan, 141, 146
Alpaca, Andean, 123
Alpha cell, 144–145
Amblonyx cinerea, 294
Ameloblast, 79
Amenorrhea, secondary, 116
Amidophosphoribosyltransferase, 181
Amine oxidase, 180
α-Aminoisobutyrate transport, 9
Aminopeptidase, 183
Ammospermophilus leucurus, 293
Anaphylatoxin, 235
Anaphylaxis, slow reacting substance of,
    241
Ancyostoma, 248
    caninum, 247–248, 252
Androgen, 112, 117, 123–125
Androstrenedione, 174
Androsterone, 123
Angiostrongylus vasorum, 248
Angiotensin, 122, 152–154
8-Anilino-1-naphthalene sulfonate
    fluorescence, 63
Anticoagulant principle, 251
Antidiuretic hormone, 105
Arachidonic acid, 15, 17, 19, 22, 37, 54
Arginase, 183
Arginine, 108–109, 145, 147
Armadillo, 173, 179
Aromatic-L-amino-acid decarboxylase,
    184
Arrhenius plot, 6–10
    lipid domains, 60–62
Artiodactyla, 289, 294, 298–299
Aryl 4-monooxygenase, 180
    sulfotransferase, 182
Arylsulfatase, 183, 242
Ascaris, 254
    suum, 249, 255, 261
Aspartate, 148
    aminotransferase, 181

329

Aspirin, 128
Ass, 282
Asthma, 246
ATP citrate-lyase, 184
  pyrophosphatase, 184
ATPase, 49, 57, 60, 62–65, 67, 121
  hibernator, 9, 14, 34–36, 39–40
Autoimmune disease, 136–138

B

Babirussa, 301
Baboon, 108, 168, 182, 273
Badger, 282
Balaenoptera musculus, 203
  physalus, 294, 295
  sibbaldi, 295
Bat, 27
Basal metabolic rate, 132–133
Bear, 296
  grizzly, 203
Bile, 80
  salt, mammalian, 287–302
Birthmark, 279
Bitis, 297
Blastocyst, 117–118, 176
Blastokinin, 186
Blastulation, 186
Bloodworm, intestinal, 247
Blue color, 273
Bone, calcification process, 86–88
  matrix, 83–85, 94
  mineral component, 84–85
    metabolism, 77–102
  resorption, 82, 88–89, 90, 91
  structural, 81–83
Bos, 295
  tarus, 203
Bovine, 64
  phospholipid, 26, 29, 31, 32
Brachiopod, 79
Bradykinin, 232, 252, 253
Brain, 62
  activity, hibernation, 33–37
  cholesterol, 58–60
  phospholipid, 50, 53
    hibernator, 17–19, 24–26, 33–37
Breast, 200
2-Br-α-ergocryptine, 216
Bromocriptine, 216

Bronchoconstriction, 241
Bubalus caffer, 295
Buffalo, 282
Bunostomum phlebotomum, 247–248

C

Calcification, 79, 86–88
Calcitonin, 90, 91–92, 94–95, 136–137
Calcium, 38, 223
  absorption, 80–81, 87, 89, 93
  carbonate, 85
  phosphate, 79, 84–85, 88
  plasma, 138, 140, 145, 146
  uptake, 63
Canine, 64
Canis familiaris, 203
  lupus, 294
Cannabinol, 125
Capra, 295
  hircus, 203
Carbamoyl-phosphate synthase, 182
Carboxylesterase, 182
Caribou, 5
Carnitine palmitoyltransferase, 181
Carnivora, 289, 294–297
Carotenemia, 284
β-Carotene, 281
Carotenoid, 281–284
Cartilage, 82–83, 97, 98
Casein, 208
  α-, 223
  β-, 212, 223
  γ-, 223
  κ-, 223
  mRNA, 213–216
Castor canadensis, 294
Castration cell, 116
Cat, 179, 203, 296
Catalase, 180
Catechol methyltransferase, 181
Catecholamine, 119, 216
Cathepsin, 241
  B, 183
  P, 183
Cavia, 293
  porcellus, 203
Cementoblast, 79
Cerebellum, 108
Cerebrospinal fluid, 150, 151

Cestode, inflammation response, 245–246, 249–250, 253, 256–258, 262–264
Cetacea, 289, 294, 295
Chenodeoxycholic acid, 287–299
Chimpanzee, 164
Chinchilla, 170
Chiroptera, 289
Cholecalciferal, 90
Cholesterol, 79, 174, 287, 288, 299
  esterase, 182
  membrane, 56–60
Cholic acid, 287–299
Choline acetyltransferase, 181
Cholinesterase, 182
Choloepus hoffmani, 291, 293
Chondroblast, 79
Chondrogenesis, 94
Choriocarcinoma, 171
Chorionic gonadotropin, 117–118
  human, 162–164
Chrysocgon brachyrus, 294, 296
Chymase, 240
Ciliatine, 288, 299
Ciliatocholic acid, 299
Circadian rhythm, 122, 128, 150
Circannual rhythm, 2
Citellus lateralis, 2, 4, 10, 17, 20–22, 24, 27, 28, 50
  tridemlineatus, 20, 136
  undulatus, 38
Citrate secretion, 217
Clot inhibition, 248–249
Coagulation factor x11, 232
  reaction, 231–233, 247–252
Collagen, 79, 83–84, 86, 96, 98, 232
Colostrum, 201, 207, 211, 216
Complement cascade, 232, 234–236
  system, 234–236, 254–256
Concanavalin A, 146
Corpus luteum, 107, 112, 116, 126, 162, 176, 283
  rubrum, 283
Correxonuclease, 183
Corticosteroids, 97, 120–123
Corticosterone, 107
Corticotropin, 107, 112, 113, 171–172
  releasing factor, 107, 110–111
Cortisol, 122–123
Cow, 163, 166, 169, 176, 179–184, 201, 203, 206, 283

Creatinine kinase, 182
Cricetus auratus, 292
CTP-cytidyltransferase, 38
3′,5′-Cyclic-AMP phosphodiesterase, 183
Cynomys ludovicianus, 293
Cystyl-aminopeptidase, 183
Cytidyltransferase, 52
Cytochrome oxidase, 180
  c oxidoreductase, 63
  P_{450} reductase, 63

D

Damaliscus albifrons, 295
Deciduoma formation, 187–188
Deer, 203
  fallow, 170
  roe, 282
7-Dehydrocholesterol, 90
Dehydroepiandrosterone sulfate, 174
Deoxycholic acid, 287–299
Deoxyribonuclease, 183
Dermoptera, 289
Dexamethasone, 122
Diabetes insipidus, 122
  mellitus, 122, 143
Diacylglycerol lipase, 182
Dichloroindophenol reductase, 65
Dicrostonyx groenlandicus, 119
Didelphis marsupialis, 203
Diethylstilbestrol, 127
Difucosyllactose, 204
Dihydrotestosterone, 124
3α, 6α-Dihydroxy-5β-cholan-24-oic acid, 298
3α,7α-Dihydroxy-5β-cholan-24-oic acid, 287
3α,7α-Dihydroxy-5β-cholestan-26-oic acid, 290, 293
3α,7β-Dihydroxy-5β-cholan-24-oic acid, 290, 296
3α,12α-Dihydroxy-5β-cholan-24-oic acid, 287
Dimyristoyl phosphatidylcholine, 36
  phosphatidylethanolamine, 36
Dioleoylphosphatidylcholine, 65
Dipalmitoylphosphatidylcholine, 56–57
Dipalmitoylphosphalidylglycerol, 67
Diphosphatidylglycerol, 27, 32, 38, 39, 50, 52

Diphyllobothrium latum, 263
Dipodomys deserti, 294
  merriami, 294
  microps, 293
Distearoylphosphatidylcholine, 36
DNA nucleotidyltransferase, 182
Docosahexaenoic acid, 17, 19, 22, 37
Docosapentaenoic acid, 22
Dog, 170, 176, 203, 282
Dogfish, phospholipid, 26
Donkey, 165
Dopachrome, 275
Dopamine, 216
Dormouse, 34
Dracunculosis, 244

E

Echidna, 203, 204
Echidna aculeatus, 290
Echinococcus granulosus, 250
Edema, facial, 256
Edentata, 289, 291, 293
Elaidic acid, 15
Elephant, 165, 203, 282, 297
Elephas maximus, 203
Endocrinology, adrenal gland, 118–123
  androgens, 123–125
  chorion, 117–118, 162–164
  kidney, 152–154
  mammalian, 103–158
  ovarian hormones, 125–128, 162, 176,
    283
  pancreas, 139–148
  parathyroid gland, 138–139
  pineal body, 148–152
  pituitary, 105–118
  sex hormones, 123–128
  thyroid gland, 128–138
Endometrial cup, 165
Endometrium, secretion, 186–188
Endoperoxides, 241
Endotoxin, 232
Enzyme activity, Arrhenius plot, 6–10
  placental, 177–185
Epinephrine, 118–119, 134, 136
Epiphyseal growth plate, 96
Epithelial growth factor, 207
Equus caballus, 203, 294, 298
Ergastoplasm, 200

Ergot, 216
Erinaceus europaeus, 290
Erythromelanin, 275, 278
Erythropoietin, 154
Escherichia coli, 15
Esterone, 174
  sulfotransferase, 182
Estradiol, 124
  17β, 174, 205
    dehydrogenase, 179
Estrogen, 112, 125–128, 201, 206
  placental, 173–176
  receptor, 126–127
Estrus, 111, 117, 187, 205
Ethanolamine, 55
Eumelanin, 274, 278
Eutheria, 201
Ewe, 179

F

Fasciola hepatica, 250, 259
Fasciolopsis buski, 253, 265
Fat, brown, 119, 280
  depot, 4
  extrusion, 217–218
Fatty acid, hibernator tissue, 14–22
  synthase, 220–221
  synthesis, 214, 220–222
Feeding control, 118–119
Felis concolor, 294
  felis, 294
  leo, 294
  pardus, 294
  tigris, 294
Felius catus, 203
Ferret, 179
Fetus, uterine environment, 185–188
Fibrinolysis, 233
Fibroblast, Arrhenius discontinuity, 64
Fish, 64
Fissipeda, 296
Fitchet, 282
Fluke, 246–247, 250–251
Fluoro-β-methasone 17,21-diacetate, 122
Follicle-stimulating hormone, 107, 112,
  116–118, 164, 205, 216
  releasing factor, 107
Fox, 282
Frog, phospholipid, 29, 31, 32

Fructokinase, 181
Fructose-1,6-diphosphatase, 53
Fuscin, 280
α-L-Fucosidase, 183
Fumarate hydratase, 184

G

Galactokinase, 181
Galactopoiesis, 204
α-Galactosidase, 183
β-Galactosidase, 183
β-Galactoside transport, 15
Galactosyltransferase, 210, 214
Gallbladder, 288
Ganglioside, 11
Gazella bennetti, 295
  rufifrons, 295
Genetics, carotenoid, 283
  lactation regulation, 211–216
  melanization, 279
Gerbil, Mongolian, 293
Gestation period, 113
Giraffe, 165, 176
Glucagon, 106, 108, 142–145
Glucocorticoids, 120, 122, 174, 175
  adrenal, 206
Gluconeogenesis, 53, 120, 222
Glucose, 119, 220
  dehydrogenase, 179
  plasma, 140–145, 147
  transport, 64
Glucose-6-phosphatase, 39, 63, 183
Glucose-6-phosphate dehydrogenase, 179
Glucosephosphate isomerase, 184
α-Glucosidase, 183
β-Glucosidase, 183
β-Glucoside transport, 15
β-Glucuronidase, 183
Glutamate dehydrogenase, 180
Glutaminase, 183
Glutathione-oxytocin transhydro-genase, 180
Glycerol release, 134, 136
Glycerol-3-phosphate dehydrogenase, 180
Glycine, 148, 288, 290–299
  aminotransferase, 181
Glycochenodeoxycholate, 298
Glycocholic acid, 300

Glycogen, 220
  synthase, 181
Glycogneolysis, 119
Glycoprotein, 79, 84, 98, 210, 220
Glycosaminoglycan, 98
Glycosyltransferase, 220
Glycyl-glycine dipeptidase, 183
Goat, 80, 170, 176, 203, 282
  Grecian wild, 295
Goiter, 128
Goldfish, phospholipid, 26
Golgi vesicle, 217–218
Gonadotropin, 110, 111, 112, 116–118, 162–166
  chorionic, 117–118
    human, 162–164
  pregnant mare serum, 165
  releasing hormone, 110, 116
Gorilla, 164
Ground squirrel, 34–35, 62, 64
  Artic, 38
  golden mantled, 2, 4, 10, 17, 20–22, 24, 27, 28
  13-lined, 136
  torpidation phase, 3–4
Growth factor, serum, 207
  hormone, 95, 106, 107, 111–113, 142, 143, 152, 206–207
    release-inhibiting factor, 107, 111
    releasing factor, 107
Guanine deaminase, 183
Guanosine monophosphate, cyclic, 113, 128
Guinea pig, 64, 166, 170, 173, 176, 179–184, 201, 203, 282, 293
  phospholipid, 24, 26, 32
  placental enzyme, 179–184
Gull, Artic herring, 5

H

Haemonchosis, 248
Haemonchus contortus, 247
Hageman factor, 231–233, 249, 251, 252
Hamster, 24, 58–60, 62, 150, 152–153, 170, 203, 292
  Syrian, 15, 34, 35, 37
Hare, 282
Heart phospholipid, hibernator, 20–22
  rate, hibernation, 4
Hedgehog, 4, 33, 282, 290

Helarctos malayanus, 294, 296
Helminth parasite, host inflammation, 229–271
    tissue reaction to, 243–247
Hemitragus jemlahicus, 295
Heparin, 240, 241
Herpestes edwardsi, 294
Hexadecenoic acid, 17
Hexokinase, 181
Hibernating cycle, 2
Hibernation, brain activity, 33–37
    membrane phospholipids, 1–47
        structure and function, 49–75
Hippopotamus amphibius, 295
Histaminase, 242
Histamine, 232, 235, 239, 249
Homo sapiens, 203, 293
Homostatic agents, interaction, 95–98
Hookworm, canine, 247
    cattle, 247
Horse, 165, 170, 174, 179, 181, 184, 203, 282
Human, 203, 206, 216, 282
    Arrhenius discontinuity, 64
    phospholipid, 26, 29, 31, 32
    placental enzymes, 179–184
        hormones, 161–177
Hydrocortisone, 206
Hydroxyapatite, 79, 84–85, 86
β-Hydroxybutyrate, 119, 222
    dehydrogenase, 53
3α-Hydroxy-5β-cholan-24-oic acid, 287
3α-Hydroxy-6-oxo-5β-cholan-24-oic acid, 294, 298
3α-Hydroxy-7-oxo-5β-cholan-24-oic acid, 290
17-Hydroxycorticosterone, 122
16α-Hydroxy-dehydroepioandrosterone sulfate, 174
6-Hydroxydopamine, 118
15-Hydroxyprostaglandin dehydrogenase, 179
Hydroxysphingomyelin, 40
3β-Hydroxysteroid dehydrogenase, 178, 179
11β-Hydroxysteroid dehydrogenase, 180
20α-Hydroxysteroid dehydrogenase, 180
21-Hydroxysteroid dehydrogenase, 180
5-Hydroxytryptamine, 239
Hylochoerus, 301

Hyocholic acid, 294, 298
Hyodeoxycholic acid, 294, 298–299
Hypercalcemia, 90
Hypertension, 232
Hypersensitivity, immediate, 236–242, 259–265
Hypocalcemia, 90
Hypoglycemia, 108, 121, 146
Hypoproteinemia, 250
Hypothalmus, 105, 108, 110, 113, 118, 122, 124, 150, 216
Hypothyroidism, 131, 133
Hyracoidea, 289

I

Imidazole methyltransferase, 181
Immunoglobin, 216
    A, 217
    E, 237–238, 242, 250, 260, 262
    G, 217, 235, 238, 242, 261
    M, 217, 235
IMP dehydrogenase, 180
Indole-5,6-quinone, 275
Indomethacin, 128
Inflammation, acute, 236–243
    cestode, 245–246, 249–250, 253, 256–258, 262–264
    complement cascade, 234–236, 254–256
    Hageman factor, 231–233
    nematode, 243–245, 247–249, 252–256, 260–262
    parasite pathway, 247–265
    pathways to, 230–242
    trematode, 246–247, 250–253, 258–259, 264–265
Insulin, 106, 108–109, 119, 121–122, 139–142, 145–148, 205–207, 213, 235
    receptor, 145, 146
    resistance, 145–146
    shock, 146
Insectivora, 289, 290
Intermedin, 111
Interstitial cell-stimulating hormone, 112, 116
Intestinal cells, mineral metabolism, 92–95
Iridescence, 274
Iron, 79
Isocitrate dehydrogenase, 179

K

Kallikrein, 232–233, 252, 255
Kangaroo, 203
  dwarf, 119–120
Kidney, endocrinology, 152–154
  mineral metabolism, 91–95
  phospholipid, 50, 53
    hibernator, 27–29, 39–41
Kinin generation, 231, 233, 249
  system, 252–253
Kininogen, 232
Koala, 290

L

α-Lactalbumin, 202, 204, 207, 208, 210,
  214, 215, 220
Lactase, 132, 204
Lactate dehydrogenase, 61, 179
Lactation, 199–228
  chemical synthesis, 219–223
  genetic regulation, 211–216
  hormonal control, 204–210
  mammary gland, 200–201
  milk ejection, 219
    secretion, 216–218
  model system, 207–208
  regression, 223–224
Lactobacillus, 202
  bafidus, 204
Lactogen, 166–170
  placental, 205
    bovine, 169
    human, 166–169, 205, 207
Lactogenesis, 206–207
β-Lactoglobulin, 202, 212
Lactose, 202–204, 217, 219–220
  intolerance, 204
  synthetase, regulation, 210
Lagomorpha, 289, 291–292, 293
Lama pacos, 123
Lemming, Artic, 119
Lemmus trimucronatus, 119
Lemur, 181
Lepus californicus, 292, 293
  timidus, 293
Leucine aminotransferase, 181
  incorporation, 63, 64

Leucocyte, basophil, 237–243, 260
  eosinophilic, 244, 246, 255, 257, 260
  mononuclear, 232, 235, 244
  neutrophilic, 232, 238
  polymorphonuclear, 235, 244, 260
Lingula, 79
Linoleic acid, 15, 17, 19, 20, 22, 36
Linolenic acid, 15
Lipid, adrenal, 283
  annular, 67
  Arrhenius plot, 60–62
  boundary, 67
  chemotactic, 241
  phase transition, 62–67
Liposomes, 41, 58
Lithocholic acid, 287, 290, 292, 294
Litocranius walleri, 295
Liver, 115, 174
  monooxygenase, 63
  phospholipid, 50
    hibernator, 26–29, 37–39
  UDP-glucuronyl transferase, 8
Lobster, 64
Lung, phospholipid, 50
  hibernator, 30–31, 41–42
Lutein, 281
Luteinizing hormone, 107, 112, 116–118,
  163–165, 205
  releasing factor, 107
Luteotropic hormone, 112, 113–115
Luteotropin, 107
  release-inhibiting factor, 107
  releasing factor, 107
Lymphocyte, 145
L-Lysine-2-oxoglutarate reductase, 184
Lysolecithin acyltransferase, 181
Lysozyme, 183

M

Macaca irus, 293
  maurus, 293
  mulatta, 142, 203
Macropus rufogriseus, 293
Mageleia rufa, 203
Magnesium, 79
Malate dehydrogenase, 179
Male antigen, 123–124
Mammalian endocrines, 103–158
  see specific glands and hormones

Mammalia, bile salt, 287–302
  pigment, 273–285
Mammary carcinoma, 115
  gland, 107, 112, 200–201, 205–206
    regression, 223–224
  tumor, 126–127
  virus, 206
Mammogenic hormone, 205
Manis pentadactyla, 291, 293
  tricuspis, 291, 293
Mannose, 119
α-Mannosidase, 183
Marijuana, 125
Marmoset, 165
Marmot, 3, 4
Marmota monax, 2
Marsupialia, 289, 290, 293
Martes martes, 294
Mast cell, 232, 235, 236–243, 245, 255,
  259–265
Maternal behavior, 126
Meconium, 204
Melanin, 274–279
  blond, 275, 277
  red, 275
Melanization, control, 279
Melanocyte, 278
Melanocyte, stimulating hormone, 172
Melanophore, 107, 111
  stimulating hormone, 111
Melanotropin, 107
  release-inhibiting factor, 107, 150–151
  releasing hormone, 107
Melatonin, 111, 148, 150–152
Melursus ursinus, 294, 296
Membrane asymmetry, 54–56
  boundary lipid, 67
  cholesterol, 56–60
  compartmentation, hibernation, 53–67
  enzymes, Arrhenius plot, 60–62
  fatty acid, hibernator, 14–22
  function, Arrhenius plot, 6–10
  low temperature function, 67–69
  lipid alteration, 67–69
    unrelated to low temperature, 51–53
  phase transition, 62–67
  phospholipids, 1–47
    fatty acids of, 10–22
    temperature, 6–10
  transport, temperature, 51

Meriones unguiculatus, 293, 294
Mesocricetus auratus, 15, 24, 50, 150, 203
Metatheria, 201
Methionine adenosyltransferase, 181
Microsome, Arrhenius discontinuity,
  62–63
  cholesterol, 58–60
  phase transition, 62–63
Microtubule, 140
Milk, 110
  composition, 202–204
  ejection, 219
  protein, 222–223
  secretion, 216–218
Milking, 216, 219
Mineral homeostasis, agent action, 92–98
  regulation, 89–92
  metabolism, 77–102
    extraskeletal, 80–81
    general theory, 78–80
    structural, 80–85
Mineralization, 79
Mineralocorticoids, 120, 174
Mitochondria, 87, 92
  Arrhenius discontinuity, 63
  lipid alteration, 52
  liver, hibernator, 19–20
Mole, hydatidiform, 170–171
Mongoose, 181
Monkey, 203, 273
  placenta, 168, 181–182
  rhesus, 142, 164, 165, 168, 172, 182
  squirrel, 122
  stump tail, 165
Monocyte, 82, 145
Monotremata, 159, 289, 290
Monotreme, 200
Mouse, 170, 179–184, 205, 206
  house, 124
  laboratory, 292
  phospholipid, 29, 31, 32
α-Muricholic acid, 292
β-Muricholic acid, 292
Mus, 292
  musculus, 124
Muscle, skeletal, phospholipid, 30–32, 41
Mustela vison, 294
Mycoplasma, 58
Myotis lucifugus, 27
Myrmecophaga tridactyla, 291, 293

N

N. brasiliensis, 261–262
N. ppustrongylus, 254
NAD kinase, 181
  transhydrogenase, 180
NADH-dehydrogenase, 63
  diaphorase, 180
  oxidase, 63
NADP-isocitrate dehydrogenase, 222
  transhydrogenase, 180
NADPH-cytochrome reductase, 180
  diaphorase, 180
Nematoda, 243
Nematode inflammation, 247–250, 252–256
  tissue reaction to, 243–245
Nematodiasis, 254
Neotoma lepida, 293, 294
Nerve, sciatic, conduction, 5
Neurohypophysis, 105–111
Noradrenaline, 118
Norepinephrine, 118–120, 150, 216
5'-Nucleotidase, 183
Nucleotide pyrophosphatase, 184

O

Obesity, 118, 133–134
Odocoileus hemionus, 203
Odontoblast, 79
Oleic acid, 15, 17, 22, 37
Oppossum, 203
Ornithine decarboxylase, 184
Ornithorhynchus anatinus, 204, 290
Orycteropus afer, 294, 297
Oryctolagus cuniculus, 291, 293
Oryx beisa, 295
Ossification, 79
Osteoblast, 79, 81, 82, 86, 87, 94
Osteoclast, 82, 92, 94
Osteocyte, 81, 82, 86, 98
Osteogenesis, 97, 98
Osteoporosis, 97
Ourebia ourebi, 295
Ovarian follicles, 107, 112
  hormones, 97, 125–128
Ovary, 205
Ovis aries, 203
  musimon, 295

Ovulation, 165
Ox, 299
3-Oxo-5α-steroid dehydrogenase, 180
Oxytocin, 105

P

Pagophilus groenlandicus, 203
Palmitic acid, 17, 19, 36, 221
Pancreas, 106
  endocrinology, 139–148
Papio ursinus, 293
Parascaris equorum, 247, 262
Parasite helminth, 229–271
Parathormone, 123, 138
Parathyroid, 91, 93–94, 123, 138–139
  hormone, 90–91
Parkinson's disease, 279
Parotid gland, 145
Parturition, 105, 123, 127, 206, 208, 216
Perissodactyla, 289, 294, 298
Perognathus formosus, 293
Peroxidase, 180
Phacochoerus, 301
  aethiopicus, 295, 299
Phagocytosis, 235
Phascolarctos cinereus, 290, 293
Pheomelanin, 275
Pheromone, 115
Phoca vitulina, 294
Phocaecholic acid, 297
α-Phocaecholic acid, 297
Pholidota, 289, 291, 293
Phosphatase, 187
Phosphate excretion, 138
Phosphatidylcholine, 15, 24, 30, 33, 36–37, 40–41, 50, 52–53, 55–58, 66
Phosphatidylethanolamine, 15, 17, 24, 27, 28, 30, 32, 36–37, 39, 41, 50, 53, 55, 59, 70
Phosphalidylinositol, 17, 27, 39, 62
Phosphatidylserine, 15, 17, 27, 30, 39, 50, 55, 66
Phosphaturia, 95
Phosphodiesterase, 129
Phosphoenolpyruvate carboxykinase, 184
6-Phosphofructokinase, 181
Phosphoglucomutase, 182
Phosphogluconate dehydrogenase, 179

Phospholipase, 182
  B, 242
  D, 241
Phospholipids, acidic, 37–38
  brain microsome, 16
  extraction and analysis, 10–14
  membrane, hibernation, 1–47
  metabolism, 22–24
  steady stale levels, 50
Phosphorylase, 181
Photoperiodicity, 148, 150
Pig, 163, 166, 170, 176, 179, 181, 184, 186, 203, 282, 298–299
Pigment, carotenoid, 281–284
  dispersion, 107
  fuscin, 280
  iridescence, 274
  mammal, 273–285
  melanin, 274–279
  red, 274
  structural color, 273–274
  Tyndall color, 273–274
Pineal gland, 111, 118, 148–152
Pinnipedia, 297
Pituitary, 223
  anterior, 205
  endocrinology, 105–118
Placenta, enzymes, 177–185
  estrogen, 173–176
  peptide hormone, 161–173
  progesterone, 173–176
  steroidogenesis, 173–177
Placental hormones, 161–177
Plasma calcium, 80
  cortisol, 122
  thromboplastin antecedent, 232
Plasmin, 232, 235
Plasminogen, 232
Platelet activating factor, 241
Platypus, 204
Polyamine, 206
Polyol dehydrogenase, 179
Polyribosome, bound, 214
Porcine, 64
Potassium elimination, 122
  uptake, 63
Potoroo, 119–120
Potorous tridactylus, 119–120
Pregnenolone, 174
Prekallikrein, 232

Preoptic nucleus, 110
Preproinsulin, 213
Primate, 289, 290–291, 293
Proboscidea, 289, 297
Procyon lotor, 294
Progesterone, 107, 112, 116, 125–128, 201, 206, 215
  placental, 173–176, 186
Proinsulin, 213
Prolactin, 107, 112, 113–115, 126, 206, 207, 215
  control, 216
  inhibiting factor, 216
  receptor, 208–209
  release-inhibiting factor, 107
  releasing factor, 107
  receptor, 115
Propanolol, 122
Properdin, 236
Prostaglandin, 54, 129–130, 241
  A_1, 129, 130
  E_1, 129–130
  $F_{2\alpha}$, 129, 131
Proteoglycan, 79
Prothiombin, 233
Prototheria, 200, 201
Pruritus, 246
Pseudoicterus, 283
Pseudopregnancy, 186
Psoriasis, 122
Puromycin, 132, 133
Pyridoxal kinase, 182
Pyrophosphatase, 187
Pyrophosphate, 232
Pyruvate kinase, 53, 182

Q

Quazodine, 129, 130

R

Rabbit, 59, 64, 166, 169, 176, 179, 181, 183, 186, 206, 215, 282, 283, 291
  cottontail, 203
  phospholipid, 26
Rat, 59, 64, 83, 91, 151, 176, 179–184, 187, 203, 205, 206, 282
  laboratory, 292
  phospholipid, 26, 29, 31, 32
  placental enzymes, 177–185

Rattlesnake, phospholipid, 26
Rattus, 292
  norvegicus, 203
Red blood cell, Arrhenius transitions, 62, 64
Relaxin, 127–128, 172
Renin, 122, 152–154, 183
Respiration, Arrhenius discontinuity, 63
Reverse transcriptase, 182
Rhinoceros, 176
Ribonuclease, 183
Ribosephosphate isomerase, 184
Ribulosephosphate 3-epimerase, 184
Rickets, 93
RNA nucteotidyltranferase, 182
RNA-dependent DNA polymerase, 182
Rodentia, 289, 292–293
Ruminant, 222, 223

S

Saccharopine dehydrogenase, 180
Saimiri sciureus, 122, 293
Salt balance, 120
Sarcoplasmic reticulum, 62–63, 67
Schistosoma mansoni, 250, 264
Schistosome, 251–252
Scleroblast, 79
Sea lion, 204
Seal, 176
  fur, 186
  harp, 203
Secretion, 106
Selenarctos thibetanus, 294
    japonicus, 296
Serine glycerophosphatide, 55
Serotonin, 150–152, 239, 241
Sex hormone, endocrinology, 123–128
Sexual differentiation, 123–124, 162
Sheep, 64, 80, 123, 163, 166, 168, 179, 181, 184, 203, 282
Shrew, Asian musk, 125–126
Siafic acid, 84
Sialoprotein, 84, 88
Simia satyrus, 293
Sirenia, 289
Skeletal homeostasis, 89
  muscle, phospholipid, 50
  structure, 81–83
Skin color, 278

Somatomedin, 96
Somatomammotropin, human chorionic, 166
Somatostatin, 106–109, 111
Somatotropin, 95–96, 106–108, 111–113
  release-inhibiting hormone, 108–109
  releasing factor, 107, 111
Sorbitol dehydrogenase, 179
Sparaganosis, 263
Spermidine, 206
Spermophilus richardsonii, 34, 62
    tereticaudus, 293
Sphingomyelin, 17, 25, 26, 28–31, 40–41, 52, 55, 70
Squalene monooxygenase, 180
Squirrel, phospholipid, hibernator, 22–33
Stearic acid, 17, 19, 20, 36
Steroid aromatase, 184
  17–20 lyase, 184
  20–22 lyase, 184
  monooxygenase, 180
Steroidogenesis, placental, 173–177
Sterol-sulfatase, 183
Strepsiceros imberbis, 295
  kuda, 295
Streptozotocin, 146
Strongyloides, 252
Strongylus, 247
Strontium, 79
Sublingual gland, 145
Succinic dehydrogenase, 15, 63, 180
  oxidase, 63
Suckling, 216, 219
Suncus murinus, 125–126
Superior cervical ganglion, 150
Superoxide dismutase, 180
Suidae, 298–299
Sus, 298, 299
  scrofa, 203
Sussus leucomystax, 294
Sylvilagus floridanus, 203
Syncitiotrophoblast, 166

T

Tachyglossus aculeatus, 203, 204, 290
Taenia taenialformis, 250, 253, 257–258, 263–264
Taeniidae, 245
Tapeworm, 245–246

Taurine, 288, 290, 291, 292–299
Taurochenodeoxycholate, 290, 298
Taurocholate, 290, 298
Taurodeoxycholate, 290
Tayassu, 301
Tegument, 245
Temperature, heart rate, 4
  membrane phospholipids, 6–10
  sciatic nerve conduction, 5
Testes, 152
  interstitial cell, 112, 123
  spermatic tubules, 112
Testosterone, 112, 117, 123–125, 126, 162, 174
  5α-reductase, 124
Tetany, 138
Tetrahydrocannabinol, 125
Tetrahydrofolate dehydrogenase, 180
3α,7α,12α,23ε-Tetrahydroxy-5β-cholan-24-oic acid, 297
Tetrahymena, 64, 68
  pyriformis, 66
Tetraiodothyronine, 128, 131
Thalarctos maritimus, 294, 296
Theophyllin, 113, 129, 130, 140
Thermogenesis, nonshivering, 4, 119
Thiaminase, 183
Thrombin, 233
Thrombocyte, 237
Thromboembolic lesion, 251
Thromboplastin, 248
Thromboxane, 241
Thymus, 91, 138
Thyroid, 91, 112, 205
  endocrinology, 128–138
  hormone, 19, 97, 110
  receptors, 132
Thyroid-stimulating hormone, 106, 112, 128–131, 171, 205
Thyrotropin, 112, 128
  chorionic, 170–171
Thyrotropin-releasing hormone, 109–110, 128
Thyroxine, 128, 131–136, 206
Tooth, 83
Torpidation, 2
Torpid state, environmental stimulus, 2
Transaldoase, 181
Transketolase, 181

Trematode, inflammation response, 246–247, 250–253, 258–259, 264–265
Triacylglycerol lipase, 182
Trichinella spiralis, 252
Trichostrongylus columbriformis, 260–262
3α,6α,7α-Trihydroxy-5β-cholan-24-oic acid, 298
3α,6β,7α-Trihydroxy-5β-cholan-24-oic acid, 292
3α,6β,7β-Trihydroxy-5β-cholan-24-oic acid, 292
3α,7α,12α-Trihydroxy-5β-cholan-24-oic acid, 287, 290
3α,7α,23ε-Trihydroxy-5β-cholan-24-oic acid, 297
Triiodothyronine, 128, 131
Tropocollagen, 83
Trypsin, 232, 235
Tryptophanyl-tRNA synthetase, 184
Tubulidentata, 289, 294
Tubulin, 140
Tyndall color, 273–274
Tyrosinase, 275, 278, 279
Tyrosine aminotransferase, 181

U

UDP glucose dehydrogenase, 179
  glucuronosyltransferase, 181
  glucuronyltransferase, 63
Ungulata, 201
Uric acid, 79, 232
Ursidae, 296–297
Ursodeoxycholic acid, 290, 296–297
Ursus americanus, 294, 296
  arctos horribilis, 203
  isabellinus, 297
Urticaria, 246
Uteroglobin, 186–187
Uterus, fetal environment, 185–188

V

Vasodilalor, 232
Vasopressin, 105
Vasotocin, 148–150
Virus, mammary tumor, 206
  New Castle, 64

Vitamin A, 98, 120, 281, 284
 C, 98, 120–121
 D, 89–90, 92–93, 146
 D₃, 90, 91
Viviparity, biochemical aspects, 159–197
 fetal environment, 185–188
 placental hormone, 161–173
  steriodogenesis, 173–177
von Recklinghausen's disease, 90

W

Wallaby, 187
Warthog, 299
Water balance, 120
Whale, 203
Woodchuck, 2

X

Xanthemia, 283
Xanthophyll, 281–282
Xanthosis diabetica, 284

Y

Yeast, 64

Z

Zalophus californiances, 204, 294
Zeitgeber, 2
Zymosan, 235–236

LATIN AMERICAN HISTORICAL DICTIONARIES SERIES
Edited by Laurence Hallewell

1. *Guatemala,* by Richard E. Moore, rev. ed. 1973.

2. *Panama,* by Basil C. & Anne K. Hedrick. 1970.

3. *Venezuela,* by Donna Keyse Rudolph & G.A. Rudolph. 1971.

4. *Bolivia,* by Dwight B. Heath. 1972.

5. *El Salvador,* by Philip F. Flemion. 1972.

6. *Nicaragua,* by Harvey K. Meyer. 1972.

7. *Chile,* 2nd ed., by Salvatore Bizzarro. 1987.

8. *Paraguay,* by Charles J. Kolinski. 1973.

9. *Puerto Rico and the U.S. Virgin Islands,* by Kenneth R. Farr. 1973.

10. *Ecuador,* by Albert W. Bork & Georg Maier. 1973.

11. *Uruguay,* by Jean L. Willis. 1974.

12. *British Caribbean,* by William Lux. 1975.

13. *Honduras,* by Harvey K. Meyer. 1976.

14. *Colombia,* by Robert H. Davis. 1977.

15. *Haiti,* by Roland I. Perusse. 1977.

16. *Costa Rica,* by Theodore S. Creedman. 1977.

17. *Argentina,* by Ione Wright & Lisa M. Nekhom. 1978.

18. *French and Netherlands Antilles,* by Albert Gastmann. 1978.

19. *Brazil,* by Robert M. Levine. 1979.

20. *Peru,* by Marvin Alisky. 1979.

21. *Mexico,* by Donald C. Briggs & Marvin Alisky. 1981.

22. *Cuba,* by Jaime Suchlicki. 1988.

Historical Dictionary of
CUBA

by
JAIME SUCHLICKI

Latin American Historical Dictionaries, No. 22

The Scarecrow Press, Inc.
Metuchen, N.J., & London
1988

Library of Congress Cataloging-in-Publication Data

Suchlicki, Jaime.
  Historical dictionary of Cuba / by Jaime Suchlicki.
      p.   cm. -- (Latin American historical dictionaries ;
  no. 22)
    Bibliography: p.
    ISBN 0-8108-2071-4
    1. Cuba--Dictionaries and encyclopedias. I. Title.
  II. Series.
  F1754.S83   1988   87-28406
  972.91'003'21--dc19

CONTENTS

Editor's Foreword   v

Introduction   viii

List of Abbreviations   ix

Chronology   xvii

THE DICTIONARY   1

Bibliography   306

Appendixes

   1.   Country Brief   360

   2.   Diplomatic Relations   362

   3.   Membership in International Organizations   366

   4.   Presidents of Cuba   368

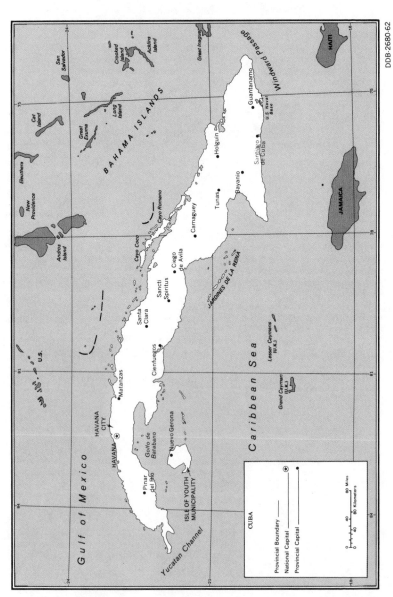

Figure 1. Map of Cuba.

EDITOR'S FOREWORD

Geography is said to dictate history. Cuba and its development gives ample evidence of this. "The Pearl of the Antilles" has been shaped by climate, soil, ocean currents, and by being the large island in a sea of more easily bypassed smaller ones. This sea provided the essential routes for Spain's discovery and colonization of the rich lands of Central and South America. From the beginning, navigators took advantage of the east to west ocean currents and the northeast trade winds prevailing off the African coast south of the Iberian Peninsula.

In 1493 in his first letter detailing his discoveries, Columbus wrote: "It is a land to be desired and once seen never to be abandoned." Further he added, "I assure Your Majesties that there is no better land or people."

The Spaniards were not the first to invade or conquer the island. The native Ciboneys had been dominated by the more advanced, non-warlike Arawaks from the South American mainland. By the time Columbus reached Cuba, the warlike Caribs also from the mainland had taken over about a third of the area, terrorizing rather than colonizing. The emissaries of Spain's feudal monarchs found the population ruled by caciques. The legacies of local and European centralized power have continued to exert strong influences.

From the days of classical Greece, Europeans had speculated upon myths of wealth and glory to be found in unknown lands beyond the sea. Columbus was propelled by Europe's need to find an alternative route to the wealth of the Orient. Although he failed to find a direct western route to Cathay (China), Columbus exploded an even more powerful myth by proving that the world was round, not flat as commonly believed.

Further, he provided an outlet for the energies of warrior Spaniards who in 1492 after seven centuries of battles succeeded in expelling the last of the Moorish conquerors. From years of daring-do in combat, few of the valiant fighters were prepared to return to former occupations--herding swine in some instances--nor were other more acceptable peaceful pursuits open to them. That same year the Spanish monarchs decreed the expulsion of Jews who refused to convert to Christianity. Thus Columbus' discovery provided outlets for the conquering spirits that might otherwise have sparked unrest or rebellion at home. It also enabled committed Jews to escape sure death at the hands of the Inquisition.

In Cuba as elsewhere in the New World the conquerors were rewarded with large grants of land complete with Indians to work them. Cuba served as the base for the discovery and conquest of Central and South America in the name of the Crown. It was from Cuba that Hernán Cortés set out on his epic conquest of the Aztec kingdom, thus giving the Crown undreamed of lands and riches.

The great wealth of Mexico, Peru, and China (which reached Acapulco via the Manila galleons) was accumulated in Veracruz and Puerto Bello (Panama) to be shipped to Spain. It was the discovery by Juan Ponce de León just north of Havana of the great water highway, the Gulf Stream. Rushing through the Straits of Florida and up the North American coast it joins the east-flowing ocean currents of Carolina that helped propel the treasure ships to the Iberian coast. Thus with its strategic location, Havana was fortified to protect these homeward-bound ships. Spain's appetite for this wealth and subsequently its huge indebtedness to European bankers became unassuageable. The Spanish fleets also attracted attacks by pirates, smugglers, buccaneers, and freebooters, some with the complicity of their European governments. Later in the eighteenth and early nineteenth centuries, Europe's squabbles spilled over into naval battles in the Caribbean.

Whereas most of Spain's American colonies achieved independence early in the nineteenth century, despite invasions, rebellions, and manifestos, Cuba remained under Spanish rule until the end of that century when the United States recognized Cuban independence and invaded the island in the war with Spain (Spanish-American War). With some U.S. strings attached, Cuba became a constitutional republic in 1902.

Geography further blessed Cuba with rich soil and a semitropical climate influenced by the trade winds, providing excellent conditions for agriculture. Indigenous crops of tobacco, corn, white and sweet potatoes, and cassava were unfamiliar to the Spaniards, who also found cotton. The tubers had the added advantage of being hurricane-proof. Early in the sixteenth century, sugar and bananas were introduced from the Canary Islands, and figs, oranges, lemons, and cattle were brought from Spain. All flourished.

Missing were iron, coal, and other minerals essential to develop industry. Early efforts to mine gold using Indian slaves were fruitless. The unaccustomed hard labor along with European diseases came close to wiping out the native population. Some blacks had accompanied the first conquerors, but with sugar, a lucrative,

large-scale crop, many hands were needed who were able to with-
stand heat and sun, and African slaves became the base of Cuba's
agricultural economy. Sugar is not a small acreage "backyard"
crop as coffee can be. It requires vast tracts of relatively flat
land. Highly prized in Europe, sugar along with tobacco--what
Fernando Ortiz called the "Cuban Counterpoint"--came from plantations with overseers supervising armies of laborers for often
absentee landlords.

Spain's mercantile policies put all its colonial trade into Spanish
hands. The monopolies set low prices for crops and sold them in
turn for high profits. The government controlled tobacco production and distribution. While Spain benefitted from Cuba's agricultural bounty, it also had to provide for the island's sustenance from
the treasures of its western and southern vice-royalties. Cuba's
geographic location and climate have indeed determined the island's
destinies.

For this much-needed compendium of Cuba's history, none is
better prepared or has the access to the extensive resources called
for as does Dr. Jaime Suchlicki of the University of Miami. He has
succeeded brilliantly. This Historical Dictionary of Cuba in the
series begun by my late husband, A. Curtis Wilgus, provides facts
and perspectives on Cuba as it developed from Columbus' first foot-
step to the complex, not always well-understood present. Dr. Suchlicki has shed much-needed light on a strategic island that suffers
from passions rather than informed judgments.

Karna S. Wilgus
New York City

INTRODUCTION

When Karna Wilgus asked me whether I was interested in doing the Historical Dictionary of Cuba, I accepted readily, confident that the task would be relatively simple. After all, I had been involved with Cuban studies for many years, the University of Miami Library had an excellent collection of materials on Cuba, and I could count on the assistance of bright and energetic research associates at the University of Miami.

As I began to organize the Dictionary my confidence began to crumble. Instead of the simple task I had envisioned, the complexities of events in pre- and post-Castro Cuba and the quantity of figures and events for both periods baffled me. What and whom should I include and why?

The final selection, unfortunately, had to be arbitrary. While an attempt has been made to include the most prominent personalities, occurrences, places, and organizations, omissions are certain to have occurred. The reader, however, should be assured that these were not intentional or the result of carelessness. Much thought and effort went into the preparation of this volume, particularly from the many people who helped me. Although it would be easier to blame others for any errors or omissions, these are necessarily my responsibility.

I am grateful to my research assistants, Marlene García, Ray Graves, and Joe Azel, who worked tirelessly and beyond the call of duty in the preparation of the Dictionary. My special gratitude is to the analysts of the University of Miami's Cuban Information System for their research assistance, in particular to its director, Graciella Cruz-Taura. I am indebted to the late Margarita Madrazo de Pelleyá who, not only typed the manuscript, but with her prodigious memory and incisive eyes caught and corrected numerous errors. Without their valuable help I could not have completed this work.

Jaime Suchlicki
Coral Gables, Florida
1987

LIST OF ABBREVIATIONS

ACC	Cuban Academy of Sciences
ACNU	Cuban United Nations Association
AIN	National Information Agency
ALIMPORT	Cuban Enterprise for Import of Foodstuffs
ANAP	National Association of Small Farmers
ANEC	National Association of Cuban Economists
ANIR	National Association of Innovators and Efficiency Experts
ANPP	National Assembly for People's Power
ATAC	Association of Sugar Technicians
AUTOIMPORT	Central Enterprise for Supply and Sales of Light Automotive Equipment
AVIAIMPORT	Cuban Enterprise for Aircraft Import
BNC	National Bank of Cuba
BRA	Aviation Repair Base
BTJ	Youth Technical Brigades
CAME	Council of Mutual Economic Assistance
CARIBEX	Caribbean Export Enterprise
CCC	Cuban Chamber of Commerce
CDCC	Committee for Development and Cooperation in the Caribbean
CDEA	Friendly Armies Sports Committee
CDR	Committee for the Defense of the Revolution
CEA	Center for American Studies
CEAC	Cuban Nuclear Energy Commission
CEATM	State Committee for Material and Technical Supply
CECE	State Committee for Economic Cooperation
CECT	State Committee for Science and Technology
CEE	State Committee for Statistics
CEEO	Center for West European Studies
CEF	State Committee for Finance
CEMSA	Agamic Seeds Improvement Center
CEN	State Committee for Standardization
CENCA	National Sugar Industry Training Center
CENDA	National Center of Author Rights

CENIC	National Scientific Research Center
CENSA	National Animal Health Center
CEP	State Committee for Prices
CEPEM	Training Center for Junior Specialists
CETSS	State Committee for Labor and Social Security
CICMA	Agricultural Machine-Building Research Center
CINAN	Information Center for Standardization
CIP	Fishery Research Center
CIPIMM	Research and Projects Center for the Mining-Metallurgical Industry
CIRP	National Section of ID Card and Population Registry
CNC	National Council for Culture
CNCU	Cuban National UNESCO Commission
CNIC	See CENIC
CNICT	National Center for Scientific Labor Research
CNUPEA	National Committee for the Peaceful Use of Atomic Energy
COC	Cuban Olympic Committee
COMARNA	National Committee for Environmental Protection and Conservation of Natural Resources
COMECON	See CAME
CONSTRUIMPORT	Cuban Enterprise for the Import of Construction Machinery and Equipment
CONSUMIMPORT	Cuban Enterprise for the Import of General Consumer Goods
CTC	Central Organization of Cuban Trade Unions
CUBACITRICOS	Citrus Fruit Enterprise
CUBACONTROL	Cuban Control Enterprise
CUBAEXPORT	Cuban Foodstuffs and Miscellaneous Products Exporting Enterprise
CUBAFRUTAS	Cuban Tropical Fruit Exporting Enterprise
CUBAINDUSTRIA	Industrial Products Exporting Enterprise
CUBALSE	Service Agency for the Diplomatic Corps and Foreign Advisers
CUBAMETALES	Cuba Metal Importing Enterprise
CUBANIQUEL	Cuban Enterprise for the Export of Minerals and Metals
CUBAPESCA	Fishing Gear and Ships Import Enterprise
CUBARTISTA	Cuban Artistic Agency
CUBATABACO	Cuban Tobacco Enterprise
CUBATECNICA	Economic Institute of Technical Assistance
CUBATEX	Cuban Enterprise for Import of Fiber, Fabrics, Hides, and Their Products
CUBATUR	National and International Tourist Enterprise
CUBAZUCAR	Cuban Enterprise for Export of Sugar and Sugar By-products
CUFLET	Cuban Freight Enterprise
DAAFAR	Antiaircraft Defense and Revolutionary Air Force

DATINSAC	Systems and Computation Techniques Integral Enterprise
DCCA	Central Directorate for Construction and Housing
DECV	Directorate for the Development of Camps and Housing
DGI	General Directorate for Intelligence
DGPEI	General Directorate for Firefighting and Prevention
DGPNR	National Directorate of the National Revolutionary Police
DIC	Directorate for Training and Cadres
DOE	Directorate of Special Operations
DOR	Department of Revolutionary Orientation
DPC	Directorate for Combat Training
DPEI	Department of Firefighting and Prevention
DSE	State Security Department
DSP	Directorate of Personal Security
ECAM	General Antonio Maceo Interservice School
ECIMACT	Commercial Enterprise for Construction Materials, Building, and Tourism
ECIMETAL	Commercial Enterprise for the Metallurgical and Metal-Working Industry
ECIQUIM	Commercial Enterprise for the Chemical Industry
ECODES	Consolidated Enterprise for Electric Household Appliance Service
ECOPREFIL	Mail, Press, and Philatelic Enterprise
EDAI	National Industrial Automation Enterprise
EGREM	Musical Recording and Publishing Enterprise
EJT	Youth Labor Army
EMCO	Enterprise for the Production of Technical Methods of Computation
EMPROMECA	National Mechanical Production Enterprise
EMTELCUBA	National Telephone Enterprise
ENC	Caribbean Navigation Enterprise
ENC	National Coastal Shipping Enterprise
ENERGIMPORT	Energy Import Enterprise
ENM	Mambisa Navigation Enterprise
ENSUME	Cuban Enterprise for the Supply of Medical Equipment
ERI	International Radio Communications
ESETC	Enterprise for Technical Computation Services of Holguín
ESICUBA	Cuban International Insurance Enterprise
EVMCC	Camilo Cienfuegos Military Vocational Schools
EXPEDIPORT	Port Expediting Enterprise
FAC	Cuban Tuna Fleet

FAR	Revolutionary Armed Forces
FCC	Cuban Shrimp Fleet
FCP	Cuban Fishing Fleet
FECUIMPORT	Cuban Railroad Importing Enterprise
FEEM	Federation of Intermediate Level Students
FERRIMPORT	Cuban Enterprise for Hardware Import
FEU	Federation of University Students
FMC	Federation of Cuban Women
FP	Omsinar Shelf Fleet
GCCAD	Antiaircraft Missile Group for the Defense of the Capital
IACC	Cuban Civil Aeronautics Institute
ICAIC	Cuban Institute of Cinematographic Art and Industry
ICAP	Cuban Institute of Friendship with the Peoples
ICGC	Cuban Geodesic and Cartographic Institute
ICH	Cuban Institute of Hydrography
ICID	Central Institute for Digital Research
ICIDCA	Cuban Institute of Research on Sugarcane By-products
ICINAZ	Cuban Sugar Research Institute
ICIODI	Cuban Institute of Research and Orientation on Consumer Demand
ICRT	Cuban Institute of Radio and Television
IDICT	Institute of Scientific and Technical Documentation and Information
IDS	Institute for Health Development
II	Children's Institute
IMACC	Institute of Mathematics, Cybernetics, and Computation
IMC	Ministry of Construction Materials Industry
IMD	Institute for Sports Medicine
IMEXIN	Commercial Enterprise for Import and Export of Infrastructure
IMEXPAL	Import and Export Enterprise for Food Processing Plants and Related Accessories
INDAF	National Institute of Forestry Development and Exploitation
INDER	National Institute of Sports, Physical Education, and Recreation
INIFAT	Institute for Basic Research in Tropical Agriculture
ININ	National Nuclear Research Institute
ININTEF	Institute for Basic Technical Research
INP	National Institute for Fishing
INRA	National Institute of Agrarian Reform
INRH	National Institute of Hydraulic Resources

INSAC	National Institute of Automated Systems and Computer Technology
INSMET	Institute of Meteorology
INT	National Institute of Tourism
INTELCUBA	International Telecommunications
INTUR	See INT
INV	National Institute for Housing
IOJ	International Organization of Journalists
IPROYAZ	Projects Institute of the Ministry of Sugar Industry
ISA	Higher Institute of Art
ISCAB	Higher Institute of Agricultural and Animal Sciences of Bayamo
ISCAH	Higher Institute of Agricultural and Animal Sciences of Havana
ISCF	Higher Institute of Physical Culture
ISCM	Higher Institute of Medical Sciences
ISDE	Higher Institute for Economic Administration
ISE	Higher Institute of Education
ISPETP	Higher Teaching Institute for Professional and Technical Training
ISPJAE	José Antonio Echeverría Polytechnical Institute
ISSE	Higher Institute of the Foreign Service
ITM	Military Technical Institute
JUCEPLAN	Central Planning Board
LIBROCUBA	Foreign Trade Enterprise for Books
LIDA	Avian Research and Diagnosis Laboratory
MAPRINTER	Cuban Enterprise for Import of Raw Materials and Intermediate Products
MAQUIMPORT	Cuban Enterprise for Import of Machinery and Equipment
MARPESCA	Cuban Enterprise for the Import of Merchant and Fishing Ships
MARPORT	Cuban Enterprise for the Import of Marine and Port Equipment
MEDICUBA	Cuban Enterprise for the Import and Export of Medical Products
MES	Ministry of Higher Education
MGR	Cuban Revolutionary Navy
MICONS	Ministry of Construction
MINAG	Ministry of Agriculture
MINAL	Ministry of the Food Industry
MINAZ	Ministry of the Sugar Industry
MINCEX	Ministry of Foreign Trade
MINCIN	Ministry of Domestic Trade

MINCOM	Ministry of Communications
MINCULT	Ministry of Culture
MINED	Ministry of Education
MINFAR	Ministry of Revolutionary Armed Forces
MINIE	Ministry of Electric Power Industry
MINIL	Ministry of Light Industry
MININT	Ministry of Interior
MINIQ	Ministry of Chemical Industry
MINJUS	Ministry of Justice
MINMG	Ministry of Mines and Geology
MINPES	Ministry of Fishing Industry
MINREX	Ministry of Foreign Relations
MINSAP	Ministry of Public Health
MISM	Ministry of Iron and Steel Industry
MITRANS	Ministry of Transportation
MMM	Ministry of Mining and Metallurgy
MPSP	Movement for the Peace and Sovereignty of the Peoples
MTT	Territorial Troop Militia
NAMUCAR	Caribbean Multinational Shipping Enterprise
OCLAE	Continental Organization of Latin American Students
OPJM	Organization of "José Martí" Pioneers
OSPAAAL	African, Asian, and Latin American Peoples Solidarity Organization
PCC	Communist Party of Cuba
PL	See PPELA
PNR	National Revolutionary Police
PRELA	Prensa Latina News Agency
QUIMIMPORT	Cuban Enterprise for Import of Chemical Products
SDPE	Economic Management and Planning System
SEPMI	Society for Patriotic-Military Education
SERVICEC	Enterprise for Calculation Centers Services
SERVITEC	Enterprise for Technical Computation Services
SIME	See MISM
SINTAE	National Arts and Show Business Workers Union
SNNMCC	National System of Standardization, Weights and Measures, and Quality Control
SNNTT	National Transportation Workers Union
SNTAF	National Farm and Forestry Workers Union

SNTAP	National Public Administration Workers Union
SNTC	National Commerce Workers Union
SNTC-FAR	National FAR Civilian Workers Union
SNTCG	National Hotel and Restaurant Workers Union
SNTEC	National Science, Education, and Sports Workers Union
SNTIA	National Food Industry Workers Union
SNTIL	National Light Industry Workers Union
SNTIQE	National Chemical and Energy Industry Workers Union
SNTIT	National Tobacco Workers Union
SNTMM	National Miners and Metalworkers Union
SNTMMPP	National Merchant Marine, Port, and Fishing Workers Union
SNTS	National Health Workers Union
TECNAPESCA	Technical Enterprise for the Fishing Industry
TECNOIMPORT	Cuban Enterprise for the Import of Technical Products
TRACTOIMPORT	Cuban Enterprise for the Import of Agricultural Machinery and Equipment
TSP	People's Supreme Court
TURIMPEX	Cuban Enterprise for the Import and Export of Tourism
UJC	Union of Young Communists
UNEAC	National Union of Cuban Writers and Artists
UNECA	Union of Caribbean Construction Enterprises
UPEC	Union of Cuban Journalists

CHRONOLOGY

1492	Christopher Columbus discovered and explored Cuba
1508	Sebastián de Ocampo circumnavigated and explored the island
1511	Diego Velázquez conquered the Indians and established various settlements
1523	Blacks were brought from Africa to work the mines and fields
1538	Havana became the seat of government
1555	French pirate Jacques de Sores captured and burnt part of Havana
1595	Cattle raisers installed sugar mills on their lands and began sugar production
1628	Dutch pirate Piet Heyn captured the Spanish fleet off the northern coast of Cuba
1662	English captured and ransacked Santiago de Cuba
1717	Veguero rebellion against Spanish tobacco monopoly
1728	University of Havana was founded
1762	English captured and occupied Havana
1763	Havana restored to Spain
1773	Real Colegio Seminario de San Carlos was founded
1790	First newspaper established
1792	Sociedad Económica de Amigos del País was founded
1808	Napoleon overthrew and captured the Spanish King, Ferdinand VII

1809   Joaquín Infante organized the first independence
                 conspiracy

1812   José Aponte organized a conspiracy of slaves and free
                 blacks

1814   Ferdinand was restored to the Spanish throne

1823   Rayos y Soles de Bolívar conspiracy was organized
                 U.S. issued the Monroe Doctrine

1828-30   Aguila Negra conspiracy was organized

1830's   Spain imposed harsher authoritarian controls

1844   La Escalera, a slave conspiracy, was suppressed

1845   Spain ended the slave trade

1850   The Ostend Manifesto calling for the purchase of Cuba
                 was issued

1848-51   Conspiracies, expeditions, and death of Narciso López

1865   Spain created the Junta de Información

1867   Spain disbanded the Junta de Información

1868   Grito de Yara began the Ten Years' War

1869   Guáimaro Constitution was drafted

1873   Rebel boat Virginius was captured and fifty-two, mostly
                 Americans and Englishmen, were shot by Spanish
                 authorities

1878   Peace of Zanjón ended the Ten Years' War
                 Protest of Baraguá--General Antonio Maceo rejected the
                 Peace and called for the abolition of slavery

1879-80   La Guerra Chiquita took place

1886   Spain abolished slavery

1892   José Martí formed the Partido Revolucionario Cubano

1895   Grito de Baire began the War of Independence
                 Martí killed on the battlefield
                 Jimaguayú Constitution was drafted

1896   Antonio Maceo killed on the battlefield in Havana

1898	U.S. battleship <u>Maine</u> was blown up in Havana harbor Spanish-American War began U.S. occupied Cuba
1899	In the Treaty of Paris, Spain relinquished Cuba
1901	Cuban Constitution was drafted, incorporating the Platt Amendment which gave the U.S. the right to inter- vene
1902	The Republic was proclaimed and U.S. intervention ended
1903	U.S.-Cuban Treaty signed whereby the U.S. obtained Guantánamo base
1906	Guerrita de Agosto, a Liberal Party uprising, hastened U.S. intervention
1906-09	Second U.S. intervention
1912	Short-lived racial uprising led by the Agrupación Independiente de Color
1917	Short-lived uprising in Oriente and Camagüey led by the Liberal Party
1920	Collapse of the sugar boom
1923	U.S. representative Enoch Crowder sent to Cuba to "reform" political process
1924	Short-lived revolt by the Association of Veterans and Patriots
1925	Gerardo Machado assumed the presidency Cuban Communist Party was founded
1927	The anti-Machado Directorio Estudiantil Universitario was founded
1929	Machado was "reelected" for a new six-year term
1930	Mass student and popular demonstrations against the regime occurred The clandestine ABC organization was established
1931	Carlos Mendieta and former President Mario G. Menocal organized a short-lived uprising in Pinar del Río Carlos Hevia and Sergio Carbó's expedition from the U.S. was crushed by the Machado army

1933	U.S. became involved in mediating between Machado and various groups seeking to overthrow his government
	Machado was ousted and Carlos Manuel de Céspedes became provisional president
	Revolt of the Sergeants led by Fulgencio Batista hastened the fall of Céspedes
	Dr. Ramón Grau San Martín became president of a revolutionary government
1934	Batista overthrew Grau's regime and appointed Mendieta as provisional president
	Platt Amendment was abrogated
	Partido Revolucionario Cubano (Auténtico) organized
1936	Miguel Mariano Gómez "elected" president
1936-40	Federico Laredo Brú became president
1939	Grau San Martín elected president of the Constitutional Assembly
1940	Constitution was drafted
	Batista elected president for a four-year term
1941	Cuba declared war on the Axis powers
1944	Communists changed their party name to Partido Socialista Popular
	Dr. Ramón Grau San Martín elected president
1947	Eduardo Chibás formed the Partido del Pueblo Cubano (Ortodoxo)
	Cayo Confite expedition against Dominican dictator Rafael L. Trujillo was thwarted by the Cuban government
1948	Fidel Castro participated in the "Bogotazo" in Colombia
	Carlos Prío Socarrás elected president
1951	Eduardo Chibás committed suicide
1952	Batista seized power through a military coup
1953	Resistance organized and led primarily by Auténticos and university students
	Castro launched the ill-fated Moncada attack
1954	Batista "reelected" president to a four-year term
1955	Attempt at political compromise organized by Sociedad de Amigos de la República failed

1956   Montecristi conspiracy within the military squashed by
                Batista
            Auténticos attacked unsuccessfully the Goicuría army
                barracks
            Castro's <u>Granma</u> expedition landed in Oriente province

1957   Members of the Directorio and the Auténticos attacked
                unsuccessfully the Presidential Palace. Directorio
                leader José Antonio Echeverría killed by police
            Strike paralyzed almost all the western provinces of
                the island
            Castro consolidated his guerrilla operations in the
                Sierra Maestra

1958   Castro-organized general strike collapsed
            Castro expanded guerrilla operations into Las Villas
                province
            Military offensive against the guerrillas failed
            U.S. gradually withdrew support for the Batista regime
            Rigged election produced the victory of Batista's candi-
                data, Andrés Rivero Agüero
            Increased demoralization and corruption led to the grad-
                ual collapse of Cuba's armed forces
            Batista and his close associates escaped to the Domini-
                can Republic

1959   Fidel Castro assumed command and began consolidation
                of power
            Castro visited Caracas, the U.S., Canada, and Buenos
                Aires
            First Agrarian Reform Law promulgated
            Castro became prime minister and replaced Manuel
                Urrutia with his hand-picked candidate, Osvaldo
                Dorticós
            Major Camilo Cienfuegos died in a plane crash
            Major Huber Matos sentenced to twenty years in jail

1960   Cuba and the Soviet Union signed a commercial treaty
                and reestablished diplomatic relations
            Major foreign businesses were nationalized by the
                government
            Transportation, banking, communications, and the
                media and educational systems were taken over by
                the government
            Central Planning Board (JUCEPLAN) created to plan
                and direct the economy
            The Committees for the Defense of the Revolution (CDR)
                were organized
            Castro issued the "Declaration of Havana" claiming Cuba's
                right to export revolution and calling for Soviet sup-
                port

The Soviet government purchased Cuban sugar that
the U.S. had refused to buy
Soviet bloc armaments began arriving in Cuba

1961   U.S. broke off diplomatic relations with Cuba
Fidel Castro proclaimed the socialist character of the
Cuban Revolution
U.S.-sponsored Bay of Pigs invasion was defeated
The U.S. declared an embargo on trade with Cuba
The Cuban government proceeded to socialize the
economy
Castro declared himself to be a Marxist-Leninist

1962   Cuba was expelled from the Organization of American
States
Castro issued the "Second Declaration of Havana"
calling for continued revolution at home and abroad
Castro formed the Integrated Revolutionary Organizations (ORI), an amalgamation of revolutionary
groups and the Communists
The Cuban government introduced rationing of most
items including food
Missile crisis brought the U.S. and the Soviet Union
to the brink of nuclear conflict

1963   Castro visited the Soviet Union for the first time
Second Agrarian Reform Law was issued
Support for revolutionary groups, particularly in
Venezuela, increased
Cuba refused to sign the Nuclear Test Ban Treaty

1964   Fidel Castro visited the Soviet Union
Cuba and the Soviet Union signed a long-term trade
agreement providing for Cuban sugar deliveries for
1965-70 of 24 million tons at a fixed price of U.S.
$.06 per pound
Marcos A. Rodríguez tried and executed
Partido Socialista Popular leaders purged
Castro announced that Cuba would produce 10 million
tons of sugar in 1970, signaling a return to depend-
ence on one agricultural crop and the abandonment
of plans for major industrialization
Conference of Latin American Communist Parties held in
Havana subscribed to the Soviet line

1965   Cuba participated in the Consultative Meeting of Com-
munist Parties held in Moscow
Ernesto "Che" Guevara initiated a series of trips to
Asia and Africa
Three-year Soviet-Cuban trade agreement signed
Fidel Castro assumed greater personal control over the
economy

Partido Unido de la Revolución Socialista (PURS) created

Cuba allowed for the exodus of tens of thousands of
Cubans

Castro rejected the Communist doctrine on "the leading
role" of Communist parties in the revolutionary struggle and criticized bitterly Latin American Communist
parties for not supporting guerrilla warfare

1966   Cuba and China criticized each other, straining relations

Tricontinental Conference held in Havana and the Organization for the Solidarity of the Peoples of Asia,
Africa, and Latin America (OSPAAL) was founded

Castro and delegates from most Latin American leftist
groups formed the Latin American Solidarity Organization (LASO)

Economic situation continued to deteriorate and Castro
proclaimed the supremacy of moral over material incentives

1967   Castro admitted to Cuban-Soviet disagreement over
Cuban support for guerrilla activities in Latin
America

First Latin American Solidarity Organization (LASO)
conference held in Havana

Flights resumed to take U.S. citizens out of Cuba after
one-year suspension

Che Guevara killed in Bolivia

Cuba refused to sign the Nuclear Non-Proliferation
Treaty

1968   The "microfaction," nine pro-Soviet members of the
Central Committee including Aníbal Escalante, were
tried as "traitors to the revolution" and received
jail terms

All remaining private business except small agricultural
plots were confiscated

The University of Havana was placed under strict military discipline and control following student demonstrations

Castro made a major address justifying the Soviet invasion of Czechoslovakia

1969   Castro committed Cuba to a long-range agricultural development plan thereby postponing the country's
move to industrialization

U.S. and Cuba signed an agreement to return passengers aboard airliners hijacked to Cuba

Cuba became the first nation to establish formal relations
with the Vietcong (National Liberation Front)

Cuba participated in the Moscow-based World Conference
of Communist Parties

Soviet Defense Minister Marshall Andrei Grechko visited Cuba

Soviet naval squadron visited the island for the first time

1970 Castro announced the capture of leaders of exile group Alpha 66 which had landed in Cuba in April

Cuba's attempt to produce 10 million tons of sugar failed and the Labor Ministry reported that productivity among sugar workers was so low that the cost of the 1970 harvest was three times higher than its value on the world market

Economic problems compelled Castro to replace several civilian ministers with military officers

The possibility of a Soviet naval base in Cienfuegos led to a diplomatic exchange between the U.S. and the Soviets and a final statement by the latter that they were not building "military bases in Cuba ... [nor] doing anything that would contradict the understanding reached between the governments of the U.S.S.R. and the U.S. in 1962"

Cuba faced economic crisis with declines in production in almost all sectors and the Labor Minister attributed the problems to "widespread passive resistance" by all workers

Castro called for closer ties with the Soviet Union and acknowledged that forms of struggle other than Castroism were possible in Latin America

A joint Inter-Governmental Soviet-Cuban Commission for Economic, Scientific, and Technological Cooperation was established and Carlos Rafael Rodríguez was appointed chairman

1971 Castro urged Cubans to work harder to increase low levels of productivity of Cuban economy

Soviet Premier Alexei Kosygin visited Cuba

Castro visited Salvador Allende in Chile

Cuba and the Soviet Union signed a long-term economic and trade agreement

1972 The Cuban government withdrew 600 million pesos from circulation (out of 3 billion total) to combat inflation

Castro toured Africa, Eastern Europe, and the Soviet Union on official visits

Cuba was formally admitted as the ninth member to COMECON, the Soviet bloc's economic alliance

Castro signed a new long-term Soviet-Cuban economic agreement whereby Cuba's large debt to the Soviets was deferred to 1986, after which it would be repaid over the next twenty-five years at no interest

Castro announced a major reorganization of administration

strengthening the structural capacity of the government to manage major economic activities

1973   U.S. and Cuba signed an agreement on hijacking
          Castro supported a pan-Latin American regional group-
              ing which would exclude the U.S.
          An experiment in local government, Poder Popular
              (People's Power), was begun in Matanzas province
          At a meeting of the Confederation of Cuban Labor
              (CTC) Castro announced the abandonment of moral
              incentives and the establishment of Soviet-type
              norms for the labor force in an attempt to increase
              productivity
          At the Conference of Non-Aligned Countries in Algiers
              Castro praised the Soviets and attacked the theory
              of two imperialisms espoused by other non-aligned
              leaders

1974   Soviet party leader Leonid Brezhnev visited Cuba on
              official mission
          Cuban-Argentine trade agreement of $1.2 billion signed--
              the largest between Cuba and any Latin American
              nation
          President Ford indicated U.S. support of Organization
              of American States' move to improve relations be-
              tween member nations and Cuba
          Cuban Foreign Minister Raúl Roa García at United Na-
              tions declared there would be no normalized moves
              with U.S. until the latter had lifted economic blockade
          Cuba stepped up its training and support of Communist
              groups in Mozambique, Guinea-Bissau, and Angola

1975   Cuban voter referendum on new constitution provided
              for National Assembly with five-year terms for those
              elected (by Cubans sixteen and older). Also involved,
              a 31-member State Council with a president, a first
              vice-president, and five second vice-presidents.
          U.S. eased restrictions on exports to Cuba by foreign
              subsidiaries of American companies; direct trade re-
              mained embargoed
          U.S. reported Cuban soldiers and advisers in Angola
              to support the Marxist group MPLA
          Castro presided over first Cuban Communist Party Con-
              gress which approved Cuba's new Socialist constitu-
              tion and five-year economic plan

1976   Cuba announced it would not withdraw troops from An-
              gola supporting the MPLA. U.S. Secretary of State
              Henry Kissinger warned Cuba about its intervention
              in Angola.
          Canadian Prime Minister Pierre Trudeau visited Cuba

Cuba informed Sweden of its intention to withdraw half
its troops from Angola by December 1976
Castro accused the U.S. of sabotage in the crash of a
Cuban passenger jet near Barbados

1977   COMECON announced decision to build a nuclear power
station in Cuba
Week of Cuban solidarity with Nicaragua held in Havana
Castro visited Africa and the Soviet Union and agreed
with Brezhnev to continue support for national liber-
ation movements
Castro and Jamaican Prime Minister Michael Manley ex-
changed official visits to each other's countries
President Jimmy Carter ended travel restrictions on U.S.
citizens to Cuba and "interest" sections of consular
officials were set up in Washington and Havana
First Congress of the Committees for the Defense of the
Revolution (CDR) held in Havana
Cuban military advisers and combat troops entered
Ethiopia
All Cuban officials were expelled from Somalia as a reac-
tion to Cuban presence in Ethiopia
Castro reasserted Cuba's determination to help black
African liberation movements, reaffirmed support for
Puerto Rican independence, and claimed that Cuba's
presence in Africa was non-negotiable with the U.S.
Granma article complained about Chinese attacks on the
Cuban Revolution and on Cuba's internationalist com-
mitments in Africa

1978   Cuba hosted the World Youth Festival
Cuban radio hailed the ratification of the Panama Canal
treaties
Cuban troop strength in Ethiopia reported by the U.S.
at 3,500-5,000
U.S. accused Cuba of supporting invading rebels in
Zaire
Tomás Borge, leader of Frente Sandinista de Liberación
Nacional (FSLN), arrived in Cuba and met with
Party leaders
Castro attacked the foreign policies of the U.S. and
China
One hundred fifty Cuban-American political prisoners
allowed to leave Cuba
Central Planning Board President Humberto Pérez ex-
plained that Cuba had not reached a rate of economic
growth allowing it to emerge from underdevelopment

1979   Cuba supported Vietnam and condemned China's "military
agression"
The Soviet Union commenced a build-up and

modernization of the Cuban military by providing the
Castro government with its first submarine and two
torpedo boats

Cuba announced the release of several hundred political
prisoners

National Assembly of People's Power met in Havana.
Castro called for higher discipline and working
standards. He criticized public services, especially
transportation.

Cuba and Jamaica signed economic, scientific, and technical cooperation agreements and continued to maintain close relations

Cuban military, technical, and economic advisers arrived
in Nicaragua following the overthrow of Somoza's government. Close relations and collaboration developed
between the two countries.

Cuba's Movement for Peace and Sovereignty of Peoples
(MPSP) appealed for solidarity with the people of El
Salvador in their struggles

The U.S. charged that a Soviet combat brigade was stationed in Cuba; Castro denied it

The Sixth Nonaligned Summit Meeting was held in Havana

Grenada admitted that it received arms from Cuba, but
only for defensive purposes. The two countries
signed a two-year technical and economic cooperation
agreement.

1980   Castro shuffled Cuban cabinet; assumed personal control
over Ministries of Defense, Interior, Public Health,
and Culture

Ten thousand Cuban refugees entered the Peruvian Embassy in Havana seeking asylum and starting a mass
exodus of Cubans to Peru and the U.S.

Cuba and Nicaragua signed an economic, scientific, and
technical agreement and established a joint intergov-
ernmental commission to set the standards of relations
in those fields

Angolan President José Eduardo Dos Santos visited Cuba
and signed an agreement establishing bilateral ex-
changes between Cuba and Angola

Cuban MIGs attacked and sunk the Bahamian patrol boat
Flamingo. The Cuban government apologized, saying
"it sincerely regretted the incident."

The M-19 guerrilla group that held diplomats hostage in
the Dominican Republic's embassy in Colombia for
sixty-one days arrived in Cuba with eleven hostages.
Hostages were released while the guerrillas remained
in the island.

Daniel Ortega, member of the Nicaraguan governing
junta, and Grenadian Prime Minister Maurice Bishop
addressed the May Day rally in Havana

Cuba and the U.S.S.R. signed an agreement for the
building of a nuclear research center in Cuba

Fidel Castro and a large delegation attended the first
anniversary celebrations of the Sandinista victory in
Nicaragua

Cuba and the U.S.S.R. signed a bilateral economic cooperation agreement for 1981-85. Trade would increase by 50% and would average over U.S. $8 bil-
lion per year

Raúl Castro announced the creation of territorial militias

President Carter accused Cuba of aiding the supply of
arms and insurgents to leftist groups trying to overthrow the government of El Salvador

The Cuban Institute for Friendship with the Peoples
(ICAP) organized a drive of solidarity with Puerto
Rico to help Puerto Rican people gain their full
sovereignty and independence

During Mexican President José López Portillo's visit to
Cuba, the two countries signed a joint communiqué
that included a call for ending the economic blockade
against the Cuban people and the cessation of violations of Cuban air space. The Mexican president
condemned the cold war and the arms buildup, noting
that nothing could be achieved without détente.

Addressing the Second Cuban Communist Party Congress,
Fidel Castro admitted failure to reach the goal of 6
percent economic growth set by the First Congress
during 1976-80

1981   Eleven new "alternative" members of the Politburo were
named. Five of them, Humberto Pérez, Vilma Espín,
Roberto Viega, José Ramírez Cruz, and Armando
Acosta would represent "mass organizations" in order
to facilitate grass-roots relations. The other six
members were all Army leaders.

The government organized militias of territorial troops
on a regional basis (Milicias de Tropas Territoriales--
MTT). Many militiamen were veterans of Angola and
Ethiopia. Militias would fight sabotage from exile
groups.

Foreign Minister Isidoro Malmierca flew to Guyana on
official visit. Affirmed Guyana's territorial integrity
in dispute with Venezuela.

Nominee for U.S. Secretary of State Alexander Haig rejected notion of improved relations with Cuba during
confirmation hearings

U.S. FBI arrested seven anti-Castro Cuban exiles in
Florida Keys. Exiles were linked to Alpha 66.

U.S. State Department failed to convince Castro to accept the return of 2,000 undesirables from 1980
Mariel boatlift

Trade agreements signed with Guyana included increased
  technical assistance
Government reported slight increase in rate of infant
  mortality from 1.93% in 1979 to 1.96% in 1980
U.S. reported that $532 million had been spent on entry
  and resettlement of 125,000 Cuban and 12,400 Haitian
  refugees who emigrated to the United States in 1980
U.S. expelled First Secretary of Cuban Interest Section, Ricardo Escortín, for illegal business deals with
  U.S. businessmen and for alleged intelligence activities
Group of Cubans seized Ecuadorean Embassy in Havana,
  holding Ecuadorean envoys hostage, demanding political asylum
Valdilev M. Vasev, Minister Counselor of the Soviet
  Embassy in Washington, D.C., denied that the Soviet
  Union had supplied arms to Salvadorean rebels, but
  said that Soviets were shipping arms to Cuba without
  restrictions on their shipment to third countries
Cuban security forces entered Ecuadorean Embassy and
  arrested would-be Cuban emigrants
Castro gave speech to Soviet Communist Party Congress;
  said that U.S. threatened a blockade and denied that
  Cuba was instigating the Salvadorean rebellion
Castro met with Leonid Brezhnev. The Soviet leader
  assured Castro of Soviet's continued support.
Colombia broke off diplomatic relations with Cuba over
  Cuban links to Colombian guerrilla insurgency
Alpha 66 reported 30 sabotage missions accomplished in
  six months inside Cuba, including attack at Regla
  power plant
Castro reaffirmed solidarity with Soviet Union in speech
  marking 20th anniversary of Bay of Pigs
Fidel Castro made overtures to Christian groups for
  revolutionary unity in country
U.S. Assistant Secretary of State Croker linked solution
  of Southwest African problem with withdrawal of
  Cuban troops from Angola
ANAP President José Ramírez launched campaign to aid
  small, private farmers, especially those belonging to
  independent cooperatives, in new agricultural policy
Costa Rica broke off diplomatic relations with Cuba over
  human rights issues
Soviet 4-ship naval squadron left Cuba after month-long
  visit and maneuvers in Caribbean
U.S. Department of State reported Cuban transfer of
  Soviet-made T-55 tanks to Nicaraguan Sandinista
  government
Vice-President Raúl Castro was awarded the USSR Order
  of the October Revolution
Aid pact with COMECON signed in Sofia. Plans made to

improve sugar production by modernizing mills and transport system. Aid package worth about $1.2 billion over ten years.

Cuba's largest cement factory, christened Karl Marx Plant, completed after six years of construction

Manuel Urrutia Lleo, first President of Revolutionary Cuba, who later became Castro foe, died in New York City at age 79

Castro voiced "profound suspicion" over the origin of crop blights and dengue fever that killed 113 Cubans and damaged harvests. He said that the epidemics had possibly been introduced by the CIA.

Castro met with President López Portillo of Mexico on the Mexican island of Cozumel. Castro accepted that his presence at the Cancún North/South summit would jeopardize meetings in light of U.S. warnings to stay away.

National Census put Cuban population at 9,706,369; Havana at 1,924,000. 97.2% of registered voters in Cuba voted in municipal elections to elect 9,763 members.

Jamaica broke diplomatic relations with Cuba, citing lack of cooperation in the extradition of Jamaican criminals

Félix Fidel Castro Díaz, Fidel Castro's only son, became Director of Atomic Energy Commission

Mexican government offered to act as "communicator" between Castro regime and U.S. government.

Fidel Castro was re-elected President of the Council of State with his brother, Raúl Castro, as first Vice-President. Other Vice-Presidents elected were Juan Almeida Bosque, Ramiro Valdés, Guillermo García, Carlos Rafael Rodríguez, and Blas Roca. Flavio Bravo replaced Roca as Council Chairman.

Humberto Pérez announced 1981 budget deficit of 785 million pesos, compared to 249 million pesos in 1980. He blamed falling commodity prices and dengue fever for shortfalls.

Government raised prices 10-30% on rationed goods and most catering services (restaurants, bars, etc.). Two ministers, including Minister of Interior Trade Serafín Rodríguez, were ousted after public outcry swelled over excessive increases in some areas. Manuel Vila Sosa replaced Fernández as Internal Trade Minister.

1982   Trade deal signed with Libya after visit by delegation headed by State Committee for Economic Cooperation President Héctor Rodríguez Llampart

Andrés Rodríguez Hernández, a stowaway aboard Miami jetliner, was first refugee to be deported to Cuba

President Ronald Reagan named F. Clifton White to lead

Radio Martí, U.S. broadcasting initiative designed to
counter anti-American broadcasts from Cuba
President Reagan said that Secretary of State Alexander
M. Haig met secretly with Cuban Vice-President Car-
los Rafael Rodríguez in Mexico in the fall of 1980.
Haig discussed arms shipments from the Soviet Union
to Cuba.
U.S. Justice Department lifted embargo on Cuban publi-
cations to the United States subscribers in effect
since mid-1981
Foreign Minister Isidoro Malmierca visited Angola; Deputy
Foreign Minister Oscar Oramas, General Abelardo
Colomé, and Politburo member Jorge Risquet visited
Ethiopia, Mozambique, Tanzania, and Zimbabwe
Cuban government liberalized restrictions on foreign investment in Cuba in effort to revive tourist industry
and gain foreign exchange
Reagan administration announced new restrictions on
travel to Cuba, banning tourist and business travel
after May 15. Only academic and family unification
exit visas would be granted.
Castro rejected Reagan administration's appeal for break-
ing off relations with the Soviet Union in speech to
Association of Small Farmers (ANAP)
ANAP President José Ramírez said that small private
farmers and cooperatives produced 70% of country's
tobacco, 67% of cocoa, 54% of coffee, 50% of vegetables,
18% of sugarcane, and owned 21% of beef and dairy
cattle stock. Castro admitted that "free market" food
policy of 1980 was a failure due to unscrupulous "mid-
dlemen."
Vice-President Carlos Rafael Rodríguez visited France
Cuban government cracked down on "capitalist activities" in free markets for food and clothing, and
made 200 arrests
Vice-President Carlos Rafael Rodríguez addressed UN
General Assembly and said Cuba had received "huge
quantities of modern and sophisticated weapons" from
the Soviet Union and allies
Torrential rains and flooding caused extensive damage
to Cuban property and agriculture
Cuban government announced that 180,000 foreigners
had visited Cuba in 1981, bringing in $80 million
U.S. expelled two Cuban envoys at Cuban mission to the
UN after finding sophisticated telecommunications
equipment in Orlando, Florida, warehouse
Former Foreign Minister Raúl Roa died.
President Fidel Castro, in Bayamo, in address during
Moncada celebrations commemorating the Cuban Revo-
lution, blamed the "capitalist crisis" for Cuba's eco-
nomic woes. He asked Cuban people to work harder

and endure, and said Cuban troops would not leave
Angola until all South African troops were out of
Namibia.

Castro announced new record sugar harvest for 1981-82;
8.21 million tons was second largest ever

Vice-President Carlos Rafael Rodríguez announced cut-
backs in Cuban imports to offset shortfalls from low
sugar prices

U.S. House of Representatives voted to authorize $7.5
million for Radio Martí

U.S. Senate voted to prevent by any means, including
the force of arms, the extension of Cuban influence
in the Western Hemisphere. Action was amendment to
a $9 billion supplemental appropriations bill.

Cuban government announced that it had asked Japanese
and Western banks to renegotiate the terms of debt
payments due in the next three years

Senate Foreign Relations Committee voted to approve
Radio Martí

Cuban authorities released poet Armando Valladares
after appeals from French President Mitterrand and
Nobel prize-winning novelist Gabriel García Márquez

NATO Report claimed that Cuba received $3 billion in
economic aid from the Soviet Union in 1981, 60% of
all aid from Soviet Union to the Third World. Cuba
was 7th in ranking of military aid recipients from the
Soviet Union, after Ethiopia, Vietnam, South Yemen,
Afghanistan, Mozambique, and Angola, with $54 mil-
lion.

Four high-ranking Cuban officials were indicted in Miami
on Federal charges of drug trafficking. Included was
Cuban Navy Vice-Admiral Aldo Santamaría Cuadrado.

Canadian University Service Overseas (SUCO), Inter-
Church Fund for International Development, and
OXFAM-Canada all signed a pact with the Cuban government pledging to promote continued scientific and
technical collaboration. There was a call for an end
to the embargo on official Canadian aid to Cuba.

Fidel Castro conferred with new Soviet leader Yuri
Andropov in Moscow

U.S. and USSR held high-level talks on Cuban troops in
Angola

1983   Cuban officials met in Panama with economic ministers
from several Western nations to renegotiate payments
on $2.6 billion in foreign debt

U.S. reported that Cuba--with Soviet backing--has expanded its amphibious fleet. Cuba also completed
building shelters for a fleet of about 225 Soviet-built
MIG fighter bombers and plans to receive four more
Foxtrot submarines from the Soviet Union.

Former Cuban agent turned federal informer revealed,
at U.S. Senate hearings, that his work as a drug
dealer for the Cuban government earned $7 million
for the Castro regime. The federal informer also
estimated that 3,000 Cuban agents entered the U.S.
during the 1980 boatlift.

Cuba rejected charges that two diplomats in their UN
delegation in New York City were guilty of spying.
The U.S. government ordered the Cuban diplomats
to leave the U.S. within 48 hours.

Cuban government charged that U.S. reconnaissance
plane violated Cuban airspace in a "deliberate and
cold provocation"

U.S. Assistant Secretary of State Thomas O. Enders met
with the head of the Cuban Interest Section in
Washington, Ramón Sánchez-Parodi, to request that
Cuba take back thousands of Cubans (who came to
the U.S. via the 1980 Mariel boatlift) because of
their criminal conduct in Cuba

U.S. Senate Foreign Relations Committee voted 13 to 4
to authorize Radio Martí

Cuban Vice-President Carlos Rafael Rodríguez claimed
that Cuba was willing to open "serious negotiations"
on re-establishing normal relations with the United
States, provided that the Reagan administration took
the first step

Osvaldo Dorticós Torrado, former President of Cuba,
committed suicide

Cuba informed the United States that it was willing to
discuss the return of some of the Cubans who came
to the U.S. illegally in 1980, but only as part of
overall negotiations on "normalization of migration"
between the two countries

President Reagan emphasized that the Soviet Union had
repeatedly violated the 1962 agreement that ended
the missile crisis by continuing to ship "offensive
weapons" into the American hemisphere

Americas Watch, a human rights organization, claimed
that at least 250 long-term political prisoners were
confined in Cuban jails under "brutal conditions" and
that up to 2,000 former prisoners were denied decent
work

French officials reported that Cuba arrested Ricardo
Bofill, former Vice-Rector of Havana University, after promises that he would be allowed to leave the
country. Bofill had sought refuge in the French
Embassy.

The United States tightened its economic embargo against
Cuba by banning imports of semifinished nickel products from the Soviet Union which is a major buyer
of Cuban nickel ore

Cuba reached a tentative agreement with its creditors to reschedule $810 million in short- and medium-term commercial debt and nearly half of its $3.5 billion debt owed to Western governments and banks

1984   President Reagan accused Cuban leaders of having betrayed the Cuban people and not telling them the truth about Cuban activities around the world. Reagan said that there were as many as 10,000 political prisoners in Cuban jails. Reagan also charged that Cuba's economy failed to provide even elementary needs.

The United States Immigration and Justice Department officials decided that 100,000 of the Cubans who came to the United States in the 1980 Mariel boatlift were eligible for legal status and citizenship opportunities under the 1966 U.S. law on earlier Cuban migration

Cuban President Castro visited Western Europe for the first time. President Castro made an unscheduled stop in Spain en route home from Yuri Andropov's funeral in Moscow.

Angolan guerrillas reported that Cuban-led Angolan forces opened a major offensive against rebel strongholds along the southeast border

South Africa labeled "unacceptable" Cuba's heavily conditioned offer to withdraw its 25,000 troops from Angola, dimming hopes for efforts to negotiate independence for Namibia

Roberto Veiga Menéndez, alternate member of the Cuban military politburo and Secretary General of the Cuban Federation of Workers, said that the Cuban government had doubled the size of its militia forces to more than 1 million men and women

Jorge Valls, prominent Cuban poet and political dissident, was released after being imprisoned for more than twenty years

Reverend Jesse Jackson arrived in Cuba for a two-day visit and in airport meeting with Fidel Castro said that the United States and Cuba "must give peace a chance." Castro said that he invited the Reverend Jackson as a "gesture of friendship to the people of the United States." Reverend Jesse Jackson returned to the U.S. with 26 freed American and Cuban prisoners.

Cuban and United States officials started discussions in New York about immigration issues, including the possible return of 1,000 Cuban refugees from the 1980 Mariel boatlift

Supreme Court reinstated Reagan administration's curbs on tourist and business travel to Cuba

Western commercial banks reportedly agreed to reschedule

about $100 million of Cuba's debts for this year on
easier terms than those of 1983

Foreign Minister Wu Xueqian said that China and Cuba
agreed to improve trade, cultural, and technological
ties despite their differences on international matters

President Fidel Castro, in speech marking 31st anniversary of the Cuban Revolution, said he would welcome
any steps to lessen bitter hostility between U.S. and
Cuba

State Department sees no change in Cuban policy reflected in President Fidel Castro's speech; rules out
comprehensive talks for now, demanding Cuba first
make fundamental foreign policy changes

1985 Cuban President Fidel Castro ordered austerity measures
expected to sharply slow the country's economic growth
and possibly cause more reliance on the Soviet Union.
The measures are aimed at saving badly needed foreign exchange.

Twenty-three Cuban 1980 Mariel refugees were returned
to Havana. These are the first of more than 2,700
unwanted Cubans who could be sent back by the
United States as part of an agreement with Fidel
Castro's government.

The Reagan administration claimed that Fidel Castro's
absence from Soviet leader Konstantin Chernenko's
funeral was evidence of friction between Cuba and
the Soviet Union over economic aid. The Reagan
administration also viewed the absence as one of
Castro's periodic urges to show his independence
from the Soviet Union.

Cuban leader Fidel Castro said that Cuba's relationship
with the Soviet Union had never been better and that
his absence from the funeral of Konstantin Chernenko was not significant

Cuba offered to withdraw 100 of its military advisors from
Nicaragua but vowed to return even more if the U.S.
continued "its dirty war" against Nicaragua

Ecuadorean President León Febres Cordero arrived in
Cuba for an official visit and was welcomed by Fidel
Castro. Talks centered on the Latin American debt
problem and unrest in Central America.

Radio Martí, the U.S. Information Agency news service
for Cuba, began broadcasts to Cuba

Havana suspended all immigration proceedings between
Cuba and the United States in response to the start-
up of Radio Martí. Cuban-Americans will be prohibited from visiting Cuba.

Fidel Castro suggested that the United States and other
industrialized nations pay the Latin American 4,360
billion debt

The Cuban government sought to improve relations with churches and synagogues and urged mutual cooperation

Cuban Central Planning Board head Humberto Pérez González was replaced by Construction Minister José López Moreno in a continuing government shakeup that is expected to lead to a more prominent role for Fidel Castro's younger brother, Raúl

Fidel Castro said that Cuba was going nuclear. A four-unit nuclear plant is already being built and two more are planned.

Fidel Castro urged Latin American nations to band together and simply refuse to pay their foreign debt

Castro refinanced Cuba's $3.5 billion debt and promised to continue making payments on loans

United States and Cuba officials agreed that relations between the two countries had reached one of the lowest points in years. A slight warming in the relationship halted abruptly when the U.S. inaugurated Radio Martí's broadcast service to Cuba, which was denounced by Fidel Castro as a U.S. attempt at subversion.

Cuba signed a pact rescheduling $90 million owed to foreign commercial banks

President Reagan imposed immediate restrictions on entry of Cuban officials into the U.S.

Soviet Foreign Minister Eduard A. Sheverdnadze arrived in Cuba for talks with President Fidel Castro

About 20,000 protesters denounced the United States, outside the U.S. Mission in Havana, after Cuba protested what it called spy flights by U.S. surveillance planes

Four Cuban Embassy employees in Madrid, Spain, tried to kidnap former Cuban official Manuel Antonio Sánchez Pérez, who had asked for political asylum. The attempt was foiled by the intervention of 30 bystanders. Four employees, including the vice-consul, were arrested.

Cuba reportedly planned to reduce its outlays for military and public security in 1986, after failing to meet production goals for leading exports in 1985

President Reagan warned that the U.S. had a legal right to defend itself against five nations that he claimed were sponsoring terrorist "acts of war" against America: Iran, Libya, North Korea, Cuba, and Nicaragua

THE DICTIONARY

ABC. A prominent clandestine organization opposing the dictatorial regime of Gerardo Machado (1925-1933) in the early 1930s. Composed of intellectuals, students, and representatives from the middle sectors of society, the ABC used the cellular concept for an underground organization. Each cell within the ABC contained seven members who had no knowledge of the other cells. The directing cell was known as A, the second tier of cells was B, then C, and so on. In December of 1932, ABC issued its program manifesto, which opposed not only the Machado regime, but also the circumstances that brought it into existence. ABC advocated the breakup of large landholdings, nationalization of public services, limitations on land ownership by U.S. interests, as well as political liberty and social justice.

In late 1932, the ABC conceived a two-phased plan to eliminate Machado. Phase I succeeded when Senate President Clemente Vásquez-Bello was assassinated in Havana. Phase II, which contemplated blowing up all top government leaders during Vásquez-Bello's funeral in Havana cemetery, failed when the Vásquez-Bello family requested that he be buried in the family plot in Santa Clara cemetery. A gardener working in the Havana cemetery later found the buried explosives. Following the ABC's fiasco, Machado's police launched a manhunt for real and suspected members of the organization, driving most of the ABC leadership into exile. In 1933, the ABC supported the mediation efforts of U.S. Ambassador Sumner Welles to ease Machado out of the Presidency and bring about a peaceful solution to the violence then prevailing on the island. Following Machado's resignation, the group participated in the short-lived regime of Carlos Miguel de Céspedes in 1933 and later opposed the more radical regime of Dr. Ramon Grau San Martín (1933-1934). The ABC ceased to operate after those turbulent years, but many of its leaders came to occupy prominent positions in later administrations and to exert considerable influence, such as Joaquín Martínez Sáenz, Carlos Saladrigas, and Jorge Mañach.

ABC RADICAL. A splinter group that broke from the ABC parent organization due to the latter's support of U.S. Ambassador Sumner Welles's mediation efforts between Cuban President Gerardo Machado's government (1925-33) and its opposition in

1933. ABC Radical opposed any U.S. interference in Cuba's
internal affairs. Led by Juan Govea and Oscar de la Torre,
the group exercised little influence in Cuba's political life.
During President Carlos Mendieta's administration (1934-35),
José Becquer, one of ABC Radical's leaders, became a Presidential Cabinet member.

ABOLITIONISM. A challenge to Cuban society in the 1800's, the
international movement to abolish slavery originated in England
as a result of opposition to Spain and France's West Indian
monopoly. A powerful humanitarian movement and religious
effort, abolitionism represented a shift in attitudes towards
the uneconomical institution. The Spanish by 1820 had accepted abolitionists' demands, and formally abolished the slave
trade. From 1816 to 1820 at least 100,000 slaves had been imported into Cuba to work in the sugar industry. Illegal trade,
however, continued into the mid-1800's as the powerful sugar
planters in Cuba sought to further their financial interests.
Several factors--among them the pressure of reformers, British
insistence on abolition, and the example of the Civil War in the
United States--gravitated in favor of abolition. A number of
large estate owners had come to realize the economic disadvan-
tages of slavery at a time of increased mechanization. Also, in
the late 1840's and 1850's a number of Chinese laborers and
Indians from Yucatan had arrived in Cuba to work in the sugar
fields and in railroad construction, providing a cheap, although
still small, labor supply whose maintenance was less problematic
than that of the slave population. A significant number of
white workers, many with skills needed to operate the new machinery, also joined the labor force in the mills. Finally, those
seeking the overthrow of Spanish power in Cuba looked at the
black population as a strong and necessary ally in any attempt
to liberate Cuba.

Despite these strong forces at work, emancipation came about
slowly. In 1865 the slave trade was partially curtailed, but it
was not until 1886, long after the end of the Ten Years' War,
Cuba's first major attempt at independence, that slavery was
completely abolished. On November 5, 1879, the Spanish government issued a law abolishing slavery in Cuba. It established
an eight-year state of tutelage (patronato) for all liberated
slaves. Patterned somewhat on the earlier apprenticeship system introduced by England for its Caribbean possessions, the
patronato guaranteed the continued labor of blacks for their
masters. Masters were required to furnish their wards with
proper food and clothing, and to furnish them monthly wages.
It proved more profitable, however, for the masters to free the
blacks and hire them as laborers, thereby avoiding the necessity of maintaining them during slack seasons. With sugar
developing as a modern, mechanized industry, without a continuous supply of new slaves, and with the availability of an
alternate labor supply composed of poor whites and Asians,

black slavery became increasingly uneconomical. The abolitionists themselves condemned the patronato as "harder servitude
than slavery itself," and on October 7, 1886, two years before
the patronato was to terminate, slavery was abolished in Cuba
by Moret's law.

ABREU, MARTA (1846-1909). Born in Santa Clara from a very
wealthy family, Marta Abreu Arencibia de Estévez became known
as Marta Abreu through her philanthropy. She believed that
great wealth implied a duty and should be used for social welfare. She paid for the construction of a great theater in Santa
Clara and donated it to the city so that the income would support schools and facilities for the poor. She not only spent all
her money on the poor but also gave it away as a patriot during
the wars of independence. She died in Paris after Cuba had
reached independence.

ACCION CATOLICA. A Catholic organization of lay persons closely
associated with the Cuban Church and advised by a priest. It
established branches in major Cuban cities. Even though its
objectives were primarily religious, Acción Católica became increasingly involved in opposition to President Fulgencio Batista
(1952-59). Some of its leaders, such as Amalio Fiallo and
Andrés Valdespino, participated directly in insurrectional activities against Batista. After Fidel Castro's victory, Acción
Católica and the Church welcomed the initial reforms of the
Castro government, but as the revolution radicalized and Marxist influence became more important, all Catholic organizations
opposed the regime. In 1961 Castro expelled a significant number of nuns and priests from Cuba and curtailed Church activities to a minimum. See also Church, Roman Catholic.

ACCION REVOLUCIONARIA GUITERAS (ARG). One of several terrorist groups operating in Cuba in the mid-1940's during the
administration of President Ramón Gran San Martín (1944-48).
ARG activists were originally linked to Grau's revolutionary
movement in 1933 and to Antonio Guiteras's anti-Batista group,
Joven Cuba. ARG contained some 800 members and enjoyed
strong links with trade unions. One ARG leader, Fabio Ruiz,
was appointed Chief of Police in Havana by Grau in return for
the organization's support. Due primarily to the establishment
of Fulgencio Batista's dictatorship (1952-59), the ARG and
other similar groups were disbanded or ceased to operate on
the island during the 1950's.

ACOSTA, AGUSTIN (1886-1979). Cuban poet and politician who
served as Secretary to President Carlos Mendieta (1934-1936),
Acosta was also a Senator from Matanzas province during Fulgencio Batista's first administration (1940-1944). He won immediate recognition in 1912 for his collection of décimas (stan-
zas of ten octosyllabic lines) entitled A la bandera cubana.

His famous "La Zafra" (1929) depicted the lives of the Cuban
sugar workers of the 1920s and portrayed the United States as
the "colossus of the North," the symbol of world capitalism.
Honored by the Cuban Congress with the title of "Cuba's National Poet" in 1955, he left Cuba in 1972. His last two books,
El apóstol y su isla and Trigo de luna, were published in exile.
He died in Miami in 1979.

AFRICAN RELIGIOUS PRACTICES. See also Slavery, Blacks. Particular Afro-Cuban religions have traditionally been associated
with specific ethnic groups and African homelands, although
smaller groups have been absorbed and people have been attracted from other groups and from the white population. The
principal religion, known as Santería, is that of the Lucumí,
descendants of the Nigerian Yoruba. In the cities of Havana
and Matanzas, and perhaps elsewhere, there are Ñañigos, whose
beliefs originated among the Efik of eastern Nigeria. In Oriente
province, which has a high percentage of blacks, Bantu and
Congo are said to be numerous; presumably their religious practices differ from those of the Lucumí and Ñañigos, but little
is known of them. All versions of Santería have incorporated
some Roman Catholic ritual and mythology. Their devotees regard themselves as Christians and believe that the names of the
saints are translations into Spanish of the Nigerian names of
African spirits; thus Roman Catholicism is viewed as the Spanish tribal version of Santería. The association between Roman
Catholic and African beliefs is based largely on the superficial
similarity between the Yoruba pantheon and the Roman Catholic
saints, particularly in the emblems pictured with each saint or
spirit as an indication of the particular field in which he has
power. The spirits are called, interchangeably, santos (Spanish word for "saints") or orishas (Yoruba word for "spirits").
The orishas are important men who have died, most of whom
were kings and founders of tribes. Each worshiper has a patron saint, chosen by divination, to suit his personality, or
because his family regards a particular orisha as its ancestor
and guardian. He belongs to a cult group or congregation
which holds regular meetings. An important feature of these
meetings is the music, consisting chiefly of singing and drumming. The songs are hymns and prayers in Yoruba. The
drumming, which is used to call the orishas, serves to raise
the emotional level of the proceedings; it is done according to
fixed patterns that are themselves hymns and prayers, employing the rising and falling tones of the Yoruba language to convey meanings.
    The santero (leader) of each cult group achieves the position
through or by being apprenticed to another santero at an early
age to secure the protection of the patron saint of that santero.
A santero maintains his position by manipulating divination procedures and dispensing advice and magical charms so as to retain the confidence of a congregation, who in return support

him through the fees they pay and the offerings they make.
The duties of the santero include consulting with members of
his congregation, visitors with special problems, and even
tourists. He must look after the shrine in his house, decorated with effigies derived equally from Nigerian and Spanish
culture. He must also take care of the sacred stones, in which
the power of the orishas is thought to reside and which must be
anointed with a potent mixture of herbs and the blood of sacrificed chickens. A similar mixture is used to anoint the heads
of devotees of the orishas. The ritual of the Santería varies
according to the inclination of the santero from an imitation of
Roman Catholic ritual (complete with candles, the Lord's Prayer,
the Hail Mary, and Ritual gestures) to the deliberate creation
of a hysterial atmosphere in which the santero and others present are possessed by the orishas. A large number of possessions in one session is regarded as a good omen. When possessed, worshipers put on ritual clothes and adopt the character of particular orishas, who are at once recognized by the
congregation. In this state of trance the worshipers speak with
the voices of the orishas, who in this manner offer advice to
their followers. The procedure lends itself to manipulation by
the santero, but probably most of the trances are genuine.

In the 1960's the most famous of the legendary secret religions, Ñañiguismo, was rediscovered by the Cuban poet Lydia
Cabrera. Ñañigos are members of the Abakuá secret society for
men only; the society, believed by most Cubans to have been
extinct since the early twentieth century, was notorious in the
nineteenth century for its reputed practice of child sacrifice.
An important feature of the Abakuá rites is the oath of secrecy
required of initiates. The beliefs of the cult emphasize the
dangerous influences that threaten people, particularly those of
African descent. Membership in the Abakuá, however, confers
protection in this world and the next. The society is organized
into ethnic groups--known as potencias (powers) or naciones
(nations)--in which the seven leaders are the descendants of
chiefs and priests in the African homeland. Rituals include a
sacrificial communion feast and liturgical reenactments of legends.
Participants in these rites must be in a state of ritual purity.
The deities are spirits of the dead whom it is necessary to
placate. They are represented by masked dancers, who at one
time appeared in public on saints' days and carnival occasions
but at present are only evident in the self-conscious folk re-
vivals of the Cuban theater. There is some incorporation of
Roman Catholic ritual comparable to that of the Santería but on
a much smaller scale. As in the Santería, drums are given
ritual significance as the voices of the spirits, but it is believed that possession is not a feature of the Abakuá ritual.

Cubans have always been disposed to regard all unknown
beliefs as sorcery or brujería (witchcraft). Brujería is the
usual upper-class term used to denote African cults and all
other practices dismissed as superstitious. It is also used by

followers of African cults to denote black magic--the deliberate
misuse of legitimate religious techniques for malicious ends.
Thus all santeros are supposed to be capable of black magic,
but nobody admits to practicing it. The rituals of all unfamiliar
cults are believed to consist exclusively of black magic. The
smallest and least known religious and ethnic groups are credited
with the most dangerous powers. They include, in approximate
order of deadliness, Ñañigos, Congo, Jamaicans, Haitians, Canary Islanders, and Chinese. Believers in Santería are likely to
attribute to sorcery such abnormal and inexplicable conditions
as insanity and any condition that fails to improve when appropriate offerings have been made to the orishas. Cubans who do
not associate with any organized cult group may still be fright-
ened at apparent indications of occult malice; for example, white
chicken feathers unaccountably discovered in the house.
Countermagic obtainable from a santero may be thought necessary. If the magic of a santero is not considered strong enough
in a particular situation, recourse may be had to a priest of one
of the more mysterious cults.

AFRO-CUBAN MUSIC. The large importation of slaves from Africa
in the eighteenth and nineteenth centuries had a definite impact
on the development of Cuban music. The African musical heritage was religious in origin and keyed to the beat of the drum.
By the twentieth century, Spanish melodies combined with the
percussion beats of Afro music and produced such well-known
non-religious rhythms as the rumba, bolero, conga, and the
son. Also see Music.

AGRAMONTE, IGNACIO (1841-1873). A leader in the Ten Years'
War (1868-1878) when planters from the eastern part of Cuba
rose up against Spain and emancipated their slaves in order to
obtain their help in the fighting. Agramonte, a Creole cattle
farmer and lawyer, rallied to the cause initiated by Carlos
Manuel de Céspedes. For five years, Agramonte led the revolt
in his native Camagüey. He participated in the drafting of the
Guáimaro Constitution of 1869 and executed numerous brave
military feats against Spanish forces. Best known of these
was the daring rescue of his colleague and friend, Col. Julio
Sanguily, whom Agramonte freed from the Spanish Army. He
achieved the rank of General of the rebel forces just prior to
his untimely death in a pitch battle with the Royalist Army in
1873. See also Guáimaro Constitution.

AGRAMONTE, ROBERTO (1909-   ). A Professor of Sociology and
author of numerous books on the subject, Agramonte became in
1951 leader of the Ortodoxo party after the death of Eduardo
Chibás. During the Presidency of Dr. Ramón Grau San Martín
(1944-1948), Agramonte became Ambassador to Mexico and later
the Ortodoxo party candidate for the Presidency in the 1952
elections canceled by General Fulgencio Batista's coup of March

10, 1952. Agramonte had the support of many middle-class
liberal Cubans and his rise to the leadership of the ortodoxos
was seen as an attempt at compromise between the various party
factions. Professor Agramonte was active against the Batista
dictatorship (1952-1959) and the Castro Revolution. During
the past few years he has lived in Puerto Rico devoted to writing and teaching.

AGRARIAN REFORM LAWS. After the Revolution of 1959 a large
portion of the nation's farmland was placed under state control
through two agrarian reform laws. Many small-scale farmers
were permitted to retain their farms, and some landless farmers
were given title to land they had previously rented or share-
cropped. Cooperatives and collectives, or communes, were also
created, but in the mid-1970's they were not numerous and did
not cover much acreage. The first major agricultural change
was wrought by the Agrarian Reform Law of 1959, which set a
limit of about 1,000 acres as the maximum land area most persons or corporate bodies could own, although unusually productive farms up to 3,300 acres were permitted. Under provisions of the 1959 law over 8.6 million acres of land were placed
under state control, and additional lands were expropriated.
Few of the large estates were split up; they were kept intact
by the government. A few years later the Agrarian Reform
Law of 1963 provided for the expropriation of virtually all private holdings over 165 acres. About 11,000 farms were taken
over under the new law, and those owners who were dispos-
sessed received monthly compensation of up to 250 pesos for a
ten-year period. Among farms escaping expropriation were
holdings being worked by brothers--provided that each brother's proportional share did not exceed the 165-acre limitation--
and a few larger farms because of the proven revolutionary
background of the owners.

As a result of the implementation of the 1963 act and further
acquisitions of private farmland by purchase from or death of
the owner, the state had acquired between 70 and 80 percent
of all farmland by the 1980's. The 1963 Agrarian Reform Law
also restructured the National Institute of Agrarian Reform
(Instituto Nacional de Reforma Agraria--INRA), which had
been created under the 1959 law to handle all matters related
to agricultural production, land reform, credit, and trade.
The 1963 law made INRA a ministry of government directed by
a cabinet minister. During the initial years following the two
agrarian reform laws, INRA experimented with setting up various kinds of state farms having different forms of administration and organization, including state-managed cooperatives.
Eventually all farms were converted into state farms, and their
previous designations were dropped. Merging of some state
farms reduced the number and all operated on the same administrative principles, farm workers receiving wages rather than
a share of the farm's profits. Several state farms located in

the same geographic region are usually formed into an
agrupación (administrative unit) administered by a director and
a six-member advisory council. The directors reported to regional directors, who were subordinate to INRA organs in
Havana. INRA disappeared in 1974 when all its functions were
absorbed by the Ministry of Agriculture (MINAG).

AGRICULTURE. Of Cuba's total area of 28 million acres, about 80%
is in farmland, of which 25.5% was cultivated and 39.3% was in
pasture in 1952. About 52% of the cultivated land is in sugarcane, much of it formerly held in large estates. Until 1959 only
16% of the land was directly cultivated by its owners. Tenant
farming, sharecropping, and hired farm labor were the norm.
Sugar companies owned about 25% of the land area of Cuba. An
agrarian reform law of June 1959 made the government proprietor of all land in Cuba, created a National Institute of Agrarian
Reform (Instituto Nacional de Reforma Agraria--INRA) as administrator, and set a general limit of 30 caballerías (999 acres) of
farmland to be held by any one owner. In payment for the expropriated property, the government agreed to issue 28-year
bonds bearing 4.5% interest annually. The largest estates were
to be made into cooperatives managed by INRA. INRA set up
2 caballerías (67 acres) as the minimum for a rural family of
five with priorities going to veterans of the revolution and victims of war crimes. The law provided that new occupants must
deliver their crops to the INRA for sale. By 1961 most farming
was done on cooperatives or state farms. Many cooperatives
were later converted into state farms. By the end of 1968 some
80% of all arable land was under state jurisdiction.
    At the beginning of 1980 the government had eliminated most
private land ownership, except for small plots, and relied heavily on state farms for agricultural production. Sugar, Cuba's
most vital crop and its largest export, is grown throughout the
island but mainly in the eastern half. The government regulates
sugar production and prices. Until 1962 the policy of industrialization and crop diversification resulted in the neglect of
sugar culture, and the size of the crop was further reduced
by drought, bad management, mechanical failures, and labor
shortages. The use of volunteers has been necessary in every
harvest since 1961. Sugar production has fluctuated radically.
Production reached 6,767,000 tons in 1961, fell to 3,821,000
tons in 1963, rose to 6,082,000 tons in 1965, and fell again to
5,315,000 tons in 1968. Production in the 1970's has averaged
about 6 million tons per year. Tobacco, the second crop in importance, is grown on small farms requiring intensive cultivation, mainly in Pinar del Río, Habana, Las Villas, and Camagüey
provinces. Production fell drastically in the late 1970's due to
a blue fungus that affected the lands. By the early 1980's
production again was on the upswing. Most Cuban coffee is
grown in the highlands of Oriente province. Domestic consumption of coffee as well as most other foodstuffs and

industrial products are rationed. Cocoa, mainly for domestic
consumption, is grown in Oriente province. Sisal and cotton
are also produced.

AGRUPACION INDEPENDIENTE DE COLOR (INDEPENDENT COLORED
ASSOCIATION). Organized by a group of radical blacks in the
early 1900's who were unhappy with the lack of political opportunities, the Agrupación developed into a political party while
Cuba was still under the second United States intervention
(1906-09). Despite their appeal to the racial consciousness of
the blacks, the party did poorly in the 1908 elections. Their
electoral fiasco increased their frustration. When the Cuban
Senate passed a law prohibiting parties along racial lines, the
Agrupación staged an uprising. Led by Evaristo Estenoz, a
former soldier during the War for Independence, several bands
of blacks roamed through the mountains of Oriente province.
The ill-organized rebellion met with much opposition and distinguished black leaders such as Senator Martín Morúa Delgado
and Juan Gualberto Gómez criticized the rebels. The United
States became alarmed over the uprising and, over the protests of President José Miguel Gómez, landed marines in several
parts of the island. Trying to avert another full-fledged intervention, Gómez moved swiftly and harshly. Estenoz and most
of the minor leaders were captured and executed and the rebellion was crushed. The Agrupación collapsed soon after these
unfortunate events. It was to be the last time that a revolt
along strictly racial lines was to develop in Cuba.

AGUAYO, ALFREDO M. (1866-1948). Cuban educator born in
Puerto Rico on March 28, 1866. While he was still a child his
family moved to Havana where he was educated, obtaining the
degrees of Licentiate in Law and Doctor in Pedagogy. He was
appointed Superintendent of Schools for the Province of Havana
and later became Professor of Pedagogical Psychology, History
of Pedagogy, and School Hygiene at the University of Havana.
Dr. Aguayo founded, and for a number of years edited, the
Revista de Educación of Havana, and in 1912 established the
Laboratory of Child Study at the University. He has written
much, both as contributor to magazines and as author of serious books. Among his works are La escuela primaria como debe
ser; Enseñanza de la lengua materna en la escuela primaria;
Pedagogía del escolar (translated from the German); Luis Vives
como educador; La pedagogía en las universidades; La pedagogía
de las escuelas secundarias; Las escuelas normales y su organización en Cuba; Desarrollo y educación del poder de observa-
ción; Ideas pedagógicas del Padre Varela; Geografía elemental;
Geografía de Cuba; Elementos de aritmética; and Estudio de la
naturaleza.

AGÜERO, JOAQUIN (1816-1851). Born in Camagüey, he lived and
studied in Havana until his father became sick and he had to

return to run the family businesses. Early in his life he showed
his liberal ideas by founding a free school for the poor in
Guáimaro and freeing his slaves. This aroused suspicion from
the Spanish who watched him closely. Nevertheless, he became
president of the Liberation Society in Camagüey, working closely with exiles in New York. He built clandestine printing offices
for the publication of revolutionary literature from New York.
These efforts were too premature to be successful and he was
taken prisoner and executed in Punta de Ganado with his con-
spirator friends José Tomás Betancourt, Fernando Zayas, and
Miguel Benavides.

AGUILA NEGRA CONSPIRACY. Ill-fated conspiracy in favor of
Cuban independence in the early 1820's. On July 4, 1825, a
group of Cuban exiles in Mexico founded the Junta Patriótica
(Patriotic Front) and tried to obtain an endorsement of Cuban
independence from Mexican President Guadalupe Victoria as well
as from the powerful General Santa Ana. The exiles received
sympathy but little material or official support. They established links with Cuba and organized a conspiracy led in Mexico
by Simón de Chávez, a former priest, known as Aguila Negra
(Black Eagle). In Cuba the movement was led by José Julián
Solís and Manuel Rojo. The conspiracy soon spread and many
prominent Cubans joined its ranks. The Spanish Minister in
Washington learned about the movement and communicated the
news to Captain-General Dionisio Vives, who captured Solís and
other conspirators and obtained information about the movement.
The Spanish government moved swiftly to crush the conspiracy,
arresting most of its leaders and sentencing them to jail terms.

AGUIRRE, MIRTA (1912-1980). Poet and writer. Born in Havana,
she became an editor of the newspaper Hoy, the organ of the
Cuban Communist Party, and in 1962 was in charge of Marxist
indoctrination in schools. She also was very active in the feminist movement but always with a Marxist orientation. She won
the coveted Justo de Lara award for literary achievement. Her
best-known works include Presencia interior, Canción antigua a
Che Guevara, and Del encausto a la sangre: Sor Juana Inés
de la Cruz. At the time of her death, she was president of
the Institute of Literature and Linguistics of Cuba's Academy
of Science.

ALA IZQUIERDA ESTUDIANTIL. An underground student group
opposed to the dictatorship of General Gerardo Machado (1924-
1933). The group was created in 1931 as a splinter organization from the Directorio Estudiantil. Ala leaders were from
more humble backgrounds and more radical than Directorio
members. Strongly influenced by Marxist ideas, Ala members
opposed any relations with Cuban political parties such as
those maintained by Directorio leaders. Ala itself moved close
to and became a tool of the Cuban Communist Party in the 1930's.

ALBEMARLE, GEORGE, 3rd Earl of (1724-1772). Commander-in-
chief of the British Expedition that captured Havana from the
Spanish during the Seven Years' War in 1762. Albemarle proclaimed himself Captain-General and Governor and broke Cuba's
links with the viceroyalty of New Spain (Mexico), delighting
those Cuban Creoles who disliked the peninsulares or Spanish-
born who controlled the politics of the island. After the establishment of peace with Spain and France in 1763, the English
relinquished their hold over Cuba and Albemarle departed for
England.

ALDAMA FAMILY. Prominent sugar planters in 19th-century Cuba
whose first members arrived from Spain at the beginning of the
century. In 1842 the Aldama family bought a major share of
the Havana Railway Company as an important capital asset for
their sugar interests. By the 1860s the Aldamas were firmly
entrenched in the hierarchy of the Cuban sugar industry. They
tried unsuccessfully to run some of their mills with only white
labor, since legalized slavery was in its final stages in Cuba.
The most famous member of this family was Miguel (1821-88),
who led the Club de la Habana in the 1840s, a group that
hoped to annex Cuba to the United States in an effort to preserve slavery. With the onset of the American Civil War, these
efforts collapsed and Miguel Aldama joined the reformists, hoping to obtain political reforms from Spain. He eventually embraced the independence movement, traveling to New York to
raise money for the Ten Years' War. This war ruined him; he
died in Havana in 1888, but not in his palace, from which
poverty kept him.

ALMEIDA BOSQUE, JUAN (1927-   ). Almeida was born in Havana.
He met Fidel Castro in the early 1950s and was a member of the
rebel group led by Castro that made the unsuccessful attack on
the Moncada Barracks in Santiago de Cuba on July 26, 1953.
He and several others, including the Castro brothers, were
captured and sentenced to prison on the Isle of Pines (now
called the Isle of Youth). In 1955 members of this group were
released from custody under an amnesty granted by Fulgencio
Batista. Almeida later joined members of Castro's 26th of July
Movement (M26J) in exile in Mexico. In late November 1956
Almeida accompanied Fidel and 80 other rebels on board the
Granma which sailed from Mexico for the southern coast of
Oriente province. Almeida was one of the few survivors to
make their way to the Sierra Maestra to form the nucleus of
the rebel army. Promoted to major in 1958, he was placed in
command of column three which operated in the Bayamo-
Santiago de Cuba region of Oriente province.
    Following Batista's defeat in January 1959, Almeida was as-
signed to the Ministry of the Revolutionary Armed Forces
(MINFAR) where he served in a number of command positions.
During 1962-65 he was a member of the national directorate of

the organization that preceded the current Cuban Communist
Party (PCC). He has been a member of the PCC Central Committee and its Politburo since October 1965. During 1963-70
he served as the Politburo delegate to the former province of
Oriente. Almeida has been a Vice-President of the Council of
State since December 1976. As a deputy to the National Assembly he has represented the municipality of Santiago de Cuba
since November 1976. At the same time the PCC Central Committee appointed Almeida chairman of its National Commission on
Party Control and Revisions, which supervises party discipline
and investigates complaints by members against the leadership.
One of Fidel Castro's oldest comrades, Almeida, a dark mulatto,
is seldom seen by Westerners except on official occasions. Besides his official persona, Almeida is an accomplished poet and
songwriter, author of the current son hit "Dame un traguito"
(Give me a little drink).

AMERICAN CHAMBER OF COMMERCE OF CUBA. Founded in mid-
1919 by American manufacturers in Cuba who worried about
European competition for Cuban goods. In the 1920's this organization lobbied for lower duties on sugar and tobacco and
for a reciprocity agreement for Cuban sugar as vital to the protection of their investments. In 1930 members of the Chamber
with interests in Cuban tobacco mounted a campaign to repeal
the U.S. statute on cigar-import limitations. In the post-World
War II period the Chamber consistently backed pro-U.S. business interests and continuously stressed the need for close relations between American business interests and those of the
Batista administration (1952-59). After Castro came to power in
1959, the Chamber was dissolved and most of its members left
Cuba.

AMERICANS IN CUBA. As early as 1818 Americans, as well as the
British, were buying interests in Cuban sugar and were establishing themselves as merchants on the island. Many Americans
owned sugar plantations, including William Stewart, who owned
La Carolina, near Cienfuegos, and J.S. Baker's San José plantation. Throughout the 19th century up to the U.S. war with
Spain (1898), these same U.S. interests exercised significant
influence over economic policy in Cuba and were behind efforts
to annex the island to the U.S. prior to the Civil War. Americans in Cuba after the U.S. occupation and up to the 1960's
continued to exercise considerable internal influence. This was
evidenced in 1933 by their support for Benjamin Sumner Welles,
the U.S. Ambassador to Cuba who mediated the resignation of
President Gerardo Machado (1925-1933) and who opposed the nationalism of the Ramón Grau San Martín administration (1933-34).
The role of the American community in Cuba, although strong,
declined in the decades of the 1940's and 1950's, and finally
received a death blow with the arrival of the Castro revolution.
Most Americans migrated back to their country following the

expropriation of American investments in the island post-1959. The Havana Post, founded in 1899, was the American community's daily newspaper.

ANARCHISM. A tendency strong in Spain in the 1870's which came to Cuba via immigration and literary influx in the 1880's. Anarchism threatened to destroy the non-political trade union movement in Cuba organized by the Reformist group. By the 1890's the anarchists, while successful with Cuban workers, were cut down by the Spanish authorities, imprisoned or deported to Spain. After the 1917 Russian Revolution, Cuban anarchists tried to organize into an efficient political force but were rivaled by the Communists whose Agrupaciones became popular and coopted their members. Cuban President Gerardo Machado (1925-33) opposed anarchism, socialism, and communism and in the mid-1920's he curtailed political activities by all three groups. From that time on, the anarchists did not play any influential political role in Cuba.

ANGOLA. African country that became the focus of Soviet and Cuban attention during the struggle for independence from Portugal in the 1970's. As independence was nearing, three pro-independence armed organizations contended for power. From the mid-1960's, Cuba and the Soviets supported Agostinho Neto, leader of the Movement for the Popular Liberation of Angola (MPLA). In 1975 Cuban troops supported by Soviet military equipment landed in Angola and made possible Neto's victory. By 1986, about 35,000 troops in addition to civilian personnel remained in Angola while thousands of Angolans were attending Cuban schools and were being trained on the island.

ANNEXATION. Since Thomas Jefferson's Administration, the United States had flirted with the idea of annexing Cuba to the U.S. In the mid-1800's, U.S. leaders saw Cuba's strategic importance in the Caribbean and feared the growth of British sea power. In the 1840's and 1850's presidents Polk, Pierce, and Buchanan tried unsuccessfully to buy Cuba from Spain. In 1854 with the Ostend Manifesto, the U.S. pushed strongly for purchase of the island. The desire for separation from Spain among the Cubans focused, particularly in the first half of the century, on annexation. Fearful that England might force Spain to abolish slavery in the island or that a Haitian-type rebellion might occur and seeing numerous commercial and security advantages in a close relation with the North, some Cubans looked toward the slave society of the United States in hopes of establishing a lasting relationship. Composed mainly of Creole planters and slaveowners, and some writers and intellectuals, the annexationists realized the dangers involved in a struggle to annex Cuba to the United States.

Threatened with losing Cuba, Spain might liberate the slaves and use them against the white planters, or the blacks themselves

might see the struggle as an opportunity for liberation. The
end result of the annexationists' attempts could thus be the
opposite of their main objective: the maintenance of the slave
system. Yet the example of Jamaica, where a slave uprising
had been crushed in the 1830's, and the awareness of their own
power encouraged the annexationists. They seriously questioned the future of Cuba under Spain and characterized the
Cuban as "a slave, politically, morally, and physically." Annexation, they emphasized, would ensure "Cuba's peace and
future success; her wealth would increase; liberty would be
given to individual action, and the system of hateful and harm-
ful restrictions which paralyzed commerce and agriculture would
be destroyed."

Several events in mid-century weakened the annexationist
movement. For one thing, the fears of the Cuban planters were
somewhat appeased when Spain stiffened its resistance to English pressure to end the slave trade and emancipate the slaves.
The lack of official United States encouragement, particularly in
the 1850's, and the violent expansion of the United States into
Texas, northern Mexico, and California somewhat discouraged
annexationist effort. The development of an incipient nationalism, particularly among Creole elements within the island, also
weakened the feeling for annexation among the already small
minority who advocated it; indeed, the great majority of Cubans
certainly did not sponsor annexation. Finally the United States
Civil War dealt a death blow to those who still hoped for a close
relation with a similar slave society. By the 1890's the U.S.
expansionists were again pressing for annexation and they were
almost successful when, during the 1898 Spanish-American War,
the U.S. occupied Cuba. Independence, however, was finally
granted to the island in 1902.

ARANGO Y PARREÑO, FRANCISCO DE (1765-1837). Creole planter
and economist, he co-founded with Spanish governor Luis de
Las Casas the influential Sociedad Económica de Amigos del País,
a center of learning and discussion devoted to promoting the
sugar interests of the Cuban plantocracy. In 1792 the two established the Junta de Fomento, an agricultural development
board and the Papel Periódico, Cuba's first newspaper. In Madrid at the time of the Haitian rebellion (1789-90), which destroyed the sugar industry in that country, Arango y Parreño
realized the opportunities that this event offered for Cuba.
Through his influential pleadings with the Spanish monarchy,
he obtained the elimination of many trade barriers and permission for the free and unlimited importation of slaves.

ARARAS. Ethnic group of blacks known as Dahomeyans who were
brought to Cuba as slaves from West and Central Africa. The
Araras were settled primarily in Matanzas province. Arara
rituals employ the Fon (Arara) language and Spanish and are
closely related to voodoo, the dominant popular religion of Haiti.
See also African religious practices.

ARAWAK. The Caribbean islands were the homeland of four neo-
lithic peoples at the time of their discovery in 1492. These
were the Lucayans of the Bahamas, the Tainian Arawaks of
the Greater Antilles, the Caribs of the Lesser Antilles, and
the Ignerian Arawaks of Barbados. All these peoples seemed
to have originated from the Orinoco basin in South America
and island-hopped throughout the Caribbean. It is estimated
that the first Arawaks reached the Lesser Antilles about 300
B.C. See also Ciboneyes, Guanahatabeyes, and Taínos.

AREITOS. A most important fiesta of the Taíno Indians in which
they sang and danced for several hours and recited their past
to the young.

ARMY, REPUBLICAN. Official title of the national Cuban Army.
After the Spanish-American War (1895-1898), the United States'
dissolution of Cuba's veteran army prevented a repetition of
the typical 19th-century Spanish-American experience, wherein
the army filled the political vacuum left by Spain. In anticipa-
tion of the American withdrawal following the Spanish-American
War, the U.S. Provisional government created the Rural Guard,
a military force trained and equipped by the United States to
assume responsibility for maintaining law and order. During
the second American intervention in Cuba (1906-1909), U.S.
Governor Charles Magoon created the permanent Army. Cuban
President José Miguel Gómez (1909-1913) initiated the politicization of the army by appointing political allies and disregarding
professional standards. Subsequently, President Gerardo
Machado (1925-1933) successfully won over the military through
bribes and threats and purged officers who opposed his regime.
Machado used the military in a variety of civilian posts, thus
increasingly militarizing society. In 1933, Sergeant Fulgencio
Batista (strongman 1933-1940 and President from 1940-1944 and
1952-1958) led the so-called Sergeants' Revolt and forced Provisional President Carlos Manuel de Céspedes (August to September 1933) to resign. From that date forward, Batista skill-
fully manipulated the Armed Forces for the next 20 years. In
1952, the military supported Fulgencio Batista's coup d'état.
Favoritism and corruption further weakened the Cuban Army in
the 1952-58 period, to the extent that it was unable to crush
the guerrilla movement led by Fidel Castro and other leaders
opposed to Batista. Shortly after taking power, Castro disbanded the army and replaced it with his own Rebel Army.

ARMY, REVOLUTIONARY (FAR--Fuerzas Armadas Revolucionarias).
The origins of the Revolutionary Army are found in the guerrilla forces that fought the military dictatorship of Fulgencio
Batista (1952-58). The most important of these groups was
the 26th of July Movement led by Fidel Castro, which numbered some 3,000 men at the end of the Civil War on December
31, 1958. Less important were the guerrilla forces of the
Directorio Revolucionario and the II Frente Revolucionario del

Escambray, each numbering about 1,000 troops. After the
Revolutionary government came to power, Raúl Castro, Fidel's
brother, was appointed Commander-in-Chief of the Armed
Forces. Raúl Castro staffed the Revolutionary Army with
trusted officers, segregating the less radical elements of the
26th of July and ignoring the members of the Directorio and
the II Frente. In the early 1960's the Revolutionary Army
included the Milicias Revolucionarias, a volunteer force created
in 1959. Today, the Cuban Army is centered around a highly
professional nucleus that is expanded by a reserve force when
necessary. A territorial militia is also part of the Armed Forces.
   In 1960, acquisition of Soviet weapons and equipment was
started. The revolutionary Army progressively adopted Soviet
structure and tactics, while receiving counsel from Soviet military advisors. A soviet brigade is currently stationed in the
vicinity of Havana. The Cuban Revolutionary Army has participated through high-ranking officers in guerrilla movements
throughout Central and South America. During the Middle East
War of 1973, Cuban troops fought on the Syrian front. In 1975,
the Revolutionary Army was instrumental in winning the Angolan
Civil War for the M.P.L.A. led by Agostinho Neto. In 1982,
over 30,000 Cuban troops remain in Angola. In the war between Ethiopia and Somalia, Cuban military participation was
essential for Ethiopia's victory. Military activities in Cuba are
not the only function of the Revolutionary soldier. Military
men actually rule over large sectors of military and civilian life.
Commissioned officers on active military duty serve also on
advisory boards, planning commissions, and as high officials
of the government and the PCC (Cuban Communist Party).
Raúl Castro, who is also the first Vice-President of Cuba, remains in charge of the Revolutionary Army.

ARMY, SPANISH (IN CUBA). The Spanish Army in Cuba was concentrated mainly in the Havana garrisons before the outbreak
   of the Ten Years' War for Independence (1868-1878). For most
   of the nineteenth century, the Spanish garrison consisted of
   25,000 to 30,000 men. A civil guard was instituted in the
   1840's, although it was primarily a police corps with some military functions. During the Ten Years' War the Spanish Army
   was dispersed throughout the island in pursuit of the 10,000-
   20,000 Cuban rebels existing in that period, although the Spanish soldiers were disorganized and unenthusiastic about the
   whole affair. Nevertheless, by early 1878, the Spanish Army
   had surrounded the Cuban rebel leaders, thus ending the War
   under terms of the Pact of Zanjón. Some years later, during
   the War of Independence (Spanish-American War after 1898),
   the Spanish Army troops were substantially increased. In addition, a corps of volunteers numbering some 25,000 soldiers
   was created. During the Spanish-American War, the Spanish
   Army suffered numerous casualties due to yellow fever and
   malaria as well as from the battles of the war itself. U.S.

intervention in the war led to the naval battle of Santiago on
July 3, 1898, in which the Spanish fleet was decimated. The
battle of San Juan Hill decided the war in favor of the American forces. Soon afterward, the Spanish Army left Cuba.

ARTEAGA, MANUEL (1879-1963). Born in Camagüey, Arteaga went
to Venezuela in 1892 to study civil law and theology at the University of Caracas. He was awarded the post of foreign curate
and vicar at Cumaná after he became a priest. Later, he was
named Canon in the Cathedral of Guayana under the title of
doctor. Arteaga was sent to the Eucharistic Congress of Madrid in 1910 to represent the Archdiocese of Caracas and to
read an ecclesiastical paper. He returned to Camagüey after
the Congress to occupy the parish priesthood at Church de la
Caridad. Later, Arteaga was Provisor and Vicar General of
the Bishopric of Havana, and after the death of the Archbishop
of Havana, Mons. Manuel Ruiz, he was appointed to this highest of the Roman Catholic hierarchy in Cuba. Years later he
was also appointed Cardinal. After the Castro revolution and
the reduction in church personnel in Cuba, old and feeble
Cardinal Arteaga tried to develop an accommodation with the
communist regime, succeeding only in maintaining a much smaller and weaker institution on the island.

ASOCIACION NACIONAL DE AGRICULTORES PEQUEÑOS--ANAP
(National Association of Small Farmers). With a membership of
about 180,000, ANAP is led by a member of the Central Committee of the PCC. Members of ANAP are independent farmers,
outside the system of collectivized state farms. Their farms
constitute about 20 percent of the total cultivated land in Cuba.
They concentrate mainly on such specialized crops as tobacco,
coffee, and tubers. ANAP provides liaison between the government and the only private sector left in Cuba. It is generally
considered the most important rural organization in terms of
size and membership, if not political influence. Although the
Cuban government recognizes the right of some small farmers
to own their lands and even to sell them, the state has a
preferential right to purchase the land, and in any case farmers must have state approval for any transaction. These
farmers receive seed, fertilizer, tools, and credit from government agencies and in turn are required to sell part of their
crops to the state. As a result of successive government regulations, the number of independent farmers and the number of
wage workers on the farms have been reduced. The government uses a combination of incentives and regulations to cause
farmers to yield their land to the state in the shortest possible
time. The PCC congress in December 1975 considered a proposal to accelerate the absorption of private farmers by per-
suading them without coercion to join state-owned projects.
ANAP had been committed since its Fourth Congress in 1971 to
educating its members as to the advantages of such absorption.

ATARES. Army barracks in the suburbs of Havana on the Atarés
hill. Atarés and other military installations were built by the
Spaniards under the direction of a Spanish soldier of Irish
descent, Marshall Alejandro O'Reilly, an expert in fortifications.
The Atarés barracks were started under the first Spanish
Captain-General (Governor) after the British relinquished their
rule over Havana in 1763. Atarés was the scene of a major battle on November 9, 1933, between a group of rebel officers supported by the ABC organization and the Cuban army under
strongman Fulgencio Batista. After heavy bombardment, Batista's troops recaptured Atarés. The barracks were rebuilt
and later used as a jail.

AUDIENCIA. Political-judicial tribunal composed of Spanish judges
who heard criminal and civil appellate cases in the Americas in
the early Spanish colonial era. The Audiencia was also an important administrative institution in the government of Spanish
America. First established in the Dominican Republic, the
Audiencia had direct influence over the governors of Cuba in
the 1500's, acting as an advisory council to them which could
supervise and even investigate their administrations. After
Cuba's independence from Spain, Audiencias were incorporated
into the Cuban judicial system with only judiciary functions.

AUTENTICO. See Partido Revolucionario Cubano (Auténtico).

AUTONOMISMO. See Partido Liberal Autonomista.

AVIATION (CIVIL). The most important civil aviation company of
the country was created on October 8, 1929 under the name of
Curtiss Cuban Aviation Company. Its operations were initially
low-level and included personnel training. At that time it had
two Curtiss Robin airplanes and the original airport was built
on the farmlands of Santiago de las Vegas, Rancho Boyeros
zone. Later on it would be expanded and renamed José Martí.
In 1930, the company dedicated its operations to the transportation of passengers and cargo. On October 30 of that same
year the first regular daily flight began between Havana and
Santiago de Cuba, with stopovers in Santa Clara, Morón, and
Camagüey. Later on, another line was opened which flew to
certain zones of Oriente province and transported passengers,
cargo, and mail. In 1932, the North American consortium Pan
American Airways acquired a major part of company stocks and
changed the name to Cuban National Aviation Company (Compañía Nacional Cubana de Aviación, S.A.). In 1933, the company expanded its services, acquiring new Lockheed planes
and opening up service to Baracoa, Antilla, and Guatánamo. In
1946, Pan American Airways sold its stocks to Cuban investors
and then the company took the name Cuban Aviation Company
(Compañía Cubana de Aviación, S.A.). The fleet was renovated
and the first daily international route Havana-Miami was estab-

lished. On June 17, 1948, it made its first flight to Europe
and thus its first transatlantic flight (Havana-Madrid). After
1954, international flight routes were expanded (Mexico, Haiti,
Dominican Republic).

In 1959, the Revolutionary Government took control of the
company. Once Cuba established relations with Communist
bloc countries, flight service began with those countries
(Czechoslovakia, the Soviet Union, GDR). There has been
construction and renovation of the airports of the main provincial capitals in order to provide national service.

AYESTERAN, JOAQUIN DE (?-1870). Ayesterán was the nephew of
the prominent Diagos, who were self-made Spanish immigrants
in the second wave of big planters in the mid-1800s. In 1850,
Ayesterán introduced the "centrifugal" machine in Cuban sugar
mills, thus eliminating a costly step in the refining process.
Ayesterán was himself a successful Cuban planter, owner of
the prosperous La Amistad plantation and Tinguaro mill.
Ayesterán's heirs used the Derosne centrifugal machine to maintain the Amistad mill at Güines in operation until 1959.

AYUNTAMIENTO. See Cabildo.

AZCARATE, NICOLAS (1828-1894). Literary figure, lawyer, and
supporter of Cuban independence, who founded the Revista de
Jurisprudencia in Havana in the mid 1860's. Azcárate organized
the Noches Literarias in 1865, a club wherein on Thursdays
literary and artistic figures met to discuss politics and intellectual ideas, including discussions relative to Cuban freedom
and the abolition of slavery.

- B -

BACARDI MOREAU, EMILIO (1844-1922). Born in Santiago de Cuba.
Entrepreneur, writer, and journalist. In 1867, at age 24, he
was honored by the Lyceum of Puerto Príncipe for his work
Conveniencia de reservar a la mujer ciertos trabajos. Because
of his revolutionary activities, he was imprisoned in 1876 and
deported in 1879. He was also jailed at Ceuta in 1895 for sending arms to the insurgents during the Independence War. His
political life led him to be elected mayor of Santiago de Cuba in
1901 and then senator of the Republic. By placing his fortune
at the disposition of the community, he encouraged the embell-
ishment of his native city and founded a museum bearing his
name as well as an adjacent library and the Municipal Academy
of Fine Arts. His published works include Via crucis, Doña
Guiomar, La condesa de Merlín, and Crónicas de Santiago de
Cuba (1516 to 1902). He was a contributor to Revista Bimestre

Cubana and El Fígaro, among others. He belonged to the
Cuban Academy of History and to the National Academy of
Arts. As an entrepreneur he founded the world-famous
Bacardí Distilleries. Juan J. Remos regarded him as "the
rare case of a capitalist who not only shared his fortune with
his people but who loved culture above business."

BACHILLER Y MORALES, ANTONIO (1812-1889). Poet, historian,
archeologist, and first Director of the Instituto de Segunda
Enseñanza de La Habana, Bachiller y Morales was persecuted
and exiled by the Spanish government because of his pro-
independence beliefs. Bachiller y Morales made several significant studies of the pre-Columbian inhabitants of Cuba and
published Cuba primitiva in 1838. He also studied, made recommendations regarding public instruction, and in 1859 published Apuntes para la historia de las letras y de la instrucción
pública en la Isla de Cuba.

BAJAREQUES. Primitive Ciboney Indian dwellings also called
barbacoas.

BALAGUER CABRERA, JOSE RAMON. José Ramón Balaguer
Cabrera is currently a member of the Secretariat of the Central
Committee of Cuba's Communist Party. He was first elected to
the Central Committee in 1975. Balaguer Cabrera also serves
as Chief of the Department of Education, Science, and Sports
of the Central Committee. He is a graduate of FAR's Graduate
School of War Studies and of the U.S.S.R.'s Graduate School
of Marxism and Political Science. He joined the Rebel Army as
a member of the guerrilla operation Frank País Second Front,
and fought under the command of Raúl Castro. After the Revolution, he was promoted to the rank of Major, and in 1965 was
named Chief of Sanitation for the Revolutionary Armed Forces
(FAR). From 1975 to 1983 he served as First Secretary of the
PCC in Santiago de Cuba and was given the rank of Colonel of
the Military Reserve. In 1985 he was officially removed, without explanation, from his PCC post, but was subsequently appointed to his current position during the Third Party Congress.

BALIÑO, CARLOS (1848-1926). As a tobacco worker in Florida
during the 1890's, Baliño represented the escogedores (tobacco
leaf selectors) in the anti-Spanish movement for Cuban inde-
pendence and directed La tribuna del pueblo, published in Tampa.
Baliño was a follower of José Martí, and in 1892 he signed the
Bases, a document outlining the ideas and program of José
Martí's Partido Revolucionario Cubano. After independence,
Baliño moved to Cuba. Although he espoused anarchist ideas,
he gradually accepted Marxism and at age 70 he founded the
Communist Association of Havana. Baliño introduced student
leader Julio Antonio Mella into the Association. In 1924 the
Association, with Baliño as its President, attempted to unite

all Cuban communists and by August 1925, a Congress was
convened in Havana with Mexican Communist Party leader
Enrique Flores Magón present to provide support and guidance.
Out of this Congress came the establishment of the Cuban Communist Party, of which Baliño was a leader until his death one
year later in 1926.

BANCO DE FOMENTO AGRICOLA E INDUSTRIAL DE CUBA--
BANFAIC (Agricultural and Industrial Development Bank).
BANFAIC was established in December 1950 as an autonomous
state institution whose purpose was to create and maintain the
necessary financial facilities that would enable small farmers
and industrialists to have access to credit, especially medium
and long-term loans. BANFAIC was set up by the Cuban government with a $10 million fund. Expenditures were divided
equally between the agricultural and industrial sectors.
BANFAIC operated successfully through 1958 until the Castro
Revolution redesigned the economic framework of Cuba.

BANCO NACIONAL. The Banco Nacional (National Bank) originally
began operations in 1950 as a central bank. Its function was
to act as the state treasury, to issue currency, and to administer the country's gold and foreign exchange reserves. It also
fixed the exchange rate and obtained credit. In November
1960, the Revolutionary government of Fidel Castro nationalized
all banks functioning in Cuba. All banks were consolidated
under the National Bank, which maintains over 250 branches
around the country. The National Bank controls loans, receives savings from the public, pays out social security pensions, and collects utility bills. It maintains subsidiaries
abroad.

BANDERAS, QUINTIN (1834-1906). Black Cuban veteran revolutionary who led an uprising in Santiago de Cuba together with
José Maceo in the "Guerra Chiquita" or second war of Cuban
independence in 1879. Banderas soon began military maneuvers
in the eastern mountains with the following of many soldiers
who had fought in the first war of independence (1868-1878).
Banderas was a leader in the 1895 insurrection against Spain
and commanded rebel forces in Matanzas province. By 1897,
Banderas was the only surviving rebel leader left in west Cuba
and was himself surrounded by Spanish General Valeriano
Weyler's troops. His prolonged holdout along with growing
disenchantment with the costly Spanish engagement contributed
to the desire on both sides for negotiated settlement. With the
establishment of the first Cuban Republic and the re-election of
Estrada Palma, a revolutionary movement was launched, due in
part to the de-commissioning of General Banderas and other
prominent military men.

BANKS. The first bank in Cuba was set up by the Spanish in 1827

(Royal Bank of Ferdinand VII). Its aim was to supply sugar
planters with funds for their operations. Spanish banks largely continued to serve in this capacity up through the Spanish-
American War (1898-1899), when U.S. influence began to formally assert itself. By the 1930's, the United States had created international mechanisms and institutions to assist Cuba.
This continued until the Revolution led by Fidel Castro broke
economic and diplomatic relations with the U.S. When Castro
took over in 1959, Cuba had over 50 domestic and foreign banks
with nearly 200 branches, as well as a Postal Savings Bank and
a National Bank. Also prominent was the Banco de Fomento
Agrícola e Industrial (BANFAIC), a government lending institution devoted to support development in the agrarian and industrial sectors. In 1959, Cuban banks owned 60% of all deposits.
By November of 1960, the Cuban government had nationalized
all banks, incorporating all banking under the Banco Nacional.

BANTUS. Black ethnic group also known as Congos, which included
slaves brought from Angola and Cameroon. The Bantus were
settled mainly in the eastern part of Cuba, particularly in the
cities of Santiago de Cuba and Guantánamo.

BARAGUA, PROTEST OF. As the Peace of Zanjón agreement between the Spanish government and the Cuban rebels was being
signed in February 1878 ending the Ten Years' War for Independence, General Antonio Maceo decided to continue the struggle in Oriente province. Maceo agreed to meet with the Spanish
General, Arsenio Martínez Campos, near the town of Baraguá
where he communicated to Martínez Campos the desire of many
rebels to continue the struggle for independence. The rebels
set up a provisional government around Baraguá, electing Maceo
Commander-in-Chief of the Eastern rebel forces and undertook
significant military operations against nearby Spanish forces.
By May 1878, the protest of Baraguá died out and the rebels
agreed to peace.

BARBACOAS. See Bajareques.

BARNET, JOSE A. (1864-1945). Born in Spain in 1864, José A.
Barnet y Vinageras served as Secretary of State under Cuban
President Carlos Mendieta's administration (1934-36). Barnet
was made Provisional President of the Republic following the
sudden resignation of Mendieta in late 1935, and promised that
elections would be held on January 10, 1936. Barnet remained
Provisional President after said election, which Miguel Mariano
Gómez won, until the May 1936 inauguration. As a sinecure for
Colonel Fulgencio Batista, Barnet approved a decree enabling
the Army Chief of Staff (Batista) to appoint army officers as
teachers in rural schools. This action was viewed by many as
another ominous sign of the Army's control over Cuban politics.
Barnet died in Cuba as World War II ended.

BATEY. Group of buildings around a sugar mill, including living
quarters In pre-Columbian times, the Batey was an open court
where Taíno Indians played a sort of soccer game called juego
de bates and assembled for areítos, a most important fiesta.

BATISTA SANTANA, SIXTO. Division General Sixto Batista
Santana has been a member of the Central Committee of Cuba's
Communist Party since the First Party Congress held in 1975.
He is currently a member of the Secretariat and Chief of the
Military Department of the Central Committee. He fought in
the Rebel Army and in 1959 held the rank of First Lieutenant.
From 1960 to 1965, he participated in the organization and train-
ing of the Occidental, Central, and Western Armies, particularly
in the Political Section of the Armed Forces. Trained in the
Graduate School of Marxism in the Soviet Union, he was later
appointed to Chief of the Central Political Directorate of the
Ministry of the Armed Forces (MINFAR). During the past 15
years, Batista Santana has moved up rapidly in the ranks of
the military establishment. He was promoted to Major in 1970,
Brigadier General in 1975, and to Division General in 1980. He
has presided over several missions to the Soviet Union to negotiate for military supplies for FAR, and has participated as an
observer in the Warsaw Pact military exercises. General Batista
Santana's principal work has consisted of directing the consolidation of political discipline within the chain of command of FAR.

BATISTA Y ZALDIVAR, FULGENCIO (1901-1973). Former President
of Cuba. Army's "strong man" in the 1930's, elected president
in the 1940's, and dictator in the 1950's, Batista's influence was
felt in Cuban politics for over two decades. Batista was born
in Banes, Oriente province, on January 16, 1901. The son of
a railroad laborer, Batista spent his early years in poverty and
attended a Quaker missionary school. Soon after leaving school
he went to work in a variety of jobs: tailor apprentice, laborer
in the canefields, grocery clerk, barber. At 20 he joined the
Cuban army. The military afforded an opportunity for Batista's
rapid upward mobility. An ambitious and energetic young man,
he studied at night and graduated from the National School of
Journalism. In 1928 he was advanced to sergeant and assigned
as stenographer at Camp Columbia in Havana. At the time Cuba
was going through a period of considerable turmoil. The deep-
ening economic depression had worsened people's misery and the
overthrow of Gerardo Machado's dictatorship in 1933 had released
a wave of uncontrolled anger and anxiety. Unhappy with both
a proposed reduction in pay and an order restricting their pro-
motions, the lower echelons of the army began to conspire. On
September 4, 1933, Batista, together with anti-Machado student
figures, assumed leadership of the movement, demoted army officers, and overthrew Carlos Manuel de Céspedes' provisional
government. Batista and the students appointed a short-lived
five-man junta to rule Cuba and on September 10 they named a

University of Havana Professor of Physiology, Ramón Grau San
Martín, as Provisional President. Batista soon became a Colonel
and Chief of Staff of the Army. Grau's nationalistic and revolutionary regime was opposed by the United States, which refused to recognize it. Different groups within Cuba conspired
against the government.

On January 14, 1934, the unique alliance between students
and the military collapsed and Batista forced Grau to resign,
thus frustrating the revolutionary process that had begun with
Machado's overthrow. Batista emerged as the arbiter of Cuba's
politics. He ruled through puppet presidents until 1940, when
he was himself elected president. Desiring to win popular support, he sponsored an impressive body of welfare legislation.
Public administration, health, education, and public works improved. He established rural hospitals, minimum wage legislation, increased salaries for public and private employees, and
started a program of rural schools under army control. He
legalized the Cuban Communist Party and in 1943 established
diplomatic relations with the Soviet Union. The Army received
higher pay, pensions, better food, and modern medical care,
thus ensuring its loyalty. On December 9, 1941, following the
Pearl Harbor attack, Batista brought Cuba into World War II on
the Allied side. Air and naval bases were made available to the
United States, which purchased all of Cuba's sugar production
and provided generous loans and grants. In 1944 Batista allowed the election of his old-time rival, Ramón Grau San Martín.
After an extensive tour of Central and South America Batista
settled at Daytona Beach, Florida. There he wrote a book,
Sombras de América published in Mexico in 1946, in which he
surveyed his life and policies. In 1948, while still in Florida,
he was elected to the Cuban Senate from Santa Clara province.
He returned to Cuba that same year, organized his own party,
and announced his presidential candidacy for the June 1952
elections. Batista, however, prevented the elections from taking place. Aware perhaps that he had little chance to win, he
and a group of army officers on March 10, 1952, overthrew the
constitutionally elected regime of President Carlos Prío Socarrás.
Batista suspended the 1940 Constitution, as well as Congress,
cancelled the elections, and dissolved all political parties. Opposition soon developed, led primarily by university students
who rioted and demonstrated frequently.

On July 26, 1953, young revolutionaries led by Fidel Castro
unsuccessfully attacked the Moncada military barracks in Oriente
province. Some of the attackers were killed, others, among
them Castro, landed in jail. Batista seemed bent on remaining
in power. In a rigged election in November 1954 he was "re-
elected" for a four-year term. Although Cuba was prosperous,
Batista neglected social and economic problems. Corruption and
graft in his administration reached unprecedented proportions.
Political parties and groups called for new elections but with
little success. As a political compromise became unlikely, the

adherents to violence grew in number. Students increased
their activism. After his release Castro went to Mexico to prepare an expedition which landed in Cuba and began guerrilla
operations. Other groups organized an urban underground.
An attack on the Presidential Palace on March 13, 1957, by the
students and followers of deposed President Prío, nearly succeeded in killing Batista. The government met terrorism with
counter-terrorism. Political prisoners were tortured and assassinated. By 1958 national revulsion against Batista had developed. Finally, defections in the army precipitated the
crumbling of the regime on December 31, 1958. Batista escaped to the Dominican Republic and later to the Portuguese
Madeira Islands, where he wrote several books, among them
Cuba Betrayed and The Growth and Decline of the Cuban Republic, both of them apologies for his divisive role in Cuban
politics. He moved to Madrid where he died in 1973.

BAY OF PIGS INVASION. By late 1960 relations between the U.S.
and Cuba had deteriorated and the United States embarked on
a more aggressive policy toward the Castro regime. Groups of
Cuban exiles were trained under the supervision of U.S. officials in Central American camps for an attack on Cuba. The
invasion took place on April 17, 1961, at Bay of Pigs, a remote
site in the southern part of the island. The internal situation
in the island then seemed propitious for an attempt to overthrow
the Cuban regime. Although Castro still counted on significant
popular support, that support had progressively decreased.
His own 26th of July Movement was badly split on the issue of
communism. Also, a substantial urban guerrilla movement existed throughout the island, composed of former Castro allies,
Batista supporters, Catholic groups, and other elements that
had been affected by the revolution, and significant unrest was
evident within the armed forces. The urban underground saw
the landing of the U.S.-sponsored invasion force as the culminating event to follow a series of uprisings and acts of sabotage
they hoped would split Castro's army throughout the island and
weaken the regime's hold over the people. This would coincide
with Castro's assassination and with a coordinated sabotage plan.
In the weeks prior to the invasion, violence increased, bombs
exploded, shops were burnt. Yet the planners in exile were
not counting on the forces inside Cuba. They placed an un-
justified faith in the invasion's success, and feared that the
underground might be infiltrated by the regime. Arms that
were to be shipped into Cuba never arrived, and communications between the exiles and underground forces were sporadic
and confused. The underground was not alerted to the date of
the invasion until April 17, the very day of the landing, when
it could only watch the Bay of Pigs disaster in confusion and
frustration.

The whole affair was a tregedy of errors. Although the
Cuban government did not know the date or the exact place

where the exile forces would land, the fact that an invasion was
in the offing was known in and out of Cuba. The weapons and
ammunitions that were to be used by the invading force were
all placed in one ship, which was sunk the first day of the invasion. The site for the invasion was sparsely populated, surrounded by swamps, and offered little access to nearby mountains where guerrilla operations could be carried out if the invasion failed. The invading forces could, therefore, all but
discount any help from the nearby population. Some of the air
raids by Cuban exiles that were intended to cripple Castro's
air force were cancelled at the last minute by a confused and
indecisive President John F. Kennedy. Perhaps trying to reassert his authority over the CIA-sponsored invasion, to stymie
possible world reaction, or to appease the Soviets, Kennedy
ordered no further U.S. involvement. Castro's Sea Furies and
T33s could, therefore, shoot down the exiles' B26s and maintain control of the air. While the invasion was in progress,
Khrushchev threatened Kennedy: "The government of the U.S.
can still prevent the flames of war from spreading into a con-
flagration which it will be impossible to cope with.... The world
political situation now is such that any so-called 'small war' can
produce a chain reaction in all parts of the world." The failure
of the invasion and the brutal repression that followed smashed
the entire Cuban underground. On the first day of the invasion, the regime arrested thousands of real and suspected op-
positionists. The resistance never recovered from that blow.
His regime strengthened and consolidated, Castro emerged victorious and boasted of having defeated a "Yankee-sponsored invasion." The disillusionment and frustration caused by the Bay
of Pigs fiasco among anti-Castro forces, both inside and out of
Cuba, prevented the growth of significant organized opposition.
Meanwhile, U.S. prestige in Latin America and throughout the
world sank to a low point.

BAYAMO. A town in southeast Cuba, now the capital of Granma
    Province, known in colonial times as a center for a large con-
    traband market where various European smugglers exchanged
    slaves, meat, dyes, etc. Later, well-known tobacco of excel-
    lent quality was grown in the area and exported to northern
    Europe. By 1867, Bayamo, a medium-sized colonial town of
    10,000 with a town hall, prison, church, barracks, and main
    plaza, had become a center for revolutionary activity against
    the Spanish, thus earning the title Ciudad Monumento Nacional
    (National Historic Site).

BAYO, ALBERTO (1892-1967). Military leader, tactician, revolu-
    tionary. Born in Camagüey Province of Cuban mother and
    Spanish father, Bayo joined the Spanish Air Force and sided
    with the Republicans during the Spanish Civil War. He di-
    rected the failed expedition to Majorca in 1936. Bayo also
    fought as a Spanish officer during the Moroccan wars. He

was the founder of the first civil air school in Spain. After
the Spanish Civil War, Bayo went to exile in Mexico where he
became director of the School of Military Aviation. He was
also involved in the politics of exiled Republicans. Bayo provided military training for Fidel Castro and his companions in
a farm close to Mexico City. He convinced Castro of the need
to adopt guerrilla tactics instead of frontal attacks. After
Castro triumphed, Bayo moved to Havana, where he died in
1967. He is the author of Mi desembarco en Mayorca and Mi
aporte a la revolución cubana.

BEHIQUE. Indian medicine man found among pre-Columbian Indian
tribes in Cuba.

BERMUDEZ, CUNDO (1914-). Leading Cuban painter living in
exile in Puerto Rico. Born in Havana, he studied in Cuba and
Mexico. He executed several murals in Cuba and in Puerto
Rico. Many of his works have been shown also in Peru, Chile,
Haiti, Venezuela, New York, and Miami. He has exhibited in
bienales at São Paulo and Venice. His works are found in many
collections in the United States, in Paris, and in South America.
He won an Honorable Mention prize at the UNESCO Graphics
Bienal, San Juan, and a prize in the "Homage to Picasso" exhibit
at the OAS, Washington, D.C.

BETANCOURT y DAVALOS, PEDRO (1858-1933). Physician graduated from medical schools in the United States and Madrid, he
practiced medicine in his native Matanzas until the 1895 War of
Independence began when he left Cuba to talk and discuss
strategies with José Martí. The revolutionary movement in
Matanzas failed and he was imprisoned and later deported to
Madrid. He managed to escape to Paris and then went to New
York to join Calixto García's expedition in 1896. He organized
the Fifth Brigade in his native province and through his involvement in battles became a Major General. He represented
the province of Matanzas in the Santa Cruz Assembly and in the
Constituent Assembly of 1901, and was President of the Veterans'
Association. He was also a Senator and Minister of Agriculture
in the Republican period, and the name of the town of Macurijes
was changed to "Pedro Betancourt" in honor of this liberator.

BISBE, MANUEL (1906-1961). Bisbé was a University of Havana
professor and intellectual who became involved in politics as a
close supporter of Ortodoxo Party leader Eduardo Chibás. Becoming an important figure in the Ortodoxo Party in the late
1940's, Bisbé was the Ortodoxo Party candidate for the mayor-
alty of Havana in June 1950. He received 11% of the votes and
came in third place. After Fulgencio Batista's coup d'état in
1952, Bisbé participated in the so-called Diálogo Cívico (Civic
Dialogue) between top Batista supporters and the main opposition leaders. The purpose of the Diálogo was to try to achieve

an understanding between Batista and his opponents and hold
free elections on the island. But the effort failed and Bisbé
joined those who supported Castro's revolutionary movement.
He was a member of the Council of National Liberation, an um-
brella anti-Batista organization. Once Castro took power in
1959, Bisbé served as Cuban Ambassador to the United Nations
in the U.S. He eventually faded from prominence in the Revolution and died in New York in 1961.

BLACK MAGIC. Or brujería, is a term used to denote any superstitious practice or African cults that misuse religious techniques
for evil purposes. Black magic is usually regarded by middle-
and upper-class Cubans as the rituals of all unfamiliar African
cults that deal with mysterious and dangerous powers or spirits.
See African religious practices.

BLACKS. See also African religious practices, Slavery. When the
indigenous population proved unsuitable for the harsh labor required by the Spaniards, the crown authorized the importation
of black slaves, beginning in 1517. By 1865, when the slave
trade ended, roughly 750,000 slaves had been brought into
Cuba, and Havana had gradually become an important international slave market. In 1817 the colored population, including
115,000 free men, slightly outnumbered the whites; and during
the following years the colored majority steadily increased. In
about 1820 Spain agreed, under pressure from Great Britain,
to a series of treaties abolishing the slave trade. The agreements, however, were not enforced. Blacks were brought to
Cuba from all parts of West Africa--from Senegal to the Congo
and from the coast to hundreds of miles inland. The largest
and most influential group in Cuba were the Yoruba from southwestern Nigeria, who were known as Lucumí and were sent
throughout the island. The second largest group, known as
Carabalíes, a name derived from the port of Calabar in southeastern Nigeria, came from various tribes. Black Haitians and
Jamaicans were also brought from their islands to work the
sugar harvest, and some settled permanently in Cuba, preserving many elements of their own cultures. Whatever their origins, the various elements of the black population evolved a
cohesive Afro-Cuban tradition that permeated not only the artistic and intellectual spheres of Cuban life but also the social
and religious aspects, even after the Revolution. The most
popular manifestation is Afro-Cuban music, which has become
internationally famous. Writers have created a school of literature called negrismo, which records the experiences of the
Afro-Cuban in both poetry and prose. In addition the survival
of African spiritualism and the existence of hundreds of societies
devoted to a syncretic worship of African saints has greatly
influenced the reglious life of the island.

BLANCK, HUBERT DE (1856-1932). Dutch by birth, De Blanck was

a renowned pianist and composer who visited Cuba in 1882 and
settled in Havana three years later. There he established the
Conservatory of Music, a school of musical education on the island. Blanck supported the Cuban independence movement
against Spain and in 1896 he was imprisoned by General Valeriano Weyler and later expelled from Cuba. After the independence of the island he returned to Havana and re-opened his
conservatory which he re-named Conservatorio Nacional to signify that it was the national center of musical education. The
Conservatorio and De Blanck had a significant influence in the
teaching and development of music in Cuba.

BOBADILLA, FRANCISCO DE (?-1502). Spanish judge and Royal
Commissioner in the Caribbean Islands in the 16th Century.
Bobadilla was sent to Santo Domingo to harmonize the volatile
struggle between colonists and Indians. Bobadilla accepted the
vehement complaints against Columbus' rule and had the latter
sent to Spain in chains in 1500.

BOHEMIA. Cuba's most influential weekly magazine until 1960,
when it was taken over by the Castro revolution and became a
semi-official organ of the Marxist Cuban government. It was
first published in 1908.

BOHIO. Rustic hut of poles thatched with palm leaves used by the
pre-Columbian inhabitants of Cuba. Bohíos were used as shelters during the colonial and early national period and even today remnants of this type of construction can be seen in rural
areas.

BOLIVAR, SIMON (1783-1830). Latin American hero and liberator.
After obtaining the independence of several Spanish-American
nations, Bolívar sought Cuban liberation from Spanish rule.
The movement of Cubans who sought Bolívar's support was led
by José Francisco Lemus. The Soles y Rayos de Bolívar, as
the conspiratorial movement was called in Cuba, appealed mainly
to the poorer whites and to blacks. The conspiracy planned an
uprising that, if successful, would have established the Republic of Cubanacán (the Indian name for Cuba), inspired by the
ideals of Bolívar. The Spanish authorities crushed the attempt
before the uprising took place, and Lemus and his main companions were imprisoned.

BONCHES. Violent student groups in the late 1930's which dominated politics and events on the University of Havana campus.
Rivalries between various political and student groups, the
Movimiento Socialista Revolucionario, the Unión Insurreccional
Revolucionaria, and the Federation of Cuban students led to the
development of these bonches. In September of 1949, after several assassinations of student leaders, Cuban police penetrated
the campus and confiscated a large cache of arms and ammuni-
tion belonging to the bonches and arrested several students.

BONSAL, PHILIP (1903-   ). U.S. Ambassador to Cuba appointed
immediately after Fidel Castro's successful revolution in January
1959. Despite Castro's criticism of U.S. interests and his refusal to meet with Bonsal in 1959, Bonsal urged U.S. restraint
with regard to the popular leader, asserting that Castro was
not a Communist. Bonsal is author of Cuba, Castro, and the
United States, a book dealing with U.S.-Cuban relations in the
1960's.

BOSCH, JOSE (1901-   ). Cuban industrialist and President of the
Bacardi Rum Company in Cuba. Bosch was sympathetic to
Fidel Castro's rebel movement during the guerrilla war against
President Fulgencio Batista (1952-1959). When Castro took
power in 1959, Bosch offered to pay his company's annual taxes
in advance, as a sign of his enthusiastic willingness to cooperate with the new government. Bosch later became disillusioned
with Castro's communistic regime and left Cuba, headquartering
his prosperous and prestigious rum company in Miami, Florida.
The Bacardi Rum Company in Cuba was nationalized by Castro.

BOTELLAS. A system of government sinecures used successfully in
Cuba by U.S. provisional governor Charles Magoon during the
second U.S. intervention (1906-1909) and by subsequent Cuban
administrations. Supporters of the government were placed on
the payroll without performing any work.

BRADEN, SPRUILLE (1894-1978). United States Ambassador to
Cuba from 1942-1945, Braden arrived in Cuba during World
War II, a few months after Cuba declared war on the Axis
powers. Braden's arrival was preceded by a satisfactory omen
in Cuban-American relations: the signing of a $25 million loan
agreement for Cuba from the United States intended for public
works projects under the Fulgencio Batista presidency (1940-44).
Braden succeeded Sumner Welles in a long line of U.S. envoys
extraordinary, or Proconsuls, to Cuba. An intelligent man,
possessing extensive Latin American experience, Braden had
begun his professional career as a mining engineer. He began
his work as a consultant on the electrification of the Chilean
railways before entering a hectic business career in New York
in the 1920's. Throughout his successful business enterprises
Braden, whose wife was Chilean, maintained a strong interest
in South America. At age forty, Braden began developing his
political ambitions. He served as U.S. representative at various congresses, and distinguished himself as the Chairman of
the U.S. delegation at the Chaco conference after the Chaco
War. From 1939 to 1942, Braden served as U.S. Ambassador
to Colombia. Braden's outspoken views on social reform consistently earned him a reputation as a radical diplomat. Many
Cubans regarded him as the best U.S. Ambassador ever assigned to Cuba.
   Braden's activities were, by necessity, mostly military.

Secretary of State Cordell Hull's first instructions told Braden
that the U.S. War Department wished to establish a "heavy
bombardment and operational training unit" under U.S. officers,
to train U.S. and British RAF personnel. The Cuban government promptly agreed, converting San Antonio and San Julián
bases in Pinar del Río province for allied training. Braden was
instructed in July of 1942 to request a U.S. land purchase at
San Julián to build a minimum of one 7,000-foot runway. In
addition, 500 men were to be stationed there, with the United
States having operational and administrative control. Cuba's
agreement was encouraged by the loss of two Cuban freighters
through German submarine fire in August of 1942. Nine military arrangements between the U.S. and Cuba employing Cuban
territory were effected through Braden's intercession by late
1942. Nevertheless, grave problems existed in the U.S.-Cuban
economic relationship. Braden assumed a critical role in the
negotiations to obtain a central bank for Cuba. The friction in
this sphere arose from Braden's stiff insistence upon "honest
and complete management" for the successful operation of the
bank. Braden charged that administrators of this sort were
extremely rare in Cuba, and as such, he was opposed to creation of the institution for fear it would become another hotbed
of corruption. Despite Sumner Welles' support for the idea of
enhancement of Cuban sovereignty through an independent currency and a central bank, Braden persisted in his opposition
and he ultimately triumphed, again postponing creation of a
Cuban National Bank. Braden continued his campaign for socio-
political reforms in Cuba until his ambassadorial term expired
in 1945, whereupon he was appointed U.S. Ambassador to Argentina (1945-46). During the 1950's, Braden evolved from a
progressive political stand to one associated with the far Right
in the U.S. This was reflected in his view of Fidel Castro in
1957 as being a long-time sympathizer, "if not a member of the
Communist Party."

BRIGADE 2506. See also Bay of Pigs invasion. The nucleus of a
Cuban exile force of 1,297 men trained by the United States in
Guatemala which sailed from Puerto Cabezas, Nicaragua, on
April 14, 1961, to overthrow Fidel Castro, landing at the Bay
of Pigs three days later. The brigade was originally composed
of six batallions of 200 men each. The designation 2506 was
adopted from the serial number of one of the members who had
died accidentally during training. Politically, the views of
brigade members ran across the spectrum from extreme Right
to Center, with virtually no supporters of the Left. Training
had been brief and varied between two weeks to two months.
Brigade 2506 landed at the Bay of Pigs without adequate air
cover and was quickly surrounded by Cuban troops with heavy
artillery and armor. After U.S. President John F. Kennedy refused to provide U.S. air strikes to save the Brigade, Castro's
forces captured 1,180 survivors who were held for public trial

and imprisoned. One and a half years later the men were ex-
changed for medical supplies, while the brigade leaders were
ransomed for several million dollars.

BROOKE, GENERAL JOHN R. (1838-1926). First U.S. military
governor of Cuba (January-December 1899) after the Spanish-
American War. Brooke's arrival marked the end of Spanish rule
on the island. Benevolent but non-assertive, Brooke surveyed
the post-war destitution in Cuba, whose population numbered
about 1.5 million, and attempted to preserve harmony and order
by introducing U.S. laws and customs into Cuban society.
Gradually, such careful moderation caused more Cuban resentment than respect as the trappings of Spanish laws remained
very much in effect. Brooke was replaced by General Leonard
Wood in December 1899.

BRUJERIA. See Black magic.

BUROCRATISMO (Bureaucratism). This concept has been considered
anathema to the Cuban Revolution. Castro has enlisted both the
party and the military in the fight against burocratismo, which
has perpetuated economic stagnation in Cuba. To be a bureau-
crat, it is felt, is to be an enemy of the revolution. Castro
and his followers take every opportunity to blame the bureaucracy for the various failures suffered by the Cuban economy.
The effect of such attacks is to reduce rather than increase
the efficiency of those responsible for implementing the goals
of the revolution. For one thing, the attacks have almost
completely destroyed the initiative of lower-echelon functionaries. Also, the demands made by the party for more sacri-
fices and more dedication are greatly taxing the resources of
higher government functionaries, resulting, among other things,
in increasing absenteeism.

BUSTAMANTE, ANTONIO SANCHEZ DE (1865-1951). Lawyer, professor, publicist, statesman, orator. Born in Havana on 13
April 1865, Bustamante was the son of a Professor and Dean of
the Faculty of Medicine in the University of Havana. After
completing his Bachelor's degree at the famous Colegio de Belén,
Bustamante traveled to Spain to study law at the University of
Madrid. He returned to Havana, however, to complete his legal
studies. Beginning his career in law before age 20, Bustamante
gained the International Law Chair at the University, which he
occupied until his death. Bustamante was elected Senator for
Pinar del Río province when the Republic was constituted in
1902. He was reelected to represent Havana in 1909. Appointed
member of the Institute of International Law in 1895, Bustamante
was selected in 1907 to be Delegate Plenipotentiary of Cuba to
the Second Peace Conference at The Hague. Bustamante's titles included: Dean of the Faculty of Law at Havana University;
President of the Academy of Arts and Letters; President of the

Proprietor's Club (Centro de Propietarios) of Havana, over
which he presided for 20 years; Dean of the Havana Bar;
Member of the Permanent Arbitration Tribunal of The Hague.
Dr. Bustamante's works included: Tratado de Derecho Inter-
nacional Privado, Havana, 1896; Informe relativo a la Segunda
Conferencia Internacional de la Paz, Havana, 1908; Programa
de las Asignaturas de Derecho Internacional Público y Derecho
Internacional Privado, Havana, 1893; Le Canal de Panama et le
Droit International, Brussels, 1895. La Segunda Conferencia
de la Paz, Madrid, 1908; La Seconde Conférence de la Paix,
Paris, 1909; La Autarquía Personal; A study of International
Private Law, Havana, 1914; Discursos (5 volumes).

BYRNE, BONIFACIO (1861-1936). Born in Pueblo Nuevo, a sub-
    urb of Matanzas, on March 3, 1861. Byrne became a famous
    poet and journalist. Byrne's first appearance in print was in
    his seventeenth year, when some of his verses were published
    in a Matanzas weekly, La Primavera. His first essay in the
    journalistic field was an analyst of the Ateneo society, being
    associated with the veteran writer Fernando Romero Fajardo
    who was its Director. He later joined the editorial staff of
    El Pueblo in Matanzas, and was director successively of La
    Mañana, La Juventud Liberal, and El Obrero; twice he was
    editor of El Diario (Matanzas) and did special articles for El
    Imparcial and La Región. In 1895, when rebellious Cuba was
    being disciplined by Spain, Byrne's caustic and enthusiastic
    pen attracted a dangerous degree of attention; his separatist
    propaganda led to the suppression of his paper, El Diario de
    Matanzas. Byrne was arrested and tried. He was sentenced
    to six months' imprisonment but escaped serving through a
    proclamation of amnesty. At this juncture Byrne emigrated to
    Tampa, Florida, where he remained three years. Here he con-
    tinued to work for Cuban independence largely through the
    columns of El expedicionario, a small paper which he directed,
    and through his contributions to the Cuban periodicals Patria,
    El Porvenir, Cacarajícara, El Continente Americano, and Cuba.
    He also collaborated in the Album Patriótico which El Fígaro
    published in 1899. Upon his return to the island of Cuba, af-
    ter the close of the Spanish-American War, Byrne acted for a
    short time as editor of La Discusión, but in April 1899 resigned
    to accept an appointment under the provincial government of
    Matanzas; later, being appointed secretary of that government,
    he retained the position for ten years, until 1912. During a
    part of this period he directed in Matanzas the periodical
    Yucayo, serving also as Secretary to the Superintendent of
    Schools of the province. He was a member of the Cuban
    Academy of Arts and Letters and an Associate (Correspondent)
    Fellow of the Academy of the Republic of San Salvador. In
    1915, during a second visit to the U.S., Byrne wrote a book
    of verses entitled La Nación maravillosa, a tribute to the coun-
    try of Washington. His writings include the volumes of poetry

Excéntricas (Philadelphia, 1893); Efigies (Philadelphia, 1897);
Lira y Espada (1901); Poemas (1903); and En medio del camino
(1914). Several of his dramatic works were produced with
success: El Anónimo; Rayo de Sol; El Legado; Varón en
puerta; and El espíritu de Martí.

- C -

CABALLERO, JOSE AGUSTIN (1771-1835). Educator. Early advocate of Cuban autonomy. Born in Havana, Caballero became a
priest and later taught philosophy at the San Carlos Seminar.
He influenced Father Félix Varela, his disciple, who was later a
major figure in Cuban education. He was active in the Sociedad
Económica de Amigos del País. Caballero was a skilled orator,
imbued with a liberal philosophy. In 1811 he wrote a document
addressed to the Spanish Parliament (Cortes) in which he advocated Cuban autonomy. He is the author of Lecciones de
filosofía electiva.

CABILDOS (Ayuntamientos). According to Spanish tradition, Cuban
Governor Diego Velázquez created cabildos (also known as
ayuntamientos or consejos municipales, in the early 1500's).
The cabildos were town councils with a measure of self government. They possessed administrative and judicial powers as
well as the function of supervising the Indians to insure that
they were well treated. The members of the cabildos were the
alcaldes (mayors) and regidores, or town councilmen. During
the 1700's the cabildos ceded most of their functions to the
central government. The Spanish Constitution of 1812 gave
new force to the cabildos, but around the 1850's, Captain-
Generals (governors in Cuba) appointed the vacant seats of
councilmen. Under the Pacto del Zenjón (Zanjón Peace Treaty)
that ended the Ten Years' War of Independence (1868-78), the
Spanish government recognized the right of Cubans to elect
their local governments. After independence was achieved
(1902) the cabildos, now called ayuntamientos, survived and
periodical elections were held to choose councilmen. Following
Fidel Castro's revolution of 1959, centralization occurred and
the ayuntamientos lost their power. The Popular Power has
replaced them but it functions both at the municipal and provincial levels. See also Popular Power.

CABRERA, LYDIA (1900-). Writer, ethnologist, and researcher
of Afro-Cuban folklore. She worked under the renowned
Fernando Ortiz and became the foremost authority on stories
and legends of the African Yoruba and Lucumí slaves brought
to the Antilles. She was born in Havana on May 20, 1900.
She began her extensive writings in literature with the

publication of Cuentos negros de Cuba, first in French (Paris, 1936), then in Spanish (Havana, 1940). There were to follow, among others: Refranes de negros viejos, La sociedad secreta Abakuá, and now in exile in the United States where she has lived since 1960, she is editing Otan Iyebiye: Las piedras preciosas; Francisco Francisca, chascarrillos de negros viejos; La regla Kimbisa de Santo Cristo del Buen Viaje, and others. Her most important work is considered to be El monte, published first in Havana in 1954 and then again in Miami. With this work, "the best book ever written in Cuba," according to Guillermo Cabrera Infante, the writer "has reaffirmed her Cuban background."

CABRERA, RAIMUNDO (1852-1923). Lawyer, politician, and author, born in Havana. Cabrera was imprisoned at an early age for his anti-Spanish activities. He was allowed to leave Cuba and travelled to Spain where he became a lawyer. Cabrera returned to Cuba and, although he did not participate in the Ten Years' War (1868-78), he was elected the Liberal Party's provincial deputy for Havana. In 1878, Cabrera acted as founding father of the Cuban Liberal or Autonomist Party, which advocated abolition of slavery and political equality for Cubans. Disillusioned with pacifist moves towards Cuban freedom from Spanish rule, Cabrera quit politics in 1893 and lived in New York during the War of Independence (1895-98). While in New York, he founded the illustrated magazine Cuba and America, which later gained wide circulation as a leading organ of the campaign for Cuban independence. Cabrera was thus established as an influential figure in this cause. After independence, Cabrera used his influence as a revived liberal in the U.S. to exert pressure on Cuban President Mario G. Menocal (1913-21) to preserve a democratic government in Cuba and allow free elections. Cabrera later founded the newspaper La Unión in Havana and was the author of numerous books, including Cuba y sus jueces; Cuentos míos; La Casa de Beneficencia y la Sociedad Económica; Los partidos coloniales; Sombras que pasan; Mis buenos tiempos.

CABRERA INFANTE, GUILLERMO (1929-). Born in Gibara, Oriente province, Cabrera Infante was the son of a Communist militant. As a young man he was involved in politics as an orator of leftist tendencies. Cabrera Infante supported Fidel Castro and after the latter took power, he was appointed director of the cultural magazine Lunes de Revolución, later censored by the regime. He was also named Cultural Attaché of the Cuban Embassy in Brussels. Eventually, Cabrera Infante abandoned his post and denounced the Cuban Revolution. He moved to London where he continues to live. Cabrera has become one of Cuba's most important novelists. His main works are Así en la paz como en la guerra, Tres tristes tigres, La Habana para un infante difunto, Vista del amanecer en el tráfico (his vision of Cuban history), and Holy Smoke.

CACIQUES. Indian chieftains. The most famous Cuban cacique
was Hatüey, who led the natives of eastern Cuba against the
Spanish conquest of the island in 1511. The peaceful Taínos of
Cuba could not resist the better armed Spanish, despite Hatüey's
and other caciques' attempts to incite a rebellion against the
Spanish conquerors.

CAFFERY, JEFFERSON (1886-1974). American Ambassador to Cuba
(1933-36). While serving as U.S. Assistant Under Secretary of
State, Caffery was appointed in 1933 to succeed Sumner Welles
as U.S. Ambassador to Cuba. Caffery is credited with having
involved himself less in Cuban politics than his predecessor.
Nevertheless, an assassination attempt on his life occurred on
May 17, 1934. The identity of the would-be killer(s) was
never established. Like Welles, Caffery was hostile to the new
Cuban President Ramón Grau San Martín's (1933-34 and 1944-
48) programs of socio-economic reform, which he saw as detrimental to U.S. interests on the island. On the other hand,
Caffery regarded Army Colonel Fulgencio Batista as a man
friendly to U.S. interests, capable of maintaining order in
Cuba. Consequently, Batista enjoyed the Ambassador's support and advice following Grau's forced resignation on January
14, 1934. Caffery left Cuba in 1938. He died in his place of
birth, Lafayette, Louisiana, in 1974.

CALDERIO, FRANCISCO (1908-1987). Calderío was born in Manzanillo, Oriente province, and later, as a Communist Party member,
changed his name to Blas Roca. A shoemaker by trade, he became active in union activities. He joined the Communist Party
in 1929 and in 1934 he was elected Secretary General of the
Party, a post he held until 1961. During 1962-65 he served on
the national directorates and secretariats of the forerunners of
the present Cuban Communist Party (PCC). He has been a member of the PCC Central Committee since October 1965. During
1965-80 he was a member of its Secretariat. He has been a
member of the Politburo since December 1975. During 1940-52
he served in the Cuban House of Representatives. Blas Roca
was editor of the Communist newspaper Hoy during 1962-65.
As President of the PCC's Constitutional Studies Commission,
Roca was instrumental in institutionalizing the Castro Government and drafting the new constitution proclaimed in February
1976. Beginning in October 1976 nationwide elections were held
and deputies chosen to the National Assembly. Roca was elected
a deputy representing the municipality of Matanzas in November
1976 and the following month he was elected President of the
National Assembly. Also in December 1976 he was elected a
Vice-President of the Cuban Council of State. Because of age
and poor health, Roca stepped down as President of the National Assembly in December 1981.

CAMACHO AGUILERA, JULIO (1924-   ). Born in Gibara, Cuba,

Camacho is currently a member of the Politburo of the Cuban
Communist Party's Central Committee (PCC). He initiated his
political career in 1960 as a Major in the Revolutionary Armed
Forces and as Minister of Transportation. During 1964-1967
he served as Chief of the Political Directorate of the Central
Army. From 1965 to the present, he has been a member of
the PCC Central Committee. Camacho was appointed Secretary
of the Provincial Committee for Pinar del Río province from
1968 to 1979. During 1979-1980 he was a member of the Secretariat of the PCC Central Committee and Party First Secretary
for Havana City since 1979. Camacho has occupied his current
Politburo position since 1980.

CAMAGÜEY (PROVINCE). Population: 541,197; area: 15,763 Km²;
municipalities: Carlos Manuel de Céspedes, Sierra de Cubitas,
Minas, Guáimaro, Camagüey, Vertientes, Najasa, Esmeralda,
Nuevitas, Sibanicú, Florida, Jimaguayú, and Santa Cruz del
Sur. The capital of the province was founded in 1514 as Santa
María del Puerto Príncipe and was one of Cuba's original seven
Spanish villas. Before 1976, the present province of Camagüey
was one province with most of its territory in what is today
Ciego de Avila and part of the territory of the province of Las
Tunas. Thus, it was prior to this date the second largest of
the island. Its fertile plains are the largest producers of cattle, and the area is also an important source of sugar. On its
southern coast, the fishing industry is being developed in the
Port of Santa Cruz del Sur with an Industrial Fishing Enterprise that produces seafood and fish preserves for export. To
the north of the province is the city of Nuevitas, industrial
and port center.

CAPABLANCA, JOSE RAUL (1888-1942). Cuba's foremost chess
player and world master. Capablanca started to play chess
when he was only five years old, amazing his father and all
the members of the Chess Club in Havana. Soon after, his
reputation spread throughout the world and after defeating
Marshall he obtained a tremendous success in San Sebastián,
Spain, winning the world chess tournament where the participants were the best chess players at the time. He held most
of the chess records during his lifetime, among others a ten-
year period without being defeated and winning simultaneously
over 350 competitors. He was such a genius at the game that
he did away with some variants that had been in the game for
years. The most famous matches he won were against masters
such as Marshall, Kostic, Lasker, and Euwe. The most important tournaments he won were those in San Sebastián in 1911,
New York in 1927, Moscow in 1937, and Nottingham in 1938.
He lost the world championship to Alekhine in 1927 who refused
to play him ever again.

CAPITULACIONES. Negotiations or surrenders ending hostilities between opposing sides in several political conflicts in Cuban history. Among these is the Pact of Zanjón ending the Ten Years' War in 1878, and the protest of Baraguá in the same year. Also regarded as a major capitulation was that of the Spanish in December 1898, when the Treaty of Paris was signed ending the Spanish-American War and signaling the transfer of government in Cuba from Spanish to American hands.

CAPTAIN-GENERAL. In the late 1500's, this title was bestowed upon the Spanish governors of Spain's Caribbean colonial possessions. The title added military powers to the extensive civil authority Cuba's governor already exercised, and rendered him equal in rank to the commanders of the Spanish fleets that called at the port of Havana.

CARABALI. The second largest group of blacks concentrated in the province of Matanzas, and especially in the cities of Matanzas and Cárdenas. The Carabalíes were brought from the port of Calabar in southeastern Nigeria and originated from various African tribes such as the Efik and the Ibibio.

CARACAS PACT. A unity pact signed in Caracas, Venezuela, on July 28, 1958, among various organizations opposing the Batista dictatorship. Fidel Castro's 26th of July movement and the Directorio Revolucionario were the principal groups participating while the Partido Socialista Popular (Communists) was excluded. Dr. Manuel Urrutia, a distinguished judge, was appointed head of the provisional government to be established after Batista's overthrow. The participants called for a common strategy of armed insurrection, a brief provisional government followed by elections, and a minimum program guaranteeing the punishment of the guilty, workers' rights, order, peace, freedom, the fulfillment of international agreements, and the economic, social, and institutional progress of the Cuban people.

CARBO, SERGIO (1892-1971). Journalist, editor, political leader. Born in Havana. As an opponent of President Gerardo Machado (1924-33), Carbó participated in an unsuccessful invasion against his regime that landed in Gibara in August 1931. Carbó founded the newspaper La Semana, and for a brief time after the fall of Machado in 1933 sponsored radical positions. For a week, in September 1933, Carbó served in the government called the Pentarquía (composed of five members) which soon crumbled, leading to the appointment of Dr. Ramón Grau San Martín as President. Carbó had been decisive in the military promotion of Colonel Fulgencio Batista in the early stages of the latter's career. Eventually, Batista became President (1940-1944 and 1952-1958). In November 1933, a bomb destroyed Carbó's La Semana. He later founded the newspaper Prensa Libre and opposed Batista's second government. In the early stages of

Fidel Castro's regime, Carbó denounced the government restrictions on the Cuban press and the Cuban Revolution's drift
toward Communism. Carbó's newspaper was finally closed down
and he fled into exile. He died in Miami, Florida, in 1971.

CARPENTIER, ALEJO (1904-1980). Born in Havana, he travelled
as a child through France, Austria, Belgium, and Russia with
his parents. He began his studies in Paris, where he showed
interest in becoming a writer and musician. He returned to
Cuba to study architecture but failed in his university studies
and left Cuba in 1921. Supposedly by this time he had already
become a Communist. He went back to Cuba in 1923 and he became active in the Grupo Minorista. He worked as a theatre
and music critic for the newspapers La Discusión and El Heraldo
and later became editor of the magazines Social and Carteles,
sharing his editorship with intellectuals and critics such as
Jorge Mañach, Juan Marinello, Francisco Ichaso, and Martí
Casanovas of the vanguardist magazine Revista de Avance. He
visited Mexico in 1926. He organized two concerts for "new
rhythms" in 1927. Amadeo Roldán, the famous conductor,
orchestrated four of Carpentier's compositions: La Rebambaramba (Cuban colonial ballet in two acts), El milagro de Anaquille (with Afro-Cuban choreography), and Matacangrejo and
Azúcar (poems with Creole choreography). He lived in Paris
from 1928, making several study trips through Spain, Italy,
Germany, England, etc. Meanwhile he sent contributions to
the principal magazines in Cuba like Chic, Musicalia, Social,
Carteles, Suplemento del Diario de la Marina, Revista de Avance,
Revista de la Habana, Imán, La Revue de Paris, Revista Cubana,
and others. In Paris and with the collaboration of French musician Marius-François Gaillard, he published Poèmes des Antilles,
La Passion noire, Yamba O, Blue, and Dos poemas afro-cubanos.
¡Ecue-Yamba-O! (Afro-Cuban history) was published by Editorial
España in Madrid in 1933. His international reknown as a novelist was established with Reino de este mundo (1949), Los pasos
perdidos (1953), El acoso (1956), and El siglo de las luces
(1962). He continued writing until the time of his death in
Paris, where he served the Castro regime as a counseling minister. During his lifetime, he was honored with the awards
Premio Internacional Alfonso Reyes (Mexico, 1975) and Miguel
de Cervantes (Spain, 1978).

CARRERA JUSTIZ, FRANCISCO (1857-1947). Lawyer, author.
Born in Guanabacoa, province of Havana, Carrera Jústiz at-
tended church schools and graduated from the University of
Havana as a lawyer. Carrera Jústiz became a professor at the
University specializing in municipal government. He founded
the magazine Revista Municipal y de intereses económicos.
Carrera Jústiz' academic publications earned him a chair in the
Academy of Sciences of Havana and honorary membership in
the Royal Academy of Jurisprudence of Spain. Carrera Jústiz

was also Minister of Cuba to Spain, the United States, Holland, and Mexico. He was a prolific author. Some of his most important works include El Canal de Panamá y sus relaciones con Cuba; Introducción a la historia de las instituciones locales de Cuba; Ayuntamientos cubanos; Una sociología municipal; and Estudios de filosofía política.

CARRILLO, JUSTO (1912-). Prominent economist of the 1930 student generation who was appointed President of the Agricultural and Development Bank by President Carlos Prío (1948-1952) in 1951. By 1953, due to his opposition to General Fulgencio Batista's coup d'état, Carrillo had left the Bank and headed Acción Libertadora, a group backed by students, trade unionists, and some professionals, hoping for Batista's ouster. By 1955, Carrillo lived in exile in Mexico where he met Fidel Castro and the two now shared their desire for Batista's downfall. Carrillo later headed a movement running parallel to Castro's July 26th Movement called the Montecristi Movement, which hoped to free Major Rafael Barquín from Batista's jails by staging a coup on the Isle of Pines, so that Barquín could act as military counterweight to Castro. In so doing, Carrillo reasoned that the Revolution might occur sooner, be less radical, and not as opposed to U.S. interests. Although Carrillo had the promised support of Venezuela and several other Latin American countries, his plan to free Barquín was thwarted by U.S. refusal to support any revolutionary movements in the Caribbean. After the triumph of the Revolution, Carrillo was briefly a candidate for the Presidency which Castro gave to Urrutia. Then, Carrillo returned to head the Development Bank from which he resigned his post as Vice-President a second time after Castro organized INRA. By 1960, Carrillo became leader of a political front of the well-organized anti-Castro Movimiento de Rescate Revolucionario (MRR) within Havana, but he was eventually forced to flee to Miami, Florida. Carrillo later served as a leader in the anti-Castro movement in the 1960's in Miami as a member of the Frente Revolucionario Democrático. Carrillo resigned from the Frente on September 30, 1960, based on his accusation that former Batista cronies were benefitting from the organization's support. He continues to live in Miami's exile community, where he published Cuba 1933: Estudiantes, yanquis, y soldados.

CASA DE CONTRATACION (House of Trade). The Casa de Contratación was created by Queen Isabella of Spain in 1503 to regulate the flow of commerce between Spain and America. Located in Seville, this clearinghouse would oversee all goods going to and coming from the New World and would ensure the Crown's interest in such commerce. All Crown subjects in Cuba paid customs duties on goods entering or leaving the island.

CASA DE LAS AMERICAS. A cultural institution which existed prior to the Revolution and which has since been converted by the Fidel Castro regime into a propaganda vehicle for its artistic expansion and influence in Latin America. Among its activities are the organizing of literary contests--inviting Latin American intellectuals to Cuba to maintain and stimulate pro-Castro feelings in the area--and publishing books, magazines, and articles supportive of the Revolution.

CASAL, JULIAN DEL (1863-1893). Leading Cuban modernist poet who knew and emulated the great Nicaraguan poet Rubén Darío. Some of the themes characterizing Casal's work include his fascination with far-off places, the exotic, and death. Some of his more famous works are Hojas al viento (1890), Nieve (1892), and Bustos y Rimas, published posthumously in 1893.

CASTELLANOS, AGUSTIN (1902-). Born in Havana, Cuba, of extremely poor parents, he graduated from the University of Havana's School of Medicine in 1925 and is recognized as an international Pediatric Cardiologist. He developed the angiographic equipment that helped in the operation of "blue babies," saving lives of many who previously died from heart ailment. For this he was honored by the Heart Institute of Mexico by being the only Cardiologist alive to be painted with the equipment in the famous mural by Diego Rivera "Immortals of Cardiology." He was also the recipient of the Gold Medal from the Hispanic-American Medical Associations, granted previously only to Madame Curie. For his work and discoveries in Angio-Cardiology, he was twice a candidate for the Nobel prize, in 1961 and 1962. Havana's prestigious Children's Hospital was named after him. After Castro's takeover, Dr. Castellanos left for Miami where he passed the Medical Board examinations in 1967. Subsequently he became Senior Scientist of the Nacional Children's Hospital, Visiting Professor of Pediatrics at the University of Miami, and Professor of Pediatrics of the International School of Medicine of the University of Miami. From 1965 to 1969 he was Researcher of Pediatric Cardiology at Variety Children's Hospital and until 1972 Chief. From 1968 to the present, he has been Professor of Clinical Pediatrics at the University of Miami. He is an Honorary Member of Cardiology societies in many countries and keeps a busy private practice in Miami.

CASTRO, FIDEL. See Castro Ruz, Fidel.

CASTRO, RAUL (1931-). Raúl was born on his father's plantation in Birán, in the northern portion of the province of Oriente, on June 3, 1931. He is the fifth of seven children of Angel Castro y Argiz and Lina Ruz González de Castro. Raúl attended the Jesuit school Colegio Dolores in Santiago de Cuba and later also Jesuit school Colegio Belén in Havana. Later as a student at the University of Havana, Raúl's interests in politics became evident. He was a member of the Juventud Socialista (Socialist Youth), an affiliate of the Moscow-oriented Partido Socialista Popular, Cuba's Communist Party. In 1953, while still a student

at the University of Havana, Raúl made his first trip behind
the Iron Curtain. He had gone to participate in the World
Youth Congress in Vienna, and visited the Soviet bloc capitals
of Bucharest, Budapest, and Prague. Upon his return, Raúl
began to get involved in his brother's struggle against Batista's government. The younger Castro, then, supposedly broke
off from the communist organization to join his brother's fight,
and did so with enthusiasm, saying that "the government has to
be overthrown so that the revolution can begin." He agreed
with Fidel's view that "reform in Cuba could not be accomplished
by constitutional means" but by overthrowing the oppressive
government. On July 26, 1953, Raúl accompanied his brother
and 160 followers in an attack on the Moncada Barracks in Santiago de Cuba. The attack was a dismal failure; Raúl and Fidel
were captured and many of their followers were executed. Both
brothers were imprisoned and Raúl was given a 13-year sentence,
but released in a general amnesty in May of 1955. Although a
disaster, the Moncada raid gave birth to Fidel's 26th of July
Movement. After his release from prison Raúl went to Mexico
with Fidel to form and organize the movement. Late in 1956
Raúl, Fidel, and 80 revolutionaries left Mexico in a yacht named
Granma and landed in Oriente province. Most of the rebels
were killed but the Castro brothers managed to escape to the
Sierra Maestra mountains along with 12 others.

In the mountains, the Castro brothers gained support and
Raúl--on February 27, 1958--gained the rank of Major. He
took some supporters and established a second front in the
Sierra de Cristal mountains in Northeastern Oriente. Named
after underground leader Frank País (who was murdered in
July 1957), Raúl's "Frank País Second Front" grew to a force
of several hundred men. During his stay in the mountains,
Raúl Castro gained a reputation for being "the most hot-headed,
impetuous and violently anti-American" of the rebels, and for
possessing a killer instinct. He reportedly matched the Cuban
dictator Batista "terror for terror." In the summer of 1958, he
kidnapped 47 Americans and 3 Canadians, ranging from engi-
neers employed at the Moa Nickel Company to American service-
men stationed at Guantánamo Base. It was reported that Fidel
disapproved of the kidnappings and ordered Raúl to release the
hostages. Raúl did not comply. He kept the captives to negotiate with the Americans, and also because he knew that Batista
would not plan any attacks while there were American citizens
present in his camp. When the U.S. agreed to Raúl's demands,
he released the prisoners, on July 18, 1958. The kidnappings
had made headlines all over the world and new accusations of
communism were hurled at Raúl. On January 1, 1959, Dictator
Batista fled Cuba for the Dominican Republic. After months of
assaults, the rebels had won. Raúl Castro then decided to
marry his fiancée, Vilma Espín, who had fought alongside him
in the mountains (under the name of Deborah), and was reported to be his "political mentor." They were married on

January 26, 1959. She was the daughter of an upper-class rum distiller, and held a chemical engineering degree from MIT. In the wake of the revolution, Raúl began to punish Batista supporters. After he became the head of the Armed Forces he directed the execution of nearly 100 officers and soldiers of the Batista Army and ordered them buried in a mass grave near Santiago de Cuba. In 1959 Raúl was named Minister of the Revolutionary Armed Forces, giving him ministerial rank and complete control in reorganizing the armed forces. He built up the army into a highly professionalized modern military establishment closely modeled on the Soviet Armed Forces and equipped with the latest Soviet equipment. The army's motto: "At your orders, commander-in-chief--for no matter what, no matter where, and under all circumstances." He also made military service for Cuba's youths mandatory and sent thousands of young officers for training in the Soviet Union. In 1969 he completed an advanced course in military studies taught by Soviet experts. In March of the same year he asked that soldiers be trained "to exhibit friendship with the sister armies of socialist countries; especially the Great Soviet Army, whose representatives work at the side of our officers and also harvest the fruits of our common efforts."

As commander of Cuba's two military intelligence organizations, Raúl directly thwarted numerous counter-revolutionary activities. It was he who led the Cuban land forces against the exile forces in the Bay of Pigs invasion of April 17, 1961. He called the presence of the Guatánamo Naval Base in Cuba a "cancer" and a permanent focus of provocation. During the 1960's, it was Raúl who played a major role in transforming the framework of the Cuban government into a "Soviet-like single political party" structure. His power was consolidated with his positions as the 2nd secretary of the Partido Comunista de Cuba (PCC) and Vice-Premier. It was Raúl who arranged for the deployment of Soviet long-range missiles in Cuba which resulted in the Cuban Missile Crisis of 1962. In the 1970's and early 1980's he has visited the Soviet Union and Eastern Europe and has been invited as observer to the Warsaw Pact maneuvers. In the early 1970's, when Raúl was promoted to First Vice-Premier, he was also given the new military rank of division commander, which is equivalent to that of General. He is also a member of the Secretariat and the Politburo of Cuba's Communist Party. He was one of the principal figures in the move toward a more Sovietized bureaucratic order and has long enjoyed Moscow's confidence as a politically reliable Cuban leader and as a competent administrator.

CASTRO RUZ, FIDEL (1926-). Cuban President, Chief of the Armed Forces, and First Secretary of the Communist Party of Cuba. A lawyer by training, a revolutionary by vocation, Castro has led the Cuban Revolution for over two decades, transforming Cuba into a Socialist State aligned with the Soviet

Union. Castro was born on August 13, 1926, on his family's
sugar plantation near Birán, Oriente province. His father was
an immigrant from Galicia, Spain. Castro was educated in
Jesuit schools in Oriente and later in Havana. One of his
teachers at Belén high school in Havana, Father Armando
Llorente, S.J., describes Fidel as "motivated, proud, different
from the others." "Fidel had a desire to distinguish himself
primarily in sports," said Llorente, "he liked to win regardless
of efforts; he was little interested in parties or socializing and
seemed alienated from Cuban society." In 1945 Castro entered
Law School at the University of Havana, where student activism,
violence, and gang fights were common occurrences. Protected
by its autonomy, the University was a sanctuary for political
agitators. Castro soon joined the activists and associated with
one of the gangs, the Unión Insurreccional Revolucionaria. Although police implicated him in the murder of a rival student
leader and in other violent actions, nothing was proved. Castro
acquired a reputation for personal ambition, forcefulness, and
fine oratory. Yet he never became a prominent student leader.
On several occasions he was defeated in student elections, an
experience that could possibly have originated his dislike for
voting evidenced after he came to power.
       In 1947 Castro left the University temporarily to enroll in an
expedition against Dominican Dictator Rafael L. Trujillo which
did not materialize. In 1948 he participated in one of the most
controversial episodes of his life, the "Bogotazo"--a series of
riots in Bogotá following the assassination of Liberal Party leader Jorge E. Gaitán. At the time, Argentine Dictator Juan D.
Perón, who favored the establishment of an anti-imperialist Latin
American Student Union under his control, encouraged four
Cuban students, including Castro, to attend a student meeting
in Bogotá. The gathering was timed to coincide with the Ninth
Inter-American Conference which Perón opposed, and which the
Communists were also bent on disrupting. When Gaitán was
assassinated, riots and chaos followed. Castro was caught up
in the violence that rocked Colombian society. Picking up a
rifle from a police station, he joined the mobs and roamed the
streets distributing anti-U.S. propaganda and inciting the populace to revolt. One of Castro's student companions denies that
Fidel was a Communist. "It was," claimed Enrique Ovares, "a
hysteric, ambitious, and uncontrollable Fidel who acted in these
events." Pursued by Colombian police, he and the other students
went to the Cuban Embassy and were later flown back to Havana
where Castro resumed his studies.
       At the University, Castro was exposed to different ideologies.
On the campus, more than anywhere else, the nation's problems
were constantly debated. Theories of all sorts flourished vigorously. The authoritarian ideas of fascism and communism were
widely discussed. But above all, the nationalistic program of
Cuba's Partido Ortodoxo--economic independence, political liberty, social justice, and an end to corruption--captured the

imagination of the students. The party's charismatic leader, Eduardo Chibás, became their idol. Castro developed into a devoted follower of Chibás, absorbing the latter's somewhat vague but puritanical ideology. While at the University, Castro married Mirta Díaz-Balart, a young philosophy student with whom he had one son. The marriage later broke up. In 1950 Castro graduated and began practicing law in Havana. Law soon gave way to politics and revolution. Castro became a Congressional candidate on the Ortodoxo party slate for the June 1952 election. The election, however, was never held. On March 10, 1952, Fulgencio Batista and a group of army conspirators overthrew President Carlos Prío's regime. For Castro violence seemed the only way to oppose the military coup. He organized a group of followers and, on July 26, 1953, attacked the Moncada military barracks in Orient province. Castro was captured, tried, and sentenced to 15 years in prison. Castro defended himself in the trial in a speech attacking Batista's regime and outlining his political and economic ideas, most of them within the mainstream of Cuba's political tradition.

After being released by an amnesty in 1955, the untiring and determined Castro traveled to Mexico and began organizing an expedition against Batista. On December 2, 1956, Fidel, his brother Raúl, and 80 men landed in Oriente province. After encounters with the army in which all but 12 of the expeditionaries were killed or captured, Castro fled to the Sierra Maestra mountains forming there a nucleus for a guerrilla operation. At the same time, urban opposition to the Batista regime increased. While Castro was in the mountains, an attack on the Presidential Palace on March 13, 1957, led by students and followers of deposed President Prío, nearly succeeded in killing Batista. On April 9, 1958, Castro called an unsuccessful national strike. The government met terrorism with counterterrorism. Political opponents were tortured and assassinated. By 1958 a movement of national revulsion against Batista had developed. Castro emerged as the undisputed leader of the anti-Batista opposition and his guerrillas increased their control over rural areas. Finally defections in the army precipitated the crumbling of the regime on December 31, 1958. On January 1, 1959 Castro and his July 26th Movement assumed power. Castro proclaimed a provisional government and began public trials and executions of "criminals" of the Batista regime. On February 15 Castro appointed his brother commander of the armed forces.

A powerful speaker and a charismatic leader, Castro exerted an almost mystical hold over the Cuban masses. As Martí had done three quarters of a century earlier and Chibás only a decade before, Fidel lectured the Cubans on morality and public virtue. His administration was almost puritanical. He emphasized his commitment to democracy and social reform, promising to hold free elections. Denying that he was a Communist,

he described his revolution as being humanistic and promised
a nationalistic government which would respect private property
and Cuba's international obligations. Attempting to consolidate
his support inside Cuba, Castro introduced several reforms.
First, he confiscated wealth "illegally" acquired by Batista's
followers. Then, he substantially reduced rents paid by tenants of houses and apartments. Finally, he passed an agrarian
reform law which confiscated landed property. Although the
avowed purpose of this law was to develop a class of independent
farmers, in reality the regime transformed the areas seized into
cooperatives managed by a National Institute of Agrarian Reform.
As time went by, cooperatives gave way to state farms, with
farmers becoming government employees. Toward the end of
1959 a radicalization of the revolution took place. This was
accompanied by the defection or purge of revolutionary leaders
and their replacement by more radical and oftentimes Communist
militants. Castro accused the United States of harboring aggressive designs against the revolution. In February 1960,
Anastas Mikoyan, Deputy Premier of the Soviet Union, visited
Havana and signed a Cuban-Soviet trade agreement, and soon
after Cuba established diplomatic relations with the Soviet Union
and most Communist countries. Castro's verbal attacks against
the United States increased.

Several months later, when the three largest American oil
refineries in Cuba refused to refine Soviet petroleum, Castro
confiscated them. The United States retaliated by cutting
Cuba's sugar quota. Castro in turn nationalized other American properties as well as many Cuban businesses. In September 1960 Castro attended the United Nations General Assembly
in New York, exchanging embraces with Soviet Premier Nikita
Khrushchev. In January 1961 President Eisenhower broke relations with Cuba. By that time anti-Castro exiles, supported
by the United States, were training for an invasion of the island. The failure of the Bay of Pigs invasion in April 1961
consolidated Castro's power and eventually led to the introduction of missiles in Cuba and to the October 1962 missile crisis.
After the invasion, Castro declared his regime to be socialist.
Economic centralization increased. Private schools fell under
government control. This was accompanied by a nation-wide
literacy campaign and by an increase of educational facilities.
Sanitation and health improved with the establishment of rural
hospitals under state control. Religious institutions were
suppressed and clergymen expelled from the island. In December 1961, Castro openly espoused Communism. "I am a Marxist-
Leninist," he said, "and shall be one until the end of my life."
Castro also organized a single party to rule Cuba. By the middle of 1961, he merged all groups that had fought against Batista into the Integrated Revolutionary Organizations, a preparatory step twoard the creation later of the United Party of the
Socialist Revolution, transformed in 1965 into the Communist
Party of Cuba--the island's present and only ruling party.

In foreign affairs Castro moved closer to the Soviet Union. The October 1962 missile crisis, however, strained Cuban-Soviet relations. By negotiating directly with the U.S., the Soviet Union humiliated Castro. Despite Castro's two visits to the Soviet Union in April 1963 and January 1964 and increased Soviet aid, uneasy relations prevailed between Havana and Moscow. At the same time, pro-Soviet Cuban communists were eliminated from positions of power. Until the end of 1963 Castro attempted to maintain a position of neutrality in the Sino-Soviet dispute. But following the 1964 Havana Conference of pro-Soviet Latin American Communist parties, the Soviet Union pressured Fidel into supporting its policies. Cuba's relations with China deteriorated, and early in 1966 Castro denounced the Peking regime. By supporting the Soviet invasion of Czechoslovakia in 1968 Castro demonstrated his dependence on the Soviet Union as well as his determination to move closer to the Soviet camp. Another source of conflict in Cuban-Soviet relations has been Castro's determination to export his revolution. After the 1964 Havana Conference of Latin American Communist parties, the Soviet Union was temporarily able to slow down Castro's support for armed struggle in Latin America. But by 1966 Castro founded in Havana the Asia-Africa-Latin America People's Solidarity Organization (AALAPSO) to promote revolution in three continents. In July 1967 he formed the Latin American Solidarity Organization (OLAS), specifically designed to impel violence in Latin America. Castro's efforts in the 1960s were unsuccessful, as evidenced by the failure of Ernesto "Che" Guevara's guerrilla campaign in Bolivia in 1967. Yet in the 1970s he could claim that the Nicaraguan revolution vindicated his commitment to violence as the Sandinistas overthrew the Somoza dictatorship. Also Cuba, with Soviet support, dispatched more than 25,000 troops to Angola to establish a pro-Soviet Marxist regime.

Castro continues to support revolutionary, terrorist, and anti-American groups throughout the world. For almost three decades now Castro has led the Cuban Revolution. Supervising projects, making decisions, traveling constantly, Castro has conducted his government in a highly personal style. A vague ideologist himself, he has transformed the island into a Socialist State aligned with the Soviet Union. He defied the United States' power and brought the world to the brink of a nuclear holocaust. A determined revolutionary, he has made the shock waves of the Cuban Revolution felt not only in Latin America but throughout the world. Within Cuba he has maintained tight political control by clamping down on enemies and by allowing potential foes to leave the island. A hero to some, a traitor to others, a criminal demagogue to still others, Fidel Castro is undoubtedly one of these decades' most controversial political leaders.

CATHOLICS. See also Church. The Catholic Church has been

prominent in Cuba since the island's discovery by Christopher Columbus. Catholic priests assisted the conquistadores in the pacification and conversion of the Indians. Catholic schools, especially the San Carlos Seminar in Havana, influenced the outlook of the high and middle classes, whereas parish priests preached to the masses. Several influential Cuban priests, such as Félix Varela, favored independence, but were opposed by the bulk of Catholic bishops and priests who supported Spain during the Independence Wars (1868-78; 1895-98). After Cuban independence was achieved in 1902, the Church emerged with little political influence. Its wealth had never been significant and its pro-Spanish attitude further weakened its support. During the 1930's, the Church tried to increase its influence through the expansion of Catholic schools and the activity of Catholic Action (Acción Católica), a lay organization closely tied to the Church. A group of professionals highly committed to the Church founded Agrupación Católica Universitaria, as well as a Catholic university, Santo Tomás de Villanueva, established in the 1940's.

During the government of President Fulgencio Batista (1952-59), the Church became increasingly concerned about social issues and human rights. In 1957 and 1958, the Catholic weekly La Quincena denounced the Batista regime's excesses. On March 1, 1958, the Cuban bishops led by Cardinal Arteaga, and the Nuncio, Monsignor Luis Centoz, asked Batista to form a government of national unity, a proposal that was rejected. After Fidel Castro took over, the Church supported his first measures, but in 1960, the bishops increasingly criticized attacks on personal freedoms and Marxist tendencies in public education in their pastoral letters. Catholics split over their support of Castro. The government eventually developed a campaign against the Church that culminated in the expulsion of several hundred priests. Catholic private schools were nationalized and Church activities were curtailed. The Cuban government has made a considerable effort to eradicate Catholic influence through use of the media and educational systems to emphasize Marxist doctrines and the incompatibility of Communism and Catholicism.

CATTLE RAISING. The livestock industry was one of the first industries developed by the Spaniards in Cuba, particularly in the areas of Santiago, Trinidad, and Baracoa. Hernán Cortés, later conqueror of Mexico, was one of Cuba's first cattle ranchers, an occupation that vied with his regular job as mayor of Santiago. Early on, cattle ranchers fought against farmers in Cuba since both competed for limited land. Prior to 1959, 68% of Cuban lands were occupied by cattle ranchers who were divided into three groups on the basis of their economic powers: the big cattle barons who obtained the greatest profits; the medium and small ranchers who concentrated on the raising and improvement of the cattle; and the sharecroppers at the

bottom who, with only limited funds, had to compete with the
more powerful ranchers. In 1952, Cuba had 9 million heads of
beef cattle, pastured on about 90,000 ranches. By 1967, numbers had decreased to 7 million, with the largest percentage in
Oriente province (2 million) and Camagüey (1.9 million). The
most serious problem with the industry in 1959 after the Revolution lay in revitalizing the antiquated feeding system.

CENSUS. Census-taking has been an irregular event in Cuba since
the island's independence from Spain in 1898. The first recorded data on Cuba's population took place in the latter half
of the 16th century on the heels of the Spanish conquest of the
island. This early information revealed the dramatic decline of
the native Indian population, along with the simultaneous increase in the white and black populations. Successive censuses
were conducted by the ruling Spaniards on a fairly regular basis until 1887: census of the white population (1569-1570); estimated census of the late 18th century; Havana census ordered
by the Count of Ricla in 1763; census by order of the Marquis
de la Torre in 1773; census by order of Captain-General Las
Casas; census of 1817; census by order of Captain-General
Vives in 1827; census of 1841; census of 1861; and census of
1887. Eight formal censuses have occurred since 1898: under
U.S. occupation in 1899, and in 1907, 1919, 1931, 1943, 1953,
1970, and 1981.

The 1953 census is considered relatively reliable. Some of
its demographic data indicate: population = 5.8 million; average family size = 4.86; urban population = 1.63 million men and
1.69 million women; rural population = 1.35 million men and
1.15 million women; density per square mile = 132; population
increase per year = 2.3%. Following the 1959 revolution, the
official census taken by the Fidel Castro government in 1970 is
referred to as the reliable source for statistical figures on the
island's population. According to the 1970 census, Cuba's total
urban population numbered (in thousands) 5,172.1, while
the total rural population (also in thousands) numbered
3,381.3, yielding a total population figure of 8,553,400. By
province, the breakdown of Cuba's population (in thousands)
according to the 1970 census was: Pinar del Río--207.2 urban;
335.2 rural; Habana--2,156.9 urban; 178.5 rural; Matanzas--
306.3 urban; 195.0 rural; Las Villas--739.2 urban; 622.9 rural;
Camagüey--498.4 urban; 314.8 rural; Oriente--1,264.1 urban;
1,734.9 rural. The last census was conducted in 1981 and the
results were to appear in Atlas Demográfico Nacional, still not
available in 1986. At the close of 1985, the population was estimated at 10,150,000.

CENTRALES. See also Ingenio. Also known as ingenios,
centrales are the plants that process sugar cane in Cuba, converting it into refined sugar. The centrales function during
the cane harvest (the zafra), and are closed for the remainder

of the year.   Black slaves staffed the sugar plantations and
the early rudimentary sugar mills.   After independence was
achieved in 1898, however, workers at the centrales progressively attained high salaries and fringe benefits, thereby en-
joying a higher standard of living than their lower-class rural
counterparts.   At the time Fidel Castro came to power in 1959,
there were 174 centrales of which only 55 were Cuban, while
some 67 were American.   Before 1959, the sugar industry operated around a central, which together with the surrounding
canefields usually represented a capital investment of $3.5 mil-
lion and up.   By 1961, however, Castro had nationalized 160
centrales.   The industry is controlled by the Ministry of Sugar
(MINAZ).

CERVANTES KAWANAGH, IGNACIO (1847-1905).   The most important
    Cuban composer of the nineteenth century.   He studied music
    in Paris, where he received numerous awards.   He composed
    contradanzas (contredanses) and experimented with all rhyth-
    mic binary combinations.   His compositions are known as Dan-
    zas, his most famous one being La danza de los tres golpes.

CESPEDES, CARLOS MANUEL DE (1819-1874).   Lawyer and revolu-
    tionary, Céspedes initiated Cuba's Ten Years' War of Independ-
    ence against Spain in 1868, and the following year became the
    first President of a provisional government organized by Cuban
    rebels.   Carlos Manuel de Céspedes was born in Bayamo, Oriente
    province, in eastern Cuba, on April 18, 1819.   The son of a
    wealthy landowner from Oriente province, the young Céspedes
    spent his childhood in his native town.   In 1834 he went to
    Havana to attend secondary school and later enrolled at the
    University of Havana.   After a short trip to Bayamo to marry
    his cousin, María del Carmen Céspedes, he travelled to Spain.
    There he received a bachelor of law degree from the University
    of Barcelona and a doctorate of law from the University of Ma-
    drid.   In Spain Céspedes had his first taste of revolution.
    The Iberian nation was undergoing a period of political turmoil
    and Céspedes joined the conspiratorial activities of army Gen-
    eral D. Juan Prim against the Espartero regime.   The failure of
    an anti-Espartero uprising in 1843 forced Céspedes to leave the
    country.   From Spain Céspedes travelled throughout Europe,
    finally returning to Cuba in 1844.   The handsome, cultured,
    and energetic Céspedes opened a law practice and engaged in
    business in Bayamo.   But law soon gave way to politics.
    Cuba was experiencing the beginning of a strong anti-Spanish
    movement.   Narciso López' unsuccessful filibuster expeditions
    against Spanish power in Cuba and his subsequent execution
    in 1851 had an impact on the young Céspedes.   Arrested be-
    cause of his anti-Spanish statements and banished from Bayamo,
    Céspedes began to organize a war for independence in Oriente
    province.   After the 1868 "Glorious Revolution" in Spain he
    saw an opportunity for revolt in Cuba and called for immediate

revolutionary action claiming that "the power of Spain is de-
crepit and wormeaten" and that if it still appeared great and
powerful to the Cubans it was because "for more than three
centuries we have looked at it from our knees."
But Céspedes' independence ideas were not shared by many
of his compatriots. Some still hoped for reforms from Spain.
Others wanted annexation to the United States. Even those
few who advocated complete independence felt that a war for
independence should be carefully organized and cautioned
against ill-prepared attempts. But Céspedes and his group
were determined to strike a blow at Spanish control of Cuba.
When they learned that their conspiratorial activities had been
discovered by the Spanish authorities they were forced to act.
On October 10, 1868, Céspedes issued the historic "Grito de
Yara" from his plantation, La Demajugua, proclaiming the independence of Cuba. He soon freed his slaves and incorporated
them into his disorganized and ill-armed force and made public
a manifesto explaining the causes of the revolt. Issued by the
newly organized Junta Revolucionaria de Cuba, the manifesto
stated that the revolt was caused by Spanish arbitrary government, excessive taxation, corruption, exclusion of Cubans from
government employment, and deprivation of political and religious
liberty, particularly the rights of assembly and petition. It
called for complete independence from Spain, for the establishment of a republic with universal suffrage, and for the indemnified emancipation of slaves. The manifesto was followed by
the organization of a provisional government with Céspedes acting as commander-in-chief of the army and head of the government.

Céspedes' almost absolute power as well as his failure to decree the immediate abolition of slavery soon caused opposition
within the revolutionary ranks. Facing mounting pressure,
Céspedes conceded in relinquishing some of his power and called
for a constitutional convention to establish a more democratic
provisional government. A Constitutional Convention met at
Guáimaro in April 1869, with delegates from several eastern towns.
A constitution was adopted which provided for a republican type
of government. The legislative power was vested on a House of
Representatives. Unhappy with Céspedes and fearful of concentrating too much power on the office of the president, a faction led by Camagüey's rebel chieftain, Ignacio Agramonte, obtained for the House a large degree of authority as well as control over presidential decisions. This group was also able to
legalize the abolition of slavery by introducing article 24 of the
Constitution, which declared "all inhabitants of the Republic to
be absolutely free." Céspedes was elected president of the new
Republic and Manuel Quesada was appointed commander-in-chief.
The war soon intensified in eastern Cuba. Céspedes decreed
the destruction of cane fields and approved the revolutionary
practice of urging the slaves to revolt and join the mambises,
as the Cuban rebels were then called. Numerous skirmishes

occurred, but Cuban forces were unable to obtain a decisive
victory against the Spanish army. Simultaneously, Céspedes
made several unsuccessful attempts to obtain U.S. recognition
of Cuban belligerancy. For the next few years the Cubans
continued to harass Spanish forces. Concentrating in eastern
Cuba and primarily in Oriente province, the war left untouched
the rich western provinces and failed to cripple Spanish power
in Cuba. Actually, by 1873 the mambises were retreating.
With most members of the House of Representatives either
dead or in hiding, Céspedes had regained almost absolute power.
Yet despite his control, he had become alienated from most revolutionary groups. The more conservative elements resented his
abolitionist stand and decrees ordering the destruction of cane
fields. The more liberal groups disliked his attempts at absolute control. The followers of General Máximo Gómez became
particularly unhappy when Céspedes began interfering in military matters and ordered the removal of the able Gómez from
command. Dissention within the revolutionary ranks and personal jealousies finally led to Céspedes' removal as President.
The remaining members of the House of Representatives called
a meeting in October 1873 and refused to invite Céspedes. It
soon became clear that the main objective of the assembly was
Céspedes' removal. This was accomplished with little opposition and the President of the House, Salvador Cisneros Betancourt, was appointed new President of the Republic. For the
next several months Céspedes sought refuge from the Spanish
forces in San Lorenzo, a farm in Oriente province, awaiting
an opportunity to leave the island. But on February 27, 1874,
a Spanish force surrounded the farm, killing Céspedes after a
brave but futile struggle.

CESPEDES, CARLOS MANUEL DE (SON) (1871-1939). The son of
Cuba's first president during the rebellion against Spain in the
1860's and 1870's, Céspedes was selected by U.S. Ambassador
Benjamin Sumner Welles and the army as provisional president
following Gerardo Machado's resignation in August 1933.
Céspedes annulled Machado's constitutional amendments of 1928
and restored the 1901 Constitution with its pro-U.S. provisions.
This aroused the ire of the reformist students and army sergeants, who overthrew Céspedes only one month after he was
inaugurated, replacing him with University Professor Ramón
Grau San Martín (1933-34 and 1944-48). Céspedes was an accomplished orator and historian. He published several historical works including a book on his father, Carlos Manuel de
Céspedes. Other works include: Un instante decisivo de la
vida de Máximo Gómez; La evolución constitucional de Cuba;
and Gonzalo de Quesada y Loynaz.

CESPEDES, CARLOS MIGUEL DE (1881-1953). Born in Matanzas
province, Céspedes graduated in law from the University of
Havana. Due to his youth, he worked as a clerk in the

Department of Justice until 1905, when he began his practice. Céspedes was appointed consulting attorney to the Department of Public Works in 1909, and Manager of the Ports Company of Cuba in 1911. A prominent leader of the Liberal Party, Céspedes participated in the uprising of 1916 directed by former President José Miguel Gómez (1909-1913) against acting President Mario García Menocal (1913-21). As a result, Céspedes was imprisoned for a short time. Under President Gerardo Machado (1925-33), Céspedes served as Minister of Public Works. He was a hard-working public official and built a number of highways and public buildings in Havana. Most prominent among these projects was the Capitol Building. Céspedes also initiated the public Central Highway of Cuba. After Machado's downfall, Céspedes went into asylum in the Embassy of Brazil. He subsequently retired from politics.

CHACON Y CALVO, JOSE MARIA (1892-1969). Writer, diplomat. José María Chacón y Calvo was born in Santa María del Rosario, province of Habana. Chacón completed his early studies under the Jesuits. He entered the University of Havana and obtained the degree of Doctor in Civil Law and Doctor of Philosophy. He was appointed second secretary of the Cuban delegation in Madrid in 1918. Forever involved in matters of education and culture, Chacón took a position in the Education Ministry overseeing matters of higher education. He was co-founder of the Sociedad Filomática Cubana and President of the section of Literature of the Ateneo of Havana. His main works are Orígenes de la poesía en Cuba; Romances tradicionales en Cuba; Gertrudis Gómez de Avellaneda; La poesía cubana; José María Heredia; Cervantes y el Romancero.

CHADBOURNE PLAN. In the face of worldwide economic depression during Gerardo Machado's (1925-1933) presidential term, Cuba joined the Chadbourne Plan in 1930 to manifest support for worldwide control of sugar production with the purpose of restoring the price of sugar to a profitable level. The plan, codified in the Cuban Sugar Stabilization Act, established a sugar export corporation to help dispose of old surplus crops. The government bought the sugar outright and secured the purchases by floating a government bond issue and by the income from taxes on future sugar production. Export quotas were attempted, but by 1931, efforts to limit sugar output elsewhere failed, and prices reached an all-time low.

CHAMBELONA. Nickname given to the February 1917 attempt by Cuban liberals to overthrow the Mario G. Menocal regime (1913-1921). Also known as the Cuartelazo de Febrero. The rebellion broke out in Oriente and Camagüey provinces and was largely successful at first. Menocal organized the army in Havana and quickly crushed the insurrection, arresting its leader, ex-President José Miguel Gómez.

CHEMICAL INDUSTRY. In the 1950's, the chemical industry was one of the most modern and efficient enterprises behind sugar. Sulfuric acid was a principal product, along with fertilizers, paints, and varnishes. Research on the uses for the chemical by-products of the sugar industry in the production of cellulose, plastics, activated charcoal, and industrial chemicals was undertaken by pre-Castro governments on a limited scale. Fidel Castro aggressively pursued research on sugar derivatives. Cuba produced fuel and drinking alcohol, invert syrups, cane wax, fertilizers, acetones, greases, glycerine, and lactic acid from raw sugar, cane juice, and molasses in the 1960's.

CHIBAS, EDUARDO (1907-1951). Political leader; founder of Partido del Pueblo Cubano (Ortodoxo). As the prominent student leader that he was, Chibás was one of the originators of the Directorio Estudiantil at the University of Havana during the mid-1920s. Graduated as a lawyer, Chibás spent some years in exile in Miami, Florida, because of his sharp criticism of General Gerardo Machado's government (1925-1933). Following the downfall of Machado's government, Chibás returned to Cuba to support the presidential candidacy of Dr. Ramón Grau San Martín, who became President of Cuba on September 10, 1933. Less than one year later, Colonel Fulgencio Batista orchestrated Grau's removal from power. Chibás again became strongly critical of the government, and particularly of Batista. In 1938, Chibás joined the Partido Revolucionario Cubano (Auténtico) and again backed Grau for the Presidency. After Grau was elected President in 1944, Chibás became disillusioned by the President's nepotism, governmental corruption, and by the violence of politically oriented gangs.

In 1946 Chibás, a Senator at the time, broke away from the Auténtico Party and founded the Ortodoxo Party. Chibás became a candidate for the Presidency in 1948 and finished third in the race. During Carlos Prío's tenure (1948-1952), Chibás used weekly radio broadcasts to attack government policies and corruption in particular. Chibás excelled in the use of radio media as a political weapon, and his following increased considerably. His programs were full of bitter condemnation and personal attacks on public figures. One of Chibás' followers and collaborators was Fidel Castro, who was nominated for a congressional seat in the 1952 elections. Chibás shot himself in the stomach at the height of an emotional radio broadcast on August 5, 1951, in an attempt to ignite the Cuban populace into action after having failed to obtain proof with which to substantiate his charges of corruption against Minister of Education Aureliano Sánchez Arango. Chibás' death three days later prompted more sadness than anger. Moreover, his death created a political vacuum and a rift in the Ortodoxo Party, facilitating Batista's coup d'état in March 1952.

CHINA. See People's Republic of China.

CHINESE. Between 1847 and 1945, some 300,000 Chinese arrived in
Cuba. Although they spread themselves throughout the island,
Havana remains the one city where the Chinese form an important distinct community. Arriving in the latter half of the nineteenth century as indentured laborers, the Chinese progressed
steadily to positions of substantial wealth and influence, principally in commerce. According to 1960 estimates, the Chinese
in Cuba amounted to 25,000 people, with 12,000 of this number
living in Havana. Chinese women are relatively scarce on the
island, thus Chinese inter-marriage with whites and blacks has
been commonplace, leading to a marked decline in the proportion
of pure-blooded Chinese. Chinese political clubs in Cuba concerned themselves mainly with Chinese rather than Cuban matters. The Chinese Nationalist Party (Kwomintang) operated an
extensive organization based in Cuba's principal cities in 1959.
The Centro Chino (national federation of Chinese associations)
was headquartered in the Taiwanese (Chinese) Consulate. In
order to be accepted by the Cubans, however, the Chinese
have emphasized the contributions of Chinese heroes in the
Cuban War of Independence against Spain. Often, Sun Yat
Sen has been called by the Chinese "the Martí of China," comparing him with Cuba's most famous patriot. Communist influence became increasingly conspicuous among Cuba's Chinese
in the 1960s. Representatives from Peking came to open a
China-Cuba People's Friendship Week, announcing a common
bond between the two nations. With Cuba's increasing dependence on the Soviet Union, the Sino-Soviet rift manifested itself
in a decline in Sino-Cuban relations. Throughout the early
1960s many Chinese, especially those with large investments
and those who had come to Cuba following the 1949 Communist
Revolution, abandoned Cuba.

CHOTEO. A popular attitude of irreverence toward the political
process and the culture and values of Cuban society. A mixture of sarcasm, biting humor, and criticism became a national
characteristic, a sort of psychological escape from social and
political reality.

CHURCH, ROMAN CATHOLIC. The Spanish colonial empire in the
Americas was carefully regulated to create societies subservient
to the Spanish crown. Under the system of patronato real
(papal grant of royal patronage), the Roman Catholic Church
was entrusted with the moral and spiritual guidance of the
people. In lieu of the king, local authorities appointed clergy
and exercised other ecclesiastical prerogatives, including the
collection of tithes. Because of this system, religious and political affairs were never clearly separated. In Cuba, however,
lack of indigenous support restricted the church to its primary
function, which was to advise and assist the governor, and it
did not accumulate great wealth. Because the colonial population was chiefly made up of Spaniards or slaves under the

authority of Spanish masters, the island was rapidly inundated with missionaries, but this immigration tapered off in the sixteenth and seventeenth centuries. Conflicts between church and state were concerned primarily with the institution of slavery, which the church was instrumental in abolishing. It not only provided the slaves with practical and religious instruction but also guaranteed them free time and opportunity to work for wages and buy freedom. A master could also free a slave by publicly acknowledging his intent while attending Mass. Statistics confirm that the church was successful in its mission; the population of freed blacks and mestizos was sizable enough as early as 1560 to elect its own constable in Havana and by 1600 to field a full company of militia. By the early 1820's the number of baptisms and marriages among slaves equaled those among Spaniards. In 1861 the free colored population accounted for nearly 40 percent of the entire black community. Nevertheless, the clergy never managed to cleave the African from his traditional background. His truancy in religious activities was often because of participation in one of various cabildos.

An eighteenth-century bishop, Pedro Agustín Morel de Santa Cruz, was primarily responsible for the eclectic nature of the African cults that still exist. Imbued with a strong sense of Roman Catholic morality, he visited the cabildos frequently, administered the sacrament of confirmation to the members, and prayed before the image of the Virgin Mary he carried with him. He named a clergyman to each of the cabildos, requiring him to visit on Sundays and holidays to teach its members Christian doctrine. No major internal differences altered the symbiotic relationship between church and state throughout the sixteenth, seventeenth, and eighteenth centuries. The opposition that separated church and state came from external forces--the introduction of nineteenth century, French-oriented rationalism, chiefly centered in the Freemasonry movement. The Freemasons and some African cabildos joined the nationalists in the struggle for independence that began in 1895; the church, however, backed the Spanish. This position of the church led to its immediate discredit and its lack of popularity after independence. The success of the independence movement in the struggle with Spain represented a triumph for rationalism and the ethos of Freemasonry. The Constitution of 1901 separated church and state and was designed to deprive the Roman Catholic Church of its tax revenue, government support, and voice in the direction of official policy. Civil marriage was made compulsory in 1918. The rationalist viewpoint in education, moreover, fostered the constitutional provision of compulsory, free, and secular education. Religious instruction in public schools was forbidden. The church was denied state subsidies and was not compensated for properties seized by the previous colonial government for its needs during the independence struggle. Although the immediate constitutional provisions that affected the church had no influence on Freemasonry, during the

republican years this movement lost its revolutionary purpose
and became increasingly more conservative.
It was during those years that Protestantism established its
foundations. Emanating chiefly from the United States, it sided
with neither the church nor the rationalists but remained generally nonpolitical. Despite the establishment and growth of a
number of Protestant missions, Roman Catholicism remained the
dominant religious influence, although its political outlook
changed from conservatism in the 1920's to reformism in the
1940's. After the foundation of the Christian Social Democracy
Movement in 1942, efforts were made to demonstrate the relevance of the church to the population, particularly to the urban
lower class, and to obtain a voice in the political forum. These
ideas were not radical for the time and were supported by conservatives. The church remained principally aligned with the
upper class and the conservative press, although its influence
spread to other classes through the expansion of such lay
groups as Acción Católica (Catholic Action) and related clubs
and health clinics. Religious history between 1952 and 1959,
with some important exceptions, was characterized more by
passivity than by activity. Although Protestants, chiefly the
Southern Baptists and such other sects as Jehovah's Witnesses
and Gideonites (white-robed street-corner preachers) were
evangelizing the lower classes, they were conspicuously absent
from any discussion of political events. The Roman Catholics
paid scant attention to the rural lower classes except for sending a few priests into the countryside. Because of the poverty
and indifference of the rural peasants, these priests were
forced to accept the financial support of local sugar mill mag-
nates, to the priests' eventual discredit. According to the historian Leslie Dewart, Batista's last regime "enjoyed an unprecedented degree of episcopal benison during six and a half of its
not quite seven years." Much of the credit for this has been
attributed to the largess bestowed on the church by Batista's
second wife and to the nature of the Roman Catholic hierarchy
and the clergy in general.
    Archbishop Manuel Arteaga, head of the Havana diocese,
preferred diplomacy to opposition; he offered congratulations
when Batista assumed power and was seen with him on cere-
monial occasions. Other leading church authorities collaborated
with the regime or believed that politics was none of the church's
business. Supported by various members of the clergy, Roman
Catholics actively opposed the government. The participation of
the Workers' Catholic Action in the unsuccessful antigovernment
strike of 1955 resulted not only in a raid on their headquarters
but also in a slander campaign against their chaplain. In late
1956, as antigovernment activities attracted more sympathizers,
government response became more direct; this, however, only
forced more clergy and laity into the Castro underground movement. In 1957 the first Roman Catholic priest, Guillermo Sardiñas,
joined Castro in the mountains. By the time Castro came to

power, six priests and one Protestant minister were directly
attached to rebel units. Relations between all churches and
the Castro government vacillated considerably in the first
fifteen years of the regime. Church-state relations shifted
from a period of initial cordiality and acceptance to open confrontation and finally in the mid-1970's into a period of cautious
détente. Both sides have contributed to the strained relations
and offer various reasons for the rift. The government has
curtailed many of the church's traditional functions and has
persecuted both laity and clergy when they stood in the path
of the revolution, and both Roman Catholic and Protestant
churches have encouraged counterrevolutionary activities on
occasion. Exile groups and traditional Roman Catholic elements
attribute repression by the government to its Marxist-Leninist
character and its theoretical emphasis on atheism, and they
consider the government anti-religious. Government officials
have accused the church of being a repository for counterrevolutionary forces linked to external agents who want the regime
overthrown. Other observers have hypothesized that the repression stemmed from the resurgence of an anti-clerical movement that had lasted during the first several decades of the
twentieth century and then fallen into abeyance around 1940.

Most scholars have seen governmental repression of organized
religion as part of a general endeavor to remove all existing or
potential organized opposition. Despite internal weaknesses and
vacillation the church still possessed the most extensive and
best organized institutional apparatus in prerevolutionary Cuba.
It was the only mass organization, and therefore its potential
as a pressure group was considerable. According to this explanation, the church was not persecuted because of antirelig-
ious sentiment within the government but because it could, and
eventually did, serve as the springboard for counterrevolutionary activities. Differences between church and state were not
immediately apparent, and there was a brief honeymoon period
after the Revolution. During this time, which lasted through
the early months of 1959, laity and clergy alike were generally
favorable to the new government and its reforms. Various
Roman Catholic leaders were named to important government
posts, and high-ranking officials openly participated in religious activities. Early opposition to the government began to
coalesce around the various religious institutions after the passage of agrarian and educational reform laws. The land reform
was more radical than had been expected, and many landholders
who mobilized against it were supported by church organizations.
Educational reform struck an even greater blow, as it confirmed
the secular character of the public schools and prohibited any
form of religious instruction in them. Members of the clergy
were outspoken in defense of the Catholic University of Havana
and its right to exist. They urged freedom of education and
expression and mercy for those sentenced to the firing squad.
At the National Catholic Conference in November 1959 about 1
million people turned out to support the stance of the church.

A period of open confrontation began in 1960 and lasted for the next four years, although tension lessened considerably after 1961. More priests and Roman Catholic laity, as well as some Protestants, including many who had initially supported the government, participated in counterrevolutionary activities. In January 1960 Castro condemned this kind of activity, particularly among the Spanish clergy--traditionally the most conservative element in the church hierarchy. The resumption of relations with the Soviet Union and dialogues with the People's Republic of China further alienated clerics and the Roman Catholic faithful. Roman Catholic publications began a campaign against the rise of communism in Cuba, and this was followed by sporadic civil disorder, verbal attacks on the government in the churches, and some arrests. In August 1960 Castro directly attacked the hierarchy of the church and followed with his own anticlerical campaign in the mass media. During the fall of that year lay organizations aligned themselves with the hierarchy, and Castro again criticized them. A pastoral letter signed by all the bishops denounced governmental attacks on the church and cited in detail repressive measures taken against the church as a whole and certain clergy specifically. Name-calling accelerated on both sides in early 1961. Some churches and Roman Catholic schools were temporarily occupied by the military, and some religious services were disrupted. The involvement of prominent Roman Catholics, including three Spanish priests, in the Bay of Pigs invasion of April 1961 brought further repression by the government. Two prelates and many priests were arrested or taken into custody, many churches were temporarily closed, and the offices of several Roman Catholic organizations were occupied in the days after the attempted overthrow. The residence permits of foreign clerics were reviewed, and some clergy were expelled from the country. All private schools were nationalized. Hundreds of priests and nuns left Cuba-- either through force or by their own volition. For example, of 3,000 nuns only about 200 remained in the early 1960s. The Church learned to operate keeping a low profile, with incidents of persecution reported. The year 1986 saw the resurgence of Church activity with the Encuentro Eclesial Cubano (ENEC), a national encounter to review the Church position and actions vis-à-vis the Castro regime.

CIBONEY. Knowledge about the early inhabitants of Cuba is generally sketchy. The Indians that inhabited the island at the time of Columbus' landing, possessed no written language and most of them, although peaceful, were annihilated, absorbed, or died out as a result of the shock of conquest. Whatever information is available comes primarily from the writings of early explorers and from later archeological discoveries and studies of village sites, burial places, and middens. These sources indicate that at least three cultures, the Guanahatabeyes, the Ciboneyes, and the Taínos swept through the island before the arrival of the Spaniards. The Ciboneyes were part of the larger South American Arawak group. They inhabited western Cuba and the southwestern peninsula of Hispaniola. It is generally agreed that the Ciboneyes, as well as the more advanced Taínos, the other Arawak group found in Cuba, originated in South America and had island-hopped along the West Indies. The Ciboneyes were a Stone Age culture and were more advanced than the Guanahatabeyes. It is believed that they migrated to Cuba already carrying with them a stone culture. They were highly skilled collectors, hunters, and fishermen, and inhabited towns particularly near rivers or the sea. Some lived in caves while others had begun to inhabit primitive dwellings called bajareques or barbacoas. The Ciboneyes practiced some form of elementary agriculture and their diet, more varied than that of the Guanahatabeyes, included turtles, fish, birds, and molluscs. Two of the more typical artifacts they developed included a stone digger (gladiolito) and a ball (esferolito), both symbols of authority or high social status also considered magical objects. The Ciboneyes fell prey to the more advanced Taínos and became their servants or naboríes. Father Bartolomé de las Casas, an early chronicler, described the Ciboneyes as "a most simple and gentle kind of people who were held like savages."

CIEGO DE AVILA (PROVINCE). Population: 272,721; area: 7,203 km²; municipalities: Chambas, Bolivia, Ciro Redondo, Majagua, Venezuela, Morón, Primero de Enero, Florencia, Ciego de Avila, and Baraguá. Its area, from North to South, includes territory that forms part of the former provinces of Las Villas and Camagüey. It is a suger province, although its pineapple, orange, and potato crops are also very important. This region has been called the Llanura de La Trocha. The capital, Ciego de Avila, emerged around 1840 in the savanna along the primitive royal road of Cuba, and sustained greater growth than Morón, the second most important city of the province.

CIENFUEGOS, CAMILO (1932-1959). Born in Jesús del Monte, a district of Havana, Cienfuegos was an early supporter of Fidel Castro who accompanied the latter on the ill-fated Granma expedition in November 1956. Following the fiasco, Cienfuegos fled to the Sierra Maestra with Castro. Cienfuegos remained in the mountains with Castro throughout 1957 and 1958, conducting guerrilla raids against the army of Dictator Fulgencio Batista (1952-1958). As a major in Castro's victorious rebel army, Cienfuegos entered Havana with Che Guevara in January 1959. He was appointed Chief of the Armed Forces. Cienfuegos' influence on Castro was slight and his post-revolutionary position was largely due to his prominence during the last months of the war. Castro, nonetheless, popularized Cienfuegos, along with himself, his brother Raúl, and Che Guevara. After Cienfuegos' disappearance in a plane crash over the ocean in

October of 1959, he became a martyr of the revolution, although some suspected that Castro had eliminated Cienfuegos as a popular and potential rival.

CIENFUEGOS, OSMANI (?-). Cienfuegos, an architect by training, initiated his political career in 1959 as Minister of Public Works, a position he occupied until 1963. Cienfuegos did not participate in revolutionary combat in Cuba. However, he is the elder brother of Camilo Cienfuegos, the Chief of Staff of the Rebel Army in early 1959 who died in an airplane accident in 1959. In 1965 he became a member of the National Directorate of the Party. He served as Secretary General of the Executive Secretariat of the Afro-Asian-Latin American Solidarity Organization (OSPAAAL) in 1966. During 1966-1973, Cienfuegos was the Chairman of the PCC Foreign Affairs Committee. In 1973 he was appointed Minister-Secretary of the Council of Ministers and continues in this post until the present. He served as Minister of Transportation in 1979, and in 1980 was elected Vice-President of the Council of Ministers. Cienfuegos currently occupies this position, wherein he is responsible for the Cuban Tourist Institute, the State Committees for Technical and Material Supply, Finance, and for Labor and Social Security. In 1986, Osmani Cienfuegos became a member of the party's Politburo. In general, it is Cienfuegos' Executive Committee of the Council of Ministers that administers Cuba on a day-to-day basis.

CIENFUEGOS. A major city on Cuba's southern coast in Cienfuegos province (prior to the Castro Revolution was Las Villas province), the population of Cienfuegos was over 100,000 in 1982. Since 1970, Cienfuegos has become strategically significant since a naval base was constructed there. The U.S. Pentagon stated that the Soviets appeared to have built a limited facility at Cienfuegos and not a submarine base on the scale of the American bases at Holy Loch in Scotland. The Cienfuegos base can, however, repair, arm, and refuel Soviet nuclear submarines.

CIENFUEGOS (PROVINCE). Population: 274,134; area: 4,150 km²; municipalities: Aguada de Pasajeros, Palmira, Cruces, Cienfuegos, Rodas, Lajas, Cumanayagua, and Abreu. The greatest activity of the province is agriculture. Its industrial activity is located in the city of Cienfuegos on the Bay of Jagua. Presently, the Nuclear Plant of Juraguá is being constructed; a cement factory and a hydroelectric plant are already operating. The capital of the province, Cienfuegos, was founded with the name of Fernandina de Jagua in 1819 as a part of a colonization project by the white emigrant families from Louisiana. Other important cities are Cruces and Cumanayagua.

CIMARRON. Those runaway black slaves in colonial Cuba who by example encouraged other poorly fed, housed, and clothed

slaves to escape captivity and to rebel. As early as 1538,
Cimarrones rioted and looted Havana while French privateers
were attacking the city from the sea.

CISNEROS BETANCOURT, GASPAR (1803-1866). Also known as
El Lugareño, Cisneros Betancourt was born in Camagüey and
wrote Escenas Cotidianas, published in the Gaceta de Puerto
Prncipe from 1838 to 1840. His simple and clear style satirized
the daily routine, the less civilized practices, and the indolence
of many Cubans. Betancourt was also interested in Cuba's political situation and in the 1850's switched from favoring Cuban
annexation to the U.S. to supporting the separatist movement
and Cuban independence.

CISNEROS BETANCOURT, SALVADOR (1828-1914). Independence
leader, public figure, and legislator, Cisneros was otherwise
known as the Marquis of Santa Lucía. Cisneros was born in
Camagüey to a wealthy family. As a young man, he became
linked to the independence movement of Joaquín de Agüero in
1851. Cisneros was subsequently imprisoned in Spain for his
anti-Spanish actions. A leader in Camagüey of the conspiracy
that culminated in the October 10, 1868, uprising against Spain
in Yara, Cisneros became very active in the direction of the
civil aspects of what evolved into the Ten Years' War for Independence. He was President of the Chamber of Representatives
during the War and succeeded Carlos Manuel de Céspedes as
President of the rebel government. He served from 1873 to
1875, having tried unsuccessfully to promote the invasion of
the western part of Cuba. After Jose Martí's death in 1895,
Cisneros was elected President until the end of the War.
Cisneros had an inclination for legislative work as shown by
his participation in the Constitutional Assembly of Guáimaro,
and especially in the Constitutional Convention of 1901 which
approved the first Constitution of Cuba as an independent
nation.

CLIMATE. Cuba is situated entirely within the torrid zone, but
the northern coastline near Havana lies only a fraction of a
degree below the Tropic of Cancer, and prevailing trade winds
combine with the warm waters of the Gulf Stream to produce a
mild climate that is described variously as border tropical,
trade wind tropical, or--because of its relatively light rainfall--
dry tropical. Located on the southern flank of the North Atlantic high-pressure zone, the island is under the influence of
the northeasterly trade winds in winter and the north-
northeasterly trade winds in summer. Despite its elongated
shape, the island's landmass is of sufficient size for some continental climatic characteristics to develop in the interior,
where diurnal and seasonal ranges in temperature are somewhat greater than along the coastlines. From year to year
there is little change in maximum temperature ranges, but

minimum ranges vary substantially in response to intermittent
winter intrusions of cold air masses from the north. Diurnal
temperature changes, even in the interior, are moderate except
at higher elevations, where the temperature drops sharply during the night. The mean temperature for the country as a whole
is about 77°F in winter and 80°F in summer, and averages range
only between 70°F and 82°F for the coldest and warmest months
respectively. Summer readings of as high as 100°F have been
recorded, and occasional freezing temperatures occur only in
mountain areas. Temperature drops to below 50°F are fairly
frequent, however, and during a severe winter cold wave in
1958 a reading of 37.4°F was recorded in the city of Camagüey.

The relative average humidity varies from 60 to 70 percent
in the daytime and from 80 to 90 percent during the night in
the summer, and from 65 to 70 percent during the day and
from 85 to 90 percent during the night in winter. An increase
in humidity is observed at higher elevations in combination with
relatively lower temperatures. Most of the country experiences
a rainy season from May through October, a period during which
three-fourths of the annual precipitation occurs. In the mountains west of Baracoa in Oriente province, however, high mountains and direct exposure to the trade winds result in a more
even distribution of rainfall. Unlike other islands of the Greater
Antilles, Cuba has no high central mountain spine, and this
absence of a strong relief barrier results in a generally adequate but lower amount of rainfall. Precipitation ranges from
40 to 70 inches per year in coastal areas and from 50 to 60
inches in the interior. For Cuba as a whole the monthly average ranges from a maximum of 7 to 8 inches during May and
June to less than 2 inches during December and January.
Pinar del Río province receives the most rainfall, and in general rainfall diminishes progressively toward the east, although
in no part of the country is the amount deficient during an
average year. The amount of rainfall varies substantially from
year to year, and serious droughts are fairly frequent. Such
periods occurred during the 1961-62, 1967-68, and 1973-74 dry
seasons. During the drought that continued from November
1973 to April 1974 precipitation was 51 percent or normal.
Some localities had no rain, and in usually well-watered Camagüey
province, the hardest hit, rainfall was 39 percent of the usual
amount. The equatorial low-pressure zone exerts influence on
the quantity and distribution of precipitation during the rainy
season, and heavy showers, thunderstorms, and occasional
flooding result.

The most violent climatic phenomenon to occur, however, is
the tropical hurricane. From Havana westward, Cuba lies square-
ly in the southern track of maximum hurricane frequency, and
the northern track of maximum frequency skirts the northern
coast from Cape Maisi as far westward as Cárdenas. Vulnerability is greatest in the far west, but all parts of the island
have experienced hurricanes. Although they occur as early

as May and as late as December, most occur from August through October, and October is the month of peak hazard. Since the year 1800 some 85 hurricanes have struck Cuba, a frequency of about one every other year. In 1944 the eye of a major hurricane passed directly over Havana, and hurricane Flora in 1963 was the most serious since that date. The five days during which it swept over Oriente province with winds of up to 110 miles an hour brought precipitation equal to the total for an average year. The storm affected 172,000 people, destroyed or severely damaged 30,000 dwellings, and caused 4,200 fatalities and countless livestock losses. On the occasion of hurricane Flora, the Cuban government accused the American Red Cross of being a political tool and the United States Weather Bureau of refusing to fly hurricane tracking planes, although Cuba had requested the flights. There had, however, been effective cooperation between the weather bureaus of the two countries during the five-day period. Cuba has an understandable fear of hurricanes, and in 1975 the government was reported to be completing a network of meteorological stations that would give the country the most effective hurricane warning system in Latin America.

CLOTHING INDUSTRY. Until the 1930's, most textiles were imported. By 1950, there were some thirty large textile mills producing flat cotton, rayon goods, synthetic cord, wool textiles, towels, and hosiery. The clothing industry developed further in the 1950's, employing some 30,000 workers. There were five plants producing rayon textiles with imported rayon fibers. Under the Castro revolution, the textile industry was nationalized and placed under government control. No significant improvement has taken place in this important industry in the past two decades: productivity is low and the industry suffers from mismanagement and inefficiency.

CLUB DE LA HABANA. Founded in 1847, the Club de La Habana was a secret organization of writers, professional people, and entrepreneurs who opposed continued Spanish domination and advocated Cuba's annexation to the United States. The plantocracy was motivated in an effort to preserve the institution of slavery. The Club's meetings, while suspicious to the Spanish authorities, were not subject to persecution. Its leaders included José Luis Alfonso, Carlos Núñez del Castillo, Miguel Aldama, and Cristóbal Madán. See Annexation.

CLUBS, SOCIAL. Social clubs or centros were cooperatively organized for mutual aid and social purposes in Cuba throughout the 19th and 20th centuries. These clubs were mainly associations of persons from provinces and regions of Spain, particularly Galicia and Asturias. Membership in the centros ranged from 10,000 to 90,000, and the largest also maintained schools, homes for the aged, mausoleums, as well as excellent hospitals and clinics. After independence several social clubs were founded, disassociated with the Spanish such as the very exclusive Havana Yacht Club, Country Club, Biltmore Yacht and Country Club, Miramar Yacht Club, and Vedado Tennis Club. The Castro Revolution took over all these clubs after 1959. Membership is determined by labor union or branch of the Armed Forces.

COARTACION. Cuban slaves in colonial times could buy their freedom or that of their children or parents by means of coartación, or obligations. Coartación was the legal right possessed by slaves to pay a sum of money to their master guaranteeing that they could not be sold except at a fixed price. Slaves thus had the right to possess or accumulate capital and property by performing extra work in agriculture. In the 18th century, slave mothers could still purchase the freedom of their unborn children for a specified sum. In the 19th century, few coartados had been purchased, due in part to the high price attached to slaves in their prime: $500.

COFFEE. Cultivation of coffee was introduced to Cuba in 1768. French immigrants who fled to Cuba from Haiti following the Haitian slave revolt of 1791 brought an interest in developing coffee production. New techniques were introduced by the French and coffee production increased. The Cuban people progressively acquired a taste for coffee, and most of its production was internally consumed. Unlike the tobacco industry, the coffee industry was centered in Eastern Cuba. The Ten Years' War (1868-1878) caused extensive damage to the Cuban coffee industry, which remained on uncertain (weak) ground for many years thereafter. Nevertheless, U.S. interest in Cuban products including coffee became increasingly active throughout the late 1800's and early 1900's, peaking in the 1920's. This helped to rehabilitate the industry. Under the revolutionary government of Fidel Castro (1959-), an effort has been made to raise coffee production substantially. According to the present government's statistics, coffee production in the early 1960's fluctuated between 36,320 and 60,615 tons, with virtually the entire crop domestically consumed. Coffee was grown on over 12,426 hectares until 1963, but since that time the land area devoted to coffee production has been expanded. The revolutionary government claimed a value of over $55 million for its 1961-62 crop. Nearly a third of the areas used for growing coffee in the Oriente mountains was destroyed by hurricane Flora in 1963. However, production has recuperated since. Traditionally, the chief coffee-growing regions in Cuba have been Palma Soriano, La Maya, and Guantánamo, in Oriente; Trinidad and Cienfuegos, in Las Villas; and the hills of Cuzco, between San Cristóbal and Guanajay, in Pinar del Río.

COGOBIERNO. A peculiar feature of University administration in

Cuba as well as other Latin American countries where students and professors share the decision-making process and the management of higher education institutions.

COLLAZO, ENRIQUE (1848-1921). Born in Santiago de Cuba, Collazo went to study in Spain at the Academia de Artillería. He returned to Cuba to fight for independence following the Ten Years' War (1868-1878) as aide to General Máximo Gómez. After the Zanjón Pact (1878), he moved to Jamaica, returning in 1895 to fight with General Calixto García. In the Cuban Republic, Collazo occupied several political posts and was elected to the House of Representatives. He founded and ran the daily La Nación in Havana and wrote several volumes on Cuban history, including La guerra de Cuba and Cuba independiente.

COLOME IBARRA, ABELARDO (1940-). Division General Abelardo Colomé Ibarra is currently a member of the Politburo--the Central Committee of the Cuban Communist Party--the Council of State, and First Substitute to the Minister of the Revolutionary Armed Forces (FAR). In addition, he holds the honorific title of "Hero of the Republic of Cuba." General Colomé Ibarra was born in Santiago de Cuba in 1940. He joined the 26th of July Movement in 1957 and became a member of the guerrilla operation "Frank País Second Front" directed by Raúl Castro. Since 1959, he has held numerous positions within the Cuban intelligence apparatus, working directly under Ramiro Valdés and Manuel Piñeiro Losada. In 1963 he was made an official of the General Intelligence Directorate (DGI). In that capacity, he is reported to have participated in support of various subversive activities in Latin America. In 1963 he joined the Cuban Communist Party and was elected a member of the Central Committee. Since 1971 he became Vice-Minister of FAR in charge of counter-intelligence. In 1977 he was chief of the Angola expeditionary corps, an activity that helped him consolidate his position in the Cuban hierarchy. At the end of the Third Party Congress (1986), Colomé Ibarra emerged as one of the principal figures in the Cuban government.

COLONO. Sugar planter and owner of a colonia (sugar farm), who sells cane--usually to a central (sugar mill).

COLUMBIA, CAMP (Campamento de Columbia). The military barracks in Havana where Sergeant Fulgencio Batista initiated the Sergeants' Conspiracy of September 1933 which brought to power the revolutionary regime of Dr. Ramón Grau San Martín. By 1936, Batista had built himself an elegant house at Camp Columbia along with recreation clubs, new barracks, and a hospital for the army. It was also at Camp Columbia that Batista launched his successful coup in 1952, overthrowing the administration of President Carlos Prío Socarrás (1948-1952). During

1959, Fidel Castro promised to turn barracks into schools, as the revolution would not need firearms. Only part of Columbia was remodeled and equipped to become a school; the rest has remained a military zone.

COLUMBUS, CHRISTOPHER (1451-1506). Genoese seaman, discoverer of the New World. On his first voyage to the Caribbean, Columbus sighted Cuba on October 27, 1492, and explored its northeast coast for several weeks. On his second voyage in 1493, he explored the southern Cuban coast sailing almost to the western tip of the island thinking that it was a peninsula of the mainland. Columbus' third and fourth voyages carried him to the northern coast of South America and the eastern coast of Central America and Jamaica. Disillusioned with internecine disputes and warfare with the Indians, he ultimately returned to Spain where he died in 1506.

COMMITTEES FOR THE DEFENSE OF THE REVOLUTION (CDR). Block-to-block vigilante committees established by the Fidel Castro revolution to ensure internal security and public control. Created in September 1960, their major function according to the government was "fighting against the counterrevolutionary class enemy." These novel neighborhood committees were created to counter the increased antigovernment activity the regime perceived in the early 1960's by acting as local vigilante committees. The CDR originally had the single purpose of serving as an adjunct to the security apparatus of the Ministry of the Interior and FAR. It was their function to identify counterrevolutionaries at the local level, to keep track of their activities, and to inform the proper governmental authorities. At the outset, the CDR became a major civil defense force during the national emergency created by the Bay of Pigs invasion. During this period they rounded up numerous actual and suspected enemies of the regime and delivered them to the revolutionary tribunals for summary disposal. They also served to keep the Cuban population mobilized for defense. During the Cuban missile crisis of 1962, the CDR heightened their mobilization and surveillance roles. In 1968, as part of the Revolutionary Offensive, the government confiscated about 55,000 small businesses, such as bars, stores, restaurants, and nightclubs.

The CDR played a vital role in the rapid nationalization process by assigning "volunteers" to guard expropriated property and "people's administrators" to operate the businesses. It took only a single day to move a great number of individual operations from private to public ownership. As the CDR evolved, they developed a number of roles in addition to their primary function of revolutionary vigilance. As a logical extension of their major function, they developed a crime-prevention role. Most of the individuals tried before the popular tribunals were apprehended by the local CDR. All CDR members stand guard duty for about four hours one night each month. They patrol the streets unarmed and are to report robberies or unusual activities to the police or militia. They have often gone so far as to capture thieves or burglars and hold them for the police. The CDR were relied upon regularly for testimony about the morals or behavior of individuals brought before the popular tribunals, and the local CDR were responsible for enforcing the tribunal's sentence in cases of banishment or house confinement. Other functions exercised by these neighborhood committees included reporting lazy workers and absentees, catching truants, organizing "click patrols" to check on the use of electricity, and listing the possessions of those individuals who had asked to leave the country. As a supplement to the internal security apparatus, the CDR has played a significant role in the indoctrination and propaganda process. Each local committee was responsible for the ideological development of its membership, holding seminars to discuss such matters as Marxist-Leninist ideology, the goals of the Revolution, Cuban history (from a Marxist perspective), the sugar harvest, or the Vietnam conflict. Ideas for topics and approaches to instruction came from committees on revolutionary instruction who advised local committees. Attendance at these seminars did not appear to be compulsory, but because of the intimacy of the local CDR, absences were easily noted.

The functions of the CDR expanded to the point that they were no longer simply supplements to the ministries of the interior and the armed forces. They had developed a number of roles as aids to other ministries acting to implement local policies. They mobilized their membership for volunteer labor in agriculture and production. They held health drives, immunization campaigns, clean-up campaigns, saving drives, and campaigns to recycle such materials as stamps, bottles, and scrap metal. They watched the quality of local services and local education. In the early 1980's, the CDR had a membership of between 4.5 and 5 million Cuban citizens over fourteen years of age. This was directed and controlled by a National Directorate. Directly subordinate to the National Directorate were 6 provincial directorates that supervised over 200 district directorates; these in turn supervised approximately 4,500 sectional directorates. There were over 30,000 base or neighborhood committees having an average membership of approximately 120 to 150 each. Almost every work center, apartment building, people's farm, or city block had a local CDR. The local CDR were directed by a president elected by the membership and were divided into a number of sub-committees (frentes) that coordinated their activities with various government organs. Such an extensive organizational structure by its very nature made it difficult, if not impossible, for citizens to avoid contact with the CDR, thus making their vigilance role that much more easily accomplished.

COMMUNIST ASSOCIATION OF HAVANA. One of the various Communist groups that sprang up in Cuba following the Russian Revolution (1917). Founded by Carlos Baliño in the early 1920's, the association brought together the various groups to form the Cuban Communist Party in August 1925. Baliño as well as a prominent student leader, Julio Antonio Mella, became important leaders in the new party which soon became affiliated with the Communist International. See also Partido Socialista Popular.

COMMUNIST CONGRESS OF 1925. Organized in August 1925 by Julio Antonio Mella, Carlos Baliño, and the Communist Association of Havana. This Congress brought together in Havana the various Communist groups in the island creating the Cuban Communist Party with José Miguel Pérez as its first Secretary General. See also Partido Socialista Popular.

COMMUNIST PARTIES OF LATIN AMERICA AND CUBA. During the early and middle sixties Fidel Castro endorsed half-heartedly the Soviet doctrine of peaceful coexistence and the peaceful road to power, but he insisted that the Cuban model of violent revolution was the only valid one for Latin America and that Cuba should exercise the leadership of the anti-imperialist movement in the area. This brought Castro in conflict with Moscow and with the pro-Soviet popular front-oriented Communist parties of Latin America, which were not now about to abandon their confortable and peaceful positions in the Latin American political arena to follow Castro's violent path. The overthrow of the leftist João Goulart regime in Brazil in 1964 and the defeat of Salvador Allende in the 1964 Chilean elections weakened the Soviet's peaceful road to power policy toward Latin America and reinforced Castro's position that violence was the best tactic. Despite a short period of harmonious Soviet-Cuban relations following Khrushchev's ouster, differences again arose, this time directly involving the Communist parties of Latin America. Castro quarreled bitterly with the leadership of these parties for not supporting guerrilla movements and denounced the Kremlin for seeking to establish diplomatic and commercial relations with "reactionary" regimes hostile to the Cuban revolution. Castro proclaimed that the duty of every revolutionary was to make revolution and rejected the Communist doctrine that the Communist Party should play the "leading role" in the national liberation struggle.

In a small book entitled Revolution Within the Revolution? by Régis Debray, a French Marxist, Castro's new line was elaborated. Not only are Communist theory and leadership--which insists on the guiding role of the party and diminishes the possibility of struggle in the countryside--a hindrance to the liberation movement, but parties and ideology are unnecessary in the initial states of the struggle. Debray explains

that the decisive contribution of Castroism to the international
revolutionary experience is that "under certain conditions the
political and the military are not separate but form one organic
whole, consisting of the people's army, where the nucleus is
the guerrilla army. The vanguard party can exist in the form
of the guerrilla foco itself. The guerrilla force is the party in
embryo." At the Tricontinental Conference in Havana (1966),
attended by revolutionary leaders from throughout the world,
Castro insisted on his independent line, seeking to gain the
undisputed leadership of a continent-wide guerrilla struggle
and offering to provide the institutional means to promote his
line. His attempts at revolution all ended in disaster, however.
The Venezuelan venture proved a real fiasco, with the majority
of the Venezuelan people rejecting Cuba's interference in their
internal affairs. The other major effort, led by Che Guevara,
to open a guerrilla front in Bolivia, ended in his capture and
death in 1967. Neither another Cuba nor "many Vietnams," as
Castro had prophesied earlier, erupted in Latin America. Castro's failures in the area weakened his leverage with the Soviets,
increased Soviet influence within Cuba, and forced Castro to
look inward to improve his faltering economy. The success of
the Cuban military interventions in Angola and Ethiopia had as
one of the results an increase in the Cuban leverage on the
Soviet Union. The victory of the Sandinista Revolution in
Nicaragua, the development of an extended guerrilla war in El
Salvador, and the presence of urban and rural guerrillas in
Guatemala and to a lesser extent in Honduras, has given new
force to the Castro position in favor of armed struggle as the
way to socialism and has added to the Cuban influence on the
Communist parties of Latin America.

COMMUNISTS. See Partido Comunista de Cuba and Partido Socialista
    Popular.

CONFEDERATION OF CUBAN WORKERS (CTC). The unions had
    been organized in Cuba since the 1920's, and they grew in
    strength and political significance during the 1930's, generally
    under Socialist or Communist leadership. Through the influence
    of Fulgencio Batista (1940-44 and 1952-59), the Confederación
    de Trabajadores de Cuba (Confederation of Cuban Workers) was
    established in 1939. It was an umbrella for Communist domina-
    tion of labor unions then. It was given official status in 1942
    and it continued under Communist influence during the first
    years of the presidency of Ramón Grau San Martín (1944-1948)
    until 1947, when President Grau broke with the Communists
    and pushed them out and the CTC became the stronghold of
    Auténtico Party workers and labor leaders. Under President
    Carlos Prío Socarrás (1948-1952) the exclusion of the Commu-
    nists from CTC was so complete that the former Secretary Gen-
    eral and one of the founders of the CTC, Lázaro Peña, left
    Cuba for Mexico, and Blas Roca, another of the strong

Communists, left for Russia. Eusebio Mujal became the Secretary General. After Batista's coup against President Prío in 1952, Mujal continued in command of CTC and remained very cooperative to the regime, denouncing the general strikes against Batista called for by Fidel Castro and other antigovernment leaders. Under the revolutionary regime of Fidel Castro, the CTC lost importance until 1970, when it was reconstituted through elections held in 37,047 local sections. These sections elected 164,367 officials to represent the 2.2 million members. The CTC Charter approved in 1973 stated that the CTC and the trade unions were neither party organizations nor part of the state apparatus but were mass organizations with voluntary membership.

CONFITES EXPEDITION. An abortive expedition in 1947 against the Dominican Republic dictator Rafael Leónidas Trujillo. The expeditionary force was allegedly financed by the Ramón Grau San Martín government and supported by the Federation of Cuban Students (FEU). Fidel Castro, Manolo Castro, Rolando Masferrer, and others, were deeply involved. A Dominican general, backed by exiled Dominican leader Juan Bosch, commanded the forces. Training was held at Cayo Confites in the northeastern coast of Cuba. Soon the Cuban government, pressured by several Latin American nations and the United States, called off the expedition.

CONGRESO DE PANAMA. Simón Bolívar, the great hero of South American independence, as well as other leaders of the newly independent nations, saw a threat to their countries in the Spanish presence in Cuba. Consequently they encouraged Cuban movements in favor of independence. They promised help and Bolívar actually wanted to participate personally in the war against Spain for Cuban freedom. The Congreso de Panamá took place in the spring of 1826. It had been called by Bolívar who had the dream of constituting a Latin American confederation. He offered to some Cubans to deal in the Congreso with the subject of Cuban independence, but only a few nations sent representatives and it was not possible to put into practice the ideals of Bolívar.

CONGRESS. Cuba's two important constitutions have specified the existence and powers of a legislative body. The 1901 Constitution vested legislative power in a Congress consisting of a Senate and a Chamber of Representatives. An equal number of senators from each province were elected for eight-year terms, half of the Senate being renewed every four years. The Chamber of Representatives had one representative for each 25,000 persons elected by universal male suffrage for four-year terms, half of the House being renewed every two years. The powers of the two houses were defined along the lines of those assigned to the U.S. Senate and House of Representatives. In the 1940

Constitution, the presidential cabinet was made responsible to the Congress instead of being removable by impeachment. The powers of the two houses of the Congress were much the same as in the 1901 Constitution, except for additional increases in control over the executive. Also, nine senators from each province were to be elected for four-year terms. Representatives were to represent 35,000 people and had four-year terms, half of the House being renewed every two years. After the Castro revolution, Congress was abolished. The Constitution of 1976 allows for poder popular (people's power), which holds national assemblies. See Poder Popular.

CONSOLIDADO. The consolidated enterprise or consolidado, is the basic administrative unit of production in the non-agricultural sector of the Cuban economy. It is a group of industrial factories and firms of the pre-socialist period, which were confiscated and merged under a common management. In mid-1961 the first regulations were enacted creating departments to take care of the functions of planning, accounting, input and output, and sales. The number of factories or firms in each consolidado varies from a few units to several hundred. By March 1961, 18,500 industrial enterprises, accounting for 80% of the total industrial production, had been combined into several dozen consolidados. In the 1980's, 90% of gross industrial output was generated by 56 consolidated enterprises.

CONSTITUTION OF 1901. The first constitution of the independent Republic of Cuba was drawn up in 1901 during the first United States occupation. The influence of the United States in the creation of this document was evident in its emphasis on the separation of executive, legislative, and judicial powers; its provisions guaranteeing individual rights and freedoms; and its establishment of an independent judiciary with the power of judicial review. The Constitution of 1901 declared Cuba a republic. It detailed such freedoms as those of speech, press, religion, and assembly; established equality before the law; and prohibited self-incrimination and ex post facto legislation. Although the functions and responsibilities of the various governmental branches were strictly defined, the 1901 Constitution did allow for the imposition of that particularly Latin American phenomenon--the state of siege. The president, in a state of emergency or serious public disturbance, could suspend the constitution and rule by means of executive orders and decrees. The most controversial part of the 1901 Constitution was the Platt Amendment, which many Cubans considered an infringement on and derogation of Cuban sovereignty. By this amendment Cuba consented that "the United States may exercise the right to intervene for the preservation of Cuban independence, the maintenance of a government adequate for the protection of life, property and individual liberty." Not only was the first Cuban constitution modeled on the United States Constitution,

it contained a provision allowing the United States to supervise
the form and actions of the government. This provision was
revised in the 1930's as one of a number of revisions the
1901 Constitution underwent in its almost forty years of existence. The first Cuban constitution, like many Latin American
constitutions, was regarded mainly as a statement of national
ideals and goals rather than as a serious plan for governing.
After the economic difficulties and political turmoil of the 1930's,
many Cubans felt a need for new goals and new policies, particularly a curb on excessive executive power. The result was
the Constitution of 1940.

CONSTITUTION OF 1940. The Constitution of 1940 was a liberal,
democratic document created by a constituent assembly convened in 1939, which included all the country's major political
factions. Even the Popular Socialist Party, a forerunner of
the PCC, was represented; certain members who later assumed
positions in the Castro government exercised influence in the
creation of this constitution. (This helps to explain why, until the 1970's, the Castro government maintained the façade of
adhering to the 1940 Constitution.) The drafters borrowed
heavily from the previous constitution with regard to respect
for individual rights. Provisions prohibiting discrimination,
self-incrimination, and ex post facto laws were carried over.
Freedom of speech, religion, assembly, and privacy and the
importance of habeas corpus proceedings were reiterated, as
was the separation of church and state. The 1940 Constitution
departed significantly from the earlier one, however, in regard
to the structure of government and the allocation of powers and
responsibilities within the government. A semiparliamentary
governmental system was created with the specific intention of
curbing past executive abuses. The government was divided
into three separate branches with considerable interdependence
between the executive and the legislature. The president was
elected by universal suffrage for a four-year term and could
not succeed himself until after an eight-year interval. His
cabinet, which was to assist him in his executive duties, was
directly responsible to the Congress and had to answer to it.
The Congress controlled the cabinet through its ability to deny
any minister or the cabinet as a whole a vote of confidence.
The cabinet was headed by a prime minister who was to represent the executive before Congress. The most important minister was the minister of the interior, who supervised local government and controlled the national police. The legislature was
bicameral, comprising a 54-member senate (nine senators from
each province) and a House of Representatives elected on the
basis of one representative for every 35,000 persons. Both
houses were elected for four-year terms. The functions of
Congress were similar to those of the United States Congress
and were relatively unchanged from the 1901 Constitution. The
judiciary was independent and headed by a supreme court of

fifteen judges. The supreme court was divided into various
chambers concerned with the constitutionality of laws (judicial
review), the supervision and governing of the lower courts,
and questions pertaining to social guarantees.

Under this constitution the legal system was founded in pre-
independence civil law; criminal codes and statutes afforded the
major source of law for the judiciary. Probably the two most
innovative aspects of the 1940 Constitution were its stress on
social and economic matters and its provisions for suspension of
constitutional rule. The Constitution could be suspended either
by congressional vote or by executive decree in case of national
emergency or serious public disturbance. This suspension--
under which all major individual rights, including habeas corpus, were restricted--could not last for more than forty-five
days. The Constitution's emphasis on economic and social concerns reflected the drafters' desire to establish goals and ideals
for the nation's future. It proclaimed the right of the individual to work and decried economic exploitation. A guaranteed
income, health protection, unemployment compensation, and social security were declared to be governmental responsibilities.
Social and economic equality were considered essential individual
rights. The gap between political reality and this espoused
idealism was evident from the Constitution's inception. The
ideals have continued to be upheld, however, and were incorporated by the Castro government into the Fundamental Law of
1959 and into the draft socialist constitution that came into
effect in 1976.

CONSTITUTIONAL LAW (1952). Issued by General Fulgencio Batista in 1952 after coming to power, these were provisional statutes that replaced the 1940 Constitution. They set up a consultative council of 80 members to replace the elected Congress
and dissolved all political parties. The law created a council
of ministers which was entrusted with powers to modify said
statutes.

CONSTITUTIONS. See Constitution (1901); Constitution (1940);
Constitutional Law; Guáimaro Constitution; Jimaguayú Constitution; Yaya Constitution.

CONSUEGRA, HUGO (1929-   ). Born in Havana, he studied art
and painting at the San Alejandro School and graduated as an
Architect from the University of Havana. He has held one-
man shows in Cuba, Venezuela, New York, Miami, and Washington, D.C. He has participated in many group shows in
Latin America, the United States, and Europe. His works were
shown in the III, VI, and VII Bienals of Brazil; in the II Bienal
of Mexico where he received an Honorable Mention; in the III
Bienal of Paris; and in the II and III Bienals of Puerto Rico.
Included in the many collections that have Consuegra's work are
The Isaac Delgado Museum in New Orleans, Louisiana; the

National Endowment for the Arts in New York; and the New
York University Collection.

CONSULADO DE AGRICULTURA Y COMERCIO DE LA HABANA.
During the tenure of Captain-General (Governor) Luis de las
Casas (1790-1796), Cuba experienced considerable economic
development. The most important advisor of de las Casas was
Francisco de Arango y Parreño, who was for many years the
force moving a new institution known as Consulado de Agricultura y Comercio de La Habana. The function of the Consulado
was to recommend and take measures in favor of development.
The directory of the Consulado was composed by merchants and
landowners. Its creation was one of the few instances in Colonial Latin America where a governor relinquished voluntarily
part of his power. Later the Consulado was re-named Junta de
Fomento.

CONTINUISMO. A Latin American tradition also prevalent in Cuba,
continuismo occurs when a leader holds the position of chief
executive for an extended period of time, usually through unconstitutional methods such as rigging elections.

COOPERATIVES. Before the 1959 Revolution cooperatives were not
important in Cuba. In his defense speech during the trial for
his attack on the Moncada Barracks in 1953, Fidel Castro stated
that land should be distributed among the landless and that the
government should provide cooperatives in farming and cattle
raising. After taking over the Cuban government, Castro
pushed for changes in economic policy including the forming of
agricultural cooperatives. In 1960 as the Agrarian Reform Law
took effect, 60% of all land was vested in the National Agrarian
Reform Institute (Instituto Nacional de la Reforma Agraria--
INRA) to be administered through cooperatives. Private owners could not buy land without INRA consent and no one could
hold more than 1,000 acres total. In practice the essence of
cooperativism, which is participation in decisions and profits
by members of the cooperatives, never took place in Cuba.
Later the name of cooperatives was changed to granjas del
pueblo (state farms), a highly centralized operation.
      At the end of the 70s, and part of its new developmental
strategy, the Government created the Agricultural Production
Cooperatives or CPA (Cooperatives de Producción Agrícola)
through the National Association of Small Farmers or ANAP
(Asociación Nacional de Agricultores Pequeños). This agricultural cooperative model was established to promote community self-sufficiency in rural areas and increase supply to the
urban areas by utilizing more advanced agricultural techniques.
The farmers who would join this type of cooperative would have
to abandon their homes in the country and move to wherever
the new farming community was located. They would also have
to turn over their lands to state-run cooperatives and work the

collective lands destined for self-sufficiency after having worked
their regular state jobs. In order to expedite the process, the
State placed priority on the supply of agricultural materials and
equipment to the farmers who were members of the CPA and increased pressure on and control over the activities of the independent farmers who refused to join such an organization. In
1986, the Government continued to pressure the independent
farmers and closed the so-called Farmer's Free Market where
the independent producers sold their products.

CORDOBA REFORM MOVEMENT. In the 1920's, the Córdoba Reform Movement in Argentina sought to reform that country's
universities to make them more accessible to the less privileged
sectors of society and to project them into the social, political,
and economic life of the country. Cuban students were greatly
influenced by the movement as well as by the Mexican and Russian revolutions and, in 1923, launched their own university
movement, which led to a series of academic and administrative
reforms, larger government subsidies, and the establishment of
a University Commission, composed of professors, students, and
alumni. The Commission drew plans to reform the University
and purge several professors accused of incompetence.

CORTES, HERNAN (1485-1547). Spanish navigator and explorer
who headed an expedition to conquer the treasures of the New
World for Spain in 1518. Cortés stopped briefly in Cuba on his
way to Mexico, thus making the island the source of support
for the conquest of nearby lands. Cortés sailed from Cuba in
1519 to conquer the Aztec Empire, entering through the province of Tabasco, Mexico, where he defeated the Tabascan and
Tlascala Indians in 1519. He was nevertheless received amicably
by the Aztec chief, Montezuma, in Tenochtitlán, the Aztec capital (which is now Mexico City). Cortés proceeded to imprison
the Aztec chief, and following Montezuma's death in 1520, he
crushed all Aztec resistance, opening Mexico's doors to Spain
by 1521. The conquest of Mexico meant temporary prosperity
and great euphoria, but it also meant the decline of Cuba's importance. Farmers and adventurers left the island in search of
Mexican riches, and for the next two centuries, Spain focused
on its continental colonies from where it obtained much-needed
mineral wealth. Cuba was thus relegated to a mere stopping
point for passing ships.

CORTES. The Spanish parliament to which Cuba was allowed to
send representatives in 1820. Two such representatives in 1823
were José María Heredia and Father Félix Varela y Morales. After the Ten Years' War (1868-1878), Cubans were again represented in the Cortes although with a very restricted franchise
such that conservatives and pro-Spanish elements won nearly
all of the regionally distributed seats.

COTTON. Scant production of cotton occurred in colonial and 19th-century Cuba. After independence Cuba grew more cotton, the state claiming some 100,000 acres under cultivation. Under the government of Fidel Castro cotton production has not increased significantly.

COUNCIL OF MINISTERS (CABINET). After independence the Cuban presidents appointed secretaries or ministers who constituted the Cabinet or Council of Ministers. It was an institution recognized in Cuban constitutions as part of the executive power and directly dependent on the president. The 1959 Revolution gave the Council of Ministers the power to change parts of the 1940 Constitution and the power to legislate. The Constitution of 1976 established at the top of the government hierarchy the executive committee of the Council of Ministers, composed of the President of the Republic, the eight vice-presidents of the Council of Ministers, and a secretary to the Council of Ministers. The President names all ministers. In 1986 the Council of Ministers totaled some 65 members including 34 ministers.

COUNCIL OF STATE. The Constitution of 1976 established a National Assembly in charge of legislation. The assembly in 1976 elected a Council of State to function when the National Assembly was not in session. The President of the Council of State became also the President of the Republic and of the Cabinet. Fidel Castro received all of these titles as well as that of Secretary-General of Cuba's Communist Party. By the provisions of the Constitution, Cuba had broken with the practice found in other socialist countries of separating the head of state from the head of the government and the Party secretary.

COURTS [This section has been reprinted from Area Handbook for Cuba (Washington, D.C.: The American University, Foreign Area Studies, 1976) with permission herein gratefully acknowledged]. Cuban courts have consistently been controlled by the executive branch either subtly through appointments and corruption--as was the case before 1959--or by being made clearly subordinate to the Council of Ministers as has been the case since 1959. Before independence the Cuban legal system had a reputation for producing first-rate attorneys and justices. Until the Revolution, however, corruption and political manipulation was sufficiently rife within the courts as to offset the value of such good legal training. The Constitution of 1940 attempted to remedy Cuba's judicial shortcomings by creating an elaborate and independent court system in which the appointment and promotion of justices was carefully insulated. Members of the Supreme Court were nominated by the law faculty of the University of Havana, appointed by the president, and approved by the Senate. These justices could be removed only by a two-thirds vote by the Senate. Appointment, promotion, and removal

of members of the lower judiciary were the function of a special
governmental section of the Supreme Court. The Supreme
Court, composed of several chambers, dealt mainly with appeals
from lower courts but had original jurisdiction over cases involving major government officials. Under the Social Defense
Code of 1938 and the Constitution of 1940, the legal system was
revised in a number of ways. Constitutional guarantees of individual liberties were stressed along with numerous broad social guarantees of such rights as collective bargaining and a
minimum wage. Judicial review was established, allowing the
Supreme Court to rule on the constitutionality of any law challenged by the lower courts, an independent organization, or a
group of twenty-five citizens. Below the Supreme Court and
under its direction were seven audiencias or courts of appeal.
Composed of a president and a number of magistrates and divided into criminal, civil, and administrative chambers, these
courts tried cases on appeal from courts of first instance, first-
class courts, police courts, and municipal courts. They had
original jurisdiction over cases referred to them by the investigative courts of arraignment. Other specialized courts, such
as the urgency courts established to try political offenders and
used sporadically since the 1920's also were part of the system.

Until 1973 the Castro regime did not change the structure of
the court system radically, but the personnel and the essence
of justice underwent revolutionary changes. As a result of
initial, hostile rulings by the courts toward the more radical
measures proposed by the revolutionaries, the new government
deemed it necessary to fill judicial positions with individuals
who would prosecute counterrevolutionaries and further the
new laws of the Revolution. During the first year in power
the Castro regime replaced almost all of the justices on the
Supreme Court, the presidents of the audiencias, and most of
the lower court justices with individuals loyal to the Revolution.
This was accomplished by suspending for forty-five days in
late 1960 the tenure law for judges in order to completely reorganize judicial personnel. Although the Supreme Court and
the audiencias continued to exist, their functions were altered.
In addition to appointing an entirely new Supreme Court, the
government brought this court and all lower courts directly
under the control of the Council of Ministers. The old regular
judiciary was decreed to have no competence in political cases,
and even in nonpolitical cases judges were often dismissed for
ruling against the interests of the Revolution. During the first
three years the Castro government was in power more than 1,000
laws were passed revising the entire political, economic, and social structures of Cuba. Laws dealing with nationalization,
agrarian reform, urban reform, and the like were felt to need
revolutionary judges to interpret them. This could be done
within the old judicial structure providing that the consciousness of the justices working within that structure was revolutionary and providing the judiciary was supervised by the
governing elite.

What could not be dealt with within the context of the old judicial structure was the determination and prosecution of counterrevolutionary activity. Originally the new government set up emergency military tribunals to deal with enemies from the second regime of Fulgencio Batista (1952-59). Because members of the old judiciary could not be trusted to deal with counterrevolutionaries and there was no time to develop a new judicial cadre, counterrevolutionaries were tried by the military. Members of the revolutionary courts were to be appointed by the Minister of the Revolutionary Armed Forces. Their authority extended to any activity deemed a threat to the state. This included actions by members of the military until 1969 when military personnel were placed under the authority of separate military tribunals. By the time the revolutionary courts were disbanded as part of the 1973 judicial reform, they had proliferated to such an extent that there existed one such court in each province. The busiest of these military tribunals-- Havana--dealt with approximately sixty cases a month. Each court was composed of a number of three-member panels consisting of a president, who was an officer, and two enlisted men (vocales). As the revolutionary courts evolved, their jurisdiction expanded from individuals charged with attempting to maintain the "Batista tyranny" to those charged with such offenses as embezzlement of state funds, sabotage of industrial facilities, armed attacks against the people, and pilfering from government enterprises. These courts were the only tribunals in Cuba that could impose capital punishment[,] which they did not hesitate to do. Bail was not permitted in any case heard by the revolutionary tribunals. Sentences were carried out immediately, although an appeal procedure was automatic in cases involving the death penalty. In non-capital punishment[,] appeal had to be lodged by the defendant before the same court within twenty-four hours. The defendant could provide his own defense counsel or the court would provide one for him. Numerous claims have been made that although such rights may have existed in form[,] the revolutionary tribunals were actually flagrant in their disregard for due process and the right to defense or appeal, particularly in their earlier years and in the period immediately after the Bay of Pigs invasion.

The piecemeal judicial system of revolutionary Cuba was institutionalized in June 1973 by the passage of the Law of the Organization of the Judicial System. This law was the outgrowth of the work of the Commission of Judicial Studies of the Central Committee of the Communist Party of Cuba (Partido Comunista de Cuba--PCC). The purpose of this commission was to recommend to the Council of Ministers a series of measures that would unify the diffuse court system and revise criminal and civil law. As a result of the 1973 law the role of the Supreme Court has been upgraded, the powers of the regular courts have been expanded to include the authority previously vested in the revolutionary tribunals that have been eliminated, and the popular tribunals have served as a basis for the entire system. Apparently military tribunals with civilian jurisdiction have been eliminated, and military offenses since 1973 have been tried by the military chamber of the Supreme Court. A notable aspect of the judicial reform is that lay or nonprofessional judges now complemented professional judges with legal training at all levels of the judicial system except in the popular tribunals where they were the sole dispensers of justice. Two lay judges sat on each bench in the Supreme Court, in the provincial courts, and in the regional courts. Each popular tribunal was composed of three lay justices. Lay judges, at all levels, were selected for three years and served two one-month terms each year. They received compensation in accordance with what they earned in their regular jobs. The logic in having lay judges on all Cuban courts is to cut through legalism and technicalities to create a kind of revolutionary justice.

At the top of the Cuban judicial hierarchy is the Supreme Court. As a result of the 1973 reorganization law, this court was divided into four chambers or sections dealing with criminal offenses, civil and administrative matters, offenses against the security of the state, and military matters. Each chamber was composed of five judges--three professional and two lay members. The judges were selected by the Council of Ministers from a list of nominees, who had to be over thirty years of age, provided by the Cuban Communist Party (PCC) and the mass organizations. Professional judges were to serve for a set five-year term but could be removed by the Council of Ministers at any time. The head of the Supreme Court was its president, who was appointed for a seven-year term by the president of Cuba with the agreement of the Council of Ministers. The president of the court presides over the Council of Ministers, directs and administers the entire court system. The jurisdiction of the Supreme Court was primarily appellate-- appeals coming from the provincial tribunals. There appeared to be some first instance jurisdiction in regard to the actions of the upper echelons of the government and the PCC: They were to be tried by the chamber of offenses against the security of the state and were immune from prosecution in all lower courts. As a result of the reorganization, directly beneath the Supreme Court were the six provincial courts. These courts were divided into three chambers--criminal, civil, and state security. As in the case of the Supreme Court, there were two lay and three professional judges, the latter serving five-year terms. Judges were selected at the provincial and regional levels by PCC commissions and were to be over twenty-five years of age. Provincial courts had appellate jurisdiction over cases from the regional courts and original jurisdiction in any case defined as grave or major by the law or in cases involving crimes committed outside the country. They also had original jurisdiction in cases of counterrevolutionary activity. Regional courts did not have jurisdiction over cases involving state security. Their original jurisdiction was limited to less serious criminal and civil offenses and to cases of precriminal states of dangerousness. Cases were referred to them on appeal from the popular tribunals. Regional courts consisted of two lay and one professional judge.

In addition to the regular court system another series of minor courts had been formed in the 1970's as a result of increased labor legislation and the government's push to motivate workers. They were similar to the popular tribunals and existed at larger work centers. Judges were elected by the workers from among workers, and court was held in open session during nonworking hours. They dealt mainly with such minor offenses as absenteeism, negligence, and minor disputes on the job. Sanctions usually consisted of reprimands before one's fellow workers but could extend to transfer to another work center or loss of vacation. A unique aspect of the Cuban judicial system has been the neighborhood people's courts. The first of these courts were organized in rural areas in 1966. As the experiment proved successful they were established in the cities and eventually in districts (zonas) composed of 2,000 to 6,000 inhabitants. The purpose of the people's courts is more than administering justice for petty crimes and disputes, although this is certainly a major function. By putting law on a layman's level within a revolutionary context the people's courts promote public order in a twofold manner: They involve the community in the process of adjudication, and they provide a means for indoctrination or consciousness-raising. Judges for the courts have been elected by members of the local PCC and CDR (Committees for the Defense of the Revolution) from the local community. A number of judges were chosen for each district although only three presided at any given time. The judges need not be members of the PCC; in fact, most have not been. They are required to be committed to the Revolution, members of a mass organization, over twenty-one years old, and to have at least a sixth-grade education. Once selected the judges are given a three-week training course in the law and are provided with a judge's manual, which cites no statutes but sets forth examples and very general guidelines. The judges serve for two years without pay in addition to their regular jobs such as factory workers or teachers. The nature of their regular jobs was not considered as important as the extent to which they are known and know the community in which they preside. Court is usually held after work in public assembly. It may be held anywhere, is quite informal, and usually is well attended. The judges on the people's courts preside in panels of three consisting of a president and two other members. If a defendant objects to one of the judges for any reason, another is chosen from those selected for that district. Any of the judges may question the accused, the accuser or the witnesses at any point during the trial. Witnesses from the audience may also testify spontaneously regarding such things as the evidence or the character of disputants.

A major function of the court is community education, and lectures by its president for the benefit of the audience have been fairly common. The judges of the people's courts are assisted in their functions by an <u>asesor</u>, who is an official of the Ministry of the Interior. The asesor has legal training and is either an attorney or an advanced law student. Each asesor is assigned a number of courts for which he has various responsibilities. He helps in the selection of judges, is largely responsible for their training period, and assists them the first few months they are in office. In the past the asesor also acted as an appellate judge and with the assistance of two other people's court judges from the same district could reconsider a particular verdict. Under these courts cases have been heard fairly quickly, usually within ten days. Bail was seldom required if an individual's work center or local CDR would vouch for him. Since the late 1960's the jurisdiction of these courts, although relegated to minor affairs, has been fairly encompassing. They dealt with torts of less than 1,000 pesos; such misdemeanors as petty theft; health and sanitary violations; juvenile delinquency; and such antisocial conduct as personal injuries, public disorders, and drunkenness. Personal injuries were the single most common offense dealt with by the people's courts. They could not deal with felonies, major torts, or counterrevolutionary activity. One category of offenses under their jurisdiction-- delicts [crimes] against the popular economy--was based solely on laws promulgated since the Revolution. The seven such delicts [crimes] were: altering prices, hoarding, selling too much of a scarce commodity to oneself or friends, exporting illicit goods, infringement of rationing regulations, false weights and measures, and clandestine business. Contraventions against good customs [manners] and public decorum, such as obscene language in public, selling pornographic material, girl pinching, or nude swimming also accounted for many cases heard by the people's courts. The people's courts have had considerable discretion in dealing with cases in spite of the limited range of sanctions available to them. Sanctions applied by the courts were intended to [be] rehabilitative and not punitive. Consequently, capital punishment and imprisonment were not within the purview of these courts. Fines, which they originally could levy, were abandoned in 1968 because of their punitive nature. The most common sanction employed by the people's courts has been public admonition. This has been administered in almost every case and it may or may not be accompanied by more severe penalties. On the surface, public admonition served to embarrass the defendant and ideally has resulted in remorse and repentance. Because the trials were well attended and the judges lectured the audiences, they served as a warning to the community and as an opportunity to educate the people about the qualities required of an individual by the Revolution.

Other sanctions imposed by the people's courts may require
individuals to pursue educational improvement by attending an
educational center to attain a certain grade level or by participating in a study group for self-improvement. They may deprive individuals of certain rights if they determine that those
rights have been abused. For example, an individual who has
obtained certain goods illegally may be prohibited from purchasing such goods for a certain period. In one case before a court
a man found guilty of drunkenness was prohibited from drinking
for two years. The courts may banish a defendant from a specific place. For example, a man found guilty of creating a row
at his mother-in-law's house may be prohibited from visiting
her, or a habitual drunkard may be banned from bars. A
somewhat more stringent sanction consists of confinement. The
courts may sentence individuals to be confined to their homes
for up to six months, allowing them to leave only to go to their
jobs. The most severe sanction imposed by the people's courts
is the confinement of a defendant to a work farm for productive
labor for a period up to six months. Additional sanctions that
the courts may apply individually or in combination are confis-
cation of illegal goods or the instruments used in an offense,
restitution for injuries, or medical or psychiatric treatment.
Court decisions are to be enforced by the National Revolutionary Police and particularly by the CDR[,] who report to the
courts regarding adherence to a sanction.

The people's courts developed a form of rapid, community-
oriented justice dispensed in a manner that educates the people
in their responsibilities to each other and to the collectivity.
They helped to clear the dockets of the upper courts for more
serious crimes. These courts made justice a community function
based on equity and compromise rather than legal technicalities.
The attempt to distinguish justice from legal formalities, however,
no matter how commendable, may result in the elimination of pro-
cedural safeguards. The participatory nature, described by
some as the circuslike atmosphere, of the courts would not seem
likely to engender a respect for the written law. The lack of
legal training for justices and their revolutionary zeal tend to
make sentencing arbitrary, based on the community esteem of
the accused and the degree of puritanism in the nature of the
court. Individuals who can establish that they are good revolutionaries receive lighter sentences, and the puritanism among
revolutionaries has meant that certain antisocial activities tend
to be punished inordinately. A girl-pincher without good revolutionary credentials might find himself working on a state form
for 180 days, but a wife-beater or petty thief with an otherwise
good revolutionary record may receive only an admonition. Finally, the attempts by the courts to make the entire community
responsible for law and its enforcement may create a tendency
by some to use the courts as a means of denouncement and to
settle grudges, and others may fear denouncement by their
neighbors for some minor action. The need to select friends

who are good revolutionaries or have some local power seems a likely element of such a system.

CRIOLLOS. Term used to denote native-born Cubans of Spanish descent during the colonial period. Spanish policy of preventing criollos from occupying positions in the governmental structure, the judiciary, and the higher echelons of the military created much resentment and tension between criollos and peninsulares (Spanish-born). Criollos confined their activities mostly to economic endeavors, such as sugar plantations, cattle farming, commerce, or the law profession. In the 19th century with the rapid expansion of the sugar industry, criollos became an active, wealthy class of hacendados and entrepreneurs that based its prosperity on sugar, coffee, land speculation, and the slave trade.

CROMBET HERNANDEZ-BANQUERO, JAIME (1940-). Jaime Crombet, an electrical engineer, was born in Santiago de Cuba in 1940. Since 1975, he has been a member of the Secretariat and the Central Committee of the Cuban Communist Party. During the late 1950s, he was a member of the 26th of July Movement in Oriente, and later joined the National Student Militia. In 1965 he was named President of Oriente's Provincial Committee of the Federation of University Students (FEU). He joined the Communist Party in 1966, and was appointed Secretary General of the National Committee of the Young Communists Union (UJC). In this capacity, he has represented the UJC in several international youth forums. He served in Angola from 1979 to 1980 as a delegate of the Cuban Politburo. Crombet Hernández-Banquero belongs to a group of so-called new leaders of the Revolution. His mother, Hortensia, was arrested in 1963 and sentenced to 15 years in prison for allegedly being a CIA agent. She was released and later died in exile, in Miami. His father was also arrested and sentenced to 6 years in prison.

CROMBET, FLOR (1851-1895). Patriot, leader of the War for Independence. Born in Oriente province, he was exiled to Spain early in his life because of anti-Spanish activities. From there, he traveled to Central America, settling in Honduras where President Marco Aurelio Soto appointed him militia commander. He returned to Cuba in 1889 only to be deported again the next year. In 1895 he returned to Cuba and joined General Antonio Maceo and others in fighting to end Spanish domination over the island. Crombet participated in various early combats of the War for Independence and was fatally wounded in 1895.

CROWDER, GENERAL ENOCH (1859-1932). U.S. envoy sent to Cuba in 1921 to supervise the peaceful settlement of political differences during President Alfredo Zayas' administration (1921-1925). Crowder had previous electoral reform experience in Cuba as architect of an electoral system which led to the untainted election of José Miguel Gómez (1909-1913). As special envoy in 1921, Crowder pressured Zayas to eliminate corrupt practices and went so far as to propose the formation of an "Honest Cabinet" to Zayas in 1922. The cabinet attacked corruption in the civil service, reduced the budget, and trimmed several questionable public works contracts. Crowder later served as U.S. Ambassador to Cuba in 1927.

CUARTELAZO. A military uprising or coup d'état to overthrow a constitutional regime, i.e., Fulgencio Batista's military coup against constitutionally elected President Carlos Prío Socarrás (1948-1952) on March 10, 1952.

CURRENCY. See Peso.

CURRO. Originally curro was the name given to blacks or mulattoes, characterized by affectation in speech and dressing similar to that of Andalusia. Eventually the term was applied to Spanish immigrants from Andalusia.

- D -

DAHOMEYANS. See Ararás.

DANZA DE LOS MILLONES. Period just after World War I when sugar interests in Cuba were prospering due to the precipitous rise in sugar prices. This period lasted only a short while for in the late 1920s the price of sugar dropped dizzily, ruining many planters as well as banks on the island.

DE LA TORRE, CARLOS (1858-1930). Born in Matanzas, de la Torre graduated from the University of Havana as a physician, specializing in pharmacology and natural sciences. In 1881 he received a doctorate in natural sciences from the Central University of Madrid. De la Torre succeeded the famous scientist Felipe Poey in his chair of Zoology at the University of Havana. During the Independence War (1895-1898) he was prosecuted by the Spanish colonial government. De la Torre went to Mexico where he worked as Professor of Natural Science in Chihuahua. After Cuban independence, he returned to the island to teach at the University of Havana. Upon his return to Cuba, de la Torre became one of the founders of the Nationalist Party. In 1902 he was elected to Congress and was later elected President of the House. In 1905 he retired from politics to pursue his scientific interests. In 1912 Harvard University conferred upon de la Torre the honorary title of Doctor of Science. In the early 1920's he became Rector of the University of Havana.

DEBRAY, REGIS (1942-). A French Marxist and author born in
Paris, Debray became a journalist with the Mexican magazine
Sucesos. Following Fidel Castro's revolution in 1959, Debray
developed a close relationship with Castro and Che Guevara,
Castro's economic minister. Debray visited Cuba on various
occasions and was invited to lecture at the University of Havana. In his book Revolution within the Revolution (1967),
Debray elaborated on Castro's theory of revolution, emphasizing the national liberation struggle. Debray articulated Castro's theme that a determined group of revolutionaries without
the support of a party structure could initiate a guerrilla foco
in the rural areas and thereby create the necessary conditions
for the defeat of the armed forces and the overthrow of an established government. In the mid-1960's Debray was arrested
in Bolivia where he had gone to interview Che Guevara, who
was leading a guerrilla movement in that country. Debray was
sentenced to 3 years in prison, but was released in 1970
through the intervention of his wealthy French parents. In
1981 Debray joined the socialist government of François Mitterrand as an advisor for Latin America in the French Foreign
Ministry. Among his works are Armed Struggle and Political
Struggle in Latin America and Strategy for Revolution.

DECLARATION OF HAVANA. The first of the two Declarations of
Havana was issued by Castro in response to the Declaration of
San José, Costa Rica, the OAS document that condemned intervention by extra-continental powers in Latin America. The
First Declaration was issued in September 1960, condemning the
exploitation of man by men and accusing the United States of
being responsible for Latin American misery and political oppression. It also accepted the Soviet Union's "symbolic" offer to
protect Cuba with missiles and announced the establishment of
diplomatic relations with the People's Republic of China. Fidel
Castro submitted the document to the people of Cuba present at
a mass meeting in Havana stating that it was an example of
pure democracy in action.

DECLARATION OF HAVANA II. In response to the OAS agreement
of January 31, 1962 by which Cuba was suspended from the
inter-American organization, Castro proclaimed the Second Declaration of Havana, on February 4, 1962, in which the growing
differences between the Soviets and the Cuban leader were made
even more evident. Fidel Castro failed to mention peaceful coexistence, "national democracy," "peaceful transition," or any
of the other recommended Soviet formulas for the "third world"
in his words to the peoples of Latin America. The Second Declaration reflects a mixture of the Chinese tendency to convert
Marxism into the ideology of the colored races, Guevara's notions about guerrilla warfare, and the ignorance and lack of
esteem of Castro and his henchmen for ideological affairs. To
the Indians, Negroes, mulattoes, mestizos, and peasants of

Latin America the example of Cuba was offered. To follow the Cuban example, a front consisting of the working class, the peasants, the intellectual workers, and the most progressive strata of the national bourgeoisie had to be formed. Since the peasants were the most numerous, their participation was decisive. But they could not be the leaders of the struggle. For this they needed the "revolutionary and political leadership of the workers and the revolutionary intellectuals." This leadership should never forget that "the national bourgeoisie can never lead the fight against feudalism and imperialism."

The declaration then stated that "the first and most important thing is to understand that it is not right or correct to distract the people with the vain and convenient illusion that they can conquer by legal means the power that the monopolies and oligarchies will defend with blood and fire." Once this is understood they should act accordingly; in other words, they should "follow Cuba's example," starting guerrilla warfare because "the armies structured and equipped for conventional war are totally impotent when faced with the irregular warfare of the peasants on their own terrain...." Finally, they must never forget the masses because in this fight to the death "with the most powerful imperialist metropolis in the world" victory is not possible if the great majorities do not take part.

DECLARATION OF SAN JOSE. Adopted by the Organization of American States (OAS) at San José, Costa Rica, on August 28, 1960, the San José Declaration (1) condemns intervention or threat of intervention from an extra-continental power in the affairs of the American republics; (2) rejects the attempt of the Sino-Soviet powers to make use of the political, economic, or social situation of any American state capable of destroying hemispheric unity; (3) reaffirms the principles of non-intervention by any American state in the internal or external affairs of the other American states; (4) reaffirms that the Inter-American system is incompatible with any form of totalitarianism and that democracy will achieve the full scope of its objectives; (5) proclaims that all member states of the regional organization are under obligation to submit to the discipline of the Inter-American system; (6) declares that all controversies between member states should be resolved peacefully; (7) reaffirms faith in their regional system and in the OAS. The resolution passed by a vote of nineteen to zero with Mexico abstaining.

DECLARATION OF SANTIAGO. A statement issued by the Foreign Ministers of the countries belonging to the Organization of American States in August 1959 in response to Cuban-sponsored expeditions against Nicaragua, the Dominican Republic, and Haiti. Although no specific mention was made of Cuba, the statement reaffirmed the principle of non-intervention in the internal affairs of Latin American countries and advocated democratic governments as the ideal type for membership in the OAS.

DEMAJAGUA, LA. The famous <u>batey</u> or sugar-mill community near Manzanillo in Oriente province which served as the location for the outbreak of rebel activity known as the Ten Years' War (1868-1878) in October 1868. Carlos Manuel de Céspedes united a large group of pro-independence followers at La Demajagua on 10 October 1868 and moved on the small town of Yara nearby with the familiar cry of "Viva Cuba Libre!" (Long Live a Free Cuba!). The famous Grito de Yara (Cry of Yara) set off the Ten Years' War.

DENGUE. Prior to World War II, dengue was recognized as a relatively mild, non-fatal infection from which most people recovered in seven days; but in the 1950's a virulent form emerged, known as dengue hemorrhagic fever (DHF). It causes severe hemorrhages, primary in children, and can develop into a potentially fatal dengue shock syndrome (DSS). The female Aedes Aegypti mosquito is the vector of dengue in the New World. Members of this species are prevalent in most of the Caribbean. In Cuba in 1976 an epidemic of dengue took place. A second epidemic, even worse than the first, took place in 1981. The government of Fidel Castro accused the American government of introducing dengue in Cuba. The other explanation is that Cuban troops in Africa or Cuban workers in Jamaica introduced the disease to Cuba.

DEPARTAMENTO DE SEGURIDAD DEL ESTADO (DSE). Major intelligence service of the Cuban government for domestic purposes. The Departamento de Seguridad del Estado is attached to the Ministry of the Interior and handles both intelligence gathering and counter-revolutionary operations. See also Dirección General de Inteligencia.

DIALOGO CIVICO (Civic Dialogue). It represented a series of meetings in 1955 and 1956 between then-dictator Fulgencio Batista (1952-1958) and leaders of the Sociedad de Amigos de la República (SAR), a civic organization attempting to bring about a compromise between Batista and opposition leaders. Cosme de la Torriente, head of the SAR, met with Batista on several occasions. In March 1956, when Batista refused to consider a proposal calling for elections that year, the dialogue collapsed and an important attempt at finding a peaceful solution to a growing civil war ended in failure.

DIARIO DE LA MARINA. One of Cuba's oldest and most conservative newspapers. Founded in the 19th century as a pro-Spanish publication, it continued being published until 1960, when it was taken over by the Castro regime. During most of the national period it espoused conservative, pro-U.S. policies, and after 1959 until it ceased publication, opposed the Castro revolution.

DIAZ DE ESPADA, BISHOP (?-1832). A director of the Economic
Society in Havana in 1802, Espada pushed for reforms and new
educational instututions, and was a leader in sponsoring the
ideas of the enlightenment in Cuba. His support for two centers of higher learning in Havana, the Real y Pontificia Universidad de San Gerónimo and the Real Colegio Seminario de San
Carlos enhanced intellectual development in Cuba during the
19th century. Espada also liberalized available readings and
abandoned the rote routines of scholasticism that had permeated Latin American and Cuban education. He sponsored a
crusade for smallpox vaccination and the building of a new
cemetery and opened his own library to students and friends.
Espada encouraged the teaching of modern philosophy and
constitutional law at the Seminary of San Carlos. He exerted
significant influence in the thinking and training of such
prominent criollos as José Agustín Caballero, José de la Luz y
Caballero, Félix Varela, and José Antonio Saco.

DIET. In post-revolutionary Cuba, around 1961, Cubans seemed
to have enough to eat such that their per capita daily calorie
intake (2,740) was higher than their estimated requirements
(2,460). Most Cubans ate two meals a day with a diet high in
carbohydrates, particularly starches, and low in proteins, minerals, and vitamins. Outside estimates in 1950 placed the percentage of the population suffering from serious nutritional
deficiencies at 60% rurally and 35% in the cities. Rice and
beans are staples in the average Cuban's diet with substantial
amounts of sugar. Meats are often fried and pork is the most
commonly eaten and preferred meat. Green vegetables play a
minor role in the diet and although fruits like oranges, guavas,
and bananas abound, little fruit is eaten. The chief use of
citrus fruits is for seasoning.

DIRECCION GENERAL DE INTELIGENCIA (DGI). Major intelligence
organization of the Cuban government attached to the Ministry
of the Interior. It operates primarily outside of Cuba and in
close collaboration with the Soviet KGB. See also Departamento
de Seguridad del Estado.

DIRECTORIO ESTUDIANTIL UNIVERSITARIO. A small but active
group of university students organized the Directorio Estudiantil
Universitario (University Student's Directorate) in mid-1927 to
oppose the regime of Gerardo Machado (1925-1933). Following
Machado's orders, the University Council expelled most of the
Directorio leaders from the University. Throughout 1929 the
expelled leaders of the 1927 Directorio renewed their contacts
with university students. In September 1930 they established
a second Directorio, agreed to issue a manifesto condemning
the regime, and planned a massive demonstration for September 30. During the demonstration policemen killed a Directorio
leader, Rafael Trejo. His death unleashed a wave of anti-
Machado feeling. The government reponded this time by

closing the University and many high schools. The University of Havana remained closed until the fall of Machado in 1933. While the principal leaders of the Directorio were in jail in 1931, a split developed and a small group formed a splinter organization, the Ala Izquierda Estudiantil (Student Left Wing). Originally the Directorio leaders had no program beyond eliminating Machado. Machado's removal was considered the panacea that would cure all of Cuba's ills, but some of the Directorio leaders advocated several reforms. They wanted not only to overthrow Machado but also to wipe out all vestiges of his regime. They called for a complete reorganization of Cuba's economic structure and they wanted the removal of the Platt Amendment. Finally, the students wanted an autonomous university, sheltered from political interference. The Directorio leaders also opposed the short-lived government of Carlos Manuel de Céspedes (August-September 1933) and demanded a constitutional convention. Later the Directorio was instrumental in forming the five-member executive commission (the Pentarquía government) from September 3-10, 1933, and in appointing Dr. Ramón Grau San Martín (September 1933-January 1934 and 1944-1948). On November 6, 1933, the Directorio, feeling that its mandate had expired, declared itself dissolved, announcing, however, that its members would continue to support President Grau.

DIRECTORIO REVOLUCIONARIO (Revolutionary Directorate). A revolutionary offshoot of the Federation of University Students (FEU) organized in December 1955 which promoted an insurrectionary response to the repression of the Batista regime. The Directorio Revolucionario was predominantly a student organization and established contact with other opponents of Batista, including Fidel Castro. In March 1957 the Directorio organized an attack on the Presidential Palace with the goal of killing Batista. The attempt failed and as a result José A. Echeverría, the leader of the Directorio, was killed. The remaining leaders went into exile and began reorganizing in late 1957 under the direction of Faure Chomón. Those Directorio leaders in Cuba shifted to guerrilla warfare in opposing Batista. By opening a guerrilla front in the center of the island, the students felt they would be closer to the capital than the other guerrilla forces in the mountains in Oriente province, and could control Havana, the key to capturing power. Early in 1958, a Directorio force sailed from Miami, landed in Cuba, and started guerrilla activities in the mountains of Escambray in Las Villas province. Leaders Faure Chomón and Eloy Gutiérrez Menoyo clashed over who should direct military operations. In mid-1958 the two factions split. Gutiérrez Menoyo forces, known later as "Second Front of the Escambray," continued to fight Batista as an independent organization opposing alliances. The remaining of the Directorio Revolucionario joined forces with the 26th of July Movement troops under Ernesto (Che) Guevara.

In July 1958 delegates of the Directorio met in Caracas with other
anti-Batista organizations and signed a new unity pact (see
Caracas Pact). After Fidel Castro took over, the Directorio
Revolucionario merged with the 26th of July Movement and the
Partido Socialista Popular (PSP) and formed the organization
presently known as the Cuban Communist Party (PCC).

DIRECTORIO REVOLUCIONARIO ESTUDIANTIL. A group of students of the University of Havana led by Juan Manuel Salvat
and Alberto Müller, both Catholic leaders, started in 1960 to
denounce communist gains within the Revolution. They used
the newspaper Trinchera, addressed to the students. The
Trinchera group sought to alert the Cuban people to the communist challenge mainly through protests, riots, and propaganda. During 1960 Salvat and the rest of the Trinchera group
formed the Directorio Revolucionario Estudiantil--DRE (Revolutionary Student Directorate) as a wing in the University of the
anti-Castro Movimiento de Recuperación Revolucionaria--MRR
(Revolutionary Recovery Movement). The activities of DRE
interfered with the government's attempts to control the University. The government repressed the anti-Castro organization through expulsions from the University, beatings, and
temporary arrests. Denied the shelter of the University, the
DRE leaders had to choose between the underground, imprisonment, or exile. In mid-1960, after several months in the underground, Müller and Salvat escaped to the United States. In
Miami the DRE broke with the MRR and joined the Democratic
Revolutionary Front, a loosely coordinated body of anti-Castro
organizations. The DRE never attained the prominence of the
anti-Batista Directorio. Between the end of 1960 and the ill-
fated invasion of Bay of Pigs in April 1961, the DRE engaged
in a variety of successful anti-Castro underground activities
such as sabotage, propaganda, and a partially successful national student strike; a small party led by Alberto Müller began guerrilla activities in Oriente province. The failure of the
Bay of Pigs invasion and the brutal repression that followed
smashed the entire Cuban underground. DRE and the rest of
the underground never recovered from the blow. They continued activities in exile for a time before disolving.

DISEASES. Immediately following the Spanish-American War in 1898,
the United States began to improve health standards in Cuba
undertaking the eradication of the mosquito-transmitted yellow
fever. Compared to some Central American countries during
the 20th century, Cuba had a relatively favorable state of
health. From 1950 to 1955 Cuba had an annual death rate of 15
per 1,000 and an annual infant mortality of 125 per 1,000 live
births. Life expectancy at birth was around 50 years. By the
1950's, malarial infection was down to 2 cases per 100,000 population, and influenza epidemics were non-existent since the
1920's. In 1956, the chief communicable diseases were syphilis

at 62 cases per 100,000, tuberculosis at 31 cases per 100,000, and typhoid fever with a rate of 16.5. Parasitical infection was endemic in rural populations both adult and juvenile, and anemia and gastrointestinal disorders were commonplace. In 1976 and 1981 epidemics of dengue took place in Cuba. President Fidel Castro attributed its introduction in Cuba to the U.S. government. Other sources believe dengue was carried by Cubans serving military or assistance activities who contracted it in Africa or the Caribbean.

DORTICOS TORRADO, OSVALDO (1919-1983). Born in Cienfuegos, Las Villas province, Dorticós attended elementary school at the Roman Catholic Instituto Champagnat de Cienfuegos and later graduated from high school at the Public Secondary Institute there. Dorticós Torrado became politically active at an early age and participated in the general strike of 1935. He graduated from the University of Havana Law School in 1941 and practiced law in Cienfuegos. He served as Dean of the Cienfuegos Bar Association and subsequently as vice-president of the National Bar Association. An opponent of President Fulgencio Batista (1952-1959), Dorticós joined the Movement of Civic Resistance, a coalition of businessmen and professionals engaged in anti-Batista clandestine activities. Dorticós was imprisoned for a short time following an anti-government uprising in Cienfuegos in 1957. After his release, Dorticós joined Fidel Castro's 26th of July Movement and soon became its coordinator in Cienfuegos. Dorticós supplied revolutionary commander Ernesto "Che" Guevara with guerrilla troops in the nearby Escambray mountains. In December of 1958, Dorticós was again arrested but soon released. Dorticós fled to Mexico, returning after Castro's revolutionary victory in 1959 as Minister of Justice. His office rendered him responsible for a series of decrees that modified the Constitution of 1940 to fit revolutionary dispositions.

When President Manuel Urrutia resigned on July 17, 1959, Dorticós ascended to the Presidency of Cuba, becoming at 40 the youngest President in the island's history. Real power remained in the hands of Prime Minister Fidel Castro, but Dorticós nevertheless remained his loyal ally receiving additional appointments that included Minister of the Economy and Director of Juceplán (the Cuban Central Planning Board). In 1965, Dorticós became a member of the Cuban Communist Party, its Politburo and Secretariat. After Aníbal Escalante was deposed from his post as Secretary of National Organization of the United Party of the Socialist Revolution (formerly the ORI) on charges of "grave error" in 1962, Dorticós temporarily assumed the position. When Castro reorganized the Cuban government in the 1970's Dorticós stepped down from the Presidency. Dorticós held positions in the National Assembly of the People's Power and Council of State. He was also Vice-President of the Council of Ministers and a member of the Political Bureau of the

Central Committee of the Cuban Communist Party. He was
married to María Caridad Molina who died in 1982 and they had
no children. Dorticós committed suicide in 1983.

DUQUE, MATIAS (1869-1941). Born in San Antonio de los Baños,
he graduated from Medical School in 1891 but when the War of
Independence began in 1895 he immediately joined, reaching
the rank of Colonel. He was appointed Secretary of Health in
the republican period and was a notable expert in leprosy. He
wrote extensively on medical as well as historical-patriotic matters. His most famous works include La vivienda del guajiro
en Cuba; Historia de la lepra en Cuba; Cómo deben ser las
leproserías; Elementos de anatomía, fisiología e higiene.
Among his historical-patriotic books the best known are the two
editions of Nuestra Patria--one for adults and the other for
children. He also wrote a two-act play: La guerra libertadora.

- E -

ECHEVERRIA, JOSE ANTONIO (1932-1957). Student leader, revolutionary. Born in Cárdenas, Matanzas province, Echeverría attended the Marist Brothers School and the Public High School
in his hometown. He entered the University of Havana's School
of Architecture and soon became involved in student and national
politics. He became one of the leading opponents of the Batista
dictatorship (1952-59) and as President of the Federation of University Students (FEU) he led numerous demonstrations against
the regime. In 1955 he traveled to Costa Rica to defend the
democratic regime of José Figueres, which was being attacked by
forces supported by Nicaraguan dictator Anastasio Somoza. After he returned to Cuba, he established the Directorio Revolucionario, an underground organization of students and others
devoted to overthrowing the Batista regime through violent means
and restoring democracy in Cuba. Under his direction, the Directorio carried out sabotage, terrorism, and attacks on prominent figures of the regime. In 1956, Echeverría led a student
delegation to Mexico where they signed a cooperation agreement
with Fidel Castro, who was organizing a landing of his own
group in Cuba. On March 13, 1957, Echeverría and followers
of former President Carlos Prío (1948-52) led an unsuccessful
attempt to assassinate Batista. While others attacked the Presidential Palace, Echeverría and the students took over a major
radio station in Havana, and unaware of the failure of the Palace attack, they broadcast an announcement that Batista had
been killed and his regime brought down. In an encounter with
government forces minutes later, the police shot and killed
Echeverría and wounded several other students. See also
Mexican letter.

ECONOMIC HISTORY. An outstanding feature of the Cuban economy has been its historical dependence on sugar exports. The
sugar industry has been a principal source of livelihood employing 25% of the labor force (pre-1959) and has accounted for 80%
of the nation's exports. Prior to 1959 large-scale foreign investment (mostly U.S.) had a heavy influence on the sugar
industry, reinforcing its dominance. Although the backbone of
the Cuban economy, sugar is also its weakest link since world
prices for sugar fluctuate and sugar cultivation is a highly
seasonal activity. A goal for Cuba in economic policy has been
to reduce this heavy dependence by expanding and diversifying
the productive base in both agriculture and industry. Over
the years, however, dependence on sugar has so pervaded the
psychology of investors and attitudes of labor, government,
and commercial interests that the great potential of resources
and capital were never exploited to their fullest potential. Historically, Cubans tended to view the factors of economic activity
as fixed quantities available to individuals and groups through
political rather than economic effort. The influence of politics
and the government's tendency to formulate economic policies
without reference to accurate factual data acted as an obstacle
to diversification of Cuban economic activity. Other agricultural
activities inside the Cuban economic structure include such crops
as coffee, cacao, tobacco, rice, fruits, vegetables, and cotton.
Livestock is raised on the grassy plains. Compared with sugar,
however, the other crops are considered frutos menores, or
lesser fruits. According to the 1946 agricultural census, of the
total land surface (28.3 million acres), 79.3% was devoted to
farmland of which only about 20% was under cultivation (about
4 million acres). The rest was wasteland or pasture area.
    Land ownership has been altered through several systems
over the years beginning with the Spanish Crown's system of
awards of land to slave-owning colonists. The more slaves the
colonist owned, the larger his title to allotments of land. A
large class of small landowners arose during colonial times.
Later systems reduced the number of landowners and in the
post-colonial period some 70% of all farms were worked by people with no ownership interest in them--renters, sharecroppers,
squatters, etc. Major industrial progress in Cuba did not
evolve until World War II when wartime shortages stimulated industrial production. Between 1950 and 1959, U.S. investments
in Cuban manufacturing (other than sugar) doubled. Most U.S.
industries were large-scale subsidiaries of U.S. corporations in
the rubber, chemical, and pharmaceutical areas. Until Castro's
revolution, private enterprise was prevalent with no industries
operated by the government. Government efforts to spark
domestic interest in industry began with the 1927 protective
tariffs. These were again utilized in 1950 along with tax reforms designed to boast private Cuban activity in industry.
Castro, after 1959, showed continued interest in this domestic
industrial activity using Cuban raw materials. The regime was

hampered by the loss of the U.S. and other markets and by the
lack of trained technicians, many of whom fled the country.
After 1959 the revolutionary government, following policies espoused by Ernesto Che Guevara, attempted to deemphasize the
sugar economy to achieve agricultural diversification and industrialization. This policy proved disastrous to the sugar crop.
In 1962, Premier Castro reversed the Guevara policy and announced a goal of a 10 million-ton crop by 1970. Despite a
disastrous drought in 1968-69, Cuba did achieve a record 8.5
million tons in 1970. This was accomplished, however, at great
material cost and to the neglect of other sectors of the economy.
Earnings from tourism have considerably diminished and tobacco
exports, suffering from the U.S. embargo, have greatly declined. Total agricultural output declined by 19% between 1961
and 1968, while per capita production suffered a decrease of
29% over the same period. In May 1969 bread rationing was introduced in Havana. Cuba relies on Eastern Europe for 98% of
its crude oil supplies, a situation that has placed considerable
strain on the economy.

During the 1970s, Cuba's relations with the Soviet Union
improved significantly, materializing in more Soviet aid, while
the rising price of sugar in the world markets led to an improvement of economic conditions in Cuba. After 1975, coin-
ciding with Cuba's military involvement in Africa, Soviet aid to
Cuba quadrupled. Cuba began to reexport Soviet crude as
naphtha, crude, or as other byproducts as a way to earn convertible currency, which it badly needs to service its debt to
Western countries. By 1985, 40 percent of Cuba's hard currency earnings came from reexports of oil and oil products.
Uncertainty as to whether the Soviet Union will continue to sub-
sidize the Cuban economy to such a high degree makes Cuba
highly vulnerable. Internal management of economic affairs has
continued to be very poor. In December 1986, at the closing of
the Third Party Congress, Castro announced more austerity
measures, going into effect on January 1, 1987.

EDUCATION [Some fragments taken from Area Handbook for Cuba
    (Washington, D.C.: The American University, Foreign Area
    Studies, 1976) with permission herein gratefully acknowledged].
    In colonial Cuba what education existed was offered within the
    Catholic Church. Elementary and religious education was pro-
    vided for Spanish as well as for criollo children and a few se-
    lected Indian and black children. The elite taught their chil-
    dren at home or sent them to institutions abroad, and the first
    formal classes were not organized in Havana until 1792 when the
    Patriotic Economic Society (Sociedad Económica de Amigos del
    País) was established. This society, which became the center
    of all cultural and educational activities on the island, relied
    principally on private donations but was eventually able to se-
    cure some public support. Bishop Espada joined the Society
    in bringing the ideas of the Enlightenment to Cuba, paving the

way for a period of intellectual flourishing spanning the 1800's.
The number of schools increased from forty in 1793 to ninety in
1817, but it was not until 1841 that Spain recognized the obligation to provide poorer children, white and black, with some
kind of education, and a board of education and committees of
public instruction were organized. The number of schools increased rapidly during the latter part of the century, but they
were both racially and sexually segregated, and in 1860 the 285
schools in existence included only two for blacks. On the eve
of the Spanish-American War the 88,000 students constituted
only about 16 percent of the island's school-aged children, and
the illiteracy rate stood at 72 percent. When the Spaniards
left the island, the United States occupation forces at once
commenced to replace the war-ravished educational system with
one modeled on that of the United States, emphasizing study of
the English language. The Constitution of 1901 declared education to be free and, somewhat unrealistically, to be compulsory
for children between the ages of six and fourteen. School enrollments increased rapidly during the early years of the republic. During the 1920's the proportion of school-aged children
enrolled was the highest in Latin America, and Cuban educators
served as advisers to several other countries of the region.
The system, however, which continued closely to resemble that
of the United States, included only a few schools in the countryside, where half of the population resided.

By the 1920's the University of Havana was still the only
center of higher learning in Cuba, having been founded in 1728.
The University was graduating too many professionals for the
economic and social needs of Cuba, and was in need of a complete overhaul of its facilities. As Cuban nationalism increased
in the 1920's, student activism and effervescence also increased.
The Student Federation was created in 1922 and it pushed for
student goals of a more modern university, administrative reforms, larger government subsidies etc. A continuing decline
in the educational program commenced after 1930, and President
Gerardo Machado (1925-1933) closed the universities and secondary schools between 1930 and 1933 because of student opposition to his regime. President Fulgencio Batista (1940-1944 and
1952-1959) made a serious effort to improve education on the
elementary and secondary levels. Under his fiat, a new kind
of rural institution called the civic-military school for primary-
level children was introduced with army sergeants as teachers,
and the Civic-Military Institute (Instituto Cívico-Militar) was
established for secondary schools. In 1959 the literacy rate
was among the highest in Latin America, and the educational
system, when rated in the general context, was not quite as
bad as revolutionary leaders have since chosen to portray it.
Under Fidel Castro the purpose of education has been to help
to create a new Cuban man (the Socialist Man) in an integrated
and classless society. Elimination of the urban/rural distinction has been sought by sending urban students to the

countryside to perform voluntary labor. Urban secondary schools were moved to rural sites, where students work during half of the school day on farms attached to the schools. The combination of work and study is considered essential by the revolutionary planners.

In 1965 Armando Hart Dávalos, then minister of education, proclaimed in a much-publicized speech the inauguration of a new phase in educational development, marked by an energetic effort to spur the ideological and social consciousness of the citizenry. Since this was what the revolutionary leaders had been trying to accomplish all along, the statement seemed to be one of dissatisfaction with the rate of progress achieved. The anticipated improvement in student diligence did not develop. High dropout, absenteeism, and grade-repetition rates continued through the late 1960s and into the early 1970s. Moreover, although the revolutionary leaders had urged students at the secondary level to enroll in the new industrial and agricultural schools, in 1971 enrollments in language schools slightly exceeded those in industrial and agricultural institutions combined. In the same year the Ministry of Education issues a statement deploring the lack of respect students were showing for socialist property and cited as an example that 50 percent of all school books were lost each year through lack of care. In 1972 Castro regretfully acknowledged that the Revolution had settled into a transitional stage in which "we still do not have the new man, and we no longer have the old one." In developing the revolutionary educational program, planners have been highly pragmatic in their policies with respect to the influence of the United States and the Soviet Union. The inadequacies of the prerevolutionary schools were attributed in large part to American imperialism, but the teaching of English was not abandoned. In the mid-1970s it was the only foreign language taught, and its study was mandatory, beginning in the seventh grade in some schools and in the ninth grade in the remainder. Animosity toward the United States was taught but not stressed in the schools, and a distinction was made between the good American people and the bad American government. Most of the new basic secondary boarding schools were given revolutionary slogan names, such as Heroic Viet Nam and Paris Commune. The name of honor, however, awarded the first of the schools to open its doors, was Martyrs of Kent (State University).

Of necessity the program of study has been much influenced by the Soviet Union and Eastern European states. In building the new curriculum the services of Soviet advisers have been used extensively and Soviet systems of study adopted. In 1974 the national Radio Rebelde commenced broadcasting daily Russian-language lessons, and a year later, the Havana press with characteristic fervor announced over 100,000 people were enrolled in the course of study, which was spreading across the country "like a highly contagious fever." Russian, however, was not a regular course of study in the schools.

Ideology is stressed in revolutionary schools both inside and outside the classroom, but it is peculiarly Cuban. Marx is mentioned in school discussions much more frequently than Lenin. A school is named for Lenin, but the name of José Martí--Cuba's hero-educator--is invoked with such frequency along with that of Marx that his name seems almost to have replaced that of Lenin in the Cuban version of the communist pantheon. For example, according to a 1971 statement by a leading Cuban educator, the establishment of basic secondary boarding schools in the countryside brought together the ideas of "two great thinkers, Marx and Martí." In revolutionary Cuba the role of education in society cannot be explained in terms of the extent to which society values and promotes education and the latter serves society's needs, for no simple relationship exists between the two. Instead there is a complicated triangular relationship that involves education, society, and the Revolution itself. Expenditures for education for the size of the country are the highest in Latin America and probably among the highest in the world. According to the well-publicized calculation for 1966, the expenditure per student was 39 pesos. This amount was said to be 6.4 times the average for Latin America. In 1973 the budget total of 700 million pesos indicated a per capita expense of about 77 pesos. During the early 1970s, educational costs were increasing because of the emphasis on boarding-school facilities. Cuban economists estimate the cost of maintaining a boarding-school student at ten to fifteen times that of a day-school youngster. In addition schools at all levels are tuition-free, and there are virtually no peripheral costs to the individual. Uniforms, meals, and classroom equipment are free, and boarding students are transported by bus to and from their homes at no cost. The free and relatively numerous textbooks available to students include many Soviet titles and photo-offset copies of English-language texts reproduced by the Instituto del Libro (Book Institute) without regard to copyright laws.

Cuba maintains one of the world's largest scholarship programs. For the 1976 school year the government offered almost 150,000 new scholarships--excluding university scholarships--60,000 for secondary schools, 50,000 for technical and vocational training, 18,000 for teacher training, and more than 20,000 to service and production organizations, such as the Revolutionary Armed Forces, the Ministry of Public Health, and the National Cultural Council. The burden of the extensive domestic scholarship program, like school costs generally, is borne by the Cuban government, whereas most scholarships for advanced schooling abroad are financed by the Soviet Union. In the mid-1970s the fundamental pattern of the teacher-training program was one in which teachers for the primary schools were prepared at teacher-training schools in five-year courses of study open to candidates who had themselves completed primary school. Secondary-school teachers were prepared in five-year

courses at pedagogical institutes attached to the universities
and open to students who had completed basic secondary schooling. In late 1975 Castro reported the capacity of teacher-
training facilities to be 78,240 places. In prerevolutionary
Cuba there had been no organized program for the preparation
of secondary-level teaching personnel, and the 1959 enrollment
in schools for the training of primary-level teachers is most
frequently reported at a little less than 9,000. In 1970 the
combined enrollment of students in training for primary- and
secondary-school assignments was a little more than 30,000.

The first regular five-year program for teachers was a three-
stage affair in which the first year was spent at a facility located
in an old sanatorium at Minas del Frío in Oriente province, the
second and third years at the Manuel Ascunce Domenech (named
for a martyred revolutionary hero) at Topes de Collantes in the
Escambray mountains, and the final two years at the Makarenko
Pedagogical Institute in Havana. All three schools were boarding institutions. The Minas del Frío site was in a remote part
of the Sierra Maestra, deliberatedly selected to isolate the students and by its wild surroundings to give the student a sense
of involvement in the Revolution. Completion of the sixth grade
was nominally required of the candidates, but many had not
progressed so far, and the year at Minas del Frío was largely
devoted to catching up. The first real teacher training came
during the two years in the Escambray mountains school where
the primitive conditions of the countryside gave students additional familiarity with the kind of rural locality in which many
would later be required to teach. The final two years at the
Makarenko Institute were devoted principally to practice teaching.
Students were quartered at a boarding facility at the former
beach resort of Tarará, a few miles east of Havana, and each
day they were bused into the city for practice teaching in more
than 250 primary schools. The student-teacher was assigned
to his own class of about twelve students, supervised by a
monitor who observed and evaluated the teaching performance
but did not participate in the teaching. The program for training secondary-level instructional personnel was offered in pedagogical institutes attached to the universities.

In addition to the University of Havana (founded in 1728),
the University of Oriente in Santiago de Cuba was established
in 1947 and the Marta Abreu Central University (later the Central University of Las Villas) in Santa Clara was founded in
1948. After the enactment of legislation in 1949 authorizing
establishment of private institutions of higher education some
private universities were opened, like Santo Tomás de Villanueva of the Catholic priests of St. Augustine. The private
universities in 1961 were either suppressed or consolidated into
the three--the University of Oriente for the east, the Central
University of Las Villas for central Cuba, and the University
of Havana for the west. Under Castro no additional universities were created until 1974, when the former branch of the

Central University of Las Villas located in the city of Camagüey was raised to university status as the Ignacio Agramonte University of Camagüey. A university reform promulgated in January 1962 placed the institutions at the service of the nation, abolishing university autonomy and faculty tenure. Under the new system admission requirements and academic standards were lowered radically. Students were subjected to stricter discipline, and attendance at lectures was made virtually compulsory. Militia membership was compulsory, performance of agricultural work during at least half of the two-month summer vacation was required, and pressure was exerted to engage in volunteer work during the limited spare time that remained. The new disciplined atmosphere left little room for dissent and, as one University of Havana professor remarked, at least for the time being progress and great intellectual freedom were not compatible in Cuba.

ENCOMIENDA. Spanish institution brought to the New World to deal with the Indian population. Encomiendas entailed assigning Indian families or other inhabitants of a town to a Spaniard who would Christianize them while at the same time extracting labor and tribute from them. Receiving an encomienda did not carry with it title to land nor ownership of the Indians; instead, the Indians were seen by the Crown as "free" subjects, who could be compelled to work for the encomendero who represented authority. Encomiendas were misused by many who exploited the Indians, extracting unreasonable amounts of labor and often overlooking the provisions about Christianization. In Cuba and other parts of Latin America, the Crown used the encomienda as a political instrument to consolidate its control over the Indian population and to organize needed labor in the first century of colonization. The Spanish monarchy itself profited from the encomienda system, using Indians as miners and taxing the encomenderos for the number of Indians they recruited. Despite attempts to regulate the functioning of the encomienda and prevent further abuses, the Crown lacked a means of enforcement. Finally, the New Laws (1542) ended the granting of new encomiendas.

ENERGY AND POWER. Cuba has no coal; hydroelectric potential is slight, and developed petroleum reserves are small. Bagazo (sugarcane waste) supplies most of the sugar industry's fuel. In 1968, Cuba had an estimated installed capacity of 940,000 kw. with another 300,000 kw. associated with sugar refining facilities. The nationalized Cuban Electric Company supplies all the power used in Cuba through some 8,000 miles of electric lines. A large hydroelectric irrigation project was constructed on the Cauto River. See also Nuclear energy.

ESCALANTE, ANIBAL (1909-). A member of the Cuban Communist Party (PSP) since his youth, Escalante was an aide to the PSP's Secretary-General, Jorge Antonio Vivó, in the early

1930's. Escalante was editor of the Communist Party's news-paper, Hoy, from 1938 until 1959. In September 1951, gunmen destroyed Hoy's printing press. Escalante immediately de-nounced President Carlos Prío (1948-52) and sued the Cuban Federation of Workers (CTC), which he charged had organized the attack. One week after the incident, an assassination at-tempt was made on Escalante's life. Again, he accused the CTC leadership. In 1953, Escalante was one of the five PSP members who directed the Party during Secretary-General Blas Roca's illness. Escalante remained at large during the ban on Commu-nist activities enforced under General Fulgencio Batista's gov-ernment (1952-59). After Fidel Castro's triumph in 1959, Es-calante continued as editor of Hoy. In June 1959 Escalante charged figurehead President Manuel Urrutia with having weak-ened revolutionary solidarity through his attacks on the Commu-nists. Shortly thereafter, Urrutia was deposed as President and placed under house arrest as a result of his differences with Fidel Castro. Like his Communist colleagues, Urrutia favored keeping the Cuban middle class and some sectors of the upper class within the Revolution.

In July 1961, the main organizations that opposed Batista (the 26th of July Movement, the Communist Party, and the Revolutionary Directorate) merged to form a new organization called Integrated Revolutionary Organizations (ORI). In No-vember 1961, Castro announced the transformation of the ORI into the United Party of the Socialist Revolution. Escalante was placed in charge of national organization, a post that en-abled him to create a bureaucratic machinery in which his Com-munist Party friends held the most important posts. However, on March 27, 1962, Fidel Castro violently and publicly accused Escalante of "sectarianism" and of having committed "grave er-rors." Escalante's "error" was to systematically purge from high posts members of Castro's 26th of July Movement and sub-stitute them with PSP cadre. Escalante was exiled to the Soviet Union and later transferred to Prague, Czechoslovakia. Relative moderation was restored to Cuban politics, ending the so-called period of sectarianism. In 1964, Escalante was allowed to re-enter Cuba as administrator of a small farm. By 1968 Escalante and his companions were again accused of organizing a "micro-faction" that conspired against the government. In an eight-hour, unpublished speech to the Central Committee, Castro de-nounced Escalante and his friends, charging that they were taking Moscow's side in the ideological polemic between Cuba and the USSR (armed struggle vs. peaceful coexistence). Cas-tro made Escalante into an example for public consumption of a "mistaken" revolutionary leader, and said that Escalante had been passing documents and information to Russia clandestinely. Escalante and his collaborators received lengthy prison sentences after a long, arduous, and mostly secret trial. The significance of the Escalante case was the position it occupied in the dialec-tic between "old" and "new" Communists (and Communism) in Cuba, which has since been resolved in favor of the "new."

ESCALERA CONSPIRACY. Name given to the Conspiracy of 1844,
climax to consecutive waves of negro slave insurrections throughout Cuba in the early 1840s.   Although introduction of black
slaves into Cuba had been formally prohibited by a charter between England and Spain in 1821, the Spanish continued to import Africans in large quantities in order to maintain production on the island's sugar, coffee, and tobacco plantations.
Meanwhile, Cuban and British abolitionists on the island continued to encourage slaves to seek their liberty. In 1840, the
British government, bitterly opposed to the slave trade, sent
the abolitionist Consul, David Turnbull, to Havana to help
emancipate the slaves. The slaves steadily gathered courage
from this support as well as from their rapidly growing numbers. By 1841, over half the island's population was black.
As their ranks swelled, many Cubans began dreading the possibility of a slave take-over such as that which had occurred
in Haiti. In response, Spain sent Lieutenant-General Leopoldo
O'Donnell to govern Cuba in the autumn of 1843 and thereby
end the slave problem. O'Donnell was informed of a slave uprising planned for Christmas Day of 1943 at a sugar plantation
in Matanzas. O'Donnell immediately ordered the execution of
all those involved in the conspiracy. This harsh retaliation
led to a massive "witch hunt" throughout Cuba of all slaves
and even free blacks suspected of involvement in any conspiracy. All those arrested would be chained to the notorious
escalera (ladder) from whence the name of the conspiracy
arises, and then whipped until they delivered the information
desired by the Spanish Military Commission. Many talented
and respected free black Cubans died in the Escalera Conspiracy such as the poet Gabriel de la Concepción Valdés (Plácido),
the dentist Andrés Dodge, and Santiago Pimienta, the musician.
All whites involved were eventually released, including José de
la Luz y Caballero, who had returned to Cuba to defend himself upon being informed of the charges against him while in
France.

ESPIN, VILMA (c. 1934-   ). An engineer with a degree from MIT,
she came from a middle-class family from Santiago de Cuba,
Oriente. She joined Fidel Castro's 26th of July movement in
1956 against the regime of Fulgencio Batista (1952-58). She
visited Castro in Mexico and helped coordinate his landing in
Oriente province in late 1956. She later joined the guerrillas
in Sierra Maestra and fought next to Raúl Castro, Fidel's
younger brother. After the triumph of the Revolution, Vilma
married Raúl and became President of the Federation of Cuban
Women, a post she still occupies. The FMC was created to
organize, mobilize, and socialize Cuban women under the revolution. Vilma is also a member of the Central Committee of the
Cuban Communist Party.

ESTANCIAS. One of the three types of land grants or mercedes

distributed by the Spanish Crown to colonists in Cuba during
the last half of the 16th century. Estancias varied in size but
were usually smaller than one league in radius and developed
in the immediate vicinity of towns and villages.

ESTANCO DEL TABACO. The popularity of Cuban tobacco in
Europe during colonial times moved the Spanish government to
purchase the tobacco on the island and later resell it abroad,
thus taking charge of this lucrative enterprise. After trying
the planned system and still not meeting the great demand, the
Spanish Crown issued a decree in 1717 ordering the estanco
(royal monopoly) of tobacco in Cuba. Especially affected was
the six-league tobacco-growing region around Havana. Spain
thereby reserved the right to purchase all tobacco grown in
Cuban fields and imposed a tariff on it. This edict outraged
many vegueros (tobacco farmers), who protested loudly. The
estanco was abolished in 1818, its disappearance authorizing the
free sale of tobacco for both domestic and export purposes. Ultimately many small farmers benefitted from this directive.

ESTENOZ, EVARISTO (?-1912). Leader of the Agrupación Independiente de Color (Independent Colored Association), a political party organized after the independence of the island from
Spain. Despite the group's appeal to the racial consciousness
of the blacks, they had an extremely poor showing in the election of 1908. In 1912, Estenoz led an uprising against the
Presidency of José Miguel Gómez (1909-13). Estenoz, a former
soldier in the War of Independence, organized several bands of
blacks and roamed around Oriente province. The ill-organized
rebellion met with much opposition and distinguished black leaders such as Senator Martín Moruá Delgado and Juan Gualberto
Gómez criticized the rebels. The U.S. became alarmed over the
uprising and landed marines in several parts of the island.
Gómez, trying to avoid a full-fledged U.S. intervention, quickly
put down this rebellion, which marked the last time that a revolt along strictly racial lines was to occur in Cuba. Estenoz
was captured and executed and the Agrupación collapsed.

ESTOPIÑAN, ROBERTO (1921-). Outstanding sculptor born in
Cuba and now living in New York. He studied at the school of
Fine Arts San Alejandro in Havana. He taught drawing in Cuba
for six years. He was a three-time winner of the National Prize
for Sculptors and took a third in an International Competition in
the Tate Gallery in London. His most famous pieces are in
bronze and marble. He has exhibited in the Museum of Modern
Art in New York; Philadelphia Civic Center; and in Cuba, Argentina, Colombia, Venezuela, France, Spain, and the U.S.A.

ESTRADA PALMA, TOMAS (1835-1908). First President of Cuba.
An honest, austere, and simple man, Estrada Palma was President of the provisional government during Cuba's War of

Independence. In 1902 he became the first President of the
Republic. Estrada Palma was born near Bayamo, Oriente province, on July 9, 1835. He attended schools in Havana and the
University of Seville in Spain, where he enrolled in the law
school but failed to obtain a degree because family matters
forced his return to Cuba. As soon as Cuba's Ten-Years' War
(1868-1878) against Spain broke out, Estrada Palma joined the
rebels. In 1876 he was selected president of the provisional
government but in 1877 was captured by Spanish forces and
exiled to Spain where he remained in jail until the end of the
war. After his release he traveled to Paris, New York, and
Honduras. In Honduras he married Genoveva Guardiola, daughter of Honduran President Santos Guardiola, and was appointed
Director of Postal Service. From Central America he moved to
Orange County, New York, where he opened a boys' school.
Teaching soon gave way to politics. The Cubans were resuming
the war against Spain, and Jose Martí visited Estrada Palma to
enlist his support for the revolutionary cause. When Martí was
killed in Cuba in the early months of the war (May 1895), Estrada Palma was named delegate-in-exile and head of the Cuban
Junta in New York, carrying out diplomatic negotiations (primarily with the United States), raising funds, and promoting
the Cuban cause. In 1902, after the end of the American intervention in Cuba, Estrada Palma became the first President of
the Republic, having been elected in 1901 by an overwhelming
majority. The new president encouraged foreign investment
and Cuba's exports. He continued and expanded public and
educational projects initiated during the American intervention.
He negotiated with the United States in 1903 a Permanent Treaty
which gave contractual form to the Platt Amendment and governed relations between the two countries. On July 16, 1903,
the United States recognized Cuba's sovereignty over the Isle
of Pines and Cuba granted the United States the right to lease
and establish naval bases at Guantánamo and Bahía Honda. In
1903 Cuba signed a Reciprocity Treaty with the United States
which gave Cuban products, particularly sugar, a preferential
rate for import duties into the U.S. and gave selected American
products preference in Cuban rates.

Estrada Palma was less successful in his domestic policies.
A popular and well-intentioned man, he remained at first above
partisan politics, conducting an honest and paternalistic government. He had little faith in the ability of his compatriots to
govern themselves and scolded them for their shortcomings.
Teacher turned politician, he lectured the Cubans on political
virtue and good government. As the elections of 1905 approached, political difficulties increased. Estrada Palma joined
the Conservative Republican Party--or Moderate Party, as it
was now called--and sought reelection. He purged unfriendly
office holders. Corruption increased. Tensions mounted. Accusing the administration of fraud, the opposition Liberal Party
boycotted the elections, thus allowing Estrada Palma's unopposed

reelection. The Liberals charged a corrupt election and resorted to violence. In August 1906 an uprising took place in Pinar del Río province which quickly spread throughout the island. Estrada Palma appealed for U.S. intervention. President Theodore Roosevelt sent Secretary of War William H. Taft to mediate between government and opposition. But when Taft proposed, with Liberal backing, that all the elections be nullified, except those of president and vice-president, Estrada Palma rejected the proposal and resigned in September 1906. Taft then ordered the landing of U.S. marines. He dissuaded the Liberals from fighting and proclaimed a provisional government led first by himself as acting governor and later by Charles E. Magoon. U.S. intervention lasted until January 1909. Estrada Palma retired quietly to his modest holding in Bayamo, where he died on November 4, 1908.

ETHNIC GROUPS. After the slave trade was abolished in 1865, an effort was made to bring in Chinese to work as laborers under conditions little better than slavery. By 1900, some 15,000 Chinese resided in Cuba, most leaving the rural sugar mills to become active in commercial and entrepreneurial activities in the cities. The earliest Jewish immigration before World War I was made up of Jews from the eastern Mediterranean region followed in the 1920's by European Jews (Russians and Poles). Other ethnic groups included Haitians, Frenchmen, Irish Catholics, Japanese, Britishers, Canadians, and Americans. The 1953 census distinguished several racial groups which remained constant into the 1970's. Of the total population 73% were white; 12% were black, 15% were mestizo; and Orientals numbered some 30,000, less than 1%. The pre-Columbian Indian population was reduced so drastically that little evidence of it remains. Over 99% of the population in the 1970's was native-born Cuban.

EXILES. See Migration.

- F -

FACTORIA. The Royal Crown office in Havana in the 18th century entrusted with buying and exporting Cuban tobacco to Europe. Branch offices were located in Santiago de Cuba, Trinidad, Sancti Spíritus, and Bayamo.

FAMILY. See also Women. Before the 1959 Revolution the family was the major integrating force in Cuban society. Upper-level families preserved traditional Hispanic elements of sexual behavior patterns, notions of noblesse oblige, and the purity of white skin. Women were supposed to be chaste and obedient and to be in charge of domestic and religious affairs. Middle-class families would follow the pattern of upper-class behavior as long as their economies would permit. In the lower level the mother's influence was determinant. Hispanic and African heritage mixed. Civil and religious marriage was sometimes replaced by common-law unions. Thus, in many lower-class families women were not only heads of households but also main bread-winners, and men were little more than infrequent visitors. Children received little schooling but were encouraged to join the work force at an early age to help support the family. Under the Revolution led by Fidel Castro the role of the family as the preserver of class status and of traditional patterns of sexual and social relations was abolished. Divorce has increased dramatically. The new-found independence of women has also contributed to a change in the character of the Cuban family. The Revolution has taken over some of the primary functions of the family in prerevolutionary society. The family remains an important institution and its existence and structure is protected by the Family Code of 1975. Its structure has been altered, however, by the changing role of women, who have entered the labor force and become increasingly independent.

FARMERS. See also Colono. Under Spanish law all land belonged ultimately to the sovereign, but cabildos were empowered to confer grazing rights which were in effect freehold grants of lands. They were known as mercedes. Even when mercedes were extensive pieces of land, in time small plots were given to poorer farmers in exchange for work or part of the crops. Toward the end of the nineteenth century a cluster of technological and other developments initiated the era of the great landed estates (latifundios) supporting large-scale industrial sugar mills. In the early 1800s some 40% of the agricultural labor force were small self-employed farmers and colonos linked to the sugar mills. By 1930 the vast expansion of the larger enterprises (partly at the expense of smaller ones) and a shift from individual to corporate ownership had rendered the rural upper class insignificant and greatly reduced the size and independence of the rural middle class. In 1953, 231,000 persons were listed as farm operators, the vast majority of whom were wageworkers (perhaps as many as one-third were owners). Farmers in Cuba have been historically disadvantaged by the widespread absence of farm-to-market roads that retarded rural development, especially in Camagüey and Oriente provinces. After the 1959 Revolution the government nationalized sugar mills, setting up highly centralized farms managed by state-appointed managers who followed the orientation of the National Institute for Agrarian Reform (INRA) (Instituto Nacional de la Reforma Agraria). Under the Agrarian Reform a few small farmers could work their land independently subject to direction by INRA. A National Association of Small Farmers (ANAP) consisting of ranchers and coffee and sugar growers was formed in the early 1960s.

FEDERACION DE MUJERES CUBANAS (FMC). Mass organization established in 1960 to mobilize, socialize, and integrate Cuban women. Led from its inception by Vilma Espín, Raúl Castro's wife, FMC enrolls about 80% of all adult women. The organization promotes voluntary work, engages in vigilance, manages child-care centers, sponsors courses on personal hygiene and pre- and post-natal care, and trains traffic policewomen and civil defense workers. FMC has also been instrumental in various aspects of the modernization of Cuban women, particularly their incorporation into the salaried work force.

FEDERACION ESTUDIANTIL UNIVERSITARIA (FEU). Cuba's first organized Student Federation created at the end of 1922 in angry response by university students to the corruption and incompetence of many professors at the University of Havana. Julio Antonio Mella was the Secretary and popular figure-head of the organization, which issued its first important manifesto on January 10, 1923. Although the demands made therein concerned university issues, the document was regarded as a challenge to the government of President Alfredo Zayas (1921-1925). From that point on, the students assumed a significant political role they would not abandon until the 1960's. Throughout the war years, the FEU remained relatively quiet and supported the Allied war effort. In July 1942, however, the FEU reorganized following the reopening of the University in 1937, and demanded that Batista form an honest, efficient war cabinet prior to registration for obligatory military service. The students also turned against the Communists and Russia's war posture, while denouncing the corruption and disorganization of the Batista administration. During the late 1940's, the FEU degenerated into a stepping stone for political prominence with the advent of the gangster-type student action groups that preyed upon faculty and administrators at the University of Havana. The turmoil surrounding the organization climaxed when in February 1948, former FEU President and MSR leader Manolo Castro was assassinated in Havana. In the 1950's, the FEU was the single organization representing the 17,000 students at the University of Havana. By 1955, the FEU had become a vocal opponent to the Batista regime calling for strikes in November to protest Batista's refusal to call for elections. Already the students had confronted the government en masse in the early 1950's, demanding the restoration of the 1940 Constitution, the reestablishment of civil government, and the holding of free elections.
    The leaders of the FEU in the mid-1950's, President José Echeverría and Vice-President Fructuoso Rodríguez, realized that efforts by third parties to mediate the struggle between the government and the student opposition were futile. The Revolutionary Directorate, a clandestine off-shoot of the FEU that advocated terrorist methods, was formed in 1956 to combat Batista's repression. During the latter part of the 1950's, the Communists sought to ally themselves with and eventually

dominate the FEU, but the union never materialized. Following the "Humboldt events" of April 20, 1957, when Batista's police massacred acting FEU President Fructuoso Rodríguez and several other student activists, the FEU became virtually inactive. In 1959, after Fidel Castro had succeeded in dethroning Batista, the students of the newly re-opened University of Havana took provisional charge of the FEU and purged all pro-Batista academicians from the University under the authority of the newly formed University Reform Commission. Castro realized that in order to mobilize and indoctrinate the students, control of the FEU was essential. He therefore advocated unity among the various student factions and successfully obtained the FEU leadership's support by installing his brother Raúl's favorite, Rolando Cubela. The FEU remained faithful to the Castro regime until well into the 1980's.

FERRARA, ORESTES (1876-1972). Lawyer, politician, congressman, and diplomat. Born in Italy, Ferrara participated in the Cuban War of Independence (1895-1898) as Secretary to General José Miguel Gómez. After the establishment of the Cuban republic, Ferrara became active in politics and joined the Liberal Party revolt against President Tomás Estrada Palma (1902-1906). He was elected to Congress and was appointed Foreign Minister by General Gerardo Machado (1925-1933). He also served as Ambassador to the United States. Elected to the 1940 Constitutional Convention, Ferrara survived an assassination attempt in 1943. Afterward, he progressively drifted away from politics. Ferrara served as the Cuban representative to UNESCO under the government of Fulgencio Batista (1952-1959). After Fidel Castro took over, in 1959, Ferrara went back to Italy where he died in 1972.

FERRER DIAZ, HORACIO (1876-1960). Born in Sabanilla del Encomendador, he fought in the Cuban War of Independence where he achieved the rank of colonel. He studied medicine and became an outstanding ophthalmologist. He developed equipment for improving cataract operations which was acclaimed throughout the world. His scientific articles and books were translated into German, French, English, Italian, and Portuguese. His most famous works include La fenometría antes y después de la operación de la catarata; Higiene Militar; and Asuntos sobre la ración alimenticia del obrero cubano. Horacio Ferrer served President Carlos M. de Céspedes (1933) as Minister of War. He consulted with Sumner Welles during the Revolution of 1933. When the Pentarchy took possession of the government and when Ramón Grau was sworn as president, he recommended to Welles that the United States intervene militarily. His memoirs as a mambí and during his years in public life were published under the title Con el rifle al hombro.

FIGUEREDO, PEDRO (1819-1970). Patriot, writer, and musician.

Born in Bayamo, Oriente. Founded the newspaper El Correo
de la Tarde. Joined the 1868 revolutionary war and fought in
the capture of Bayamo where the song that he composed, "La
Bayamesa" was sung for the first time. The hymn--known also
as "Himno Bayamés"--was adopted later as the Cuban national
anthem. In 1870 while acting as the Minister of War for the
revolutionary army, he was captured and executed in Santiago
de Cuba.

FINLAY, CARLOS J. (1833-1915). Cuban physician and epidemiolo-
gist, discoverer of the mosquito vector of yellow fever. Finlay
was born in Camagüey, Cuba, on December 3, 1833, of a Scot-
tish father and a French mother. He spent his early years in
his father's coffee plantation but soon was sent to school in
France and England. From there he traveled to the United
States where he received a degree in medicine at Jefferson
Medical College, Philadelphia, in 1855. He returned to Cuba
and began to practice medicine after validating his degree at
the University of Havana. From Cuba he traveled to Peru,
Trinidad, and France, working in various hospitals. In 1870
he settled permanently in Cuba, developing an interest in the
island's sanitary and health problems. When in 1879 an American mission arrived in Cuba to study the causes of yellow fever,
the Spanish government designated Finlay to work with the
group. He developed the idea that the transmission of yellow
fever required a vector. At the International Sanitary Conference held in Washington in February 1881, he explained his
theory. A few months later, in August 1881, Finlay read before the Academy of Sciences of Havana his historic work showing a mosquito, Culex fasciatus or Stegomyia fasciata (later
known as Aedes aegypti) to be the vector of the yellow fever
organism. Although he advanced numerous experiments and observations to support his conclusions, his theory was not accepted by the scientific world for almost two decades. Finlay
continued his experiments and published the results in articles
and pamphlets. These were received with indifference by his
colleagues. When he presented a paper to the Convention of
Hygiene and Demography in Budapest in 1894 he only found
contempt and disbelief.

In a report to the International Sanitary Conference held in
Havana in 1901, Walter Reed confirmed Finlay's discovery.
When United States troops landed in Cuba in 1898 during the
Spanish-American War, Finlay worked with the American army
in Santiago de Cuba. He further tested his theories in practice and advocated a campaign against the mosquito vector. As
a result of his urgings, W.C. Gorgas, United States health chief
in Cuba, began a program (later extended to Panama) to exter-
minate the mosquito vector, thus putting an end to a sickness
that had plagued the Caribbean for many years. In addition to
his work in the epidemiology of yellow fever, Finlay wrote extensively on ophthalmology, tuberculosis, tetanus, trichinosis,

filariasis, leprosy, beriberi, cholera, and exophtalmic goiter.
After the establishment of the Cuban Republic in 1902, he
was appointed public health chief, and the Cuban government
created in his honor the Finlay Institute for Investigations in
Tropical Medicine. Finlay died in Havana on August 20, 1915.

FISHING. Since 1959 the fishing industry has expanded greatly as
a result of the Revolution's concentration on this sector. Despite the presence of many varieties of edible fish in Cuban
waters, little effort was made to develop a modern fishing industry before 1959. After the Revolution all commercial fishing
ventures were either nationalized or organized into cooperatives
and placed under the control of the National Institute for Agrarian Reform (INRA). Later the National Fishing Institute was
created to take over responsibility for all aspects of fishing including research. The Institute has five oceangoing fleets and
two coastal fleets in operation with over 14,000 fishermen and
numerous onshore installations employing almost 20,000 additional
people. Total catches increased steadily to over 165,000 tons
in the late 1970's--many times the 1958 level of 22,000 tons. In
volume, mackerel, bonito, and different varieties of snapper,
are the more important fish caught, but the more valuable
catches, such as lobsters and shrimp, are exported to earn
foreign exchange. Very few lobsters were used domestically--
most were exported fresh or frozen to France, Italy, the United
Kingdom, and Canada. The Cuban Fishing Fleet and the Cuban
Tuna Fleet are the most far-reaching of the various fleets, each
of which specializes in an area or variety and fishes in flotillas.
Although the tuna fleet operates worldwide, it remains mainly in
the Caribbean and Atlantic. Crews of both fleets are rotated
back to Cuba every six months, and fresh crews are assigned.
The vessels seldom return home; their catches are brought back
by the refrigerated mother ships. Three fleets concentrated on
shrimping, although some of the other fleets also landed shrimp
occasionally. About 6 percent of total annual catches are
shrimp.

The fleet having the largest number of vessels is the Shelf
Fleet, composed of between 2,000 and 3,000 small wooden vessels based at many ports and generally fishing on the continental shelf not far from shore. The Shelf Fleet was grouped into
thirty-two units, each specializing in certain species and areas.
The small craft were organized into cooperatives, which included
both state and private fishermen. State fishermen work for
wages and operate state-owned boats, while the private fishermen own their pre-revolution boats and sell their catch to the
state at fixed prices. They purchase all their supplies from
the state except for replacement vessels, which they are not
allowed to buy. Eventually all fishermen will be salaried employees of the state when the remaining private vessels go out
of commission. Some fishermen operating on the Isle of Pines
specialize in the sea turtles, frogs, and sponges that abound

in that area. The catch is processed on the Isle of Pines and shipped to Havana for further disposition. Several thousand crocodiles are bred commercially in captivity in the Zapata Swamp, in the south of Matanzas province. Fishmeal is processed at sea by special vessels acquired from the German Democratic Republic (East Germany) in the early 1970's. The fishmeal is used as supplementary food for poultry and cattle. Such freshwater fish as trout and carp are stocked in lakes and lagoons by the Fish Culture Department of the National Fishing Institute. The institute also operated about twelve freezer plants, four canneries, a fishmeal plant, a broad network of cold-storage facilities throughout the island, and more than 200 refrigerated trucks to distribute the frozen fish to outlying population centers. The cold-storage plants had a total capacity of over 60,000. The largest fishing port is Havana, where a 33-acre zone has been set aside for fishing vessels, complete with cold-storage facilities, ice plant, repair shops, and miscellaneous facilities. Research centers, fishing schools, and several specialized laboratories are also maintained by the institute. The research centers are responsible for technological improvements in the industry, the provision of data on species of commercial value, and the oceanological conditions of the fishing areas. Two foreign trade firms also belonged to the institute: the Caribbean Export Enterprise handles all exports of fish products, and CUBAPESCA imports all fishing vessels and equipment. The Cuban Merchant Marine comprises more than 94 ships, with a gross tonnage of several million tons.

FLORA AND FAUNA. Cuba has approximately 5,000 different species of flowers, plants, and trees, including yucca, tobacco, pineapple, sugar cane, and sweet potato, all native to the island. Cuba is a natural botanical garden, being situated in the path of hurricanes and of north and south bird migrations, both of which encourage the dispersal of seeds. The mountain areas are covered by tropical forest, but Cuba is essentially a palm-studded grassland. Pines like those in southeastern U.S. grow on the slopes of the Sierra de los Organos and on the Isle of Youth. The lower coastal areas, especially in the south, have mangrove swamps. Desert-type plants grow in a small area around Guantánamo Bay. Only small animals inhabit Cuba. Tropical bats, rodents, birds, and some non-poisonous species of reptiles and insects are commonly found.

FORDNEY-McCUMBER TARIFF. The post-World War I slump in the economy of the United States produced an increased demand for an upward revision of the tariff on sugar importation. The beet-sugar producers clamored protection from Cuban sugar which was produced at lower costs. In 1921 the tax on Cuban sugar was 1.0048 cents a pound or 20% less than the full duty. The chairman of the Senate Finance Committee and the chairman

of the House Ways and Means Committee were in favor of the
beet-sugar interests. During the discussion in Congress that
laster over a year and a half, the Harding Administration proposed and received approval of the Emergency Tariff Act which
raised the duty on Cuban sugar to 1.6 cents a pound. After
considerable lobbying efforts by beet-sugar interests on the
one hand and pro-Cuban interests on the other, the Fordney-
McCumber Treaty Act of September 1922 was approved at
1.7640 cents.

FORESTRY. Much of the natural forest cover was removed in colonial times to plant sugar cane, but cutting since the end of
World War I reduced the woodland area to about 10% of the total
area and led to soil erosion. A reforestation program is in progress. New nurseries are being established and, since 1960,
large plantings of pine, eucalyptus, cedar, majagua, dyewoods,
and teak have been carried out. Lumber for construction constitutes the major part of production; mahogany and other hard-
woods are cut, but may not be exported.

FREEMASONRY. Rationalism, in the form of Freemasonry, arrived
in Cuba during the 18th century, challenging Catholicism.
Freemasonry was introduced by the British during their occupation of Havana in 1763 and tolerated by the Spanish authorities in Cuba until 1811. Masonic doctrine asserted that universal moral truths existed that were accessible to the individual
without the intervention of ecclesiastical tradition. Up to 1900
most Masonic lodges with their secret structure were centers of
revolutionary pro-independence sentiment. Many were linked
to lodges inside the United States which offered moral support
and money to aid the anti-colonialist cause. Most of the independence leaders, including José Martí, were Masons. After
independence, an English-speaking lodge was formed in Havana.
Freemasonry after independence became increasingly conservative and was noticeable mainly by its anti-Catholicism. Before
Castro came to power Freemasons supported a university in Havana. Under Castro some Mason leaders were persecuted and
have become anti-Castro activists in exile. Others remain in
Cuba supporting mildly the Revolution in order to preserve
some measure of freedom for their activities.

FRIAS JACOTT, FRANCISCO DE (1809-1877). Better known for
his title of Conde de Pozos Dulces, Frías was born in Havana
of a very wealthy family and throughout his life he advocated
reforms and later independence for the island. He studied in
Baltimore and returned to Cuba in 1829. He went to Spain in
1832 but from 1833 to 1842 he dedicated himself to agricultural
studies receiving a prize for his work La industria pecuaria en
Cuba. He enriched the Cuban economic and cultural scene
through works such as Informe sobre el Instituto de Investigaciones químicas and El trabajo agrícola y la población en

Cuba, but principally as a member of the Junta de Fomento
(Development Board). He founded the magazine El Porvenir
del Carmelo, devoted to agriculture, and the newspaper El
Siglo, which voiced Cuban aspirations for reform. He was
exiled by the Spaniards and his property confiscated, but he
never relented his stand in favor of Cuba, dying in poverty
in Paris.

FUEL. See also Petroleum. Petroleum was produced commercially
in Cuba in small quantities beginning in 1914. Lacking substantial coal deposits, Cuba had to look to petroleum for its
principal fuel needs. From 1914 to 1955 only 2.5 million barrels of oil were produced, barely enough for two months'
consumption. Between 1945 and 1959 some $25 million were
spent by large American oil companies in search of Cuban
petroleum, a process revived in 1980-1981 when Pemex, the
Mexican oil company, offered to explore for oil resources.

FUNDAMENTAL LAW (1959). The Fundamental Law, issued by the
Castro regime in 1959, replaced the 1940 Constitution. It incorporated all the modifications the revolutionary government
had made affecting individual rights; it extended the authority
and powers of the prime minister; and it established that no
indemnification was necessary in cases of expropriation. The
minimum age for the presidency was reduced to 30 years. The
Fundamental Law did not contain the customary references to
God and was replaced by the Constitution of 1976.

                                - G -

GACETA OFICIAL. Official publication of the Cuban government
where laws, decrees, international agreements, etc., are reprinted.

GAMBLING. Common practice among a vast portion of the Cuban
population reflecting the traditionally weak popular confidence
in economic ventures. Usually the enthusiasm for obtaining
large economic gains in the shortest time possible was expressed through mass participation in the weekly lottery; one
entered the lottery by purchasing billetes (tickets). The national lottery was also an effective vehicle for distributing
patronage by various powerful political leaders. It was prohibited by the U.S. government of occupation at the time of
independence but reinstated during the government of José
Miguel Gómez (1909-1913) to serve as a powerful mechanism of
political corruption. Cockfights were another popular form of
gambling, particularly in the countryside, where in some rural
areas, almost every household would raise its own fighting

cocks. Other significant forms of gambling included Havana's casinos and jai-alai. During the 1940s and 50s, one of the regular functions of the Cuban police in the major cities was to protect gambling houses and prostitution, as a result of their manipulation for political purposes. Under the Castro regime, an effort was made to wipe out the small entrepreneurship of Cuba's many gamblers, but cockfighting remains a popular sport, of sorts, among the people. The national lottery was abolished by the revolutionary government in the early 1960s.

GARAY, SINDO (1868-1968). Cuban composer born in Havana and best known for his work La Bayamesa. Garay's music is sentimental and based upon Cuban themes. Other works include Guarina; La tarde; El Erial; A Maceo.

GARCIA, CALIXTO (1839-1898). Born in Holguín, García was the second of eight children. García served as an apprentice to his uncle in Havana as a youth, but was unable to realize his envisioned education in law there because he had to return home to tend the family business. Subsequent to his marriage to Isabel Vélez Cabrera, Calixto became involved in a conspiratorial movement organized by Aguilera and Céspedes directed against the Spanish. Under the leadership of Donato Mármol, García joined 100 other men in an uprising at the Santa Teresa ranch near Río Cautillo in the Jiguaní zone. García was recognized as a man of military skill and strategy when he participated in the seige of Bayamo. Later, when his forces and those of Mármol attacked the town of Santa Rita, taking Jiguaní, whose governor was arrested, García won the accolades of "Lion of Santa Rita" and "Hero of Jiguaní." Calixto García fought numerous other battles successfully and Máximo Gómez, Commander of the Cuban rebel forces, appointed him Brigade General. Calixto García was subsequently made Chief of the Eastern Forces, replacing General Máximo Gómez, who was displaced through an error of Rebel President Céspedes. Near the end of the Ten Years' War (1868-1878), García was surrounded by superior Spanish forces and attempted to take his life by shooting himself. The Spaniards took him prisoner and he convalesced. He was set free at the end of the war, whereupon he travelled to the United States.

In New York, he immediately founded the Cuban Revolutionary Committee. Supporters of his cause launched the "Guerra Chiquita" in Cuba. In his efforts to join the short-lived rebellion, he sailed to Cuba's eastern coast but was captured barely after landing. Once again, he was exiled to Spain where he was eventually set free and worked as a banker and English teacher. Some years later, in 1895, he returned to Cuba to join the rebel uprising of February 24, and was named Military Chief of Oriente province. An event of major significance occurred in 1897, when U.S. President William MacKinley sent Navy Lieutenant Rowan to Calixto García's camp, El Aserradero, to determine what the posture of the Cuban rebel forces would be toward the impending Spanish-American conflict. U.S. Generals Sampson and Shafter organized a military plan to attack Santiago de Cuba, which was swiftly and efficiently taken by June 16, 1898, largely through the admirable cooperation of Calixto García and his troops. On December 11, a few hours after the signing of the Treaty of Paris officially ending the Cuban-Spanish-American War, Calixto García died in Washington.

GARCIA BARCENA, RAFAEL (1907-1961). Early in his political career, García Bárcena became active in the Ala Izquierda Estudiantil, a leftist student group in opposition to the government of Cuban dictator, Gerardo Machado (1925-1933). García Bárcena later became a university professor, and joined the Ortodoxo Party. When General Fulgencio Batista (1952-1958) took over the government by military coup, García Bárcena launched a campaign of bitter criticism and insurrection against him. Known for his strong anti-Communist and Nationalist positions, García Bárcena founded the Movimiento Nacional Revolucionario (National Revolutionary Movement) under these ideals. In 1953, García Bárcena planned an attack upon the main Cuban military base that had given Batista command of the army, Camp Columbia. Expecting his attack to coincide with a military coup as coordinated with army officers, García was duped when the plot was averted by Batista's intelligence. García Bárcena was arrested in April 1953, tortured, and sentenced to two years in prison. He abandoned politics thereafter and died in exile in 1961.

GARCIA FRIAS, GUILLERMO (?-   ). A member of the Cuban government's top leadership, García was the Minister of Transportation until 1986 when he was deposed. García initiated his political career in the Revolutionary government of Fidel Castro in 1962 as a Major in the Cuban Revolutionary Armed Forces. He occupied this military position until 1966. From 1963 to 1965, he served as member of the National Dictorate of the National Party of the Socialist Revolution. In 1965, he was appointed to the select thirteen-member Political Bureau (Politburo) of the Cuban Communist Party (PCC), a seat he still occupies today. Between 1967 and 1970 García was chosen as Party First Secretary for Oriente Province. García was subsequently appointed Vice-President to the Council of Ministers (responsible for transportation and communications) in 1972. In 1976 he was elevated to Vice-President of the Thirty-member Council of State. García served in this position until 1979, when he was appointed to the Ministry of Transportation.

GARCIA MENOCAL, MARIO (1866-1941). Third president of Cuba. Engineer, Major General of the army during Cuba's war for independence, Mario García Menocal became a prominent public

figure during the first decades of the Cuban Republic. García
Menocal was born in the Province of Matanzas on December 17,
1866. When Cuba's Ten Years' War (1868-1878) against Spain
broke out two years later, he was taken to the United States
and then to Mexico, where his father settled as a sugar planter.
When Menocal was thirteen his father sent him to the United
States to attend several schools, obtaining a degree in engineer-
ing from Cornell University in 1888. Soon after graduation,
Menocal began practicing his profession. He joined an uncle in
Nicaragua who was studying the feasibility of a Nicaraguan canal
route. In 1891 he returned to Cuba and was employed by a
French company. He surveyed a proposed railway in Camagüey
province, but soon got involved in Cuba's political problems.
When Cubans resumed the war against Spain in 1895, Menocal
joined. He fought under Máximo Gómez, Antonio Maceo, and
Calixto García, the leading generals of the war. As a soldier
he exhibited talent for military affairs and a definite aptitude
for strategy, achieving the rank of General. When the United
States declared war on Spain and intervened in Cuba, Menocal
was promoted to Major General in charge of Havana and Matanzas provinces. During the United States military government
of Cuba Menocal was made Havana's Chief of Police, a post he
held only briefly. He soon returned to engineering, building
the Chaparra sugar plantation for the American Sugar Company.
Under his able management, Chaparra became one of the largest
sugar-producing estates of its kind in the world.

   A successful businessman, a veteran of the war for independence, his popularity increasing, Menocal turned to politics
after Cuba became independent in 1902. In 1908 he ran unsuccessfully for the presidency on the Conservative party ticket
and in 1912 he was elected, becoming Cuba's third president.
Menocal served two terms (1913-1921). During his first administration, education, public health, and agricultural production
improved. He introduced administrative and financial reforms,
particularly the establishment of a Cuban monetary system.
He strengthened relations with the United States and exposed
the corruption of the previous administration. As time went
by, however, his administration deteriorated. Graft and corruption became widespread. Opposition and violence increased.
The regime resorted to repressive measures. Menocal's reelection in 1917 caused much discontent. Opponents complained of
fraud. Despite U.S. warnings that revolution would not be
tolerated and that the Menocal administration would be supported, the Liberal party, led by former President José Miguel
Gómez, staged an unsuccessful revolt that was harshly suppressed. American supervision of Cuban affairs and American
economic influence grew during Menocal's administration. On
April 7, 1917, one day after the U.S. declared war on Germany, Menocal took Cuba into the war. Cuba floated loans in
the United States and marines landed on the island, supposedly
for training purposes. Although Cuba's contribution to the

war effort was slight, Menocal collaborated with the United
States and sold Cuba's sugar production to the Allies. This
arrangement resulted in a short-lived period of great prosperity called "The Dance of the Millions."

Prosperity brought corruption, speculation, and inflation.
When sugar prices collapsed in 1920, Cuba plunged into depression and misery. Properties were foreclosed and sold at bargain prices, many to American investors. As the 1920 elections
approached, the United States sent Major General Enoch E.
Crowder to prepare an electoral code. American control of
Cuban affairs was growing and with it nationalism and anti-
Americanism. After turning over the Presidency to Alfredo
Zayas in 1921, Menocal returned to business, but always kept
close to politics. He was defeated in the 1924 presidential
elections. When President Gerardo Machado extended his presidential term, Menocal participated in an ill-fated expedition and
uprising against the regime in August 1931. Later, he made
another unsuccessful bid for the presidency in 1936 and was a
member of the convention that drafted the 1940 Constitution.
Shortly after, in 1941, he died in Havana.

GENERATION OF 1930. Term applied to the students around whom
coalesced the opposition to Cuban President Gerardo Machado's
regime (1925-1933) in 1930, which was particularly heightened
following the violence that occurred on September 30 during a
mass anti-government demonstration led by the Directorio
Estudiantil. Most prominent among this generation of student
idealists were young leaders such as Carlos Prío Socarrás,
Ruben de León, Grau San Martín, Joaquín Martínez Sáenz,
Juan Antonio Rubio Padilla, Jorge Mañach, and Francisco
Ichaso. Militant student "action groups" vocalizing the demands of the Generation of 1930 for political as well as social
reforms sprung up throughout Cuba after 1931. Several of
the young intellectuals of this Generation joined under the
leadership of Dr. Martínez Sáenz in a militant student political
organization known as the ABC, which was principally concerned with punishing the Machado regime for atrocities committed against the opposition. This secret organization became
the anti-government spearhead for the new revolutionary sentiment of the members of the Generation of 1930. Ultimately,
these student leaders succeeded in their aim of ousting Machado
on August 12, 1933.

In keeping with their alma mater of nationalism and freedom
from foreign dominance of Cuba, the Student Directory leaders
of the Generation of 1930 joined in a sergeants' coup against
the provisional government of Carlos Manuel de Céspedes on
September 4, 1933, whom they regarded as a puppet of the
United States. Thus, the Generation of 1930 ignited the ideal-
istic Revolution of 1933, assuming political power through a
Revolutionary Junta whose military muscle was vested in Sergeant Fulgencio Batista. For the students of the Generation of

1930, the decade to follow represented an opportunity to implement the many notions of revolution and radical reform they
had been striving for since 1927. The Generation of 1930 enjoyed immense popularity among the Cuban people. Its members were considered noble, disinterested idealists who represented the universal desires of the people. These young student
leaders represented the hopes and aspirations for true reform,
as well as the sentiments of nationalism and patriotism of all
Cubans. Unfortunately, the utopian ideal of government envisioned by the Generation of 1930 disintegrated shortly after its
revolutionary triumph under the presidency of the popular
Ramon Grau San Martín in 1934. Ten years of puppet presidents under Batista's control followed.

GEOGRAPHY. The long, narrow island of Cuba extends some 746
miles from Cape Maisi on the east to Cape San Antonio on the
west. It has a maximum breadth of 120 miles and a median of
about 62 miles. The largest of the West Indies islands, Cuba
has a territorial extent about matching that of all the other
islands combined. In addition to the main island, the Cuban
archipelago includes the Isle of Pines now renamed Isle of
Youth, near the south coast in the Gulf of Batabano and some
1,600 coastal cays and islets. Cuba's landmass is 46,300
square miles. The main island occupies 94.7 percent of the
national territory, and the Isle of Pines and the other cays
and islets occupy respectively 2.0 percent and 3.3 percent of
the total. Cuba was raised from the seafloor by geological action occurring about 20 million years ago and at one time was
connected with other Antillean islands. The mountains of
southeastern Cuba are related to those of southern Mexico,
Jamaica, and Hispaniola; and the limestone formations that
make up much of the island resemble those of Florida, Jamaica,
and the Yucatán Peninsula. Cuba's topography has resulted
from the interaction of constructive forces that determined the
basic structure and alignment of landforms and destructive
forces of wind and water that sculpted the structure into its
present configurations. Soil erosion, however, has been less
severe than on most other Antillean islands. The island is
still subject to some crustal instability, and its history has
been marked by earthquakes of varying intensity. The zone
of maximum instability occurs in the southeastern mountains.
Light quakes are recorded frequently in the southeast. Havana recorded several strong disturbances during the nineteenth century, and a severe tremor struck Pinar del Río
province in 1880. None, however, has occurred in the twentieth century.
　　The least mountainous of the Greater Antilles, Cuba has a
median elevation of no more than 300 feet above sea level, and
its three principal mountainous zones are isolated and separated by plains. The most extensive highland zone occupies
much of the island's eastern extremity, the second rises near

the center of the island, and the third rises in the extreme west. The loftiest mountain system is the Sierra Maestra, which skirts the southeastern coastline west of Guantánamo Bay except where it is broken by the small lowland depression on which Santiago de Cuba is located. It is the most heavily dissected and steepest of the Cuban ranges, and its peaks include Pico Turquino (6,540 feet), which is the country's highest elevation. On the east, the Sierra Maestra terminates in a low area around the United States Naval Base at Guantánamo Bay. The lowlands around Guantánamo mark the termination of the Central Valley, which is some 60 miles in length and merges with plains to the west. Most of the island east of a line from north to south between Nipe Bay and Santiago de Cuba is mountainous, however, and includes such ranges as the Sierra de Nipe, the Sierra de Nicaro, the Sierra del Cristal, and the Cuchillas de Toa. The port of Baracoa on the northeast coast is the most isolated urban center on the island. The mountains of central Cuba, less extensive and of lower elevation than those of the east, occupy the southern portion of Las Villas province. The two principal systems, known collectively as the Escambray mountains and separated by the Agabama River, are the Sierra de Trinidad in the west and the Sierra de Sancti Spíritus in the east. The principal ranges of the western highlands are the Sierra del Rosario, which commences near the town of Guanajay west of Havana and extends southwestward along the spine of the island for about sixty miles, and the Sierra de los Organos, which continues in the same direction almost to the tip of the island. These western highlands are limestone formations weathered into strange shapes. Ranks of tall erosion-resistant limestone columns resembling organ pipes gave the Sierra de los Organos its name. The numerous shapes, sinkholes, and underground caverns and streams are limestone developments known as karst. Karst landscape is most characteristic of the western highlands but is widely distributed about the island.

Most of the country's more than 200 rivers originate in the interior near the island's watershed and flow northward or southward to the sea. Smaller streams and arroyos that remain dry during most of the year are also numerous. River levels rise significantly during the rainy season, when 80 percent of their flow occurs, and seasonal flooding is common. The watercourses for the most part are not navigable, and their potential as sources of hydroelectric power has yet to be realized. The longest and heaviest flowing is the Cauto River of Oriente province, which rises in the Sierra Maestra near Santiago de Cuba and flows westward to the Gulf of Guanacayabo. Rivers are most numerous in Oriente province and are fewest in La Habana province and in Camagüey province, and in the west of Cuba. Seven subterranean river basins constitute the sources of many surface rivers, and there are extensive reservoirs of fresh and brackish groundwater. Sulfide mineral springs are located in

Pinar del Río and Matanzas provinces, and there are radioactive thermal springs in Las Villas province and on the Isle of Pines. There are no large lakes, but coastal swamplands are numerous and extensive; the largest covers more than 1,700 square miles on the Zapata Peninsula. Others occur on both the northern and the southern coasts and on the Isle of Pines. Almost two-thirds of the Cuban landscape consists of flatlands and rolling plains. With hills and the lower and gentler slopes of the mountains, they make up as much as three-fourths of the national territory. The generally easy gradients minimize the hazards of land erosion and facilitate both development of the transportation network and land tillage, including the use of mechanized equipment. Limestone, usually with a high clay content, is the basic ingredient of most Cuban soils. In particular a red limestone earth known as Matanzas clay extends in a wide and continuous belt from a point west of Havana to near Cienfuegos on the southern coast and reappears in extensive patches in western Camagüey province. The dark red material, which undergoes little chemical or physical alternation to depths of as much as 20 feet, permits a good downward percolation of moisture that minimizes runoff and has so little stickiness that land can be plowed within a few hours after a heavy rain. The rich Matanzas clay would be suitable for a wide variety of crops, but the traditional Cuban practice, somewhat modified since 1959, has been to plant sugarcane on the better soils and to plant other crops on less productive lands.

A second zone of fertile soil occurs on the plains north of Cienfuegos between the Sierra de Sancti Spíritus and the Caribbean coast. Scattered alluvial soils also have a high degree of fertility but occur only in the narrow floodplains of watercourses. Certain soils of Camagüey province and those of the Guantanamo Basin are suitable for cane production but are of lower fertility. Extensive areas of sandy soil in Pinar del Río, western Las Villas, and portions of Camagüey are characterized by an inability to hold moisture. This soil can support only palm trees, xerophytic shrubs and grasses suitable for pasture. Soils suitable for coffee cultivation occur in all three major mountain zones; tobacco grows well in eastern Las Villas and in portions of Pinar del Río and Oriente. The mangrove-dotted coastal swamps and cays have soils of limited fertility made up of organic silt, organic clay, and peat, each underlain with clay of high plasticity. The main island of Cuba rests on a subsurface shelf from which the numerous cays, coral islets, and reefs rise. Submerged about 300 to 600 feet below sea level, the shelf varies in breadth off the north coast, is almost nonexistent off the southeast coast, and attains its maximum breadth off the remainder of the southern coastline, where it extends to the limits of the gulfs of Batabanó, Ana María, and Guacanayabo. Its outer rim is flanked on the southeast by the deep Bartlett Trough, which separates Cuba and Jamaica, and on the southwest by the Cayman or Yucatán depression. The two troughs are separated by the shallows of the submerged Cayman ridge, a continuation of the Sierra Maestra range that reemerges as the Cayman Islands. Off the northern coast of Camagüey province the sea-lane of the Old Bahama Channel at some points is only 10 miles wide as it passes between the Cuban shelf and the shallows of the Great Bahama Bank. Except where the precipitous cliffs of the Sierra Maestra plunge into the sea, most of the Cuban shoreline is fringed with coral reefs and archipelagoes of cays. In the north an almost unbroken chain of cays extends from Cárdenas to Nuevitas. In the south the Isle of Pines is the largest member of the Canarreos Archipelago, and the Jardines de la Reina chain flanks the Gulf of Ana María. Coral reefs are interlaced with many of the cays, clog the gulfs of Ana María and Guacanayabo, and form a chain of unborn islets off the western extremity of the island. Cuba's approximately 2,200-mile coastline is indented by some of the world's finest natural harbors, namely, Havana, Guantánamo, and Cienfuegos. There are about 200 in all, and many are of the pouch or bottleneck variety with narrow entrances that broaden into spacious deepwater anchorages. Among the ports on the north coast that have harbors of this kind are Mariel, Havana, Nuevitas, Manatí, Puerto Padre, Gíbara, and Antilla. South coast bottleneck ports include Guantánamo, Santiago de Cuba, and Cienfuegos. The principal open bay ports, Cárdenas and Matanzas, are located close to one another on the north coast of Matanzas province. Most of these ports were developed primarily for the export of sugar. Nuevitas ranked first and Havana second in volume of sugar handled, but all handled commercially important quantities. There are no important harbors west of Cienfuegos on the south coast or Mariel on the north. Shallow waters, coral formations, and a lack of good natural harbors are characteristic of the coasts of western Cuba, but since sugarcane is not grown in the west the need for ports is reduced accordingly. Elsewhere along the coastline good ports are lacking only along the 150 miles between Caibarién and Nuevitas, where cays are numerous and there is extensive coral development.

GLADIOLITO. Typical Ciboney Indian artifact; symbol of authority or high social status. Also considered a magical object.

GOICURIA CABRERA, DOMINGO (1804-1870). Merchant, patriot, anti-Spanish leader. A rich merchant who increased his fortune by running supplies from Mexico to Texas during the American Civil War, Goicuría favored slavery and opposed Spanish rule in Cuba. He supported the idea of annexing Cuba to the United States. In 1855, he supported a U.S.-backed expedition to Cuba which failed. In 1856, Goicuría participated in an invasion of Nicaragua directed by the American adventurer William Walker. Goicuría expected an eventual invasion of Cuba in return for his support. After

taking over the Nicaraguan government, Walker appointed
Goicuría Minister to London. When Walker was deposed,
Goicuría increased his anti-Spanish activities. He invaded
Cuba in 1870 but was defeated by the Spanish and executed.

GOMEZ, JOSE MIGUEL (1858-1921). Major General in the Independence Army, President of Cuba (1909-1913). Barely out of high
school, Gómez joined the anti-Spanish forces in the Ten Years'
War for Independence (1868-78). Gómez also participated in
the unsuccessful rebellion against Spain known as La Guerra
Chiquita. During the War of Independence (1895-98), Gómez
achieved the rank of Major General and although he fought
numerous battles he was never defeated. Having been born in
Sancti Spíritus, province of Las Villas, he became a legislator
and Governor of the province after independence. In 1905 he
was an unsuccessful candidate to the Presidency for the Liberal
Party. He was later elected President in the campaign of
1908. His main achievements were the reorganization of the
army, increase of the navy, the construction of railways, and
the expansion of the public school system. Gómez led a rebellion in 1913 contesting the election of General Mario García
Menocal as President. He was seized and imprisoned briefly.
Soon thereafter he retired to the United States where he died
but was buried in Havana, Cuba, with all honors.

GOMEZ, JUAN GUALBERTO (1854-1933). Journalist, public man.
Born in Matanzas, July 12, 1854, Gómez became, at an early
age, active in the campaigns for the liberty of Cuba, writing
on the subject for various papers. He went to Paris and entered the Escuela Central de Ingenieros, but having to earn
his living, he became a journalist. He travelled as a teacher
through the French Antilles and from there to Mexico; then,
going to Havana, he went into the offices of La Discusión as
editor. He spent afterwards nearly ten years in Madrid.
There he published: La cuestión de Cuba en 1884 (1885); La
isla de Puerto Rico (1891); Las islas Carolina y Marianas
(1885). He directed in Madrid the dailies El Pueblo and El
Progreso, and was editor of La Tribuna. He was an ardent
advocate of the abolition of slavery and the betterment of the
negro race. He was secretary of the abolitionist society in
Madrid. He returned to Cuba to join the War of Independence
and his part in the insurrection of Ibarra caused the Spanish
Government to deport him in 1897. He constantly took an active part in political life after Cuba became independent. He
was secretary of the committee of Consultations which edited
the organic law of the Cuban Republic. He was one of the
editors of the daily paper La Lucha and in the 1910's was
elected to the national Senate.

GOMEZ, MAXIMO (1836-1905). A Dominican by birth, Gómez became a general in Cuba's Independence army and a hero of

the struggle which ended Spanish domination over "the ever-faithful island" in 1898. Gómez was born in the small town of Baní in the Dominican Republic on November 18, 1836. The son of a lower-middle-class family, he completed primary school in his home town and then entered a religious seminary. His religion instruction was soon interrupted by a Haitian invasion of the Dominican Republic in the mid-1850's. Joining the forces of Dominican patriots, he fought bravely in the battle of Santomé in 1856 and in numerous subsequent battles against the Haitian invaders. When Dominican General Pedro Santana invited Spain to reestablish control over the Dominican Republic in 1861, Gómez accepted a commission as captain in the Dominican army reserve. He retained that post until the end of the Spanish domination in 1865 when he moved to Santiago de Cuba. Cuba was then experiencing revolutionary turmoil, as Cuban patriots conspired to rid the island of Spanish control. Unhappy with the treatment he and other Dominicans had received from Spain and horrified by the exploitation of the black slaves, he started to conspire with Cuban revolutionaries. When on October 10, 1968, Carlos Manuel de Céspedes and other leaders began Cuba's Ten Years' War for Independence, Gómez joined the rebellion. His experience in military strategy was of significant importance to the revolutionary cause and he was soon promoted to the rank of General and later to Commander-in-Chief of the Rebel Army. However, disappointed and disillusioned, he left Cuba just prior to the signing of the Peace of Zanjón in February 1878 which ended the Ten Years' War.

From Cuba, Gómez travelled to Jamaica and then to Honduras where he was appointed army General. From Honduras he supported the ill-fated Guerra Chiquita (Little War, 1879–1880), an attempt by several Cuban rebel leaders, led by General Calixto García, to continue the war against Spain. In 1884 Gómez left Honduras for the United States to organize and collect funds for a new rebellion in Cuba. In New York he met with veteran General Antonio Maceo and with José Martí, then engaged in mobilizing the Cuban exiles in support of the war against Spain. But Gómez and Martí soon clashed when the latter, fearful of Gómez' authoritarian attitude and insistence that his orders be obeyed without question, withdrew from the movement. In a letter to Gómez, Martí chastised the Dominican: "A nation cannot be founded in the same manner as an army camp is commanded. What are we, General, the modest and heroic servants of an idea or the brave and lucky caudillos that are getting ready to take the war to a people for the purpose of later subjugating them?" Gómez' withdrawal, desertion from the revolutionary ranks, disillusionment among exiled Cubans, lack of capital and weapons, and poor organization all doomed the new movement to failure. Gómez travelled to Panama for a short stay and then settled back in the Dominican Republic. There he received a new call from Martí in 1892 for a final effort to liberate Cuba. Martí had organized a revolutionary party in exile and now offered Gómez the post of military chief. Forgetting old differences Gómez accepted promptly and joined Martí and Maceo in their revolutionary endeavors.

For the next few years the three men worked tirelessly, organizing Cubans in and out of the island until finally on February 24, 1895, the War of Independence began. In April of that same year, Gómez, Martí, and other leaders landed in Cuba and joined Maceo, who was already in the battlefield. Now Gómez finally was able to implement his invasion plan. Although Martí's tragic death in Dos Ríos, on May 19, 1895, in one of the early combats of the war, dealt a strong blow to the morale of the rebellion, Gómez and Maceo did not waver. In repeated attacks the two generals undermined and defeated the Spanish troops and carried the war to the western provinces. By 1896, the Cubans seemed victorious throughout the island. Then came a change in the Spanish command and the more conciliatory Spanish Marshal Arsenio Martínez Campos was replaced by General Valeriano Weyler, a tough and harsh disciplinarian. Weyler's policy of concentrating the rural population in garrisoned towns, increasing numbers of Spanish troops, and Maceo's death, allowed the Spaniards to regain the initiative. Yet they were unable to defeat the Cuban rebels or even engage them in a major battle. Gómez retreated to the eastern provinces and from there carried on guerrilla operations. He rejected any compromise with Spain and in January 1898, when the Spanish monarchy introduced an autonomy plan that would have made Cuba a self-governing province within the Spanish empire, Gómez categorically opposed it.

This was the existing condition in Cuba when the United States declared war on Spain on April 25, 1898, following the explosion of the battleship Maine in Havana's harbor earlier that year. The Cuban forces collaborated with the U.S. army in the short campaign against Spain. By August hostilities had ceased and Spain agreed to relinquish sovereignty over the island. Gómez and his troops retired to the sugar mill Narcisa in Las Villas province and there awaited the departure of Spanish troops. After the withdrawal, Gómez made a triumphant tour of the island and amidst general joy entered Havana on February 24, 1899. But Gómez, the most popular hero of the war soon got into trouble. He requested that the Americans pay the Cuban army for the aid that American soldiers had received. The United States refused and offered $3 million or an estimated $75.00 for each soldier who turned in his weapons. Gómez also clashed with the Cuban Assembly, composed of army delegates, over a proposed U.S. loan. Gómez opposed the loan as well as its onerous terms and criticized the Assembly for considering it. The Assembly also resented Gómez' highhanded manner and the secret conversations he held with representatives of the U.S. Government to secure payment for the war veterans. ·The Assembly finally dismissed Gómez as Commander-in-Chief of the Army. His dismissal only increased his popularity.

As the end of the American occupation approached and candidates emerged for the presidential election of 1901, Gómez was the most popular figure. Yet the old general refused to be considered, claiming that "I would much rather liberate men than govern them." Instead, he campaigned for and helped elect Tomás Estrada Palma, former rebel President and delegate in exile of the Cuban Republic in Arms. Gómez supported Es-Trada Palma's administration, but when the President announced his intention to reelect himself, he met with Gómez' stiff opposition. Old and sick, General Gómez went on a speaking tour but could do little for he died on June 17, 1905.

GOMEZ, MIGUEL MARIANO (1889-1950). Born in Sancti Spíritus, the son of Cuban Independence War General José Miguel Gómez, who later became President of Cuba (1909-1913), the younger Gómez was active in politics at an early age and became an opponent of President Gerardo Machado (1925-1933). After Machado's downfall in 1933, Gómez was appointed Mayor of Havana, a job he retained under President Carlos Mendieta (1934-1936). On January 1, 1936, Gómez, backed by Cuban strongman, General Fulgencio Batista, was elected President and took office in May. Gómez opposed Batista's growing influence and vetoed a bill providing a tax to create new military rural schools, a measure designed to increase Batista's political power. An impeachment trial followed and finally Gómez was voted out of office. Afterwards he never regained influence in Cuban public life.

GOMEZ CARBONELL, MARIA (1903-   ). Born in Havana and a literary figure since her early years. Her Ph.D. dissertation was on "La numismática y su significado," and after graduating from the University of Havana she delivered lectures on "El mío Cid"; "William Shakespeare"; "Simón Bolívar, cumbre de América"; and "La lírica de Martí." She founded the Alianza Nacional Feminista and was twice its president. Her most famous speech on the feminist movement was "La mujer, camino del mañana." As a poet her most famous compositions are the elegy "Se fue papá" and also "Viejecita blanca," and patriotic poems such as "Mi tierra." She was elected twice to the House of Representatives and was Senator at the time of Castro's takeover in Cuba. She immediately went into exile. As a legislator she helped to advance the status of women, not only in Cuba but throughout Latin America. She also worked to improve education in Cuba; she founded a school for elementary and secondary education and helped establish the higher schools for commercial education through her work as director of Escuelas de Comercio y Normales (for teachers) at the Ministry of Education. She lives in Miami, Florida.

GOMEZ DE AVELLANEDA, GERTRUDIS (1814-1873). Poet. Born in Camagüey, Gómez de Avellaneda achieved recognition as one

of the greatest poets of the Spanish language. She used the
pseudonym "La Peregrina" in most of her writings. She excelled in lyric poetry but was also the author of plays and
novels. Gómez de Avellaneda moved to Spain where she spent
a good part of her life. Her main poetic compositions were
Llanto de la cruz; Soledad de alma; Poesías; La vuelta a mi
patria.

GOOD NEIGHBOR POLICY. Foreshadowed by President Coolidge
and advocated by President Hoover, the good neighbor ideal
for U.S.-Latin American relations became President Franklin
Roosevelt's major proposal to inaugurate a new U.S. policy
with regard to the Western Hemisphere. He renounced U.S.
armed intervention, proclaiming respect for the sovereignty of
Latin American countries. Important U.S. business interests
were exhorting Roosevelt to increase export trade with nations
such as Cuba, which many Roosevelt appointees saw as a testing ground for the Good Neighbor Policy. Roosevelt's deeds
confirmed the policy when he sent Ambassador Sumner Welles to
Cuba to mediate the friction caused by President Gerardo Machado's dictatorship (1925-33) and the vocal opposition to it,
rather than sending in U.S. troops. After Welles successfully
engineered Machado's resignation, the subsequent Sergeant's
Revolt, led by Fulgencio Batista, put some stress on the policy
since the U.S. refused to recognize the government of Ramón
Grau San Martín (1933-34) and the threat of armed intervention
loomed greatly. Roosevelt's ultimate decision to avoid armed
intervention in Cuba was seen as demonstrable proof of the
value and importance of the Good Neighbor Policy in the 1930's
when Latin America was suspicious of U.S. motives. A further
manifestation of Roosevelt's policy was Cuba's release from the
terms of the Platt Amendment.

GRAJALES, MARIANA (1808-1893). Born in Santiago de Cuba of
Dominican parents on July 16, 1808. She married Francisco
Regüeiferos with whom she had four children before becoming
a widow. She was remarried to Marcos Maceo and had nine
children with him between 1845 and 1860, the oldest of whom
was Antonio Maceo. All her children fought in Cuba's War of
Independence, including her husband, Marcos, who while dying
on the battlefield on May 14, 1869, twenty-four years before
her death, was said to have murmured: "I have done all I can
for Mariana," meaning Mariana's mambí cause. Martí wrote
about her to Antonio Maceo: "I will now see again one of the
women who has moved my heart: your mother." On another
occasion, "I will remember her with love all my life." Mariana
Grajales died on November 28, 1893. In Havana there is a monu-
ment in honor of this exemplary mother, dedicated to her by
the people of Cuba and built by Teodoro Ramos Blanco.

GRANJAS DEL PUEBLO. State farms established in the 1960's by

the Castro Revolution after cooperatives failed to perform according to government expectations. State farms became the
principal form of organization in the rural areas providing
stable employment throughout the year to many peasants and
laborers previously affected by seasonal unemployment. Workers in these enterprises receive minimum wages and state services, but lacking incentive, they have reduced their labor
efforts; productivity declined and absenteeism increased.
Yields of the few farms in the private sector were consistently
higher than in the state sector in the late 1970's.

GRANMA (PROVINCE). Population: 641,757; Area: 8,461 Km²;
Municipalities: Río Cauto, Jiguaní, Yara, Campechuela, Niquero,
Bartolomé Masó, Guisa, Cauto Cristo, Bayamo, Manzanillo,
Media Luna, Pilón, Buey Arriba. Province of the new political-
administrative division of 1976. It received its name from the
boat carrying the expedition led by Fidel Castro that landed on
Las Coloradas beach in 1956. Its capital city is Bayamo.
Forest resources, sugarcane, and fishing. Manzanillo and
Niquero are the two other important cities in the province.
See also Bayamo.

GRANMA. The 58-foot yacht purchased by Fidel Castro with money
from ex-President Carlos Prío (1948-1952) in Mexico to launch
his invasion against the dictatorship of Batista. Castro and 82
men sailed the Granma out of Tuxpán in late November 1956,
and headed toward Oriente province in Cuba. Castro and his
men hoped to initiate a rebellion throughout the island. The
Granma reached Cuba on December 2, 1956, landing in swampy
terrain. After several days of skirmishing with Batista's soldiers, the invasion collapsed and Castro fled to the Sierra
Maestra with a few followers, beginning his famous guerrilla
campaign.

GRANMA. Since 1965, the official daily newspaper of the Cuban
Communist Party, also appearing in weekly international editions
in various languages. See also Hoy, Revolución.

GRAU SAN MARTIN, RAMON (1887-1969). Former President of
Cuba. Physician and Professor of Physiology at the University
of Havana, Grau was appointed Provisional President of Cuba
in 1933 and was later elected to the presidency in 1944. Grau
was born in Pinar del Río province, September 13, 1887. Although his father, a prosperous tobacco grower, wanted his
son to continue in his business, Grau dreamed of becoming a
doctor. Despite family opposition he entered the University of
Havana, receiving his degree of Doctor of Medicine in 1908.
From Cuba he traveled to France, Italy, and Spain to round
out his medical training. He returned to Cuba and in 1921
became Professor of Physiology at the University of Havana.
He wrote extensively on medical subjects including a university

textbook on physiology. Grau's reputation, however, rests not
on his medical achievement but on his political involvement. In
the late 1920's he supported student protests against dictator
Gerardo Machado (1925-33), and in 1931 was imprisoned. After
his release he went into exile in the U.S. With the overthrow
of the Machado regime, Grau was catapulted into national prominence. When on September 4, 1933, students and the military
led by Sergeant Fulgencio Batista deposed the provisional government of President Carlos Manuel de Céspedes and appointed
a five-man junta to rule Cuba, Grau was selected as one of its
members. This pentarchy, however, was short-lived and the
students soon chose their old professor as provisional president.

Grau's regime (September 10, 1933-January 14, 1934) was
the high water mark of a revolutionary process that began with
Machado's overthrow. In a unique alliance, students and the
military ruled. The government was pro-labor and nationalistic
opposing the dominance of foreign capital. Grau issued decrees
establishing a maximum working day of eight hours, and requiring all business enterprises to employ native Cubans for at
least half of their total working force. Another decree was also
issued which sought to "Cubanize" the labor movement and re-
strict Communist influence by limiting the number of foreign
leaders. Grau denounced the Platt Amendment and began negotiations for its abrogation. These measures aroused American
hostility and the United States refused to recognize Grau.
Since recognition was considered by Cuban political leaders as
a key factor for the existence of any Cuban government, United
States policy in effect condemned the Grau regime and encouraged opposition. On January 14, 1934, Army Chief Fulgencio
Batista forced Grau to resign. Grau went into exile. There
he was soon appointed president of a newly created nationalist
party, the Partido Revolucionario Cubano (Auténtico). Grau
returned to Cuba in time to be elected to the convention that
drafted the 1940 Constitution. In the presidential elections
of that year he lost against his rival, Fulgencio Batista. In
1944 he tried again, this time successfully.

Grau's administration coincided with the end of World War II.
He inherited an economic boom as sugar production and prices
rose. Grau inaugurated a program of public works and school
construction. Social security benefits were increased. Economic development and agricultural production was encouraged.
Increased prosperity brought increased corruption. Nepotism
and favoritism flourished. Urban violence, a legacy of the
early 1930's reappeared now with tragic proportions. Groups
employing violence organized not only to mobilize political power
but also to obtain government privileges and subsidies. The
reformist zeal evident during Grau's first administration had
diminished considerably in the intervening decade. Grau himself seemed softened after years of exile and frustration. He
faced, furthermore, determined opposition in Congress and from
conservative elements in his own party. For many Cubans Grau

failed to fulfill the aspirations of the anti-Machado revolution.
After turning over the presidency to his protégé, Carlos Prío,
in 1948, Grau virtually withdrew from public life. He emerged
again in 1952 to oppose Batista's coup d'état. He ran for president in the 1954 and 1958 Batista-sponsored elections, but
withdrew just prior to each election day claiming government
fraud. After Castro came to power in 1959 Grau retired to his
home in Havana.

GRENET, ELISEO (1893-1950). Cuban composer of popular folk
music. Among his best known songs are "Mamá Inés"; "Papá
Montero"; "Allá en la Siria."

GRITO DE BAIRE. The historic call to rebellion on February 24,
1895, in the town of Baire, Oriente, initiating the War of Independence against Spain (1895-1898). See Independence War.

GRITO DE YARA. The call to rebellion by Carlos Manuel de
Céspedes on October 10, 1868, initiating the Ten Years' War
against Spanish rule in Cuba, and proclaiming independence.
In the Grito de Yara, Céspedes called for the gradual emancipation of slaves, which had the immediate effect of attracting
large numbers of abolitionist followers to his cause. See Ten
Years' War.

GROBART, FABIO (1901-   ). Abraham Simkovitz, known as Fabio
Grobart, was born in Poland. He arrived in Cuba about 1922,
and since then has been a leading Communist organizer and
ideologist. During most of the forty years the Cuban Communist Party has been in existence, he has worked behind the
scenes, frequently using aliases and avoiding publicity. Since
Castro's announcement of his affiliation with Marxism-Leninism
in 1961, Grobart has been the editor of the official journal,
Cuba Socialista. On October 5, 1965, he was appointed to the
Central Committee of the new Cuban Communist Party. Little
is known about his activities in Cuba other than what has been
sporadically published by the newspaper Hoy or by Cuba Socialista. He seemed to have been a trusted pro-Soviet Communist
who enjoyed close relations with the Kremlin leadership and
became, together with Carlos Rafael Rodríguez, key liaison people between Castro and Eastern European leaders.

GRUPO MINORISTA. A group of young writers bent on reforming
Cuba's intellectual and political life. Organized in 1923, they
published a document that attracted national attention condemning political graft and corruption and calling for social and political reforms. This militant and defiant document was signed
by several young writers who later became prominent and influential Cuban intellectuals, such as Jorge Mañach, Rubén
Martínez Villena, and Juan Marinello.

GUAIMARO. Town in Camagüey province where an assembly presided over by Carlos Manuel de Céspedes representing the
various insurrectionary groups of the independence struggle,
met in 1869 to draw up a constitution establishing the first
revolutionary government. Céspedes was appointed President
of the rebel republic.

GUAIMARO CONSTITUTION. At the Constituent Assembly of
Guáimaro, on April 10, 1869, the first President of the Republic, Carlos Manuel de Céspedes, was elected, and this
Constitution was declared the first of the Republic of Cuba.
It contained 29 brief articles that recognized three powers:
legislative (House of Representatives, in permanent session
until the end of the war), the executive (president named by
the House, with a General in Chief subordinated to the executive), and an independent judicial power. The decisions of the
House needed the President's sanction in order to be compulsory. The Constitution could only undergo reforms whenever
the House unanimously decided to do so. According to this
document, the citizens of the Republic were entirely free and
slavery was abolished. In addition, all citizens were considered soldiers of the Republic. Among its fifteen signatories
were Carlos M. de Céspedes, Salvador Cisneros Betancourt,
and Ignacio Agramonte.

GUAJIRO. A Cuban peasant who lives away from any major urban
center.

GUANAHATABEY. Oldest Indian culture in Cuba, before the arrival
of the Spaniards, it has been described as a shell culture, characterized by the use of shell gouge and spoon as its principal
artifacts. They lived mostly in caves in the western part of
the island, primarily the Peninsula of Guanahacabibes. They
quickly disappeared or were killed following the Spanish colonization. See also Ciboneyes; Taínos.

GUANTANAMO (PROVINCE). Population: 416,115; Area: 6,370
km²; Municipalities: El Salvador, Yateras, Baracoa, Imías,
Manuel Tames, Niceto Pérez, Guantánamo, Maisí, San Antonio del
Sur, Caimanera. It emerged during the first years of the
nineteenth century alongside the shores of the Guaso River with
a city plan similar to that of the City of Cienfuegos, characteristic of the age. It also felt French influence derived from the
immigration of colonists during the Haitian revolution. Its capital city has the same name as the province and is the easternmost city of Cuba. Before 1976 it was part of the Province of
Oriente. In Guantánamo Bay there is a North American air-
naval base. The Province of Guantánamo is second in territorial
size of the eastern provinces. The zone is dedicated mostly to
fishing and agriculture, sugar, coffee, and less important fruits.

GUANTANAMO BASE. U.S. naval facility, 45 square miles in area,
in Guantánamo province (formerly part of Oriente) on Cuba's
southeastern coast, which was granted to the United States by
Article 7 of the Platt Amendment in 1901. By 1912 Guantánamo
Base had become a leased property of the United States at
$2,000 per year. After the 1959 Revolution and the break be-
tween the United States and Fidel Castro, the U.S. naval base
at Guantánamo has been a controversial issue with Castro demanding its return and the United States refusing to yield its
strategic military outpost in the Caribbean. As a form of keeping pressure on the Castro regime, the United States has
staged military maneuvers in Guantánamo during the Carter
and Reagan administrations.

GUERRA CHIQUITA (Little War). A short-lived rebellion against
Spain in 1878 following the conclusion of the Ten Years' War for
Independence (1868-1878). Rebel General Antonio Maceo and
Major-General Calixto García tried to re-start the war against
the Spanish but with the Cuban forces exhausted after years
of bloodshed, this effort failed. Maceo was exiled and García
was temporarily imprisoned.

GUERRA Y SANCHEZ, RAMIRO (1880-1970). Historian and political
advisor. Significant Cuban intellectual and historian, Guerra
influenced post-independence Cuba with his writings. In his
books Guerra criticizes American policies toward Cuba. He was
also influential as a professor, having been director of the Normal School of Havana and Superindendent of Schools. During
the government of General Gerardo Machado, Guerra was advisor
to the President. His main books are Historia de Cuba; La
lección de la escuela primaria; La patria en la escuela; El
Padre Varela; Ecuador; Azúcar y población en las Antillas; El
Cardenal Cisneros y el principio de autoridad; Fines de la
educación nacional; La expansión territorial de los Estados
Unidos a expensas de España y de los países hispanoamericanos;
and Guerra de los diez años, 1868-78.

GUERRILLA WARFARE. After the failure of his attack to the Moncada barracks in 1953, Fidel Castro was sentenced to jail and
released under amnesty in May 1955. Castro went to Mexico
where he gathered a group of followers, among them Ernesto
(Che) Guevara. Castro and his followers were trained in guerrilla warfare by retired General Alberto Bayo, a well-known
veteran of the Spanish Civil War. After landing in Cuba on
December 2, 1956, Castro started a guerrilla war that evolved
in a civil war that put him in power in January 1959. Soon
after the triumph of the Cuban Revolution Ernesto (Che)
Guevara published his book, La guerra de guerrillas (Guerrilla Warfare), in which he states that conventional armed forces
can be overpowered by rural guerrilla forces, as demonstrated

by the Cuban experience. French intellectual and revolutionary
Régis Debray would adopt the same position in his 1967 book
Revolution Within the Revolution. Both authors stress the importance of creating a revolutionary foco (a focus or starting
point) to spread revolution from that point even when all conditions for revolution would not be appropriate. Guevara's ideas
would become the official position of Revolutionary Cuba on the
subject of the way to achieve power in Latin America by leftist
movements. During the 1960's, Fidel Castro supported guerrilla
movements in Latin America in contradiction to Soviet theory
stressing the peaceful way to achieve power. Castro-backed
guerrilla attempts in Venezuela (1963) and Bolivia (1967) ended
in disaster and guerrilla warfare moved into the background in
the 1970's. Political instability and civil war erupted in Central
America: Nicaragua (1979-1980), El Salvador, and Guatemala
(1980-1981-1982). Guerrilla activity has been utilized in those
areas to destabilize governments in power. Castro's support to
those movements, as disclosed by the United States, suggests a
resurgence of this activity.

GUERRITA DE AGOSTO. A short-lived uprising in 1906 led by
   José Miguel Gómez and his Liberal Party followers to prevent
   the re-election of President Tomás Estrada Palma (1902-1906) to
   a second term in office. The uprising was crushed partially
   with the help of U.S. marines, who stayed until 1909, marking
   the second U.S. intervention in Cuba.

GUEVARA, ERNESTO (CHE) (1928-1967). Argentine revolutionary.
   Cuba's former Minister of Industries, guerrilla theoretician, and
   Fidel Castro's trusted adviser, Guevara died in Bolivia while
   attempting to lead an "anti-imperialist" revolution. Guevara
   was born on June 14, 1928, in Rosario, Argentina. Of Spanish
   and Irish descent, he suffered from asthma, spending his childhood in a mountain town near Rosario. At an early age he read
   history and sociology books in the family library and was particularly influenced by the writings of the Chilean Communist
   poet Pablo Neruda. He finished high school, and at nineteen
   entered the medical school of the University of Buenos Aires.
   In 1952 "Che" Guevara ("Che" is an Argentine equivalent of
   "pal") broke off his studies in order to set out with a friend
   on a transcontinental trip that included motorcycling to Chile,
   riding a raft on the Amazon, and taking a plane to Florida.
   He returned to Argentina to resume his studies, graduating
   with a degree of doctor of medicine and surgery in 1953. Late
   in 1953 Guevara left Argentina, this time for good. He moved
   to Guatemala, where he had his first experience of a country at
   war. He supported the Jacobo Arbenz regime, and when it was
   overthrown in 1954 Guevara sought asylum in the Argentine Embassy, remaining there until he could travel to Mexico. It was
   here that Guevara met the Castro brothers, Raúl and Fidel. At
   the time Fidel Castro was planning an expedition against Cuban

Dictator Fulgencio Batista, and Guevara agreed to go along as
a doctor. On December 2, 1956, the expeditionaries landed in
eastern Cuba, becoming the nucleus of a guerrilla force which
operated in the Sierra Maestra mountains. The guerrillas, together with a loosely connected underground movement, undermined the government. Finally a wave of popular discontent
and desertions in the army precipitated the crumbling of the
Batista regime on December 31, 1958. In January 1959, Guevara was one of the first rebel commanders to enter Havana and
take control of the capital.

He held several posts in the Castro government: commander
of La Cabaña fortress, President of the National Bank, Minister
of Industries--and always, most important of all, one of Castro's
most influential advisers. Guevara visited Communist countries
in the fall of 1960 to build up trade relations with the Soviet
bloc and criticized U.S. policy toward Cuba. He also directed
an unsuccessful plan to bring rapid industrialization to Cuba
and advocated the supremacy of moral over material incentives
to increase production. Guevara also masterminded Cuba's subversive program in Latin America and wrote extensively on this
subject. In his first book, Guerrilla Warfare (1960), he provided basic instructions on this type of conflict. Guevara's official tasks did not cure him of his restlessness. He continued
to travel. In December 1964 he set out on a long journey to
Europe, Africa, and Asia. After his return to Havana he sur-
prisingly disappeared from public view. His wanderings took
him to Africa to lead a guerrilla movement which failed. He returned to Cuba, preparing a team of Cuban Army Officers who
would accompany him to his next fighting area, Bolivia. Guevara expected that a spreading guerrilla operation in Bolivia
would force United States intervention, thus creating "two,
three, or many Vietnams." Instead, the United States provided the Bolivian army with needed weapons and training in
counter-insurgency techniques. Although the guerrillas enjoyed initial successes, they failed to gain much support.
Peasant assistance was lacking. The Bolivian army tracked
down and annihilated the guerrillas and captured Guevara on
October 8, 1967. The next day Ernesto "Che" Guevara was
executed.

GUILLEN, NICOLAS (1902-). Born in Camagüey, Guillén became
a prominent revolutionary poet and President of the National
Union of Writers and Artists of Cuba after Fidel Castro came to
power in 1959. Long before the Revolution, Guillén protested
American domination and social injustice. A mulatto himself,
Guillén emphasized the African contribution to Cuban society.
In 1920, he began publishing Camagüey Gráfico in Camagüey,
and in 1922 he finished his book of verses Cerebro y Corazón.
In 1926, he moved to Havana as a civil service employee and
joined the Vanguardista movement, publishing poetry in the
magazine Orto and in the section titled "Ideales de una raza"

in Diario de la Marina. This section of the most conservative newspaper in Cuba was paradoxically dedicated to enhance the Cuban blacks' social position. Guillén published strong articles against racism, criticizing Cuban society. He wrote Versos de ayer y de hoy, a collection of early works and of his vanguardistic poems, and Al margen de un libro de estudio. But his most acclaimed work of that time was Motivos de Son, which caused praise in literary circles when it appeared in 1930. In 1931, Guillén published Songoro Cosongo. With irony and grace, he described vignettes of Cuban society, mixed with Afro-Cuban folklore. In West Indies Ltd. (1934), without excluding the presence of Afro-Cuban elements, Guillén broadens his concerns and attacks the U.S. as well as Cuban politicians. In 1937 he published Cantos para soldados y sones para turistas and España: Poema en cuatro angustias y una esperanza.

In 1938 he joined the Cuban Communist Party, having previously been a collaborator of the Communist press as editor of the weekly Resumen and co-editor of the magazine Mediodía. Guillén became an activist and traveled through South America and Europe. In 1947 he published El son entero, a collection of his works. During the government of Fulgencio Batista (1952-1959), Guillén lived in exile. In 1956, he received the Lenin Peace Prize. In 1958 his book La paloma de vuelo popular: Elegías, appeared, wherein Guillén prophesied the triumph of a revolution in Cuba. In 1961 he was appointed President of the Cuban Union of Writers and Artists (UNEAC). Since then, Guillén has published Tengo (1964); El Gran Zoo (1967); La rueda dentada; and El diario que a diario. Guillén is currently revered in Cuba as one of its greatest living native poets.

GUITERAS HOLMES, ANTONIO (1906-1935). Young Cuban revolutionary of the Generation of 1930, he was the founder and leader of Joven Cuba, a nationalistic and anti-U.S. group committed to profound social, political, and economic reform. A strong activist in the struggle against the dictatorship of Gerardo Machado (1924-33), Guiteras had considerable influence in the choosing of Ramón Grau San Martín as provisional President in 1933. He held the post of Minister of the Interior during Grau's administration and was responsible for most of the social laws enacted during that period. An anti-imperialist with strong nationalistic and pro-labor feelings, he was killed in 1935 by the Batista-led army while trying to leave Cuba for Mexico.

- H -

HACIENDAS COMUNERAS. Early collective cattle ranches in Cuba. During the 17th and part of the 18th centuries, most of Cuba

was covered by large cattle ranches. Obtained originally by
individual grants, these were later transformed into collective
properties of haciendas comuneras via property transfers and
sales or inheritances. Each comunero received, not a piece of
land, but a share or peso de posesión of the farm's appraised
value, and branded his livestock within the collective farm.
Over a period of time, this encouraged small-scale land ownership cutting down large cattle estates.

HART, ARMANDO (1928-). Revolutionary leader under Castro;
Minister of Education and Culture; Cuban Communist Party official. Son of a prominent judge, Hart graduated from the University of Havana Law School and soon became involved in politics. He was one of the leaders of the National Revolutionary
Movement (MNR), a nationalistic group that opposed the government of General Fulgencio Batista (1952-58). After a failed attempt by his group to overthrow Batista through a violent uprising, Hart joined Fidel Castro's 26th of July Movement. He
became an important clandestine leader and organized the Re-
sistencia Cívica, an underground group of professionals and
businessmen opposed to the Batista regime. Hart also organized the supplying of Castro's rebel forces in the Sierra Maestra. From January 1958 to the triumph of Castro's revolution
in January 1959, Hart was in jail, having been captured by Batista's police for his revolutionary activities. In 1956 he married Haydée Santamaría, one of the top female leaders of the
26th of July Movement. He later divorced her. She committed
suicide in 1980. Under Castro, Hart served as Minister of Education, Secretary of the Cuban Communist Party, and Organizational Secretary of the Party. A close ally of Fidel, Hart is
currently Minister of Culture and a member of the Political
Bureau of the Cuban Communist Party.

HATUEY. Arawak cacique who fled from Hispaniola to escape Spanish conquistadores. He was captured and burned alive in Cuba.
The Spaniards attempted to convert him to Catholicism prior to
his death, but he refused, and legend has it that he insisted
that he did not want to go to heaven where he would find the
Catholic Spaniards who were about to kill him.

HAVANA (CITY, PROVINCE). Population: 1,786,552; Area: 739
Km²; Municipalities: Playa, Centro Habana, Regla, Guanabacoa,
Diez de Octubre, Marianao, Boyero, Cotorro, Plaza de la Revolución, Cerro, La Habana Vieja, La Habana del Este, La Lisa,
San Miguel del Padrón, Arroyo Naranjo. It is the capital of
the Republic of Cuba. In 1976 a new political-administrative
division was enforced nationally and what was originally the
Province of Havana was divided in two: City of Havana and
Havana (with 19 municipalities). The City of Havana was
founded around 1519 by Spanish conquistadores and became
an important city at that time because every Spanish ship came

to its port on route to the American colonies and then returned
to Spain loaded with bullion. In order to provide it protection,
forts were built around the city, many of which still stand.
Because of this, the City of Havana was the target of corsair
and pirate attacks and on one occasion (1762) it was occupied
by the English during the Seven Years' War. The City of Havana has always been the center of political and cultural activity of the nation. The University of Havana was founded in
1728. Presently, the province has the highest population and
population density. The Port of Havana, located in the bay of
the same name, is the most important of the country and is surrounded by a wide industrial belt, refineries, and a fishing port.
The beaches that extend toward the East are an important tourist attraction. The oldest section of the city (Old Havana) was
declared Humanity's Heritage by UNESCO and its conservation
has been promoted. UN grants have facilitated restoration
projects, to be carried out until 1990.

HAVANA (PROVINCE). Population: 524,001; Area: 5,828 Km^2;
Municipalities: Mariel, Caimito, San Antonio de Los Baños,
Bauta, San José de las Lajas, Santa Cruz del Norte, Guines,
Nueva Paz, Batabanó, Güira de Melena, Artemisa, Guanajay,
Bejucal, Jaruco, Madruga, Alquízar, San Nicolás, Melena del
Sur, Quivicán. One of Cuba's original six provinces, during
the administrative changes made by the revolutionary government in 1976 it lost the City of Havana, now a new province,
while gaining territory previously part of Pinar del Río province. The City of Havana is the provincial capital of this
province, even though it lies outside its limits. The Cuban
sugar industry had its birth and experienced tremendous growth
during the nineteenth century in this area. Into the twentieth
century, it was still the most important industry. As the century progressed, sugar was displaced as the leading industry
by the development of manufacturing and other light industry.
Although tobacco cultivation is minimal, cigar and cigarette
processing is mostly done in this province. Fruits, vegetables,
and sisal production are also important as well as fishing.

HAWLEY-SMOOT TARIFF. By 1929, the battle over tariffs and
quotas on Cuban sugar coming into the U.S. had become intense. The domestic sugar growers in the U.S. sought to raise
the tariff on Cuban sugar above the 1.7 cent per pound level
of the 1921 Fordney-McCumber Tariff. These interests prevailed over U.S. investors in Cuban sugar who would be damaged by an increased tariff. In January 1930, the Hawley-
Smoot Tariff raised the surcharge on Cuban sugar to 2 cents
per pound. This increase, along with the depression in the
United States, struck a blow to the Cuban sugar industry.
See also Fordney-McCumber Tariff.

HEALTH. Health services were advanced in Cuba prior to the

Castro Revolution in 1959. Statistics show that Cuba's general
rate of mortality was the lowest in Latin America. One of the
features of the health system was the Clínicas. Clínicas were
special private facilities run by nonprofit organizations, such as
immigrant mutual aid associations, which collected a monthly fee
from their associates and provided complete medical attention including medications. Clínicas were directed by representatives
of the associates and had modern medical equipment. Doctors
were hired as employees in the Clínicas. In some cases, physicians established medical cooperatives (also called Clínicas).
Their main distinction was that in the medical cooperatives, doctors would direct administrative operations. It was estimated
that over a third of the population received attention from
these institutions. Most of these were established in urban
centers to the neglect of the rural population. With the advent of the Castro revolution, the Clínicas were taken over by
the government or abandoned by physicians who left the country. The Cuban government centralized and expanded health
care throughout the island. All services are administered by
the Ministry of Public Health (MINSAP).

HEREDIA, JOSE MARIA DE (1803-1839). Prominent poet, best
known for his ode Al Niágara. Heredia graduated from the
University of Havana Law School and joined the Soles y Rayos
de Bolívar, an anti-Spanish movement in Cuba to gain independence for the island and annex it to Simón Bolívar's Republic in South America. Heredia fled Cuba and lived in Mexico,
where he continued to be active against Spanish rule. He returned to his homeland only to be forced into exile. He died
in Mexico in 1839. His main poems also include En el Teocalli
de Cholula; El himno del desterrado; A Emilia.

HERNANDEZ CATA, ALFONSO (1885-1940). Cuban novelist and
playwright; author of Cuentos pastorales, Los frutos ácidos,
El bebedor de lágrimas, La juventud de Aurelio Zaldívar.

HEVIA, CARLOS (1904-1964). Minister of Agriculture under the
Revolutionary Government of 1933; President of the Republic
for 48 hours after the resignation of Grau San Martín. A
graduate in engineering of the Naval Academy at Annapolis
(1919), Hevia participated in World War I aboard the U.S. war-
ship Missouri. Following his return to Cuba, he became active
in politics, expressing strong opposition to the dictatorship of
President Gerardo Machado (1925-1933). Hevia was appointed
Minister of Agriculture under President Ramón Grau San Martín
(1933-34) in 1933. Upon Grau's resignation on January 16,
1934, Hevia assumed the Presidency. However, on January 17,
he was forced to resign in favor of Colonel Carlos Mendieta.
Hevia was a founder of the Partido Revolucionario Cubano
(Auténtico) shortly thereafter in February 1934. Under President Carlos Prío Socarrás (1948-52), Hevia served as Secretary

of State, Minister of Agriculture, and President of the Development Commission. He was the favored Auténtico candidate in the 1952 presidential elections, which were thwarted by General Fulgencio Batista's coup of March 10, 1952. The government of General Batista (1952-58) imprisoned Hevia for a short time. After Fidel Castro took power in 1959, Hevia left Cuba and lived in exile until his death in 1964.

HISPANIDAD. Movement initiated by Ramiro de Maetzu which was very popular in Spain during the 1940's. Hispanidad claimed that the independence of Latin America had been frustrated due to lack of social reforms, and lamented that Anglo-Saxon values had supplanted Spanish cultural domination. The movement called for closer identity among the Latin American nations as well as with Spain, and emphasized that the new Spain had been liberated from both Marxism-Leninism and Anglo-Saxon materialism. Liberal democracy was criticized as decadent; the supremacy of spiritual over material values was hailed. Hispanidad was widely acclaimed in Cuba, particularly by exiled Spanish intellectuals. As a student at Belén Jesuit School in Havana, Fidel Castro seemed to have been greatly influenced by the ideas of Hispanidad, as by several of his teachers who discussed the movement and taught its philosophy at the school.

HOLGUIN (PROVINCE). Population: 777,262; Area: 9,119 Km^2; Municipalities: Gíbara, Frank País, Holguín, Banes, Sagua de Tánamo, Urbano Noris, Báguanos, Mayau, Calixto García, Rafael Freire, Moa, Cacocum, Antilla, Cueto. Capital of the Province, Holguín was founded in 1523 by a land grant to Francisco García Holguín by Diego de Velázquez. In the 1900s, Holguín was famous for its industrial and commercial activity and because of it was a transportation center for the Oriente province, of which it was part before 1976. Its proximity to the Port of Gíbara was favorable to its development. Great mining richness found in deposits in Moa and Nicaro (iron and nickel). The bay of Nipe, located within the territory of this province, is one of the largest enclosed bays of the world measuring 120 Km^2. Crops are sugarcane, corn, plantains, cacao; its forests are rich in Mayarí pines. It has been called the Cuban granary. Other cities are Antilla and Mayari.

HOUSING. Like most Latin American countries, Cuba in the 1970's was in the grip of a serious housing shortage. Its intensity had been alleviated by the exodus of an estimated 600,000 refugees who had fled Castro's Revolution, while a relatively low rate of population growth kept in bounds the rate at which it was worsening. There had been a housing boom of sorts during the years after World War II, but it had been limited largely to the construction of homes for well-to-do families, and the housing situation inherited by the revolutionary government was poor. The 1953 census classified the condition of 15 percent

of the dwellings as ruinous or worthless and 31.6 percent as
bad. In the countryside some 75.2 percent were classified as
ruinous. Between 1953 and 1958 only 25,000 housing units
were constructed, and during 1943-58 almost all of the 143,000
units constructed had been in urban localities with the exception of a few hundred that had been built along the Central
Highway for political display purposes. In 1959 well under one-
half of the population lived in the countryside, but an estimated
two-thirds of the housing shortage was rural. In Havana the
6 percent of the population that lived in communities of jerry-
built shacks fringing the city was a smaller proportion than that
in similar communities in many other Latin American capitals, but
a large proportion of the population lived in inner-city tenements.

   In the country as a whole the availability of public services
in urban localities compared favorably with that elsewhere in
Latin America. Before 1959 well-to-do families had lived in old
masonry or wooden houses with tiled floors and roofs, built in
the Spanish style around interior patios in good residential districts, such as the Miramar section of Havana, and in comfor-
table modern houses in the suburbs. There were a few luxury,
garden-style apartment buildings and in the 1950's a number of
high-rise apartments were built. The typical dwelling of the
countryside was the bohío, constructed in a style that had been
developed by the Taíno and Ciboney Indians. Made entirely of
local materials, it consisted of a rectangular skeleton of poles,
walls of rough lumber or royal palm bark, earthen floor, and
a roof constructed of cane or palm fronds or of a palm thatch
called guano. The more elaborate had interior partitions and a
connected cooking area. The furniture was limited to a few
stools, one or more cots, and a kerosene stove. In the small
towns and villages long rows of wood or stucco houses with
one or two floors fronted directly on the unpaved streets. On
the sugar estates workers were housed in barracks (barracones),
long buildings usually partitioned off into rooms in which ham-
mocks were hung for sleeping. A separate structure nearby
served for cooking. These buildings were often built on stilts,
raising them about four feet above the ground.

   Among the first measures undertaken by the revolutionary
government was a 1959 law reducing rents by 50 percent for
housing units renting for 100 pesos or less per month. Smaller
reductions were prescribed for higher rentals and for units
owned by private savings and retirement programs. The 1959
law was subject to extensive evasion, however, and it was replaced by the comprehensive and important Urban Reform Law
of October 1960. The Urban Reform Law established the right
of all families to decent housing by means of a three-stage program. In the first stage the state assumed responsibility for
ensuring each tenant household full amortization of the premises
occupied during a period ranging from five to twenty years,
through payments to be made in lieu of rent. The second
stage called for a massive construction program giving occupants

permanent usufruct of the units occupied through monthly payments not exceeding 10 percent of the family income. The third
stage was to involve the cession of permanent usufruct of
dwellings without cost to the occupant. Implementation of
stages one and two commenced promptly, but in the late 1970's
full implementation of stage three appeared unlikely in the fore-
seeable future. Rent-free housing had been provided to rural
families resettled in new housing units, to inhabitants of tenements who had made at least sixty monthly rental payments
since 1959, to victims of violence during the regime of Fulgenio
Batista, and to households with monthly incomes of twenty-five
pesos or less. Moving from one dwelling to another became a
difficult matter. Rental advertisements were not accepted by
newspapers and radio stations, and the growing family in search
of more spacious living quarters was forced to rely largely on
word of mouth and messages placed on public bulletin boards.
It was necessary to obtain moving permits approved by a government agency, and special controls were placed on moving
articles of furniture in order to prevent their disposal by pro-
spective refugees.

HOY. Official daily newspaper of the Partido Socialista Popular,
former name of the Communist Party. Founded in the 1930s,
the newspaper was marged with Revolución in 1965 to create
Granma, official daily newspaper of the Castro Revolution. See
also Granma; Revolución.

HULL, CORDELL (1871-1955). U.S. Secretary of State in the early
Franklin Roosevelt administration who instructed Sumner Welles
to mediate the violent confrontation in Cuba between the dictator Gerardo Machado (1925-1933) and his opposition. Later in
1933, Hull resisted Ambassador Welles' recommendations that
U.S. armed intervention was necessary to protect U.S. commercial interests. Hull opposed the presidency of Dr. Ramón Grau
San Martín in 1933 and refused to extend recognition of the
Cuban government because of Grau's nationalization of U.S.
properties and his nationalistic and anti-American posture.

HUMBOLDT, BARON FRIEDRICH ALEXANDER VON (1769-1859).
Scientist, prolific writer, and traveler who visited Cuba in
1800 and 1801 taking copious notes of his observations which
he later published as Personal Narrative of Travels to the
Equinoctial Regions of the New Continent (1799-1804) (3 volumes).

- I -

IMMIGRATION, JAMAICAN. Since the development of the sugar

industry at the beginning of the 19th century, Jamaicans were
brought to Cuba, mostly as slaves, to work the sugar harvest.
Some settled permanently, primarily in Oriente province, bringing with them their language and culture, but ultimately adapt-
ing to Cuban values and customs. Until 1959, many workers
joined the sugar harvest annually but would return to Jamaica
when the slack season set in.

IMMIGRATION, SPANISH. The increased wealth of Cuba as a result of the growth of the sugar industry brought a great influx
of Spaniards during the second half of the 18th century. Immigrants also came from Martinique, Jamaica, and Haiti as a result
of political turmoil and revolution in these islands. Spaniards
arrived in Cuba in increasing numbers from South America following the independence of those countries in the 1820's. The
development of the tobacco industry as well as the growth of
agricultural production created a demand for new Spanish immigrants throughout the 19th century and even after independence
from Spain was achieved in 1902, Spaniards continued to migrate
to the island. Some of the late-comers brought syndicalist experiences and were influential in the organization of the first
Cuban labor unions. The fall of the Spanish Republic at the
end of the Spanish Civil War (1936-39) also brought political
refugees to Cuba.

INDEPENDENCE WAR (1895-1898). A three-year struggle that led
to U.S. intervention and eventually to independence in 1902.
Organized by José Martí and led militarily by Antonio Maceo
and Máximo Gómez, the war began in Oriente province. Martí
was killed in action and the leadership of the war fell to Gómez
and Maceo who soon began an invasion of the western provinces.
In repeated attacks they undermined and defeated the Spanish
troops and carried the war to the sugar heart of the island.
From January to March 1896 Maceo waged a bitter but successful campaign against larger Spanish forces in the provinces of
Pinar del Río and Havana. By mid-1896 the Spanish troops were
on the retreat and the Cubans seemed victorious throughout the
island. Then came a change in the Spanish command: the more
conciliatory Marshal Arsenio Martínez Campos was replaced by
General Valeriano Weyler, a tough and harsh disciplinarian.
Weyler's policy of concentrating the rural population in garrisoned towns and increasing numbers of Spanish troops allowed
the Spaniards to regain the initiative after Maceo's death on
December 7, 1896, in a minor battle. Yet they were unable to
defeat the Cuban rebels or even engage them in a major battle.
Gómez retreated to the eastern provinces and from there carried on guerrilla operations. He rejected any compromise with
Spain. In January 1898 when the Spanish monarchy introduced
a plan that would have made Cuba a self-governing province
within the Spanish empire, Gómez categorically opposed it. In
1898, following the explosion of the S.S. Maine outside the har-

bor of Havana, the United States declared war on Spain. The Cuban-Spanish-American War, as Cubans patriotically refer to it, led to Cuba's independence from Spain that same year and to American intervention until 1902.

INDIANS. The Indians that inhabited Cuba at the time of Columbus' landing, in numbers anywhere from 16,000 to 60,000 up to a high of 600,000, possessed no written language and most of them, although peaceful, were annihilated or assimilated as a result of the shock of conquest. At least three cultures, the Guanahatabeyes, the Ciboneyes, and the Taínos, swept through the island before the Spaniard's arrival. Of these, the first two groups were fairly primitive, living in caves and practicing very crude agriculture. The Taínos, an Arawak group of Indians, entered Cuba some two centuries before the conquest and developed a rather advanced economic system based on agriculture with commonly cultivated fields of yuca, tobacco, cotton, corn, and white and sweet potatoes. In terms of economic development, social organization, technological advances, and art, the Indians of Cuba were far inferior to mainland Indians such as the Maya, Aztec, and Inca. There was little mingling of races between Spaniards and Indians and the new society of Spaniards and subsequently of Spaniards and blacks supplanted the Indian society. New institutions, new values, and a new culture replaced the old ones. For the most part, the Cuban Indians' contribution to the development of a Cuban nationality must be considered minor. After the conquest Indian warriors like Hatuey, who fought the Spanish conquest in eastern Cuba, were glorified in the pages of Cuban history books and raised to folk hero status. They represented for Cuban children a symbol of the native resistance against the oppressive Spanish conquistador. However, whereas Mexicans searching for a national identity could look back to their Indian past as a rallying point for revolution, Cubans could not. See Ciboney, Guanahatabey, Taíno.

INDUSTRY. Major industrial progress in Cuba was not realized until World War II when domestic industrial production in Cuba was stimulated by wartime demands. In the decade of the 1950's U.S. investment in Cuban manufacturing doubled over pre-1950 levels. Most of the American industries were large-scale subsidiaries of U.S. corporations in the rubber, chemical, and pharmaceutical sectors. Other non-food industries included fertilizers, textiles, leather products, building materials, glass, lumber, furniture, metal products, and machinery. Overall industrial activity by 1950 included between 12,000 and 13,000 industries with total assets equalling approximately $300 million. Investments in the sugar industry in 1950 far outdistanced all other forms of industrial investment. The Province of Havana contained 50% of all Cuban manufacturing and non-sugar industrial establishments in 1950. American investment was chiefly responsible for making Cuban industry heavily capitalized until 1959. Until the Castro Revolution, private enterprise controlled industrial activity where no industry was state-dominated. In 1953 some 327,000 persons, or 16.6% of the labor force, were employed in the industrial sector. Sugar, rum, and industrial alcohol manufactures accounted for well over the total industrial output of 1 billion pesos. In 1958 there were 161 sugar mills (centrales), 36 of which were U.S.-owned, and the latter accounted for 36% of total output (121 mills were Cuban-owned). By 1970 virtually all Cuban industrial production had been nationalized. Production is now under the direction of the industries ministry.

The chemical industry was in 1950 perhaps the most modern and efficient enterprise after sugar. Sulphuric acid, chlorine, and chemical fertilizers were main products. The domestic textile and clothing industries were small but expanding in the 1950's. The construction industry has been a major contribution to the economy since World War II as public works projects were stimulated with the 1950 Economic Development Plan. It suffered a setback with the Revolution but recovered somewhat with a government-sponsored public housing program. Meat and meat products were produced before 1959 in ample enough quantities ($75 million worth a year) to supply domestic requirements. Processed meats, sausage, canned meat, hams, bacon, and jerked beef were produced by some thirty plants. Today, Cuban food requirements are not supplied by domestic food production and processing. The result is rationing and importation of food products. In December 1981 the Cuban government doubled prices of some 1,500 food staples due to supply shortages. More austerity measures were announced by Castro himself in January 1987.

INGENIO. Usually a large, steam-powered sugar mill. See also Sugar; Centrales.

INSTITUTIONALIZATION. Refers to the consolidation of power and control by Fidel Castro and his on-going Revolution begun in the 1960's. Castro, as a strong personalist figure, is in absolute control of the Cuban government, with the party and other organs of the state acting merely as instruments carrying out his orders. Castro's political style and his institutionalization of the Cuban Revolution has led to a vertical structure of political power where control is in the hands of a small elite. The stability of the regime is based primarily on the strength of the armed forces and of the Cuban Communist Party, the two most powerful institutions in Cuba today.

INSTITUTO CUBANO DEL ARTE E INDUSTRIA CINEMATOGRAFICA (ICAIC). From a modest beginning, ICAIC grew to become a vast organization which controls all phases of the country's film industry. Under the Castro regime it owns all Cuban movie

theatres, is the sole importer and distributor of foreign-made
films and materials, and produces newsreels and full-length films
at its own studios and laboratories. An impressive number of
awards have been given to Cuban feature films and documentaries at international film festivals.

INSTITUTO NACIONAL DE LA REFORMA AGRARIA (INRA). Established in 1959, INRA was entrusted to carry out and administer all the provisions of the Agrarian Reform Law of May 17,
1959. The institute was given sweeping authority to organize
collective cultivation of the land and to regulate all agricultural
production. The country was divided into twenty-eight agrarian
development zones, each to be governed by an appointee of
INRA. All private lands, with the exception of several thousand
small farms, were expropriated and placed under the control of
INRA. The institute proceeded to organize agricultural production through state farms, which remain today the principal organizational structures in the agrarian sector. In 1974 INRA
was absorbed by the Ministry of Agriculture (MINAG). See
also Agrarian Reform Laws.

INTEGRATED REVOLUTIONARY ORGANIZATIONS (ORI). See
Organizaciones Revolucionarias Integradas.

INTEGRISTAS. Supporters among the peninsular (Spanish-born)
class in Cuba during the early 1800's of a movement known as
Integridad Nacional (National Integrity). They supported continuation of Spanish control in Cuba against the rising tide of
nationalism and independence movements. During the second
War of Independence in 1895, the Integristas were strong supporters of General Valeriano Weyler's efforts at suppressing the
Cuban rebels.

INTENDENCIA. The political system introduced by Spain in Cuba
in the late 1700's as part of the Bourbon reforms. It sought
to improve the colonial administrative system and to curb the
excesses that drained revenue from the Royal treasury. The
Intendant in Havana ranked equal to the Captain-General (Governor) and was appointed to oversee the Departments of Treasury and War. His primary function was to collect revenues, en-
force trade laws, and encourage any agricultural or commercial
activities that would increase prosperity. José Pablo Valiente
served as one of the first Spanish Intendants in Cuba, selected
by Captain-General Luis de Las Casas for his exceptional fiscal
abilities.

INTER-GOVERNMENTAL SOVIET-CUBAN COMMISSION (Coordinating
Committee). The Inter-Governmental Soviet-Cuban Commission
was set up in December 1970 to expand cooperation between
Cuba and the Soviet Union in areas of economic relations as
well as scientific and technological exchange. The Commission

was significant as a symbol of the new closeness between the
two nations and of the integration of Cuba to the Soviet political and economic spheres of influence.

INTERNATIONAL COOPERATION. Cuba is a member of the United
Nations and of all the specialized agencies except the IBRD.
Its membership in the OAS was suspended at the second Punta
del Este meeting, in February 1962, through U.S. initiative.
The isolation of Cuba from the inter-American community was
made almost complete when at Caracas, on July 26, 1964, the
OAS voted 15-4 for mandatory termination of all trade with the
Castro government. Mexico, citing its adherence to the principle of nonintervention, has refused to adhere to the Caracas
decisions and was the first OAS member to maintain diplomatic
relations with Cuba. Chile extended full recognition in November 1970. Jamaica and Trinidad and Tobago maintain relations
with Cuba. Other OAS members, including Bolivia, Brazil,
Nicaragua, and Colombia, have recently renewed relations with
Cuba. In 1981, Cuba was elected Chairman of the Conference
of Non-aligned Nations. Cuba is very active in a number of
other Western Hemisphere organizations.

INTERNATIONAL SUGAR AGREEMENT (1937). In 1937, Cuban sugar
interests were still trying to recover from the disastrous effects
of the Depression, after which Cuban sugar exports to the
United States were substantially reduced. Prices on the world
sugar market remained low and the United States, through the
International Sugar Agreement in 1937, offered a premium price
above the cost of producing sugar in Cuba, thus keeping the
industry afloat.

INTERNATIONAL SUGAR AGREEMENT (1957). During the 1950's,
political events helped to create a roller-coaster price for
Cuban sugar. The Korean War (1952), the Suez Crisis (1956),
windfall sugar purchases by the Soviet Union, and aggressive
efforts by the Cubans themselves helped maintain an acceptable
picture for the Cuban sugar industry. In 1957, the negotiation
of the International Sugar Agreement setting export quotas was
also helpful in stabilizing sugar prices.

INTERNATIONAL SUGAR AGREEMENT (1968). In 1960, the Soviets
had taken over U.S. responsibility for purchasing Cuban sugar
and had to supplement their purchases with extensive credits.
Cuba could not produce even half the sugar for the Russian
market that it contracted to supply. More price fluctuations
in the 1960's forced Castro in the Fall of 1968 to sign an international agreement limiting the amounts of sugar he would sell
on the world market in an attempt to fulfill his commitments to
the Soviet Union.

INTERVENTION (1906). The second period of U.S. intervention in

Cuba which lasted from 1906 to 1909 differed significantly from
the first (1898-1902). A civilian, Charles Magoon, instead of a
military man, was the provisional governor. Magoon dispensed
government sinecures to pacify the various quarreling factions
and drew up an electoral law, along with an organic body of
law for the executive, judiciary, provincial, and municipal
branches of government. The second intervention had as its
purpose the enactment of fair legislation that would prevent
civil wars and enhance the democratic process. In 1908, the
U.S. called for municipal and national elections. José Miguel
Gómez was elected to the presidency in 1909, thus paving the
way for withdrawal of U.S. interventionist forces.

IRON. Only two of Cuba's original six provinces, Oriente and
Pinar del Río, have significant deposits of iron. The eastern
Santiago zone of iron was under active exploitation by American
companies from 1880 to 1930, when 27 million tons were exported
to the U.S. The iron in the Oriente zone is also rich in nickel
and cobalt and reserves indicate a figure of 813 million tons of
ore.

ISLE OF PINES. See Isle of Youth.

ISLE OF PINES TREATY. Negotiated in 1904, Ratified in 1925.
(U.S. Senate, Papers Relating to the Adjustment of Title to
the Ownership of the Isle of Pines. Senate Document No. 166.
68th Cong., 2nd Sess. Washington, D.C., 1924. p. 232.)
The United States of America and the Republic of Cuba, being
desirous to give full effect to the sixth article of the Provision
in regard to the relations to exist between the United States
and Cuba, contained in the Act of the Congress of the United
States of America, approved March second, nineteen hundred
and one, which sixth Article aforesaid is included in the Appendix to the Constitution of the Republic of Cuba, promulgated
on the 20th day of May, nineteen hundred and two and provides that "The Island of Pines shall be omitted from the boundaries of Cuba specified in the Constitution, the title of ownership thereof being left to future adjustment by treaty" have for
that purpose appointed as their Plenipotentiaries to conclude a
treaty to that end: The President of the United States of
America, John Hay, Secretary of State of the United States of
America; and The President of the Republic of Cuba, Gonzalo
de Quesada, Envoy Extraordinary and Minister Plenipotentiary
of Cuba to the United States of America: Who, after communi-
cating to each other their full powers, found in good and due
form, have agreed upon the following articles: Article I. The
United States of America relinquishes in favor of the Republic
of Cuba all claim of title to the Island of Pines situated in the
Caribbean Sea near the southwestern part of the Island of Cuba,
which has been or may be made in virtue of Articles I and II
of the Treaty of Peace between the United States and Spain,

signed at Paris on the tenth day of December eighteen hundred and ninety-eight. Article II. This relinquishment, on the part of the United States of America, of claim of title to the said Island of Pines, is in consideration of the grants of coaling and naval stations in the Island of Cuba heretofore made to the United States of America by the Republic of Cuba. Article III. Citizens of the United States of America who, at the time of the exchange of ratifications of this treaty, shall be residing or holding property in the Island of Pines shall suffer no diminution of the rights and privileges which they have acquired prior to the date of exchange of ratifications of this treaty; they may remain there or may remove therefrom, retaining in either event all their rights of property, including the right to sell or dispose of such property or of its proceeds and they shall also have the right to carry on their industry, commerce and professions being subject in respect thereof to such laws as are applicable to other foreigners.

ISLE OF YOUTH. Population: 30,103; Area: 2,231 Km². Until 1976, the Isle of Pines with its main city at Nueva Gerona. This territory is made up of an island located to the south of the present Havana province (59 miles away). It was always under the administrative auspices of Havana but after the new political-administrative division of 1976 it became a Special Municipality with an administrative structure similar to that of a province. The island was discovered by Columbus in 1494. During the sixteenth century and on, it was visited by corsairs and pirates who razed the Caribbean. At the beginning of the twentieth century, its autochthonous population was small and was made up of a notable number of North Americans, most of whom left following the Hay-Quesada Treaty (1925). Its economic activity was basically in the free port to the East of the island. Later on in 1926, a modern prison was built there (Presidio Modelo) which was always shrouded by stories of mistreatment until it was closed in 1975. Afterwards, the island was populated by worker migrations from other provinces. The Isle of Youth is the largest producer of citrus fruits for export. Fishing and marble quarries are also important activities. With the ascent to power of the Revolutionary Government, schools were built along the length of the island where a large number of students from Africa and other parts of the world have received scholarships to study mainly at the primary and secondary levels. To the southwest of the island there is an important naval airbase.

- J -

JESUITS. In 1704, the first Jesuits founded a school in Cuba in

Puerto Príncipe under the auspices of Bishop Jerónimo Valdés.
The Jesuits, a religious order dedicated primarily to the teaching profession, were aided by Bishop Compostela in establishing
a site for their school in Havana. In 1854, the Jesuits opened
a second school and installed the first observatory in Cuba in
the old convent of Belén. The scientist at the head of the
observatory was Father Gutiérrez Lanza. Later, the Jesuits
established the famous Belén School in Havana.

JEWS. In 1959, the Jewish population of Havana numbered some
8,000 with approximately 3,000 more distributed throughout
the island, principally in Las Villas, Camagüey, and Oriente
provinces. In the 1950's, about 75% of the Jewish population
were engaged in small-scale retail trade, while 15% were owners
of larger stores and 10% were engaged in the production of consumer goods. Jewish commercial and professional associations
evolved as did a Jewish Chamber of Commerce. After the Castro revolution, most Jews migrated to the United States or
Puerto Rico as their businesses were confiscated by the Castro
regime. Several hundred Jews, mostly aged, still remain in the
island.

JIMAGUAYU CONSTITUTION. The rebel constitution approved on
September 16, 1895, established a centralized revolutionary
government. The Constitution joined the executive and legis-
lative powers into one united Consejo de Gobierno (Government
Council) composed of a President, Vice-President, and four
Ministries: War, Interior, External Relations, and the Treasury.
The Jimaguayú Constitution improved upon an earlier revolution-
ary constitution (Guáimaro) by placing control over the military
in the hands of the Consejo de Gobierno and a General-in-Chief.
Salvador Cisneros Betancourt was elected President, Bartolomé
Masó, Vice-President, and Máximo Gómez became General-in-
Chief.

JOVEN CUBA. A clandestine revolutionary organization founded in
1934 by Antonio Guiteras, Dr. Ramón Grau San Martín's (1933-
34) former Minister of the Interior. Joven Cuba violently opposed Fulgencio Batista's control. Batista's most militant opponents joined Joven Cuba, utilizing tactics of urban violence in
an attempt to cripple the administration of Col. Carlos Mendieta
(1934-36) and thus Batista's power. See Guiteras, Antonio.

JUDICIARY. See also Courts; Tribunals. During the first U.S.
intervention in 1902, after Cuban independence had been granted, General Wood, the U.S. provisional governor, reorganized
the judicial system, placing Cuban judges on a payroll for the
first time in Cuba's history. The judiciary was declared independent of the other branches of government and a Supreme
Court was organized to be appointed by the President. The
Supreme Court had the power to declare laws unconstitutional.

In the 1940 Constitution, a Chamber of constitutional and social guarantees was to be established within the Supreme Court to consider constitutional questions. The Court members were still to be appointed by the President, and the Supreme Court would appoint, promote, and regulate the members of the lower judiciary. Entry into the judiciary was to be by examination and no judge could be removed without cause. Judicial reorganization was again carried out after the January 1959 revolution when the Cabinet appointed a new Supreme Court. In late 1960, as Fidel Castro consolidated his power, most Supreme Court justices were dismissed and their number reduced from thirty-two to fifteen.

JUNTA CENTRAL DE PLANIFICACION (Central Planning Board)-- JUCEPLAN. Set up in February 1960 by the Castro regime, the Board's function is to plan and direct Cuba's economic development. Swift transformation of Cuba's private enterprise system into a centralized state-controlled economy was a rough procedure, causing growing inflation, disorganization, bureaucratic chaos, and inefficiency. As the planning focus escalated into the decade of the 1980s, the Central Planning Board as an institution became heavily controlled by the military and by Party bureaucrats.

JUNTA DE FOMENTO. The agricultural and development board established by Spanish Governor Don Luis de las Casas and Francisco Arango y Parreño in 1792. This board was a sort of embryonic ministry of industries for Cuba and was significantly led by criollo (Cuban-born) entrepreneurs instead of peninsular (Spanish-born) merchants.

JUNTA DE INFORMACION (1866-1867). The reform commission composed of 12 criollo (Cuban-born) reformers and 4 peninsulars (Spanish-born) established in 1866 by the Spanish monarchy. The Junta met in Madrid in late 1866 and early 1867 adopting a number of political reforms including representation in the Cortes, equality of access to civil employment, freedom from arbitrary arrest and search, and civil-criminal codes. The commissioners also favored gradual emancipation of slaves. In early 1867, the Narváez government disbanded the commission, dismissing all of its recommendations and imposed new and irritating taxes. The failure of the Junta gave new impetus to the independence movement.

JUVENTUD REBELDE. Since the mid-1960's, the official daily newspaper of the youth section of the Cuban Communist Party.

JUVENTUD SOCIALISTA. See also Partido Socialista Popular. In 1953, widespread disillusion among Ortodoxo members, due to the party's ineffective leadership, caused many young ortodoxos to join the Juventud Socialista or young communists. Members

Kennedy 150

of Juventud Socialista were not supporters of Fidel Castro as
were many young ortodoxos in 1953. By late 1953, Juventud
Socialista had been banned by president Fulgencio Batista
(1952-1958). By 1957-1958, some members had been absorbed
by Castro's revolutionary movement and were with him in the
Sierra Maestra engaged in revolutionary activities.

- K -

KENNEDY, JOHN F. (1917-1963). Assumed the U.S. Presidency on
January 20, 1961, and inherited the Bay of Pigs invasion plans
for overthrowing Fidel Castro's government in Cuba. Kennedy
sanctioned the invasion but was more concerned with his general plan for Latin American development through an Alliance
for Progress. The Cuban issue gave rise once again to the
controversy over U.S. intervention. Kennedy was undecided
about whether to carry through with the invasion plans, which
he had left almost entirely in the hands of the CIA. Kennedy
publicly announced that the United States would not intervene
in a Cuban conflict, which angered the Cuban Revolutionary
Council. On April 17, 1961, Kennedy did not permit U.S. air
support after the invaders landed at Bay of Pigs. Prior to the
landing, Kennedy had cancelled follow-up B-26 air strikes intended to finish off the small Cuban Air Force, and therefore
Castro had air power with which to repel the invaders. After
the failure of the invasion Kennedy pursued a vigorous policy
of isolating the Cuban regime in order to strangle it economical-
ly, and in January of 1962 Kennedy was successful in forcing
Cuba's suspension from the OAS by a slim majority. However,
Kennedy had provided a bland performance during his summit
meeting with Nikita Khrushchev in Vienna in June 1961. Kennedy had been apologetic and deferential, leading the Soviet
Premier to evaluate him as a weak President. Thus, by mid-
1962, the Soviets felt bold enough to introduce nuclear missiles
into Cuba.

In September, President Kennedy solicited Congressional approval to mobilize 150,000 reserves. One month later, Kennedy
publicly denounced the Soviet transports of ICBM missiles into
Cuba and reacted strongly by blockading the island and demanding the withdrawal of all offensive weapons. After the
Soviets backed down, Kennedy admitted that the chances for
nuclear war with the Soviets had been quite high (perhaps
50/50) and that on October 22 he had had all U.S. missile
crews placed on maximum alert. On December 29, 1962, Kennedy welcomed back Brigade 2506 at the Orange Bowl in Miami,
Florida, promising a free Cuba, but he later sought a less con-
frontal stance vis-à-vis Castro. Kennedy eventually accepted
the fact of a Cuban anti-American Communist regime which had

been powered by hostility toward the U.S. On November 22, 1963, John F. Kennedy was shot and killed during a parade in Texas as he rode beside his wife, Jacqueline, in a convertible automobile.

KEPPEL, GENERAL WILLIAM (1727-1782). Younger brother of Lord Albermarle, the English commander who captured Havana in August 1762, Keppel himself took El Morro during the battle. In January 1763 Keppel was appointed military governor of Havana when Albermarle and another Keppel brother (a commodore) left Cuba to secure Jamaica for the British Crown.

KHRUSHCHEV, NIKITA (1894-1971). Born in Kalinovka, Kursk, in Russia. Soviet Premier and Party Chairman who met with U.S. President John F. Kennedy in Vienna in June 1961 and evaluated the latter as a weak President due to Kennedy's Bay of Pigs fiasco earlier that year. Khrushchev installed offensive nuclear weapons in Cuba (ICBM's) in 1962, prompting the October 1962 missile crisis to which Kennedy responded by ordering a blockade of Cuba and demanding an immediate withdrawal of the Russian missiles. Khrushchev backed away from the possibility of nuclear conflict with the U.S. by ordering the dismantling of all the missile sites without even consulting Fidel Castro, who did not permit United Nations inspection of the sites. Nevertheless, the U.S. was able to check that the Russians had indeed fulfilled their pledge by using U-2 plane radar. Russian foreign policy was restrained and cautious for several years following the Missile Crisis. In internal Russian policy, the Cuban missile crisis was a turning point. As a result of Kennedy's tough stance at the Summit, Khrushchev lost much international prestige. He soon faced domestic difficulties in 1963 with the decline of destalinization, and by October 1964 was replaced as the Soviet leader. However, despite Castro's arrogance with Russia during the 1962 missile crisis, Cuba and Russia maintained close relations following the event, largely because of Castro's fast friendship with Khrushchev, and his economic and military dependence on the Soviets. Khrushchev died in Moscow.

- L -

LA COUBRE. French freighter with a cargo of Belgian arms for the Cuban government that blew up in Havana harbor on March 4, 1960. Castro blamed the U.S. government for the explosion.

LABOR. Labor organizations in Cuba served as forums for the exchange of ideas between management--in this case the state-- and the workers. They were more than a vehicle for the implementation of governmental labor policy, although that was

their most important function. They conveyed to policymakers
the views of workers concerning proposed labor legislation.
They also provided voluntary labor, coordinated cultural events
and other benefits and services for union members, and encouraged workers to use educational facilities in the work place.
The most important organization of workers in Cuba was the
Confederation of Cuban Workers (Confederación de Trabajadores
Cubanos). Although the labor movement as a whole had little
to do with the armed struggle against Batista, a number of
labor leaders were involved in the underground, and these
leaders gained control of the movement after 1959. A survey of
worker attitudes conducted in 1962 indicated that most workers
identified strongly with the revolutionary government, although
only 29 percent of the sample had supported the Communist
Party before the Revolution. Jorge Risquet Valdés, the Minister of Labor, admitted in a 1970 speech that the government
had allowed the union structure to atrophy during the 1960's
because it had mistakenly assumed that in a revolutionary system the workers rights would be automatically protected and
promoted by the government. But, added Risquet, the existence of labor laws enacted by the revolutionary government did
not guarantee that workers knew what their rights were or that
employers (that is, managers of state enterprises) knew what
their obligations were toward workers. Thus the government
came to appreciate the need for a strong and active labor movement to provide the link between statutory guarantees and actual
practices. Risquet said that the first duty of the unions should
be to ensure that labor legislation be applied and workers' rights
protected. After more than a decade of eclipse, the Confederation of Cuban workers (CTC) was reconstituted in late 1970
through elections held in 37,047 local sections. These sections
elected 164,367 officials to represent the approximately 2.2 million members. Nationwide about 27 percent of those elected
were incumbent union officials, but in Havana 40 percent of
those elected were incumbents.

   Between 1970 and 1973 the national union structure was reorganized. Whereas some of the workers had been organized
by trade, after the reorganization all workers employed by a
central ministry or agency were unionized together, regardless
of their trades. The Thirteenth Congress of the CTC, held in
November 1973, was reportedly attended by 88 percent of all
workers in the state sector. The congress elected Lázaro Peña,
a founding member of the CTC, as secretary general for the
1973-1977 term. After his death in 1974 the office was assumed
by Roberto Veiga González. The CTC charter approved in 1973
stated that the CTC and the trade unions were neither party
organizations nor part of the state apparatus but were mass organizations with voluntary membership. Participants in the
congress reiterated the obligation of the labor movement to
protect the rights of workers, to monitor the enforcement of
labor legislation and work safety regulations, and to act on

the complaints of affiliates. It also listed among the movement's objectives: supporting the government, cooperating in improving managerial performance, strengthening labor discipline, and raising the political consciousness of workers. The evaluation of drafts of proposed labor legislation by the unions, a common practice since the early 1960's, was again endorsed, and detailed guidelines were provided as to how that should be carried out. It was also agreed that unions should participate in the administration of enterprises through "production assemblies" and "management councils" and, in his speech closing the congress, Castro suggested that the CTC should be represented at meetings of the Executive Committee of the Council of Ministers. The extent to which these policies and proposals had been implemented remained unclear to most observers in late 1985. The draft socialist constitution does not expressly recognize the right to strike. Chapter 6, entitled "Fundamental Rights, Obligations, and Guarantees," states "the rights to assembly, demonstration, and association are exercised by manual and intellectual workers in the city and in the countryside, and they have all the facilities they need to carry out those activities in which the members have full freedom of speech and opinion based on the unlimited right of initiative and criticism."

LABOR MOVEMENTS. The structure of labor in Cuba was altered radically as the 20th century dawned. The 18th- and 19th-century criollo aristocracy during sugar's golden age had been displaced by ambitious Spanish capitalist immigrants with their syndicalist ideas. The wars of independence and the challenge of North American technology and capital in turn affected the labor situation in Cuba in the early 1900's. Sugar workers remained at the bottom of the social scale. By the 1920's, the chief labor movement in Cuba was still led by the Spanish anarcho-syndicalists. They founded the Confederación Nacional Obrera Cubana (CNOC) in 1925, some of whose members simultaneously joined the Cuban Communist Party. By 1939, CNOC had disappeared and a new labor organization was established, the Confederación de Trabajadores de Cuba (CTC), directed by the Communists. This quickly became closely associated with General Fulgencio Batista's Ministry of Labor and in effect was a state-supported trade Union. The CTC was present at the Latin American Labor Congress in Mexico in September 1938, and joined the Confederación de Trabajadores de América Latina which was created there. The CTC was eclipsed by the 1959 revolution and Castro's Frente Obrero Nacional de Unidad, but was reconstituted in the 1970's through "elections" held at some 37,000 locals. Some 167,000 officials were "elected" to posts inside the CTC, representing the 2.2 million members. The structure of the CTC was reorganized and all workers were unionized together, regardless of their trades. Rights of workers, enforcement of labor legislation, and work safety regulations were reinstated. The CTC works closely with the government supporting and promoting policies developed by the Cuban Communist Party. The acronym CTC now stands for Central de Trabajadores de Cuba.

LAM, WIFREDO (1902-1982). Cuban painter born in Sagua La Grande, province of Las Villas, of Chinese and black parentage. His contact with African, Chinese, and Spanish cultures in his birthplace were elements recreated in his paintings such as La Jungla, where Western details combined with exotic and mystic tones of Africa and China. He studied in the Academy of Fine Arts of San Alejandro in Havana. Between 1920 and 1923 he held several exhibits in the capital where his works, Puente del castillo del príncipe, Mediodía, Quinta de los molinos, and others appeared. He left Cuba for the first time in 1926 when he went to Spain. There he signed up with the Fifth Regiment which fought in the Civil War in defense of the Spanish Republic. In 1938, he went to Paris where he stayed until the Second World War began, after which he returned to Cuba in 1942. While in Cuba he painted, among others, La jungla (1943), La silla (1943), and also took frequent trips to New York and Paris, where he finally stayed in 1952. In 1951 he won first prize in the National Exhibition of Havana and in 1953 he received the Lissone prize (Italy). After this date, he exhibited his works in Paris, Havana (1955), Venezuela, and Switzerland. In 1960 he established a study in Italy and stayed there indefinitely. His last visit to Cuba was in 1980. He died in Paris, on September 12, 1982, at the age of eighty. He is the only Cuban representative at New York's Latin American Art Exposition for 1987.

LAMAR, HORTENSIA (1888-1967). Born in Matanzas of an upper-middle-class family, she devoted her whole life to the defense of women and children. A good writer, she urged for equal political and social rights for women. She founded and directed the magazine La Mujer Moderna. She was also the Cuban delegate at the First International Feminist Congress in Chile in 1926 where she introduced a proposal on women suffrage that was approved unanimously. A beautiful and interesting woman, her lectures on "La mujer ciudadana" and "Las obras sociales de protección a la infancia," delivered at the Lyceum in 1929 and 1932, had a positive impact. She also wrote Cuentos and was honored in her native Matanzas in 1967 shortly before she died.

LAREDO BRU, FEDERICO (1875-1946). Lawyer and Cuban President from 1936 to 1940. Laredo Brú served as Colonel in the Cuban War for Independence or Spanish-American War (1895-1898). Later in 1923, Laredo Brú led an uprising of angry veterans in Las Villas against President Alfredo Zayas. Laredo Brú was elected Vice-President of Cuba in January 1936 while

Miguel Mariano Gómez was elected President. Gómez was impeached by a pro-Fulgencio Batista Senate and Laredo Brú was installed as the new figurehead President, with a cabinet composed totally of Batista supporters. From 1936 to 1940, Laredo Brú maintained a compliant stance regarding Batista's control, and signed a decree in 1938 whereby any newspaper could be suspended at the government's command. In 1939, Laredo Brú ensured that the Constituent Assembly would make Batista a Presidential Candidate in the 1940 elections. Before leaving office in July of 1940, Laredo Brú was consulted by U.S. Diplomatic Envoy Cordell Hull, who was in Cuba for the conference on the consequences of World War II for the Western Hemisphere. Hull hinted to Laredo Brú about the need to maintain orderly governments that fulfill their obligations. As a result of this interaction, Laredo Brú signed the Casanova Bill, granting the Cubans further U.S. loans with which to pay large debts. Laredo Brú died in Cuba in 1946.

LAS CASAS, FATHER BARTOLOME DE (1474-1466). Born in Seville, Spain, Father Bartolomé de Las Casas was a Dominican friar recognized as an early chronicler of Cuban history and as the "protector" of the Cuban Indians. Las Casas was one of the pioneer conquistadores of the island between 1511 and 1514. He would lead the Spaniards into Indian villages and succeed in convincing the natives to cooperate with the conquistadores. After witnessing the harsh subjugation and even massacres of the Indians by Governor Diego Velázquez' lieutenants, Las Casas became an outspoken critic of the Spanish conquest of Cuba. His History of the Destruction of the Indies, which condemned the Spaniards' treatment of the Indians, whom Las Casas argued were rational, free people entitled to retain their lands, received widespread attention in Europe. Las Casas disputed the conquistadores' idea that Spaniards were naturally superior and had divine authorization to use force to convert the natives to Christianity. His writings, some of which may have exaggerated accounts of Spanish brutality, gave rise to the "Black Legend" of Spanish cruelty. Toward the later years of his life, Las Casas became a strong advocate of importing black slaves to the Caribbean as a chief source of labor and as a way to preserve the fast-disappearing Indian population.

LAS CASAS, DON LUIS DE (1745-1807). Born in Sopuerto, Vizcaya, in Spain, Las Casas served as aide to his brother, the Spanish ambassador to Russia, and fought in several wars before going to Havana. He was the brother-in-law of Marshall Alejandro O'Reilly, fortifier of Havana, whom he accompanied in the field of battle. Las Casas became Captain-General (Governor) of Cuba in 1790. He acquired a sugar mill in the old tobacco zone of Güines, and had interests similar to those of the criollo plantation class. Together with Arango y Parreño, Las Casas founded the Sociedad Económica de Amigos del País, the Junta

de Fomento, and the Papel Periódico in 1792, three institutions
that represented the incipient stage of Cuban self-determination.
Their leaders were anxious to create a rich sugar colony through
these organizations. Las Casas died in Cuba, and is recognized
as a great promoter of Cuba's economic development, particularly the phenomenal growth of the sugar industry in the nineteenth century.

LAS TUNAS (PROVINCE). Population: 331,831; Area: 6,378 Km²;
Municipalities: Manatí, Jesús Menéndez, Las Tunas, Colombia,
Puerto Padre, Majibacoa, Jobabo, Amancio. Province in the new
political-administrative division, it is one of the eastern provinces. Before 1976, it was part of the provinces of Oriente
and Camagüey. Its capital city is Las Tunas (originally Victoria de Las Tunas). During the War of Independence of 1895-
1898, it was an important military station for the Spanish colonial forces and was attacked and taken by mambí troops under
the command of Major General Calixto García Iñiguez. Sugar
industry is the main economic activity.

LAWS. The legal system in colonial Cuba was characterized by a
lack of moral restraint and injustice toward both criollos and
poor black slaves. The town municipality and the cabildo system, as well as the audiencia, remained strong institutions well
into the 1800's under continued Spanish rule. Litigation in 19th-
century Cuba was an endless process due to confusion over
land titles after the collapse of the old system of circular grants
(mercedes). The law was neither clear nor ethical and many
verdicts based on bribes or graft. Judges were paid little and
by the hour, as were witnesses and court officials. Legal proceedings were long, drawn out, and usually conducted in writing, thus favoring the rich who could afford them and who were
literate. After the wars of independence and the Spanish-
American War (1898), the U.S. decided to reorganize Cuban
law. In 1899, the U.S. judge-advocate found that the Cuban
laws as modified prescribed correct remedies for injustices, but
that the court procedure and administration of the laws were
very poor. He also found that the court system was corrupt.
The U.S. revised the Spanish laws on taxation, issued decrees
on property, rents, past-due taxes and monetary claims, court
procedure, and appeals. A Supreme Court was established with
a President and six associate justices. Beneath this Court were
six audiencias or courts based on the six old provinces with
only Havana to divide criminal and civil jurisdictions. During
the second U.S. intervention (1906-1909), Charles Magoon, the
U.S. provisional governor, set up a commission to revise the
whole body of Cuban law which was still a confusing mélange of
Spanish statutes, laws of the first U.S. occupation, and laws
set up by the 1902 Constitution. Electoral, judicial, municipal,
and civil service law were all rewritten. A Supreme Court was
redefined and reconstituted. Electoral law was the most ambi-

tious reform offered, since a stable, democratic government was
what the U.S. most wanted for an independent Cuba.

Law in post-1959 revolutionary Cuba has undergone some
transformations. The law treats conventional crimes or misdemeanors normally, quickly, and relatively free of bribery, if
occasionally brutally or arbitrarily. For "political crimes" there
is no rule of law established save "revolutionary law"--which
defines these crimes as perpetrations against the revolution and
the state. The legal principle of habeas corpus has not existed
since 1959; Cuban citizens may therefore be detained and inter-
rogated by the government indefinitely without trial. Appeals
for political crimes do not exist. Under the revolutionary government of Fidel Castro, the Supreme Court, subject to the
jurisdiction of the Council of Ministers, directs the judiciary.
The judiciary is divided into provincial and regional courts
with original and appellate jurisdiction. People's courts deal
with minor offenses at the local level. All courts have some
lay judges. Laws and courts became an integral part of the
centralized governmental structure created by the leadership
of the 1959 Revolution. Crime, in a society geared to development, redistribution, and government control of the economy,
had been redefined to coincide with the evolving socialist nature
of the regime. Statutes and laws in Cuba have been intended
largely for the good of the collective or the state. This has
meant that individual liberties have been restricted and treated
arbitrarily. Such rights as habeas corpus, appeal, and self-
defense were for a time severely limited. A number of sources
have indicated that these individual rights had been reaffirmed
in principle since the early 1970's, but it was still too difficult
a decade later to ascertain whether these rights would be honored. Most of the laws that have been in force since 1959 have
dealt with the economy and individual responsibilities to society.
Laws such as the antiloafing law, which makes absenteeism a
crime, the law requiring individuals to inform on others who
commit a crime, or the law that requires individuals to register
with the Ministry of the Interior and carry an identification
card have enhanced the ability of the state to maintain public
order.

Since the Revolution an entirely new category of offenses
against the economy and the well-being of the state have been
designated crimes; these made up a majority of the crimes being prosecuted in the mid-1980's. There is considerable evidence that such crimes as homicide, prostitution, dealing in
drugs, and gambling have been drastically reduced. When
crimes of this nature have occurred, the penalty has been severe because they are considered bourgeois crimes carried over
from the days when Cuba was exploited not only by United
States commercial interests but also by United States criminal
interests. Minor crimes and misdemeanors--fights, drunkenness, slander, and minor thefts--have been prosecuted expedi-
tiously with considerable justice at the local levels by the

community-oriented popular tribunals. The zealousness of the
revolutionaries and the thoroughness with which mass organizations have involved the population in the establishment of a
new social order have also tended to reduce certain crimes by
making everyone a guardian of the Revolution and as such responsible for the conduct of others. In certain other areas,
however, "criminal" activity has increased as a result of regulations against, and a decrease in community acceptance of,
deviant behavior. The prudishness of the Revolution resulted
for a time in the persecution and prosecution of such non-
conformists as homosexuals, minor sex offenders, drunks, and
hippies--youths wearing long hair or miniskirts.

By the mid-1970's, however, long hair was commonplace, and
even highly placed female government officials were wearing
miniskirts. Intellectuals occasionally found that work not considered consistent with revolutionary ideals was punishable as
counterrevolutionary activity. The economic shortages and social tensions created by the government's push for rapid economic development and redistribution in the late 1960's resulted
in numerous crimes against state property ranging from minor
thefts to sabotage and arson. The strict rationing of the 1960's
resulted in a considerable black market, and shortages contributed to labor-related crimes of loafing and absenteeism. Juvenile
delinquency remained a very serious problem for authorities.
Since 1973 public order in Cuba has been further enhanced by
the institutionalization of the judicial system. The revolutionary
tribunals of the 1960's as well as their instantaneous justice and
arbitrary sentences have been eliminated. The contemporary
judicial system is structured hierarchically, having an appellate
procedure that culminates in the Supreme Court (Tribunal Supremo Popular). This court and the provincial courts, as the
highest local tribunals, hear all cases involving counterrevolutionary activity. Forming the base of the entire judicial system
are the people's courts (tribunales populares de base) which
are neighborhood-oriented and deal with almost all minor offenses by prescribing corrective measures appropriate to the
particular offense and giving lectures on the nature of what
constitutes good revolutionaries.

LAZO HERNANDEZ, ESTEBAN. Elected to the Cuban Politburo during the Third Party Congress, he is its youngest member. Born
on February 26, 1944, he was 15 years old at the time of revolutionary victory in 1959. He joined the militia in 1961 and the
Communist Party in 1963. He was named Provincial Delegate of
the Agricultural Ministry in 1979, after having worked in the
Department of Construction, Transport, and Communications.
Currently, he holds the position of First Secretary of the Provincial Committee of the Cuban Communist Party in Matanzas.

LECUONA, ERNESTO (1895-1963). A child prodigy who is known
to have played at age five. Famous pianist and composer, he

won national and international acclaim. He popularized Cuban and Afro-Cuban melodies throughout the world. He had piano recitals in North and South America, France, and Spain. As a pianist he had a superb technique and he was a composer of songs, boleros, operettas, and zarzuelas. He composed albums of over 200 songs and Hollywood used one of his compositions, "With a Song in My Heart," as the theme for the famous movie of the same title. His operetta María la O is the most famous of his compositions and others include El cafetal, El batey, Rosa la china, La flor del sitio. His songs "Siboney," "La comparsa," and "Malagueña" have attained universal recognition. He died in Spain, where he was exiled following the Revolution of 1959.

LEON, RUBEN DE (1907-1971). Born in Manzanillo, Oriente province, De León was a militant student leader of the Generation of 1930 and member of the Directorio Estudiantil who acted with other student leaders in 1933 to force the election of Ramón Grau San Martín as President. De León denounced strongman General Fulgencio Batista as a traitor to the revolution in 1934, when Batista forced Grau's resignation in favor of Carlos Hevia. De León was one of the founders of Grau's Partido Revolucionario Cubano and was elected President of the House of Representatives and Senator. He became Minister of Defense and Minister of the Interior under the government of Carlos Prío Socarrás (1948-52). De León opposed both the Batista and Fidel Castro regimes and lived mostly in exile beginning in 1952. He is the author of El origen del mal (1964).

LERSUNDI, FRANCISCO (1817-1874). Reactionary captain-general (1867-1869) sent to Cuba after the Junta de Información (Reform Commission) was disbanded. Lersundi prohibited public meetings and clamped tight political censure over reformist literature. Lersundi's iron despotism and harsh brand of administration greatly alienated and disillusioned many Cubans, contributing to mounting tension.

LEZAMA LIMA, JOSE (1910-1976). Poet, essayist, and lawyer. Born in Havana, as a young man he was attracted to writing. During his first years of study at the University of Havana, he participated in the founding and publication of several literary magazines (Verbum, 1937; Espuela de Plata, 1939; Orígenes, 1944). Possessing an extensive culture, he accumulated an immense vocabulary. His poetry is an attempt at creating a universal poetic system. He wrote a large amount of poetry grouped in volumes (Muerte de Narciso, 1937; La fijeza, 1949; Dador, 1960). His novel Paradiso (1966) is considered one of the most important "boom" novels of the 1960s in Latin America. The novel as well as his essays require a profound ability on the part of the reader to be able to read his intricate prose. His most important essays are Analecta del reloj

(1953), <u>Las eras imaginarias</u> (1971), and <u>La cantidad hechizada</u>
(1970). Despite being one of the most important novelists and
poets in Cuban literature, Lezama Lima spent his later years in
almost total isolation because of pressure from the authorities
over his lack of identification with the Communist regime. He
died on August 9, 1976, from pneumonia in the presence of his
wife and a couple of friends.

LIBERAL PARTY. <u>See</u> Partido Liberal.

LIBORIO. Symbolic nationalistic name for a Cuban, corresponding
to the United States' Uncle Sam or Great Britain's John Bull.
The personification of Liborio is that of a humble <u>guajiro</u> (peasant) with a straw hat and a <u>guayabera</u>.

LITERACY CAMPAIGN. Illiteracy in Cuba had long been a major
concern of responsible leaders. Rapid strides towards its
eradication had been made in the early years of the Republic,
and the percentage of literate Cubans increased from 43.2 in
1899 to 71.7 in 1931. Despite the political rhetoric that filled
the air during every election campaign, little progress was
made thereafter. The percentage of literates in 1953 was 76.4,
only a slight improvement over 1931. The problem was mainly
rural, for only 58.3 percent of the population in the countryside was literate in 1953. With power in his hands to do so,
Fidel Castro announced in the Fall of 1960 that the following
year would be the Year of Education. Teaching materials were
prepared, along with instruction manuals for literacy tutors.
In order to gauge the magnitude of the job, a census of illiterates was undertaken with the aid of teachers and volunteers.
The aim was to discover all the illiterates who were fourteen
years of age or older, especially those in the countryside. By
August 1961, a reported 985,000 illiterates had been located.
The regular school term was cut short in April 1961 and teachers as well as pupils in primary and secondary grades were
given two weeks of instruction before being sent out into the
country. Professional schoolteachers served mainly in technical and organizational positions.

The student volunteers were called <u>brigadistas</u> and usually
lived with the families they were assigned to instruct. They
were organized in units of twenty-five to fifty under a local
peasant leader, with a regular teacher to supervise their techniques of instruction. In addition to the brigadistas, thousands of literate adults volunteered to instruct illiterates in their
local neighborhoods. These volunteer teachers were called
<u>alfabetizadores populares</u>. A third group, the Patria o Muerte
Brigade, was set up in August, when it appeared to the Second
National Congress of Education that the work was not proceed-
ing as well as it should. This brigade was composed of 13,882
workers who were paid their regular salaries and sent into rural
areas to assist in the organization of the student brigadistas,

provide them with supplies, and generally help in instruction.
When the campaign ended on December 21, the regime was able
to report that of the 979,207 illiterates in the country, "707,212
adults have been taught to read and write." Although the government has repeatedly claimed that the illiteracy rate is less
than 3 percent, the accuracy of this number may be questioned.
For one thing, the figure for the total population of Cuba in
1961, used as a base of computation, was only an estimate. No
true census had been taken since 1953. One may further raise
questions about the completeness of the census of illiterates
taken by volunteers and the degree of literacy achieved by the
707,000 new literates, but there is no doubt that the massive
effort should have reduced illiteracy significantly.

LITERATURE. See Novel, the; Poetry; Theatre.

LOBO, JULIO (1898-1983). Powerful sugar baron in Oriente prov-
    ince whose sugar plantation, Niquero, served as a rendez-vous
    point for bands of Fidel Castro's forces. Lobo gave President
    Fulgencio Batista's (1952-58) opposition $50,000 in 1957, al-
    though he did not specifically offer the money to Castro him-
    self. At least $25,000 of this amount was employed by Castro-
    ist opposition leader Justo Carrillo to stage an uprising in
    Cienfuegos which was quickly dissolved by Batista's army. By
    1959 Lobo, as the most important sugar baron, had 14 ingenios
    (mills) totalling 1 million acres and possessed a reputation for
    honesty and good treatment of his workers. Lobo had been one
    of the first to try to mechanize the work of cutting cane, but
    his experimental machine was sent back to the United States by
    Cuban customs officials under Batista. He was also the most im-
    portant art collector in Cuba. His collection of Napoleon's relics
    is now on exhibit at Havana's Napoleonic museum. After Castro
    took over the government in 1959, Lobo suffered government
    confiscation of his lands and sugar mills. In that same year
    Lobo together with former Cuban President Ramón Grau San
    Martín refused the position of Minister of the Treasury, using
    the excuse that he had to attend his businesses. Later, Che
    Guevara, Castro's Economic Minister, extended Lobo the oppor-
    tunity to exchange his capitalist stripes for Communist ones by
    offering to make Lobo general manager of the Cuban sugar in-
    dustry under the revolutionary government. Lobo refused on
    October 11, 1960, and left for Miami on October 13, leaving be-
    hind his vast enterprises and holdings. He moved to Spain
    and lived in Madrid until his death in January 1983.

LOCATION AND SIZE. The Republic of Cuba consists of one large
    island and several small ones situated on the northern rim of
    the Caribbean Sea, about 100 miles south of Florida. It has an
    area of 44,218 square miles. About 760 miles long and with an
    average width of 50 to 60 miles, Cuba lies between 74°8' and
    84°57'W and 19°48' and 23°12'N. It is separated from Florida

by the Florida Strait, from the Bahamas and Jamaica by various channels, from Haiti by the Windward Passage, and from Mexico by the Yucatán Channel and the Gulf of Mexico. The coastline is about 2,170 miles long. The largest offshore island (1,182 square miles), the Isle of Youth (Isla de la Juventud) lies southward of the main island. Cuba is the largest island in the West Indies, accounting for more than one-half of West Indian land area. See also Geography.

LOPEZ, ANTONIO (1918-). Born in Havana, Tony López studied at Escuela Técnica Industrial where his father was a professor and where in 1936 he was appointed assistant professor of sculpture. He has been awarded several prizes, among them Círculo de Bellas Artes 1948 and 1956; Colegio Nacional de Arquitectos; Concurso de Artes Plásticas 1948 and 1950; Museo Nacional de Cuba; IV, VII and VIII Salón de Escultura y Pintura; II Bienal Hispano Americana; Acquisition Prize 1954. His works are in the permanent collection of the Palacio de Bellas Artes of Havana since 1955. He has also participated in several individual and group exhibits, among them Sociedad Lyceum, Havana, 1948; 16 Sculptors at the VII Pan American Architects Congress, Havana, 1950; Museo Nacional de Cuba, 1954 and 1956. In exile, he has participated in exhibits by Sculptors of Florida, Re-Encuentro Cubano, 1977 and 1978, and several others.

LOPEZ, NARCISO (1798-1851). Venezuelan by birth, he became a general in the Spanish army and later led filibustering expeditions against Spanish power in Cuba in 1850-1851. Considered by some historians as a precursor of Cuba's independence movement, he has been accused by others of attempting to annex Cuba as a slave state to the United States. López was born in Venezuela on September 13, 1798. At an early age, he joined the Spanish army fighting against Simón Bolívar, rising rapidly in the ranks. When Spanish troops withdrew from Venezuela to Cuba in 1823, López accompanied them and settled in the island. He married a sister of a high Spanish official, the Conde de Pozos Dulces. The marriage, however, soon broke up. An adventurer and warrior, López moved to Spain, where he served the crown against the Carlist rebels and was rewarded with high honors, being promoted to the rank of General. In 1841 he returned to Cuba, and during the administration of his personal friend, Captain-General Gerónimo Valdés, he occupied important posts, among them governor of the town of Trinidad. López also acted as President of a Military Tribunal, becoming notorious for the severity of the sentences he imposed on political dissenters. When a new captain-general was appointed, López lost his post. He turned to business, engaging in several unsuccessful ventures.

By then López had grown unhappy with Spanish rule. In 1848 he began to conspire with Cubans who advocated the annexation of the island to the United States. López' conspiracy,

known as "The Cuban Rose Mine" contemplated an uprising in
several parts of Cuba to coincide with the landing of an expedition of American allies. The scheme failed when Spanish authorities learned of the conspiracy and when the U.S. government,
at the time considering the purchase of Cuba from Spain, moved
against the expedition. Many of the conspirators in the island
were arrested and López fled to the United States. There he
resumed his conspiratorial activities and organized an expedition
with the support of southern leaders. In 1849 he sailed from
New Orleans with a force of 600 men, mostly American veterans
of the Mexican War, and landed in Cárdenas, Matanzas. The
flag carried by López was to be later adopted as the Cuban national emblem. The expeditionaries overwhelmed the small Spanish force and captured the town. But finding little support
from the population and faced with Spanish reinforcements,
López retreated and escaped to the United States. The untiring and daring Venezuelan sailed again in 1851 with over 400
men, mostly Southerners, some Hungarians and Germans, and
a few Cubans. He planned to join with conspirators inside the
island, but these rebelled prematurely and were rapidly annihilated. López landed in Pinar del Río in a desolate area far
from the uprisings. He found little support and was soon defeated and captured by the Spanish Army. As he was publicly
garroted in Havana on September 1, 1851, he insisted: "My
death will not change the destiny of Cuba." Historians still
disagree as to López' real objectives. While some point out
that he wanted the island's independence, others insist that he
desired Cuba's annexation to the United States. Perhaps he
wanted a free Cuba but one where slavery could be preserved.
Whatever his motivations, López' actions helped arouse anti-
Spanish sentiment in the island and paved the way for later
uprisings.

LOTTERY. See Gambling.

LOVEIRA CHIRINO, CARLOS (1882-1928). Railroad worker and
self-taught writer. Journalist. Loveira acquired such a cultural status that he was appointed a corresponding member of
the Real Academia Española and the Academia Nacional de Artes
y Letras. He was also a militant socialist and a brilliant orator.
Among his novels, the most famous are Los inmorales; Generales
y doctores; Los ciegos; Juan Criollo; and Ultima lección. On
workers' problems he wrote Lecciones de la experiencia en la
lucha obrera; La Confederación Panamericana del Trabajo; El
movimiento obrero de los Estados Unidos; and El Socialismo en
Yucatán. He was very young when he joined the efforts for
independence in Cuba and he was one of the expeditionaries of
General Lacret at the landing of Banes during the War of Independence.

LUCUMI. The largest and most influential African negro group in

Cuba, originally the Yoruba of Nigeria. See also African religious practices.

LUZ Y CABALLERO, JOSE DE LA (1800-1862). Educator. Born in Havana, Caballero was one of the three most prominent Cuban intellectuals of the mid-1800s, together with Father Félix Varela and José Antonio Saco. Caballero taught philosophy at the San Carlos Seminary, the most prominent institution of higher learning in Cuba at that time. A world traveller, Caballero visited Europe and the U.S. and became acquainted with Goethe, Walter Scott, Longfellow, Humboldt, Michele, and all important writers of the period. In Cuba, Caballero became a respected and beloved teacher at San Carlos as well as at the Colegio El Salvador. Through his lectures and writings he influenced an entire generation of Cubans with his strong nationalistic sentiment. In education Caballero was a supporter of the empirical scientific method. He wrote Texto de lecturas graduadas and Aformos. Caballero preached austerity, justice, and freedom, while strongly opposing slavery, and favored Cuban independence from Spain. Nevertheless, he returned to Cuba to defend himself of the charges that he had plotted in the Conspiración de la Escalera. His work Impugnación al examen de Consin sobre el ensayo del entendimiento humano de Locke is one of Caballero's best-known achievements.

- M -

MACEO, ANTONIO (1845-1896). Cuban mulatto patriot who rose to the rank of general in Cuba's independence army becoming a hero of the wars that ended Spanish domination over "the ever-faithful island" in 1898. Maceo was born in Santiago de Cuba, Oriente, on June 14, 1845. He was the son of Marcos Maceo, a Venezuelan mulatto émigré, and of a free Cuban black, Mariana Grajales, one of the outstanding women in Cuba's history. His brothers also played patriotic roles in the independence efforts. The young Maceo spent his early years on his father's small farm in Oriente, and received most of his education at home from private tutors. He also worked at his father's farm making occasional trips to Santiago de Cuba to sell agricultural products. The island was then experiencing revolutionary turmoil, as Cuban patriots conspired to rid themselves of Spanish control. Unhappy with Spanish domination and horrified by the exploitation of the black slaves, Maceo entered the Masonic Lodge of Santiago in 1864 and started to conspire with Cuban revolutionaries. When on October 19, 1868, Carlos Manuel de Céspedes and other leaders began Cuba's Ten Years' War, Maceo joined the rebellion. He soon showed superior ability in guerrilla fighting. Under the instructions of Máximo Gómez, a Dominican guerrilla

expert who had joined the Cuban forces, Maceo developed into
one of the most daring fighters of the Cuban army. Showing
extraordinary leadership and tactical capabilities, he defeated
the Spanish forces in numerous battles and was soon promoted
to the rank of captain. By January 1869 he was made lieutenant colonel. Maceo won respect and admiration from his men
as well as fear and scorn from the Spanish troops. He kept
tight discipline in his encampment, constantly planning and organizing future battles. Maceo enjoyed outsmarting and out-
maneuvering the Spanish generals and on successive occasions
he inflicted heavy losses on them. His incursions into the
sugar zones not only helped to disrupt the sugar harvest but
principally led to the freedom of the slaves who soon joined
the ranks of the Cuban army.

By 1872 Maceo had achieved the rank of general. His
prominent position among revolutionary leaders soon gave rise
to intrigue and suspicion. Conservative elements who supported
the war efforts began to fear the possibility of the establishment
of a Negro republic with Maceo at its head. The example of
Haiti still loomed in the minds of many, and when General Gómez
advocated an invasion of the West to cripple sugar production
and liberate the slaves, he met determined opposition. Maceo
was ordered to remain in Oriente province and the invasion of
the West had to be postponed until 1875. Even after the invasion got underway, it only reached Las Villas province in central Cuba. The destruction of the sugar estates increased the
opposition from landed and sugar interests. Supplies, weapons,
and money failed to arrive from exiles in the United States.
Dissension in the revolutionary ranks and fear of the Blacks
again slowed down the revolutionary efforts. After a prolonged
silence, Maceo finally answered those who accused him of attempting to establish a Black republic: "In planting these seeds of
distrust and dissension," he wrote on May 16, 1876, "they do
not seem to realize that it is the country that will suffer ... I
must protest energetically that neither now nor at any other
time am I to be regarded as an advocate of a Negro Republic....
This concept is a deadly thing to this democratic Republic which
is founded on the basis of liberty and fraternity." The war
dragged on with neither the Cubans nor Spaniards able to win
a decisive victory. Finally, on February 11, 1878, the Peace
of Zanjón was signed which ended the Ten Years' War. Most of
the generals of the Cuban army accepted the pact. Yet Maceo
refused to capitulate and continued to fight with his now depleted army. He held a historic meeting, known as the "Protest of Baraguá," with the head of the Spanish forces, Marshal
Arsenio Martínez Campos, requesting independence for Cuba
and complete abolition of slavery. When these two conditions
were rejected, he again resumed fighting. It was, however, a
futile effort. Years of bloodshed and war had left the Cuban
forces exhausted. Exile aid decreased and Maceo now faced the
bulk of the Spanish forces. Realizing the hopeless situation,

Maceo left for Jamaica. From there he traveled to New York to raise money and weapons necessary to continue fighting.

He soon joined the activities of Major General Calixto García, then organizing a new rebellion. This uprising, known as La Guerra Chiquita (Little War, 1879-80), ended in disaster. Maceo was retained in exile for fear of antagonizing the conservative elements in Cuba and García was captured soon after he landed on the island. Disappointed and disillusioned, Maceo traveled to the Dominican Republic and finally settled in Honduras. There he joined General Gómez and was appointed to an army post in Tegucigalpa. But Maceo saw his exile as only a temporary interruption in the struggle to liberate Cuba. He and Gómez soon began to organize a new rebellion. Maceo visited different exile centers in the United States seeking support. This movement, however, was doomed to failure. The leadership was totally in military hands, thus alienating such revolutionary leaders as José Martí. Then, weapons that were to be used for the uprising were either confiscated in the Dominican Republic or lost in Jamaica when the captain of the ship Morning Star, which was transporting weapons, fearful of being arrested, dumped them in the sea. Finally, dissensions, mistrust, and prejudice among the revolutionary leaders dealt a mortal blow to this new effort. For the next several years, Maceo wandered throughout the Caribbean and Central America. He lived in Panama, visited Santiago de Chile and Jamaica in 1890, and finally settled in Costa Rica where he engaged successfully in tobacco and sugar production. There he received a call from Martí, in 1893, for a final effort to liberate Cuba. Martí had organized a revolutionary party in exile and now offered Maceo an important position in the movement. Maceo joined Martí and Gómez in organizing the Cubans in and out of the island until finally on February 24, 1895, the War of Independence began.

One month later, Maceo and a group of expeditionaries landed in Oriente province to join the rebellion. Now Gómez and Maceo were able to implement their plan to invade the western provinces and carry the war to the other extreme of the island. The two generals and Martí met on Cuban soil to map the war strategy. Maceo advocated a strong military junta rather than civilian control to direct the effort. He argued that dissension and incompetence of the civilian government during the Ten Years' War had led to the failure of the rebellion. Although the question of civilian versus military control was not resolved, Gómez was made Commander-in-Chief of the army, Maceo military commander of Oriente, and Martí head of the revolution abroad and in non-military matters. Martí's tragic death only days after the meeting, on May 19, 1895, dealt a strong blow to the morale of the Cuban forces. Yet Maceo and Gómez did not waver. In repeated attacks the two generals undermined and defeated the Spanish troops. For the next three months Maceo and Gómez carried the war to the western provinces.

From January to March 1896 Maceo waged a bitter but successful campaign against larger Spanish forces in the provinces of
Pinar del Río and Havana. On December 7, 1896, while preparing their next campaign, near the small town of San Pedro del
Cacahual in Havana, Maceo's troops were attacked and the
courageous general was killed in a minor battle against Spanish
forces.

MACHADO VENTURA, JOSE RAMON (1930-   ). Machado was born
in Las Villas province in 1930. He graduated from the University of Havana's Medical School in 1954. During the following
year he became involved in revolutionary activity against the
regime of Fulgencio Batista and eventually joined the rebel
forces of the 26th of July Movement, serving under Raúl Castro's Second Front in northern Orient province. While assigned
as Raúl's chief of military sanitation, Machado obtained the rank
of major. Following the collapse of the Batista forces in January 1959, Machado served for a year as director of medical services for the city of Havana. He served as Minister of Public
Health during 1960-68. He was appointed to the Central Committee of the Cuban Communist Party (PCC) in October 1965, to
its Politburo in December 1975, and to its Secretariat in November 1976. Machado served as the Politburo's delegate in Matanzas province during 1968-71. He was PCC First Secretary of
Havana from 1971 to 1976 when he was assigned to the PCC Secretariat. Since November 1976 he has been a deputy to the National Assembly representing the municipality of Playa in the
City of Havana province. Machado has been a member of the
Council of State since December 1976. Today Machado is the
Political Bureau member in charge of Communist Party organization. During crises, Machado sets up the massive pro-
government demonstrations in the Plaza de la Revolución. He
has helped build Cuba's much-praised medical system.

MACHADO Y MORALES, GERARDO (1871-1937). Fifth President of
Cuba. A General of Cuba's War of Independence (1895-1898)
against Spain and later an elected President, he developed into
a harsh dictator. Machado was born in Santa Clara, Las Villas
province, September 28, 1871. He spent his childhood in his
family's cattle estate, attended private schools and in his early
twenties engaged in growing and selling tobacco. During Cuba's
Ten Years' War (1868-1878) against Spain, Machado's father had
joined the Cuban rebels, attaining the rank of major. Machado
followed in his father's steps, and when the Cubans resumed
the War in 1895 he enrolled, rising to the rank of brigadier
general. After the war ended, Machado turned to politics and
business. He became mayor of Santa Clara and during José
Miguel Gómez' administration (1909-1913) was appointed Inspector of the Armed Forces and later Secretary of Interior. Soon
after he engaged in farming and in business, and together with
American capitalists invested in public utilities. He grew wealthy,

returning to politics in the early 1920's. He won control of the
Liberal Party and, with his slogan "water, roads, and schools,"
was elected president in 1924. Machado's first administration
coincided with a period of prosperity. Sugar production expanded and the U.S. provided a close and ready market. Machado embarked on an ambitious public works program which
included the completion of the Central Highway, the construction of the National Capitol, the enlargement of the University
of Havana, and the expansion of health facilities. He also sponsored a tariff reform bill in 1927 providing protection to certain
Cuban industries.

Despite these accomplishments, Cuba's dependence on sugar
continued and U.S. influence and investments increased. Before his first administration ended, Machado sought reelection.
Claiming that his economic program could not be completed within his four-year term and that only he could carry it out, Machado announced his decision to have himself reelected and to
extend the presidential term to six years. He prevented the
growth of political opposition by controlling the Conservative
Party and small Popular Party. Through bribes and threats he
subordinated Congress and the Judiciary to the Executive will
and in 1928 he was reelected over virtually no opposition. Machado's second term was wrought with problems. Affected by
the shock waves of the world depression and oppressed by an
increasingly ruthless dictator, many Cubans, led primarily by
university students, organized resistance to the regime. In
1931 former President Mario G. Menocal led a short-lived uprising in Pinar del Río province. That same year an anti-Machado
expedition landed in Oriente province, only to be crushed by
the army. One of the most prominent of these opposition groups,
the ABC, a clandestine organization composed of intellectuals,
students, and the middle sectors of society, undermined Machado's position through sabotage and terrorism. As urban
violence increased, so did repression. Machado's police raided
secret meeting places, arresting students and ABC leaders whom
they tortured or killed. This was the existing condition in the
island when the United States, attempting to find a peaceful
solution to Cuba's political situation, sent Special Envoy Benjamin Sumner Welles to mediate between government and opposition. The mediation was supported by most political factions
and leaders with the exception of the Conservatives, and particularly the students. Sumner Welles' efforts finally led to a
general strike and an army revolt which forced Machado to
leave the country on August 12, 1933. Machado settled in the
United States and died in Miami Beach on March 29, 1939.

McKINLEY, WILLIAM (1834-1901). Elected President in 1896 in the
midst of the Second Cuban Independence War (1895-1898), Mc-
Kinley wanted to preserve peace between the U.S. and Spain if
possible. In early 1898, McKinley dispatched the USS Maine to
Havana from Key West to show his determination to protect U.S.

property and citizens in Cuba. After it was sunk in Havana harbor by unknown persons, McKinley tried vainly to resist war with Spain even offering to buy Cuba from Spain for $300 million. On April 11, 1898, he asked Congress for a declaration of war on Spain. Earlier, as a Congressman in 1890, McKinley had sponsored a tariff on sugar that carried his name. Raw sugar was placed on the free list leading to expansion of U.S.-Cuban trade of sugar.

McKINLEY TARIFF (1890). After the Ten Years' War for Independence (1868-1878), American investment, particularly in Cuban sugar, grew. Taking advantage of the bankruptcy of many Spanish and Cuban enterprises, U.S. capital acquired sugar estates. European sugar beet expansion in the 1880's hurt this new market and depressed the world price of raw sugar. This depression in price ruined many Cuban producers and facilitated U.S. economic penetration. The McKinley Tariff in 1890 placed raw sugar on the free list and led to increased U.S.-Cuban sugar trade and the expansion of sugar production.

MAGAZINES. Magazines in pre- and post-revolutionary Cuba have been important sources of information and entertainment for the general population. One of the most popular weekly news magazines, Bohemia, edited by liberal journalist Miguel Angel Quevedo, had a circulation of 250,000 in the 1950's. Founded in Havana in 1908, Bohemia was a consistent critic of Fulgencio Batista (1952-58) and survived the revolution. Bohemia's survival was due primarily to its portrayal of Fidel Castro during the struggle to overthrow Batista as a true Cuban hero molded after the U.S.'s Thomas Jefferson. In volume, critical acclaim, and international interest and impact, poetry continued to dominate the literary scene. Most of it was generated and published in Havana, frequently in the magazine Orígenes (1944-57). This magazine published the major literary figures of the day and its circulation and quality made its associates the most important literary group of the period. In fact, until the Revolution, Orígenes' thirteen years of uninterrupted publication was a record for Cuban literary magazines. Between 1937 and 1938 the ten major poets of the Orígenes group published over thirty books of poetry. Poets of the Orígenes group were considered evasive and elitist and the magazine reflected the sophistication and lack of social concern expressed by certain Cuban intellectuals. Revista de Avance (Forward Review) was another literary magazine founded in pre-revolutionary Cuba by Jorge Mañach in the year 1927. Avance articulated the concern of the new generation of poet-politicians. Many of this group, originally formed around the magazine Avance and possibly reacting against the color prejudices of Cuban society, became Marxists. Nicolas Guillén, one of Cuba's most prominent poets and allied with Avance, is credited with originating the negrismo (negritude) school in poetry. Socially oriented poets such as

Guillén and José Tallet associated themselves with other activist publications, such as the magazine Ciclón (1955-59), which was started by a cofounder of Orígenes, and Nuestro Tiempo, published from 1951 to 1959 by a cultural society of the same name. A popular news and information national weekly established in 1924 was the magazine Carteles, also popular in the pre-revolutionary period.

In the post-revolutionary period, a number of new magazines were established, all of them reflecting the government's ideology of the Revolution. The literary magazine El Caimán Barbudo published the material of the second generation of post-revolutionary writers who tended to adopt a middle-of-the-road stance. The Caimán Barbudo group tried to adhere to Castro's 1961 maxim by claiming not to write for or against the Revolution but from within it. Nevertheless, this group has also been involved in several severe conflicts with the government including the arrest of the poet Heberto Padilla in 1967 and 1968, which eventually led to the resignation of the entire El Caimán Barbudo editorial board and to the temporary suspension of its publication. Among the periodicals designed specifically to serve the government's purposes was Verde Olivo, the organ of the Revolutionary Armed Forces, and INRA (later replaced by Cuba), a glossy, general-interest magazine published by the National Institute of Agrarian Reform (Instituto Nacional de Reforma Agraria--INRA). The third important periodical of the early 1960's was Bohemia. In the mid-1970's numerous periodicals devoted to a wide range of interests were available. Nevertheless, many magazines ceased publication, including those that had previously been successful, due in large part to political circumstances. Among the periodicals that closed were many covering economic subjects, some of which were the official organs of various groups. They included: Comercio Exterior, published until 1966 by the Ministry of Foreign Trade; Vanguardia Obrera and Trabajo--closed in 1966 and 1968 respectively--both published by the Confederation of Cuban Workers; Agro, published until 1966 by the agricultural trade unions; and Teoría y Práctica, edited by the shools for revolutionary instruction until 1967. Cuba Socialista, a very popular periodical according to a study done in 1964, ceased publication in 1967. The reason offered for this action was that Cubans had not had sufficient political education and that debate on the subjects covered in the monthly should be postponed. Cuba Socialista carried news of the PCC and published analyses of political, economic, and social conditions. It was replaced in 1967 by Pensamiento Crítico, edited by the Department of Philosophy of the University of Havana.

General interest magazines were among the most popular in the 1980's. Bohemia, Cuba, and Verde Olivo contained a variety of news and politically oriented feature articles in addition to human interest material. Verde Olivo also carried military information. Cuba was published in Russian as well as in Spanish,

and issues were sent to the Soviet Union in addition to being
sold to the resident Soviet community. The monthly publication URSS, covering aspects of Soviet life, was first put out by
the embassy of the Soviet Union in 1961 and by the 1970's had
achieved considerable popularity. Other Soviet publications
were available, but they were not as well received. Mujeres
and Romances were women's magazines. Mujeres, published by
the Federation of Cuban Women, carried articles on women's
place in the Revolution under such sections as "Women in Agriculture" as well as stories about fashion, cooking, and other
traditional female pursuits. In addition to being sold, Mujeres
was distributed free by the federation. Palante, a satirical
weekly founded in 1961, also enjoyed considerable success.
Casa de Las Américas, a bi-monthly magazine published by the
organization of the same name and edited by the poet Roberto
Fernández Retamar, was considered one of the best publications
in Latin America. Funded by contributions from Cuban and
other Latin American writers, it featured articles on literature
and the social sciences. Numerous other periodicals, less widely
circulated, were also published, many for smaller specialized
readerships. Among these were Revista de Agricultura, In-
geniería Civil, Mar y Pesca, Boletín de Higiene y Epidemiología,
Signos (a cultural magazine founded in 1969), Cine Cubano (a
film review), and Gazeta de Cuba and Unión, literary journals
published by the National Union of Cuban Writers and Artists
(Unión Nacional de Escritores y Artistas Cubanos--UNEAC).
The University of Havana published Universidad de la Habana,
which contained historical articles. Periodicals of international
political groups were also published in Cuba, among them Tri-
Continental, put out by the Afro-Asian-Latin American People's
Solidarity Organization and OCLAE, published by the Continental
Organization of Latin American Students.

MAGOON, CHARLES E. (1861-1920). United States provisional gov-
    ernor in Cuba during the second U.S. intervention (1906-1909).
    Criticized as avaricious and unprincipled by many Cuban writ-
    ers, Magoon, a civilian, had the task of pacifying the various
    political factions. Magoon undertook the using of botellas, or
    government sinecures. At the same time, he embarked on an
    extensive public works program, gave Havana a new sewage
    system and established the outlines of a responsible civil ser-
    vice bureaucracy for Cuba. Magoon's electoral reform allowing
    for new elections (and thus U.S. withdrawal) was ambitious as
    was his reorganization of a modern Cuban Army. Although
    somewhat incomplete, these contributions to Cuban development
    were mostly positive.

MAINE, USS. The U.S. battleship sent to Key West, Florida,
    in 1898, as a symbol of strong U.S. naval force ready to counter
    any anti-American conspiracies in Havana. With the approval of
    the Spanish, the Maine arrived in Havana in January 1898, and

on the evening of February 15, 1898, was blown up in the harbor, killing 260 of the 355 Americans on board. Responsibility for the sinking of the Maine was not conclusively established, although most Americans believed the Spanish to be the perpetrators. An alternative theory was that Cuban revolutionaries who sought U.S. help, and American business interests also seeking a U.S. declaration of war on Spain and subsequent annexation of Cuba sabotaged the Maine. The explanation which seems most likely, however, is that the Maine's gunpowder ignited and caused the tremendous explosion which in the end caused the McKinley Administration to declare war on Spain. Yellow journalism fanned U.S. popular feeling after the Maine sinking and intense pressure later forced U.S. President McKinley's hand, despite the lack of conclusive proof as to Spanish responsibility.

MAMBISES. Plural for mambí or Cuban rebel joining the Cuban independence effort during the Ten Years' War (1868-1878) and later during the Independence War of 1895. These Creole mambises were regarded as symbols of unselfish sacrifice for Cuba, having abandoned position and comfort to fight Spanish power. The mambises also urged the black slaves to revolt and to join the cause for independence of Cuba.

MAÑACH, JORGE (1898-1961). One of the island's most distinguished writers and intellectuals, who criticized the Cuban people in his writings for the civic indolence and passive acceptance of U.S. tutelage over political and economic life. In 1923 he joined in the Protesta de los Trece, denouncing graft and corruption in public life. A Harvard graduate, he joined the ABC, a middle-class clandestine group opposing Gerardo Machado's dictatorship (1924-1933). Mañach drafted the group's program calling for profound structural changes in Cuban society. In 1947 he supported populist leader Eduardo Chibás in founding the Partido del Pueblo Cubano (Ortodoxo). Following Chibás' death, Mañach broke with the party. In the 1950s he opposed the dictatorship of Fulgencio Batista (1952-1958) and criticized the U.S. for supporting Batista. In 1955 he headed a group calling for Constitutional reform and urged the Diálogo Cívico, a meeting between Batista and opposition leaders trying to find a compromise and prevent the escalating violence then taking place. The Diálogo failed. Mañach was above all a writer and humanist much influenced in his style by the Spaniard José Ortega y Gassett. Some of his works include Indagación al choteo (1928); Historia y estilo (1944); Hacia una filosofía de la vida (1951); Estampas de San Cristóbal (1925); La pintura en Cuba (1924); El pensamiento político y social de Martí (1941); Examen del Quijotismo (1950); Imagen de Ortega y Gassett (1956); Dualidad y síntesis de Ortega (1957).

MANIGUA. Term for the open country or remote rural areas where

Cuban mambises initiated the Wars of Independence against
Spain.

MARINELLO, JUAN (1898-1977). Studied Law at the University of
Havana and became Doctor of Civil and Public Law. In 1923
he participated in the so-called Protest of the Academy or Pro-
test of the Thirteen and was elected first Vice-President of the
Falange de Acción Cubana. Marinello was arrested for his ac-
tion there in May 1923. Marinello edited the magazine Venezuela
Libre, which became América Libre. He was elected Deputy of
the School of Lawyers of Havana and designated member of the
National Code Commission. He was a founder of the Hispanic-
American Cultural Institute, of which he was Vice-President.
In 1927 he married María Josefa Vidaurreta. Marinello published
his first book of poetry, Liberación, in Madrid. Later, he became Professor at the Institute of Modern Languages at the
University of Havana, where he participated in the student
demonstrations against the Gerardo Machado regime (1924-1933)
and was detained by police awaiting trial for two months in the
Castillo del Príncipe. There he published his essay Sobre la
inquietud cubana and joined the directorate of Revista Política.
He continued his struggle against Machado and was sentenced
to six months in prison. Marinello went to Mexico as a political
exile until the downfall of Machado, but was exiled a second
time shortly after returning due to his role as director of
Diario La Palabra.
    In 1937 he participated in the First Congress of Revolutionary
Artists and writers of Mexico. By then he was an avowed Marxist and had joined the Partido Socialista Popular (PSP), Cuba's
Communist Party. In 1938-1948 he was designated to represent
the Party as candidate to the mayorship of Havana. By 1957
he became head of the Partido Socialista Popular, articulating
the Cuban Communist Party's disapproval of barracks assaults
as a method to dislodge the Batista regime (1952-1958). Marinello described the PSP's position advocating the mass struggle
and mobilization of the proletariat, both purportedly leading
toward national elections. In 1962, Marinello was named Rector
of the University of Havana. One year later, Marinello, considered an "Old Guard" PSP leader and thus a potential threat
to Fidel Castro, was replaced. Between 1962 and 1977, Marinello
was the recipient of many awards and honors. He received the
investiture of Doctor Honoris Causa in Philosophical Sciences
at the University of Carolina in Czechoslovakia; he was Cuba's
Ambassador and Permanent Delegate before UNESCO; he received
the title of Professor Emeritus at the University of Havana,
and the May First distinction instituted by the Confederation of
Cuban Workers, in recognition of his life's work and struggle
for socialism. Marinello is the author of Literatura Hispano-
americana, Maceo: líder y masa, and Momento Español.

MARQUEZ-STERLING, MANUEL (1872-1934). Born in the Cuban

Legation in Lima, Peru, of distinguished parents from a
Camagüey family. Journalist, diplomat, and author, Márquez-
Sterling was secretary to Cuban patriot Gonzalo de Quesada in
Washington in 1900 during the negotiations for establishment of
the Cuban Republic. In the same year, he was sent to the Paris
Exposition by Cuba, and afterwards devoted himself to journalism
until 1907 when he was sent to Buenos Aires as Chargé d'Af-
faires. He became Resident Minister in Rio de Janeiro in 1909;
Minister Plenipotentiary in Peru, 1911; and Minister Plenipotentiary in Mexico in 1912. He was in Mexico during the outbreak
of the Revolution and failed in an attempt to save the life of
President Francisco Madero. Retired from career diplomacy,
Sterling founded the journal Heraldo de Cuba, which achieved
success, and later founded the organ La Nación, supporting the
Liberal Party. He became Ramón Grau San Martín's Secretary
for Foreign Affairs in 1933. Following Grau's overthrow,
Márquez-Sterling was asked by supporters to accept the Presidency but this was not endorsed by strongman Fulgencio Batista
and Carlos Mendieta became President. Márquez-Sterling wrote
many books. Some of the best known are Ideas y sensaciones;
Hombres de pro; Alrededor de nuestra psicología; Los últimos
días del Presidente Madero; and Las conferencias del Shoreham:
El cesarismo en Cuba.

MARQUEZ-STERLING Y GUIRAL, CARLOS (1898-). Born in
Camagüey on September 8, 1898. He graduated in Public Law
and Philosophy and Letters from the University of Havana and
taught the same area studies at the School of Business Administration from 1937 to 1959. He was Director of the Panamerican
Office in Cuba with the rank of Minister (1929-1939) and president of the Cuban delegation to the IV Panamerican Congress in
Washington in 1931. (In 1942 he founded the Professional
School of Journalism Manuel Márquez-Sterling.) In Cuban public life he held a seat in the House of Representatives, was
president of the Constitutional Assembly of 1940, Minister of State
(1941-42), and Minister of Education (1942-43). In November
1958, he ran for president against Andrés Rivero Agüero (Batista's sponsored candidate) and former President Grau San Martín. Both Márquez-Sterling and Grau asked the Supreme Court
to annul the rigged elections. A member of the Academy of History of Cuba, his intellectual production includes biographies,
historical works, essays, and journalistic articles. Among his
works about Cuban patriots are Agramonte: El bayardo de la
Revolución and Nueva y humana visión de Martí (one of various
works on José Martí). Once out of Cuba he continued as a historian with such books as Historia de Cuba desde Cristóbal
Colón a Fidel Castro (1963-1969); Martí, ciudadano de América
(1965); Historia de los Estados Unidos de Norteamérica (1983).
His journalistic production is endless. He went into exile in
1959, taking up residence first in New York, where he was
Professor at Columbia University (1962-64) and at C.W. Post

College (1964-1979). He now lives in Miami where he continues to write and lecture.

MARQUITO AFFAIR. Refers to the trial in March 1964 of Marco Armando Rodríguez ("Marquito"), a Communist informer who had denounced the leading students involved in the attack on the Presidential Palace on March 13, 1957, during the presidency of Fulgencio Batista. Marquito's trial was an episode in the power struggle raging between Fidel Castro and old-guard Communists throughout 1961 and part of 1962 which revealed Communist maneuvers in the University of Havana student body. After exposing the student Directorio leaders, who were assassinated by Batista's policy in a Humboldt Street apartment on April 20, 1957, Marquito fled to Mexico where he received aid and protection from Joaquín Ordoqui and his wife, two leading Partido Socialista Popular (PSP) members. Ordoqui had obtained a scholarship for Marquito to study in Czechoslovakia and had helped him become a member of Juventud Socialista, the youth wing of the PSP. In 1961, at the insistence of the Directorio, which had discovered Marquito's guilt, Castro requested his arrest and extradition from Prague to Havana. Deported shortly thereafter, Marquito remained in jail in Cuba until his trial in March 1964.

   The trial--especially Marquito's testimony--shed light on an important and previously unclear matter: the tactics of the PSP in dealing with the student movement during 1955-57. Finally, Marquito explained the refusal of the Directorio to allow his participation in the plans for the Presidential Palace attack in 1957 and admitted having revealed the students' hideout to Batista's police. Castro used the Marquito Affair to reassert his own power. He tried Marquito and took advantage of the opportunity to put the Communist PSP on the bench with the accused. All the top leaders of the PSP testified; they took pains to deny that Marquito had been a member of Juventud Socialista or had been connected with their party. Toward the end of the trial--perhaps thinking of his economic dependence on the Soviet Union and fearful of provoking Moscow's wrath-- Fidel exonerated the PSP from guilt in the Humboldt Street killing and ordered Marquito's execution. Whether Marquito had acted on his own initiative or followed the directives of the party when he betrayed the four students is still unclear. The trial did, however, reveal that the Communists had used Marquito in their efforts to undermine the non-Communists revolutionary forces, especially the Directorio.

MARRERO Y ARTILES, LEVI (1911-   ). Born in Santa Clara, former Province of Las Villas, he is Cuba's most distinguished living historian and geographer. Professor of Economic History of Cuba at the University of Havana, until he was exiled in 1960, Leví Marrero's Geografía de Cuba and La tierra y sus recursos were required textbooks in Cuban secondary schools. He taught

in Venezuela until 1965; he left that year for the University of
Puerto Rico, where he taught until his retirement in 1972.
Since then, he has devoted most of his writing time to Cuba:
economía y sociedad, a monumental history of Cuba, still un-
finished, with twelve volumes published by 1986.

MARTI, JOSE (1853-1895). Cuba's greatest hero and most influential writer. Revolutionist, poet, journalist, and the principal
organizer of Cuba's war against Spain, Martí was the apostle
of Cuba's independence. Born in Havana, January 28, 1853, of
Spanish parents, Martí spent his early years as an eager student. His environment and teachers aroused in him a devotion
to the cause of freedom. He enrolled at the Instituto de Segunda
Enseñanza de La Habana but was soon arrested for political reasons. After serving several months of hard labor, he was deported to Spain in January 1871. By then Martí was already
achieving recognition as a writer. At the age of 15 he had composed several poems, and at 16 he published in Havana a newspaper, La Patria Libre, and wrote a dramatic poem, Abdala.
In Spain, Martí published a political essay, El presidio político
en Cuba, an indictment of Spanish oppression and conditions in
Cuban jails. In Spain, the young revolutionary resumed his
studies. In 1874 he received a degree in philosophy and law
from the University of Zaragoza. From Spain, Martí traveled
through Europe and in 1875 went to Mexico, where he worked
as a journalist. After a short visit to Cuba in 1877 he settled
in Guatemala, where he taught literature and philosophy. That
same year he married Carmen Zayas Bazán, daughter of a Cuban
exile, publishing shortly afterwards his first book, Guatemala.
Unhappy with Guatemala's political conditions, Martí returned to
Cuba in December 1878. The Peace of Zanjón, which ended the
Ten Years' War (1868-1878) against Spain, had just been signed,
and Martí felt that conditions in the island would be propitious
for his return. Spanish authorities, however, soon discovered
his revolutionary activities and again deported him to Spain.
He escaped to France and from there moved to the United States
and Venezuela.
     Finally in 1881 he made New York the center of his activities,
although he continued to travel and write about the many problems of Latin American nations. He wrote a regular column for
La Opinión Nacional of Caracas and for La Nación of Buenos
Aires, gaining recognition throughout Latin America. Not only
his journalistic articles, but also his poetry and prose became
popular. Martí became a precursor of the Modernist movement
in literature. In 1882 his most significant poems, composed for
his son, were published in a book called Ismaelillo. Martí's
best known poems are his Versos sencillos (1891), which emphasize such themes as friendship, sincerity, love, justice, and
freedom. Martí also won the hearts of many Latin American
youngsters with his Edad de oro (1889), a magazine especially
devoted to children. Martí's greatest contribution to Spanish-

American letters were his essays. Written in a highly personal
style, the Modernist renovation of language characterized his
writing and marked the beginning of the new Latin American
prose.

Martí realized very early that independence from Spain was
the only solution for Cuba and that this could only be achieved
through a fast war that would at the same time prevent United
States intervention in Cuba. Martí's fear of a military dictatorship after independence led in 1884 to a break with Máximo Gómez
and Antonio Maceo, two generals who were at the time engaged
in conspiratorial activities. Martí withdrew from the movement
temporarily, but by 1887 the three men were working together
with Martí assuming political leadership. In 1892 Martí formed
the Partido Revolucionario Cubano in the United States and directed his efforts toward organizing the war against Spain.
What distinguished Martí was his ability to organize and harmonize. His oratory inspired his listeners, his honesty and
sincerity inspired faith, and his conviction in the ideas he was
pursuing gained for him respect and loyalty. His writings were
not mere rhetorical exercises but moral teachings aimed at making a better man. Martí's importance transcended Cuba. Like
Bolívar he thought in terms of a continent and advocated the
unity of Latin America. His writings and ideas had impact not
only in his homeland but throughout Latin America. When in
1895 he gave the order for the resumption of hostilities against
Spain, Martí landed in Cuba to lead the war. Shortly after,
however, on May 19, 1895, he was killed in a skirmish with
Spanish troops at Dos Ríos, Oriente.

MARTINEZ CAMPOS, ARSENIO (1831-1900). Spanish general in
charge of royal forces in Cuba from 1874 to 1878, who vigorously prosecuted the end of the Ten Years' War in 1878. Martínez Campos negotiated with dissident rebel General Antonio
Maceo in March 1878, after Maceo refused to accept the terms
of the Zanjón Peace. After the Protest of Baraguá, as the
negotiations were labeled, fighting continued until May, when
Maceo capitulated. Martínez Campos became captain-general
(governor) of Cuba and pressed for a program of rights for
Cubans still under his authority. The program included political equality for Cubans and Spaniards, guarantees of freedom of speech, press, and assembly, and freedom for the
slaves. The program was not enacted by the Spanish government. During the War of 1895, Martínez Campos returned to
Cuba again as Commander-in-Chief, and Captain-General and
soon recognized the depth of the rebel resistance. He hoped
for a negotiated settlement with reforms granted to the Cubans.
When this was not agreed to by the Spanish Prime Minister,
Martínez Campos resigned in January 1896. He was succeeded
by General Valeriano Weyler. See also Moret Law.

MARTINEZ SANCHEZ, AUGUSTO (?-). Lawyer appointed to the

first cabinet after Fidel Castro's revolution in January 1959. As
Minister of Labor and with the hold of being in Fidel Castro's
very good graces he was able to keep his job through 1963 without being reshuffled. The convincing proof of this was that
when Fidel Castro left for his highly publicized visit to the
United States in 1959 it was Martínez Sánchez, not Che Guevara,
who was designated to act as prime minister. Martínez Sánchez'
key insider status became most visible on that fateful day of
July 1959 when Castro was deposing Urrutia, the president.
It was he who was put in charge of quarantining the entire
cabinet in the council chamber of the presidential palace. But
Martínez Sánchez' demise was as spectacular as his rise to
prominence. Apparently beset by personal problems, he attempted suicide in December 1964. This was most displeasing
to Fidel Castro and resulted in his political decline. A communiqué issued by Castro and president Dorticós stated: "According to fundamental revolutionary principles, we think that this
conduct is unjustifiable and improper for a revolutionary, and
we believe that comrade Augusto Martínez Sánchez could not
have been fully conscious when he engaged in such a deed, because every revolutionary knows that he does not have a right
to deprive his cause of a life which does not belong to him, and
which can only be legitimately sacrificed facing the enemy."

MARTINEZ VILLENA, RUBEN (1899-1934). While still in his early
twenties Martínez showed an interest in political and social issues. In 1923, he denounced the government's illegal sale of a
convent in the city of Santa Clara. A short time later, he co-
founded the Grupo Minorista, an association of young men concerned with reforms. Martínez also became associated with the
Anti-Clergy League whose aim was to eradicate Catholic influence
in Cuba. At that time he began to emerge as a poet of quality
with anti-American inclinations. Martínez was for a while secretary to the famous intellectual Fernando Ortiz. In 1923, he
was jailed for his participation in the Veteran's Revolt against
President Alfredo Zayas (1921-1925). He was also an active
opponent of President Gerardo Machado (1925-1933), whom he
personally insulted. He joined the Communist Party in 1927 and
by 1929 was its main leader. Martínez was a charismatic man at-
tractive to the Communist intellectuals and young people. In
1930, Martínez traveled to the Soviet Union to recover from a
bout of tuberculosis. He returned secretly to Cuba in May
1933 and became involved in the struggle against Machado.
During the last few days of Machado's rule, in August of 1933,
the Communists made a deal with the government which weakened
the party's image. Martínez relinquished his position to Blas
Roca. Despite his continuing illness, Martínez remained active
in politics until his death in January 1934. Martínez's closest
associates were subsequently drawn into the Communist Party
by his qualities and idealism. They were, however, purged
from the party in 1934 by representatives of the Comintern in
favor of Secretary-General Blas Roca and his followers.

MASFERRER, ROLANDO (1914-1975). Founder and leader of the
Movimiento Socialista Revolucionario (MSR), one of the three
most prominent violent urban groups operating in Cuba under
the government of Ramon Grau San Martín (1944-48). Masferrer
had been an early opponent of the Gerardo Machado regime
(1925-33), and had also fought on the Communist side during
the Spanish Civil War in 1937-38. Wounded in battle, he was
bestowed the nickname "El Cojo" ("Lame Leg"). Masferrer split
with the Communists in the 1940's and formed the MSR to challenge them and other groups by violent means. In 1947, Masferrer became heavily involved in an abortive expedition, allegedly financed by the Grau government, against Rafael
Leónidas Trujillo, dictator of the Dominican Republic. Fidel
Castro, then a student leader, participated in this venture.
Under the Carlos Prío Socarrás administration (1948-52), Masferrer became a Senator for the Auténticos and carried on his
illegal activities as head of a notorious gangster ring. Following Fulgencio Batista's coup d'état in 1952, Masferrer joined the
dictator's camp, thus retaining his position as Senator although
he kept a large private army called "los tigres" (the tigers) in
Oriente province. Subsequently, Masferrer and his group became a pillar of the Batista regime. In 1958, Masferrer commanded a group of his "tigers" in an attempt to wipe out Fidel
Castro's guerrillas in the mountains of Oriente province. Masferrer also published a journal, Tiempo en Cuba, which disclosed many of the political underworld intrigues in Cuba of
the 1940's-50's. During the electoral campaign of 1958, Masferrer pitted himself against Castro by staging a mass demonstration in support of Batista in the Sierra Maestra coast between
Santiago and El Uvero, during which he even promised to give
land to the peasants. On December 31, 1958, Masferrer, warned
of Batista's resignation and flight from Cuba, quickly left Santiago by yacht. He arrived in Florida on January 6. Masferrer
continued to live in Miami until 1975, when he was killed by a
bomb that exploded in his car.

MASO, BARTOLOME (1830-1907). Born in Manzanillo, Oriente
province, Masó participated in the Cuban rebel uprising against
Spain on October 10, 1968. He served as assistant to General
Carlos Manuel de Céspedes during the Independence War of
1868-1878 (Ten Years' War). By the end of the war, he held
the title of Secretary of War in the illegal rebel government.
Masó later participated in the Guerra Chiquita (1878-79). He
was captured and sent to Spain as a prisoner but was sent back
to Cuba after one year. Masó led a well-organized uprising in
Bayate on February 24, 1895, lending support to the patriots
who subsequently disembarked in Cuba. During the Independence War of 1895-98 and the Spanish-American War, Masó became
Major-General of the Independence Army and President of the
Rebel Republic. In 1901, he was a candidate against Tomás Estrada
Palma (1902-1906) for the Presidency. Retiring before the

election on claims of irregularities, Masó showed his civic character by supporting his rival's peaceful ascension to power.
Masó subsequently retired from public life.

MASONS. See Freemasonry.

MASS MEDIA. See also Press; Radio; Television. The Revolution
inherited a well if unevenly developed mass communications system based on the ideas of free enterprise and competition.
Dozens of newspapers and magazines owned by private, usually
middle-class, individuals were active in disseminating information.
Radio and television stations, also privately owned, competed
for an audience in the entertainment field. Criticism of the
government was, at times, expressed. Newspapers, especially,
were kept on a short leash through subtle but stringent controls, although during times of crisis journalists assumed a
more independent and critical stance with regard to government
policy. By 1962 the government had attained both ownership
and control of the media and had begun to shape content to
conform to socialist goals. The media were designated as important to the process of national development and integration
of all Cubans into the new society and consequently began to
reflect this in the content and kind of information released.
Moreover in the government's view the important functions of
the media were education and information, not entertainment for
its own sake and certainly not independent critical analysis of
government actions. Generally media officials have tried to follow Lenin's directives about the purposes of the mass media in
a communist state: to help organize the Communist Party to
educate party members and the public about the aims of the
government, and to agitate the masses to sustain morale and
support. Monopolization of the media allowed officials to eliminate potential vehicles of open opposition and promote greater
homogeneity of public opinion. It also allowed officials to develop those media and potential audiences, such as rural
Cubans, that they felt had been neglected under the previous
government.

The government's central role in opinion formation has been
enhanced by its role as the sole purveyor of news. Various
directorates and ministries--including the General Directorate
for Intelligence and the Ministry of Communications--set the
policy concerning newsworthiness; and the National Information
Agency and Prensa Latina (Latin Press), the state-owned international news agency, released information to the various
media channels. Prensa Latina was also charged with building
solidarity with the rest of Latin America, and in 1975 the agency
began publication of a magazine of general interst to Latin Americans. In 1986 two newspapers with nationwide circulation,
Granma and Juventud Rebelde, were published in Havana. Both
were official organs of the Communist Party of Cuba (Partido
Comunista de Cuba--PCC) organizations. This constituted a

considerable decline in the number and diversity of newspapers since the early 1960's. In addition each province supported a locally circulated daily. In the late 1970's the Cuban press faced both technical and quality problems. A shortage of spare parts and newsprint contributed to the technical difficulties, and the lack of trained journalists resulted in writing of indifferent quality. There are over 15,000 journalists in Cuba, of whom less than 100 could be considered professionals. The rest are volunteers. The number of magazines and journals was not reduced as drastically as that of newspapers, and by 1986 numerous periodicals were available, many of them having been published continuously since before the Revolution. All are controlled by the government and ranged from general interest to women's magazines and specialized trade publications. Some, such as Bohemia and Casa de las Américas, were distributed throughout Latin America, where they enjoyed considerable popularity. Many magazines tried to appeal to special-interest groups and provided more in-depth coverage of certain critical issues than newspapers. Because other magazines were not as closely tied to the PCC as Granma and Juventud Rebelde, they enjoyed greater latitude of political comment. Publications from Communist nations were available on the newsstands, but Western European and United States periodicals could be read only in the José Martí National Library, which subscribed to them. Book publishing soared in the 1960's and 1970's. The most important publishing organization was the Book Institute (Instituto del Libro), which took care of all phases of production and distribution. About two-third of the books published were free textbooks, and the rest were low-cost works of general appeal.

Film production was not well developed before the Revolution. Since 1959, however, the cinema has become one of the most important cultural and artistic branches of the revolutionary apparatus. Films are used to inform Cubans of the events and direction of the Revolution and to garner support for its changes. Film production is under the direction of the Cuban Institute of Cinema Art and Industry (Instituto Cubano del Arte e Industria Cinematográfica--ICAIC), established in 1959 to develop a film industry responsive to the needs of the Cuban people and the desires of the new government. Most films were political, cultural, or educational documentaries and short subjects, but some feature films have also been made that have drawn international attention. In the mid-1980's radio and television were under the direction of the Cuban Broadcasting Institute (Instituto Cubano de Radiodifusión--ICR). In the decade after the Revolution service was expanded and the number of receivers increased so that most Cubans had access to either radio or television and many to both. Most Cubans expected these media to continue to provide entertainment as they had previously, and in response to this popular demand old--frequently foreign--films and serials were still shown, and popular music

was still heard. In 1975, however, one of the three television
stations and two of the five national radio networks were devoted to education and cultural development, which were the
government's preferred uses of the media.
The decade of the 1960's saw dynamic changes in the direction and thrust of mass media development. All communications
channels became branches of the government, and new agencies
were formed to administer them. Because many trained professionals in the communications field either left Cuba or were ignored by the new government, a new group of official media
professionals began to form under socialist tutelage. The government and its new administrators placed emphasis on developing media that would not only reach but also politicize the Cuban
masses. Inasmuch as the basic apparatus for administration and
development had been set up by the late 1960's, the government
devoted its attention to stabilization of these changes in the
1970's. There was some indication that their efforts to institu-
tionalize the new apparatus had met with success among the
young and the dedicated, who expressed the importance of the
media in economic and cultural development even at the cost of
independent analysis and quality. Other Cubans, however, did
not support the government's role in directing the media and
preferred to ignore the more propagandistic elements in newspapers, television, and films. See also Telecommunciations.

MATANZAS (PROVINCE). Population: 454,486; area: 11,882 Km²;
Municipalities: Matanzas, Varadero, Colón, Jovellanos, Limonar,
Ciénaga de Zapata, Jagüey Grande, Los Arabos, Cárdenas, Martí,
Perico, Pedro Betancourt, Unión de Reyes, Calimete. Located
in the western part of Cuba, it shares borders with the provinces of Havana, Cienfuegos, and Villa Clara. Its territory began to expand in 1976 when it acquired Ciénaga de Zapata after
the famous April 1961 invasion of the Bay of Pigs. During the
colonial age, it was a major producer of sugar and many black
slaves worked there, although around 1920, it was surpassed in
this productive area by the provinces of Las Villas and Camagüey. Most of the soils of the so-called Llanura Habana Matanzas were the most valuable on the island and were located
in this province. The capital, Matanzas, was called "the Athens
of Cuba" because of its cultural houses and the fact that it is
located only 100 km from the City of Havana. It is divided by
two rivers, the San Juan and Yumurí, having been founded in
1693 on a beautiful bay at the site of the Indian community
Yucayo. Other cities of this province are Cárdenas, an important industrial port to the north with much historial tradition; and Colón, in the center of Matanzas, with mostly agricultural development.

MEDEROS DE GONZALEZ, ELENA (1900-1981). Pioneer of the
feminist movement in Cuba, founded the Alianza Nacional
Feminista in 1920 and the first women's club, the Lyceum, in

1929 with co-founder Berta Arocena de Martínez Márquez. Four
years later, during Grau San Martín's government, women were
granted the right to vote. Her aims were not only political but
combined encouragement of intellectual pursuits with programs
of social reform. From 1928-1949 she was Cuban delegate to
five Inter-American conferences of the Organization of American
States and president of the Hispanic-Cuban Cultural Institute,
the Cuban Foundation for Social Services, Cuban-American Allied Relief Fund, and the Cuban Good Neighbor Foundation (a
civil social welfare organization). Founder of the School of Social Work, Elena Mederos succeeded in having the school incorporated into the University of Havana, where she was instructor and Supervisor of Case Work and Director of Theses. She
was also Member of the Board of the National Public Assistance
Corporation, the Franklin D. Roosevelt Institute for the Rehabilitation of the Handicapped, and the Foundation for Medical Research. Elena Mederos was the Cuban Delegate to the United
Nations Commission on the Status of Women; she worked closely
with Cosme de la Torriente as Vice-President of the Society of
Friends of the Republic, mediators for elections during Fulgencio
Batista's regime to avoid revolutionary violence. When Fidel
Castro came to power, under the presidency of Manuel Urrutia,
Elena Mederos was appointed Minister of Social Welfare, but,
disillusioned with the political direction of the government, she
went into exile working first in Bogotá, Colombia, and later in
Washington, D.C., where she died.

MEDIACION, LA. Refers to the attempt by the United States to
find a peaceful solution to the situation in Cuba in 1933 where
opposition to the Gerardo Machado regime (1925-33) was creating
violent political turmoil. Franklin D. Roosevelt sent Ambassador
Benjamin Sumner Welles to Cuba in 1933 to act as a mediator between the government and the opposition. The mediation was
opposed by Conservative followers of ex-President Mario García
Menocal (1913-25) and the Directorio Estudiantil (Student Directorate) who wanted more than just a regime substitution. Welles'
efforts at mediation culminated in a general strike, dissension
within the armed forces, and small army revolts, forcing Machado
to resign and leave Cuba in August 1933.

MELLA, JULIO ANTONIO (1905-1929). One of the principal leaders
in the Cuban reform movement of the 1920s, Mella later became
a powerful speaker and anti-American agitator at the University
of Havana. There, he was elected Secretary-General of the
Student Federation. Mella advanced the reform movement beyond the academic locale, seeing it as part of the social struggle to improve the lot of the "have nots" in Cuban society.
Influenced by Víctor Raúl Haya de la Torre, Peruvian leader
and founder of the APRA movement, Mella established the short-
lived José Martí Popular University in Havana, a leftist institution devoted to education of the workers. By 1925, Mella had

become a leader in the Cuban Communist Party as well as its foremost propagandist. Mella led a group of students in opposition to President-elect Gerardo Machado (1925-33), whose authoritarian tinge alienated the militant Cuban communists, as well as other sectors of society. Machado had Mella jailed, but released him after the latter staged a hunger strike, whereupon Mella travelled to Mexico and the Soviet Union. He was assassinated in 1929 in Mexico shortly after shedding his affiliation with the Communist Party. It is not clear who had him killed-- Machado or his former Communist comrades. Mella shared with his University colleagues a desire to improve the educational and political conditions of Cuba and to oppose U.S. supervision of Cuban affairs. He differed from them, however, in that he rejected his generation's romantic nationalism and vague ideological conceptions to embrace an international movement devoted to the overthrow of the existing order and to the establishment of a proleterian dictatorship.

MENDIETA, CARLOS (1873-1960). Journalist, soldier, and statesman, Carlos Mendieta was born on his family's sugar plantation, La Matilde, in Santa Clara, Cuba. He studied at Belén and El Mesías schools in Havana. He received his Bachelor's Degree from the Institute of Santa Clara in 1895, and entered the University of Havana to study medicine. In early 1896 he went to his family's sugar plantation in Pinar del Río to join the revolutionary movement, abandoning his studies. Mendieta joined the Army of Liberation with a contingent of 125 men which he personally armed and equipped. During the war, Mendieta ascended to the rank of Colonel. He resigned his post as Captain of the Rural Guard during the first American Intervention and resumed his studies at the University of Havana, from which he graduated in 1901 as a Doctor of Medicine. After graduating, he was elected Congressional Representative from Santa Clara province. Under the Provisional government he was appointed Inspector of Health of the Republic. He was elected once again to the lower House of Congress by Santa Clara in 1908, and again in 1912. In 1916, he was nominated for the Vice-Presidency of the Liberal Party, and in 1919 became Director of the Heraldo de Cuba newspaper. In 1931 Mendieta organized the Unión Nacionalista, a political party opposed to Machado which condemned the regime in newspapers and in public demonstrations. In that same year Mendieta and former President Mario García Menocal (1913-21) organized a short-lived uprising in Pinar del Río province.

On January 14, 1934, Army Chief Fulgencio Batista forced President Grau San Martín (1933-35 and 1944-48) to resign; after a two-day rule by engineer Carlos Hevia, Batista appointed Carlos Mendieta as Cuba's Provisional President. Within five days after Mendieta's accession to power, the United States recognized the new Cuban government, since Batista and Mendieta represented stability to the United States, which had refused to recognize Grau the year before. Mendieta gained the reputation of a "puppet President." The frustrated students and members of the Generation of 1930 who had supported Grau, were particularly active against the Mendieta government, although the membership was split by the resort of some to violence. One student, Antonio Guiteras, founded the Joven Cuba, a clandestine revolutionary organization opposing Batista and Mendieta. Mendieta did nevertheless grant the University autonomy. Labor unrest, strikes, widespread discontent, and opposition from the newly formed Auténtico Party culminated in a general strike against Mendieta and Batista in March 1935. Mendieta criticized the strike as being unpatriotic and later Batista and the military moved to subdue its leaders and consolidate the government's power. Unions were dissolved; the University was closed or occupied by the military. Repression continued, and the leader of Joven Cuba was murdered. Mendieta proved to be a weak President who confessed his own failure to unite Cuban political parties. He was even forced to invite Dr. Harold Willis Dodds, President of Princeton University, to Havana to act as constitutional counsel to the government, a curious gesture on the part of a sovereign nation. Mendieta was, however, an honest president among many corrupt government officials. Batista replaced Mendieta as president with José A. Barnet (1935-36).

MENDIVE, RAFAEL MARIA (1821-1886). A committed supporter of Cuban independence, Mendive was a teacher as well as a poet. He was exiled to New York after being accused of attending a demonstration against Spanish rule in Cuba. Mendive especially exercised his influence by spreading separatist ideas. José Martí (1853-1895), his most prominent disciple, became the most renowned figure of Cuban independence. Mendive counseled Martí throughout the youth's high-school years, indoctrinating the future patriot in the committment to independence and love of literature. Also to Mendive's credit is his contribution as founder of the pro-independence Revista de la Havana.

MENENDEZ DE AVILES, PEDRO (1519-1574). First Adelantado (Royal deputy entrusted to found a colony) of Florida as appointed by Phillip II, and later governor of Cuba. Menéndez was responsible for coordinating land and sea defenses in the Caribbean for the Spanish in reaction to English expansion of trade there in the 1560's. Menéndez secured safe passage for Spanish fleets in convoys and fortified Havana, making it an impregnable fortress. His efforts to protect Spanish fleets and ports in and around Cuba succeeded until 1628, when the Dutch plundered a Spanish fleet off Cuba's north coast. Menéndez returned to Spain where he died, in Santander.

MENOCAL, ARMANDO (1861-1942). Born in Havana where he studied until sent to Madrid to study painting with masters such as Francisco Jover and Francisco Domingo. His painting

Generosidad Castellana when he was only 21 years of age won
him a second prize at the Madrid Exposition of 1876. His most
notable portrait was that of Bishop Santander. His Embarque
de Colón por orden de Bobadilla was exhibited at the Chicago
World Fair. His painting El Derecho feudal also won a prize in
1893. By 1895 he was fighting in the War of Independence
where he did sketches of the leaders in the field. After independence, he became a Professor of Landscaping at the Academia
de San Alejandro. Two of his most famous paintings were done
for the home of Rosalía Abreu, patriot and philanthropist--La
Invasión and Coliseo. He decorated the Presidential Palace
where his historic painting La toma de Guáimaro has President
Menocal in the forefront. His painting La muerte de Maceo was
done for Havana City Hall, and he also did the murals of the
Aula Magna (Main Lecture Hall) of the University of Havana.
In addition to his historical paintings, portraits, landscapes,
still lifes, and seascapes, Menocal was also a poet, famous for
his sonnets to Maceo and Máximo Gómez.

MENOCAL, MARIO. See García Menocal, Mario.

MENOCAL, PEDRO (1928-). Self-taught painter. Born in Havana
he attended Jesuit schools and studied architecture at the University of Havana but did not finish, going into private business
in agriculture (rice). After being imprisoned twice by the Castro government he left Cuba in 1961 and worked in New York as
a fashion illustrator for Saks and Lord and Taylor. He began
painting in New York and continued later in Mexico until 1977
when he went to Miami. He had two successful exhibitions at
the Bacardi Art Gallery in 1966 and 1972, but most of his exhibitions have been in private homes in New York, Mexico, and
Europe. His paintings are all privately owned and include the
White House, having painted two portraits of First Lady Nancy
Reagan. His work is in oils, pastels, and crayons, specializing
in portraiture but also horses and animals (pets). He has also
painted still lifes which he does not sell.

MERCEDES. The principal type of land grant by the Spanish crown
in Cuba in the 16th century. Mercedes were made in the form
of circles marking the distances or radius from a given point.
The recipient was bound to furnish the nearby town or settlement with an inn and all food stocks needed.

MEXICAN LETTER. An agreement known as La Carta de México,
signed by Fidel Castro and José A. Echeverría, the Cuban student leader, in Mexico in early 1956. The students pledged a
series of diversionary riots in Havana to coincide with Castro's
landing from Mexico on the other end of the island and the
coordination of activities to overthrow the Batista regime.

MIGRATION. What began as a trickle in early 1959, became a

veritable flood of human migrants through the following years
until 1962. Restrictions were placed on the exodus of Cubans
from that time on except for the 1965 massive airlift agreed on
by the Cuban and U.S. government and the 1980 Mariel exodus,
when the Cuban government opened, unilaterally, that port for
Cubans outside the island to come and collect relatives. It is
estimated that, since the Revolution, over 600,000 Cubans have
left the island. The majority came to the U.S. where the Federal Government established a reception center and made provisions for temporary care while encouraging relocation throughout the U.S. Since Miami became the destination of the bulk of
the refugees, special Federal subsidies were provided for health
care and for the education of Cuban children through the local
school system. Other major refugee centers include New Jersey,
New York, and Puerto Rico.

MIJARES, JOSE MARIA (1922-). Cuban-born painter. He studied
at San Alejandro, Havana, where he won two national first prizes,
1944 and 1950, and one second in 1952. His work was shown in
the São Paulo Bienal in 1953 and in the Venice Bienal in 1956.
He has held many one-man shows and has participated in important group exhibitions in Paris in the Modern Museum of Art, in
Japan, Venezuela, Puerto Rico, and the United States. He left
Cuba after Castro's takeover and now lives in Miami. He was
awarded the Oscar Cintas Fellowship. He painted many murals
in Havana in buildings and private homes which have been carefully preserved by the Castro Revolution covering them with
glass.

MILANES, JOSE JACINTO (1814-1863). Cuban poet born in Matanzas, distinguished by his melancholy and compassion, moods
evident in his poems. Milanés also wrote patriotic poems and
one famous play, El Conde Alarcus, shown in Havana in 1838.
His more famous works include La fuga de la tórtola; El beso;
El mendigo; Los dormidos.

MILIAN CASTRO, ARNALDO (1913-1983). As Vice-President of the
Council of Ministers and Minister of Agriculture, Milián was one
of the top members in the Cuban government leadership. Milián
initiated his political career in 1944 as a delegate from Matanzas
province to the Popular Socialist Party in Havana. In 1959,
Milián ascended to membership in the Popular Socialist Party
Central Committee. Following the Revolution Milián was elected
Secretary General of the Cuban Communist Party (PCC) for Las
Villas province in 1962. By 1965 he was appointed Supplementary
Member of the PCC's Central Committee. Milián eventually became
a member of the PCC Central Committee and its Political Bureau
(Politburo) in 1975. One year later, in 1976, he was appointed
Deputy from the new Province of Villa Clara to the National As-
sembly of Poder Popular as well as to the thirty-member Council
of State. Milián was appointed to the powerful eleven-member

Secretariat of the PCC Central Committee in 1977 and retained
that position until 1980. He died in 1983.

MILITIA. Conceived by Castro in October 1959, a voluntary army
of workers was the forerunner of the famous revolutionary militia. By February 1960, Captain Acevedo was leading this national militia which had replaced the corroded rebel army. Men
and women--150,000 strong--who had supported the revolution
put on uniforms and took up weapons after their daily work.
They served in the militia about eight hours per week, guard-
ing public buildings and other important installations against
attack by counterrevolutionaries. By 1961, the presence of the
militia was pervasive, organized and run by army officers who
had been loyal to the revolution and to Fidel Castro. During
the Bay of Pigs invasion in April 1961, Castro mobilized the
militia and moved south in Matanzas province to be used if necessary to repel the invasion. Militia service became compulsory
for all young men and women later in the 1960s, a development
that many students disliked. The organization was originally
called Milicias Nacionales Revolucionaries (MNR). In 1981, Castro felt the need to mobilize the Cuban population to reestablish
control in the aftermath of the Mariel exodus. The MNR was
replaced by Milicias de Tropas Territoriales (MTT). By 1986,
over 1 million civilians were participating in the MTT, ready to
protect the island from imperialist invaders, in what is referred
to as Guerra de Todo el Pueblo (all the people's war).

MINING. The early Spanish explorers hoped to mine gold and
copper in Cuba, but were disappointed at the small amounts
encountered. Until the early 20th century, mining remained
dormant. From 1902 to 1950, Cuba's mineral resources became
a focus for exploitation and total mineral production totalled
$467 million, equivalent to one sugar crop. U.S. interests pro-
vided the driving force behind the little mining activity occurring in Cuba during the 1950's. Until 1959, the Cuban government had a small mining department under the jurisdiction of
the Agriculture Ministry. Minerals of significance in Cuba today
include only nickel and its by-products. Some iron, chromite,
manganese, and copper deposits have also been mined in smaller
amounts.

MIRET PRIETO, PEDRO (1927-). Miret was born in Santiago de
Cuba, Oriente province. He attended the University of Havana
for a time but did not complete his studies because of his involvement in anti-government activities against the regime of
Fulgencio Batista (1952-1958). Miret joined Fidel Castro's rebel
group and participated in the abortive attack on the Moncada
Barracks in Santiago de Cuba on July 26, 1953. He was captured and sent to prison along with other survivors. Released
from prison under Batista's general amnesty of 1955, Miret joined
Castro in Mexico where he helped Castro organize the 26th of

July Movement (M26J). Arrested by the Mexican authorities for
arms violations, Miret did not take part in the Granma expedition
that left Mexico in late November 1956 for Cuba. Subsequently
released by the Mexican authorities, Miret eventually returned
to Cuba and joined the Castro brothers in the mountains where
he acquired the rank of major as head of one of the rebel columns operating in Oriente. Following the collapse of the Batista
regime in January 1959, Miret served for six months as Vice-
Minister of Defense. He was Minister of Agriculture from June
1959 until early 1961 when the position was abolished. In late
1961 Miret joined the Ministry of the Revolutionary Armed Forces
(MINFAR) where he served in the artillery directorate until
1966. In late 1966 he was promoted to MINFAR Vice-Minister
and during 1968-69 he was First Vice-Minister of MINFAR.
Miret has been a member of the Central Committee of the Cuban
Communist Party since October 1965, its Secretariat since January 1974, and its Politburo since December 1975. He served as
Minister of Mines and Metallurgy during 1969-73. From November 1972 to January 1974 he was Vice-Prime Minister for Basic
Industries, which had, among others, the responsibility for the
Ministry of Mines and Metallurgy. Miret has been a deputy to
the National Assembly representing the municipality of Camagüey
since November 1976 and a member of the Council of State since
December 1976.

MIRO CARDONA, JOSE (1902-1974). President of the Cuban Bar
   Association, Miró Cardona supported the call for President Fulgencio Batista (1952-58) to step down in 1956. He was also influential in the so-called Civic Dialogue intended to seek a
   peaceful solution to the Cuban governmental crisis. He was
   forced to seek asylum in the Uruguayan Embassy in Havana in
   1958 after a document circulated by the president of the Cuban
   Medical Association calling for Batista's resignation was attributed to him. While in exile, Miró acted as Secretary-General in
   charge of coordinating the forces comprising the Frente Cívico
   Revolucionario Democrático (Civic-Democratic Revolutionary
   Front). Miró returned to Cuba as Prime Minister of the revolu-
   tionary government following Fidel Castro's triumph in 1959.
   After only one month as Prime Minister, Miró resigned in the
   realization that true power was really in Fidel Castro's hands.
   He was appointed Ambassador to Spain and then recalled to Cuba
   after an incident between the Spanish Ambassador and Fidel
   Castro. Miró opposed Castro's revolutionary reforms in the
   University of Havana, thus losing favor with the Cuban leader.
   In June 1960 he went into exile. Miró was chosen President of
   the Cuban Revolutionary Council in March 1961. He would have
   become President of Cuba had the April 1961 Bay of Pigs invasion been successful. Afterwards Miró continued his political
   work against Castro's government from Puerto Rico, where he
   settled and later died.

MIRO Y ARGENTER, JOSE (1857-1925). Soldier and journalist, Miró was born in Barcelona and served as General Antonio Maceo's Chief of Staff during the War of Independence (1895-98) in 1895. He later reached the rank of General and after the war became Director of the Archives of the Army of Liberation. After the war, Miró took an active interest in journalism, founding several journals including La Doctrina, El Liberal, and Vida Militar in 1898. He authored several pamphlets on the Independence War, including Muerte del General Maceo and Crónicas de la Guerra.

MISSILE CRISIS (1962). After the embarrassment to the U.S. of the failure in April 1961 of the Bay of Pigs invasion, the Soviet Union felt it could upgrade its economic and military involvement in Cuba without receiving a strong U.S. reaction. By mid-1962, the Soviets had introduced nuclear-capacity bombers and nuclear missiles into Cuba. On October 22, President Kennedy publicly denounced the existence of these offensive weapons in Cuba and demanded their withdrawal. He also instituted a naval blockade of the island saying that the U.S. would intercept Soviet ships bringing ICBM's into Cuba. After twelve days of hectic posturing both in public at the U.S. and in private "hot-line" correspondence, Khrushchev agreed to remove the missiles and bombers and to allow U.N.-supervised inspection of this removal in exchange for a U.S. pledge not to invade Cuba. Castro never allowed the U.N. inspection, having been angered at being excluded from the Kennedy/Khrushchev dialogue during the wind-down of the crisis. The missiles and bombers were removed as verified by U.S. aerial surveillance and the crisis ended in early November. The U.S. never publicly acknowledged its pledge not to invade Cuba although until the early 1980's a U.S.-Soviet understanding was extant including a U.S. "hands-off" policy toward the island. The Cuban missile crisis had a significant impact on the countries involved. It did lead to a thaw in U.S.-Soviet relations but conversely strained Cuban-Soviet relations. Castro was humiliated over being reduced to a pawn during the crisis where nuclear conflict was the ultimate specter.

MONCADA ATTACK. Fidel Castro's ill-fated assault against the Fulgencio Batista regime (1952-58) of July 26, 1953, was directed against the Moncada military barracks in Oriente province. Castro planned the attack to coincide with the Santiago de Cuba carnival in eastern Cuba, expecting army discipline to be lax at this time. The attack was to have been supported by pro-Orthodox Party army officers in order to paralyze the army which would then be unable to act against the rebels. To Castro's surprise, the garrison at Moncada was not relaxed as the assault began, and the attackers were routed. Castro, escaping to the mountains, was later captured and sentenced to years in prison. Although the assault failed, it gave Castro

and his following national prestige and imbued in him the
knowledge that publicity and the mass media were tools to be
utilized later in the struggle against Batista.

MONTE, DOMINGO DEL (1804-1853). Born in Venezuela, del Monte
went to Cuba in 1810 where he remained most of his life.
Del Monte was an inspiration for the Cuban students of his era,
conducting social gatherings in his home, where young Cuban
writers and poets would gather to exchange ideas. From such
tertulias (as these gatherings were called) the Romantic movement was born in Cuba. Del Monte was a contemporary and
friend of such notable Cuban intellectuals as José A. Saco and
José de la Luz y Caballero. He was also an abolitionist and opposed the movement for annexation of Cuba to the United
States. Del Monte was a strong defender of the liberal reforms which were popularly expressed in Cuba during the 1840's.
He was later exiled to Spain and died in Madrid. Del Monte
published one well-known novel, Clementina o los recuerdos de
un gentil hombre, and later published a book of poems called
Romances cubanos.

MONTECRISTI CONSPIRACY. One of many plots against the Fulgencio Batista regime (1952-1958) during the mid 1950's. The
Montecristi group plotted with Army officers to overthrow the
regime. Batista uncovered the conspiracy and arrested its in-
stigators in April 1956.

MONTORI, ARTURO (1878-1932). Notable educator in early 20th-
century Cuba, Montori was Director and Professor of the Escuela Normal de La Habana, circa 1913. His works include:
La fatiga intelectual (1913); Ideales de los niños cubanos;
and the well-known El feminismo contemporáneo (1922). His
most famous work of fiction was the novel El tormento de vivir
(1923), which reflects his ideas on truth and ethics.

MONTORO, RAFAEL (1852-1933). Rafael Montoro was born in
Havana, scion of an old, wealthy family. Montoro graduated
as a lawyer in Madrid, where he became well versed in philosophy and literature. An eloquent orator, Montoro served as
Cuban Representative to the Spanish Cortes (Parliament).
Along with José Martí (1853-1895), he is considered one of the
best orators of his period. Beginning his career as an editori-
alist, Montoro became Chief Editor of Revista Contemporánea
and also contributed to the Revista Europea, both Spanish
publications. In 1877, he was appointed Vice-President of the
Political and Moral Sciences section of the Atheneum of Madrid,
as well as Second Secretary of the Association of Spanish Writers and Artists. Returning to Cuba just before the end of the
Ten Years' War (1868-1878), Montoro founded the jouranl El
Triunfo, of which he was editor, with a group of literary
friends. Politically, Montoro favored autonomy from Spain

rather than independence, and joined the Liberal Autonomist
Party. In 1879 he was elected to the party's Central Committee;
Montoro was also made Havana Province's Deputy to the Spanish
Cortes. He was reelected to this post in 1886 and again in 1893
to represent the Province of Puerto Príncipe. A firm believer
in autonomy rather than total separation from Spain, Montoro
refused to participate in the Independence (Spanish-American)
War (1895-98).

When Spain finally granted autonomy in 1897, Montoro was
appointed Secretary of the Treasury of the new Cuban government. However, the new government had no support and fell
apart with the U.S. intervention. Montoro subsequently retired
from public life. With the end of the American occupation in
1902, he again became politically active and was appointed Envoy
Extraordinary, Minister Plenipotentiary to England and Minister
to Germany. In 1906, he was appointed Delegate to the Pan
American Conference in Rio de Janeiro and Buenos Aires. Montoro was a significant member of the Consultative Commission
during the second U.S. intervention (1906-1909). In 1912 he
was appointed Secretary of the President and was later a candidate for Vice-President of the Republic. Montoro was member
and Director of the National Academy of Arts and Letters and
member of the Advisory Law Commission. His main works include Principios de moral e instrucción cívica (1903) and
Nociones de instrucción moral y cívica (1908).

MORAL INCENTIVES. Moral incentives were extolled by Ernesto
Guevara after Castro's consolidation of power in the early
1960s, as those means of shaping the new man and the new
society in Cuba. High consciousness of social duty and intolerance of violation of social interest would predominate. The
new characteristics of the socialist morality included: abnegation, a spirit of sacrifice, courage, and discipline. The new
society would be abundant in material wealth but each individual
would not seek personal wealth; he would rather work to produce for the whole society. Material incentives, like money,
would be eliminated, and via a gigantic effort to organize the
productive, social, educational, and cultural activity of the
Cuban people, the new Communist ideology would be imposed.
During the Third Party Congress (1986), Castro complained of
the Revolution's failure to create a new society. He reaffirmed
that socialism could not be built with capitalist incentives and
launched a program of austerity measures relying on the socialist man's motivation to only require moral incentives to create a
new society.

MORALES LEMUS, JOSE (1808-1870). Pioneer of Cuban independence
and of the anti-slavery movement, Morales was raised by Spaniards from the Canary Islands after his father abandoned him.
He became a lawyer and inherited a fortune from his benefac-
tors. His sense of justice made him free his slaves and led him

to conspiratorial activities for the independence of Cuba. He
contributed to El Siglo, the publication that defended Cuban aspirations for independence. He was elected for the Junta de
Información to the Cortes in Madrid. Back in Cuba, he actively participated in the Ten Years' War and managed to escape to
New York but all his properties were confiscated. He dedicated
his exile to patriotic endeavors, first as president of the Junta
Cubana and later as Minister of the Republic in Arms to the
U.S. Government. His efforts were premature but helped
arouse support for the war effort against Spain. He died penniless in New York.

MORENO FRAGINALS, MANUEL (1920-   ). Cuban historian, whose
publication in 1964 of El ingenio: el complejo económico social
cubano del azúcar, 1760-1860 presented a new interpretation of
Cuban historical development in terms of the technological transformations suffered by the sugar industry at the turn of the
nineteenth century, as led by a capitalist-minded Creole elite.
A three-volume expanded edition of El ingenio appeared in 1978.
Moreno lives in Cuba and occasionally travels abroad to attend
conferences related to slavery and black culture in the Americas.

MORET LAW (1870). Law that abolished slavery in Cuba, named after
Segismundo Moret, who became Spanish Minister of the Colonies
during the Ten Years' War. It was Spain's attempt to counter-
act the article in the Guáimaro Constitution of 1869, by the
Cuban insurgents, declaring all men free and equal. By Moret's
Law, all children would be born free, and the freedom of those under eighteen months of age would be bought by the state. These
as well as old slaves would stay until age 18, under their masters'
tutelage, patronato. All other slaves would be freed, with compensation if they had served Spain during the war. In 1879,
Arsenio Martínez Campos, then prime minister in Spain, announced that the patronato would end in 1888, but it officially
ended two years earlier, given that there were only 26,000
slaves in the island.

MORGAN, SIR HENRY (1635-1688). Henry Morgan was a 17th-
century British buccaneer; a mercenary with no political alliance. Buccaneers served European nations in times of war and
plundered for themselves in times of peace. Morgan was one
of the most notorious of these privateers. He terrorized Spanish settlers in Cuba and throughout the Caribbean. Morgan
also ransacked Porto Bello and northern South America in general. In Cuba, the ruthless buccaneer attacked Camagüey and
Oriente provinces, torturing or killing those Spaniards not for-
tunate enough to escape him. He died in Lawrencefield, Jamaica.

MOVIMIENTO SOCIALISTA REVOLUCIONARIO (MSR). One of three
urban semi-gangster groups in Cuba in the 1940's. MSR

numbered some 300. Founded by Rolando Masferrer, MSR articulated a genuine social purpose which remained secondary to murder and gangster tactics that infested the movement.

MUJAL, EUSEBIO (1915-1985). Son of a Catalan banker from Guantánamo, Mujal joined the Cuban Communist Party in 1930. He later severed his communist affiliation and became Secretary General of Guiteras' Joven Cuba in 1934. In 1938, Mujal joined Ramón Grau San Martín's Auténtico Party. With the formation of the Cuban Confederation of Labor (CTC), Mujal became a prominent agitator in the Cuban labor movement. Under the administration of Cuban President Carlos Prío (1948-52), Mujal expelled all communists from important positions in the CTC with Prío's backing. When Fulgencio Batista (1952-58) overthrew Prío in March of 1952, Mujal managed to arrive at an accommodation with Batista as leader at the CTC, thus averting a general strike. Mujal remained head of the CTC until 1959, gradually moving closer to identification with Batista although he retained autonomous control over the confederation. As soon as Fidel Castro took over the government in 1959, Mujal left Cuba and dedicated his life to anti-Castro activities in the United States, especially those connected with labor.

MUSIC. Music received early impetus in Cuba from both Spanish colonials and African slaves. As early as the mid-sixteenth century Miguel de Velazco, an organist and professor of grammar, gained recognition as the first Cuban-born musician. He was followed by the first Cuban composer, Esteban Salas, around the end of the eighteenth century. Salas was the leading cultural figure of his day, and his compositions reflect his transitional position from a late baroque style to the beginning of neoclassicism. Despite the basic division of Cuban musical styles into Euro-Cuban, Afro-Cuban, popular (or mulatto) and concert music, the traditions are not clearly disparate. Classification of the various kinds and schools depends upon the degree of mixture rather than the purity of form and expression. Euro-Cuban and Afro-Cuban traditions did develop somewhat separately until the twentieth century. The former was confined largely to white groups in the countryside and the upper class in the cities. Their music, played at formal balls and elaborate social gatherings, was based on Spanish forms and melodies. Instruments used were the small guitar and occasionally the violin; voice accompaniment was also used. The Spanish forms were modified: the tempo was slowed and the beat shifted until new Cuban rhythms emerged--the punto, the guajira, and the zapateo. The meter of the Spanish bolero was changed also from three-four to two-four time and became the Cuban bolero. Perhaps the best known of the Cuban-developed rhythms is the habanera, although it is a more stately dance than is suggested by Georges Bizet's use of it in the opera Carmen. While the habanera was being danced by the

white upper class of Cuba's sharply stratified society, the lower
class, almost entirely black, was maintaining and elaborating its
own varied African musical heritage, emphasizing the drum and
other percussion instruments.

Initially Afro-Cuban music had a primarily religious function
and was an integral part of popular rituals. The most widespread ritual was that practiced by the Lucumí (Cuban descendants of the Nigerian Yoruba). The Yoruba drum (bata) is
hourglass-shaped and is held across the knees while the player
hits both ends. The drum body is made from a tree trunk hol-
lowed by fire, and the skins are permanently attached so that
tension or pitch is not adjustable. Early in the nineteenth century a secular Afro-Cuban music developed that gradually came
to the attention of the white population. Lower-class benefit
societies, largely black, began to appear in public to drum,
sing, and dance in masks and costumes on such feast days as
the Epiphany. Some of the religious societies also developed a
secular tradition and began to play an important part in the
pre-Lenten carnival celebrations. The many societies that participated came to have their own distinctive songs, dances, and
costumes. Spanish melodies were superimposed on the secular
Afro-Cuban music, and the guitar was added to the drums, producing, in the words of the writer Ortiz, "love affairs of the
Spanish guitar with the African drum." The dances and rhythms
that resulted from the mingling--the rumba, conga, son (perhaps
the oldest of these blends) and bolero--spread to much of the
world. The rumba that is known outside Cuba actually is more
closely related to the son and the Cuban bolero than to the
rumba, which is a much faster, more dramatic dance usually
confined to exhibition dancing. After independence, blacks increasingly moved to the major urban centers, and their music
was performed and assimilated by professional musicians, composers, and poets. It was scored for additional instruments--
piano, trumpet, and trombone; the resulting popular music came
to override completely the Euro-Cuban forms that had previously flourished. Many of the more stately dances of the rural
and urban upper class are now found only among poorer groups
in the countryside.

At the beginning of the twentieth century several diverse
elements appeared on the musical scene. Some Cuban composers
such as Ignacio Cervantes Kawanagh (1847-1905) went through
a cosmopolitan period in which they reflected the influence of
foreign styles ranging from the impression of Claude Debussy
to the atonality of Igor Stravinsky and Arnold Schönberg. During the 1920 and 1930s many composers found inspiration in romantic and sensual Afro-Cuban elements. Ernesto Lecuona
(1896-1963) produced such representative songs as "Malagueña,"
"Siboney," and "Siempre en mi corazón" (Always in My Heart),
which were enthusiastically received both in Cuba and abroad
and have retained their popularity for nearly half a century.
Cuban concert music has not received the recognition accorded

to other kinds of music either inside or outside Cuba. Formal
compositions that antedate independence either drew upon and
entered the Euro-Cuban tradition or have been completely for-
gotten and have not served as a source for later composers.
The two great composers of the period who were given considerable recognition after 1959 were the composer-violinist Amadeo
Roldán (1900-1939) and Alejandro García Caturla (1906-1940).
Roldán admired Guillén and consciously wove African rhythms
into his music. Like some other composers, he wrote overtures,
ballets, and pieces for chorus, piano, and percussion. García
Caturla, like Roldán, was an admirer of Guillén and was caught
up in the rediscovery of Afro-Cuban music. García Caturla finally turned from the black influence and sought inspiration
from the Euro-Cuban music of the colonial past. In this regard he followed in the footsteps of such earlier composers as
Eduardo Sánchez de Fuentes (1874-1944), who composed many
popular habaneras, and Joaquín Nin (1879-1944). In the 1930's
there was a movement away from both cosmopolitanism and Afro-
Cubanism, led by José Ardévol (1911-   ). From 1932 to 1945
Ardévol led a group of composers known as the Renovation
Group, which returned to the rigidity of classical and neoclas-
sical forms. Few of the group made any attempt to draw upon
the Cuban idiom.

The turning point came in 1945 when many composers, eventually including Ardévol, returned to Cuban sources for inspiration. Gonzalo Roig (1900-70), the founder of Havana's symphony
orchestra and composer of many popular songs, also mixed many
folkloric elements in his musical work. His chief contribution,
an adaptation of Villaverde's novel Cecilia Valdés, was performed
in 1932, precisely at the zenith of the Afro-Cuban vogue. It
combined such typical dances as the habanera, tango, conga,
and guaracha with Spanish and African musical themes. The
timely appearance of the work gave respectability to Afro-Cuban
motifs as well as gaining international recognition for them.
During the first part of the 1960s, Cuban popular music con-
served the vitality derived from the popularity of the mambo
and the cha-cha-chá of the 1950s with such figures as Dámaso
Pérez Prado, Enrique Jorrín, and Benny Moré. Cuban popular
music was enriched by the works of filin (feeling) music, a
style at the beginning of the 1950s, characterized by a smooth
harmonious personal style as seen in the music of José Antonio
Méndez, César Portillo de la Luz, and Frank Domínguez. Popular music found itself in the middle of a crisis initiated during
the first years of the Revolution by a breakdown in the broadcasting structure and the implementation of rigid cultural policy.
Beginning in 1966, the State undertook an isolationist attitude
censuring all foreign and nationally produced music as a way to
avoid foreign cultural influence and any ideological deviation.
Official policy was that the works of Cuban authors must reflect the social commitment of the artist to a revolutionary system and the necessity of recovering lost musical values was

emphasized. Several short-lived rhythms emerged at this time, such as the pilón, the paca, and the mozambique, which never left Cuba.

At the end of 1968, popular music reawakened with the appearance of a new musical style, the Nueva Trova. Soloists accompanied by guitar played songs generally of a social nature which conformed to the demands of cultural policy. From that moment on, official steps were taken to improve the state of popular music and new music schools were established. Foreign music began to be heard more as well as the important figures of Cuban traditional music. During the seventies, there was more emphasis on culture; and jazz and experimental music groups appeared with a notable increase in musical compositions. In order to encourage the development of music, national and international festivals were held, such as the Festival of the Popular Music of Varadero, the Benny Moré Festival, and the Adolfo Guzmán Music Contest. See also Nueva Trova.

- N -

NARVAEZ, PANFILO DE (1470-1528). Born in Valladolid, Spain. A Lieutenant under the famous Spanish conquistador Diego Velázquez, Narváez subdued the Indians in Cuba in 1511. Possessing neither Velázquez's courage nor wisdom, Narváez dealt harshly with the Indians, allowing his soldiers to plunder Indian villages. In 1512, Narváez ultimately occupied the Indian center of Bayamo. Narváez returned to Spain in 1527 and led an ill-fated expedition to conquer Florida. Unsuccessful in their search for gold, Narváez and his men sailed from Florida toward Mexico on September 22, 1528. All but two members of the company perished when Narváez's ship was wrecked during a storm.

NATIONAL ASSEMBLY OF PEOPLE'S POWER. See Poder popular.

NATIONAL COUNCIL OF UNIVERSITIES. In January 1962, the Castro government proclaimed a comprehensive university reform plan. In this new program, the task of directing the universities in Cuba (and only three remained: in Havana, Las Villas, and Oriente) fell to the National Council of Universities. This body was composed of faculty and student representatives from the three universities and of government representatives. Presided over by the Education Ministry, the Council guides all administrative and academic decisions as well as planning changes in higher educational programs. The Council works with the university rectors who are appointed by the government. The rectors implement the Council's directives and handle the university's academic and administrative affairs.

NATIONALISM. After the First War for Cuban independence (the
    Ten Years' War of 1868-1878), the first stirrings of nationalism
    took form especially within the heterogeneous ranks of the rebel
    army. Rich and poor, black and white, Chinese and mulatto,
    peasant and worker--all were thrown together in a nationalist
    cause that never died out. After Cuban independence had been
    achieved, U.S. intervention and the "Platt Amendment mentality"
    tended to difuse this sense of national purpose and fostered a
    continuing sense of civic indolence. By the 1920's nationalism
    had reasserted itself with an economic focus when U.S. sugar
    interests seemed to dominate the affairs of the country. Anti-
    U.S. feelings, xenophobia, and the retrieval of economic wealth
    became the main themes of the new nationalism. Nationalism in
    the 1920's was perhaps most significantly captured by the
    cubanismo theme so prevalent in the works of Cuban literary
    figures like Trelles, Ortiz, and Mañach. The Reformist spirit
    once again surfaced in 1933 with the successor of Gerardo Ma-
    chado (1925-1933), Carlos Manuel de Céspedes, who was seen
    by the reformers of the Generation of 1930 as a tool of the U.S.
    and a continuation of his predecessor's policies.
        With the arrival of Dr. Ramón Grau San Martín in 1933, the
    nationalist spirit of the Generation of 1930 was catapulted into
    power. The government was pro-labor and opposed to an econ-
    omy dominated by foreign (U.S.) capital. Grau abrogated the
    1901 Constitution and asked for abrogation of the U.S.-designed
    Platt Amendment. Grau implemented many reforms which in gen-
    eral sought to "Cubanize" the labor movement and were seen as
    threats by U.S. business interests. The nationalist mood sub-
    sided after Batista assumed a leadership role in Cuba and re-
    mained moribund for 25 years. Castro and the 1959 Revolution
    began to reemphasize certain aspects of the past especially the
    Cuban sacrifices at the time of the wars of independence. The
    cult of José Martí has continued unabated, though his writings
    have been screened to select those which reflect his anti-American
    and pro-Cuban sentiments. Nationalism, as well as Marxism, guided
    the revolution's path at locating Cuba's new identity. Castro
    and the Cuban leadership present the Revolution as the embodi-
    ment of the ideas of the independence movement and the frus-
    trated 1933 revolutionary process begun by Grau. As proclaimed
    by Castro himself in 1968, the centennial thesis states that the
    Cuban people fought for 100 years for their independence, i.e.,
    since 1868, year when the Ten Years' War began.

NATIONALIZATION. In March 1959, Castro publicly declared that
    he had no intention of nationalizing any foreign-owned industries.
    By late 1959, however, many nationalizations were beginning to
    occur. Land belonging to American companies were seized. By
    July 1960, as U.S.-Cuban relations steadily deteriorated, Castro
    ordered 600 U.S.-owned industries to inventory their stocks, in
    preparation for a complete nationalization of U.S. property. On
    August 6, 1960, all oil refineries and sugar mills, along with the

Cuban Telephone and Electric Companies were expropriated.
By October, INRA (National Institute of Agrarian Reform) had
taken over 382 large private enterprises including all the banks
(except for 2 Canadian banks), all the remaining private sugar
mills, 18 distilleries, 61 textile mills, 16 rice mills, 11 cinemas,
and 13 large stores. Later in October, Castro nationalized 166
more U.S. enterprises, including all major U.S. enterprises
such as Woolworth, Sears Roebuck, General Electric, Westing-
house, International Harvester, and Coca-Cola. Soon afterward
the U.S. Ambassador was withdrawn from Havana, never to return.

NAVY. In 1933, the Cuban navy consisted of only 13 ships (11
gunboats, 1 training ship, and a 1,500-ton "cruiser"). All but
1 gunboat and the cruiser Cuba were in poor condition. During
World War II the Cuban navy owned one submarine bought from
the U.S. which did manage to sink a German submarine; however,
the Cuban role throughout the war was more passively one of
Cuban territory being used for training purposes by U.S. and
RAF personnel. The naval base at Cienfuegos was the site of
an abortive effort in 1957 to generate opposition to Batista with
a naval mutiny using the cruiser Cuba to force the dictator to
yield. This attempt ultimately failed after the base landed in
rebel hands for several days. Following the revolution of 1959,
Fidel Castro revamped Cuba's Armed Forces. Conscription is
for a three-year period starting at age 17; conscripts also work
the land. In July 1985, the navy numbered 13,500, constituting
the smallest branch of the Cuban Armed Forces (the Army is
the largest by far). Estimated defense expenditure for 1980
was 811 million pesos, most of that amount going to the Army.
Cuba's Navy receives considerable aid from communist countries,
notably the U.S.S.R. In 1986 the Cuban Navy possessed 3 ex-
Soviet submarines: 2 F-, 1 W-class; 14 ex-Soviet large patrol
craft: 10 SO-1, 4 Kronshtadt; 27 ex-Soviet FAC (M) with Styx
SSM: 6 Osa-I, 5 Osa-II, 16 Komar; 26 ex-Soviet FAC (T):
2 Turya, 12 P-6, 12 P-4; 12 ex-Soviet Zhuk FAC (P); 12
coastal patrol craft; 14 minesweepers: 2 ex-Soviet Yevgenya,
1 ex-Pol K-8; 10 T-4 LCM; and some 50 Samlet coast-defense
SSM. Naval bases are located at Cienfuegos, Havana, Mariel,
Punta Ballenatos, and Canasí.

NEWSPAPERS. See also Magazines; Mass Media; Press. In the
period from 1935 to 1958, the Cuban government influenced
many newspapers through an intricate system of official bribery. Fifty-eight daily newspapers were in circulation in 1956,
with no more than six or seven of these meeting their costs
through subscription and advertising alone. According to
former government officials, Cuban dictator Fulgencio Batista
(1952-58) paid some $217,000 monthly to Cuban periodicals in
exchange for minimal press criticism in the mid-1950s. Cuban
journalism maintained high standards in the 1940s and 1950s in

comparison with the Latin American press in general. Journalists had to be graduates of an Education Ministry-sponsored
4-year program in journalism, and belong to the Newspapermen's
Association, the disciplinary organ of the profession. Cuba
ranked fourth in 1956 among Western Hemisphere nations in
ratio of newspaper circulation to total population, below the
U.S., Canada, and Brazil. Their circulation was largely urban. Havana in 1956 had only two fewer newspapers than
London, a city nine times its size. The majority of Cuba's
newspapers and radio stations were confiscated between October 1959 and July 1960 by Fidel Castro. In February 1961,
only six general newspapers were being published daily in
Havana; 10 fewer than on January 1, 1959. Five of these:
Revolución, La Calle, El Combate, El Mundo, and Prensa Libre
were semi-official, while the sixth, Noticias de Hoy, was an
organ of the PSP (Partido Socialista Popular), edited by Carlos
Rafael Rodríguez. By the mid-sixties all of these disappeared
and were replaced by Granma, the official newspaper of the
Cuban Communist Party.

NICKEL INDUSTRY. In 1946, Cuba supplied 9 percent of the
world's production of nickel. During World War II, the U.S.
subsidized mining installations to mine Cuban nickel, installing
the big Nicaro Nickel Co. plant in Moa, Holguín, which operated from 1943 to 1947, producing some 37,700 metric tons of
unrefined nickel worth about $16.6 million. This plant alone,
working at maximum capacity, could produce 5 percent of the
total international demand for nickel. In 1948, the Nicaro
plant was inactive and did not re-open until 1952 when the
U.S. was preparing for the Korean War. In 1977, Cuba's
nickel reserves were re-evaluated as being the fourth largest
in the world. Nickel could provide the country with an opportunity to diversify its export base away from sugar,
strengthen its economy, earn hard currency, and reduce its
dependence on the Soviet Union. The industry is government owned.

NOVAS CALVO, LINO (1905-1983). Although he was born in Spain,
Novás Calvo is considered a Cuban writer. He became known
in 1933 with the publication of a book on a slave trader, El Neg-
rero. He is best known for his short stories, such as La luna
nona y otros cuentos (1942), Cayo Canas (1946), and En los
traspatios (1946), and the novelettes Un experimento en el
barrio chino (1936) and No sé quién soy (1945). He left Cuba
in 1960 and served as editor of Bohemia Libre.

NOVEL, THE. The nineteenth century saw in Latin America the
influence of romanticism and the birth of the costumbrista
(manners) novel. The first important Latin American costumbrista novelist was the Cuban Cirilo Villaverde (1812-1894),
whose most famous work Cecilia Valdés (1839, 1882) is possibly
the best representation of life in nineteenth-century Cuba. The

first abolitionist novel in Latin America was the work of Anselmo
Suárez y Romero (1818-1878), Francisco. Villaverde and Suárez
were among the participants in Domingo del Monte's tertulias,
where the romantic spirit of freedom was embedded into a class
of young Cubans. Also worthy of mention is Sab, the romantic
novel by Gertrudis Gómez de Avellaneda. José Martí's Amistad
funesta (1885), although considered romantic by some, has the
modernist elements that characterized his modernist prose. By
the end of the century, Cuban fiction had developed characteristics of its own independent of the literary movements then affecting Spain. As with the rest of Latin America, Cuban literature will be difficult to classify as literary movements and
schools converge in a mestizo literature and socioeconomic conditions will make the novel, in particular, highly politicized.

The social and political life of the early republic are best
captured by Carlos Loveira (1882-1928) in Generales y doctores
(1920) and Juan Criollo (1928). The author's indignation at
social injustice and political corruption characterized his work.
In Alejo Carpentier (1904-1980) the Cuban novel would find its
finest expression. Carpentier brought music to his baroque
prose, along with history, geography, and black mythology,
the magic realism of the new novel. Another major exponent
of magic realism is Paradiso, the only novel by poet José Lezama Lima (1912-1976). Paradiso is considered by many critics
the most outstanding work of the boom novel period. Virgilio
Piñera (1914-1979) is another important member of this generation, although best remembered for his short stories. Other
outstanding Cuban novelists are Guillermo Cabrera Infante
(1914-), Severo Sarduy (1937-), and Reynaldo Arenas
(1943-), all living in exile now. Edmundo Desnoes (1930-)
and Lisandro Otero (1932-) are the only major novelists who
have remained loyal to the Castro regime, but their best-
acclaimed novels were published in the early 1960s. No novels
of the stature of Paradiso or Carpentier's El siglo de las luces
has appeared since the prolific early years of the Revolution,
with Arenas's El mundo alucinante closing that period in 1969.

NUCLEAR ENERGY. A program for nuclear development in Cuba
    began in 1969 with the creation of the Institute for Nuclear
    Research, part of the Cuban Academy of Sciences (ACC), for
    the formation of qualified personnel and the gradual creation of
    the necessary conditions for the subsequent introduction of
    nuclear energy into the country. In 1976 the Cuban Govern-
    ment signed an agreement with the USSR for the construction
    of the first electronuclear plant in Juraguá, Cienfuegos Bay,
    on the Southern coast of Cuba. Cuba then began to partici-
    pate in the COMECON Permanent Commission to collaborate on
    the peaceful use of atomic energy. At the beginning of 1980,
    a decision was made to restructure the nuclear sphere and thus
    the Cuban Atomic Energy Commission (CEAC), adjoined to the
    Counsel of Ministers, was established as well as the Executive

Secretariat for Nuclear Matters (SEAN) which would oversee
the use of nuclear energy in the country. Both organizations
are directed by Fidel Castro Díaz-Balart, the president's son.
The Juraguá Electronuclear Plant is of Soviet design and has
four VVER-type reactors pressurized by water. The complex
is earthquake-proof. The capacity of each reactor is 417 Mw
and the entire plant is of 1600 Mw. It is expected that the first
425 Mw bloc will be operative around 1990.

NUEVA TROVA. See also Music. This musical genre emerged at
the end of 1968 in the midst of a creativity crisis in Cuban
popular music. Although the traditional trova of the beginning
of the twentieth century is said to be a direct antecedent of
the Nueva Trova, there is little apparent relationship between
the two. Thus we find that the themes, tone, and melodies of
their songs only widen the distance between them. As far as
theme in songs, the traditional trova exalted nature and indigenous figures as being national symbols, women's esthetic qualities, while in every song a certain lyricism stood out. On the
other hand, the Nueva Trova sings of social and political themes
that reflected the artist's position within the new political structure. This musical genre acquires the character of a movement
within the country--from which is derived its name, New Trova
Movement. It is promoted by the State and by its interplay
with other forms of musical expression--Protest Song or Com-
promising Song--which appeared in Latin America around the
1970s, an age of political awakening in many nations of the
area. Beginning with the I Congress of Education and Culture
(Havana, 1971), politicization of culture and other artistic expressions permitted the consolidation of this movement and the
appearance of more complex musical groups concerned with what
could be called the "rescue" of national and Latin American musical values, one of the proposals of the I Congress of Education and Culture. The most outstanding figures of this movement have been Pablo Milanés (said to be its founder), Silvio
Rodríguez, Martín Rojas, Noel Nicola, Eduardo Ramos (Tatica),
and Vicente Feliú, in addition to groups such as Moncada,
ICAIC's Sonora Experimental Group, and Manguaré.

NUÑEZ PORTUONDO, EMILIO (1898-1978). Son of the Independence
War General Emilio Núñez. He became a lawyer and served as
legal advisor to the Cuban railways. Núñez eventually served
as Representative and Senator in the Cuban Congress. Under
President Fulgencio Batista (1952-58), Núñez was appointed
Cuba's Ambassador to the UN, in which capacity he curried
favor with the U.S. State Department. In March 1958 Batista
offered Núñez the Cuban Government Premiership, but Núñez
refused in light of the deterioration of Cuba's political situation.
Immediately following Batista's departure from Cuba, Núñez attempted to assist Supreme Court Judge Carlos Manuel Piedra in
forming a transition government. The effort was a failure and
eventually Núñez went into exile.

- O -

OCHOA, EMILIO (1907-). Nicknamed "Millo" as a youth, Ochoa,
    born in Holguín, Province of Oriente, was an early opponent
    of President Gerardo Machado (1925-33) and later became a
    founder of Eduardo Chibás' Ortodoxo party. Chibás' death in
    1951 and Fulgencio Batista's military coup of 1952 transformed
    Ochoa into a fierce opponent of the Batista government (1952-
    58). He was tried on charges of using the television media to
    incite the public to revolt in August 1952. Found guilty, he
    refused to pay a fine and spent a short time in prison. Mean-
    while, the Ortodoxo party split between Ochoa and Roberto Ag-
    ramonte, another party leader. Exiled in Montreal, Canada,
    Ochoa signed an accord with former Cuban President Carlos
    Prío Socarrás (1948-52) by which they committed themselves to
    restoration of the 1940 Constitution. In 1957, Ochoa's branch
    of Ortodoxos and four other political groups formed a coalition
    that proposed elections to be held in a period of 90 days with
    Batista resigning in favor of the current senior Supreme Court
    judge. Batista did not accept the plan, but did agree to par-
    ticipate in the government elections to be held in June of 1958.
    Batista never entered the race. Ochoa remained in Cuba for a
    time although he eventually moved to the U.S. as an exile.

OFICIALES REALES. In the colonial governmental system these
    bureaucrats were responsible for the collection and expenditure
    of revenues and all financial affairs as delegated to them by the
    governor. They were appointed by the Crown and worked with
    the governor and the Casa de Contratación in Seville to increase
    royal revenue. The oficiales consisted of a treasurer, an ac-
    countant, and sometimes a factor, who was appointed to admin-
    ister the King's property. By the 1700's, with Cuba's popula-
    tion growth and increasing wealth, which the Crown actively
    sought to enlarge, the oficiales were forced to appoint deputies
    to represent them throughout the island.

OIL. See Petroleum.

ORGANIZACION DE PIONEROS DE JOSE MARTI (OPJM). Although
    it has been compared to the U.S. Boy Scouts, the OPJM's pri-
    mary goal is to train and educate children in the norms and
    views of Marxist-Leninist society. Up to 1966, the Pioneros
    (Pioneers) were somewhat selective, but then the decision was
    made to turn the group into a mass youth organization. It
    was originally named Unión de Pioneros de Cuba (UPC), but
    the organization was later renamed Organización de Pioneros
    José Martí (OPJM). Today, 98% of all primary-school children
    are Pioneers. The Organization is led, tutored, and controlled
    by Communist cadres who instill in children "a sense of honor,
    modesty, courage, comradeship, love of both physical and intellec-
    tual work, respect for workers, and love for our Revolutionary

Armed Forces and the Ministry of the Interior." OPJM is an autonomous organization with its own administration, and works very closely with the Unión de Jóvenes Comunistas (UJC), a selective organization of the Cuban Communist Party.

ORGANIZACION DE SOLIDARIDAD CON LOS PUEBLOS DE ASIA, AFRICA Y AMERICA LATINA (OSPAAAL). A Cuban-based organization created in September 1966 which evidenced Castro's desire to identify with young students and political ideologues world-wide who embraced the "national liberation" credo. The organization declared in 1968 that revolutionary violence was inevitable even in the United States and that the black's struggle there for equality and against oppression was supported by Cuba.

ORGANIZACIONES REVOLUCIONARIAS INTEGRADAS (ORI). Body created in 1961 by the merging of the Partido Socialista Popular, the Cuban Communist Party, Castro's July 26th Movement, and the Directorio Revolucionario, the last two being the principal organizations that had opposed Batista in the 1950s. The government's objective was to unify these major groups into a united party controlled by the Castro brothers. It was directed by Aníbal Escalante, an old-time communist leader. In 1963, the ORI was transformed into the United Party of the Socialist Revolution (PURS), later renamed the Cuban Communist Party, the island's present ruling party.

ORI. See Organizaciones Revolucionarias Integradas.

ORIENTE (PROVINCE). One of Cuba's six original provinces, it was divided into Las Tunas, Holguín, Granma, Santiago de Cuba, and Guantánamo provinces following the new administrative geographic divisions established in 1976 by the revolutionary government. See related entries for each new province.

ORTIZ Y FERNANDEZ, FERNANDO (1881-1969). Lawyer, anthropologist, historian, and author, Ortiz was born in Havana and educated in Minorca, Spain. In 1901 he received his Doctorate of Laws from the Central University of Madrid. He became Doctor of Civil Law in 1902 and Doctor of Public Law in 1906, both titles conferred by the University of Havana. Ortiz served as Cuba's consular representative in Italy and Spain between 1902 and 1906, then became Public Prosecutor in Havana for two years. He was Professor of Public Law at the University of Havana, but resigned his post after being elected to the Cuban House of Representatives for a 7-year term in 1916. Dr. Ortiz, a fervent student of Cuban history, investigated the Afro-Negro influence in Cuban culture. He was a member of many prominent scientific organizations, including the American Institute of Criminal Law and Criminology, American Academy of Political and Social Sciences, Academia de la Historia de Cuba, and

Sociedad Económica de Amigos del País. From 1910 on, he
was editor of Revista Bimestre Cubana. Ortiz published several studies of Cuban folklore specifically related to the Afro-
Cuban influence including Los negros brujos and Los negros
esclavos, the two best-known studies of the history and culture of blacks in Cuba. Ortiz is most recognized outside of
Cuba for his study Cuban Counterpoint: Tobacco and Sugar.
Ortiz established the Sociedad del Folklore Cubana in 1923 and
the Society for Afro-Cuban Studies in 1926.

ORTODOXO. See Partido del Pueblo Cubano.

OSTEND MANIFESTO (1854). Conceived in 1854, the Ostend Manifesto was a peaceful attempt by the United States to acquire
Cuba from Spain. U.S. ministers to Europe signed a secret
report, later known as the "Ostend Manifesto," calling for U.S.
purchase of Cuba from Spain, and failing this, a forced wrest-
ing of the island from Spanish rule.

OTERO, LISANDRO (1932-). Cuban writer and novelist, Otero is
author of La situación (1963), considered the most experimental
of the new Cuban novels in the early 1960's; En ciudad seme-
jante, which deals with the dictatorship of Fulgencio Batista
(1952-58); and of Pasión de Urbano. He was an active participant in the struggle against Batista and later became one of the
high-ranking spokesmen for cultural affairs under Fidel Castro's
new revolutionary regime after 1959. He had advocated impos-
ing restrictions on the freedom of expression of Cuban intellectuals who, according to the Cuban government must support
fully the Revolution and desist from anti-government criticism.
In 1971 Otero was involved in the condemnation of poet Heberto
Padilla who had previously criticized Otero's novel Pasión de
Urbano. Otero was also made director of the Castro regime's
magazine Cuba.

OVANDO, NICOLAS DE (1451-1511). Born in Brozas, Spain.
Governor of the Spanish island-colony of Hispaniola in 1508 who
sent an expedition headed by Sebastián de Ocampo to circum-
navigate Cuba. This exploration brought back tales of wealth
and a more detailed picture of the island's fine terrain and in-
lets. Ovando was instrumental in the assignment of his lieutenant, Diego Velázques, as conquistador (conqueror) of Cuba
in 1511. Ovando died in Cuba.

- P -

PADILLA, HEBERTO (1932-). Cuban poet who became a cause
célèbre in 1968 when he was awarded the annual prize for

poetry by the Union of Cuban Writers. However, Padilla was
not allowed to accept his prize because one of his poems, "Fuera
del fuego," was not considered sufficiently supportive of the
Communist regime. Already Padilla had been denounced by
Lisandro Otero, novelist and director of the journal Cuba, for
having published criticisms of Cuba's repressive cultural atmosphere in the newspaper El Caimán Barbudo. Padilla was also
attacked for his "anti-revolutionary" works in Verde Olivo,
organ of the Cuban armed forces. "Fuera del fuego" brought
Padilla the support and sympathy of important European and
Latin American intellectuals, as well as the condemnation of the
Executive Committee of the Union of Cuban Writers. Arrested
in 1971, Padilla "repented" and "confessed" his sins against the
Revolution shortly thereafter. The Padilla Case, as the whole
affair was termed, revealed the restraints placed by the Cuban
government on the creativity of intellectuals as well as its intolerance of dissent. Padilla left Cuba in 1980 and currently
lives in exile. He is the editor of Linden Lane Magazine, a
bilingual literary periodical published in the United States.

PAIS, FRANK (1934-1957). Born in Oriente, País became a student
leader at the University of Santiago de Cuba in the mid-1950's,
where he became associated with two minor movements: Acción
Libertadora and Accion Nacional Revolucionaria. País became a
sympathizer of Fidel Castro and reached an agreement with him
in Havana in 1955. After joining the 26th of July Movement,
País was selected by Castro to head the Action Groups in Cuba
against Batista while Castro organized the movement from exile
in Mexico. País traveled to Mexico in November 1956 to help
coordinate an uprising in Santiago de Cuba scheduled for the
end of November, a plan to which he was opposed. Nevertheless, as national underground coordinator of the 26th of July
Movement, he led a group against Batista forces in late 1956,
and by December 1, controlled the city of Santiago de Cuba
for a short while. Later, País organized rebels, provided reinforcements from Santiago de Cuba to Castro's small group of
rebels which had landed in Oriente province and were carrying
on guerrilla operations. On July 30, 1957, País was shot and
killed by Batista's police in Santiago de Cuba; his death sparked
a spontaneous strike in all three of Cuba's easternmost provinces, strengthening Castro's conviction that a general strike
could topple the Batista regime.

PAPEL PERIODICO. Cuba's first newspaper founded by Don Luis
de las Casas, Spanish Governor of Cuba, in 1790. This newspaper counted among its contributors the most significant literary intellects of the time, including Arango y Parreño, Tomás
Romay, Félix Varela, José A. Caballero, and others. Published
weekly from 1791-1793, Papel Periódico became a daily newspaper
in 1793.

PARTIDO COMUNISTA DE CUBA (Cuban Communist Party). In
1961 Fidel Castro, through the merger of Castro's 26th of July
Movement with the Partido Socialista Popular (PSP) and the
Directorio Revolucionario, established a new structure called
Organizaciones Revolucionarias Integradas--ORI--(Integrated
Revolutionary Organizations), a preparatory step toward the
creation of the United Party of the Socialist Revolution--
Partido Unido de la Revolución Socialista (PURS) which in 1965
was transformed into the Partido Comunista de Cuba (PCC),
the Communist Party of Cuba which is the island's ruling and
only party. During the early period the party remained small,
disorganized, and relegated to a secondary position vis-à-vis
the military. It lacked a clear and defined role. Internal
leadership and coordination was poor and meetings were scarce
and of questionable value. Castro saw little need for a well-
developed party structure which would have reduced or at least
rivaled his personalist style of leadership. Conflict between
old-guard Communists and Fidelistas also created tension and
prevented the development of a strong organization. Competition from the military or the bureaucracy took away from the
party the best talents. These cadres saw better opportunities
for advancement in those other sectors than in a party riddled
with factionalism and not warmly supported by the líder máximo.

The decade of the 1970's was one of expansion and consolidation for the Party. During the first half membership expanded
from some 55,000 members in 1969 to 202,807 at the time of the
First Party Congress in 1975. During the second half the rapid
rate of expansion slowed down somewhat. Secretariat member
José R. Machado Ventura disclosed that by 1981 full members
and candidates numbered 434,143. Recently greater emphasis
has been placed on candidates that work on production, teaching, and services. Since many of the earlier Party members
had been promoted rapidly within the ranks and had become
party bureaucrats, the need was for cadres that worked in industry and agriculture and, therefore hopefully, were more
aware of production problems and in closer contact with the
reality of the economy. Also an attempt has been made to bring
more women into the Party. From the time of its organization,
women had been under-represented in the Party's rank and
leadership organs. Even when more women were entering the
labor force few were attaining leadership status. Since 1975,
attempts have been made to correct this situation. Yet women
today are still relegated to secondary positions in the Party
hierarchy and have not achieved representation commensurate
with their participation in the labor force. The First Party
Congress was a watershed in legitimizing the position of the
Party as the guiding and controlling force in society. It re-
assured the Soviets of Cuba's loyalty and friendship, extolling
the former's continuous military and economic aid to the Cuban
Revolution and rehabilitating old-guard Communists, some of
whom had been mistrusted and persecuted by the Castroites.

Three old-guard Communists, Carlos Rafael Rodríguez, Blas Roca, and Arnaldo Milián, were elected to the Political Bureau; these in addition to Fidel and Raúl Castro, Juan Almeida, Guillermo García, Ramiro Valdés, Armando Hart, Osvaldo Dorticós, Sergio del Valle, Pedro Miret, and José Ramón Machado.

During the Second Party Congress in 1980 the Political Bureau was expanded to 16 members. The new members were Osmani Cienfuegos, Julio Camacho, and Jorge Risquet. The First Congress of the Communist Party also expanded the Party's Central Committee from 19 to 112, increased the Political Bureau from 8 to 13, and maintained the Secretariat at 11 members with Fidel Castro and his brother Raúl as First and Second Secretaries. In his report to the Congress, Castro attempted to reconcile the adoption of Soviet-style institutions in the island with a renewed emphasis on nationalism and on the historical roots of the Cuban Revolution. He emphasized that Cuban socialism was the culmination of a struggle against Spanish colonialism and U.S. neo-colonial involvement in Cuban affairs. With total disregard for Martí's ideas, Castro linked the Cuban independence leader with Lenin in order to justify Cuba's move into the Communist camp. The First Party Congress adopted a Five-Year Plan calling for closer economic integration with the Soviet Union and an economic system modeled on other socialist states. The approval of the Party's platform stressing "Marxist-Leninist principles and the leading role of the Party" was further evidence of the impact of Soviet-style orthodoxy in the island.

The second PCC Congress convened in December 1980 and solidified the main tenets of the 1975 Congress report while presenting new dictums for the 1980s. At the international level, the Congress reaffirmed Cuba's strong ties with the Soviet Union, defended such internationalist principles as support for revolutionary movements abroad (with special reference to Nicaragua and Grenada). Within this context, the FAR's external as well as internal roles of assistance to national liberation struggles and national defense buildups were exalted. Regarding U.S.-Cuban relations, the Party anticipated an escalation of tensions between the two countries as a result of Ronald Reagan's election. In the economic sphere, the Congress emphasized the need for more state planning and warned of a possible decline in production and exports due to the world economic situation. The 1980 Congress strengthened the PCC structure and function in the political sphere. Following traditional Marxist-Leninist principles, the PCC was envisioned as the vanguard of the people.

As preparations were made in 1985 for the Third Congress of the Cuban Communist Party, many senior economic planners hoped for a liberalization of the economy and a new policy of decentralization. The Congress, however, was delayed until February 1986, presumably because Castro was formulating an economic plan for the next quinquennial. At the Congress, in

a speech lasting nearly six hours, Castro chastised his fellow
citizens for the rate of absenteeism among Cuban workers, low
productivity, shoddy product quality, waste of materials and
resources, and similar concerns. Some of the most important
work of the Congress was postponed until the end of 1986.
According to Castro, the most significant achievement of this
first part of the Congress was to rejuvenate the party leadership. Almost immediately after the first session of the Party
Congress ended, Castro ordered the closing of the farmers'
market on the grounds of illicit enrichment and corruption
charges.

PARTIDO CONSERVADOR. Created in 1908 during the second U.S.
intervention from the dissolution of the moderate party of Estrada Palma, the Partido Conservador shared the political power
of the nation with the Liberal Party until 1933. Its first presidential candidate elected was General Mario García Menocal, who
won the election of 1912 and was re-elected in 1916 amid out-
cries of fraud from the liberals. By 1924 the party was suffering internal deterioration and weakness primarily from its
long association with the corruption of the Zayas government.
Their candidate for the 1924 election (García Menocal) lost to
Gerardo Machado. The Conservatives' support of the dictatorship of Machado brought them the antipathy of the Directorio
Estudiantil, who in their proclamation of 1933 advocated that
the Conservative Party, as well as the Liberal and Popular
parties be banned due to their support of Machado (1925-33).

PARTIDO DEL PUEBLO CUBANO (ORTODOXO). The Ortodoxo
Party was created in 1946 by a faction that broke with the
auténticos, disillusioned with the public corruption of the
Ramón Grau San Martín administration (1944-48). It was led
by Eduardo Chibás, adopting a broom as its emblem and proclaiming Vergüenza contra dinero (Dignity versus Money), the
ortodoxos aimed to "clean up" Cuban politics and rescue the
revolution of 1933. Chibás was elected to the Senate and in
the 1948 election he ran for the presidency. The ortodoxos
pulled 20% of the votes but lost the presidency to the
auténticos' candidate, Carlos Prío Socarrás (1948-52). In the
1952 election, Chibás again campaigned in a three-way race
against the auténtico candidate, Carlos Hevia, and the other
candidate, Fulgencio Batista. Yet, a few months prior to the
election, Chibás committed suicide. After his death the party
lost much of its dynamism and momentum. Nevertheless,
Chibás' candidacy was replaced by that of Roberto Agramonte,
but three months before the election, Fulgencio Batista staged
a military coup and annulled the election, putting a permanent
end to the political aspirations of the Ortodoxo Party. Members
of the Partido Ortodoxo joined the opposition to Batista and
many became involved in the revolutionary struggle that culminated in the Castro revolution of 1959.

PARTIDO LIBERAL. One of the three principal political parties of
the pre-World War II era. It was established as part of a political alliance organized in Cuba during the second U.S. intervention. Under a platform of "Cubanism" and reform, its candidate, General José Miguel Gómez, was elected President in
1908. After the death of Gómez in 1921, the party suffered
because of internal rivalries. Two potential leaders were Colonel Carlos Mendieta and General Gerardo Machado. Though
Mendieta had considerable popular support, Machado had the
support of powerful allies, including the backing of U.S. interests, which contributed to his nomination as official party
candidate for the election of 1924, and later to his election as
President of Cuba. Under the leadership of Machado, the
Liberal Party dominated the political life of Cuba until the
Revolution of 1933. The Party was characterized by a policy
of moderate economic nationalism, merging with powerful American interests, and widespread corruption and violence.

PARTIDO LIBERAL AUTONOMISTA (Autonomismo). Autonomismo
was the movement that advocated autonomous rule for Cuba
from Spain and differed little from Reformism. After the Ten
Years' War (1868-1878), the movement coalesced into the Partido
Liberal Autonomista, which called for a system of local self-
government patterned after the English colonial model. By
1892, with no changes forthcoming from Spain, disillusionment
and frustration gripped the Party, many of whose members had
hoped for a continuous association with Spain. The Party
warned Spain that without changes and an easing of repression, another rebellion would be inevitable. The Independence
War of 1895-98 soon followed.

PARTIDO REFORMISTA (Reformismo). The reformist movement,
begun in the early 19th century, received new impetus by mid-
century, with both annexation faltering and independence still
nascent. The reformers sought numerous political and economic
concessions from Spain: more equitable taxation, freer trade,
and political representation at the Spanish Cortes. In 1865,
the reform movement was strong enough to evolve into the
Partido Reformista which, although not cohesive, advocated
equal rights for Cubans and peninsulars, limitation on the
powers of the captain-general (governor), and greater political freedom in the island. It also supported freer trade and
gradual abolition of slavery, and called for an increase of
white immigrants into Cuba. Following the successful independence movement in Santo Domingo in 1865, along with Reform
Party pronouncements in its newspaper El Siglo, the Spanish
monarchy called for the election of a reform commission to discuss changes for Cuba. This commission was later known as
the Junta de Información.

PARTIDO REVOLUCIONARIO CUBANO. The Cuban Revolutionary

Party was conceived from the set of resolutions or bases that evolved out of an exile meeting in Key West on January 5, 1892, attended by José Martí. These bases represented a pragmatic approach to the Cuban situation, appealed to all classes and races, and discussed Cuba as a factor in the international political scenario. Among other things, the bases, as the ideological statement of the Partido Revolucionario Cubano (PRC), called for complete independence for Cuba and Puerto Rico and for another war of independence to seek the first point. The revolutionary party was given a democratic organization by Martí based on the civilian command over the military. The party was active for about three years, during which time it (and Martí) argued for "continuous intervention of the Cuban people in the control of its own affairs."

PARTIDO REVOLUCIONARIO CUBANO (Auténtico). After Fulgencio Batista had engineered Dr. Ramón Grau San Martín's downfall and the thwarting of the 1933 Revolution, the students who had supported Grau were frustrated and disillusioned. Many desired to continue fighting for their frustrated revolution, forming the Auténtico Party in 1934. The party's model was the earlier party of José Martí (the Partido Revolucionario Cubano of 1892). Directorio leaders joined the new Auténtico Party and Grau San Martín was appointed President. The program of the auténticos called for economic and political nationalism, social justice, and civil liberties. Through the 1930's and into the mid-1940's, the auténticos consistently opposed Batista and his puppet presidents. In 1944, Grau and the auténticos were elected at the same time that organized use of violence to achieve political power increased. To many Cubans in the late 1940's, the auténticos and Grau had failed to fulfill the aspirations of the anti-Machado revolution, especially with regard to administrative honesty. Eduardo Chibás and other auténticos split from the party in 1947, forming the Partido del Pueblo Cubano (Ortodoxo). The failure of the auténticos to bring profound structural economic and political changes to Cuba was perhaps the single most important factor contributing to the 1952 coup d'état by Fulgencio Batista.

PARTIDO SOCIALISTA POPULAR (PSP). Original name of the Cuban Communist Party. Carlos Baliño, a prestigious figure of Cuba's war of independence and later founder of the Communist Association of Havana, collaborated closely with Julio Antonio Mella in the magazine Juventud, bringing the young student leader into the association. Encouraged by the Mexican Communist Party and supported by its envoy in Havana, Enrique Flores Magón, Baliño and Mella called a congress of all Communist groups in the island for August 1925. The number of militant Communists in Cuba was small, and of the nine Communist groups only four sent delegates. From this 1925 congress emerged the PSP. José Miguel Pérez, a Spanish

Communist, was appointed Secretary General, and the party was
soon affiliated with the Communist International. Mella became
one of its most important leaders, entrusted to propagandize
the creation of the party, edit its newspaper, Lucha de Clases,
and direct the education of new party members. Although the
party was small and disorganized, the Communists soon formed
a Youth League, using Mella and other student leaders to agitate and gain followers within the University. The Communists
adopted the name Unión Revolucionaria Comunista and dedicated
themselves during the 1930s to the development of labor unions.
These efforts brought about the Cuban Confederation of Workers (CTC--Confederación de Trabajadores de Cuba), established
in 1939 and given official status in 1942. The Communists opposed the government of Gerardo Machaco (1925-1933) but at
the end of this regime promised him cooperation in opposing the
general strike that forced him to resign.

This final deal with Machado gave a bad reputation to the
Communists. They supported Fulgencio Batista (1940-44 and
1952-59) when they joined in the forming of the Coalición
Socialista Democrática for the candidacy of Batista in the 1940
elections. During Batista's presidency they had a Minister
without Portfolio, Carlos Rafael Rodríguez, and were very influential in the government. The Communists changed the name
of the party to Partido Socialista Popular in 1940. President
Ramón Grau San Martín (1944-1948) permitted Communist activities until 1947 when he broke with them and sought the control
of labor unions by his Auténtico party. President Carlos Prío
Socarrás (1948-1952) attacked the Communists and curbed their
remaining influence in labor unions. After Batista's coup in
1952, ousting President Prío, he cooled off with the Communists,
nevertheless the PSP leaders condemned Fidel Castro's violent
opposition to the regime. Only at the end of the struggle
against Batista some Communists joined the fighting. Former
Batista cabinet member Carlos Rafael Rodríguez went to the
Sierra Maestra mountains and spent over a year with Fidel
Castro until his victory in 1959. In 1961 the revolutionary
government merged the three organizations that had opposed
Batista--the Partido Socialista Popular, Castro's 26th of July
Movement, and the Directorio Revolucionario--and remained supportive of Castro. The organization was called Organizaciones
Revolucionarias Integradas (ORI). In 1962 its name was changed
to Partido Unido de la Revolución Socialista (PURS) and ulti-
mately to Cuban Communist Party in 1965.

PARTIDO UNIDO DE LA REVOLUCION SOCIALISTA (PURS). In
1963, ORI (Organizaciones Revolucionarias Integradas) disappeared without fanfare and its successor, PURS (Partido Unido
de la Revolución Socialista) appeared without inauguration. As
part of the evolutionary process by Castro of weeding out figures unacceptable to him, PURS itself gave way to the new Communist Party of Cuba (PCC) in 1965.

PATRONATO REAL. A body of rights and privileges granted by
the Pope to the Spanish monarchy during the colonial period
which permitted the kings to nominate all higher church digni-
taries coming to the New World as well as to control the administration of ecclesiastic taxation. In practice, the King and his
officials in the New World became the secular heads of the
Church. The Church became a political arm of the state and
was to dominate and influence Indians and blacks as well as
colonists.

PEDRERO PACT. A unity pact signed in November 1958 in Las
Villas province between the Guevara forces and the Directorio
Revolucionario when Che Guevara arrived from Oriente in Las
Villas and took command of anti-Batista operations in that area.

PELAEZ, AMELIA (1897-1968). A famous modern Cuban artist born
in Yaguajay, Las Villas province. After studying art in Cuba,
she won a scholarship to continue her studies at La Grande
Chamière in Paris. As a teacher in San Alejandro School in
Havana, she greatly influenced the younger generation and
was a pioneer in changing the "classical" school of painting.
Her work in ceramics was also outstanding. Most of her work
is in collections throughout the world and she held exhibits
in Havana, New York, and Paris (Galerie Zak). She was a
master in oil and water color.

PEÑA, LAZARO (1911-1974). Born in Havana, Peña left school at
age 13 to support his family and eventually became a tobacco
worker. In 1932 he was imprisoned for his participation in the
Torcedores (manual tobacco-manufacturing) strike. He was a
member of the Tobacco Workers' Union of Havana and a member
of the Communist Party since 1930. Peña denounced the
Gerardo Machado regime (1924-33) and was jailed for three
months. Upon release, Peña began to organize the workers
of the principal industries in Havana as well as in the interior;
his activities cost him jail sentences under various regimes. In
January 1939, a new Cuban Labor Federation, the Confederación
de Trabajadores de Cuba (CTC) was established through Peña's
efforts. Peña was unanimously elected the CTC's first
Secretary-General. The CTC was directed by the Cuban Communists for some eight years. In 1947, the anti-Communist
Minister of Labor, Carlos Prío Socarrás, began a campaign
against Peña and the Communists under the government of
Ramón Grau San Martín (1944-48) to end Peña's control over
the Cuban labor movement. Peña fled to Mexico until 1953,
traveling later to Moscow until Fidel Castro took power in
1959. He returned to become once again the leader of the CTC.
Peña was an open supporter of Castro. He advised and worked
closely with Castro regarding Cuban labor.

PENINSULARES. Name given to Spaniards living in America during
the colonial period, usually the main upholders of power.

PENTARCHY (Pentarquía). After the overthrow of the short-lived
Carlos Manuel de Céspedes presidency in September 1933, Fulgencio Batista, the army, and the Directorio Estudiantil, named
a 5-man commission to lead a provisional government. The five
civilian members of the Pentarchy were: Dr. Ramón Grau San
Martín, Sergio Carbó, Porfirio Franca, Guillermo Portela, and
José M. Irisarri. The power behind the Pentarchy was based
on a shaky alliance of several factions bound together by common revolutionary ideals. The pentarchy lacked U.S. and
Cuban political parties' support and thus collapsed in September 1933 over bickering as to selection of the next president.
The Directorio, with Batista's consent, selected from the
pentarchy members Grau San Martín as the new President,
ending the Pentarchy's function.

PEOPLE'S POWER. With the advent of the Revolution of 1959 and
the subsequent radicalization of the system, the traditional
forms of local government disappeared (municipalities, municipal
committees, etc.) and were replaced by other state administrative organizations with more limited functions under the JUCEI
of the sixties (Coordination, Implementation, and Inspection
Committee). Later on the Poder Local (Local Power) acquired
the functions of the JUCEI, i.e., the control of purely public
duties. The authority of the Fidel Castro regime was sub-
stantively enlarged and to some extent modified with the new
political structures added during the institution-building phase
of the 1970s. The National Assembly of the People's Power is
such a structure, resulting from popular approval of the new
Constitution through a national referendum held on February
24, 1975. Under the new Socialist Constitution, the regime
sought to increase the institutional cohesiveness of the state
by integrating it further into the newly established political
structures. In agreement with the constitutional mandate, on
December 2, 1976, the National Assembly of the People's Power--
with its powerful Council of State--was called to session to hold
its first historic meeting and exercise legislative power. The
National Assembly is headed by a president, a vice-president,
and a secretary, 14 politico-administrative committees of 7 to 21
members, and 169 politico-administrative Municipal Assemblies
with executive committees of 5 to 15 members. The delegates
to the Municipal Assemblies are elected for a 2 1/2-year term
by the neighborhood Comité de Defensa de la Revolución (CDR)
by secret, popular vote on the basis of one municipal delegate
for each electoral circumscription, and it is not necessary to be
a party member to be elected. Following an indirect, electionary
process, the municipal delegates elect their provincial counterparts for a term similar to their own. There is one Provincial
Assembly for each of the 14 provinces.
   The number of provisional delegates is determined by the
population of each province on the basis of one delegate for
every 10,000 citizens, or fraction greater than 5,000. Also,

the municipal delegates elect the deputies to the National Assembly with one national deputy for every 20,000 citizens. Approximately 55.5% of all national deputies come from the ranks of the municipal delegates. At the national level, approximately 90% of the delegates are members of the PCC; at the municipal level approximately 50% are party members. The 1976 (present) Constitution established the jurisdiction of the Organs of People's Power at its three different levels: national, provincial, and municipal. The Provincial and Municipal Assemblies also have authority over the operation of schools, hospitals, and transportation. Administrative functions are exercised at both municipal and provincial levels. They include: (1) enforcing laws and regulations coming from superior state organs; (2) electing the members of their executive committees and revoking their mandate if the need arises; (3) determining the organization, functioning, and tasks of the administrative units in charge of the socio-economic activities under their control; (4) appointing and recalling the judges of the people's courts in their jurisdictions in agreement with the 1977 reorganization of the judicial system; and (5) enhancing socialist legality and internal order. The OPP provincial level works as an intermediate unit between the National and Municipal Assemblies and as a centralizing agency between different municipalities. However, in some instances, as when national deputies are rendering to their Municipal Assemblies, the provincial level is circumvented. As expected, the National Assembly elected Fidel Castro as the first president under the new Constitution. Blas Roca was appointed first President of the National Assembly. According to Article 67 of the 1976 Constitution, the National Assembly stands at the top of the pyramidal structure of the Organs of the People's Power, representing the "sovereign will of all the working people."

Besides being invested with legislative and constitutional reform authority, its powers include revoking the decree laws issued by the Council of State (the functioning body when the National Assembly is not in session); approving the national budget; declaring a state of war in the face of military aggression; appointing the members of such organs as the Council of State, the president, vice-president, other judges of the Supreme Court, and the attorney general of the Republic; and exercising control over the organs of state and government. The National Assembly holds regular sessions, usually for two days, twice a year, in July and December; also, whenever the need arises, it is called to hold an extraordinary session. During its first three years, from 1976 to 1979, the National Assembly approved approximately 30 laws, reached more than 35 agreements, and held several elections. This heavy legislative load includes the annual state budgets, yearly economic and social development plans, and civil, penal, administrative, public, military, and other laws. The laws are drafted by the Commission on Judicial and Constitutional Affairs of the National

Assembly, but they are put together in coordination with any
of the 20 standing commissions that function throughout the
year conducting business in preparation for the regular sessions. The deputies are normally assigned to one or more
commissions, which are staffed with advisory and clerical personnel. Regardless of how demanding working as the people's
representative could be, the national deputies continue holding
their former jobs with no additional pay other than being reim-
bursed for their expenses while discharging their responsibilities; however, they are freed from their daily jobs whenever
they need to attend Assembly sessions or commission meetings,
which are held in different cities throughout the year. The
record of the National Assembly seems to indicate that the
deputies take in earnest their political representative roles,
discussing their constituency's concerns as well as their own.
Laws can be proposed not only by the national deputies, but
by the Council of State, the Council of Ministers, the commissions of the National Assembly, the trade unions and mass organizations, the People's Supreme Court, the Attorney General,
and through a petition signed by at least 10,000 citizens who
are eligible to vote. In spite of such advances, it does not yet
seem that national deputies can freely demonstrate their desire
to reject a major policy or propose new alternatives aside from
what is officially recommended.

PEOPLE'S REPUBLIC OF CHINA. Diplomatic relations between the
PRC and Cuba were established in September 1960. By November, 1961, with Cuba comfortably within the Communist bloc,
China sent an ambassador to Havana. Cuba recognized and
tried to maintain neutrality towards the Sino-Soviet dispute in
the mid 1960's, concluding trade agreements with the Chinese
in November 1961 and exchanging sugar for equipment and
technical aid. During the October 1962 missile crisis, China
backed Castro's rejection of the private bargain between Kennedy and Khrushchev and supported Castro's call for no UN
inspection of dismantled sites and U.S. withdrawal from Guantánamo. Castro's assertion of Cuban aid to revolutionary
movements around the world after the Tricontinental Conference in January 1966, along with his attack on Chinese perfidy
with regard to Communism, signalled a Cuban movement towards
the Soviet Union. This re-alignment with Russia lasted into the
decade of the 1980's, as Sino-Cuban and Sino-Soviet relations
cooled.

PERIODICALS. See also Mass Media; Newspapers; Press. Cuba in
the 1950s had 60 to 70 well-established dailies, most of which
were in Havana. Cuban periodicals were read throughout the
island thanks to good transportation service. The 28 main
newspapers claimed a circulation of 580,000. Periodicals were
exempted from paying income tax and from duties on import
of raw materials. Notable newspapers in Cuban history include:

Diario de la Marina, a conservative paper founded in the late
19th century; El Mundo, founded in 1901. El País, Prensa
Libre, and Excelsior, were three more recent (mid 1950s) newspapers. After 1959 these newspapers were confiscated by the
government and were replaced by an official daily, Revolución,
later renamed Granma, and Hoy, the Communist daily. Cuba
also enjoyed two fine weekly magazines, Bohemia and Carteles,
patterned after U.S. magazines such as Time or Newsweek.
They too were taken over by the Cuban government which has
continued to publish Bohemia.

PERSONALISMO. A Latin American tradition in which the personality
of a single leader is the major unifying force in the government,
political party, or movement. Personalismo has led to a reduced
importance of institutions and to a growing allegiance to one man
or group.

PESO. Cuban monetary unit. Prior to 1959, one peso equalled one
U.S. dollar. After the revolution the peso lost value and could
be purchased in the black market during the 1960's at the rate
of 15 pesos for one U.S. dollar. The Cuban government maintains an official exchange rate of one U.S. dollar for 70 Cuban
cents (centavos).

PETROLEUM. See also Fuel. Before the Castro Revolution, most of
Cuba's production was in the hands of the United States, although
the British did have one significant distributive organ, the Compañía Petrolera Shell de Cuba in the late 1950's. Petroleum refining capacity underwent a sixfold increase from 9,000 barrels
per day in 1954 to 59,000 barrels in 1957, due to government
encouragement of refining of imported crude. Daily consumption of refined petroleum at 60,000 barrels was therefore met
adequately. After Castro nationalized U.S. assets, including
petroleum in 1960 and 1961, the industry encountered difficulty
in refining the different grades of the new Soviet crude and production declined into the early 1960s. Cuba became dependent
on the Soviet crude. In 1968, during a difficult period in Soviet-
Cuban relations, deliveries experienced a slowdown. Since then,
the USSR has met Cuban oil needs, often at subsidized prices
and bartered for Cuban products. This has caused Cuba to run
into enormous deficits. Since 1980, Cuban domestic oil production has increased significantly to 740,000 MT in 1983. Cuba
began in the 1970s--and on a major scale since 1980--reexporting
Soviet crude and its refined products, accounting for 40% of
Cuba's hard currency earnings for 1983-85. Cuba needs these
earnings but depends on Soviet willingness to maintain itself as
an oil exporter.

PEZUELA, JUAN MANUEL DE LA (1810-1875). Pezuela was the son
of the last Spanish Viceroy of Peru and was thus a marquis.
Anti-slavery and incorruptible, Pezuela was also ultra-conservative

and politically motivated. He was appointed the first Captain-General (governor) of Cuba under the Spanish government in 1853. Upon arriving on the island, Pezuela immediately published a series of articles in the local newspapers which called for emancipation of slaves and enforcement of treaties prohibiting the slave trade. Pezuela was accused by Cuban planters of conspiring to "Africanize" Cuba by sending large quantities of negroes to the island, only to free them after a short period of forced labor. On May 3, 1854, Pezuela issued a decree in Havana ordering the annual registration of all slaves, which provided that any owner who could not present a document of title for his slave must emancipate that slave. This measure created fear and hostility among Cuban hacendados. By November 1854, after the new Spanish government had refused a U.S. offer to purchase Cuba, Captain-General Pezuela was de-commissioned and replaced by General Gutiérrez de la Concha. Within a few weeks, Pezuela's crusade against slave trafficking was abandoned, and the "Africanization" issue was no longer considered a threat.

PINAR DEL RIO (PROVINCE). Population: 547,288; Area: 10,901 Km²; Municipalities: Sandino, Minas de Matahambre, La Palma, Candelaria, Los Palacios, Pinar del Río, San Juan y Martínez, Mantua, Viñales, Bahía Honda, San Cristóbal, Consolación del Sur, San Luis, Guane. This is the westernmost of the provinces. With few changes it maintains the same boundaries it did before the political-administrative division of 1976. It possesses the most appropriate lands for the cultivation of tobacco (Vueltabajo) and it is the most important producer in the country. It maintains mining activity in Matahambre and fishing on both coasts. It cultivates minor fruits and is also involved in some forest exploration. Its main mountainous area is located in the Organos mountain range where one can find the beautiful Viñales Valley. The capital of the province is the city of Pinar del Río which got its name from the settlement in 1699 of a village along the nearby pine shores of the Guama River. It became a city officially in 1880 under the name of Nueva Filipina. Other important cities are Guane, Consolación del Sur, and Bahía Honda.

PIÑEIRO, MANUEL (1933-   ). Born in Matanzas on March 13, 1933, to a middle-class family and received his secondary education in his home town. In September 1953 he entered the American Language Center, School of General Studies, of Columbia University in New York. He attended classes there until mid-1955. After returning to Cuba late in 1955, Piñeiro joined the underground movement against the late President and Dictator (1940-44; 1952-58) Fulgencio Batista and became chief of the 26th of July Movement in Matanzas. In June 1957 he joined Castro's rebel forces in the Sierra Maestra. When Raúl Castro opened up the "Frank País Second Front" in the Sierra del Cristal in northern Oriente in March 1958, Piñeiro accompanied him as his aide. Piñeiro subsequently attained the rank of comandante

(major) and was appointed the territorial chief of the Sierra del
Cristal. Following the collapse of the Batista regime on January
1, 1959, he became commander of the Moncada Barracks in Santiago and chief of the rebel forces in Oriente province. He allegedly exercised extreme cruelty while presiding over the revolutionary tribunals that tried former batistianos. Reports indicate that friction developed between Piñeiro and some of the
military troops under his command because of his alleged Communist sympathies. The local people felt that he was involving
the military in strictly civil affairs. In mid-June 1959 he was
transferred from Oriente to the Army General Staff in Havana,
where he became involved in the Directorate of Intelligence of
the Revolutionary Army, an agency later known as G-2, under
the command of Major Ramiro Valdés Menéndez. Piñeiro served
as assistant to Valdés. When the Interior Ministry was established in June 1961 under Valdés, all the Cuban intelligence
agencies were incorporated under it.

In late 1961 Piñeiro was identified as the nominal chief of a
MININT section known as El Aparato (The Apparatus), which
was allegedly responsible for checking on high government and
military officials as well as foreigners working for the government. Soviet personnel allegedly controlled that section. Shortly afterward the DGI was created. Piñeiro became its director
and probably acquired the title of Technical Vice-Minister at
this time. Internal and external intelligence operations were
separated, the DGI being given the responsibility for foreign
intelligence collection and clandestine operations. In 1962
Piñeiro served as a member of the five-man military tribunal
that presided over the trial of the prisoners captured during
the April 1961 Bay of Pigs invasion of Cuba. In January 1966
he was a delegate to the Tricontinental Conference held in Havana. After the failure of several Cuban operations, the collapse of national liberation movements in various Latin American
countries, and defections within its own ranks, the DGI and
Piñeiro found themselves in serious trouble for about a year
during 1968 to 1969. In 1969 the Soviets persuaded Cuba, particularly through Raúl Castro, to reorganize the DGI into an
intelligence collection agency responsive to Soviet needs as well
as those of Cuba. Under these circumstances Piñeiro's projects
to promote revolutions abroad, then unacceptable to the Soviets
but a personal interest of Fidel's were removed from the DGI
and made a separate organization, staffed with officers loyal to
Piñeiro, which subsequently evolved into the Americas Department of the Cuban Communist Party. In July 1980, Piñeiro accompanied Fidel Castro and other top-ranking Cuban leaders as
a member of the Cuban delegation to the first anniversary of the
FSLN victory in Nicaragua. Piñeiro is a member of the Central
Committee of the Party. He is readily recognizable by his red-
dish thick mustache and full beard, whence his nickname "Bar-
baroja" (Redbeard) is derived.

PIÑEYRO BARBI, ENRIQUE (1839-1911). Prolific writer and superior

literary critic and active participant in the struggle for independence from Spain, Piñeyro was also a teacher at Colegio del
Salvador in Cuba. While in exile in the United States, he was
a member of the Junta Revolucionaria and Secretary of the Diplomatic Delegation (República en Armas) presided by Morales
Lemus. He was editor of La Revolución (1870) and founder of
El Mundo Nuevo (1871), both pro-independence newspapers
edited in the U.S. His intense revolutionary campaign did not
divert his attention from his literary work. He was the author
of Historia de la Literatura Española, Cómo acabó la dominación
de España en América, and Biografías americanas, among others.
He returned to Cuba after the Pact of Zanjón (1878) and left
for Paris in 1882, where he resided until his death.

PIONEROS. See Organización de Pioneros José Martí.

PIRATES AND CORSAIRS. In the 16th century, the colonial powers
did not maintain large navies as offensive or defensive security
forces for their colonies. When war broke out, explorers and
traders and their men continued commercial activities and simply
added on the extra task of periodically attacking enemy colonies.
These men were the corsairs. The English coveted the new
wealth emanating from Spanish America and, by the 1570's or
1580's, Francis Drake, the most renowned corsair, began ham-
mering away at the possessions of Spain. Later, Dutch corsairs
plundered Caribbean islands, including Cuba into the early 17th
century. Pirates differed from corsairs in the sense that their
allegiance to any one nation was less obvious. Early pirates in
the late 16th and early 17th centuries were mostly British and
indeed in 1655 they helped capture Jamaica from the Spanish.
This made Cuba vulnerable to further attacks by pirates and
smugglers. Toward the latter part of the 17th century, a new
type of international pirates emerged who were mercenaries with
no political allegiance. They plundered for whichever nation
paid them more in times of war and for themselves in times of
peace. These pirates or buccaneers lived on various Caribbean
islands and preyed upon Spanish, Dutch, French, and Portuguese colonies.

PLATT AMENDMENT, 1901. U.S. Congressional resolution included
in the 1901 Cuban Constitution, the Amendment was defined in a
declaration entitled "For the recognition of the independence of
the people of Cuba, demanding that the government of Spain
relinquish its authority and government in the island of Cuba,
and to withdraw its ... forces from Cuba ... and directing the
U.S. President to carry these resolutions into effect." The
amendment also specified that the United States would leave the
government and control of Cuba to its people as soon as Cuba's
government could establish a constitution, which would define
its relations with the United States. Provisions of the Platt
Amendment include: "I. That the government of Cuba shall

never enter into any treaty or compact with any foreign power
or powers which will impair or tend to impair the independence
of Cuba.... II. That said government shall not assume or contract any public debt.... III. That the government of Cuba
consents that the U.S. may exercise the right to intervene for
the preservation of Cuban independence.... IV. That all Acts
of the United States in Cuba during its military occupancy
thereof are ratified and validated, and all lawful rights acquired thereunder shall be maintained and protected.... VII.
That to enable the United States to maintain the independence
of Cuba, and to protect the people thereof, as well as for its
own defense, the government of Cuba will sell or lease to the
United States land necessary for coaling or naval stations...."

PLATT AMENDMENT, ABROGATED. The Treaty of Relations with
Cuba abrogating the Platt Amendment, which was concluded on
May 22, 1934, was signed on May 29, 1934. Under the Treaty,
all of the acts effected in Cuba by the United States during its
military occupation of the island remained valid, as did all the
rights legally acquired by virtue of those acts. The treaty also
provided that any agreements in regard to the lease to the
United States of lands in Cuba for coaling and naval stations
at Guantánamo Bay would continue in effect until abrogated by
both parties.

PODER LOCAL (Local Power). Grass-roots governing units created
by the revolutionary government of Fidel Castro to supervise
the small businesses nationalized in 1968. With the beginning of
the Revolutionary Offensive, however, the poder local units,
lacking coordination and organization, were replaced by poder
popular. See People's power.

PODER POPULAR. See People's power.

POETRY. With the spread of romanticism to Cuba, poetry became
the main vehicle of the artist. Themes included slavery, pre-
Columbian Indians, and local customs. Much Cuban poetry was
written in exile, however, and the forms were European.
    Colonial Period (neo-classical style). Until the late eighteenth
century Spain devoted few of its resources to Cuba, and life was
difficult for the colonists. The few educated persons among
them endeavored to create an intellectual and artistic life modeled
in the image of Spain. Cuban poets of the colonial period created a body of literature memorized by succeeding generations
that passionately told of their love of country. A characteristic common to poetry of this and future periods was the identification of the poet's personal well-being with the welfare of
his homeland. Much of it, written in exile, had a quality of
unreality and sang of a country only dimly remembered, whose
unity and peasants were idealized and whose slaves would be
freed. One of the most frequently quoted poets of this period,

the last to write in the neo-classical style, replete with classical allusions and verse styles but nonetheless lyrical, was José María Heredia y Heredia (1803-1839). The young Heredia wrote and published most of his work in the United States (although he is also known for his works in French), and is best known for his "En el Teocalli de Cholula" (In the Temple of Cholula), written in 1820, and "El Niágara," written in 1824 after a visit to the waterfall. Another poet of this genre was the mulatto Gabriel de la Concepción Valdés, known as Plácido (1809-1844). An early romantic poet, he favored such themes as medieval legends, Moorish Spain, idealized pre-Columbian Indians, and political freedom. He was a forerunner of the specifically Cuban brand of romanticism called ciboneyismo. Gertrudis Gómez de Avellaneda (1814-1873) is also recognized for her neo-classical poetic style.

Pre-Independence Period (modernism). Poets of this period sought to develop a distinctive, regional Latin American style flavored with nationalistic sentiment. One of the best known is José Martí (1853-1895). Martí was the personification of revolutionary romanticism, but he was also one of the forerunners of the Latin American modernist movement. Martí reserved his poetry primarily for the expression of his innermost thoughts, his loves, and his increasing preoccupation with death. In Ismaelillo (1882) he recorded his tender feelings for his son and homeland, expressed in regular meters but in a style that presaged the coming of the modernists. His Versos Libres (Free Verses), written around 1882 but published posthumously, was a collection of compact, introspective poems. Although the modernist movement never attracted the following in Cuba that it did elsewhere in Latin America, Julián del Casal (1863-1893) and Martí are widely respected throughout the hemisphere as being among its originators and most influential figures. Casal's poetry was increasingly affected by his ill health and reflected his preoccupation with death. He wrote three books of poetry: Hojas al viento (Leaves to the Wind), Nieve (Snow), and Bustos y Rimas (Busts and Rhymes), written in 1890, 1892, and 1893 respectively. By the end of the nineteenth century he, along with Martí and others, had established poetry as the dominant genre of the republican period. Federico Urbach (1873-1932) was the principal exponent of the related symbolist school. Imitating luxurious verbal textures of contemporary "decadent" poets in Europe and the United States--nonpolitical, introspective, fond of the exotic in verse form and subject matter--Urbach sought to develop a distinctive, regional Latin American style.

Republican Period. With the success of the independence movement, the literary effort lost its chief inspiration. From about 1925 on, however, growing discontent among middle-class intellectuals was accompanied by an increased literary output. Greater concern with social conditions of the nation was expressed by the poets of this period. Agustín Acosta (1886-

1979) wrote "La Zafra" (Cane Harvest), a poem depicting the
life of the sugar workers in 1926. Still widely known in Cuba,
this poem presaged the emergence of the 1930's self-consciously
proletarian literature. Many poets of the period, possibly reacting against the color prejudices of Cuban society, became Marxists. Nicolás Guillén is credited with originating the negrismo
(negritude) school of poetry, using Afro-Cuban forms and set-
tings to write strongly proletarian literature, portraying the
bitter life of the lower classes, both black and white, and their
exploitation by those in authority. In Guillén's Motivos de son
(Motives of Dance), written in 1930, he translated the musical
form of the son dance rhythm (originating in Haiti) into a literary form. In this and later poems he imitated African drums,
using repetition and alliteration in a style reminiscent of the
American poet Vachel Lindsay. In his West Indies, Ltd. (1934)
and Cantos para soldados y sones para turistas (Songs for
Soldiers and Dances for Tourists), written in 1937, the hopes
and frustrations of the exploited classes are expressed. Guillén became head of the cultural department of the Ministry of
Education under Castro. Typical of the negrismo group is the
work of Regino Pedroso, a Havana factory worker. Juan Marinello, who was subsequently a leader of the Popular Socialist
Party was another poet of the negrismo school. Eugenio Florit
(born 1903 in Madrid) was a poet of the republican period who
wrote fastidious, contemplative verse, collected in Poema mío
(1947) and Antología poética (1956). Florit was on the faculty
of Columbia University in New York for over twenty years.

    Pre-Revolutionary Period. Poetry continued to dominate the
literary scene throughout the 1940's and 1950's. Most of it was
generated and published in Havana, frequently in the magazine
Orígenes (1944-1957) (see Magazines). Poets in the Orígenes
group were characterized as transcendentalist, evasive, and
elitist. The more socially oriented poets such as Guillén and
José Z. Tallet (1893-1985) associated themselves with more
activist publications, such as Ciclón (1955-1959) (see Magazines). Most of Cuba's leftist intellectuals--who included only
a few poets--belonged to this society. Another school of poets
evolved in the province of Las Villas whose work resembled that
of the Orígenes group stylistically and thematically but whose
chief concerns were the incorporation of folkloric elements and
the simplification of language. Samuel Feijóo (1914-), Alcides
Iznaga (1914-), and Aldo Menéndez (1918-) were the chief
representatives of the Las Villas group. These poets were very
prolific, publishing over fifteen books of poetry between 1944
and 1958, as well as books of criticism and folkloric research.

    Post-Revolutionary Period. Following the Revolution, there
was a decided tightening of the license allowed artists and writers. The final stage in the controversy between the revolutionary government and Cuban intellectuals over freedom of expression was ignited and executed largely by the poet Heberto
Padilla (1932-), considered the most overtly dissident writer

in Cuba at the time. Padilla lost his job with the government-
line newspaper Granma and was denied permission to travel to
Italy, but he appealed to Fidel Castro and was given a job at
the University of Havana. Padilla was awarded the UNEAC
prize by an international jury in 1968 for his book of critical
verse, Fuera del juego (Outside the Game). The poem's title
implied Padilla's hostility to the Revolution, as it alluded to
Castro's statement, "Outside the Revolution, nothing." Nevertheless, Padilla's book was published anyway, with the political
disclaimer of UNEAC. A series of articles in Verde Olivo, the
official magazine of the armed forces, began by censuring Padilla and ended by criticizing Cuban writers in general. Padilla's jailing and subsequent public self-flagellation aroused international anger against the Castro regime. European intellectuals
wrote letters protesting his imprisonment and confession and comparing them with events of the Stalinist era in the Soviet Union.
After 1959 most poets continued to be active to one degree or
another, depending on their support for the new socialist ideas.
Those who were classified as "social" poets--Guillén and Iznaga--
have since become the greatest promoters of the Revolution,
whereas the Revolution has little affected the themes or style
of the Orígenes group.

Although several of the original group have since left the
country, those remaining have continued to be prolific and influential. Other poets of the Batista era have written sporadically, several adopting other writing styles. The first generation
of the Revolution in poetry consists of those born around 1930
who began publishing during the Batista era but whose works
matured after 1959. This group includes Roberto Fernández
Retamar (1930-), Fayad Jamís (1930-), Pablo Armando Fernández (1932-1962), and Padilla. This generation shares the
experience of having lived abroad during the social and political upheavals of the 1950's so that they missed an important
part in the formation of the revolutionary culture. Consequently, these poets are occasionally out of step with the Revolution,
as Padilla was in the late 1960's. Although this group lacks cohesiveness in attitude, themes, and style, they still represent
the most prominent voice in literary circles, largely because of
their prolific output and excellent quality. During the Revolution's first decade, these poets dominated literary prizes, cultural institutions, and major journals. The power and universality of their verse led them to be recognized and acclaimed
outside of Cuba as well.

Those born between 1940 and 1946 constitute the second generation of poets who reflect the internal conflicts posed by the
attempt by Cuba's cultural leaders to institute praxis. One
sector of the second generation was recognized through the
establishment of a private publishing house, El Puente (The
Bridge), which operated from 1960 to 1965. The El Puente
group has been accused of common aesthetic and moral and
political weakness despite the heterogeneity of its poets. The

purge and labor camps of the mid-1960's were directed largely against this group, when the founder of El Puente was himself sent to a labor camp, the group falling apart. After this incident many went into exile. This group constituted a temporary flowering of Cuban literature, regardless of their commitment to the Revolution. Another segment of the second generation that tended to take a moderate stance, although politically committed, associated themselves with the literary magazine El Caimán Barbudo. Victor Casaús, Luis Rogelio Nogueras, and Jesús Díaz Rodríguez, all formed part of this group. Critical of both the propagandists and the liberals, these poets rejected the didactic, populist stance of the former and the transcendental, escapist tendencies of the latter. This group tried to adhere to Castro's 1961 mandate by claiming not to write for or against the Revolution from within it. This group has nevertheless been involved in several severe conflicts with the government including the Padilla case in 1967 and 1968. Eventually, these problems led to the resignation of the entire El Caimán Barbudo editorial board and to the temporary suspension of its publication. All original members of this group have, however, remained in Cuba and have continued to be energetic contributors in the field.

A third generation of poets in their thirties during the mid-1970's is still being formed. These poets belong neither to the El Puente liberals nor to the El Caimán Barbudo moderates, although Lina de Feria (1945-   ), one third-generation poet, became editor of El Caimán Barbudo after the publication's temporary demise. Another promising third-generation poet is Excilia Saldeña (1964-   ). Irrespective of age or political conviction, Cuban writers seem to have found poetry the most expressive way of describing the introspective, interpersonal, and societal conflicts generated by the Revolution. Themes have evolved from the metaphysical concerns of the Orígenes group to the practical and physical problems presented by the new society, such as the changing of old traditions and values, international conflict, and the pressures for conformity. Attempts at purely lyrical poetry have been made throughout. New poetic forms have emerged to suit the new realism, and there is a definite trend away from sophisticated language and construction toward a much more simplified and colloquial expression. Notable Cuban dissident poets who have been imprisoned by the Castro regime include Armando Valladares, Enrique Salas, and Ernesto Díaz.

POEY Y ALOY, FELIPE (1799-1891). Born in Havana, Poey y Aloy was a brilliant scholar of the natural sciences and professor at the University of Havana. He specialized in the study of fishes. Although Poey had graduated as a lawyer from the University of Madrid, he never practiced the profession, choosing instead to write poetry and literary criticism. Much of his work is gathered in his volume Obras literarias (1888). But it was in the field of zoological study and investigation that Poey excelled,

gaining wide recognition by European scientists, such as Cuvier
and Valenciennes, for his scientific observations, which in some
aspects predated those of Darwin. Poey was a founding member
of the Sociedad Económica de Amigos del País, the Academy of
Sciences, and the Society of the Island of Cuba. Among Poey's
writings are Ictología cubana; Memorias sobre la historia natural
de la isla de Cuba; Sinopsis o catálogo razonado de los peces
cubanos; Curso elemental de minerología. Poey died in Havana.

POLAVIEJA, GENERAL CAMILO GARCIA DE (1838-1914). The
Spaniard Polavieja was an aide to Spanish General Arsenio
Martínez Campos and was instrumental in drafting the Treaty of
Zanjón, which put an end to the Ten Years' Independence War
(1868-1878). He also helped to crush the Guerra Chiquita
(1878-1879) in Oriente province. Later, Polavieja reversed his
anti-independence posture and favored concession of independence to Cuba by the Spanish crown. From 1890-1892 Polavieja
was governor of Cuba. He resigned and later became governor
of the Philippines and conducted anti-independence campaigns
there. In 1897 Polavieja was considered for the position of
Spanish Prime Minister but never attained the position despite
support from the Catholic Church.

POLITICAL PRISONERS. During colonial times, Spain confronted
sporadic opposition to its rule and crushed these efforts by
executing and imprisoning Cuban patriots for short periods of
time as well as by sending them into exile. It was expected
that this would discourage opposition. Cubans, however, infiltrated back into the island or sought support for the revolutionary cause in foreign countries. The father of Cuban independence, José Martí, was imprisoned briefly, forced to work
in a quarry, and finally exiled to Spain. During the Cuban
republican period before the Castro Revolution, no government
was free from strong opposition but only twice, in the 1930's
and 1950's, were there significant numbers of political prisoners. During those periods opposition was of revolutionary proportions and developed after dictatorial regimes assumed power
and became increasingly repressive. In 1906 and because of the
Platt Amendment--which gave the United States the right to intervene if political turmoil threatened stability in Cuba--the
United States intervened, but there were no political prisoners.
The first presidents of Cuba solved political opposition by power-
sharing or sinecures. During Alfredo Zayas' presidency (1921-
1925), there was an uprising against the regime in Cienfuegos,
Las Villas province, and the president himself intervened to end
opposition. When asked whether he wanted soldiers to accompany
him he replied that he had his checkbook and fountain pen.
Dictator Gerardo Machado (1925-1933) crushed opposition by offering sinecures to politicians or by short imprisonment. Political prisoners were usually released at the request of the president's friends and were sent into exile. Torturing and even

killing of prisoners "while trying to escape" (Ley de fuga) became a method to deal with opposition. The repressive measures
of Machado were followed by the strong-arm tactics of Fulgencio
Batista in the late 1930's and by the execution of opponents of
his regime. When opposition escalated after his 1952 coup, Batista became increasingly repressive, jailing and torturing opponents. Batista, however, solved the accumulation of political
prisoners by periodic amnesties, his belief being that a short
period of imprisonment would ameliorate anti-government feelings.
Even Fidel Castro, imprisoned after the attack on the Moncada
Barracks on July 26, 1953, was released by an amnesty in 1955.

When Castro's revolution triumphed, executions of Batista
supporters were followed by massive political imprisonment--
characteristic of the Castro regime--the largest Cuba had ever
witnessed. Within the first two years of the revolution large
numbers of real or suspected opponents had been jailed. Sources
indicate that as many as 150,000 persons were arrested during
the Bay of Pigs invasion, without due process of law. A number of these remained in jail until the 1970's. Cuba's prisons,
estimated to be about fifty scattered throughout the island, are
mostly maximum security institutions. Until the 1981 Mariel exodus these were overcrowded, harshly administered, and the
scene of torture and deprivation. The most notorious of these
are Castillo del Príncipe and La Cabaña fortress, Boniato prison,
San Severino castle, and América Libre, one of the three women's
penitentiaries. Before its conversion to the Island of Youth
where students from foreign countries are trained, the Isle of
Pines was the largest prison in Cuba. Forced labor is imposed
on male political prisoners; women are forced to work in the
fields. The government has consistently refused to allow inspection of Cuban jails or to provide information regarding the
treatment and numbers of political prisoners in Cuba. Despite
attempts by groups such as the Inter-American Human Rights
Commission, the International Commission of Jurists, Amnesty
International, and the International Red Cross to gain information, estimates remain highly speculative and vary greatly.
Castro told U.S. journalist Lee Lockwood in 1965 that there
were 20,000 political prisoners and not the 100,000 claimed by
exile sources. Jaime Caldevilla, a Spanish diplomat in Havana,
gave these figures in 1969: 24,000 in concentration camps,
7,000 in jails, 7,200 on penal farms, and 17,000 held by the
security police.

Castro has used the "ransom" technique to release political
prisoners for hard currency or foreign products. After the
Bay of Pigs invasion, $100,000 ransom was placed on the heads
of prisoners of latifundium families. A few were released on
this basis until a deal was negotiated between the United
States and the Castro Government for the release of some
2,000 captured invaders by paying $35 million in medicines and
agricultural equipment. After 1976 a group of exiled Cubans
initiated conversations with Castro concerning the release of

prisoners as well as other matters. Success on the release of
prisoners was achieved. During the Mariel exodus in 1980 some
political prisoners who had served their sentences but had not
been allowed to leave the island were sent over to the United
States along with numerous common criminals and mental patients
released by the Castro regime. Cuba's political prisoners are
divided into two groups, those who have submitted to what is
referred to as "Rehabilitation," and those who have refused to
do so. The former are removed from maximum security prisons
to labor camps from which they have a chance of being released
upon the completion of their terms. Those who refuse to cooperate with the government, or so-called plantados (the unyield-
ing ones), are held in maximum security prisons with limited
visits and food packages. Hunger strikes have been common.
Former Rebel Army Commander Eloy Gutiérrez Menoyo, who had
been a plantado for twenty-one years, was released in December
1986, following Spanish Primer Minister Felipe González's visit
to Cuba.

POPULATION. With 10 million inhabitants by 1985, Cuba was fairly
densely populated, but the density was considerably lower than
in the other major islands of the Antilles. A large proportion
of the land was arable, and the population crowding that was
beginning to plague other islands had yet to become a matter of
immediate concern to Cubans. Between 1959 and 1971, moreover, as many as 7% of the country's citizens chose not to live
in revolutionary Cuba and engaged in a massive emigration,
principally to the United States. During the period that immediately preceded the Revolution, the birthrate had been somewhat lower than the average for Latin America, but a low and
declining rate of mortality had been extending the span of life
expectancy. During the first years under the revolutionary
government there was a sharp increase in the birthrate, which
some observers attributed at least in part to the excitement and
sense of euphoria that accompanied the revolutionary takeover.
The rate commenced a decline in 1965, however, and dropped to
under 3% annually in the late 1960's and early 1970's. Although
the Cuban government's 1971 suspension of further refugee
flights and the termination of those flights after a short-lived
resumption between December 1972 and April 1973 caused an up-
turn in the growth rate, the rate anticipated for the late 1970's
and the 1980's was a moderate one well below the Latin American
average.

     In the mid 1970's, the population of Cuba was young but not
excessively so in comparison with other countries of Latin
America. The 1970 census, showing the median age to be some-
where between 22 and 23 years, and in many other countries of
the region half or more of the population, has considerable political significance. About 40% had been born after the revolutionary government's assumption of power, and more than half had
no real memory of any other government. In the mid-1970's the

Cuban government had no strong feelings about family planning
and population control. Its international posture was one of
support for the position of the Third World, which related over-
population to imperialism and an unjust distribution of economic
wealth rather than a function of unchecked population growth
in underdeveloped countries. With respect to its own population, however, it recognized the right of parents to limit the
size of their families and their right to be informed on population problems and planning techniques. It did not actively engage in family-planning activities but permitted both the distribution of contraceptive devices and the use of medical facilities
for family-planning clinics and studies. Before 1959 the Cuban
government had adopted a policy of official encouragement of
family planning. The revolutionary government never officially
abandoned that position, and in the mid-1970's international organizations concerned with the family-planning program continued
to list Cuba in that category. Cuba was one of the few countries of Latin America not to have joined the International Planned
Parenthood Federation (IPPF), but, because of its moderate
growth rate and relatively large amount of arable land, population pressure remained a distant problem in the mid-1980's.

Although it had a population about one-tenth of Japan's,
Cuba had about the same amount of land under cultivation. In
one important aspect the country's position on population was
unique. While all other Latin American countries were experiencing unemployment in varying degrees and in a majority the
rate gave promise only of increasing, Cuba had what it regarded
as a shortage of workers. The socialist economy had converted
the fairly heavy unemployment of the late 1950's into a full-
employment situation, and, although there was some doubt about
the reality of the labor shortage, economic and social planners
were not required to concern themselves immediately with consideration of population controls as a means of combating future
unemployment. According to the government 1971 census, for
every 1,000 inhabitants, the crude birthrate was 29.5, the
crude death rate was 6.0, the rate of natural increase was 23.5,
the net emigration rate was 5.8, and the net growth rate was
17.7. According to the same census, the total population numbered 8,569,121. Structurally, there were 4,392,970 males and
4,176,151 females. The urban population was 5,169,420 and the
rural population 3,399,701. In 1981, Cuba's total population
numbered 9,796,000; the average annual rate of growth in 1979-
80 was 1.1%; the birth rate (1977) was 17.6 per 1,000 persons;
the population density (1981) was 86 per square km. or 223 per
square mile; and the ethnic division of the population was 51%
of mixed origin, 37% European, 11% African, and 1% Chinese.

PORRA. Special secret police during Gerardo Machado's regime
(1925-33) which was outside the control of either the army or
the police. The Porra was organized to combat street demonstrations and to intimidate enemies of Machado.

PORTELL VILA, HERMINIO (1901-   ). Historian, journalist, and
professor. Born in Cárdenas, he was a Professor of History
at the University of Havana but also taught at universities in
Washington and California. An adversary of the Cuban Communist regime of Fidel Castro, he went into exile in 1959 and
taught and worked in Washington, D.C., as a newspaper and
radio commentator on international events. His most famous
historical books include Historia de Cárdenas; Narciso López y
su época; El ideario político del Padre Varela; Evolución de la
política y la democracia en Estados Unidos; Historia de Cuba en
sus relaciones con Estados Unidos y España (4 volumes).

PORTOCARRERO, RENE (1912-   ). Painter. Portocarrero is the
most famous exponent of expressionist painting. He is magnificent in his murals and during his early period his paintings expressed the nostalgia of the colonial patio, the Havana rooftops,
the corners of solitary streets. His paintings have been shown
in Europe and North and South America, and he has been acclaimed internationally. He lives in Cuba where he has accepted
honors and support from the Castro government.

PORTS. The leading Cuban ports are Havana, Santiago de Cuba,
Matanzas, Nuevitas, Cienfuegos, and Mariel. In addition, there
are some other twenty-five serviceable ports of lesser importance.

POZOS DULCES, CONDE DE (FRANCISCO DE FRIAS) (1809-1877).
Multi-faceted aristocrat involved in scientific research, journalism, and politics, he has been called the Cuban most representative of his generation. The Conde de Pozos Dulces was a reformist, probably the first to suggest the need for agrarian reform in Cuba. He was editor of El Siglo, the voice of Cuban
Creole aspirations. As a founding member of the Junta de Información (1865) in Spain, he called for political reforms for
Cuba. He was responsible for writing the protest of the commissioners when the hopes for reform faded. The Conde de
Pozos Dulces conceived and designed the Vedado de La Habana,
a well-known neighborhood in Havana. He was the author of
several works on Cuban agriculture, and a member of the Economic Society, the Academy of Sciences, and many others. The
Conde de Pozos Dulces advocated Cuban independence through
evolution, not revolution, and became so disillusioned by the War
of 1868, that he emigrated to Paris despite financial difficulties.
He remained in Paris, and nearing death, requested that his remains not be transported to Cuba until the island was free.

PRENSA LATINA. New agency created by Fidel Castro after the
triumph of the revolution to disseminate the views of the Cuban
regime.

PRESIDENTIAL PALACE ATTACK. Attempt to kill President Fulgencio Batista (1952-58) on March 13, 1957, in a combined effort

of the Directorio Estudiantil and the Auténticos, led by Carlos
Gutiérrez Menoyo, by assaulting the Presidential Palace. The
failure of reinforcements to arrive turned the attack into a
costly defeat for the revolutionaries. Most of the would-be
assassins were killed, including Gutiérrez Menoyo and José
Antonio Echeverría, leader of the Directorio, after storming a
Havana radio station to announce the fall of Batista, as had
been planned.

PRESNO, JOSE ANTONIO (1876-1950). Physician, surgeon, and
teacher, Presno was born in Regla, Cuba, and was educated in
the Church Schools of Guanabacoa, where he graduated with a
Bachelor's degree. He later graduated from the Medical School
of the University of Havana, and in 1893 was named Assistant
in Dissection in the Mercedes Hospital. Presno was founder of
the Revista de Medicina y Cirugía de la Habana and attained
the degree of Licentiate in Medicine. He became Doctor in
Medicine and Assistant Instructor in Surgical Anatomy and
Operations in 1897. Dr. Francisco Zayas, Secretary of Public
Instruction, appointed Presno Assistant Professor Extraordinary
in 1898. Presno was the recipient of many more prestigious
awards, titles, and recognitions between 1898 and 1904, when
he was chosen to represent Cuba by the organizing Committee
of the Second Latin-American Medical Congress, and in 1905, he
was elected General Secretary of the First National Medical Congress of Cuba. Presno was suggested as a Presidential candidate by some students after the overthrow of President Gerardo
Machado in 1933. He was Rector of the University of Havana
during the first administration of President Ramón Grau San
Martín (1933-34) and Minister of Health from 1944-48 during the
second Grau administration. Presno was a contributor to Cuba's
medical literature, publishing among other works, Tratamiento de
los aneurismos externos; and La situación topográfica del apéndice
cecal.

PRESS. The Cuban press has, since the 1930's and 1940's, been at
least partially subsidized by several governments including those
of Grau and Batista. Journalists were often directly involved in
political life since many newspapers were published by political
leaders or dependent upon the subsidies already mentioned. As
opposition to Batista mounted in the late 1950's, many newspapers loyal to him tried to garner support and labeled the activities of Castro's forces as "repressive terrorism." Castro in
turn solicited foreign press correspondents to visit him in the
Sierra Maestra, to relate the story of the guerrillas' struggle.
After Fidel Castro assumed power in January 1959, several
papers did attack the Revolution (including Diario de la Marina,
Prensa Libre, and Avance). Castro ignored these protests,
considering the sources. Castro relied on his mouthpieces,
Revolución and Hoy, to counter and discredit the opposition.
Castro used radio and TV stations to denounce the U.S. and

Spanish ambassadors (Bonsal and Lojendio) as war criminals,
thieves, and oligarchs as early as January 1959. After consolidating his position and that of the revolution, Castro, in
the 1960's, used censorship of all forms of the media to control dissemination of information. TV, radio, and the print
media all have been used to extol the virtues of the revolution
and to warn of possible future tyrannies, should the U.S. gain
any foothold. Newspapers have become rather tedious, repetitive organs of the Central Committee of the Communist Party of
Cuba. Granma, for example, seems less a newspaper than an
exhortation sheet, reporting Castro's speeches in their entirety,
with little or controlled news.

PRIO SOCARRAS, CARLOS (1903-1977). President of Cuba; Lawyer.
Born in Bahía Honda, Pinar del Río province, Prío became active
in politics at an early age. While studying law at the University
of Havana, he became an active member of the Directorio Estudiantil in the 1930's. In 1939, he was a founding member of the
Constitutional Assembly that drafted the Constitution of 1940.
Prío was elected Senator (1940-48) and also served as Prime
Minister (1945) and Minister of Labor (1947-48) in the administration of President Ramón Grau San Martín (1944-1948). By
1948 Prío was the presidential candidate of the Auténticos and
won decisively. Unlike his predecessors, Prío was an experienced statesman who emphasized that the Executive and Legislative branches of government had to work together. Prío's
administration adhered closely to the Constitution of 1940, re-
specting civil liberties and achieving major accomplishments in
the areas of social and labor legislation such as the social security benefits plan for sugar and wheat industry workers
(Retiro Azucarero and Retiro Harinero) and the law to stabilize
rent payments (Rebaja de Alquileres). Prío established the
Banco Nacional de Cuba, subjecting currency and banking to
strict government regulation. Public works legislation provided
for the extension of major highways and construction of the
National Library. Also, the sugar export agreements negotiated by the government with the United States, Great Britain,
and West Germany brought an economic boom to Cuba. Under
Prío, the Tratado Interamericano de Asistencia Recíproca (Rio
Treaty) and the Tratado de Asilo Político (Treaty of Political
Asylum) were ratified by the Senate.
    As President, Prío espoused democratic ideals while enrich-
ing himself on the spoils of corrupt friends and crime figures.
His administration was beleaguered by the problems of gangster-
ismo or pandillismo inherited from his predecessor, wherein
gangs used terror and violence to extract political favors from
government and business. Prío facilitated the existence of
these gangs by refusing to assume an aggressive stance against
their tactics. Many of the gangsters were protected by Prío's
own cabinet. Corruption among government officials reached
alarming proportions during his administration. Prío attempted

a reorganization of government in response to accusations of official graft and corruption in a program known as Los nuevos rumbos (the new routes), whereby he purged his administration of the worst perpetrators of corruption, including his own brother. By 1950, however, the Ortodoxo Party of Senator Eduardo Chibás had become greatly popular and vocal in demanding an end to corruption. Despite Prío's achievements, politics came to be regarded with disrespect by the Cuban people. To become a politican was to enter an elite, a new class apart from the interests of the people. Political figures were the objects of popular mockery. In particular, the image of the Presidency was ridiculed and abused. Chibás' criticism helped to undermine not only the authority of the Auténticos but the stability of Cuba's already fragile political institutions.

On March 10, 1952, less than three months before national elections were held, Fulgencio Batista staged a coup d'état with heavy military backing and overthrew the Prío administration, arresting key supporters and assuming control over the mass media. Batista faced virtually no opposition. Prío sought asylum in the Mexican Embassy and later left for Mexico. Prío moved to Florida in 1953 and actively worked against the Batista regime. He was arrested temporarily on charges of violating the U.S. Neutrality Law of 1939. Prío returned to Cuba in August 1955 under the terms of Batista's general amnesty. Prío began organizing an opposition movement to Batista and also supported Fidel Castro. He was forced to return to Miami in May 1956 when Batista charged him with conspiring to overthrow the government. From the United States, Prío continued to conspire against the Batista regime, and was instrumental in the organization of a Council of National Liberation which included representatives of all opposition groups except the Communists. Prío financed most of these opposition groups. He was indicted before a Federal Jury in Miami for planning an arms delivery to rebels in Cuba from Miami, and for a short time went to jail. In July 1958, by now free, Prío participated together with Castro in forming in Caracas a "Junta of Unity" against the Batista regime. With the triumph of the revolution in 1959, Prío returned to Cuba and vocalized his support of the new Castro government. Eventually, however, he became disillusioned and departed for Miami, where he lived until he committed suicide in 1977.

PRISONERS. See Political prisoners.

PROCURADOR. A solicitor chosen by the Cabildo during the early colonial period to represent the interests and desires of the community. He also served as liaison between the settlers and the Spanish crown. The procuradores met annually to discuss the island's needs and to choose a general procurador to carry their grievances and requests to the King. By 1532, the governor was suspicious of this embryo of a representative government,

and the meetings of the procuradores as well as their powers
were abolished.

PROTESTANTISM. See also Church; Religion. Spanish ancestry
and Catholicism have dominated in Cuba since the 16th century--
thus Cuban attitudes and value were those of a Catholic country
prior to 1959. Competing moral systems arrived in Cuba by
means of Protestant missionaries in the late 19th century. Rationalism in the form of Freemasonry also challenged the Catholic dominance among the middle and upper classes. Protestantism, however, remained outside the debate between rationalism
and Catholicism, both of which it opposed. Protestantism was
introduced in 1884 in a small fashion and remained largely
insignificant until the first occupation of Cuba in 1898. Thereafter, some 40 denominations established missions, schools,
clinics, and hospitals. The leading Protestant denominations
were Southern Baptist, Methodist, Presbyterian, and Episcopal.
Protestant sects were supported financially by U.S. sources
enabling them to offer "free religion" so they did not have to
charge fees for marriages and baptisms. Most Protestant missionary work was concentrated in lower and middle income
groups in the urban areas. Protestantism never successfully
competed with Catholicism for the devout Cuban.

PROVINCES. When the Republic was established in 1902, the island
was divided into six provinces: Pinar del Río, La Habana,
Matanzas, Las Villas, Camagüey, and Oriente, following the
divisions made by the colonial government in 1879. The revolutionary government changed the administrative structure of Cuba
in 1976 and created the following divisions: provinces of Ciudad
de La Habana (base of the central government), Pinar del Río,
La Habana, Matanzas, Cienfuegos, Villa Clara, Sancti Spíritus,
Ciego de Avila, Camagüey, Las Tunas, Granma, Holguín, Santiago de Cuba, and Guantánamo. The municipality of Isla de la
Juventud (Isle of Youth) encompasses the former Isle of Pines
and is administered by the central government. See related
entries under each new province.

PURS. See Partido Unido de la Revolución Socialista.

- Q -

QUESADA AROSTEGUI, GONZALO (1868-1915). Born in Havana.
Writer, orator, and collaborator of Martí and his closest friend.
Quesada was the first to publish the works of Martí and the
executor of his will. In the works of Martí, Quesada included
his own notes, comments, and a prologue to each volume. He
was Cuban Minister to Washington and Germany where he also

published Martí's writings. A friend of Emperor William II, he
died in Germany in 1915.

QUINTAS. Community recreational and health service centers established by Spaniards during the colonial period. Quintas
offered cultural, social, and educational activities as well as
medical services for its Spanish members who paid regular
monthly dues. The various regional provinces of Spain were
represented by their respective quintas (e.g., Asturias, Galicia,
Andalucía). These mutual aid societies were the early precursors of the modern-day clínicas (clinics), which are more health
services-oriented in nature.

- R -

RADIO. Historically, the most significant means of transmitting political information as well as entertainment was the broadcast
media. After the 1959 Revolution, all radio stations were confiscated to ensure that no dissenting views would be aired to
"impair the independence of the revolution." The leading radio
station was WCMQ, which the government changed to Radio
Liberación. Radio Rebelde is considered the official government
station for domestic broadcasting, while Radio Habana Cuba,
with broadcasts in various languages, is designed strictly for
an international audience. In addition, local and regional stations also operate throughout the island, but all national and
international newsbroadcasts are controlled by the National Information Agency (AIN).

RAILROADS. By 1837, the first railroad in Latin America was operating in Cuba and the decades that followed saw railways rapidly built, 400 miles by 1860 and 1300 miles by the turn of the
century. Investments were at first mostly British and as the
century progressed, American. Railway development cannot be
separated from the growth that the sugar industry experienced
at the time. It facilitated and cheapened transportation to
warehouses and ports. Access to railroads also marked the difference between those small planters who could not afford to
build railways and a new class of sugar kings, increasingly of
U.S. origins. The nation's extensive railroad system deteriorated rapidly in the first years after the Revolution, but service
was slowly being restored. According to the Cuban government,
about 12 million tons of general freight was being carried annually in the early 1970s compared with an annual average of
about 25 million tons in the 1950s. The figures do not include
the large amounts of sugarcane and milled sugar transported by
rail; about 70 percent of cane and sugar travels by railroad. In
1958 there were fourteen railroad companies. After nationalization

they were placed under one enterprise, Cuban Railroads, which is part of the Ministry of Transportation. Cuban Railroads, employing 20,000 workers in 1984, is composed of four divisions and several subdivisions. Total trackage of Cuban Railroads in the early 1980s was about 2,800 miles, although some uneconomic lines were being phased out. An additional 5,600 miles of track between sugar fields and the mills, formerly belonging to the sugar plantations, are operated by the Ministry of the Sugar Industry, which has its own rolling stock. Some of the sugar ministry's rail cars use Cuban Railroads track, and some of the rolling stock of Cuban Railroads is loaned to the sugar ministry during the harvest season, the only time that the ministry's railroad operate. Most of the track gauge, including the sugar lines, was standard gauge (56.5 inches).

The Cuban Railroads operated about 11,000 freight cars and the Ministry of the Sugar Industry about 29,000 in the early 1980s. Locomotives totaled 300 and 800 respectively during the same period. There were nearly 300 railroad stations along the lines of Cuban Railroads, some of them only a few miles apart. The Ministry of the Sugar Industry's lines operated between the sugar mills and 444 cane collection centers. After nationalization major maintenance problems were aggravated by the shortage of replacements. The government showed little inclination to assign high priority to railroad maintenance. The consequent deterioration of rail lines and rolling stock were quickly felt, because the major lines had been making little or no profit since the 1930s and hence had done no more than keep the equipment marginally operable. In 1972, for example, half of the Cuban Railroads' locomotives were out of use for considerable periods of time. With the help of the Soviet Union, a long-range transportation improvement program was developed during the 1960s and was being implemented in the early 1970s. The policy included modernizing the more important rail lines, rebuilding tracks, repairing key branches, and constructing larger stations. Some lines have been completely relocated, particularly along the Havana-Santiago de Cuba route. Older wooden bridges were replaced by concrete spans, and new diesel locomotives have been acquired from abroad. The older steam locomotives of the Ministry of the Sugar Industry are also to be replaced by diesel locomotives. Railroad piggyback service began in 1975 between Havana and Santiago de Cuba, whereby containers loaded with cargo traveled on flat cars and then were reloaded on trucks or ships.

RAINFALL. In Cuba there are two distinct seasons based on rainfall. The rainy season runs from May to October, with almost daily afternoon storms. The dry season runs from November to April. Average rainfall ranges from 8.2 inches in June to 1.5 inches in February. Heaviest rainfall accompanies the hurricane season--August, September, and October. The provinces of Pinar del Río and Havana suffer most of the brunt of this potentially destructive period.

RAMIREZ, ALEJANDRO (1777-1821). Spanish intendant; economist.
Ramírez served as intendant in Cuba from 1816 to 1821 under
the Spanish governors José Cienfuegos and Juan M. Cagigal.
He had previously been stationed in Guatemala and Puerto Rico.
Ramírez was an able administrator and his tenure was characterized by prosperity and significant development in the island,
particularly in the sugar industry. He supported the Economic
Society, abolished in 1817 the tobacco monopoly, and encouraged small farming and diversification. He obtained the Crown's
approval in 1819 to consider landowners all those who could
prove that they had been on their land for the past forty years.
This process facilitated the breakdown of large estates, contributed to the growth of the sugar industry, and benefitted a
new class of proprietors who could either sell their land at a
profit, become sugar producers themselves, or lease their land
to others.

RAMOS ENRIQUEZ, DOMINGO (1894-1957). Foremost Cuban and internationally renowned painter, outstanding for his landscapes.
His paintings of the Valle de Viñales in Pinar del Río province
have not been surpassed either by Cuban or foreign painters.
As a professor and for many years director of the Academia
de San Alejandro he influenced many Cuban painters. He held
more than 20 exhibits in Europe and in America. Among his
many works that won prizes are Valle de viñales; El coloso en
la cumbre; Abril florido; El hato de Caiguanabo; El Tajo de
Ronda; Mediodía; Lebena.

RATIONING. See also Economic development. Rationing was introduced in April 1962 after the accumulated reserves of the old
regime had been exhausted. Increased demands brought increased consumption of food and consumer goods and when
stocks were not replaced, the store shelves became empty.
Rationing came as a shock where rationing had not existed
even during World War II. Years of affluence were replaced
by years of scarcity (continuing into 1980s) and declining productivity of workers has only accentuated the scarcities.
Shortages continued to be both widespread and severe into the
decade of the 1980s. By 1971 the ration list included: 6 lbs.
of rice per person per month, 1 1/4 ounces of coffee per week,
5 lbs. of sugar per month, and 1/2 lb. of lard and 1 lb. of
meat per person per month when available. Only one bar of
laundry soap and one of hand soap per month, and clothing
distributed only every 6 months. In addition, electricity was
rationed by means of brownouts in Havana. In December 1986, the regime announced more austerity measures.
Clothing was further rationed, as were milk and dairy products;
and meals were reduced to one per day at daycare centers.

REAL COMPAÑIA DEL COMERCIO DE LA HABANA. The commercial
outfit that in 1740 obtained from the Spanish Crown the exclusive privilege of transporting products between Cuba and Spain.

For some twenty years, the company purchased products such
as flour and fabrics at bargain prices and sold them in Spain
while importing Spanish merchandise and selling it in Cuba for
high prices. In addition, the monopoly company introduced
over 5,000 black slaves into Cuba, enriching its directors while
entrepreneurs outside the island shared no similar benefits.

REALENGOS. Under Spanish colonial rule the realengos were par-
cels of land retained by the Crown, which fell between the official circular land grants known as mercedes. The town councils or cabildos in Spanish-ruled colonial Cuba were authorized
to grant the various realengos to citizens who paid the appropriate tax to the Crown.

REBEL ARMY. See Army, revolutionary.

RECIPROCAL TRADE AGREEMENT OF 1934. Signed in August
1933 between the U.S. and Cuba, the reciprocal agreement
reduced the tariff on Cuban sugar another 9 cents per pound
over the 1.5-cent decrease in the 1930 tariff act. The new
treaty gave Cuba certain benefits for sugar, rum, tobacco,
and vegetables. In turn, the U.S. received several concessions from Cuba, including lower import duties on some U.S.
goods and abolition or reduction of internal taxes on a large
number of American products. The reciprocal trade agreement was one example of Franklin D. Roosevelt's New Deal
approach to reconcile the international economic interests of
the country with domestic interests.

RECIPROCITY TREATY OF 1903. President Theodore Roosevelt in
1901 agreed with the concept of reciprocity between Cuba, a
newly formed nation, and the United States. Reciprocity meant
a reduction of duties upon Cuban sugar and tobacco, with a
simultaneous lowering of import duties in Cuba on U.S. products. Roosevelt believed in the U.S.'s destiny to control the
Cuban sugar market and indeed the economic activity of all
Caribbean and South American lands. A heated U.S. Congressional debate over the Reciprocity Treaty occurred throughout
1902 and most of 1903 due to U.S. domestic sugar interests'
lobbying to defeat the bill. It finally passed in mid-1903 and
was signed in December 1903. The Treaty provided a 20%
preferential reduction in the American tariff on Cuban sugar
(and other exports) while various American products received
from 20% to 40% reductions in the Cuban tariff.

RECONCENTRACION. See also Weyler, General Valeriano. The
most controversial policy undertaken by the Spanish general
Valeriano Weyler, who was sent to Cuba to put down the 1895
insurrection against Spain. Weyler ordered entire populations
of towns and villages to be concentrated in "military" areas in
specific outposts which would be served by special zones of

cultivation. Most of Cuba in 1896 was such a "military area,"
and so Weyler's policy of concentration was designed to suppress popular uprising throughout the island. Treason was
punishable by death and military commanders had extraordinary
powers to try, punish, and execute "traitors." Significantly,
the policy caused extensive destruction of the agricultural wealth
with an estimated loss of $40 million, since sugar mills and to-
bacco factories were at a standstill throughout 1896.

REFORMISM. See also Partido Reformista. Existing in Cuba since
the early 1800's, reformism took new impetus in mid-century
due to failures of the violent opponents to Spanish control and
the growth of a feeling of nationalism. The reformists called
upon Spain for numerous political and economic concessions such
as equitable taxation and freer trade along with representation
in the Spanish Cortes. In 1865, the Partido Reformista (Reformist Party) was organized. Its members generally favored eventual
independence for Cuba, equal rights for Cubans and Peninsulars,
freer trade, and gradual abolition of slavery. Due to Spanish
fears of losing Cuba to the clashes between Reformists and
Peninsulars, Spain moderated its reaction to reformism in 1865,
and for two years tried to mediate its demands.

RELIGIONS. See also African religious practices; Catholics; Jews;
Protestantism. The Spanish Catholic Church has dominated
Cuban life since the 16th century, yet the Church has not been
as powerful in Cuban national life as it has elsewhere in Latin
America. The Catholic Church has repeatedly been a target of
co-optation by leaders of revolutionary movements throughout
Cuban history, and this has partly accounted for the Church's
poor following. Protestantism and African traditions competed
with Catholicism as they were brought into Cuba by middle-
class European intellectuals and slaves. Catholicism and Protestantism were middle-class concerns and only half the total
population participated in any organized religious exercises up
to the 1950s. In rural areas, churches were scarce and those
in existence were usually built by local sugar planters on whom
the community was dependent. Churches, therefore, were not
like social clubs, the focal points of social and cultural activity
on the island. In colonial Cuba, the absence of any large Indian
populations eliminated the Church's special powers over indigenous populations as in Mexico and Peru. Negro slaves practiced
their African cults despite sporadic oppression by the authorities, thus observance of these Afro-Cuban religions into the
20th century were cloaked in secrecy mandated by the need to
avoid persecution. The 1901 Constitution ordered separation of
Church and State to deprive the Church of its tax revenues
and the political support of the government. In the 1920s, a
union of the Church, upper class, and conservative press gave
rise to a conservative, moral atmosphere in Cuba. Catholic
schools burgeoned in the 1930s, and efforts were made to

demonstrate the relevance of the Church to the populace into
the 1940s. In the middle 1950s, the sense of national crisis as
the excesses of the Batista regime multiplied pushed many upper-
and middle-class people to look to the Church for salvation. In
1959 there were 200 parishes served by 700 priests (mostly Spanish) and a secular clergy of 1,000 men and 2,400 women in religious orders, most of these also Spanish. The highest officials
of the Church in 1959 were Cuban, but only one-fifth of the
priests were Cuban nationals.

REVOLUCION. Underground newspaper of Castro's 26th of July
movement. After 1959, it became the official newspaper of the
revolution until 1965, when it was renamed Granma.

RICE. Rice began to be grown in Italy, France, and Spain some 25
years prior to the discovery of the New World. It was introduced into the United States by a chance event, in 1694, where
it quickly gained popularity. No exact date exists for when rice
was first cultivated in Cuba. Antonio Bachiller y Morales, in
his Compendium of General Agriculture, dated 1856, refers to
its planting in the 1850's "in unwatered land." There is indication
that rice was grown in 1862 "on low terrain in Pinar del Río
and Santa Clara, and on the banks of the Ariguanabo River in
Güines." As a precedent for the consumption of rice in Cuba,
it should be noted that the first vessel carrying Chinese contract laborers arrived in Cuba in 1847, and it is known that
rice constitutes the staple diet of the Chinese. Rice has traditionally been one of Cuba's basic food imports, bought at an
average price rate of 30 million pesos per year. The United
States, which had controlled the Cuban rice market since 1937,
was the only supplier of Cuba's imported rice up until the 1959
Revolution. After that, rice was a most important crop entirely in state hands. After August 1961, new tracts of land
totalling 440,000 acres had been reclaimed from idleness and
much of it planted with rice which, under Castro, became the
principal crop of diversification away from sugar and tobacco.
Three types of rice have been grown in Cuba, taking into consideration the size and shape of the grain: short grain, medium
grain, and long grain. The varieties of short grain are Bolo
Prieto and Grano de Oro; of medium grain they are Cristalino,
Grano de Oro Largo, and Barbudo; and the long grain varieties
are Fortuna, Edith, Rexora, and Nira.
    In recent years the Honduras (Zayas Bazán) varieties have
been introduced, from which the strains Paquita and Alba have
been derived. Also, by crossing the Honduras with the Buffalo,
the variety Patiprieto has been produced, one that occupied
first place in domestic rice production by the late 1960's. From
the United States, the varieties Blue Bonnet and Blue Bonnet 50,
Century Patna 231 and Rosa del Golfo have been introduced. Of
the latter varieties, the Century Patna 231, which has a short
vegetation cycle, is preferred, because the plants are resistant

to most of the diseases common to rice in Cuba, because very
good yields per planted surface have been obtained, and because of its culinary qualities. The others produce good yields,
but they are more prone to diseases. By early 1970, behavioral
tests had been performed on over 70 foreign varieties introduced
in Cuba. According to the records of the pre-revolution Administration of Rice Stabilization, in 1956-57 there were in Cuba
268 rice mills; by 1968 there were 314. In 1956, there were
7,046 harvesters, 225 traders-wholesalers, 132 rice packaging
plants, and 1,704 retailers distributing rice. In his speech of
July 17, 1968, at the opening of the 150 houses of the state
Rice Plan on Cuba's southern coast, Fidel Castro declared with
regard to the growing of rice in Cuba that by 1970 the country
would have 17,000 physical caballerías available for rice cultivation. However, in comparison with large sugarcane and cattle
holdings, individual rice farms in Cuba remain usually quite
small. Nevertheless, the area planted in rice more than quadrupled between 1966 and 1970, mainly in low-lying coastal
areas of Oriente and Camagüey provinces and in the south of
Habana province. Irrigation and mechanization have also helped
increase rice production, as has the introduction of newer,
high-yielding varieties. Most of the rice comes from state
farms, but some small quantities are grown by private farmers.
By 1983, production of rice in thousands of metric tons in Cuba
reached 227.

RISQUET, JORGE (?-). Minister of Labor under the government
of Fidel Castro (1959-), Risquet attended the Communist Youth
Congress held in Czechoslovakia as a sympathizer in the 1940s.
In the mid 1950s Risquet was involved in subversive revolutionary activities on the Communist side in Guatemala. He subsequently returned to Cuba and joined the revolutionary movement
against Fulgencio Batista (1952-58). Risquet was by then an
acknowledged "old guard" Communist and had been a member of
the Communist Party for some time. Risquet joined the ranks
of Frank País's guerrilla operation, Segundo Frente, in Santiago
de Cuba in the late 1950s. Raúl Castro put him in charge of
political indoctrination of the local peasantry in the Sierra Cristal; therefore, Risquet was never active in combat duty. Following the triumph of Castro's Revolution in 1959, Risquet was
promoted to the position of Captain of the rebel Army and was
assigned to Santiago de Cuba as Chief of the Department of Culture. The Province of Oriente was under his charge, and his
office was involved in executing basically the same type of indoctrination efforts he had while in the Sierra Cristal, only on
a much larger scale. Risquet was appointed Minister of Labor
in 1967 replacing former Minister Basilio Rodríguez. He organized a number of commissions against excessive bureaucracy,
integrating young radicals from the Schools of Revolutionary
Instruction (EIR). Risquet was able to overcome many of the
administrative difficulties facing Cuba's labor force and succeeded

in the government's campaign to correct the inadequate distribution of the labor force by "rationality" (transferring large numbers of surplus urban workers to rural areas for agricultural work).

Toward the end of 1967 Risquet announced future plans for women to take over the factory jobs abandoned by men who went to cut 'cane in the countryside, thus launching an ambitious campaign to incorporate one million women into Cuba's economy. In 1970, Risquet initiated a plan of labor reform designed to "democratize" labor, integrating the workers into the decision-making process of the trade union movement by holding union elections and passing legislation to protect workers' rights. Nevertheless, Risquet ensured tight government control over the labor force by eliminating the non-agricultural private sector and taking a national population and housing census in September 1970. Risquet was responsible for a law against loafing passed in 1971 obligating all capable men between ages 17 and 60 to work, primarily for purposes of curtailing absenteeism, and incorporating idle men into the work force. Risquet has been a frequent contributor to the Cuban revolutionary magazines Verde Olivo and Moncada, as well as an eloquent public speaker. He is a member of the Central Committee of the Communist Party of Cuba and of the National Assembly of the People's Power. Risquet has traveled to Africa, where he has led high-ranking Cuban delegations' visits to various African nations.

RIVERO AGÜERO, ANDRES (1905-   ). Born in Burene, Oriente province, of humble origin, Rivero Agüero struggled to become a lawyer, graduating from the University of Havana in the early 1930's. He became a close associate of strongman Fulgencio Batista, becoming the latter's political secretary. In 1940, Batista appointed him Minister of Agriculture. When Batista left office in 1944, Rivero Agüero faded from public life. He returned after Batista's coup in 1952 to become Minister of Education and later Prime Minister. He was the successful presidential candidate in the 1958 fraudulent elections, but never took office, as he and Batista escaped from Cuba on December 31, 1958. He and his family have lived modestly in Miami since that time.

RIVERON, ENRIQUE (1906-   ). Born in Cienfuegos, Las Villas province, he began his studies in Havana, winning his first scholarship award in 1924. He studied in Spain, Italy, France, and Belgium. Later he worked in Havana as a cartoonist and illustrator until he went to New York City. There he worked for the New York Times, the New Yorker, and Cine Mundial. He moved to South America, doing cartooning and illustrations in Chile, Argentina, and Brazil. Returning to the United States he became a staff artist for the Walt Disney Studios. In 1941 he became dedicated to painting and sculpture. In 1964

he moved to Miami, where he continues to work. He has held
one-man shows in Paris, Havana, New York, Wichita, and Miami.
He has shown his work in numerous juried group exhibitions,
having received several awards.

RIZO ALVAREZ, JULIAN (1930-   ). Born in Matanzas in 1930,
Julián Rizo Alvarez has been a member of the Central Committee of the Cuban Communist Party since 1970, and a Secretariat
member since 1980. During the Third Party Congress of 1986,
he was also elected as an alternate member of the Politburo.
Rizo Alvarez is a graduate in Political Science from the University of Matanzas. He fought under Raúl Castro in the guerrilla
operation "Frank País Second Front," where he reached the
grade of captain. During the 1960s and 1970s, he held a number of Provincial Party assignments in Pinar del Río and Matanzas. Reportedly, he has been involved in various political and
personal scandals regarding misuse of government funds, but
has retained his posts as a protégé of Raúl Castro.

ROA, RAUL (1908-1982). Prominent intellectual, writer, and Minister of Foreign Relations in the Castro Revolution during the
1960's and early 1970's. Roa adopted Marxism-Leninism after
1959 and became the official spokesman for the Castro line in
Western capitals. Roa was part of the Generation of 1930 and
member of the student leadership, writing and distributing
propaganda directed against the Machado regime (1925-33). Although temporarily, he joined the Communist party in 1927 and
was jailed in 1930 for his opposition to Machado. Roa graduated
with a doctorate in public and civil law from the University of
Havana in 1934. Philosophically, he was devoted to the Hispanic-
Americanism of José Martí and Simón Bolívar. Roa was awarded
the chair of History of Social Doctrines in the University of Havana's School of Social Science and Public Law in 1940, where
he taught while engaging in journalistic writings. Roa received
a Guggenheim Fellowship in 1942, and spent one year in the
United States, at Columbia University, writing a book on Franklin D. Roosevelt's New Deal. Roa returned to Cuba and later
served as Director of the Department of Culture in the Ministry
of Education, although he expressed disgust over the corruption
and political gangsterism during the Auténticos' administration.
In 1948 he was elected Dean of the School of Social Sciences at
the University of Havana. Roa wrote articles sharply critical
of General Fulgencio Batista's takeover in 1952, and in 1953
fled to Mexico, fearful of political persecution.
    Roa returned to Cuba in 1955 and resumed teaching at the
University. He did not participate in insurrectionary activities
against the regime until 1956, when he joined the Civic Resistance Movement, a group of prominent professionals and university professors who supported various rebel groups, principally
Fidel Castro's 26th of July movement. Following Castro's victory, Roa was designated Ambassador to the OAS in Washington.

A short while later, Castro appointed him Minister of Foreign
Affairs due in part to Roa's influence in two key countries,
Mexico and Venezuela. Throughout the 1960's Roa became
Cuba's spokesman before the UN Security Council, repeatedly
denouncing the United States. Roa gained the reputation of
"unconventional diplomat" due to his lack of diplomatic protocol
with regard to his manners and violent tirades against other
nations. In 1972 Roa negotiated an anti-hijacking treaty with
the United States. He was replaced as Minister of Foreign
Relations in 1976. Roa was a member of the Central Committee
of the Cuban Communist Party, member of the Council of State
and, Vice-President of the National Assembly of the People's
Government. He was the author of many books, mostly of a
political nature, including José Martí y el destino americano
(1938), Pablo de la Torriente Brau y la revolución española
(1939), Retorno a la alborada (1964), as well as the author of
political essays, the most significant of which is En pie (1959),
a collection of articles that documents his earlier disenchantment
with Marxism.

ROADS. Cuba has one of the most extensive road networks in Latin
America. The revolutionary government has undertaken further
expansion, even though road maintenance has decreased significantly. Motor vehicles, mainly buses, carry over 98 percent of
all passengers. There are about 12,420 miles of highways of
which 5,280 miles are paved with concrete or asphalt. These
figures do not include the vast number of rural access roads,
mainly in the sugarcane fields, which total about 87,600 miles.
Almost all unpaved, these minor roads traditionally comprised
the most neglected part of the nation's road system, especially
in Oriente province. The most important roadway and the chief
connecting link between major cities is the Central Highway,
which runs from Pinar del Río through Havana to Santiago de
Cuba. In general this road follows the island's central axis,
except for one portion that skirts northward to Havana and another section that follows a Z-shaped pattern in Oriente province. Although other roads parallel the Central Highway in
some areas, especially in the western part of the country, it
is still the most convenient route for long-distance traffic.
Long parts of the Central Highway were expanded from
its original two lanes to six (eight lanes near Havana) and re-
surfaced during the 1970's. The stretch along the south near
Santiago de Cuba had been renamed National Southern Highway.

ROCA, BLAS. See Francisco Calderío.

RODRIGUEZ, CARLOS RAFAEL (1913-). Communist Party leader,
lawyer, economist, Vice-President of Cuba. Born in Cienfuegos,
Las Villas province, Rodríguez attended the University of Havana, obtained his law degree, and served intermittently on the
University faculty. In 1936, he joined the Partido Socialista

Popular, Cuba's pro-Soviet Communist Party. In 1939, Rodríguez was elected to the Party's Central Committee. He married
Communist leader Edith García Buchaca, but the marriage soon
ended when she left him for another Communist leader, Joaquín
Ordoqui. After General Gerardo Machado was overthrown in
1933, Rodríguez served as mayor of Cienfuegos. From 1936 to
1939, he was editor of the magazine Mediodía, and in 1938 he
became coeditor of the magazine Universidad. He was Court
Attorney for the City of Havana and a civil service employee
from 1943 to 1944. During Fulgencio Batista's administration
(1940-44), Rodríguez became Minister without Portfolio and a
member of the National Committee for Post-War Studies. He
became interested in economics and wrote many articles on
Marxist theory. He opposed Batista's dictatorship (1952-58)
and in early 1958 travelled to the Sierra Maestra mountains, in
Oriente province, where Fidel Castro had his headquarters, in
order to form an alliance between the Communists and Castro's
26th of July Movement. Rodríguez remained with the guerrillas
during the rest of the war and established a close relationship
with Castro and his top followers.

After Castro's victory in 1959, Rodríguez became the editor
of the Communist Party newspaper Hoy. He helped to draw up
government decrees and in 1960-61 was involved in University
reform. In 1962 he was appointed President of the National
Institute of Agrarian Reform (INRA). In the power struggle
between Castro and the Old Guard Communists during the early
years of the Revolution, Rodríguez supported Castro. In 1963,
INRA was upgraded to a ministry and, as its President, Rodríguez received the rank of Minister. Under his direction, lands
were confiscated and transformed into state farms. Production
decreased and food shortages increased. In 1965, Fidel Castro
assumed the presidency of INRA and Rodríguez was appointed
to the Cuban Communist Party Secretariat. In 1970, he became
President of the National Commission on Economic, Scientific,
and Technical Collaboration of the Council of Ministers. During
the 1970s, Rodríguez's role in the Cuban government became
more important. In 1971, he became Vice-President of the Council of Ministers. When Castro reorganized the Cuban government in 1972, Rodríguez was named one of Cuba's seven Vice-
Presidents, a post he still holds today. He is also a member
of the Political Bureau and Secretariat of the Communist Party,
a member of the National Assembly of the People's government,
and a Vice-President of both the Council of State and the Council of Ministers. Rodríguez is considered one of the most important figures in the Cuban government after Fidel and Raúl Castro. He has remained closely associated with Castro rather than
with the old communists and has served as Fidel Castro's principal liaison man with the Soviet Union and Eastern European
nations.

ROIG, GONZALO (1890-1970). Composer, conductor, and music

professor. Born in Havana where he founded the Symphony
Orchestra in 1922. He composed the popular zarzuela "Cecilia
Valdés," a classic, and popular songs like "Quiéreme mucho,"
"Aquellos ojos brujos," "Nadie se muere de amor," and "Me
gustas."

ROIG DE LEUCHSENRING, EMILIO (1889-1964). Lawyer, journalist,
historian. Born in Havana, Roig was educated in the Jesuit
Colegio de Belén. In 1916 he graduated in Law from the University of Havana. Shortly after graduation, Roig became secretary and editor of the proceedings of the First National Law
Congress held in Havana in 1916 and one of the editors of the
magazines Gráfico, El Fígaro, and the Revista Jurídica. He
wrote and lectured on such topics as the Negro and Indian contributions to Cuban culture and the social and economic evolution of Cuba. Roig became a historian and with Fernando Ortiz
and others reinterpreted the events that led to the War of Independence and the problems of establishing a Cuban nation in
the shadow of the United States. Officially appointed Historiador de la Ciudad de La Habana in 1933, Roig was general editor
of Cuaderno de historia habanera and led a group of revisionist
and Marxist historians in the establishment of the Sociedad Cubana de Estudios Históricos e Internacionales in 1946. Roig also
organized a series of National Historic Congresses that emphasized themes such as anti-imperialism and nationalism. His main
works are: La Habana: apuntes históricos; Los Estados Unidos
contra Cuba libre; Martí, antimperialista; Vida y pensamiento de
Félix Varela; Curso de introducción a la historia de Cuba; El
Sesquicentenario del Papel Periódico de la Havana; Historia de
la Enmienda Platt: una interpretación de la realidad cubana;
Tradición antimperialista de nuestra historia; Cuba no debe su
independencia a los Estados Unidos.

ROLDAN, AMADEO (1900-1939). Composer, violinist. One of the
first to use Afro-Cuban rhythms, Roldán was born in Paris
but moved to Cuba at age 19. He became director of the Havana Philharmonic Orchestra. His music successfully integrated
African folkloric influences and Cuban themes. Roldán wrote
La rebambaramba, of great rhythmic boldness and intensity;
Obertura sobre temas cubanos; Los tres pequeños poemas; El
milagro de Anaquille; Danza negra; and Motivos de son. Roldán
continued to write songs almost until the time of his cancer-
caused death in 1939.

ROLOFF, CARLOS (1842-1907). Carlos Roloff Mialofski was born in
Warsaw in 1842. A Polish Jew by birth, Roloff was a Cuban rebel
leader and general in 1896. Roloff became a U.S. citizen and
subsequently went to live in Cuba as an employee of Bishop and
Co. in Caibarién. By 1895, Roloff was the Secretary for War
inside the political organization of the rebel cause under the
leadership of Máximo Gómez and Antonio Maceo. He became one

of the prime nationalist followers of José Martí. General Roloff
launched a campaign in 1896 to raise money for the rebel cause,
soliciting contributions from Cuban as well as American sources.
Roloff was appointed Treasurer of Cuba under the government
of Estrada Palma (1902-6) after the War of Independence. He
died in Guanabacoa, Cuba.

ROMAÑACH, LEOPOLDO (1869-1951). Born in Sierra Morena, Santa
Clara province, Romañach was educated in Spain. Upon returning to Cuba, young Romañach decided to become a self-taught
painter. Santa Clara's Provincial Assembly later granted him a
pension for study in Italy where he studied for five years under
the renowned master Professor Filippi Prosperi, Director of the
Institute of Fine Arts of Rome. In 1895 Romañach's pension was
retracted due to the War of Independence. Romañach moved to
the United States where he stayed until 1900, earning a living
by painting. He returned to Havana and mounted an exhibition
of his works, after which he was appointed Professor of Color
in the Academy of Havana, a position he held for over 18 years.
Exhibitions of Romañach's paintings have been held in Paris at
the 1902 World's Fair, Buffalo (1903), St. Louis (1907), Havana
(1911) and San Francisco (1916). He won First Prize at the
Havana Expo and received a Medal of Honor in San Francisco.
In the 1920's Romañach led the artistic movement that broke
with the academic tradition of Spanish and colonial culture, as
head of the San Alejandro Academy (Cuba's first art academy
established in 1818). He was joined by many students who returned from Paris in developing colorful, musical, vibrant art.
Some of Romañach's best known works are: Convaleciente,
Abandonada, Un nido de miseria, La promesa, Juventud.

ROMAY, TOMAS (1764-1849). Born in Havana, Romay was the emi-
nent physician in Cuba who first introduced the vaccine against
smallpox and who initiated early investigations into the causes
of yellow fever. Romay was also a Professor of Philosophy at
the University of Havana and a strong ally to Las Casas and
Arango y Parreño in their efforts to shape the intellectual and
cultural climate of Cuba. A founding member of the Sociedad
Patriótica, Romay wrote many articles and monographs of public
interest, as well as articles of a literary nature.

ROOSEVELT, FRANKLIN D. (1882-1945). Born in Hyde Park, N.Y.,
Roosevelt was the 32nd President of the United States (1933-45).
He graduated from Harvard in 1904, and served as Democratic
governor of New York in 1929-33. Roosevelt's legislative program, known as the New Deal, helped lift the United States out
of the Great Depression of the 1930's. His Good Neighbor Policy advocated non-intervention in Latin America in general and
Cuba in particular. This policy had great ramifications for
Cuba, since the United States had always been heavily involved
in the island's affairs. In 1933, Roosevelt sent Benjamin Sumner

Welles as Ambassador Plenipotentiary and mediator to settle the
struggle between Cuban dictator Gerardo Machado (1925-33)
and his opposition. Following Machado's resignation, the Roosevelt administration moved in support of the short-lived Carlos
Manuel de Céspedes government (1933). When the more nationalistic and revolutionary leader, Ramón Grau San Martín, assumed the presidency, Roosevelt refused to recognize the new
Cuban government after receiving information from Ambassador
Welles depicting Grau as unfavorable to U.S. interests. After
Fulgencio Batista forced Grau to resign on January 15, 1934,
the Roosevelt administration quickly recognized the puppet
presidency of Col. Carlos Mendieta, hoping for a government
friendly toward the United States under Batista's fiat.

From 1934 until 1940, when Batista formally assumed the
presidency, Roosevelt maintained a friendly stand toward Batista and visited with the Cuban dictator when the latter
traveled to Washington in September 1938 on a friendship tour.
Roosevelt was personally responsible for the decision to recognize Soviet Russia in November 1933, and was a strong advocate
of formal neutrality at the outset of World War II. However, in
June 1940, he authorized the transfer of arms to Britain, and
by March 1941, he had approved the Lend Lease Act, by which
the United States could lease, sell, or exchange war supplies
with any country considered vital to its defense. Roosevelt
led the United States through World War II and joined Winston
Churchill (1941) in drawing the Atlantic Charter. Roosevelt
attended meetings with Allied leaders at Casablanca, Quebec,
Teheran, and Yalta, and inspired the creation of the United
Nations. Roosevelt was elected to a 4th term in 1944, and
shortly thereafter he died of a stroke on April 12, 1945.

ROOSEVELT, THEODORE (1858-1919). Twenty-sixth President of
the United States, Roosevelt was born in New York and graduated from Harvard in 1880. He devoted time to historical writing: The Naval War of 1812 (1882) and The Winning of the West
(1889-96), and spent two years leading a tough life on a North
Dakota cattle ranch after a term in the New York State Assembly (1882-84). He assisted in New York State administration
(1889-97), after running unsuccessfully for mayor in 1886. As
Assistant Secretary of the Navy (1897-98), he prepared the
U.S. Fleet for the Spanish-American War of 1898. Roosevelt
was thus in the forefront of the movement to assert U.S. dominance in the Western Hemisphere and to oust the Spanish from
Cuba. During the war, Roosevelt gained world acclaim as a
colonel commanding the volunteer "Rough Riders" in the island,
in Santiago de Cuba and San Juan Hill. He was elected governor of New York upon his return to the United States and became McKinley's Vice-President in 1901. Following McKinley's
assassination several months later, Roosevelt ascended to the
Presidency. As President (1901-1909), he advocated a strong
executive branch and passed strict anti-trust legislation. (He

made famous the saying that to go far one must "speak softly
and carry a big stick.")

The demise of the Spanish proved of vital strategic importance for Roosevelt, since it opened a new era of outward-
oriented vision and policy for the United States; Roosevelt was
favorable to American governments in Cuba following the war,
giving full support to McKinley's appointed military governors,
generals John Brooke and Leonard Wood. Roosevelt supported
the Platt Amendment to the 1901 Cuban Constitution, which gave
the United States the right to intervene on the island. He
called a special session of Congress for negotiation of a commercial reciprocity treaty with Cuba in 1902, since the United
States was responsible for Cuba's economic direction due to its
occupation of the island. Although Wood terminated his rule in
1902, Roosevelt later sent his Secretary of War, William Howard
Taft, to Cuba in 1906 as "Provisional Governor," thus presiding
over a second period of U.S. occupation which lasted until
1909. Roosevelt did not seek to annex Cuba, but in keeping
with his corollary to the Monroe Doctrine, he saw the U.S.
responsibility as that of moral teacher to preserve order and
prevent any insurrection on the island. Roosevelt is credited
with many political accomplishments. In 1903, he secured the
right to build the Panama Canal, and in 1906, was awarded the
Nobel Peace Prize for his mediation efforts in the Russo-Japanese
War of 1904-05. Roosevelt withdrew from politics in 1908 in
support of Taft, whom he had recalled from Cuba and made
Secretary of State. At the end of a two-year world tour, he
tried to gain the Republican presidential nomination in 1912, but
failed and ran as a Progressive ("Bull Moose") candidate against
Taft and Woodrow Wilson. Wilson won and Roosevelt went on an
expedition to explore Brazil's "River of Doubt," later named
"Rîo Teodoro" in his honor, in 1914, while remaining critical of
Wilson's policy of neutrality and non-intervention.

RUBENS, HORATIO (1869-1941). A New York lawyer and a friend
    of José Martí. In 1893, he proved that Cuba's exportation of
    labor to Florida in the face of a strike by revolutionary Cuban
    tobacco workers in Florida was against the U.S. Law of Labor
    Contracts of 1885. Thus, the Cuban strikers triumphed against
    the Spanish colonial government in Cuba. Rubens was the law-
    yer for the Junta de Nueva York, a pro-Independence group of
    Cuban exiles who negotiated in 1897-98 with American bankers
    to purchase Cuban independence. In April 1898, following
    Spain's declaration of cessation of hostilities in Cuba. U.S.
    Senator Foraker introduced an amendment to recognize the
    rebel Cuban government of Masó. Rubens, who was very sen-
    sitive to Cuba's independence, demanded that the amendment
    include a declaration of the U.S.'s intention to make Cuba a
    fully independent nation, thus authoring the introduction of the
    Teller amendment, by which the U.S. renounced all interest to
    exercise sovereignty of any kind over the island. Rubens

later became President of Consolidated Railways of Cuba, President of the Council of Administration of the Cuban Railroad, Director of Cuba Company, and President of the Council of Cuba Railways Company. He received the titles of "Havana's adopted son" and "amigo de La Habana." In 1922, he represented the North American Committee before the U.S. Senate Sessions regarding the reduction of tariffs on Cuban sugar. Rubens became president of the newly created Consolidated Railways of Cuba in 1933. He died in New York on April 3, 1941.

RYDER, JEANNETTE (1866-1931). Born in Wisconsin, she came to Cuba in 1899 and founded the Bando de Piedad de Cuba (Piety Association of Cuba), devoting herself to helping handicapped or wayward women, the poor, children, and also to the protection of animals. Her whole life was devoted to charity and she worked tirelessly for all the unfortunate people.

- S -

SACO, JOSE ANTONIO (1797-1879). Born in Bayamo, Saco was a prominent statesman, Creole writer, and historian. He was an early advocate of annexation, a position he later rejected, and a critic of slavery. Saco was a Professor of Philosophy at the San Carlos Seminary and a distinguished member of the Patriotic Society (Sociedad Patriótica), director of the Revista Bimestre Cubana, and collaborator with El Siglo. Saco's efforts to establish the Cuban Academy of Literature were opposed by the Spanish, and he was forced to go to New York where he began publishing El Mensajero Semanal together with his mentor, Father Varela. Thrice elected as Deputy to the Cortes (Spanish parliament), Saco was never able to assume the office. In 1866, he represented Santiago de Cuba before the Junta de Información for overseas reforms in Madrid, where he formulated his famous Voto Particular, opposing election of deputies and proposing his criteria for improvement of Spanish colonies. Saco strongly defended his pro-independence posture, and understood that in order to enjoy sovereignty, the Cuban people must be educated. Saco was the central figure of Reformism and is credited with being the father of the Cuban Encyclopedia. A great critic and analyst, Saco published his Colección de papeles científicos, históricos, políticos y de otros ramos, sobre la Isla de Cuba in 1858-59. He is remembered for his unparalleled work Historia de la esclavitud (3 vols.). Saco died in Barcelona, Spain.

SAINT LOUIS. Luxury cruise ship of the Hamburg-America line which sailed to Cuba from Germany on May 13, 1939, with 936 Jewish refugees on board. This group of Jews was among the

last to escape from the Nazis during World War II. Each refugee was required to pay an exorbitant passage fare plus an additional fee to guarantee return passage in the case that Cuba
would not accept them. Cuba was to be a temporary sanctuary
for 734 of the Jewish passengers; the United States would be
their ultimate destination. Cuba and the U.S. refused entry
and the ship returned to Europe where most of the passengers
were killed by the Nazis.

SALABARRIA, MARIO (1913-   ). Important figure in the Movimiento
    Socialista Revolucionario (MSR) in 1944 and a friend of President
    Ramón Grau San Martín's, who appointed him Chief of the Secret
    Police during his administration (1944-48). Salabarría became
    head of the MSR and through his position began the persecution
    of all his enemies as a truly gangster-type vendetta. In 1947
    Salabarría took part in one of Cuba's more notorious gangster
    wars during the Grau period waged between the MSR and its
    arch-rival, the Unión Insurreccional Revolucionaria (UIR).
    Salabarría had become involved in a plot sponsored by the
    Grau government to overthrow Rafael Leónidas Trujillo, dictator
    of the Dominican Republic. While the planned invasionary ex-
    pedition was still organizing on Cayo Confites off Camagüey
    province, Salabarría went to Orfila, a suburb of Marianao, seek-
    ing to force gangster Emilio Tro to withdraw from his role in
    the invasion. Tro, who was the head of Unión Insurreccional
    Revolucionaria, refused, and a bloody gangster war ensued
    during which Tro and 4 others were killed. For his part in
    the senseless violence, Salabarría was sentenced to 30 years in
    prison. Following the Castro revolution in 1959, Salabarría went
    into exile.

SALINAS, BARUJ (1938-   ). Born in Havana, he studied architec-
    ture at Kent State University in Ohio but soon showed outstand-
    ing painting abilities. He has exhibited in one-man shows
    throughout the United States and in France, Switzerland, Spain,
    Israel, Canada, and Mexico. Among many of the awards re-
    ceived he was given an Excellence Prize in the 7th Grand Prix
    International de Peinture in Cannes, and was awarded a First
    by the Cintas Foundation in the watercolor category and the
    Best Watercolor in the Annual Hortt Memorial Exhibit at the Fort
    Lauderdale Museum in 1969. Since then his work has been in
    many galleries throughout the world.

SAN ALEJANDRO, ACADEMIA DE. Founded in 1818 by the Sociedad
    Económica de Amigos del País, at the recommendation of the
    Spanish intendent Alejandro Ramírez as an art and drafting
    school. Its first director was Jean Baptiste Vermay, a Bona-
    partist exile, who had opened an art school in Havana the pre-
    vious year. The Academy did not prosper financially through-
    out the 1800s and became a government institution in 1863. It
    was revitalized and expanded during the educational reorganization

program of Enrique José Varona in the early 1900s, when it was renamed Escuela Nacional de Bellas Artes, although it continued to be popularly referred to as San Alejandro. The Escuela Elemental de Artes Plásticas was annexed to San Alejandro in the 1940s as a further addition. Camilo Cuyás was San Alejandro's first Cuban-born director (1833). After 1878, when Miguel Melero became director, the direction of the school was always in Cuban hands. Leopoldo Romañach and Esteban Valderrama were among its most distinguished directors, and Sicre, Gelabert, and Domingo Ramos were among the faculty. Many Cuban artists received training at San Alejandro, such as Wilfredo Lam and Carlos Enríquez.

SAN AMBROSIO, SEMINARIO DE. Established during the late 1600's in Cuba by Bishop Compostela, San Ambrosio's function was the "preliminary preparation of future ecclesiastics." In 1773, the Seminary of San Carlos was founded and annexed to San Ambrosio which was in a moribund state. San Carlos was specifically designed for the sons of the aristocracy and no blacks, mulattoes, or mestizos were admitted.

SAN CARLOS, REAL COLEGIO Y SEMINARIO DE. Founded in 1773 by Bishop José de Hechavarría in the house of the Jesuits' school, San Carlos was annexed to the seminary of San Ambrosio. San Carlos was set up for the sons of the aristocracy to prepare only this select group for ecclesiastical careers. In 1820, San Carlos served as the site for the creation of the cátedra de constitución by the liberal movement. Bishop Félix Varela revolutionized instruction in Cuba and at the seminary in the 1800's with his pedagogical innovations, discussion of concepts, dialogic form of exposition, and deductive reasoning. In addition, new studies were introduced by Bishop Varela at San Carlos, including physics, chemistry, botany, and political economics. San Carlos had the strongest impact on intellectual developments in Cuba during the 19th century. Out of its classrooms came prominent members of the Sociedad Económica, including José Agustín Caballero, Francisco de Arango y Parreño, and José Antonio Saco.

SAN GERONIMO, REAL Y PONTIFICIA UNIVERSIDAD. One of two centers of higher learning in Havana in the early 1800's, San Gerónimo had been founded in 1728. Bishop Espada imposed many educational reforms in both San Gerónimo and the Seminary of San Carlos in 1802 as he directed the Sociedad Económica.

SANCHEZ ARANGO, AURELIANO (1907-1976). Founding member of the small but active Directorio Estudiantil Universitario organized in mid-1927 to oppose Gerardo Machado's regime (1925-33). Arango was arrested in 1931, accused of plotting against the government along with the principal Directorio leaders. A

member in 1933 of the Ala Izquierda Estudiantil, Arango broke
away from the movement following Machado's downfall due to his
disillusion with communism. He was later appointed Minister of
Education by President Carlos Prío (1948-52) in 1948. Arango
subsequently claimed that his efforts to correct widespread
government corruption were unsuccessful. In June 1951,
Arango attempted to defend the Prío government from attacks
by the popular radio broadcaster and Ortodoxo Party leader
Eduardo Chibás, who had mistakenly included Arango in his
list of corruption accusations. Arango was made Prío's Foreign
Minister in 1952, and fled into exile with President Prío after
General Fulgencio Batista's coup d'état in March. Sánchez
Arango then founded the Triple A (Amigos de Aureliano Arango)
movement, which opposed Batista (1952-58) from exile throughout
the dictator's rule. Staunchly loyal to Prío, Sánchez Arango
supported Fidel Castro's guerrillas while financing and organizing many revolutionary acts against Batista's regime together
with Prío. Following the Revolution's success, he lived in 1959
as a private citizen in Havana but was later publicly identified
as a counterrevolutionary due to his contacts with the exiles'
political organization (MRR). Arango escaped to Miami through
the Ecuadorean Embassy. He died in Miami in 1976.

SANCHEZ DE BUSTAMANTE Y SIRVEN, ANTONIO (1865-1951).
Born in Havana, he studied law in Madrid and after competitive
exercises was appointed professor of International, Public, and
Private Law at the University of Havana. He was a great ora-
tor, writer, and teacher, and founder and director of the In-
ternational Law Review and director for seven years of the
Forum Review. He was president of the Permanent Tribunal of
the Hague and the League of Nations as well as president of the
Constitutive Assembly that extended the presidential term of
Gerardo Machado (1925-1933) and also granted the dictator an
Honoris Causa degree. He wrote extensively and his Discursos
(Speeches) were published in many volumes. His most famous
contribution to law was the Code of International, Public, and
Private Law, which made him internationally famous and was
used as a textbook in several foreign universities.

SANCHEZ DE FUENTES, EDUARDO (1874-1944). One of the most
renowned Cuban composers whose exquisite melodies won him
international acclaim. Sánchez de Fuente is the author of the
famous patriotic song Tú, written in habanera rhythm and
especially popular among the Cuban Independence rebels of the
1890's. Always in the process of musical evolution, Sánchez de
Fuentes is also remembered for the versatility of his many op-
eras, among them Yumurí, El Náufrago, and Dolorosa. He was
also president of Cuba's National Academy of Arts and Letters.

SANCHEZ GALARRAGA, GUSTAVO (1893-1934). Lawyer by profession and poet by vocation, he mastered all literary styles, and

the music master Ernesto Lecuona used his lyrics in a duet considered to be the Cuban Rogers and Hammerstein. The most famous Lecuona operettas, such as María la O, Rosa la China, Flor del Sito, and El Cafetal, used Sánchez Galarraga's lyrics. His most famous poetic compositions include "Canto a la mujer cubana," "Música triste," and "Senderos de luna." For the theater he wrote La vida falsa, La princesa buena, La máscara de anoche, and more. In the historical field, Sangre mambisa, and as a novelist his most famous is La expulsada, among others.

SANCTI SPIRITUS (CITY). One of seven municipalities into which Cuba was divided in the 16th century by the Spanish crown. Each municipality was the seat of government for the surrounding territory with a cabildo, three regidores, and two alcaldes (mayors). Sancti Spíritus was in the modern Province of Las Villas and in 1536 was the site of the first merced or freehold land grant where grazing rights were conferred. The city remained a center of the cattle industry as well as of traditional Spanish attitudes for many decades.

SANCTI SPIRITUS (PROVINCE). Population: 366,601; area: 6,752 Km²; municipalities: Taguasco, Fomento, Sancti Spíritus, Jatibonico, Yaguajay, La Sierpe, Trinidad, Cabaiguán. Preeminently a cattle province, this territory consists of--beginning with the new political-administrative division--what was once part of Las Villas and Camagüey. Its agricultural importance is derived mainly from tobacco and sugar. Its capital, Sancti Spíritus, was founded in 1516 and was one of the first seven towns colonized by Spain in Cuba. It was also the site of the first merced or freehold land grant where grazing rights were conferred (1536).

SANGUILY, JULIO (1846-1906). Like his brother Manuel, he was a student of José de la Luz y Caballero at El Salvador School and later taught at his alma mater. During the Ten Years' War (1868-1878), he held the rank of major general. In 1871, he was rescued from the Spanish by Ignacio Agramonte in one of the most heroic deeds of the war. At the end of the Ten Years' War, he left Cuba for the United States, where he became a U.S. citizen. In 1895 he was back in Cuba conspiring for independence, when the Spanish sentenced him to life in prison. His American citizenship spared him from serving. In 1897 he joined an expeditionary force in the U.S. commanded by José Lacret, and fought with the mambises.

SANGUILY, MANUEL (1848-1925). Soldier, author, and administrator, Sanguily was born in Havana and studied at El Salvador School under the direction of the famous José de la Luz y Caballero. When the Ten Years' War broke out (1868-1878) he quit his studies of law at the University of Havana to join in the fighting. In 1897, the Revolutionary government sent

Sanguily to New York with the rank of Colonel to act as Secretary to his brother, Major General Julio Sanguily. He traveled to Spain when the Peace of Zanjón (1878) was signed, to finish his law studies at the University of Madrid. He returned to Cuba in 1880, but was unable to practice law because he refused to take the oath of loyalty required by the Spanish courts. Turning to literature for a living, Sanguily contributed to Dr. Enrique J. Varona's magazine, Revista Cubana; and later founded Hojas Literarias in 1893. Colonel Sanguily emigrated to the U.S. with his family during the War of Independence (1895-1898) and returned near the end of the war as a member of the Assembly of Santa Cruz. He was then appointed to a commission under Major General Calixto García to ask for U.S. aid for the veteran Cuban soldiers. Sanguily was most notably a strong supporter of Cuban sovereignty before as well as after the Spanish-American War of 1898. He was elected a member of the Constitutional Convention for the Province of Havana in 1901, wherein he acted as a voice for patriotism whose wisdom and personal judgment were invaluable. He was a spokesman for the diversification of Cuba's international relations, away from U.S. dependence, and opposed the sale of Cuban land to foreigners. Sanguily later served as Director of the Institute of Havana, President of the Senate, Secretary of State in the Cabinet of President José Miguel Gómez, and Inspector-General of the Armed Forces of Cuba under President Mario G. Menocal. He resigned from the latter post over disagreement with Menocal's re-election schemes. Member of Cuba's Academy of History, his published works include Los caribes de la isla; Céspedes y Martí; Cuba y la furia española; and Un insurrecto cubano en la corte.

SANTAMARIA CUADRADO DE HART, HAYDEE (1927-1980). Born in a small village in Santa Clara province of a middle-class family, she lived with her brother Abel when he attended the University of Havana. Concerned with farmers' plight she soon became a devotee of the political philosophy of Eduardo Chibás and she and her brother became active members of the Ortodoxo Party. She joined Fidel Castro in the early 1950's and from the Santamaría apartment in Havana the attack on the Moncada barracks (July 26, 1953) was planned. During the attack she lost her brother and her fiancé, Boris Santa Coloma, who were captured, tortured, and killed. Haydée was captured but after less than six months of imprisonment in Guanajay prison she was released. She began working with the anti-Batista underground and participated along with Frank País in an uprising in 1956 to coincide with Castro's landing on Oriente province. During this period she met Armando Hart, a young lawyer and later Castro's minister, who became her husband. They both joined Castro in the mountains. In spite of chronic asthma she lived the difficult life of a guerrilla. From the Sierra Maestra she was sent on complex missions abroad and represented the

26th of July Movement in the United States until Castro's victory when she returned to Cuba. She became a member of the Council of State and one of the five women members of the Central Committee of the Cuban Communist Party. In 1967 she presided over the Conference of the Latin American Solidarity Organization (OLAS) in Havana. Until her death she was the head of Casa de las Américas, Cuba's cultural center and state publishing house. She committed suicide on July 26, 1980.

SANTERIA. A syncretic cult--in which Catholic saints are equated with African deities--widely practiced in Cuba, particularly among the poorer elements of the population, black and white, and also among some of the white middle class. The practice of santería has diminished somewhat with the advent of the Castro revolution. See African religious practices.

SANTIAGO DE CUBA (PROVINCE). Population: 792,519; area: 6,343 Km²; municipalities: Contramaestre, San Luis, Songo La Maya, Santiago de Cuba, Tercer Frente, Guamá, Mella, Palma, Soriano, Segundo Frente. One of the new provinces established in 1976, it was originally part of Oriente province. The provincial capital bears the same name, Santiago de Cuba. It was one of the seven original villas in Cuba designated by the Spanish Crown in the 1500s and thus the first seat of government for the territory. Santiago was the early commercial center in Cuba until Havana supplanted it in the 17th century. Santiago remained the main urban site in eastern Cuba and was chosen by Fidel Castro as the location for his July 26, 1953, unsuccessful attack on the Moncada barracks in an attempt to overthrow the Batista regime. The City of Santiago de Cuba is still the second in the country. During the Cuban-Spanish-American War a naval battle took place in Santiago bay between the Spanish and North American fleets which was decisive in the defeat of the colonial troops that fought against the Cuban independent forces. Agricultural activity is low, principally fruits prepared for preserves. There are also distilleries. The port is important for foreign trade and the city is a crossroads for highways and railroads.

SANTOVENIA, EMETERIO (1889-1968). Journalist, author, and historian, Santovenia was born in the village of Mantua, Pinar del Río, where he studied as a boy. He obtained his Bachelor of Letters and Sciences at the Institute of Pinar del Río and his degree of Doctor of Civil Law at the University of Havana, where he was made Professor of Public Instruction. After one year, Santovenia abandoned this post and devoted himself entirely to his literary pursuits as a journalist and author. Santovenia contributed to several prominent Havana newspapers including El Comercio, El Triunfo, La Prensa, and La Nación. Fulgencio Batista approached Santovenia about assuming the Presidency following Batista's coup against the government of

Prío Socarrás (1948-52) in 1952. Santovenia declined although
he remained a loyal supporter of Batista throughout the 1950s.
By 1958, however, he was ruined politically in the aftermath of
the civil war between Batista and Fidel Castro, and was soon
arrested in 1959, having remained in Cuba following the Revolution. Santovenia's chief literary contributions are in the field
of history; he was a corresponding member of the Academy of
History of Cuba, being one of the editors of Historia de la nación cubana. Included among Santovenia's many published works
are: Tranquilino Sandalio de Noda, Cirilo Villaverde, El municipio de Ramón Lazo, Gonzalo de Quesada, and Una heroína cubana.

SEGURA BUSTAMANTE, INES (1909-   ). Psychologist who was expelled from the University of Havana in 1928 because of her activities opposing the Gerardo Machado (1925-1933) regime. She
was imprisoned with her mother at the Guanabacoa Prison first
and later at Nueva Gerona in the Isle of Pines, then Isle of
Youth, until finally released in 1932. She graduated as Doctor
of Philosophy and Letters from the University of Havana in 1934
and later became a professor of logic and civics and married
Manuel Antonio de Varona, a lawyer who later became Prime
Minister during the Carlos Prío (1948-1952) government. She
divorced him after bearing a son and graduated as Psychologist
from Columbia University. She is an active feminist, working
and using her political influence to support all causes that
would advance women's status. Exiled in the United States,
she has continued to write.

SEGURIDAD DEL ESTADO. See Departamento de Seguridad del
Estado.

SERRANO, GENERAL FRANCISCO (1810-1885). Half-Cuban and one-
time lover of Spanish Queen Isabella II, Serrano, a General in
the Spanish Army, encouraged the importation of Chinese laborers into Cuba in 1859. General Serrano was sent to
Cuba in that same year by his colleague and former Captain-
General of Cuba, Leopoldo O'Donnell. Cuban reformists in the
1860's tried to maintain good relations with Serrano, who was
made Captain-General of Cuba, especially after he endorsed reforms before the Spanish Cortes (Parliament). These reforms
included stimulation of trade with the U.S.; Cuban representation in the Cortes, and abolition of the slave trade. Cuban
planters opposed this and later in 1866, following a shake-up
in the Spanish Cortes culminating in the ouster of O'Donnell's
government, Serrano was exiled to the Canary Islands. In
1967, Serrano led a revolt in Spain expelling Queen Isabella II
from the country. Forcing his way into Madrid in October,
Serrano formed a new government with himself as Prime Minister on October 8. In November, Serrano appointed his abolitionist friend, Domingo Dulce, Captain-General of Cuba. Dulce
arrived in Cuba in January 1869 with a modest program of

reforms. At the outbreak of the Cuban Ten Years' War in 1868, Serrano stepped aside as Prime Minister yielding to Juan Prim, and became regent to the throne.

SERVICIO DE INTELIGENCIA MILITAR (SIM). SIM was Cuba's military intelligence and special police organization created by strongman president Fulgencio Batista during the 1950's (after his 1952 coup against president Carlos Prío). SIM headquarters were located in Santiago de Cuba under the direction of Ramón Cruz Vidal, formerly an ordinary soldier who had taken part in the "Sergeants' Revolt" of 1933 led by Batista. President Batista employed SIM to identify the rebels who took part in the attack on the Moncada Barracks led by Fidel Castro in 1953. Later, SIM was able to seize the rebels' secret radio station when several of its operators defected to the government's side. SIM was known for its violent techniques and brutal mass murders of anti-Batista elements throughout the 1950's.

SEVEN YEARS' WAR (1756-1763). Spain joined a conflict between England and France on the French side due to its discomfort with English colonies in the Caribbean and Central America. Spain did not want to see English preeminence in the Western Hemisphere, yet it lacked the naval power to confront the English or to prevent them from capturing its possessions. In August 1762 the English destroyed a large Spanish naval force and captured Manila and Havana, only to trade the latter back to Spain in exchange for Florida in the Treaty of Paris in 1763.

SHELTON VILLALON, RITA (1901-). She studied medicine at the University of Havana where she graduated in 1923. She practiced pediatrics and specialized in tuberculosis. She became a professor of anatomy and pathological histology at the University of Havana. She was an active member of the Feminist Club of Cuba and organized the establishment of the "Gota de Leche" (Drop of Milk) program whereby expectant mothers and children were supplied with milk to help combat tuberculosis. She was active in the struggle against the dictatorial regime of Gerardo Machado (1925-1933) and went into exile in Madrid. She published extensively on scientific matters; some of her works include Amor a la Infancia (Love for Children) and Eugenesia, themes that she lectured on at the Lyceum and the Hispano-Cubana de Cultura (Hispanic-Cuban Culture). She now lives in exile in Miami.

SIBONEY. See Ciboney.

SIERRA DEL ESCAMBRAY. Mountain range in southern Cuba, near Trinidad and Cienfuegos. During the Revolution to overthrow Fulgencio Batista (1952-58) from 1956 to 1959, the Sierra del Escambray was a focal point for several guerrilla groups, particularly those belonging to the Directorio Revolucionario.

Castro succeeded in asserting his authority over the guerrillas
in the Escambray in late 1958.   After Castro assumed power,
late in 1959, approximately 1,000 rebels belonging to his Revolution assembled in the Sierra del Escambray and were later
starved into surrender after Castro evacuated all peasants who
could otherwise have fed them.

SIERRA MAESTRA. The longest (150 miles) mountain range in Cuba,
   located in southern Oriente province, west of Santiago de Cuba.
   The highest mountain in Cuba, Pico Turquino, is in the Sierra
   Maestra, rising to 6,000 feet.   Up to the early 20th century,
   much of the eastern section of the Sierra Maestra consisted of
   latifundia where coffee was exclusively grown.   Many regarded
   the area as a refuge for fugitive criminals and it resembled the
   lawlessness of the wild West in the United States, until the
   1950's when Castro used it as a base for his guerrilla activities.
   Castro and his small band of invaders reassembled in the Sierra
   Maestra after landing the Granma in southern Oriente on December 2, 1956.   During much of 1957 the Sierra Maestra served as
   an area of training and maneuvers by the guerrillas and in July
   1957 Castro issued a manifesto calling on Cubans to join his
   revolutionary front to force free elections and a democratic
   government.   By 1958 Castro had about 300 armed men in the
   Sierra Maestra and controlled a 2,000 square mile area, aided
   by friendly peasants who helped him set up an elaborate supply
   and maintenance system complete with factories and hospitals.
   From the territorio libre, as the Sierra Maestra was called,
   Castro throughout 1958 attacked, in guerrilla fashion, Batista's
   nearby garrisons and conducted extensive sabotage of railways,
   harbors, and warehouses.   After the Revolution's success in
   1959, a group of anti-Castro rebels grouped in the Sierra Maestra under the command of ex-rebel Captain Manuel Beatón but
   they were easily defeated.

SIMONS, MOISES (1884-1944).   Cuban composer and musician who
   spent many years in Paris working on typically Cuban musical
   themes.   His most famous composition, El Manisero, brought him
   worldwide fame.   Simons is also the author of zarzuelas and
   operettas; La Consigna is his best work.

SLAVERY.   During most of Cuba's colonial period (16th, 17th, and
   18th centuries) uninterrupted slavery decisively influenced the
   development of society.   African slavery existed in Spain and
   the first slaves had come to Cuba with the early conquistadores.
   Later more slaves were imported to work in the washing of gold.
   As in the U.S., black slaves were considered stronger and better suited for hard labor, so they replaced the weaker indigenous workers.   Slave importation slowed as gold reserves were
   depleted and the cost for slaves ($500 in the 17th century for
   a healthy adult male slave) increased.   Only after the full-scale
   development of sugar in Cuba did the demand for slaves resurge.

As early as 1538 black slaves rioted and looted Havana while
French privateers attacked from the sea. Many blacks worked
in urban industries and some were able to obtain their own
earnings and free themselves by paying the price of their (or
their children's) emancipation. Some were freed after they had
performed services for their masters. By 1774 the number of
slaves decreased continuously until reaching 39,000 out of a
total population of 171,000. In the late 18th and 19th centuries
as the sugar industry took off, an economic revolution occurred
in Cuba. The large sugar estate evolved and thousands of
black slaves entered Cuba, such that by 1825 the black population had surpassed the white one. Many slaves, known as
cimarrones, escaped to the mountains rather than face harsh
treatment by their owners. Spanish law, Catholic religion, and
the economic condition in Cuba all facilitated the integration of
the freed blacks into Cuban society to a much greater extent
than their counterparts in the U.S. Slavery was abolished in
Cuba in 1886 by the Moret Law. See also Moret Law.

SOCIAL STRUCTURE. Before 1959 the class system was characterized by four strata, and the gap between the lowest class and
those above it was emphasized by rural/urban differences. Members of the upper, middle, and working classes had a modern,
urban orientation toward life and enjoyed a near monopoly of
economic and political power. Their political outlook was international, and their Hispanic culture was often overlaid with
North American traits. Members of the lower class, largely
rural dwellers, regarded themselves as an integral part of Cuban society but were not fully accepted as such by other Cubans, in part because of the infusion of African elements into
their basically Hispanic culture. Because economic, educational,
and cultural opportunities were available only in the cities, the
lower class often sought upward mobility through rural-urban
migration. Although some who migrated succeeded in bettering
their status, many remained on the periphery of city life, ex-
acerbating already serious urban problems of overcrowding, unemployment, and intense competition for goods and services.
Migration further depressed rural areas, which remained sparsely settled, backward, and without even the most basic goods
and services. The unique circumstances of Cuban history led
to the development of a social structure that was quite different
from that of most other Latin American countries. Most Spanish
colonies acquired independence early in the nineteenth century
and developed a patrimonial agrarian order consisting of a landed
elite and the peasants who served them. The relationship that
developed between these classes was characterized by mutual obligations and benefits.
    In Cuba, however, the development of such an indigenous
patron-client relationship was circumvented by Cuba's protracted
colonial status under the Spanish and thereafter by its economic
tutelage under the United States. This meant that the criollo

upper class came to be defined more in economic terms--by
money and occupation--than by inherited social status. The
upper class consisted of urban businessmen, bankers, merchants, and professionals but not of a landed elite that commanded traditional loyalty and respect because of its patriarchal
role. Because upper-class status could be bought--with a high
salary, a position of prestige, education, and a cultured lifestyle
--the prerevolutionary class structure was more fluid than elsewhere in Latin America, where power, wealth, and status were
in the hands of a hereditary and frequently rural oligarchy.
This fluidity meant more mobility for all Cubans in a relatively
open and modern structure. Other distinctive features of the
upper class were its economic ties to foreign interests and its
adoption of foreign cultural traits--specifically, North American.
The accessibility of upper-class status coupled with upper-class
dependence on foreign sources of capital and culture left the
class in a relatively vulnerable position. Since the status of the
upper class was based on materialistic and occupational criteria
rather than on traditional notions of noblesse oblige and the
justness of the social order, the rest of the population was less
in awe of the upper class. This lack of moral legitimacy was a
factor in the eventual deposal of the upper class.

The middle class was an amorphous group whose existence as
a class has been questioned by Cuban scholars. By purely
economic standards the class constituted a sizable sector of the
urban, and to a lesser extent the rural, population. It lacked,
however, the collective consciousness and shared values on which
political cohesion might have been built. University-educated
middle-class men--a sector characterized by intense competition
for individual gain--were foremost among those engaged in politics, from conservatism to advocacy of radical reform or socialism. Castro and many of those originally involved in the Revolution are examples of middle-class revolutionaries. Like the
upper class, much of the middle class was absorbed in United
States-funded commercial activities or influenced by United
States economic policies. This sector included sugar growers,
merchants who relied on constant supplies of United States equipment and spare parts, and employees and managers of United
States-based multinational corporations. The only element of
the class unaffected by external ties were a few professionals,
intellectuals, and members of the government bureaucracy.
Thus, except for those committed to change, the fragmented
and dependent character of much of the middle class left it in
a position similar to that of the upper class on the eve of the
Revolution. That is, their socioeconomic positions were viewed
as indefensible and their roles as expendable by the less afflu-
ent majority of Cubans. Because Cuba became highly industrialized soon after independence, a nucleus of organized, cohesive, industrial and agricultural workers was formed which became politically active. The working class expanded in the first
half of the twentieth century in keeping with an increasingly

differentiated and relatively fluid socioeconomic structure.
Unions were not an isolated or uniquely urban phenomenon.
They were also found in the mines and on the sugar mills
(centrales) in the rural areas.

By 1955 about one-fifth of the Cuban population claimed
union membership. The power and cohesion of organized labor
--as well as its high level of integration into the society--was
demonstrated repeatedly, especially in the downfall of President Gerardo Machado in 1933. The Communist Party was active in organized labor for much of this period, stimulating the
desire for change and building the feelings of collective strength.
Not only the mentality but also the lifestyle of working-class
Cubans was different from that of their lower class compatriots.
They had access to public education; their standard of living
was higher and their job security greater; and they had a
chance for upward mobility. The lower class was unskilled,
uneducated, and frequently unemployed. Although many lived
in the slums of Havana and other cities, the vast majority lived
in pockets of poverty throughout the countryside. Status in
the rural areas was based on one's relationship to the land.
At the lowest level and constituting the largest occupational
group were the nonunionized farm laborers, who had no permanent claim to the land. Many of these were migratory, their
labor was seasonal, and their existence precarious. They were
the most deprived of health services and education. A little
above them were the squatters, sharecroppers, and subrenters,
sometimes distinguished from the migratory laborers by their
geographic stability rather than by greater prosperity. The
size of the piece of land owned, rented, or subrented was vital
in determining status and wealth among this segment of the
rural lower class. See also Society.

SOCIALIST MAN. Concept envisioned by Fidel Castro, influenced
mainly by Ernesto Guevara, which attempts to redefine Cuban
society and the role of Cuban citizens inside it. The new Socialist Man should be devoted to the causes of the Fatherland
and to Communism. The Socialist Man should constantly labor
for the welfare of society such that the collective interest will
supersede the individual one. Racial prejudice will be elminated,
and honesty will guide everyone's life. Party discipline will be
adhered to and enforced. The ideals of sacrifice, abnegation,
courage, and discipline as demonstrated by Castro's rebel army
in the mountains will be the new predominant virtues in Cuba.
See also Moral incentives.

SOCIEDAD DE AMIGOS DE LA REPUBLICA (SAR). A nonpartisan
organization that emerged in Cuba in 1955 as a final attempt at
compromise between President Fulgencio Batista (1952-58) and
his opponents. SAR's head, Colonel Cosme de la Torriente, a
respected jurist and veteran of the Cuban War of Independence,
sought to persuade Batista to hold new elections. SAR's

sponsorship of mass public meetings to show support for the
need for new elections initially compelled Batista to enter into
negotiations with Cosme de la Torriente. These became known
as el diálogo cívico, which tried to forge a compromise formula
for new elections. By March 1956, Batista had terminated the
civic dialogue and no new elections were scheduled.

SOCIEDAD ECONOMICA DE AMIGOS DEL PAIS. Founded in 1792
by Captain-General Don Luis de las Casas and Bishop Díaz de
Espada as a center of learning and discussion, the Sociedad
Económica de Amigos del País became the headquarters of a
well-organized and influential group of Creoles from Havana
and the site of numerous meetings where these men discussed
science, the arts, culture and education, and commerce and
industry. Of primary interest to Sociedad members was agriculture, especially sugar and its economic trappings, and by
1815 the group had convinced the Spanish government to permit outright ownership of land formerly held in usufruct. The
power of the 200 members of the society was economic in nature,
and given the corrupt and inefficient nature of Spanish bureaucracy in Cuba these sugar planters succeeded in forcing the
Crown to sell them lands for their cultivation of sugar.

SOCIETY. The Cuban Revolution was not an ordinary rebellion but
instead a drastic overturn of the existing order in Cuba. What
started out as an armed protest against a dictatorial regime
ended up as a revolution along the Socialist-Communist pattern
which sought the complete abolition of the traditional Cuban social system based on private property, democratic institutions,
and constitutional guarantees of individual rights. Castro's
goal became identical to the classical Marxist one of the ultimate
classless society where all are equal with no "exploiters" nor
"exploited." The subtle criterion that has historically been important in determining class structure in Cuba has been the
conception derived from the Spanish heritage that physical
labor was degrading. Thus the lower-class individual engaged
in physical work while upper- and middle-class Cubans did not.
In addition, Cuban society has traditionally been drawn along
the two lines of the urban (specifically Havana) and rural sectors. The wealthy elite resided and worked in the urban areas
and their suburbs while the laborers and agricultural workers
naturally worked in the rural areas. Commercial and financial
activities in the urban sectors delienated the upper classes while
land ownership did the same for rural dwellers. The Revolution
of 1959 changed the societal givens drastically when property was
seized and nationalized for the benefit of all. No wealthy pro-
pertied class can exist in revolutionary Cuba. Similarly, the
middle classes in Cuba largely were separated from their few
luxury possessions, and many joined the upper class in exodus to Miami, Florida. The new elite group in Cuba emerged
as the coterie around Fidel Castro in the mid-1960's whose

strength is derived from their dictatorial and sweeping powers. The symbiotic relationship of the military and the Party with the former holding true power is the new criterion for leadership in Cuba.

SOLES Y RAYOS DE BOLIVAR. Founded in 1821, the Soles y Rayos de Bolívar was a secret society whose purpose was to create a Cuban Republic which they would name Cubanacán. The society survived two years, dying out as its leaders, including José Francisco Lemus, were arrested by the Spanish. The name "Soles y Rayos" came from the cellular nature of the society where each new recruit would in turn enlist six others. The recruiter would receive the rank of sol (sun) and his recruits would become rayos (rays).

SOLIDARITY ORGANIZATION OF THE PEOPLE OF ASIA, AFRICA, AND LATIN AMERICA. See Organización de Solidaridad con los Pueblos de Asia, Africa, y América Latina (OSPAAAL).

SOMERUELOS, MARQUES DE (1754-1813). Salvador de Muro is known in Cuban history as the Marqués de Someruelos, Captain-General (Governor) of Cuba from 1799 to 1812. Someruelos was sent by Spain to govern during one of Cuba's most turbulent historic periods. Someruelos proved his keen political savvy by acknowledging the importance of the wealthy Cuban Creole class, which he was able to win over. This helped the governor to crush spawning revolutionary and annexationist plots on the island. During his tenure, Someruelos received U.S. General James Wilkinson, an envoy of U.S. President Thomas Jefferson. Wilkinson tried to convince Someruelos of the convenience of selling Cuba to the United States--just as Napoleon had sold the Louisiana Territory--in order to prevent Cuba's falling under British or French control. The bright, energetic Someruelos confronted the considerable foreign pressures of his day, succeeding in his endeavor to keep Cuba for Spain.

SORES, JACQUES DE (c. early 1500's-c. mid-1500's). Huguenot sailing under the French flag who, in 1555, attacked, captured, and burned Havana. From the middle of the sixteenth century on, the Spanish were threatened by foreign interlopers--mainly pirates and privateers sailing under the flag of a particular European nation--who visited what Spain considered its exclusive waters. It was the French privateers, however, who became particularly active in the Caribbean and threatened Spanish commerce. The outbreak of hostilities between Spain and France and the Treaty of Lyon between France and Portugal, which prohibited France from attacking Portuguese shipping, focused French attention on Cuba and other Spanish possessions in the Caribbean, leading to numerous forays on towns and harbors. The most notorious of these was the attack by Jacques de Sores. For nearly one month, Sores laid seige on Havana, seeking ransom for captives and bounty, as well as the life of the governor, Gonzalo Pérez de Angulo. In the end, Sores and his men released their Cuban prisoners and abandoned the island after having burned all they could not carry with them.

SOTO, HERNANDO DE (c. 1496-1542). Born in Extremadura, Spain, between 1496 and 1500. Spanish conquistador and early governor of Cuba. De Soto was the first governor of royal title to govern the island, and was responsible for initiating the fortification of Havana. Scarcely eleven months after his arrival in Cuba, De Soto set out in the Spring of 1539 to conquer Florida despite the failure of Pánfilo de Narváez's similar expedition twelve years earlier. In his stead, De Soto left his wife, Isabel de Bobadilla, the first and only woman in Cuba's history, to govern the island. De Soto's ill-fated expedition led to his death in the wilds of Mississippi in 1542, after his discovery of the river that bears the same name. During his lifetime, De Soto had earned a reputation as a brave, honest, and just captain as a result of his participation in the conquest of Perú. De Soto closed the century of Spanish expeditions to unexplored continents that used Cuba as a base or stopping point.

SOTO PRIETO, LIONEL. A member of the Secretariat of the Cuban Communist Party, Lionel Soto Prieto has also been a member of the Central Committee since 1975. A personal friend of Fidel Castro since 1946, he helped Castro considerably during the struggle for leadership of the Federation of University Students (FEU) during 1948-51. A long-time militant member of the Popular Socialist Party (PSP), Soto Prieto became International Secretary of the PSP in 1945. As a member of the PSP, he did not participate in the revolutionary activities of the 26th of July Movement, but is known to have favored support for the armed struggle movement. During the early 1960s he was director of the magazine Pensamiento Crítico, an orthodox pro-Soviet publication, and held various positions responsible for the dissemination of Marxist-Leninist ideology. During Castro's 1968 ideological purges, he was removed from his assignments. Accused of conspiring with the old-guard of the Communist Party, he was assigned to perform agricultural work as punishment. Soto Prieto was later reinstated and in 1973 was named Cuban Ambassador to Great Britain, a position he held until 1976. In 1978, he was appointed to direct the foreign relations of the Central Committee, and in 1981 was named as Chief of the Economic Commission of the Central Committee. Lionel Soto Prieto was named Ambassador to the Soviet Union in 1983, a position he held until mid-1986.

SOVIET-CUBAN RELATIONS. Following the Bay of Pigs fiasco in April 1961, the United States together with several Latin American nations isolated the Cuban regime as Fidel Castro moved fully into the socialist camp. Initially, Moscow maintained a careful stand towards Castro as the latter sought allies to defy the United States. Moscow was also uneasy with regard to Castro's promotion of violent revolution in Latin America. Castro's policies were at odds with Moscow's philosophy of "peaceful coexistence" and support for the popular front-oriented Communist parties in Latin America. By 1962 Castro had curbed the power of the Cuban Communist Party (PSP) and firmly consolidated his own personalist control. By the early 1960s, the Soviets began perceiving U.S.-Cuban tensions as an opportunity to offset earlier failures to obtain U.S. concessions over Berlin as well as to embarrass the U.S. in exploiting its "loss" of Cuba. Soviet leverage increased in Cuba, where the economy was plagued by low productivity, mismanagement, poor planning, and shortages of many items. Long-term Soviet-Cuban trade agreements perpetuated Cuba's role as sugar producer, with the Soviet Union supplying up to $1.5 million per day in aid throughout the 1980s for a total of $12 billion beginning in 1960. Since mid-1968, the Soviet-Cuban relationship has been one of close collaboration and friendliness. Castro's support of the August 1968 Soviet invasion of Czechoslovakia and the late 1979 Soviet incursion into Afghanistan illustrate this point. The failure of Castro's guerrilla activities in Latin America (Venezuela and Bolivia) have removed an important irritant in Soviet-Cuban relations and facilitated rapprochement between the two. Cuba's involvement in Angola and Ethiopia, stationing 50,000 Cuban troops in Africa, would have been impossible without Soviet logistic support. Likewise, the chaotic state of the Cuban economy into the 1980s requires Soviet subsidies, particularly in the form of Soviet crude, which Cuba reexports as naphtha, other refined products, and as crude to earn hard currency. Thus, Cuba remains vulnerable to Soviet ability and willingness to transfer petroleum to Cuba.

SPAIN. Spain was the first European nation to explore the New World in the 15th century, particularly the Caribbean Basin area. Spanish explorations led to conquest and colonization. After Hispaniola, Cuba was valued as an early site for gold extraction and as a source of indigenous labor. Following Diego Velázquez's pacification of the Indians in Cuba and the firm entrenchment of the encomienda system, Spanish rule in Cuba was established, enduring for four centuries. As elsewhere in the New World, colonial Cuba was characterized by the web of political and bureaucratic government devices implemented in Spain. The audiencia, residencia, cabildo, alcaldes, and later the intendencia and captaincy-general were all utilized effectively by Spain to structure and control all areas of Cuban society--economic, social, religious, and political. As Latin America achieved independence in the early part of the 19th century, Cuban intellectuals failed in similar efforts to debilitate Spanish power. Lacking widespread support, the early independence conspiracies of the 1820's and 1830's failed, especially due to the fact that many Spanish royalists and troops settled in Cuba following their defeat in Latin America. Cuba became a heavily fortified garrison, the last significant bastion of Spanish power in the New World. Independence for Cuba was therefore late in arriving and was achieved only after the second Independence War (1895-98), coupled with the Spanish defeat by the United States in the Spanish-American War (1898). Through the Monroe Doctrine, the United States had preferred Cuba under a weak Spain during the 1800's so that it might later be annexed to the United States rather than become a British, Dutch, or French colony.

SPANIARDS IN CUBA. After Cuban independence in 1902, Spanish immigration to the island began on a grand scale. Between 1902 and 1910, some 200,000 Spaniards, mostly Galicians and Asturians, emigrated to Cuba, attracted by high wages in Cuban urban areas, especially Havana. Spanish immigration was particularly significant as it helped to articulate the benefits and power of organized labor in Cuba during the early 1900's. Spanish and other immigration to Cuba dropped off during the Great Depression of the 1930's and thereafter. Many Spaniards returned to Spain at that time, and later few could return due to Cuba's new nationalistic laws, restricting the number of foreigners an employer could hire. Nevertheless, many Spaniards sought political asylum in Cuba following the fall of the Spanish Republic in 1939.

SPANISH-AMERICAN WAR. The Cuban wars of independence (1868-78 and 1895-98) became an international issue in 1898 when the United States, angered by Spanish excesses and prodded by those who saw a possibility of acquisition of greater American influence in the island, presented an utlimatum threatening war if Spain did not relinquish its authority over the island. Spain naturally refused, and American troops invaded Cuba and Puerto Rico. Naval battles were also fought in the Philippines. The Spanish were no match for the American armies. The years of struggling against the Cuban mambises had also weakened the strength as well as willingness of the Spaniards. At the Treaty of Paris in 1898 the Spanish Crown relinquished Cuba to the United States. Cubans refer to the Independence War of 1895 which ended with American intervention as the Cuban-Spanish-American War. See also Independence War; Maine, USS; McKinley, William.

SPANISH CIVIL WAR REFUGEES. Many refugees from the Spanish Civil War (1936-39) were violence-prone and they extended their activism and rivalries to Cuba. Their anti-Communist character was strong and in part contributed to their rise in the early 1940's. Rolando Masferrer had fought on the Communist side

during the war and upon his return to Cuba he founded the
Movimiento Socialista Revolucionario (MSR), one of three prominent action groups in Cuba. Many Spanish refugees joined
groups like MSR, UIR, and ARG in the 1940's and were active
in the sort of "gangsterism" practiced by these organizations.

STUDENTS. See Federación Estudiantil Universitaria.

SUGAR. Sugar is and always has been Cuba's principal crop.
Production of sugar was first stimulated by loans given to
early settlers by the Casa de Contratación in 1523. Up to the
early 1800's, the sugar industry grew slowly in Cuba, primarily
due to lack of markets. By 1805, annual production reached
34,000 tons and the number of sugar mills was 478, more than
twice as many as there were in 1761. By 1810, Cuba enjoyed
substantial sugar trade with the United States and other foreign countries. After independence at the beginning of the
twentieth century, the Cuban sugar industry was expanded by
American capital investments, which accelerated the concentration of land but, more importantly, cemented Cuba's dependence
on sugar as a single-export crop. Cuba's economic fortunes in
the 1920's and 1930's were tied to the fluctuations of the price
of Cuban sugar. During World War II, sugar production was
paralyzed in many areas of Europe and Asia, thus accelerating
further the expansion of the Cuban sugar industry. Simultaneously, the process of nationalizing the sugar industry was enhanced, such that by 1958 there were 121 Cuban-owned sugar
mills out of a total 161 (the remainder being 36 American, 3
Spanish, and 1 French). From 1949 to 1958 about 30% of Cuba's
GNP was generated by the sugar sector, and during this same
period sugar accounted for 85% of exports. Cuba continued to
be highly vulnerable to the effects of price fluctuations in the
international sugar market as well as dependent upon the quotas
and prices set by the U.S. Congress. At the time of the Revolution (1959), sugar plantations occupied 10% of all the national
territory and 53% of Cuba's cultivated soil. Efforts to diversify
agriculture failed during the 1960's and, by 1966, a return to
concentrate on the sugar industry became policy. Better
prices on the world market improved conditions temporarily.
Into the 1980's Cuba continues to depend on sugarcane exports
and sugar is planted throughout the island.

SUGAR ACTS (1937, 1948, 1952, 1956, 1960). A series of U.S.
legislative acts designed primarily to protect U.S. sugar interests. The Sugar Act of 1937 created an excise tax on domestic
sugar and fixed a quota system that allotted Cuba 28.6% of the
total consumption requirements of the United States. The Sugar
Act of 1948 changed the method of quota allocation in favor of
U.S. domestic beet and sugarcane producers (including Hawaii,
Puerto Rico, and the Virgin Islands) by giving them fixed rather
than percentage quotas. Cuba and other sugar-producing areas

received percentage quotas based on the remainder or "leftover" annual consumption requirement after deduction of the fixed quotas. In addition, domestic producers received subsidy payments based on unplanted acreage or crop deficiency. The 1952 and 1956 Sugar Acts basically renewed the system set up in 1948. President Eisenhower changed the system in 1960 by increasing the domestic allotment by 200,000 tons and tying the Cuban quota to Puerto Rican production. This contributed to future instability of the Cuban sugar producer whose production was far down the priority list of the Acts.

SUGAR AGREEMENT (1953). The Sugar Agreement of 1953 launched an era of permanent regulation of Cuba's primary product. Some changes in ownership of the sugar mills occurred but sugar remained the leader in the economy.

SUGAR COORDINATION LAWS. The two edicts issued by Batista, one in 1936, the other in 1937, to protect the rights of the small Cuban sugar farmer or colono. The first law guaranteed the tenants of small sugar plantations that they would not be easily evicted. The second, in addition to restating this, also allowed the colono to lease his land in perpetuity so long as he kept his land cultivated with sugarcane, produced the minimum required amount, and delivered it to the mill on time.

SUGAR DIFFERENTIAL. The "sugar differential" was compensation for increased price levels received from the U.S. government as primary purchasers of the sugar crop. In the 1940's this compensation was supposed to be divided among the producers but more often was impounded by the government.

SUGAR QUOTAS. During the 20th century, due to the heavy influence of American business interests inside and out of the Cuban sugar industry, the quota for Cuban sugar entering the United States was an important issue. In 1939 President Roosevelt suspended the quotas on sugar, raising the tariff on it to 1.5 cents per pound. Later, the Sugar Acts of 1948, 1952, 1956, and 1960 would similarly shift the quotas for Cuban sugar into the United States based largely on the success in Congress of the domestic sugar producers' lobby. Cuba, throughout all these irregular fluctuations of quotas and tariffs, continued to experience insecurity with its one-crop economy.

SUGAR STABILIZATION ACT (1930). As sugar prices worldwide dropped rapidly in 1929, North American interests in the Cuban sugar industry sought to stabilize prices to avoid financial disaster. In November 1930, a Sugar Stabilization Act was passed which allowed the newly formed New Sugar Export Corporation to purchase much of the 1929 crop at $4.00 a bag. Export quotas on Cuban sugar were imposed and in the context of the worldwide depression, the Cuban share of the U.S. market dropped from 50% to 25% in 1933.

SUGAR STABILIZATION INSTITUTE. Established in Washington,
D.C., in the 1930's, the Cuban Sugar Stabilization Institute
functioned as an organization that transmitted American economic and political views to Cuba. The representatives of the
Institute would communicate to the Cuban government the shift-
ing position of the many interests involved regarding the sugar
quota. The battle for the quota was acrimonious and the Insti-
tude tried to serve the interests of both U.S. businessmen and
the Cuban government, which were usually incompatible.

- T -

TACON, MIGUEL (1775-1855). Spanish Captain-General of Cuba
(1834-38), Tacón was a general in the Spanish Army who fought
against independence movements in order to preserve Spanish
rule in South America. When he arrived in Cuba as Captain-
General (Governor) in 1834, Tacón was a widower and a firm
believer in strong rule. Tacón initiated a policy of administrative reform designed to establish order in the countryside
and law in the city. He launched a program of public works,
including street cleaning, which improved health conditions in
Havana. A theater, a prison, streets, roads, and a broad
avenue called El Paseo were constructed under General Tacón.
He permitted the slave trade, from which he profited significantly. Cuban Creoles desiring a more benign form of rule or
abolitionism were exiled by Tacón, which radicalized many of
them toward the independentist position. The intellectual José
Antonio Saco was among those exiled. Tacón also prevented
election of Cubans to the Spanish Cortes (Parliament), thus
wounding Cuban nationalist pride and increasing opposition to
his rule. Tacón practiced iron despotism in administering Cuba,
which progressively alienated and disillusioned those Cubans
seeking independence. In 1838, Tacón's resignation from the
governorship of Cuba was accepted and upon return to Spain
he was named Vizconde de Bayamo and Marqués de la Unión de
Cuba. Hefty profits garnered from his involvement in the
slave trade allowed Tacón to later retire to the palace he built
in Majorca.

TAINO. The second and more advanced Arawak Indian group (over
the Ciboneyes) to enter Cuba about two centuries before the
Spanish conquest. The Tainos occupied the eastern and central
parts of the island, as well as most of Hispaniola, Jamaica, and
Puerto Rico. The short, olive-skinned people subjected their
children to artificial cranium changes by binding the frontal
or occipital regions of the head. A Spanish eyewitness described
them as "meek, humble, obedient, and very hospitable." A
favorite Taino pastime (and an irritant to the Spaniards) was

bathing frequently. The Taino developed a fairly advanced economic system based on agriculture with commonly cultivated fields of yuca, tobacco, cotton, corn, and white and sweet potato. Taino society was organized along distinct class lines with the cacique at the top, managing the affairs of the community. The line of inheritance went to the cacique's eldest nephew, if any, otherwise to his eldest son. Taino religion conceived of a supreme invisible being and a series of lesser gods represented by idols. In 1492, the South American Caribs challenged the Tainos in Cuba and terrorized the more peaceful Tainos with their cannibalistic practices. The arrival of the Spanish abbreviated this potential conflict and, proving superior to all Indians, the Spanish pushed the Caribs to the Lesser Antilles and away from Cuba.

TALLET, JOSE ZACARIAS (1893-1985). Cuban poet; leftist leader. A poet whose works first appeared in Social (1923) and El Heraldo (1924), Tallet developed an early interest in Marxism. In 1923, Tallet participated in an attack on the administrative corruption surrounding the sale of a convent in the city of Santa Clara, Protesta de los Trece. During the early 1920's, Tallet together with Julio Antonio Mella founded the Universidad Popular José Martí, dedicated to the education of the poor. Tallet served as its first president, while giving lessons in Marxism to audiences of workers, thereby adding a revolutionary flavor to his literary aspirations. He was a founding member of the Grupo Minorista and Falange de Acción Cubana, and with Rubén Martínez Villena participated in Movimiento de Veteranos y Patriotas. The Marxist poet-educator emerged as an opponent to the government of President Gerardo Machado (1925-1933). Former editor of Venezuela Libre and contributor to Revista de Avance, Tallet remained loyal to the Castro regime until his death in 1985. A frequent contributor to Bohemia, he continued writing prose and poetry until his death.

TANGANAS. These were impromptu protest gatherings by University of Havana students while the university was closed by President Gerardo Machado (1925-33) from 1933, which led to clashes with the police and subsequently violence and terrorism. Tánganas formed after the violent demonstration by the Directorio Estudiantil against Machado, whose suppression of the student organization caused many students to conduct activities clandestinely.

TAX REFORM LAW. Early after the 1959 Revolution, the Tax Reform Law was passed as a signal to the private sector in Cuba that it would continue to be protected. The Tax Reform Law acted as a counterbalance to the Agrarian and Urban Reform laws, which were more revolutionary in scope, promising more lands and reduced rents to the lower classes.

TELECOMMUNICATIONS. Most postal, telegraph, and telephone service is operated by the various entities belonging to the Ministry of Communications, although some international telegraph
service is still being provided by British-owned Cable and Wire-
less (West Indies). Between 19,000 and 20,000 persons, 40 percent of whom were women, worked in the telecommunications
sector in the early 1980's. Some difficulties were being experienced by employees either unfamiliar or not complying with technical procedures, and telephone and telegraph systems suffered
frequent interruptions. There were nearly 700 joint postal and
telegraph offices in the country; the offices also sold magazines
and newspapers. The level of mail decreased during the 1970's,
as compared with the 1960's, and the government placed the
blame on a shortage of writing paper and envelopes and on
transportation deficiencies. The country has 12,000 miles of
telegraph lines plus radio telegraph service. Some modernization of the telegraph service was slowly being accomplished by
the installation of a teleprinter system obtained from East Germany and known as "TGX." The new teletypewriters were replacing manually operated lines.

  The transcriber system known internationally as Telex was
also in use, but was reserved for communications between government agencies. Telex instruments were being installed in
ministries, major enterprises, and factories. There were over
280,000 telephones in use, including thousands of free public
booths, although many of the free telephones were frequently
out of order. Priority for new telephone service is being given
to out-lying rural areas where service either did not exist or
was insufficient to the needs of the area. A number of localities, such as the Isle of Youth, have had direct dialing to Havana since mid-1974. Operators, however, must be used to
connect Havana with those same localities, as direct dialing is
not reciprocal. Telephone, telex, and television reception between Cuba and the Soviet Union and East European countries
has existed since 1974 via a Soviet satellite.

TELEVISION. Television was introduced into Cuba at approximately the same time it appeared on the U.S. market. According to
the 1953 Census, there were 78,931 television sets in Cuba then.
By 1958 there were approximately 400,000--more than in any
other country in Latin America. Early after the 1959 Revolution, the Cuban government assumed operation of the island's
radio and television networks.

TELLER AMENDMENT. The amendment proposed by Senator Henry
M. Teller, was added to the U.S. war resolution against Spain in
April 1898. The Teller Amendment provided that the United
States would disclaim interest in exercising sovereignty over
Cuba after it achieved independence from Spanish rule. A
qualifying statement, however, did permit the United States to
conduct "pacification" of Cuba after hostilities had ceased until
the island could be governed by Cubans themselves.

TEN YEARS' WAR (1868-1878). The Ten Years' War broke out in
1868 and was directed by radical Creole landowners in the
Oriente region along with a group of lawyers and professionals.
The leader of the movement was Carlos Manuel de Céspedes,
whose freed slaves formed the early vanguard of the rebel
group. Céspedes' manifesto, the Grito de Yara, called for
complete independence from Spain, the establishment of a republic with universal suffrage, and the indemnified emancipation of slaves. The war intensified in eastern Cuba in 1870
with no decisive victories on either side and significantly, no
U.S. recognition of the Cuban belligerency. By 1871 the
rebels had been pushed back into Oriente and the rebellion
was contained there. Máximo Gómez and Antonio Maceo conducted the military campaigns for the rebels, who languished
in Oriente unaided by either Cuban sugar interests or exiles
in the United States. The war dragged on with neither the
Cubans nor the Spaniards able to win a decisive victory. Finally, on February 11, 1878, the Peace of Zanjón ended the
Ten Years' War with most of the rebel Cuban generals accept-
ing the pact. One notable exception was General Antonio
Maceo, who continued to fight for a short while with his depleted army. During the centennial celebration in 1968 Fidel
Castro proclaimed the Grito de Yara as marking the beginning
of a Hundred Years' War, thus including himself and the Revo-
lution of 1959 in Cuba's liberation process. See also Agramonte,
Ignacio; Céspedes, Carlos Manuel; Maceo, Antonio; Guáimaro;
Guáimaro Constitution; Zanjón, Peace of.

THEATER. The first vestiges of theater in the island date back to
1520 with the presentations of farces in the Christian festivities
of Corpus Christi, in Santiago de Cuba, produced by Pedro
Santiago. Later on, in 1577, Juan Pérez de Bargas presented
his work, whose content was reviewed by the ecclesiastical and
political authorities of the colony. From that moment on there
was a development of the stagework ranging from simple dances
and comedies to religious plays, short farces, and so-called good
works. There is also evidence of plays in the churches themselves, but religious orders objected to them since they were
considered profane. The first Cuban work, El príncipe jardinero
y fingido Cloridano, by Santiago Pito y Borroto from Havana, appeared around 1730. With this work the choteo (jeering) was
born in the Cuban theater. On January 20, 1775, the first
theater was inaugurated in Havana, El Coliseo, through the initiative of the Governor of the Island, Marques de la Torre, and
which served as stage for Spanish and Italian companies (beginning in 1834). It was destroyed by a hurricane and rein-
augurated as El Principal (also destroyed by a hurricane in
1846). During that time local theaters were emerging all over
the city which were highly inadequate and uncomfortable, and
which were constructed in the middle of the capital (presently
restored and called García Lorca). It was considered one of

the most magnificent in America at the time. The construction
of theaters slowly developed outside the capital and increased
with the development of the sugar industry. In 1823 the first
theater in Santiago de Cuba was established (Coliseum or Marine Theater); in 1850 El Principal, in Camagüey; the Sauto
Theater in Matanzas (1863); and another in Pinar del Río (1845).
The first theater companies emerged (1810-1811), and with them
the first actor, Francisco Covarrubias, who inaugurated the
vernacular genre. The one-act farce (sainete) and melodrama
were developed, and the country became a first-class theater
setting, with Spanish and Italian companies visiting the island.
    The works of José María de Heredia (1803-1839) gave new
strength to the Cuban theater and a wave of political criticism
was initiated, disguised in parables. Thus, the theater became
the setting for Creole conflicts with the metropolis. José Jacinto Milanés (1814-1863), "El Conde Alarcos," and Gertrudis
Gómez de Avellaneda (1814-1873) brought romanticism to the
Cuban theater. The birth of the national comedy in 1842 whose
first performance was Una aventura o el camino más corto, by
José Agustín Millán, opened a new era. Prior to 1868, melodrama or "serious" drama appeared little by little on the scene,
and never achieved the popularity it had in England or France.
Drama reached its highest point with Joaquín Lorenzo Luaces
and closed the triad of romantic writers with Milanés and Avellaneda. During the insurrectional era (1868-1896), patriotic
works appeared, although not very numerous, such as Abdala
by José Martí, La muerte de Plácido by Diego Vicente Tejera,
and others. A genre with popular roots was the comic theater
(teatro bufo), begun in 1868 upon the debut of the Havana
Comedians, although this genre had its antecedents in Covarrubias, Creto Ganga, Zafra, Guerrero, and Socorro León. The
second stage of this genre, which began with the creation of
the Miguel Salas Company, was much richer in authors and ac-
tors. The 20th century began with the absence on playbills of
the most important figures, some having retired and others having died. During this lapse, dramatic tradition was broken and
only the teatro bufo stayed alive through the presentations of
the Alhambra Theater in Havana until its closure in 1935.
    José Antonio Ramos (1885-1946), author of works with strong
social content, was one of the most outstanding figures in the
first three decades of 1900, with works such as Tembladera
(1918), together with Federico Villoch (1868-1954--La chambelona), Francisco Robreño (1871-1921), and Gustavo Robreño
(1873-1957). These three decades saw such figures of the
comic theater as Aníbal de Mar, Arquímides Pous, Regino López, Candita Quintana, and José Sanabria. The lyrical theater
had a strong popular base with the zarzuela. Authors such as
Jorge Anckermann and Ernesto Lecuona caused this genre to
become one of the most important artistic expressions of the
beginning of the century. In this genre the outstanding artists were Rita Montaner, Candita Quintana, Rosita Fornes, and

Esther Borja. Other important authors were Moisés Simons,
Gonzalo Roig, Rodrigo Prats, and Eliseo Grenet. The crisis in
the dramatic theater lasted until 1940, when the Academy of
Dramatic Arts (ADAD) was created and through which several
theater groups were founded, namely, The Comedians and Free
Stage Group. Some were avant-garde, such as the Experimental Theatre of Art (TEDA), in which Vicente and Raquel Revuelta, Paco Alfonso, and Marcos Behmara appeared. In 1959,
the Cuban theater was successful again, and the most important
topics were presented in the Study Theatre group (Vicente and
Raquel Revuelta, Hector García Mesa, Sergio Corrieri). There
were the works of talented dramatists such as Virgilio Piñera.
But the theater was not immune to the new political structures
and lost its independence. All cultural activity was placed under the control of the Ministry of Culture. The theater began
to express ideology, and included works in socialist realism of
authors such as Bertolt Brecht, Gorki, and Vasiliev, wherein
also the revolutionary epic and communist thought were exalted.
The works of Cuban authors contained social topics concerned
with the renewal of a classless society. In order to achieve
the objectives of the official policy utilizing art as a medium of
mass consciousness theater groups were created (Escambray
Theater, Rita Montaner Group, People's Art Theater, Guiñol
Theater, Cabildo de Santiago, La Yaya, Cubana de Acero).
Among the most outstanding authors of this period are Antón
Arrufat, Eugenio Hernández Espinosa, Abelardo Estorino,
Nicolas Dorr, Albio Paz, and José Triana.

TOBACCO. Tobacco figured prominently in pre-Columbian Indian
society in Cuba. It was used for smoking as well as for religious ceremonies and for curing the ill. After the Spanish took
over the island, tobacco became a popular export as it gained
acceptance for smoking and snuff in Europe. By the 17th and
18th centuries, increasing European demand sparked the growth
of tobacco farms or vegas, especially around Havana and Pinar
del Río. Spain tightened control of the burgeoning tobacco industry, setting up a government monopoly in which the Crown
purchased Cuban tobacco at a cheap price and sold it in Europe
at a considerable profit. By 1717 all tobacco production was
placed under such a monopoly, which lasted until 1812, when
tobacco planters revolted at such oppressive mercantilistic practices. Well into the 20th century, tobacco held second place in
the Cuban economy, behind sugar. In the 1950's tobacco occupied fifth place in agricultural output, and as an industry
was ranked third. The industrial phase--the manufacture of
cigars and cigarettes--employed some 35,000 workers in 1,000
establishments in the 1950's. In addition, some 1,300 individual
cigar makers added to the total production output.

    As opposed to sugarcane, which is an imported plant, tobac-
co is indigenous. Whereas sugarcane is grown over vast ex-
panses of land, giving rise to large estates, tobacco is raised

on small farms, nearly always less than 1 caballería (13.42 hectares) in size. In recent years, the tobacco harvest has exceeded 100 million pounds (46 million kilograms). In 1962, tobacco plantations covered 66,843 hectares, with an estimated 2,624 plants per hectare. Over half of the crop is exported, mainly in the form of unstripped branches; next, as stripped branches; and finally, already industrialized as twisted tobacco, the world-famous Havana cigar. Exports of cigarettes and cut tobacco are very meager. Nearly half of the crop is consumed in the country, in the form of cigarettes and twists, and, to a lesser extent, as cut tobacco. The value of the domestic consumption of these products exceeds 100 million pesos, and exports have an annual value of between 40 and 55 million pesos. This makes tobacco one of Cuba's main resources, with its production, industry, and trade exceeding 150 million pesos per year.

TOPOGRAPHY, See Geography.

TORRIENTE, COSME DE LA (1872-1956). Independence War colonel. Conservative Party leader. Mediator between President Batista and opposition. Born on his family's estate in Matanzas province, Torriente attended elementary and secondary school in Matanzas. He initiated his law studies at the University of Havana, which he was to complete after the Independence War (1895-1898). In 1895 he went into exile in the United States where he joined a military expedition headed by Cuban General Calixto García which landed in Baracoa, Oriente province. Torriente served in the war under generals Calixto García and Máximo Gómez, among others. By the end of the war, Torriente had been made a colonel. During the American occupation following the Spanish defeat in Cuba, Torriente was appointed Magistrate of the Provincial Courts. After Cuba achieved total independence, he served as a diplomat in Spain. Later in Cuba, Torriente became dedicated to the Conservative Party and held the offices of General Secretary, Vice-President, and President of the party. In 1918 he was elected Senator for an eight-year term. Torriente became the first Cuban Secretary of State under President Mario García Menocal's administration (1913-1921). In 1955 Torriente, at the age of 83, headed the Society of Friends of the Republic (SAR). As leader of this nonpartisan organization, he attempted to discuss new elections with strongman President Fulgencio Batista (1952-1958), who refused to believe Torriente's allegations that he was not the leader of another political faction. Batista's refusal brought Torriente national prominence. The SAR held a public meeting to demonstrate mass support for the opposition's cause and thereby compel the government to allow new elections. Batista indeed reversed his attitude and initiated a series of lengthy negotiations--known as El Diálogo Cívico (the civil dialogue)--with Torriente and other opposition leaders to devise a compromise

formula for elections. The negotiations collapsed when in
March 1956 Batista's delegates to the talks refused to consider
a proposal advocating elections that same year. Following this
episode Torriente retired permanently from public life.

TOURISM. Prior to 1959, tourism generated an ample amount of
foreign exchange but was often offset by the amount of money
Cubans spent abroad. After World War I, the huge fortunes
made by U.S. foreign investors in Cuban sugar stimulated
tourism, especially during the winter months. Tourism by
wealthy Americans became almost an institution in the 1950's,
enduring until the anti-American turn of the revolutionary
government of Fidel Castro in 1960 and 1961. Tourism virtually
disappeared in Cuba, at least as far as the United States and
Western Europe were concerned, and consequently the foreign
exchange it had generated also disappeared. A few Eastern
Europeans came to Cuba, but in general, tourism as an industry stagnated. In 1973 the Castro government, seeking renewed sources of foreign exchange, started to encourage international tourism and improve tourist facilities. Older hotels
were refurbished, and new hotels and motels were built. This
stimulus was effective, as more than 100,000 foreign tourists
visited Cuba in 1985, nearly half of them Canadian. Cuba's
tourism industry has continued to expand in the 1980's.

TRADE AND COMMERCE. Domestic trade. Before Fidel Castro
assumed power in 1959, the domestic market in Cuba fluctuated
as a direct response to the sugar harvest or zafra. During
the zafra, railway traffic, banking transcations, and wholesale
and retail trade all increased measurably. After the zafra,
wages dropped with the fall in consumer purchases and domestic trade stagnated. This yearly fluctuation was especially evident prior to World War II. During the late 1950's, strongman
President Fulgencio Batista tried to regulate the awkward cycle
of boom/bust tied to sugar. Before 1959, distribution of domestic products was carried out by Cuba's many wholesalers, retailers, and commission merchants with the help of Cuba's adequate highway and railway systems emanating from Havana.
The distribution of imported goods was similarly well developed.
Of the total value of wholesale trade and services in 1955, imported products accounted for 55% and domestic products for
45%. Local markets were generally large buildings inside of
which wholesale and retail trade was carried on simultaneously.
No price lists were published and no indices existed to give
the farmer a reliable estimate of the worth of his products.
This situation provided the merchant an opportunity to exploit
both farmer and consumer. Goods were not subject to any
grading method, so that merchants could sell as first-quality
goods items purchased from the producer at second-quality
prices. Farmers made daily or weekly trips to nearby towns
to market their produce and receive their mail. In more

remote areas, farmers resorted to bartering, although some cash exchange also occurred.

Before 1959, retail prices were assigned inconsistently with mark-ups on U.S. items ranging frequently between 20% and 200% above the U.S. retail price. Prices often varied according to the avenues of trade from manufacturer to consumer. Some manufacturers sold directly to small retail outfits, while others preferred bulk sales to wholesalers. Hoarding to corner the market and to speculate on prices was common, and retail merchants often took advantage of the confusion resulting from the use of both Spanish and English standards of measure. After 1959, direct government control of domestic trade, both wholesale and retail, has evolved. Domestically manufactured goods are controlled at the source, since factories and plants have been nationalized. As of 1961 shortages of consumer goods and difficulty with maintaining controls over prices to reflect wages have emerged as significant problems. After the U.S. embargo on non-food products which had furnished 70% of Cuba's imports, shortages in manufactured goods became acute. Cuban industry could not fill the gap, nor could the goods be supplied by the Soviet bloc. The resultant rationing in the 1960's has become institutionalized in the 1980's. Since Castro came to power, most wholesale and retail trade and distribution have been controlled by the Ministry of Domestic Trade, which also operates most service establishments. The Ministry maintains warehouses for food and non-food products and is also responsible for setting wholesale and retail prices. Its outlets include department stores and specialty shops in the cities, general stores in rural areas, and urban street-vending kiosks. In the 1960's some 2,000 general stores were opened in rural areas to attend to the needs of peasants who previously had to travel to urban areas to purchase most non-food items.

Foreign trade. Cuba's primary products, sugar and tobacco, and much of its pre-1959 industrial production insured the nation at least a minimal export economy and, consequently, foreign economic commerce. From 1900 to 1960, Cuba's foreign trade was almost exclusively dependent on the United States. The close relationship was bolstered by a myriad commercial treaties giving preference to U.S. interests. The firm U.S. market for sugar obtained under multiple quota arrangements throughout the first half of the 20th century brought prosperity during times of war and economic distress when world sugar supply was up. As early as 1896, U.S. investment in Cuba was significant, ranging between $30 and $50 million. From 1920 to 1945, the U.S. and other foreign interests influenced the economy and wielded significant control over both the sugar and banking industries. Before 1950, Cuba maintained a positive balance of payments as trade surpluses usually offset losses in services and freight and remitted profits on investments. High sugar earnings offset deficits except in 1921, when the sudden drop in sugar prices gave Cuba little time to adjust to the decrease in export earnings. After 1950 and until 1959, sugar earnings remained high, but as their exports fluctuated, revenues did not cover the prices for foreign goods, and unfavorable account balances occurred in every year except 1953.

Since the end of World War II, exports have represented about 30% of Cuba's GNP and 40% of its national income. Sugar has accounted for 80-90% of export earnings throughout the period and thus even a slight drop in sugar prices has had a marked effect on the entire Cuban economy. Under Castro all foreign trade is conducted by the government and since 1960 no sugar has been sold to the United States. In addition, all U.S. business interests, valued at about $1 billion total, were nationalized by the end of 1960. Foreign trade since 1961 has largely been conducted with the Soviet Union and East-bloc nations via some 24 foreign trade enterprises inside the Ministry of Foreign Trade. Throughout the 1960's and early 1970's, exports fell far behind imports every year. The first surplus in Cuba's balance of trade since the Castro takeover occurred in 1974 and was caused by soaring world sugar prices. During the 1960's, Cuba depended heavily on the willingness of the East-bloc nations to purchase most of the sugar crop at prices above the world market price. This subsidization did not provide Cuba with foreign exchange surpluses but did keep the economy afloat. The Cuban economy throughout the 1960's was dependent on a high level of imports from the Eastern-bloc nations more than it had been on the U.S. before. Cuba's ratio of imports to GNP was quite high during most of the 1960's at over 25%. The hundreds of items imported included entire industrial plants and equipment, agricultural machinery, fertilizers, paper, chemicals, vehicles, foodstuffs, raw materials, and petroleum.

TRANSPORTATION. Cuba has 8,291 miles of roads. The first-class Central Highway (Carretera Central) extends for 760 miles from Pinar del Río to Guantánamo, connecting all major cities. An extensive truck and bus network transports passengers and freight. In 1968 there were an estimated 150,000 motor vehicles, many originally imported from the U.S. Since 1962, Cuba has imported almost all its motor vehicles from Communist nations. Some buses have been imported from England and Spain. Nationalized railways connect the eastern and western extremities of the island with 3,714 miles of trunk lines and 7,542 miles of feeder lines. Before Castro, Cuba had a small merchant marine, but the revolutionary government's first steps were to develop it with the help of the Soviet Union. The USSR has supplied oceangoing vessels and fishing boats and, in the mid-1960's, built a huge fishing port in Havana Bay to service Cuban and Soviet vessels. By 1986 the Cuban merchant fleet had some 80 vessels with about

300,000 gross registered tons. Much Western shipping was curtailed after 1960 owing to pressure from the U.S. government
and to Cuba's increasing trade with the Soviet bloc. Today
Cuba's major ports--Havana, Santiago de Cuba, Nuevitas, Matanzas, Cienfuegos, Mariel, and 21 minor ports--are serviced
mainly by Soviet ships, with Spain, the United Kingdom, China,
and Eastern Europe making up the bulk of the remainder. Traffic at Havana's José Martí International Airport, once a major
international crossroads, is now limited to weekly flights to
Spain and Mexico and regular service with Moscow and Prague.
The state airline, Compañía Cubana de Aviación (CUBANA) services both internal and international routes. See also Roads;
Railroads.

TREATY OF PARIS. Ended the Spanish-American War. Concluded
December 10, 1898. Proclaimed April 11, 1899. Article I.
Spain relinquishes all claim of sovereignty over and title to
Cuba. And as the island is, upon its evacuation by Spain, to
be occupied by the United States, the United States will, so
long as such occupation shall last, assume and discharge the
obligations that may under international law result from the fact
of its occupation, for the protection of life and property.
Article VII. The United States and Spain mutually relinquish
all claims for indemnity, national and individual, of every kind
of either Government, or of its citizens or subjects, against the
other Government, that may have arisen since the beginning of
the later insurrection in Cuba and prior to the exchange of ratifications of the present treaty, including all claims for indemnity for the cost of the war. The United States will adjudicate
and settle the claims of its citizens against Spain relinquished
in this article. Article XVI. It is understood that any obligations assumed in this treaty by the United States with respect
to Cuba are limited to the time of its occupancy thereof; but it
will upon the termination of such occupancy, advise any Government established in the island to assume the same obligations.

TREJO, RAFAEL (1905-1930). Student leader killed by President
Machado's police who became a symbol of rebellion. Trejo was
an engaging leader of the Student Union of Havana University
who vigorously opposed Cuban President Gerardo Machado (1925-
33). Trejo was elevated to President of the Law School Student
Union, and became a leader of the Directorio Estudiantil Universitario (University Student Directorate), a small but active
group of students organized in 1929 to oppose Machado's authoritarian regime. In September 1930, the students established a
second Directorio, agreed to issue a manifesto condemning the
regime, and planned a massive demonstration for September 30.
When police attempted to disperse the gathering, a riot developed and Trejo was fatally wounded by a policeman. Trejo's
death was the turning point in the struggle against Machado,
creating a new image of the student generation of 1930 as

courageous underdogs. Trejo later evolved into a symbol of the Cuban Communists' struggle.

TRELLES, CARLOS M. (1866-1951). Scholar, patriot. Born in Matanzas province, Trelles received his elementary education in his native city, attending secondary school in Havana. Trelles did not complete his medical studies and instead pursued a commercial career which did not lessen his devotion to literary and bibliographic studies. In 1895, Trelles published a pamphlet entitled "Cuba and America" in which he revealed his support for Cuba's independence from Spain, and later became involved in subversion against Spain. In 1896, Trelles had to go into exile in the United States, returning to his native Matanzas after independence was won. There, he organized the Public Library of Matanzas. Trelles went to Paris in 1900 to organize the products characteristic of Matanzas province in the Cuban section of the Paris Exposition. He later served briefly in the Council of Matanzas, but politics was not his vocation. Between 1907 and 1917 he published twelve volumes of his well-known Bibliografia Cubana. Trelles is also the author of Biblioteca científica cubana and America as an Intellectual Power, among others.

TRIBUNALES POPULARES. Lower level or "grass roots" courts in Cuba during the 1960's which met in the evenings to deal with minor, non-political offenses including brawls, labor disputes, and problems of public order. Several hundred of these Tribunales Populares were convened before judges whose legal training was limited to a ten-day course and whose main qualification was that they were loyal Communists.

TRICONTINENTAL CONFERENCE (1966). Held in Havana in January 1966 and attended by revolutionary leaders from Latin America, Asia, and Africa. At the Conference, Fidel Castro insisted on following an independent course, seeking to gain the undisputed leadership of the worldwide guerrilla movement. He gave an unqualified promise that any revolutionary movement anywhere in the world could count on Cuba's support. Castro further launched a scathing attack against the Chinese, who under a "senile" Mao Ze-dong, had confused communism with fascism and had subverted the true ideals of the socialist purpose.

TRINIDAD. An old smugglers' port on Cuba's southern coast located near a large number of sugar mills built during the 18th and 19th centuries. In the 1820's, the region around Trinidad contained several of the largest sugar mills on the island, many built on the profits of the slave trade. One example was Guinea de Soto, which manifested prodigious production throughout the early 1800's.

TRO, EMILIO (?-1947). Chief of political-gangster group; civilian
employed by Police of Marianao City. Tro fought with the U.S.
forces during World War II in the Pacific, returning to Cuba
where he was appointed as trainer of the National Police Academy by Dr. Ramon Grau San Martín. Tro was leader of the
Unión Insurreccional Revolucionaria (UIR), one of the three
major urban gangster groups operating in Cuba throughout the
1940's and oriented toward violent political action. Thus, his
job connection with the Police provided him with the protection
necessary to conduct the UIR's activities. UIR became the
political-gangster organization closest to President Grau himself. Tro and his UIR were repeatedly accused of assassinations and other violent actions. Tro was killed after being
surrounded by the forces of a rival group led by Tro's mortal
enemy, Mario Salabarría.

26TH OF JULY MOVEMENT. Name given to Fidel Castro's revolutionary cause adopted from the date of his ill-fated attack on
the Moncada barracks in Santiago de Cuba in 1953. After the
attack and after Castro was freed from prison, he articulated
the movement and for over a year traveled to the U.S. and
Mexico seeking funds for his cause. The 26th of July forces
sailed from Mexico to Oriente province in 1956 aboard the yacht
Granma, again hoping to defeat Batista's forces and touch off
an island-wide insurrection. The uprising was crushed and
most of the movement leaders were either killed or imprisoned.
Castro and his 26th of July movement followers fled to the
Sierra Maestra mountains, where for three years the movement
grew in prestige, strength, and importance as it successfully
waged guerrilla warfare on Batista's military. The 26th of July
movement's underground organized a general strike in Cuba on
April 9, 1958, which fizzled but did serve to illustrate the
movement's organizational strength. Underground cells of the
movement conducted bombings, sabotage, and kidnappings
throughout 1958, as well as distributing propaganda that undermined the foundations of the government, helping to create
the somber atmosphere of a civil war. Following the Revolution's
success, the 26th of July movement became an integral part of
the governing body, merging with the PSP in the early 1960's
to form a new Cuban Communist Party, which became Castro's
ruling organ. The Party, the 26th of July movement, and the
Directorio Revolucionario were all merged into ORI (Organizaciones Revolucionarias Integradas) in 1961.

- U -

UNIDADES MILITARES DE AYUDA A LA PRODUCCION (UMAP).
One of the devices used by Castro to help shape the minds of

the new generations. UMAP were forced labor camps where the
Union of Young Communists (UJC) and the Army provided political instruction for Cuban youth. The UMAP were disbanded in
the late 1960's, given the international protest over human
rights violations in these concentration camps.

UNION DE JOVENES COMUNISTAS (Young Communist League). In
1962 the country's youth groups were merged into a single
giant organization, the Young Communist League (Unión de Jóvenes Comunistas--UJC), formerly known as the Union of Young
Rebels. Among the groups it absorbed was the Federation of
University Students, which played a leading role in Cuban student affairs before the Revolution. An objective of the UJC
was to raise the consciousness and self-discipline of young
Cubans; its members were responsible for the mobilization of
all youth. In 1972 the Second Congress of the UJC acknowledged serious weaknesses in the organization's program and
launched a campaign to extend its influence and enhance its
recruitment. It was also agreed that the organization would
work closely with the Ministry of Education to reduce the student dropout rate and direct students into the occupations in
which they were most needed. In 1987 the UJC had a membership of 604,457. It was headed by Roberto Robaina, who
was directly responsible to the Central Committee of the PCC.
The 1975 draft of the Cuban Constitution defines this organization's mission: "The Young Communist League, the organization
of the vanguard youth, is under the direction of the Party and
contributes to the education of the new generations along the
ideals of communism, by means of their participation in a program of studies and in patriotic, labor, military, and scientific
activities." Cubans may join the UJC at fourteen years of age
and may become members of the PCC at age twenty-seven. Despite a partially competitive selection process, the UJC seeks
the widest possible membership. In some specialized schools
its membership reaches 90 percent of the student body.

UNION DE PIONEROS DE CUBA (UPC). See Organización de
Pioneros José Martí (OPJM).

UNION INSURRECCIONAL REVOLUCIONARIA (UIR). One of three
prominent urban "gangster" groups operating in Cuba throughout the 1940's. UIR was headed by Emilio Tró, who was chief
of police in Marianao (a city close to Havana) and whose position afforded a protective shield for the unethical activities of
the organization. UIR and two other groups, MSR (Movimiento
Socialista Revolucionario) and ARG (Asociación Revolucionaria
Guiteras) were rivals in attracting students' support for their
activities.

UNION NACIONAL DE ESCRITORES Y ARTISTAS DE CUBA
(UNEAC). Established in 1961, UNEAC became the official

and sole organization of intellectuals, writers, poets, and artists in Cuba. Through UNEAC the Cuban government centered authority over cultural matters and established a series of cultural guidelines within the framework of Socialist Cuba. As Castro emphasized, "artistic creation will always be judged through a revolutionary prisma ... within the Revolution--everything, outside the Revolution--nothing."

UNION NACIONALISTA (National Union). A short-lived political party opposed to the Gerardo Machado (1925-33) regime in the 1930's. The National Union was founded by Carlos Mendieta, later to become President of Cuba. In 1928, Machado's Emergency Law prohibited presidential candidates to be drawn from parties other than the Liberal, Conservative, and Popular, thus barring Mendieta from running as the National Union candidate. After Mendieta's assumption of the presidency in 1934, the party waned as a political force.

UNIVERSIDAD POPULAR JOSE MARTI. Founded by the student leader of national fame Julio Antonio Mella, as an adult education institute. Mella became its Secretary-General in 1921 and José Tallet its president. The Universidad Popular had a short lifespan, since it was closed in 1927 by Cuban dictator Gerardo Machado (1925-33), who was aware of the leftist, Marxist instruction practiced by Mella and Tallet. Later, in the 1950's, the Universidad José Martí was founded in Havana as a technical college, but it was not a revised version of the original.

UNIVERSITIES. Until the early 20th century, the University of Havana, founded in 1728, was the only center of higher learning in Cuba. As Cuban university students became a cohesive political force in the 1920's, they pushed for a modern, democratic, participative university, and espoused reforms that paralleled the social reforms widely discussed in Cuba at that time. By the 1950's, other universities evolved, including the Universidad José Martí, a large technical college in Havana. In addition, several hundred special colleges (or colegios) existed that were roughly akin to the post-secondary junior college institutions found in the United States. These colegios provided certification for middle-class young adults, enabling them to secure professional employment later. After the revolution, education inside Cuba was reformed. In 1961, the universities were reorganized with many former professors exiled, and new department heads, possessing solid revolutionary credentials in place. In practice, by 1961 the universities were Marxist-Leninist and student discipline was far stricter than before. Attendance at lectures was compulsory and as in other sectors of society, membership in the militia was mandatory. By 1969, the universities in Cuba were totally structured around the revolutionary needs of the Cuban economy. To this day, students have little autonomy in selecting courses of study, as

the technical needs of Cuba's agricultural organization determine
the nature of the curricula. In 1969, universities had over
40,000 students, constituting an increase of 15,000 in 11 years.
Most available textbooks today are translations of Russian texts.
Castro, in the late 1960's, professed a desire to see the universities abolished and replaced by a system wherein normal education would include technical education in the final years and
graduates from the secondary system would enter agriculture
or industry as trained technicians.

UNIVERSITY OF HAVANA. Founded in 1728, the University of
Havana remained the only center of higher education in Cuba
for nearly a century and a half. By 1900, under the first
United States intervention, new professors were added, modern
scientific equipment was introduced, and the campus was moved
from its old location in the Santo Domingo Convent to its present
location in the center of the city. Students at the university
became politically articulate as the 1920's unfolded, seeking to
utilize the university as a national asset with which to explore
the politics, culture, and society of Cuba. They pushed for
and received academic reforms including more freedom from
governmental interference, but did not acquire autonomy. By
the 1940's a strong student organization, the Federation of
University Students (FEU), had lobbied for and obtained a
large measure of autonomy and was fully participative in the
stormy political life of Cuba. During the 1950's, the FEU represented the 17,000 students at the university which, located
in the heart of the capital city, was exposed to the continuous
shock waves of the tumultuous political scene. The university's
autonomy had converted it into a sanctuary for political agitation
--a safety zone for myriad activities--against the Batista regime.
After University of Havana student leaders, such as José Antonio Echeverría, began to openly support Fidel Castro's 26th
of July movement in the mid-1950's, Batista began restricting
university autonomy. The university was closed for several
months after November 1956 when Castro attempted to foment
insurrection with his Granma invasion of Oriente province.
After the Revolution, Castro began to reshape the university by
placing supporters inside the student federation who proselytized
for change, and he incorporated the university into the state.

URBAN REFORM. In 1960 the Urban Reform Act reduced rents and
guaranteed the urban worker that he could become the owner of
the house he was living in. This act, along with the Agrarian
Reform Act of 1960, served notice on foreign interests in Cuba
(particularly U.S. interests) that the Revolution was moving
toward nationalization of foreign assets.

URRUTIA, MANUEL (1901-1981). Judge. First Cuban President
following the Revolution of 1959. Born in Yaguajay, Las Villas
province, Urrutia graduated from law school in 1923. He was

appointed judge in Santiago de Cuba, where he presided over
the tribunal that tried a group of Fidel Castro's supporters.
In May 1957, Urrutia determined that the "Fidelistas" should be
acquitted. This decision brought him strong criticism and a
judicial reprimand from the government of President Fulgencio
Batista (1952-58). In December 1957, Urrutia went into exile
in Miami as Fidel Castro's favored candidate for the presidency
once the civil war was over. Urrutia flew to Castro's strong-
hold in the Sierra Maestra in November 1958 to consult with the
revolutionaries. In January 1959, following the triumph of the
Revolution, Urrutia took over the Presidency and appointed a
civilian cabinet composed mainly of anti-Batista political figures.
Urrutia then proceeded to tear down Batista's governmental
structure. Through a series of decrees, he dissolved Congress;
removed from office all Congressmen, provincial governors,
mayors, and municipal councilmen; abolished all of Batista's
censorship and martial law restrictions; and initiated a widespread purge of Batista supporters in the bureaucracy. It
soon became clear, however, that real power resided in the
person of Fidel Castro and his band of young rebel army officers. Castro publicly announced major policy changes without
consulting Urrutia's cabinet and complained about the slowness
of reforms. On his part, Urrutia publicly attacked the Communists. In his speech of July 17, 1959, Fidel Castro accused
Urrutia of posing an obstacle to the Revolution and of fabri-
cating the legend of a Communist danger to Cuba. Urrutia
resigned and was placed under house arrest. He was able to
take refuge in the Venezuelan Embassy and subsequently go
into exile, whereupon he retired from public life. Castro replaced Urrutia with Osvaldo Dorticós as President of Cuba.

U.S. CONGRESSIONAL RESOLUTION ON CUBAN BELLIGERENCY.
Resolution passed by the U.S. 54th Congress on April 6, 1896,
wherein both Houses expressed the opinion that the U.S. should
maintain a position of "strict neutrality" between Cuba and
Spain, which were recognized as belligerents by the Resolution.
It was resolved further that the U.S. would guarantee its friendship to Spain if the latter would decide to recognize the independence of Cuba.

U.S. JOINT CONGRESSIONAL RESOLUTION ON CUBA (September,
1962). "Whereas President James Monroe, announcing the Monroe Doctrine in 1823, declared that the United States would consider any attempt on the part of European powers 'to extend
their system to any portion of this hemisphere as dangerous to
our peace and safety'; and Whereas in the Rio Treaty of 1947
the parties agreed that 'an armed attack by any state against
an American state shall be considered as an attack against all
the American states, and, consequently, each one of the said
contracting parties undertakes to assist in meeting the attack
in the exercise of the inherent right of individual or collective

self-defense recognized by Article 51 of the Charter of the United Nations'; and Whereas the foreign ministers of the Organization of American States at Punta del Este in January, 1962, declared: 'The present Government of Cuba has identified itself with the principles of Marxist-Leninist ideology, has established a political, economic, and social system based on that doctrine, and accepts military assistance from extracontinental Communist powers, including even the threat of military intervention in America on the part of the Soviet Union'; and Whereas the international Communist movement has increasingly extended into Cuba its political, economic, and military sphere of influence: now, therefore, be it Resolved by the Senate and House of Representatives of the United States of America in Congress assembled, that the United States is determined (a) To prevent by whatever means may be necessary, including the use of arms, the Marxist-Leninist regime in Cuba from extending by force or threat of force its aggressive or subversive activities to any part of this hemisphere; (b) To prevent in Cuba the creation or use of an externally supported military capability endangering the security of the United States; and (c) To work with the Organization of American States and with freedom-loving Cubans to support the aspirations of the Cuban people for self-determination."

U.S. JOINT RESOLUTION OF CONGRESS RECOGNIZING THE INDEPENDENCE OF CUBA. The Joint Resolution of April 20, 1898, expressed: "Whereas the abhorrent conditions which have existed for more than three years in the island of Cuba, so near our own borders, have shocked the moral sense of the people of the United States, have been a disgrace to Christian civilization, culminating, as they have, in the destruction of a United States battleship, with 266 of its officers and crew, while on a friendly visit in the harbor of Havana, and cannot longer be endured, as has been set forth by the President of the United States in his message to Congress of April 11, 1898, upon which the action of Congress was invited: Therefore, Resolved by the Senate and House of Representatives of the United States of America in Congress assembled, First. That the people of the island of Cuba are and of right ought to be free and independent. Second. That it is the duty of the United States to demand, and the Government of the United States does hereby demand that the Government of Spain at once relinquish its authority and government in the island of Cuba and withdraw its land and naval forces from Cuba and Cuban waters. Third. That the President of the United States, be and hereby is, directed and empowered to use the entire land and naval forces of the United States and to call into the actual service of the United States the militia of the several States to such extent as may be necessary to carry these resolutions into effect. Fourth. That the United States hereby disclaims any disposition or intention to exercise sovereignty, jurisdiction, or control over said island except for the pacification thereof, and asserts its determination, when that is accomplished, to leave the government and control of the island to its people."

USSR-CUBAN RELATIONS. See Soviet-Cuban relations.

- V -

VALDES, GABRIEL DE LA CONCEPCION (1809-1844). Poet martyr of Cuban independence, Valdés was born in Havana, where he acquired the nickname "Plácido." He was the illegitimate mulatto child of a Cuban barber and a Spanish dancer; consequently, Valdés was raised in the Havana orphanage (Casa de Beneficencia). An avid reader, Valdés was a man of remarkable literary talent who garnered a reputation as a prolific poet. He conspired against harsh Spanish colonial rule in the Escalera (Ladder) Conspiracy, was discovered, and executed. Valdés's most important poems include "Xicotencal," "Plegaria a Dios," and "La flor de la caña."

VALDES MENENDEZ, RAMIRO (1932-   ). Valdés was born in Artemisa, Pinar del Río. He participated in the unsuccessful assault on the Moncada Barracks in Santiago de Cuba on July 26, 1953. As a member of Fidel Castro's 26th of July movement, Valdés also participated in the Granma expedition in late November 1956. He has been a member of the Cuban Communist Party Central Committee since October 1965 as well as of its Politburo. In November 1976 he was elected a deputy to the National Assembly representing the municipality of Artemisa. Valdés has served as Minister of Interior (1961-68); Vice-Minister of the Revolutionary Armed Forces (1970-71); Chief of the Construction Sector (1971-72); Vice-Prime Minister (1972-76); Vice-President of the Council of Ministers (December 1976-   ). In 1981 he was reappointed Minister of Interior, but was removed in 1986.

VALDES RODRIGUEZ, MANUEL (1849-1917). Born in Matanzas province, Valdés Rodríguez was a Cuban educator concerned with modernizing the Cuban school system. He was influential through his books announcing a reshaping of the educational system at the outset of the 20th century. Valdés Rodríguez's main works include La escuela y la sociedad; Ensayo sobre educación teórica; and La educación popular en Cuba.

VALLADARES, ARMANDO. Born in 1937, Armando Valladares graduated from the Instituto de la Habana and completed a year of law school at the University of Havana. Valladares, who was a civil servant in Castro's revolutionary government, vocally opposed the Marxists taking control of power. He was arrested in early 1961 and sentenced to thirty years in prison. While imprisoned, he wrote From My Wheelchair. Pressure from Amnesty International and several governments led to his release in 1981. In the years since his release, he has denounced Cuban human rights violations at the United Nations. Valladares is also the founder of the Paris-based organization International Resistance, which fights human rights violations worldwide. He is the author of The heart with Which I Live, a volume of poetry, and, most recently, Against All Hope (Contra Toda Esperanza), a critically acclaimed account of his years in Cuban prisons.

VALLE JIMENEZ, SERGIO DEL (1927-). Sergio del Valle Jiménez was born in Pinar del Río province. He is a graduate of the Medical School of the University of Havana. In the late 1950's he joined Fidel Castro's 26th of July movement and fought with the rebel group in the Sierra Maestra, obtaining the rank of captain. Assigned to the Ministry of the Revolutionary Armed Forces, del Valle held various posts in that ministry including Chief of the Air Force (1959-61); Vice-Minister (1961-65); and Vice-Minister for Special Affairs (1965-68). From July 1968 to December 1979 he served as Minister of Interior. Since the latter date he has been Minister of Public Health. Del Valle has been a member of the Cuban Communist Party Central Committee and its Politburo since October 1965. In November 1976 del Valle was elected a deputy to the National Assembly representing the Municipality of Pinar del Río in the province of the same name.

VARADERO. Beautiful beach in the peninsula of Hicacos which juts out in the northern part of Matanzas and is the closest place to the U.S. The peninsula protects the ports of Cárdenas and the Bay of Matanzas. Varadero also has one of Cuba's international airports used as a connecting point between Havana and Cienfuegos.

VARELA, FELIX (1787-1853). Born in Havana, Varela studied at San Carlos Seminar where he became a priest and taught philosophy. He was influential with a group of students who later distinguished themselves as intellectuals, including men such as José Antonio Saco and José de la Luz y Caballero. As a philosopher and promoter of cultural activities, Varela enjoyed the support of the prominent Bishop Espada and became the latter's protégé. Varela favored modern philosophical doctrines and pioneered in introducing the explicative teaching method, thus replacing the rote memorization method of Scholasticism. Varela was also instrumental in the acceptance of Spanish over Latin as the language of education. Popular with his students, Varela sympathized with their liberal ideas and dislike for the absolutism

of the Spanish king, Ferdinand VII. Although a Creole intellectual liberal, Varela was unique in his vocal opposition to
slavery and its retention in Cuba. Varela distinguished himself as a politician as well as a professor and philosopher. He
was a reformist and was sent as such to the Spanish Cortes
(Parliament) as a Cuban representative in 1823. While serving,
Varela advocated a more benign rule of Spain over its colonies.
Due to the political situation in Spain, Varela was forced into
exile in the United States, where he became a strong advocate
of Cuban independence from Spain. He is considered the first
Cuban to advocate complete separation from Spain. While living
in the United States, Varela published the pro-Cuban independence magazine El Habanero. Philosophical works by Varela include Instituciones de Filosofía ecléctica (1814), Apuntes filo-
sóficos (1818), and Lecciones de Filosofía (1820). Varela's non-
philosophical works are Cartas a Elpidio, Cartas eruditas, and
Teatro crítico universal.

VARONA, ENRIQUE JOSE (1849-1933). Born in Camagüey province,
Varona moved to Havana after completing his early education to
attend the University of Havana from which he graduated with
the degree of Doctor of Philosophy and Letters. A student
activist and poet in his youth, Varona completed his Odas
anacreónticas before reaching the age of twenty. In later
years at the University of Havana, Varona held the chairs of
Psychology and Sociology, and exerted influence over student
leader Julio Antonio Mella. Varona became a polemicist and
served briefly as a Cuban Deputy to the Spanish Cortes (Parliament). The poet-politician came to support Cuban independence from Spain, making his appeal Contra España (1896) to the
peoples of Latin America in favor of the Cuban Independence
War (1895-1898). Under the American Military Governor Leonard Wood (1899-1902), Varona was appointed Secretary of Finance, and later served as Secretary of Public Instruction. In
1912 he was elected Vice-President of Cuba under President
Mario García Menocal (1913-1921). After completing his term,
Varona retired to a private academic and literary life. Other
published works of his include Las reformas de la enseñanza
superior (1904); La instrucción pública en Cuba (1901); Mirando
en torno (1910); El fundamento de la moral (1903); and Por
Cuba: Discursos (1918).

VAZQUEZ BELLO, CLEMENTE (1885-1932). Born in Santa Clara,
Las Villas province, Vázquez Bello was electoral manager in
the victorious 1925 campaign of President Gerardo Machado
(1925-1933). A close friend of Machado's, Vázquez Bello was
President of the Cuban Senate in 1932 when he was shot to
death by members of the militant anti-Machado organization
called the ABC. The ABC's plot was to assassinate an important official in the Machado government and then blow up
the burial site during that official's funeral at the Havana

cemetery, thus killing all top government officials, including Machado. Although the first phase of the plan to eliminate the Machado regime was completed through the murder of Vázquez Bello, the second failed when it was decided that Bello's funeral would take place in his hometown and not in Havana. A gardener working at the cemetery subsequently discovered the buried explosives. Vázquez Bello's murder cost the ABC dearly since Machado clamped down on his opposition.

VEGETATION. Cuba's approximately 8,000 species of trees and plants render the island a natural botanical garden. Propagation of a great variety of seeds has been facilitated by Cuba's position in the routes of annual bird migration and tropical storms. Subtropical and tropical species, common to Florida, the West Indies, Mexico, and Central America are found in Cuba, as well as many species introduced by man. Cuba's forests have been reduced from 60% to 10% of its total land area, corresponding to the need for timber and wood products for agricultural uses. Cuba's lowland savannas are dotted with the country's national tree--the royal palm--and a few huge ceibas (silk-cotton trees). In the mountains of Oriente, Las Villas, and Pinar del Río provinces, broadleaf evergreen, coniferous, and semi-deciduous forests still exist. Only near Baracoa can small sectors of true tropical broadleaf forest be found due to the nature of the island's seasonal rainfall. Pine forests are relatively large in the Sierra de Nipe (Oriente province), in Pinar del Río province, and on the Isle of Youth, although in this last area the pine forests have been drastically decimated. Over forty prime woods are among Cuba's forest species, including Cuban oak and mahogany. Many woods and plants used for their medicinal properties may also be found. Over 30 species of palm can be found in the lowland areas, as well as extensive mangrove forests, valuable for their charcoal deposits. Pre-Castro administrations in Cuba made few efforts to implement reforestation programs in the wake of forest-clearing projects. More than 600 million trees have been planted, however, since 1959, according to the Castro government's report on its reforestation program in 1974. Vegetation native to Cuba includes manioc, tobacco, guava, and pineapple. Many tropical fruits grow wild on the island, including avocados, mangos, prickly pears, papayas, bananas, plantains, sweetsops, soursops, custard apples, and star apples. Some of the better-known flowering plants and trees include the beautiful eastern highland orchids, jacaranda, frangipani, magnolia, jasmine, gardenia, and red and purple bougainvillea.

VEGUERO. Tobacco farmer located primarily in the province of Pinar del Río in the western part of the island. The best tobacco farming land was in Vuelta Abajo in that province.

VEIGA MENENDEZ, ROBERTO. Roberto Veiga Menéndez has been

a member of the Central Committee of the Cuban Communist Party since 1975. During the Third Party Congress that took place in February 1986, he was elected as an active member of the Politburo and ratified as General Secretary of the Confederation of Cuban Workers (CTC). Veiga Menéndez did not participate actively in the revolutionary process prior to 1959, and his role until the 1970s was fundamentally that of a minor government labor leader. In 1976 he was named Secretary General of the National Council of the CTC. Veiga Menéndez is one of the few members of the upper echelon of the Cuban government who does not have a history of revolutionary activity.

VELAZQUEZ, DIEGO (1465-1524). Conqueror and Governor of Cuba. An able administrator and former lieutenant to Nicolás de Ovando, Governor of Hispaniola, Velázquez was one of the wealthiest Spaniards on that island and achieved a reputation for courage and sagacity because of his role in subduing Indian caciques in Hispaniola. Christopher Columbus's son and governor of the Indies, Diego, appointed Velázquez governor of Cuba. Velázquez recruited over 100 men and prepared for the conquest. In early 1511 he landed near Baracoa on the eastern tip of the island where he established the first permanent settlement and prepared for the colonization of the rest of the island. He fought his way through hostile Indian attacks, eventually capturing and burning their leader, Chief Hatuey. To some extent, Velázquez tried to reduce abuses against the Indians. He founded several towns: Baracoa, Bayamo, La Habana, Puerto Príncipe, Trinidad, and Santiago de Cuba. Velázquez issued encomiendas, a practice that entailed assigning Indian families to a Spaniard who would extract labor and tribute from them while providing for their Christianization. Velázquez remained Governor of Cuba from the time of his landing in 1511 until his death in 1524.

VENCEREMOS BRIGADE. In the late 1960's and early 1970's, the Venceremos (We Shall Win) brigades consisted of some 1,000 youths, mostly Americans, who visited Cuba to help in the sugar harvest (zafra). Their participation expressed solidarity with Cuba's criticism of U.S. involvement in Southeast Asia. Venceremos youth received liberal doses of Viet Cong information about the Vietnam War and generally were educated in the ideology of the Cuban leadership concerning revolution in general. As a propaganda tool, Venceremos brigades aided Fidel Castro in his criticism of the United States, since American youths were depicted cutting cane in Cuba, violating the U.S. blockade of the island.

VERDE OLIVO. Official weekly magazine of the Ministry of Armed Forces in Cuba.

VETERANS' AND PATRIOTS' ASSOCIATION. In 1923, under the

Alfredo Zayas (1921-25) administration in Cuba, the Veterans'
and Patriots' Association began to agitate for political reforms.
The Association threatened to initiate a revolution and did in
fact stage a minor uprising in Cienfuegos on April 30, 1924.
The poorly organized movement was easily broken up and the
U.S. administration of Calvin Coolidge exercised restraint by
not intervening a third time.

VILLA CLARA (PROVINCE). Population: 700,006; area: 8,782 Km^2;
municipalities: Corralillo, Sagua La Grande, Camajuaní, Remedios, Santa Clara, Santo Domingo, Manicaragua, Quemado de
Güines, Encrucijada, Caibarién, Placetas, Cifuentes, Ranchuelo.
Province in the central part of Cuba. Prior to the political-
administrative division of 1976, the Province of Villa Clara was
just one province (Las Villas) together with the present provinces of Cienfuegos and Sancti Spíritus (part of what would become part of Matanzas). The capital is the city of Santa Clara,
a commercial and communications center. Its population increased
in 1933, basically because of the construction of the Central
Highway. It was founded in 1689 by families coming from
neighboring Remedios, a provincial tobacco-growing city. Besides Remedios, the cities of Sagua la Grande and Placetas are
important because of their location above sea level (185 meters)
and because they are the coldest cities in Cuba.

VILLAS, LAS. One of pre-revolutionary Cuba's six provinces.
Las Villas disappeared when the new administrative divisions
were established in 1976, roughly into the new provinces of
Cienfuegos, Villa Clara, and Sancti Spíritus. In early colonial
Cuba, Las Villas province was the center of the cattle industry,
having been the location of the first Merced (land grant) by
the Cabildo (Spanish town council). Las Villas in the 18th and
19th centuries had a large concentration of sugar mills, and by
the 20th century had the highest percentage of land in farms
(90%). By 1950, Las Villas was second to La Habana province
in number of Cuban manufacturing establishments. In 1953,
the ethnic breakdown of Las Villas showed 82% white, 9% black,
and 9% mestizo, making it the least heterogeneous province in
Cuba. See also Villa Clara (province).

VILLAVERDE, CIRILO (1812-1894). Born in Pinar del Río province,
Villaverde was initially a teacher and journalist who developed
an interest in literature. In 1837 he published his short stories
"La peña blanca"; "El ave muerta"; "La Cueva de Taganana";
"El perjurio"; "El espetón de oro"; "Engañar con la verdad."
In these brief narrations, Villaverde demonstrates great power
of description and the capacity to transmit the cultural atmosphere of his time. In 1839 Villaverde published the first part
of what is considered the best folkloric novel of Cuban literature of manners: Cecilia Valdés o La loma del Angel.
The second part appeared in 1882. Through his

love story about a mulatto girl and a slave trader's son who
turns out to be her illegitimate brother, Villaverde creates a
splendid panorama of Cuba in the early 1800s describing the
haciendas, coffee plantations, and sugar mills of 19th-century
Cuba, as well as the prevailing customs and social condition of
its people. Villaverde became involved in the struggle for
Cuban independence from Spain. He fled the island in 1848
following his participation in the annexationist conspiracy of
General Narciso López, with whom Villaverde helped to design
the Cuban flag. After establishing himself in the United
States, Villaverde directed the pro-independence newspaper
La Verdad from New York. Other novels by Villaverde are
El penitente (1844) and Dos amores (1858).

VIÑALES. City in Pinar del Río province numbering some 16,000
inhabitants by 1980. Viñales is a beautiful valley within a very
mountainous territory known for its excellent tobacco land.

VIRGINIUS. During the Ten Years' War (1868-1878), a group of
armed Cuban rebels on board the Virginius was forced into
Santiago de Cuba by a Spanish naval vessel. The Spanish
governor of Oriente province sentenced the entire crew to
death as pirates. The Virginius had been a U.S. ship and had
a number of British subjects on board. Fifty of the 150 crew
members had been executed by the Spanish when a British war-
ship, the Niobe, sailed into the harbor at Santiago and intervened in the name of Great Britain to spare the remaining captives.

VIVES, GENERAL FRANCISCO DIONISIO (1735-1840). Spanish
Captain-General (Governor) of Cuba from 1822 to 1832 under
whose autocratic rule Cuban liberalism was crushed. Vives
abrogated the constitutional liberty that had existed in Cuba
for three years prior to his arrival. Vives was described as
being stiff in manners and dissolute in habits, always seeking
to enrich himself as Captain-General. A popular refrain evolved
about Vives: "Si vives como Vives, vivirás" (If you live like
Vives, you will live well). Vives enriched his friends as well.
Under his administration, general corruption and a wild west
type of atmosphere pervaded Cuban society, both rural and
urban.

VOLUNTARY LABOR. Established by Fidel Castro as a way of increasing productivity without cost. Since political and even
economic advancement in Cuba is reserved for those who sacrifice harder for the revolution, voluntary work is a way of
showing devotion and loyalty to the party and the revolutionary
cause. Women's involvement in voluntary labor is surprising.
They fill a variety of jobs in and out of the government bureaucracy, many working overnight on armed-guard duty.
Although some of this work is voluntary, there is a great

amount of direct and indirect coercion. The former takes the
the form of directives by the Party, the Federation of Women
under the leadership of Vilma Espín, Raúl Castro's wife, and
other organizations to engage women voluntarily in tasks such
as cane cutting. The latter is more subtle. The more militant
exert a sort of social pressure on the less militant to participate.
Undoubtedly, Castro has achieved success in his calls for "voluntary" work, especially when in crisis to meet the goals of
sugar harvests. See also Moral incentives.

- W -

WAR OF SPANISH SUCCESSION. Ended in 1713 by the Treaty of
Utrecht, when England obtained from Spain an asiento (privilege) to bring African slaves to the Spanish colonies.

WELFARE. During the prerevolutionary era the wealthy upper
class supported many charitable undertakings, recognizing a
social and religious obligation that was also an important means
of expressing social status. Gift giving was done conspicuously, focusing attention on the social distinction between those
giving and those receiving aid. Preeminent among these organizations was the Royal Alms House (Real Casa de Beneficencia) founded in 1795 and supported by the Roman Catholic
Church and private donors. Small groups of wealthy Spanish
expatriates had formed charitable organizations to extend assistance to newcomers so that they would not become public
charges. Each organization concerned itself with immigrants
from a particular region of Spain. Their clubs, many of which
continued operating until the early 1960's, were cooperatively
organized for mutual aid and social purposes. The largest clubs
maintained schools, homes for the aged, mausoleums, and some
of the finest hospitals and clinics in the country; most members
were interested chiefly in the free medical privileges to which
they were entitled. Many other secular organizations sponsored
welfare services. Upper-class women in Havana were particularly active in charities. The government operated various
health and welfare institutions including hospitals, dental and
maternity clinics, and detention homes for juveniles. The level
of care in the institutions appears to have been fairly good, although many apparently suffered from poor administration and
lack of operating funds. There was also the problem of over-
concentration of such facilities in Havana. Members of the urban middle class and the organized wage earners in the lower
class were the principal beneficiaries of the social insurance
plans that existed. Through their unions and professional associations these groups were able to wrest special concessions
and guarantees from the government. Laws and decrees

providing for retirement, disability, work injury, and maternity
insurance programs were enacted between 1916 and 1958, but
health insurance was available only under private schemes, and
there was no unemployment insurance.

The most significant of the programs were the highly diver-
sified pension and retirement funds. From 1919 on, public employees in general were protected by a civil service retirement
scheme, but preexisting schemes for the armed forces, the
Ministry of Communications, and members of the judiciary remained in effect, and during subsequent years special laws
favoring specific public occupational categories continued to be
enacted. The same trend toward diversification was evident in
the private sector, where separate funds for specific occupations
were brought into being by vested interests backed by political
and union pressures. Concern over the proliferation of pension
plans had been expressed as early as 1943, when an attempt
was made to establish a unified social security program. This
effort failed, however, and on the eve of the Revolution more
than fifty entities provided pension coverage for as many small
groups. Mismanagement and inefficient handling of reserves
caused nearly all the funds to encounter financial difficulties,
and many ineligible persons succeeded in collecting benefits.
Between 40 and 60 percent of the labor force was protected by
the pension plans, however, and even the lower proportion
would have been sufficiently high to place Cuba in the top rank
of Latin American countries in pension coverage during the late
1950's. Immediately after assuming power in 1959, the revolutionary government enacted legislation consolidating twenty funds
for wage and salary earners in the private sector and placing
them under the direction of the Social Security Bank, an institution created at the same time. The following year the public
funds were merged and consolidated with the private funds, and
the Ministry of Labor assumed the functions of the Social Security Bank, which was abolished. In 1962 the several funds for
professional workers were taken over by the Ministry of Labor,
which at the same time assumed responsibility for administering
the programs for protection against occupational diseases and
industrial accidents, previously administered by private employers. Enactment of the Social Security Act of 1963 completed
the standardization of worker coverage in the public sector and
what remained of the private sector of unemployment. Finally,
in 1964, these unified standards were extended to include the
self-employed and certain other categories.

The Social Security Board of the Ministry of Labor was responsible for the administration of the system; the commissions
of claims were the basic administrative units. Situated at work
centers, each was composed of about twenty-five workers, who
were supervised by three titular members: one chosen by a
general assembly of the workers, another by the enterprise itself, and the third by agreement between the trade union section and the enterprise. The commissioners were responsible

for investigating the validity of all claims made by the workers.
An individual could be denied social security benefits by an assembly of workers if his attitude toward work was not considered sufficiently "revolutionary." With minor changes the Social Security Act remained in effect in the mid-1980's, and under it the government bore all costs; no charges were assessed
against the worker's paychecks. Old-age pensions were payable
regularly to men at the age of sixty and to women at the age of
fifty-five after twenty-five years of employment or after twelve
years of employment in work classified as dangerous or arduous.
If employment had been for shorter periods of time, specified
minimum pensions were payable at the ages of sixty-five for
men and sixty for women. Full disability pensions were payable
on loss of two-thirds of working capacity, and survivor pensions were payable when the deceased was employed or a pen-
sioner at the time of death. The old-age pension amounted to
50 percent of average earnings during the last five years of
employment, plus an increment of 1 percent for each year of
employment after twenty-five years or 1.5 percent if the work
was arduous or dangerous. The amounts payable ranged from
a minimum of sixty pesos a month to a maximum of 250 pesos.
Full disability pensions were payable in a similar manner and in
similar amounts; partial disability was paid in proportion to the
degree of disability.

The survivor pension amounted to 60 to 100 percent of the
pension of the insured, the amount depending on the number of
survivors. Eligible dependents might be widows, needy widow-
ers aged sixty or over, children under the age of eighteen, or
disabled or needy parents. Sickness benefits amounted to 50
percent of earnings of the worker or 40 percent if he or she
was hospitalized, payable after a three-day waiting period during a period of up to twenty-six weeks or up to fifty-two weeks
if recovery seemed probable. On conclusion of the maximum
period the worker became eligible for disability retirement if he
had not recovered. For tuberculosis patients the payment was
100 percent of earnings for up to two years plus six months of
rehabilitation. Maternity benefits amounted to 100 percent of
earnings for a period of up to six weeks before and six weeks
after confinement on the basis of a minimum of 1.5 pesos and a
maximum of eight pesos per day. In addition a lump sum of
twenty-five pesos was payable if no hospital facilities were
available and the birth took place in the home. Continuance of
employment during the period of pregnancy and extending to
one year after the birth was guaranteed. Work injury benefits
were payable without a minimum qualifying period. Temporary
disability resulting from a work injury was payable at a rate of
70 percent of current earnings, or 60 percent if hospitalization
occurred, and was payable for twenty-six weeks if recovery
seemed probable. Permanent work-injury disability was payable
at a rate of 55 percent of earnings plus 1.1 percent per year
of employment in excess of twenty-five years of 1.65 percent if

the work was deemed dangerous or arduous. Benefits ranged
from a minimum of sixty-six pesos to a maximum of 275 pesos
monthly. Partial disability resulting from work accidents was
payable in proportion to the degree of disability. When death
resulted from the work accident, payment to survivors was
made under terms similar to those for survivors of old-age and
disability retirees. In all cases a lump-sum funeral grant of
fifty pesos was provided when the earnings of the deceased
had been 350 pesos or less per month.

There were a few special categories. Workers who had received the Heroes of Moncada Flag award for surpassing production quotas were awarded full pay on retirement. The same
benefit was conferred on some 170,000 workers who in 1968 had
announced their voluntary renunciation of overtime wages, a
benefit that Castro was reported later to have recognized as a
mistake. In 1973 the Thirteenth Congress of the Confederation
of Cuban Workers adopted a proposal that 100-percent pensions
were no longer to be granted, and at the end of the same year
the government moved to eliminate this overly generous policy.
Early in 1986 the Ministry of Labor was reported to be studying
the possible amendment or replacement of the Social Security
system. The budget of the social security program, more
than 500 million pesos in early 1976, had doubled since 1966, and
it had been suggested that all pensions in effect should be honored but that a new law should prescribe a schedule of payments more in keeping with the resources the country had at
its disposal. The minimum monthly retirement benefit of sixty
pesos had been in effect since 1969, and it had been widely
rumored that the new law might raise the minimum to 100 pesos.
In squelching the rumor the Ministry of Labor pointed out that
the minimum agricultural wage in Cuba was only seventy-five
pesos per month and that accordingly a worker receiving a minimum of 100 pesos on retirement might find his income substantially increased. Instead the studies in progress in 1976 pointed
to the advisability of replacing the minimum retirement income
by a flexible schedule. Among the other possible changes under study was a revision of the status of workers desiring to
continue employment after reaching retirement age. Under the
law in effect in early 1976 a pension was payable only on actual
retirement, but consideration was being given to a change that
would permit at least partial pension payment to the worker
desiring to remain active, particularly when his skills were in
short supply.

The Revolution had come as a crippling blow to well-to-do
Cubans, and those who left the country departed with little
more than the clothes on their backs, their houses and possessions were taken over by the government, and refugees who
had not provided themselves with bank accounts abroad arrived
at their destinations destitute. Well-to-do people who remained
behind, however, were by no means reduced to penury. Employed people who continued to work were assigned a "historic

wage" based on the amount previously paid to them rather than
the usually lower wage for similar work currently in effect.
Owners of expropriated farms were permitted to keep their
farmhouses and were given lifetime pensions of as much as 600
pesos monthly. Similarly, urban landlords lost their rental
properties to the government but were given direct government
payments for them. Owner-occupants of urban mansions were
permitted to remain in them, and those who were widows or unable to work were given pensions. Many conditions that in
other countries would have made public or private welfare assistance necessary did not exist in Cuba of the 1970's. There
was no unemployment benefits program but, at least by official
pronouncement, there was no unemployment. All medical and
dental care was free. Education was free at all levels, and
housing rentals were kept artificially low. Most food items were
rationed, but rationed food was so inexpensive that even the
family of the worker receiving minimum wages did not need to
go hungry. Organized recreational activities were numerous
and available either free or at moderate cost. The CDR and
other mass organizations provided a variety of such peripheral
welfare services as visits to the elderly and care of invalids
and children. For those in need of institutional care there
were nearly 6,000 homes for the aged and 900 homes for or-
phaned and abandoned children, and in 1985 Castro announced
a program for the construction of additional homes.

WELLES, SUMNER (1892-1962). U.S. career diplomat appointed Assistant Secretary of State in early 1933 under Franklin D.
Roosevelt's administration. Roosevelt and U.S. Secretary of
State Cordell Hull decided to send Welles to Cuba as U.S. Ambassador to mediate the violent confrontations between Cuban
dictator Gerardo Machado (1925-33) and his opposition. Welles
arrived in Cuba to stabilize internal conditions on June 1, 1933,
and advised Machado to allow political party pluralism and to
restore the office of Vice-President, but to no avail. Welles'
mediation efforts were supported by most politcal factions excluding the conservative followers of former President Mario G.
Menocal (1903-21). The mediation culminated in a general strike
and ultimately in defection of the armed forces, whose lack of
support induced Machado to heed Welles' advice. Machado thus
resigned on August 12, 1933, and fled the country. A coalition
government under Carlos Manuel de Céspedes was rapidly organized under Welles' direction, but proved unpopular and incapable of gaining military support. Welles proposed a general
election to quell strikes and public disorder. The Céspedes
government, however, was overthrown by an army coup in
September 4 under Sergeant Fulgencio Batista, who installed
Grau San Martín as President. Welles had advocated U.S. military intervention to protect the government of Céspedes but
was opposed by President Roosevelt and Secretary of State
Cordell Hull. Welles soon found that Grau's rule-by-decree

and nationalistic fervor aroused anti-Americanism, and thus he
persuaded the U.S. not to recognize Grau's government. Welles
departed Cuba and was replaced by Jefferson Caffery, who also
distrusted Grau's revolutionary ideas and programs. On January 20, 1934, Grau was overthrown and replaced by Carlos
Mendieta (1934-35) as Provisional President, favored by the
U.S., Batista, and Cuban conservatives. Due to Welles' urging,
the U.S. recognized the Mendieta government within five days.
With Mendieta in power, Welles was instrumental in negotiating
a treaty abrogating the Platt Amendment and removing other
limitations previously imposed on Cuban sovereignty. Welles
also helped to draft a reciprocal trade agreement signed by
Cuba and the U.S. that reduced duties on sugar.

WEYLER, GENERAL VALERIANO, MARQUES DE TENERIFE (1838-
   1930). Of German descent, Weyler was austere in his private
   life and authoritarian in character. Officers and soldiers under
   his command admired him because of his total commitment to his
   military vocation. During the American Civil War, Weyler served
   as the Spanish military attaché to Washington. He later participated on the Spanish side in the Cuban Independence War
   (1868-1878), and in the Carlista War of Spain. In 1896 he was
   appointed Governor (Captain General) of Cuba. Ruling with a
   severe hand, Weyler dedicated himself to the ruthless extermina-
   tion of the pro-independence Cuban rebel. He reduced the size
   of the Spanish cavalry to afford it greater mobility, and reorganized battalions to form self-sufficient columns for each area.
   Pro-Spanish rule Cubans were recruited to form military units.
   Weyler was able to generate a sense of unity among the Spanish
   army's high command, thus discouraging defection to the rebel
   side, but by far his most transcending measure was the policy
   of concentrating the rural population in garrisoned towns. Although this step gave Weyler's military commanders extraordinary
   judicial powers to try and punish any Cuban who defied his
   edicts while depriving the rebels of their support base, the
   General's policy also caused a decrease in agricultural production as well as grave sanitary problems for the population.
   Weyler's repressive measures won him the upper hand in the
   war, but the demand for Spanish troops to put down the rebellion in the Philippines and attacks against him by the liberal
   press in Madrid began to undermine his efforts. The American
   press, especially the Hearst chain of newspapers, also denounced
   Weyler's real and exaggerated atrocities. In early 1897, an unsuccessful assassination attempt on Weyler's life took place.
   Shortly thereafter in June 1897, Spanish Prime Minister Antonio
   Cánovas was murdered. General Weyler resigned a short time
   later, since Cánovas had been his main political supporter in
   Spain.

WHITE LAFITTE, JOSE SILVESTRE (1836-1918). World-famous
   violinist. White ("Joseíto," as he was called) was only eighteen

years old when he played with the famous pianist Gottschalk.
He held the Allard chair at the Imperial Conservatory of Paris
and traveled throughout the world, playing in front of royalty.
He mastered up to sixteen musical instruments and was highly
praised by critics and composers in all French and European
newspapers.

WOMEN. See also Federación de Mujeres Cubanas. After the 1959
Revolution, Fidel Castro attempted to redefine the role and position of women in Cuban society. Since the Revolution, over
half a million women have joined the labor force and a Federation of Cuban Women was organized. Party directives have
specified that women fill a variety of jobs inside and out of the
government bureaucracy. So-called voluntary work assignments
to men and women such as cane cutting have been organized to
give each person an opportunity to demonstrate devotion and
loyalty to the Party. Many more women than before the Revolution are university-educated and participate in politics and
sports. The new roles for women have contributed to under-
mining the traditional Cuban family structure. Large numbers
of children attend free boarding schools and live apart from
their parents permanently. In addition, work demands often
differ for husband and wife such that they too are separated
for extended periods. Overall, women's role has undergone a
drastic metamorphosis from its former subdued, subservient
profile in Cuban society to its present integrated, participant
character under the Revolution.

WOOD, LEONARD (1860-1927). Medical doctor to President William
McKinley, Wood participated in military campaigns against American Indians. During the Spanish-American War, he was promoted to General. Following the Spanish surrender, Wood was
appointed Governor of the city of Santiago de Cuba, and a
short time later was promoted to Governor of Oriente province.
After the Cuban Army was demobilized, Wood set up a rural
guard that provided employment to many soldiers while avoiding
U.S. soldiers interfering to repress Cubans. Wood is credited
with having improved the sanitation system and administration
in Oriente province. In December 1899, Wood took over as Military Governor of Cuba. He tried to improve Cuba's domestic
conditions with public works programs and promoted local elections. Wood was criticized by Cuban leaders as favoring annexation. In fact, he promoted a compromise between annexation
and independence that resulted in the notorious Platt Amendment.
This addition to the Cuban Constitution which so frustrated
Cuban nationalist sentiment guaranteed the United States special
rights to intervene in Cuba under circumstances of political
turmoil. Nevertheless, Wood is credited with improving the
Cuban school system and with promulgating an electoral reform
law enfranchising adult males who were literate, owned land,
or had served in the Independence army. Under Wood's

administration particular attention was given to the eradication of yellow fever. He was also known to punish those guilty of administrative corruption. On May 20, 1902, the American occupation in Cuba ended. That same day, General Wood turned over the presidency to Tomás Estrada Palma, first elected President of Cuba.

- Y -

YAYA CONSTITUTION. One of several constitutions created during the War of Independence. Convened in 1895, marking the final stages in the Cuban revolutionary process of the nineteenth century under the leadership of José Martí. Due to the negative experience of the Ten Years' War (1868-1878), the Yaya Constitution as well as the Jimaguayú Constitution were called specifically to launch the Revolution for independence from Spain. The Yaya Constitution dispensed with the creation of a Chamber of Representatives, a body that caused numerous problems during the Ten Years' War, providing instead for a government organized under an Executive Power (President of the Republic) and four Legislative Secretaries. This facilitated designation of a general-in-chief in charge of directing the war, providing for greater control and expediency than was the case during the Ten Years' War.

YORUBA. See Lucumí.

YUCA (MANIOC). Staple food for Taíno indians and still an important item in Cuba and other Caribbean islands. For the Taínos the cultivation and preparation of yuca, a sturdy tuber, played a significant role. By means of a coa or stick they opened a hole in the ground dropping the slip of the yuca. They also developed the more modern technique of cultivation in beds. They piled the soil around the yuca into mounds or montones, thus concentrating the nourishing power of the soil around the slip and preventing soil erosion. After the yuca-- which had a growth period of over a year--had been harvested, it was grated, drained of its poisonous juice, and baked into unleavened bread called cassava, which the Spaniards labeled "bread of the earth." This bread was both nutritious and tasty and kept for several months, even in humid weather.

YUMURI. A small town in Matanzas province containing some 5,000 inhabitants in 1940 and situated in a small but beautiful and fertile valley by the same name. The Yumurí River runs south to north in Matanzas province, functioning as a minor means of transport ending at the port city of Matanzas.

- Z -

ZAFRA. Sugar harvest.

ZANJON, PEACE OF (February 11, 1878). Armistice that ended the
    Ten Years' War in 1878. The terms of the Peace did not encom-
    pass the abolition of slavery nor Cuban independence, but did
    allow for political equalization of Cuba with Puerto Rico, freedom
    for all rebel leaders who left Cuba, and the liberation of those
    slaves who had fought with the rebels. Rebel General Antonio
    Maceo nevertheless refused to accept the Zanjón terms and un-
    dertook the Guerra Chiquita (Little War) to continue the fight,
    but without success.

ZAPATA, CIENAGA DE. A swamp stretching for miles to the east
    and west along Cuba's southwestern coast, bordering the Bahía
    de Cochinos (Bay of Pigs). The terrain in this area is rocky
    and hard near the beach and up to 6 miles inland, where it be-
    comes impassable. Zapata was thought by the CIA to be the
    ideal location for the Cuban exile counterrevolutionary forces to
    land in 1961, since there were only two highways leading through
    the swamps to the landing area. From a defensive point of view,
    the beach would be easy to hold but difficult to leave. The
    CIA's paramilitary organization believed that Fidel Castro would
    be completely fooled, since the Zapata swamp represented such
    an unlikely site for the Bay of Pigs invasion. The CIA, how-
    ever, was unaware of the fact that the Zapata region was where
    Castro's revolutionary forces had most concentrated their efforts
    in the two previous years, completely winning over the surround-
    ing population. For this and other military-strategic reasons,
    the Bay of Pigs landing in the Zapata swamp area was a fiasco.
    Ciénaga de Zapata's name was later renamed Guamá Recreational
    Center.

ZAYAS, ALFREDO (1861-1934). Born in Havana, he received his
    law degree from the University of Havana at age 22 and was
    also a writer and orator. Like most Cubans of his generation,
    he shared in the revolutionary activities of the time and was a
    member of the Autonomist Party, a contributor of propagandist
    articles to various newspapers, editor of a literary magazine,
    and representative of the Revolutionary Party in Havana. His
    part in the revolutionary program resulted in his arrest and
    imprisonment in 1896 and his exile in 1897. With the triumph of
    the Revolution in 1898 he became active in political life. He
    held many offices: Prosecuting Attorney as early as 1889, Mu-
    nicipal Judge in 1891, and in 1901 acting Mayor of Havana and
    Member of the Council. In 1905 he was made Senator for the
    Province of Havana and chosen President of the Senate; in 1906
    he was President of the Revolutionary Committee; in 1907 he was

chosen member of the Committee of Consultation to deal with
matters affecting Cuba and the United States. He was Vice-
President during José Miguel Gómez's term (1909-1913) and
was elected President of Cuba (1921-24). Zayas recognized
the resurgence of economic nationalism sweeping Cuba in the
1920's, which was also particularly anti-American in character.
Zayas cooperated with U.S. special envoy Enoch Crowder in
1922 when the latter attempted, on U.S. President Harding's
orders, to institute fiscal and electoral reforms in Cuba.
Crowder's "Honest Cabinet" in 1922 was composed of many
distinguished Cubans who did enact reforms to reduce the
budget, trim the redundant bureaucracy, and annul corrupt
public works projects. Zayas later disbanded this quasi commission, much to the displeasure of the U.S. government.

Despite his mismanagement, Zayas succeeded in reestablish-
ing his country's credit, preventing overt U.S. intervention,
and securing title to the Isle of Pines for Cuba after a 20-year
delay imposed by the Platt Amendment. The graft and corruption in Zayas' administration overshadowed his achievements and
by 1924 opposition in the form of the Cuban Council of Civic
Renovation was criticizing him and pushing for social reforms.
Zayas ceded the Cuban presidency to General Gerardo Machado
following elections and retired from political life. The literary
interests of Zayas were many and varied. He was a frequent
contributor to the press and to magazines on political and historical subjects; and he was for six years Librarian of the
Sociedad Económica de Amigos del País and for many years its
President. His published works include El presbítero don José
Agustín Caballero y su vida y sus obras (1891); Cuba Auto-
nómica; Lexicografía Antillana (1914).

ZENEA, JUAN CLEMENTE (1832-1871). Born in Bayamo, Zenea was
a poet, professor, and journalist who was also a fervent believer
in Cuban independence. In New York, in exile, Zenea founded
La Revolución, a journal that became the mouthpiece for the
Cuban rebels during the Ten Years' War. Sent to Cuba as a
go-between to try and secure from the Spanish a Constitution
for a semi-autonomous Cuba, Zenea was shot after the Spanish
refused to honor his safe-conduct pass. Zenea's greatest poem
was "Fidelia."

ZULUETA, JULIAN DE (1814-1878). A native of Alava, Spain,
Zulueta arrived in Cuba penniless in 1832 and later inherited
an uncle's fortune. With this he increased his capital by investing in and successfully operating large sugar mills, Habana
and Vizcaya, both in the Macagua region. Zulueta was one of
the last operators in 1845 of the triangular slave trade, and
used over 6,000 slaves on his Alava mill, which in 1860 produced over 100,000 tons of sugar turning over a profit of over
$200,000 yearly. Zulueta was considered a big planter of the
second level who made his fortune via investments and the

slave trade rather than from original grants from the Crown.
He was an important planter nonetheless, controlling 10,000
acres with his three mills in western Cuba. During the Ten
Years' War for Independence, Zulueta backed the Spanish
Captain General Lersundi and opposed the Cuban rebels whose
triumph could severely limit Zulueta's power and wealth.

BIBLIOGRAPHY

The Cuban Revolution has produced a voluminous literature and has stimulated investigation into a variety of topics, particularly historical subjects which could help explain the reasons for and the course of the Castro Revolution. The following bibliography, without being exhaustive, is quite comprehensive. It includes entries in English and Spanish only.

ABC. El ABC ante la mediación. La Habana: Maza, Caso y Cía., 1934.

ABC. Hacia la Cuba nueva. La Habana, 1934.

Academia de Ciencias de Cuba and Academia de Ciencias de la URSS. Atlas nacional de Cuba. La Habana, 1970.

Academia de la Historia de Cuba. Constituciones de la República de Cuba. La Habana: Artes Gráficas, 1952.

Acosta, Maruja, and Hardoy, Jorge. Urban Reform in Revolutionary Cuba. New Haven: Yale University, Antilles Research Program, 1973.

Acosta Rubio, Raoul. Batista ante la historia (relato de un civilista). La Habana: Jesús Montero, 1938.

Adam y Silva, Ricardo. Cuba: raíces del desastre. Jerez de la Frontera: Gráficas del Exportador, 1971.

_____. La gran mentira: 4 de septiembre de 1933. La Habana: Editorial Lex, 1947.

Agramonte, Roberto. Félix Varela, el primero que nos enseñó a pensar. La Habana: Imprenta Molina y Cía., 1937.

_____. José Agustín Caballero y los orígenes de la conciencia cubana. La Habana: Ucar García, 1952.

Aguila, Juan M. del. Cuba: Dilemmas of a Revolution. Boulder, Colo., and London: Westview Press, 1984.

306

Aguilar, Luis E. Cuba, 1933: Prologue to Revolution. Ithaca: Cornell University Press, 1972.

_____. Pasado y ambiente en el proceso cubano. La Habana: Ediciones Insula, 1957.

Ahumada y Centurión, José. Memoria histórico-política de la isla de Cuba redactada de orden del señor ministro de Ultramar. La Habana, 1874.

Aldereguía, Gustavo. Don Ramón Genio y Figura. La Habana: "La Ultima Hora," 1952.

Alienes y Urosa, Julián. Características fundamentales de la economía cubana. La Habana: Banco Nacional de Cuba, 1950.

Alloza, Fernando. Noventa entrevistas políticas. La Habana: Imp. Cuba Intelectual, 1953.

Alonso, Luis Ricardo. Territorio Libre. Oviedo: Editorial Richard Grandio, 1967.

Alvarez Acevedo, Julio M. La colonia española en la economía cubana, 1902-1936: un balance histórico. La Habana: Ucar García y Cía., 1936.

Alvarez del Real, Evelio. Patrias opacas y caudillos fulgurantes. La Habana: Imp. "La Verónica," 1942.

Alvarez Díaz, José. La destrucción de la ganadería cubana. Miami: Editorial AIP, 1965.

_____, Arredondo, A., Shelton, R.M., and Vizcaino, J.F. Cuba: Geopolítica y pensamiento económico. Miami: Duplex Paper Products of Miami, Inc., 1964.

_____. The Road to Nowhere: Castro's Rise and Fall. Miami: Editorial AIP, 1965.

Alvarez Tabío, Fernando. Evolución constitucional de Cuba, 1928-1940. La Habana: Tall. Gráfica, 1953.

Amaral Agramonte, Raúl. Al margen de la revolución. La Habana: Cultural, 1935.

Amaro, Nelson. La revolución cubana ... ¿por qué? Guatemala: Talleres Gráficos Rosales, 1967.

American University, Foreign Area Studies Division. Special Warfare Area Handbook for Cuba. Washington, D.C.: U.S. Government Printing Office, 1971.

Angulo y Pérez, Andrés. Curso de historia de las instituciones locales de Cuba, 1943.

Annino, Antonio. Dall'insurrezione al regime. Milan: Franco Angeli, 1984.

Aranda, Sergio. La revolución agraria en Cuba. Mexico City: Siglo XXI, 1965.

Araquistaín, Luis. La agonía antillana. Madrid: Espasa-Calpe, S.A. 1928.

Arboleya, José G. de. Manual de la isla de Cuba. La Habana: Imprenta del Gobierno y Capitanía General por S.M., 1852.

Arce, Luis A. de. Emilio Núñez (1875-1922): historiografía. La Habana: Editorial "Niños," 1943.

Arrate, José M. Llave del Nuevo Mundo. La Habana: 1830.

Arredondo, Alberto. Batista: un año de gobierno. La Habana: Editorial Ucacia, 1942.

_____. Cuba: tierra indefensa. La Habana: Editorial Lex, 1945.

_____. El negro en Cuba. La Habana: Editorial "Alfa," 1939.

_____. Estudio comparativo de las reformas agrarias en América. Caracas: Talleres de Reproducciones del Instituto Agrario Nacional, 1965.

_____. Reforma agraria: la experiencia cubana. Río Piedras: Editorial San Juan, 1969.

Arrom, J.J. Historia de la literatura dramática cubana, 1944.

Artime, Manuel. ¡Traición! Gritan 20,000 tumbas cubanas. Mexico City: Editorial Jus, 1960.

Arvelo, Perina. Revolución de los barbudos. Caracas: Editorial Landi, 1961.

Asociación Nacional de Hacendados de Cuba. El tratado de reciprocidad de 1934. La Habana: Asociación Nacional de Hacendados de Cuba, 1939.

Ateneo de La Habana. Los maestros de la cultura cubana. La Habana: Ateneo de La Habana, 1940.

Atkins, Edward F. Sixty Years in Cuba. Cambridge, Mass.: Riverside Press, 1926.

Azcuy Alón, Fanny. Psicografía y supervivencias de los aborígenes de Cuba, 1941.

Azcuy y Cruz, Aracelio. Cuba: campo de concentración. Mexico City: Ediciones Humanismo, 1954.

Azicri, Max. "The Governing Strategies of Mass Mobilization: The Foundations of Cuban Revolutionary Politics." Latin American Monograph Series, No. 2. Erie, Pa.: Northwestern Pennsylvania Institute for Latin American Studies, 1977.

Bachiller y Morales, Antonio. Apuntes para la historia de las letras y de la Instrucción Pública en la isla de Cuba. 3 vols. La Habana, 1859-1861.

_____. Cuba primitiva. La Habana, 1883.

Baciu, Stefan. Cortina de ferro sôbre Cuba. Rio de Janeiro: Editorâ Grafica Tupy, 1961.

Baeza Flores, Alberto. Las cadenas vienen de lejos. Mexico City: Ed. Letras, 1960.

Banco Nacional de Cuba. Economic Development Program, Progress Report, no. 1, 1956.

_____. La economía cubana en 1956-1957. La Habana: Editorial Lex, 1958.

Baralt, Blanca Z. de. El Martí que yo conocí. La Habana: Editorial Trópico, 1945.

Baran, Paul A. Reflections on the Cuban Revolution. New York: Monthly Review Press, 1961.

Barbarrosa, Enrique. El proceso de la república: análisis de la situación económica bajo el gobierno presidencial de Tomás Estrada Palma y José Miguel Gómez. La Habana: Imp. Militar, 1911.

Barbeito, José. Realidad y masificación. Caracas: Ediciones "Nuevo Orden," 1964.

Barbour, Thomas. A Naturalist in Cuba. Boston: Little, Brown, and Company, 1945.

Barnet, Miguel, ed. The Autobiography of a Runaway Slave: Esteban Montejo. New York, 1968.

Baroni, Aldo. Cuba, país de poca memoria. Mexico City: Ediciones Botas, 1944.

Bibliography   310

Barquín, Ramón. Las luchas guerrilleras en Cuba: de la colonia a la Sierra Maestra. Madrid: Editorial Playor, 1975.

Barras y Prado, Antonio de las. La Habana a mediados del siglo XIX. Madrid, 1926.

Barro y Segura, Antonio. The Truth About Sugar in Cuba. La Habana: Ucar, García y Cía., 1943.

Batista, Fulgencio. Alocución del 4 de septiembre de 1939. La Habana: Burgay y Cía., 1939.

_____. Cuba Betrayed. New York: Vantage Press, 1962.

_____. Cuba: su política interna y sus relaciones exteriores. La Habana: Prensa Indoamericana, 1939.

_____. Ideario de Batista. Edited by M. Franco Varona. La Habana: Prensa Indoamericana, 1940.

_____. Paradojas. Mexico City: Ediciones Botas, 1963.

_____. Piedras y leyes. Mexico City: Ediciones Botas, 1961.

_____. Respuesta.... Mexico City: Imp. "Manuel León Sánchez," 1960.

_____. Revolución social o política reformista (once aniversarios). La Habana: Prensa Indoamericana, 1944.

_____. Sombras de América. Mexico City: Ediapsa, 1946.

_____. The Growth and Decline of the Cuban Republic. The Devin-Adair Company, 1964.

Bayo, Alberto. Mi aporte a la Revolución Cubana. La Habana: Imp. Ejército Rebelde, 1960.

Beals, Carleton. The Crime of Cuba. Philadelphia: Lippincott, 1933.

Beato, Jorge J., and Garrido, Miguel F. Cuba en 1830. Miami: Ed. Universal, 1973.

Bekarevich, Anatolii, Bondarchuk, Vladimir N., and Kukharev, N.M. Cuba in Statistics. Moscow: Academy of Sciences of the USSR, Institute of Latin America, 1972.

_____, and Kukharev, N.M. The Soviet Union and Cuba. Moscow: Nauka, 1973.

Benítez, Fernando. La batalla de Cuba. Mexico, 1960.

Benjamin, Jules R. The United States and Cuba: Hegemony and Dependent Development. Pittsburgh, 1977.

Bernardo, Robert M. The Theory of Moral Incentive in Cuba. Alabama: The University of Alabama Press, 1971.

Bethel, Paul D. The Losers. New Rochelle, N.Y.: Arlington House, 1969.

Blutstein, Howard I., et al. Area Handbook for Cuba. Washington, D.C.: United States Printing Office, 1971.

Bonachea, Ramón, and San Martín, Marta. The Cuban Insurrection, 1952-1959. New Brunswick, N.J.: Transaction Books, 1974.

Bonachea, Rolando, and Valdés, Nelson P., eds. Cuba in Revolution. Garden City, N.Y.: Anchor Books, 1972.

_____. Che: Selected Works of Ernesto Che Guevara. Cambridge: MIT Press, 1969.

Bonsal, Philip W. Cuba, Castro, and the United States. Pittsburgh: University of Pittsburgh Press, 1971.

Boorstein, Edward. The Economic Transformation of Cuba. New York: Monthly Review Press, 1968.

Borges, Milo A. Compilación ordenada y completa de la legislación cubana de 1899 a 1950. Vols. 1-3. 2nd ed. La Habana: Editorial Lex, 1952.

Bosch, Juan. Cuba, la isla fascinante. Santiago de Chile: Editorial Universitaria, 1955.

Boza Domínguez, Luis. La situación universitaria en Cuba. Santiago: Editorial del Pacífico, 1962.

Braña Chansuolme, Manuel. El aparato. Coral Gables: Service Offset Printers, 1964.

Brennan, Ray. Castro, Cuba, and Justice. Garden City, N.Y.: Doubleday, 1959.

Brigada de Asalto 2506: History of an Aggression, Testimony, and Documents from the Trial of the Mercenary Brigade Organized by the U.S. Imperialists that Invaded Cuba on 17 April 1961. La Habana: Ediciones Venceremos, 1964.

Brimelow, T. Cuba: Economic and Commercial Conditions in Cuba. London: His Majesty's Stationery Office, 1950.

Brown, Charles H. The Correspondents' War: Journalists in the Spanish-American War. New York: Scribner's, 1967.

Brown Castillo, Gerardo. Cuba colonial. La Habana: Jesús Montero, Editor, 1952.

Brundenius, Claes. Revolutionary Cuba: The Challenge of Economic Growth with Equity. Boulder, Colo., and London: Westview Press, 1984.

Buch López, Ernesto. Historia de Santiago de Cuba. La Habana: Editorial Lex, 1947.

Buell, Raymond L., et al. Problems of the New Cuba. New York: Foreign Policy Associates, Inc., 1935.

Bueno, Salvador. Medio siglo de literatura cubana, 1902-1952. La Habana: Comisión Nacional Cubana de la UNESCO, 1953.

Burks, David D. Cuba Under Castro. New York: Foreign Policy Association, 1964.

Bustamante, José Angel. Raíces psicológicas del cubano. La Habana: Ed. Librerías Unidas, 1960.

Bustamante y Montoro, Antonio S. La ideología autonomista. La Habana: Imp. Molina y Cía., 1933.

Buttari y Gaunaurd, J. Boceto crítico histórico. La Habana: Editorial Lex, 1954.

Caballero, José Agustín. Escritos varios. La Habana: Editorial de la Universidad de la Habana, 1956.

_____. Filosofía electiva. La Habana: Editorial de la Universidad de La Habana, 1944.

Cabrera, Lydia. Anagó: vocabulario lucumí. La Habana, 1957.

_____. El Monte. Miami, 1968.

_____. La laguna sagrada de San Joaquín. Madrid: Ed. Erre. 1973.

_____. La sociedad secreta Abakuá. La Habana: Ediciones C.R., 1958.

_____. Refranes de negros viejos. La Habana, 1955.

Cabrera, Raimundo. Cuba y sus jueces. La Habana, 1922.

_____. Mis malos tiempos. La Habana: Imp. "El Siglo XX,"
1920.

Cabrera Saqui, Mario. Julián del Casal: poesías completas. La
Habana: Publicaciones del Ministerio de Educación, 1945.

Calcagno, Francisco. Diccionario biográfico cubano. New York,
1878.

Camacho, Pánfilo D. Estrada Palma, el gobernante honrado. La
Habana: Editorial Tropical, 1938.

_____. José Antonio Saco: estudio biográfico. La Habana:
Imprenta Molina, 1936.

Campoamor, Fernando G. La tragedia de Cuba. La Habana:
Editorial Hermes, 1934.

Canelas O., Amado. Cuba: socialismo en español. La Paz:
Empresa Industrial Gráfica E. Burillo, 1964.

Canet, Gerardo, and Erwin Raisz, Atlas de Cuba, 1949.

Carballal, Rodolfo A. Estudio sobre la administración del general
José M. Gómez (1909-1913). La Habana: Imp. Rambla, Bouza
y Cía., 1915.

Carbonell, José Manuel. Evolución de la cultura cubana. 18 vols.
La Habana, 1928.

Carbonell y Rivero, Néstor. Martí: sus últimos días. La Habana:
"El Siglo XX," 1950.

Cárdenas, Angel G. Soga y sangre: una página de horror del
machadato y su acusación pública. La Habana: Ediciones Montero, 1945.

Carrera Jústiz, Francisco. Introducción a la historia de las instituciones locales de Cuba. La Habana, 1905.

Carrillo, Justo. Cuba, 1933. Miami: North-South Center for the
Institute of Interamerican Studies, University of Miami, 1985.

Casa de las Américas. Cuba: transformación del hombre. La
Habana, 1961.

Casal, Lourdes, ed. El caso Padilla: literatura y revolución en
Cuba. Miami: Ediciones Universal, 1971.

Casas, Bartolomé de las. Historia de Las Indias. Mexico, 1953.

Castellanos G., Gerardo. Hacia Gibara. La Habana: Seoane y
Fernández, 1933.

_____. Panorama histórico. La Habana: Ucar, García y Cía.,
1934.

Castro, Fidel. Cartas del presidio. Edited by Luis Conte Agüero.
La Habana: Editorial Lex, 1959.

_____. Discursos de Fidel en los aniversarios de los CDR, 1960-
1967. La Habana: Instituto del Libro, 1968.

_____. Discursos del Dr. Fidel Castro Ruz, Comandante en Jefe
del Ejército Rebelde 26 de Julio y Primer Ministro del Gobierno
Provisional. Edited by Emilio Roig de Leuchsenring. La
Habana: Oficina del Historiador de la Ciudad, 1959.

_____. Discursos para la historia. La Habana, 1959.

_____. El partido marxista-leninista. Buenos Aires: Ediciones
La Rosa Blindada, 1960.

_____. El pensamiento de Fidel Castro. Bogotá: Ediciones Paz
y Socialismo, 1963.

_____. Guía del pensamiento económico de Fidel. La Habana:
Editorial Lex, 1959.

_____. History Will Absolve Me. New York: Liberal Press,
1959.

_____. Humanismo revolucionario. La Habana: Editorial Tierra
Nueva, 1959.

_____. Palabras a los intelectuales. La Habana: Ediciones del
Consejo Nacional de Cultura, 1961.

_____. Palabras para la historia. La Habana: Cooperativa
Obrera de Publicidad, 1960.

_____. Playa Girón. La Habana: Comisión Nacional del Monu-
mento a los Caídos en Playa Girón, 1961.

_____. Political, Economic, and Social Thought of Fidel Castro.
La Habana: Editorial Lex, 1959.

Castro, Martha de. El Arte en Cuba. Miami: Ediciones Universal.

Castro, Raúl. Informe. Montevideo: Documentos Latinoamericanos,
1968.

Castro Hidalgo, Orlando. Spy for Cuba. Miami: Seemann, 1971.

Casuso, Teresa. Cuba and Castro. New York: Random House, 1961.

Catá, Alvaro. De guerra a guerra. La Habana: Imp. "La Razón," 1906.

Central Intelligence Agency. Cuba: Foreign Trade. Washington, D.C., 1975.

_____. Directory of Personalities of the Cuban Government, Official Organizations, and Mass Organizations. Washington, D.C., 1974.

_____. The Cuban Economy: A Statistical Review, 1968-1976. Washington, D.C.: Library of Congress, 1976.

Cepero Bonilla, Raúl. Cuba y el Mundo. La Habana: Editorial Echevarría, 1960.

Chacón y Calvo, José María. Estudios heredianos. La Habana, 1939.

Chapman, Charles E. A History of the Cuban Republic: A Study in Hispanic American Politics. New York: Macmillan, 1927.

Chase National Bank of New York. Cuban Public Works Financing: Reply of the Report of Special Committee Created by Decree Law 140 of April 16, 1934. New York: Chase National Bank, 1934, 1935.

Chester, Edmund A. A Sergeant Named Batista. New York: Henry Holt, 1954.

Chilcote, Ronald H. Cuba, 1953-1978: A Bibliographical Guide to the Literature. 2 vols. White Plains, N.Y.: Kraus International Publications, 1986.

Childs, James B., ed. Cuba: A Guide to the Official Publications of the Other American Republics. Vol. 7. Washington, D.C.: Government Printing Office, 1939.

Clark, Juan. Religious Repression in Cuba. Miami: North-South Center for the Institute of Interamerican Studies, University of Miami, 1986.

Clavijo Aguilera, Fausto. Los sindicatos en Cuba. La Habana: Ed. Lex, 1954.

Clytus, John, with Rieker, Jane. Black Man in Red Cuba. Coral Gables, Fla.: University of Miami Press, 1970.

Collazo, Enrique. Cuba intervenida. La Habana: Imp. C. Martínez y Cía., 1910.

_____. Desde Yara hasta el Zanjón. La Habana, 1917.

_____. La revolución de agosto de 1906. La Habana: Imp. C. Martínez y Cía., 1907.

_____. Los americanos en Cuba. La Habana: Imp. C. Martínez y Cía., 1905.

Comisión Económica para América Latina. El desarrollo industrial de Cuba. ST/ECLA/Conf. 23/L. 63, 1966.

Comité Central del Partido Comunista de Cuba. Comisión de Estudios Jurídicos. Organos del sistema judicial. La Habana: Empresa de Medios de Propaganda.

_____. Departamento de Orientación Revolucionaria. Estatutos del partido comunista de Cuba. La Habana: Imprenta Federico Engels, 1976.

_____. Departamento de Orientación Revolucionaria. Política educacional. La Habana: Instituto Cubano del Libro, 1976.

_____. Departamento de Orientación Revolucionaria. Sobre el pleno ejercicio de la igualdad de la mujer. La Habana: Imprenta Federico Engels, 1976.

_____. Departamento de Orientación Revolucionaria. Sobre la cuestión agraria y las relaciones con el campesinado. La Habana: Instituto Cubano del Libro, 1976.

_____. Departamento de Orientación Revolucionaria. Sobre la lucha ideológica. La Habana: Instituto Cubano del Libro, 1976.

_____. Departamento de Orientación Revolucionaria. Sobre los órganos del poder popular. La Habana: Imprenta Federico Engels, 1976.

Comité Central del Sindicato General de Empleados del Comercio de Cuba. El Sindicato General de Empleados del Comercio de Cuba frente al IV Congreso Obrero Nacional, Federación Obrera de Cuba. La Habana, 1934.

Comité de Jóvenes Revolucionarios Cubanos. El terror en Cuba. Madrid: Imp. Editorial Castro, 1933.

Commission on Cuban Affairs. Problems of the New Cuba. New York: J.J. Little and Ives Company, 1935.

Conangla Fontanilla, José. Tomás Gener: del hispanismo a la cubanía práctica. La Habana: Academia de la Historia de Cuba, 1950.

Concheso, Aurelio Fernández. Cuba en la vida internacional. Jana and Leipzig: Wilhelm Gronau, 1935.

Confederación Nacional Obrera de Cuba. IV Congreso Nacional de Unión Sindicalista: resoluciones y acuerdos. La Habana, 1934.

Consuegra, W.I. Hechos y comentarios: la revolución de febrero de 1917 en Las Villas. La Habana: La Comercial, 1920.

Conte, Rafael, and Capmany, José. Guerra de razas: negros contra blancos en Cuba. La Habana: Imprenta Militar de Antonio Pérez, 1912.

Conte Agüero, Luis. América contra el comunismo. Miami: TA-Cuba, 1961.

_____. Cartas del presidio: anticipo de una biografía de Fidel Castro. La Habana: Editorial Lex, 1959.

_____. Cuba, la OEA y la Fuerza Interamericana de Paz. Coral Gables, Fla.: Service Offset Printing, 1965.

_____. Doctrina de la contra intervención, sovietización de la economía cubana y otros escritos. Montevideo: Ediciones Cruz del Sur, 1962.

_____. Eduardo Chibás, el adalid de Cuba. Mexico City: Editorial Jus, 1955.

_____. Fidel Castro: vida y obra. La Habana: Editorial Lex, 1959.

_____. Los dos rostros de Fidel Castro. Mexico City: Editorial Jus, 1960.

Corbitt, Duvon C. The Colonial Government of Cuba, 1938.

Corwin, Arthur. Spain and the Abolition of Slavery in Cuba, 1817-1886. Austin: The University of Texas Press, 1967.

Costa, Octavio R. Juan Gualberto Gómez. La Habana: Imprenta "El Siglo XX," 1950.

Cotera O'Bourke, Margarita de la. ¿Quién es quién en Cuba? La Habana: Julián Martín, 1925.

Cotoño, Manuel. Primera conferencia nacional del Ala Izquierda

Estudiantil, octubre de 1933: tesis política. La Habana: Imp. Molina y Cía., 1934.

Cova, J.A. Páez y la Independencia de Cuba. La Habana: "El Siglo XX," 1949.

Crassweller, Robert D. Cuba and the U.S.: The Tangled Relationship. New York: Foreign Policy Association, 1971.

Cruz, Carlos Manuel de la. Proceso histórico del machadato. La Habana: Imp. "La Milagrosa," 1935.

Cuba bajo la administración del Mayor General José Miguel Gómez. La Habana: Imp. Rambla y Bouza, 1911.

Cuba. Consejo Superior de Universidades. La reforma de la enseñanza superior en Cuba. La Habana, 1962.

Cuba: ejemplo revolucionario de Latinoamérica. Lima: Editorial Libertad, 1961.

Cuba. Ministerio de Relaciones Exteriores. The Position of Sugar in the United States. La Habana, 1960.

Cuba. Ministerio de Relaciones Exteriores. The Revolution and Cultural Problems in Cuba. La Habana, 1962.

Cuban Economic Research Project. A Study on Cuba. Coral Gables, Fla: University of Miami Press, 1965. Essentially a translation from the Spanish Un estudio sobre Cuba (University of Miami Press, 1963).

_____. Cuba: Agriculture and Planning, 1963-1964. Miami: Editorial Press, 1965.

_____. Social Security in Cuba. Mimeo. University of Miami, 1964.

_____. Stages and Problems of Industrial Development in Cuba. Mimeo. University of Miami, 1965.

_____. Sugar in Cuba. Mimeo. University of Miami, 1966.

Cuban Information Bureau. Ambassador Guggenheim and the Cuban Revolt. Washington, D.C., 1931.

Cuban Planters Association. Proclamation to the People of Cuba and General Program. La Habana: Díaz y Peredes, 1934.

Cuesta, Leonel Antonio de la, ed. Constituciones cubanas. New York: Ediciones Exilio, 1974.

Debray, Régis. Revolution in the Revolution? Translated by
Bobbye Ortiz. New York: Grove Press, 1967.

Declaraciones de La Habana y de Santiago. La Habana: Editora
Política, 1965.

Defensa de Cuba. Conferencias por Raúl Castro et al. La Habana:
Universidad Popular, 1960.

Defensa Institucional Cubana. Tres años. Mexico City: Ediciones
Botas, 1962.

Del Río, Germán Wolter. Aportaciones para una política económica
cubana. La Habana: Ucar, García y Cía., 1936.

Dewart, Leslie. Christianity and Revolution: The Lesson of Cuba.
New York: Herder and Herder, 1963.

Díaz, Emilio. Essay of the Cuban History. Coral Gables, Fla:
Service Offset Printers, 1964.

Díaz-Verson, Salvador. Caníbales del siglo XX. Miami: Edit.
Libertad, 1962.

_____. Cuando la razón se vuelve inútil. Mexico City: Ediciones
Botas, 1962.

Dinnerstein, Herbert. The Making of a Missile Crisis, October 1962.
Baltimore: Johns Hopkins University Press, 1976.

Dirección Nacional de los Tribunales Populares. Manual de los
Tribunales Populares. La Habana: Ministerio de Justicia, 1966.

Documentos de la Revolución Cubana. Montevideo: Nativa Libros,
1967.

Documentos escogidos de la Revolución Cubana. Bogotá: Ediciones
Paz y Socialismo, 1963.

Dolz, Ricardo. El proceso electoral de 1916. La Habana: Imprenta
y Papelería La Universal, 1917.

Domínguez, Jorge I. Cuba: Order and Revolution. Cambridge:
Belking Press, 1978.

Dorta Duque, Francisco. Justificando una reforma agraria. Madrid:
Raycar, 1960.

Draper, Theodore. Castroism: Theory and Practice. New York:
Frederick A. Praeger, Inc. 1965.

_____. Castro's Revolution: Myths and Realities. New York: Praeger, 1962.

Duarte Oropesa, José A. Historiología cubana. Hollywood, Ca.: privately printed, 1969.

Dubois, Jules. Fidel Castro: Rebel, Liberator, or Dictator? Indianapolis, Ind.: Bobbs Merrill, 1959.

Dumont, René. Socialism and Development. New York: Grove Press, 1970.

Dumpierre, Erasmo. Mella: esbozo biográfico. La Habana: Instituto de la Historia, 1965.

Duque, Matías. Nuestra patria. La Habana: Imp. y Librería Nueva, 1928.

_____. Ocios del presidio. La Habana: Imp. "Avisador Comercial," 1919.

Durch, William J. "The Cuban Military in Africa and the Middle East: From Algeria to Angola." Professional paper no. 201. Arlington, Va.: Center for Naval Analyses, 1977.

Echeverría Salvat, Oscar. La agricultura cubana, 1934-1966. Miami: Ediciones Universal, 1971.

Ely, Roland T. Comerciantes cubanos del siglo XX. Bogotá: Aedita Editores Ltd., 1961.

_____. Cuando reinaba Su Majestad el Azúcar. Buenos Aires: Editorial Sudamérica, 1963.

_____. La economía cubana entre las dos Isabeles, 1492-1832. Bogotá: Aedita Editores, 1962.

Entralgo, Elías. Apuntes caracteriológicos sobre el léxico cubano. La Habana, 1941.

_____. Esquema de sociografía indocubana. La Habana, 1936.

_____. La insurrección de los Diez Años. La Habana, 1950.

_____. La liberación étnica cubana. La Habana, 1953.

_____. Períoca sociográfica de la cubanidad. La Habana: J. Montero, 1947.

_____, Vitier, Medardo, and Agramonte, Roberto. Enrique José Varona: su vida, su obra y su influencia. La Habana, Cultural, 1937.

Estenger, Rafael, El pulpo de oro. Mexico City: Editorial "Olimpo," 1954.

Estep, Raymond. The Latin American Nations Today. Alabama: Air University Documentary Research Study, 1964.

Estévez y Romero, Luis. Desde el Zanjón hasta Baire. La Habana, 1899.

Facultad de Ciencias Médicas, Universidad de La Habana. Docencia de las ciencias médicas en Cuba. La Habana: Ministerio de Salud Pública, 1966.

Fagen, Richard R. Cuba: The Political Content of Adult Education. Stanford California: Hoover Institution, 1968.

_____. The Transformation of Political Culture in Cuba. Stanford, California: Stanford University Press, 1969.

_____, Richard A. Brody, and Thomas J. O'Leary. Cubans in Exile. Stanford, Ca.: Stanford University Press, 1968.

Fagg, John Edwin. Cuba, Haiti, and the Dominican Republic. Englewood Cliffs, N.J.: Prentice-Hall, 1965.

Farber, Samuel. Revolution and Reaction in Cuba. Connecticut, 1976.

Fergusson, Erna. Cuba. New York: Alfred A. Knopf, 1946.

Fermoselle, Rafael. Política y color en Cuba. Montevideo: Ed. Geminis, 1974.

Fernández, Julio César. En defensa de la revolución. La Habana: Editorial Juventud, 1936.

_____. Yo acuso a Batista. La Habana, 1940.

Fernández Caubí, Luis. Cuba, sociedad cerrada. Miami: Impreso Editorial, 1968.

Fernández de Castro, José Antonio. Esquema histórico de las letras en Cuba (1548-1902). La Habana: Universidad de La Habana, 1949.

_____. Medio siglo de historia colonial en Cuba; cartas a José Antonio Saco ordenadas y comentadas (de 1823 a 1879). La Habana: R. Veloso, 1923.

Fernández de Oviedo y Valdés, Gonzalo. Historia General y Natural de las Indias. Madrid: Imprenta de la Real Academia de la Historia, 1851-55.

Fernández Retamar, Roberto, ed. Ernesto Che Guevara: obra revolucionaria. Mexico City: Ediciones ERA, 1967.

Fernández Santos, Francisco. Cuba: una revolución en marcha. Paris: Ediciones Ruedo Ibérico, 1967.

Ferrer, Horacio. Con el rifle al hombro. La Habana: El Siglo XX, 1950.

Figueres, Francisco. Cuba y su evolución colonial. La Habana, 1907.

Fiterre, Rafael. Por la redención de Cuba. La Habana: Compañía Editora de Libros y Folletos, 1943.

Fitzgibbon, Russel H. Cuba and the United States, 1900-1935. Menasha, Wis.: George Banta Publishing Company, 1935.

Foner, Philip. A History of Cuba and Its Relations with the U.S. 2 vols. New York: International Publishers Co., 1963.

_____. The Spanish-Cuban-American War. 2 vols. New York: Monthly Review Press, 1972.

Foraker, Joseph Benson. Notes of a Busy Life. 3d ed. 2 vols. Cincinnati: Stewart and Kidd, 1917.

Foreign Policy Association, "The Cuban Crisis: a documentary record," Headline Series. New York: Foreign Policy Association, Inc., 1963.

_____, Commission on Cuban Affairs. Problems of the New Cuba. New York: Little and Ives, 1935.

Fort y Roldán, Nicolás. Cuba indígena. Madrid, 1881.

Franco, José L. Folklore criollo y afrocubano. La Habana: Junta Nacional de Arqueología y Etnología, 1959.

_____. Política continental americana de España en Cuba, 1812-1830. La Habana, 1947.

Franco Varona, M. La revolución del 4 de septiembre. La Habana: Editorial Varona-Cruz, 1934.

Frank, Waldo. Cuba: Prophetic Island. New York: Marzan and Munsell, 1961.

Franqui, Carlos. The Twelve. New York: Lyle Stuart, 1968.

Free, Lloyd A. Attitudes of the Cuban People toward the Castro

Regime. Princeton: Institute for International Social Research, 1960.

Friedlander, Heinrich E. Historia económica de Cuba. La Habana: Jesús Montero, 1944.

Frondizi, Silvio. La revolución cubana, su significación histórica. Montevideo: Editorial Ciencias Políticas, 1960.

Funston, Frederick. Memories of Two Wars: Cuban and Philippine Experience. New York: Scribner's, 1911.

Furrazola-Bermúdez, Gustavo, Judoley, Constantino M., Mijailovskaya, Marina S., Miroliubov, Yuri S., Novojatsky, Ivan P., Núñez Jiménez, Antonio, and Solsona, Juan B. Geología de Cuba. La Habana: Editora Nacional de Cuba, 1964.

García Kohly, Mario. El problema constitucional de Cuba y de nuestra América. Paris: Imprimerie Labor, 1930.

García Montes, Jorge, and Avila, Antonio Alonso. Historia del Partido Comunista de Cuba. Miami: Editorial Universal (Rema), 1970.

García Oliveras, Julio A. José Antonio Echeverría: la lucha estudiantil contra Batista. Havana: Editora Política, 1979.

García Regueiro, Ovidio. Cuba: raíces y frutos de una revolución. Madrid: IEPAL, 1970.

García Valdés, Pedro. Cuba. La Habana: Primer Congreso Nacional de Historia, 1943.

_____. La civilización taína en Pinar del Río. La Habana, 1930.

Garrigó, Roque E. Historia documentada de la Conspiración de los Soles y Rayos de Bolívar. 2 vols. La Habana, 1929.

Gay-Calbó, Enrique. El Padre Varela en las Cortes Españolas de 1822-23. La Habana: Rambla, Bouza y Cía., 1937.

Gellman, Irwin F. Roosevelt and Batista: Good Neighbor Diplomacy in Cuba, 1933-1945. Albuquerque: University of New Mexico Press, 1973.

Gerassi, John, ed. Venceremos! The speeches and writings of Che Guevara. New York: The Macmillan Co., 1964.

Gilly, Adolfo. Inside the Cuban Revolution. Translated by Félix Gutiérrez. New York: Monthly Review Press, 1964.

Giménez, Armando. Sierra Mestra: la revolución de Fidel Castro. Translated by Carmen Alfaya. Buenos Aires: Editorial Lautaro, 1959.

Godoy, Gastón. El caso cubano y la Organización de Estados Americanos. Madrid: Aldus, S.A., 1961.

Goldenberg, Boris. The Cuban Revolution and Latin America. New York: Praeger, 1965.

Gómez, Gabriel A. De la dictadura a la liberación (interpretación política y social de la revolución cubana). La Habana: Publicaciones G.A. Gómez, 1959.

Góngora Echenique, Manuel. Lo que he visto en Cuba. Madrid: Editorial Góngora, 1929.

González, Diego. Historia documentada de los movimientos revolucionarios por la independencia de Cuba de 1852 a 1867. 2 vols. La Habana, 1939.

González, Edelmira. Batista: Septiembre 4, 1933; Junio 1, 1940; Marzo 12, 1952. La Habana, 1952.

_____. La revolución en Cuba: memorias del coronel Rosendo Collazo. La Habana: Editorial "Hermes," 1934.

González, Edward. Cuba Under Castro: The Limits of Charisma. Boston: Houghton Mifflin, 1974.

_____. Partners in Deadlock: The United States and Castro, 1959-1972. California: The Southern California Arms Control and Foreign Policy Seminar, 1972.

Gonzalez, Edward, and Ronfeldt, David. Castro, Cuba, and the World. Santa Monica, Ca: Rand, 1986.

González, Manuel Pedro. José Martí. Chapel Hill: The University of North Carolina Press, 1953.

González, N.G. In Darkest Cuba. Columbia, S.C.: The State Co., 1922.

González Palacios, Alberto. El alzamiento del Ocujal. Santiago de Cuba, 1934.

González Palacios, Carlos. Revolución y seudo-revolución en Cuba. La Habana, 1948.

González Pedrero, Enrique. La revolución cubana. Mexico: Talleres Gráficos de Mexico, 1959.

González Carbajal, Ladislao. Julio Antonio Mella. La Habana:
Editorial Berea, 1941.

González del Valle, Francisco. Cronología herediana (1803-1839).
La Habana: 1938.

_____. Documentos para la vida de Heredia. La Habana, 1938.

_____. Vida política de José María Heredia. La Habana, 1936.

González Muñoz, Rafael. Doctrina Grau. La Habana: Publicaciones
del Ministerio de Estado, 1948.

Grau San Martín, Ramón. "Cuba ante la guerra." Cuadernos de
Cultura Auténtica, No. 3. La Habana: Comisión Nacional de
Propaganda del Partido Revolucionario Cubano, 1940.

_____. La revolución cubana ante América. Mexico City:
Ediciones del Partido Revolucionario Cubano (Auténtico), 1936.

Gray, Richard B. José Martí: Cuban Patriot. Gainesville:
University of Florida Press, 1962.

Green, Gil. Revolution, Cuban Style. New York: International
Publishers, 1970.

Grupos de Propaganda Doctrinal Ortodoxa. Doctrina del Partido
Ortodoxo. La Habana: Fernández, 1951.

Guas Inclán, Rafael. El general Gerardo Machado y Morales. La
Habana, 1956.

Guerra y Sánchez, Ramiro. Azúcar y población en las Antillas.
La Habana: Cultural, 1927.

_____. Cuba en la vida internacional. La Habana: "El Siglo
XX," 1923.

_____. Guerra de los Diez Años. La Habana: Cultural, S.A.,
1950.

_____. La industria azucarera de Cuba. La Habana: Cultural,
1940.

_____. Manual de historia de Cuba. La Habana, 1938.

_____. Sugar and Society in the Caribbean. New Haven: Yale
University Press, 1964.

_____, Cabrera, José M., Remos, Juan J., and Santovenia,
Emeterio S. Historia de la nación cubana. 10 vols. La Habana:
Cultural, 1952.

Guevara, Ernesto. Cartas inéditas por Ernesto Guevara. Editorial Sandino, 1967.

_____. Che Guevara on Guerrilla Warfare. Ed. Harries-Clichy Peterson. New York: Praeger, 1961.

_____. Che Guevara on Revolution: A Documentary Overview. Ed. Jay Mallin. Coral Gables, Fla.: University of Miami Press, 1969.

_____. Che Guevara Speaks: Selected Speeches and Writings. Ed. George Lavan. New York: Grove Press, 1968.

_____. Che: Selected Works of Ernesto Guevara. Ed. Rolando E. Bonachea and Nelson P. Valdés. Cambridge, Mass.: MIT Press, 1969.

_____. Economic Planning in Cuba. New York, 1961.

_____. Guerra de guerrillas. La Habana, 1960.

_____. "Man and Socialism in Cuba." In Man and Socialism in Cuba. Ed. Bertram Silverman. New York: Atheneum, 1971.

_____. Obra revolucionaria. Ed. Roberto Fernández Retamar. 2d ed. Mexico City: Ediciones Era, 1968.

_____. Pasajes de la guerra revolucionaria. La Habana, 1963.

_____. Reminiscences of the Cuban Revolutionary War. Translated by Victoria Ortiz. New York: Monthly Review Press, 1968.

Guggenheim, Harry F. The United States and Cuba. New York: Macmillan, 1934.

Guiteras, Pedro José. Historia de la conquista de la Habana. Philadelphia, 1856.

_____. Historia de la Isla de Cuba. 3 vols. La Habana: Colección de Libros Cubanos, 1928.

Gutiérrez, Gustavo. El desarrollo económico de Cuba. La Habana: Ruperto Hernández, Ed., 1952.

_____. El empleo, el subempleo y el desempleo en Cuba. La Habana: Consejo Nacional de Economía, 1958.

Gutiérrez de la Concha, José. Memorias sobre el estado político, gobierno y administración de la isla de Cuba por el teniente general don José de la Concha. Madrid, 1853.

Habana, Universidad de la. Crítica y reforma universitarias. La Habana: Imprenta de la Universidad de La Habana, 1959.

Hagedorn, Hermann. Leonard Wood, a Biography. 2 vols. New York: Harper and Bros. 1931.

Hageman, Alice, and Wheaton, Philip, eds. Religion in Cuba Today. New York: Association Press, 1971.

Halebsky, Sandor, and Kirk, John M. Cuba: Twenty-Five Years of Revolution, 1959-1984. New York: Praeger, 1985.

Halperin, Ernst. Castro and Latin American Communism. Cambridge: Center for International Studies, Massachusetts Institute of Technology, 1963.

_____. The Ideology of Castroism and Its Impact on the Communist Parties of Latin America. Cambridge: Center for International Studies, Massachusetts Institute of Technology, 1961.

Halperin, Maurice. Rise and Decline of Fidel Castro. Berkeley: University of California Press, 1972.

_____. The Taming of Fidel Castro. Berkeley: University of California Press, 1981.

Harnecker, Marta. Cuba: ¿dictadura o democracia? Mexico City: Siglo XXI, 1975.

Harrington, Mark R. Cuba antes de Colón. 2 vols. La Habana: Cultural, 1935.

Healy, David F. The United States in Cuba, 1898-1902. Madison: University of Wisconsin Press, 1963.

Henríquez Ureña, Max. Panorama histórico de la literatura cubana. New York: Las Américas Publishing Co., 1963.

Hernández, José M. The Role of the Military in the Making of the Cuban Republic. Baton Rouge, La., 1978.

Hernández-Catá, Alfonso. Un cementerio en las Antillas. Madrid: Imp. de Galo Sáez, 1933.

Hernández Corujo, Enrique. Historia constitucional de Cuba. La Habana: Cía. Editora de Libros y Folletos, 1960.

Herrera Fritot, René. Cooperativas agrícolas de los indios cubanos. La Habana, 1960.

_____. Culturas aborígenes de las Antillas. La Habana, 1936.

_____. Los complejos culturales indocubanos basados en la arqueología. La Habana, 1956.

Horowitz, Irving Louis, ed. Cuban Communism. New Brunswick: Transaction Books, 1987.

Horrego Estuch, Leopoldo. Emilia Casanova: la vehemencia del separatismo. La Habana: "El Siglo XX," 1951.

————. Legislación social de Cuba. La Habana: Editorial Librería Selecta, 1948-1949.

————. Martín Morúa Delgado: vida y mensaje. La Habana: Editorial Sánchez, 1957.

Hortsmann, Jorge A. Aguacero. La Habana: Empresa Editorial de Publicaciones, S.A., 1956.

Huberman, Leo. Socialism in Cuba. New York: Monthly Review Press, 1969.

————, and Sweezy, Paul M. Cuba: Anatomy of a Revolution. New York: Monthly Review Press, 1960.

Hull, Cordell. The Memoirs of Cordell Hull. 2 vols. New York: Macmillan, 1948.

Humboldt, Alejandro de. Ensayo político sobre la Isla de Cuba. La Habana: Talleres del Archivo Nacional, 1960.

Iglesia, Alvaro de la. Cosas de antaño. La Habana, 1927.

Iglesias, Abelardo. Revolución y dictadura en Cuba. Buenos Aires: Editorial Reconstruir, 1963.

Infiesta, Ramón. Martí constitucionalista. La Habana: Academia de la Historia de Cuba, 1951.

————. Máximo Gómez. La Habana: Imp. "El Siglo XX," 1937.

Instituto de Investigaciones Científicas de la Economía Pesquera Marina y de la Oceanografía de Cuba. Investigaciones pesqueras soviético-cubanas. Moscow: Editorial Pischevaja Promyshlennost, 1965-1971.

International Bank for Reconstruction and Development. Report on Cuba. Baltimore: Johns Hopkins University Press, 1951.

International Commission of Jurists. Cuba and the Rule of Law. Geneva: H. Studer, S.A. 1962.

International Missionary Council. The Cuban Church in a Sugar Economy. London and New York: Department of Social and Economic Research and Counsel, 1942.

Iraizoz, Antonio. Ideología de José Martí. Lisboa: La Bécarre, 1925.

_____. Outline of Education Systems and School Conditions in the Republic of Cuba-1924. La Habana: Montalvo, Cárdenas and Co., 1924.

Iznaga, J.M. Por Cuba. La Habana: Imp. Rambla y Bouza, 1907.

Iznaga, R. Tres años de República. La Habana: Imp. Rambla y Bouza, 1905.

Jackson, D. Bruce. Castro, the Kremlin, and Communism in Latin America. Baltimore: The Johns Hopkins Press, 1969.

James, Daniel. Che Guevara: A Biography. New York: Stein and Day, 1969.

_____, ed. The Complete Bolivian Diaries of Che Guevara. New York: Stein and Day, 1968.

Jenks, Leland. Our Cuban Colony. A Study in Sugar. New York: Vanguard Press, 1928.

Johnson, Haynes. The Bay of Pigs. New York: W.W. Norton, 1964.

Johnson, Willis Fletcher. The History of Cuba. 5 vols. New York: B.F. Buck, 1920.

Joven Cuba: programa. La Habana: Ed. Siglo XX, 1934.

Judson, C. Fred. Cuba and the Revolutionary Myth: The Political Education of the Cuban Rebel Army, 1953-1963. Boulder, Colo., and London: Westview Press, 1984.

Julien, Claude. La revolución cubana. Translated by Mario Trajtenberg. Montevideo: Ediciones Marcha, 1961.

Karol, K.S. Guerrillas in Power. New York: Hill and Wang, 1970.

Kaye, Martin, and Perry, Louise. Who Fights for a Free Cuba? New York: Workers Library Publishers, September 1933.

Kennan, George. Campaigning in Cuba. New York: Century Co., 1899.

Kiple, Kenneth T. Blacks in Colonial Cuba. Gainesville, Florida, 1976.

Klein, Herbert S. Slavery in the Americas. Illinois: The University of Chicago Press, 1967.

Knight, Franklin W. Slave Society in Cuba During the Nineteenth Century. Madison: The University of Wisconsin Press, 1970.

Kohly, Mario García. El problema constitucional de Cuba y nuestra América. Paris: Imprimerie Labor, 1931.

Kozolchyk, Boris. The Political Biographies of Three Castro Officials. Santa Monica, Ca.: The Rand Corporation, 1966.

Krieger, H.W. The Early Indian Culture of Cuba. Washington, D.C.: Explorations and Field works of the Smithsonian Institution, 1932.

Krishna Iyer, V.R. Cuban Panorama. Trivandrum: Prabatham, 1967.

La constitución de los órganos del poder popular. La Habana: Imprenta Federico Engels, 1974.

Labra, Rafael M. de. La brutalidad de los negros. La Habana: Imprenta de la Universidad de La Habana, 1961.

Lachatañeré, Rómulo. Manual de santería: el sistema de cultos lucumís, 1942.

Lagas, Jacques. Memorias de un capitán rebelde. Santiago de Chile: Editorial del Pacífico, 1964.

Lamar Schweyer, Alberto. Cómo cayó el presidente Machado. Madrid: Espasa-Calpe, 1941.

_____. La crisis del patriotismo. La Habana: Ed. Martí, 1929.

Langley, Lester D. The Cuban Policy of the U.S. New York: John Wiley and Sons, 1968.

Larson, David L., ed. The Cuban Crisis of 1962: Selected Documents and Chronology. Boston: Houghton Mifflin Co., 1963.

Las insurrecciones en Cuba. 2 vols. Madrid, 1972-73.

Lataste Hoffer, Albán. Cuba: hacia una nueva economía política del socialismo. Santiago: Editorial Universitaria, 1968.

Laurent, Emilio O. De oficial a revolucionario. La Habana: Imp. Ucar García y Cía., 1941.

Lavan, George, ed. Che Guevara Speaks. New York: Merit Publishers, 1967.

Lázaro, Angel. Retratos familiares. La Habana: Ucar, García y Cía., 1945.

Lazo, Mario. Dagger in the Heart: American Policy Failures in Cuba. New York: Funk and Wagnalls, 1968.

Lazo, Raimundo. La teoría de las generaciones y su aplicación al estudio histórico de la literatura cubana. La Habana, 1954.

Le Riverend Brusone, Julio. See Riverend Brusone, Julio Le.

Leiner, Marvin, with Ubell, Robert. Children Are the Revolution: Day Care in Cuba. New York: Viking Press, 1974.

Leiseca, Juan Martín. Apuntes para la historia eclesiástica de Cuba, 1938.

León, Rubén de. El origen del mal. Miami: Service Offset, 1964.

Lequerica Vélez, Fulgencio. 600 días con Fidel: tres misiones en La Habana. Ediciones Mito, 1961.

Leroy, Luis Felipe, and Gaztelu, Mons. A. Fray Gerónimo Valdés, obispo de Cuba: La iglesia parroquial del Espíritu Santo de La Habana, Reseña histórica. La Habana, 1963.

Levine, Barry B. The New Cuban Presence in the Caribbean. Boulder, Colo.: Westview Press, 1983.

Lewis, Oscar, Lewis, Ruth M., and Rigdon, Susan M. Four Men: Living the Revolution. Urbana: University of Illinois Press, 1977.

_____. Four Women: Living the Revolution. Urbana: University of Illinois Press, 1977.

_____. Neighbors: Living the Revolution. Urbana: University of Illinois Press, 1978.

Ley de organización del sistema judicial. La Habana: Ministerio de Justicia, 1973.

Lima, Alfredo. La odisea de Río verde. La Habana: Cultural, 1934.

Lizaso, Félix. Panorama de la cultura cubana. Mexico City: Fondo de Cultura Económica.

Llaverías, Joaquín. Miguel Aldama o la dignidad patriótica. La Habana. 1937.

Llerena, Mario. The Unsuspected Revolution: The Birth and Rise of Castroism. Ithaca: Cornell University Press, 1978.

Lobo, Julio. El plan Chadbourne: nuestro cancer social. La Habana: Maza, Caso y Cía., 1933.

Bibliography 332

Lockmiller, David A. Enoch H. Crowder. Columbia: University of Missouri Press, 1955.

_____. Magoon in Cuba. Chapel Hill: The University of North Carolina, 1938.

Lockwood, Lee. Castro's Cuba, Cuba's Fidel. New York: The Macmillan Company, 1967.

López Fresquet, Rufo. My Fourteen Months with Castro. New York: World Publishing Co., 1966.

López Goicochea, Francisco. El desahucio: doctrina del Tribunal Supremo de Cuba, 1947-1952. La Habana: Empresa Editora de Publicaciones, 1954.

López Segrera, Francisco. Cuba: capitalismo dependiente y subdesarrollo (1510-1959). La Habana: Casa de las Américas, 1972.

Loveira, Carlos. Generales y doctores. Edited by Shasta M. Bryant and J. Riis Owre. New York: Oxford University Press, 1965.

Loven, Sven. Origins of the Tainan Culture, West Indies. Göteborg, 1935.

Lozano Casado, F. La personalidad del General José Miguel Gómez. La Habana: Imprenta y Papelería de Rambla, Bouza, 1913.

Ludwig, Emil. Biografía de una isla. Mexico: Ed. Centauro, 1948.

Lumen, Enrique. La revolución cubana: crónica de nuestro tiempo, 1902-1934. Mexico City: Ed. Botas, 1934.

Luz y Caballero, José de la. Aforismos. La Habana: Editorial de la Universidad de La Habana, 1962.

_____. De la vida íntima: epistolario y diarios. La Habana: Editorial de la Universidad de La Habana, 1945.

_____. Elencos y discursos académicos. La Habana: Editorial de la Universidad de La Habana, 1950.

_____. Escritos educativos. 2 vols. La Habana: Editorial de la Universidad de La Habana, 1952.

_____. Escritos sociales y científicos. La Habana: Editorial de la Universidad de La Habana, 1955.

_____. La polémica filosófica. 5 vols. La Habana: Editorial de la Universidad de La Habana, 1946.

Macauley, Neill. A Rebel in Cuba: An American's Memoir. Chicago: Quadrangle, 1970.

MacGaffey, Wyatt, and Barnett, Clifford. Cuba: Its People, Society, and Culture. New Haven: Human Relations Area Files, 1962.

Machado y Morales, Gerardo. Declarations of General Gerardo Machado y Morales Regarding His Electoral Platform. La Habana: National Press Bureau, 1928.

_____. Sus discursos y su obra de gobierno. La Habana: Rambla, Bouza y Cía., 1927.

Madden, Richard R. La isla de Cuba. La Habana, 1964.

Maestri, Raúl. El latifundismo en la economía cubana. La Habana: Ed. Rev. de Avance, 1929.

Magill, Roswell, and Shoup, Carl. The Cuban Fiscal System, 1939. New York, 1939.

Mallin, Jay, ed. Che Guevara on Revolution. Coral Gables, Fla.: University of Miami Press, 1969.

_____. Fortress Cuba: Russia's American Base. Chicago: Henry Regnery, 1965.

Mañach, Jorge. Doctrina de ABC. La Habana: Pubs. de Partido ABC, Ed. Cenit, 1942.

_____. El militarismo en Cuba. La Habana: Seoane, Fernández y Cía., 1939.

_____. Historia y estilo. La Habana, 1945.

_____. Indagación del choteo. La Habana: Ed. Lex, 1936.

_____. La crisis de la alta cultura en Cuba. La Habana: Imprenta La Universal, 1925.

_____. Martí: Apostle of Freedom. New York: Devin Adair, 1950.

Mankiewicz, Frank, and Jones, Kirby. With Fidel. Chicago: Playboy Press, 1975.

Marinello, Juan. Sobre la inquietud cubana. La Habana: Rev. de Avance, 1930.

Márquez Sterling, Carlos. Don Tomás. La Habana: Editorial Lex, 1953.

_____. Historia de Cuba, desde Colón hasta Castro. New York: Las Américas, 1963.

Márquez Sterling, Manuel. Las conferencias del Shoreham: el cesarismo en Cuba. Mexico City: Ed. Botas, 1933.

_____. Proceso histórico de la Enmienda Platt (1897-1934). La Habana: Imp. "El Siglo XX," 1941.

Marrero, Levi. Cuba: economía y sociedad. Puerto Rico, 1973.

_____. Cuba: la forja de un pueblo. Puerto Rico, 1971.

Martí, Carlos. Forjando patria. La Habana: Maza y Cía., 1917.

Martínez, Marcial. Cuba, la verdad de su tragedia. Mexico City: Talleres Gráficos "Galeza," 1958.

Martínez Alier, Juan, and Martínez Alier, Verena. Cuba: economía y sociedad. Paris: Ruedo Ibérico, 1972.

Martínez Alier, Verena. Marriage, Class, and Color in Nineteenth Century Cuba: A Study of Racial Attitudes and Sexual Values in a Slave Society. London: Cambridge University Press, 1974.

Martínez Escobar, Manuel. El desahucio y su jurisprudencia. 2nd ed. La Habana: Cultural, 1942.

Martínez Fraga, Pedro. El general Menocal: apuntes para su biografía. La Habana: Talleres de Editorial "Tiempo," 1911.

Martínez Márquez, Guillermo. "Itinerario político de Ramón Grau San Martín." Cuadernos de Cultura Auténtica. La Habana: Jorge Lauderman, Impresor, 1939.

Martínez Moles, Manuel. Contribución al folklore. 3 vols. La Habana: Imprenta de "El Fígaro," 1926.

Martínez Ortiz, Rafael. Cuba: los primeros años de independencia. 2 vols. Paris: Editorial "Le Livre Libre," 1929.

_____. General Leonard Wood's Government in Cuba. Paris: Imprimerie Dubois et Bauer, 1920.

Masetti, Jorge Ricardo. Los que luchan y los que lloran (El Fidel Castro que yo vi). La Habana: Editorial Adiedo, 1959.

Masferrer Landa, Rafael. El pensamiento político del Dr. Guiteras. Manzanillo: Ed. El Arte, 1944.

Masó y Vázquez, Calixto. Consideraciones en torno a las revoluciones. La Habana: Editorial Lex, 1959.

Massip, Salvador. Introducción a la geografía de Cuba. La Habana, 1942.

Matthews, Franklin. The New-Born Cuba. New York: Harper and Bros., 1899.

Matthews, Herbert. Fidel Castro. New York: Simon and Schuster, 1969.

_____. The Cuba Story. New York: George Braziller, Inc., 1961.

McClatchy, C.K. Cuba, 1965. [A reporter's observations as published in the Sacramento Bee; several articles bound together.] Sacramento, Ca.: McClatchy Newspapers, 1965.

Medel, José Antonio. La guerra hispano-americana y sus resultados. La Habana: Imp. P. Fernández y Cía., 1932.

Medina, José Toribio. La imprenta en La Habana. Santiago de Chile, 1904.

Mella, Julio Antonio. Documentos para su vida. La Habana, 1964.

_____. La lucha revolucionaria contra el imperialismo. La Habana: Ed. Sociales, 1940.

Méndez, M. Isidro. Martí. La Habana: P. Fernández y Cía., 1941.

Meneses, Enrique. Fidel Castro: siete años de poder. Madrid: Afrodiso Afuado, S.A., 1966.

Menocal y Cueto, Raimundo. Origen y desarrollo del pensamiento cubano. La Habana: Editorial Lex, 1945.

Menton, Seymour. Prose Fiction of the Cuban Revolution. Austin: University of Texas Press, 1975.

Merino, Adolfo G. Nacimiento de un estado vasallo. Mexico City: B. Costa-Amic, 1966.

Merino, Bernardo, and Ibarzábal, F. de. La revolución de febrero: datos para la historia. 2d ed. La Habana: Librería "Cervantes," 1918.

Merino Brito, Eloy C. José Antonio Saco: su influencia en la cultura y en las ideas políticas de Cuba. La Habana: Molina, 1950.

Mesa Lago, Carmelo. Cuba in the 1970's. Albuquerque: University of New Mexico Press, 1978.

_____. Cuba: teoría y práctica de los incentivos. Latin American Studies Series, occasional paper no. 7. Pittsburgh: University of Pittsburgh, 1971.

_____, ed. Revolutionary Change in Cuba. Pittsburgh: University of Pittsburgh Press, 1971.

_____. The Labor Sector and Socialist Distribution in Cuba. New York: Praeger, for the Hoover Institution on War, Revolution and Peace, 1968.

_____, and Roberto E. Hernández. Labor Conditions in Communist Cuba. Cuban Economic Research Project. Coral Gables, Fla.: University of Miami Press, 1963.

Mesa Rodríguez, Manuel I. Los hombres de "La Demajagua." La Habana: "El Siglo XX," 1951.

_____. Tres retratos de Luz y Caballero. la Habana: "El Siglo XX," 1925.

Mestre, Arístides. La antropología y el estudio de nuestros indios. La Habana, 1925.

Mestre y Amabile, Vicente. Cuba, un año de república: hechos y notas. Paris, 1903.

Mezerik, Avraham G., ed. Cuba and the United States. New York: International Review Service 140, 1960.

Miller, Warren. 90 Miles from Home. Greenwich, Conn.: Fawcett Publications, 1961.

Millett, Allan Reed. The Politics of Intervention: The Military Occupation of Cuba, 1906-1909. Columbus: Ohio State University Press, 1968.

Millis, Walter. The Martial Spirit: A Study of Our War with Spain. Boston: Houghton Mifflin, 1931.

Mills, C. Wright. Listen, Yankee. New York: Ballantine Books, 1960.

Minniman, Paul G. The Agriculture of Cuba. Foreign Agriculture Bulletin no. 2. Washington, D.C.: Government Printing Office, 1942.

Miranda, Luis Rodolfo. Reminiscencias cubanas de la guerra y de la paz. La Habana: Imp. P. Fernández y Cía., 1941.

_____. Temas cubanos. La Habana: Imp. P. Fernández y Cía., 1942.

Miranda, Raúl. Siluetas de candidatos. Matanzas: Imprenta y
Librería La Pluma de Oro, 1910.

Miró Argenter, José. Cuba: Crónicas de la guerra, las campañas
de invasión y de occidente, 1895-1896. La Habana: Editorial
Lex, 1945.

Mitjans, Aurelio. Estudios sobre el movimiento científico y literario
de Cuba. La Habana, 1890.

_____. Historia de la literatura cubana, 1918.

Monahan, James, and Gilmore, Kenneth O. The Great Deception.
New York: Farrar, Straus, 1963.

Montaner, Carlos Alberto. Cuba, Castro, and the Caribbean.
New Brunswick, N.J., and Oxford: Transaction Books, 1985.

Montero, Tomás. Caras y Caretas. La Habana: Editorial "Montiel,"
1951.

_____. Grandezas y miserias. La Habana: Editorial ALFA, 1944.

Morales Coello, Julio. La importancia del poder naval--positivo y
negativo--en el desarrollo y en la independencia de Cuba. La
Habana: Imprenta "El Siglo," 1950.

_____. Las ciencias antropológicas en Cuba. La Habana, 1942.

Morales y Morales, Vidal. Iniciadores y primeros mártires de la
revolución cubana. La Habana, 1901.

_____. Nociones de Historia de Cuba. La Habana, 1904.

Morales Patiño, Oswaldo. Glosario terminológico de arqueología
indo-cubana. La Habana, 1943.

_____. La religión de los indios antillanos. Actas y Documentos
del Primer Congreso Histórico Municipal Interamericano. La
Habana, 1943.

Moreno Fraginals, Manuel. José Antonio Saco: estudio y bibliografía.
La Habana, 1960.

_____. The Sugarmill. New York: Monthly Review Press, 1976.

Mustelier, Gustavo. La extinción del negro. La Habana: Imprenta
de Rambla, Bouza, 1912.

Navas, José. La convulsión de febrero. Matanzas: Imp. y
Monotipo "El Escritorio," 1917.

Nelson, Lowry. Cuba: The Measure of a Revolution. Minneapolis: University of Minnesota Press, 1972.

_____. Rural Cuba. Minneapolis: University of Minnesota Press, 1950.

Newman, Phillip C. Cuba Before Castro: An Economic Appraisal. Ridgewood, N.J.: Foreign Studies Institute, 1965.

_____. Joint International Business Ventures--Cuba. New York: Columbia University Press, 1958.

Núñez Jiménez, Antonio. Cuba, con la mochila al hombro. La Habana: Ediciones Unión, 1963.

_____. Geografía de Cuba. La Habana: Editorial Lex, 1959.

O'Connor, James. The Origins of Socialism in Cuba. Ithaca: Cornell University Press, 1970.

Oliva Pulgarón, Luis. Apuntes históricos sobre la masonería cubana. La Habana. 1934.

Onís, Federico de. The America of José Martí. New York: The Noonday Press, 1953.

Ordoqui, Joaquín. Elementos para la historia del movimiento obrero en Cuba. La Habana: Imprenta Nacional de Cuba, 1961.

Organización Nacional de Bibliotecas Ambulantes y Populares. 13 documentos de la insurrección. La Habana: ONBAP, 1959.

Organization of American States. Inter-American Commission on Human Rights. Fifth Report on the Status of Human Rights in Cuba. OEA/Ser. G/CP/INF. 872/76, June 1, 1976.

_____. Report Regarding the Situation of Human Rights in Cuba. OEA/Ser. L/V/II. 17, Doc. 4 rev., June 13, 1967.

_____. Second Report on the Situation of Political Prisoners in Cuba. OEA/Ser. L/V/II. 23, Doc. 6, rev. 1, November 17, 1970.

Ortiz, Fernando. Contrapunteo cubano del azúcar y el tabaco. La Habana, 1940.

_____. Cuba primitiva: Las razas indias. La Habana, 1937.

_____. De la música afrocubana. La Habana, 1934.

_____. Hampa afrocubana: los negros brujos. La Habana, 1905.

_____. Historia de la arqueología indocubana. La Habana, 1922.

_____. Historia de una pelea cubana contra los demonios, Erre. San Juan, P.R., 1973.

_____. La africanía en la música folklórica de Cuba. La Habana, 1950.

_____. La crisis política cubana. La Habana: Imprenta y Papelería La Universal, 1919.

_____. La decadencia cubana. La Habana: Imp. y Papelería Universitaria, 1920.

_____. Las cuatro culturas indias de Cuba. La Habana, 1943.

_____. Las relaciones económicas entre Cuba y los Estados Unidos. La Habana, 1927.

_____. Las responsabilidades de los EE.UU. en los males de Cuba. Washington, D.C., 1932.

_____. Los bailes y el teatro de los negros en el folklore de Cuba. La Habana, 1951.

_____. Los cabildos afrocubanos. La Habana, 1923.

_____. Los instrumentos de la música afrocubana. La Habana, 1953.

_____. Los negros curros. La Habana, 1934.

_____. Los negros esclavos. La Habana, 1916.

Osgood, Cornelius. The Ciboney Culture of Cayo Redondo, Cuba. New Haven: Yale University Press, 1942.

Otero, Lisandro, with Martínez Hinojosa, Francisco. Cultural Policy in Cuba. Paris: UNESCO, 1972.

Otero Echeverría, Rafael. Reportaje a una revolución: de Batista a Fidel Castro. Santiago de Chile: Editorial del Pacífico, 1959.

Palacios, Alfredo L. Una revolución auténtica en Nuestra América. Mexico City: Editorial Cultura, 1960.

Pardeiro, Francisco A. Historia de la economía de Cuba. La Habana: Ed. Universidad de La Habana, 1966.

Pardo Llada, José. Memorias de la Sierra Maestra. La Habana: Editorial Tierra Nueva, 1960.

Partido Aprista Cubano. El aprismo ante la realidad cubana. La Habana: Ed. Apra, 1934.

Partido Comunista de Cuba. El Partido Comunista y los problemas de la revolución cubana. La Habana, 1933.

_____. Segundo Congreso del Partido Comunista de Cuba: resoluciones. La Habana, 1934.

Partido Revolucionario Cubano. Al pueblo de Cuba. La Habana, 1935.

_____. Programa constitucional de gobierno. La Habana, 1934.

Partido Unido de la Revolución Socialista de Cuba. Relatos del asalto al Moncada. La Habana: Empresa Consolidada de Artes Gráficas, 1964.

Peralta, Víctor M. de. Conmonitorio de intervención a intervención. La Habana: Imp. La Prueba, 1907.

Peraza, Carlos G. Machado: crímenes y horrores de un régimen. La Habana: Cultural, 1933.

Peraza Sarausa, Fermín. Anuario Bibliográfico Cubano. 1937-1959 published in Havana; 1960 in Medellín, Colombia; 1961-1966 in Gainesville, Florida.

_____. Diccionario biográfico cubano. La Habana: Ediciones Anuario Bibliográfico Cubano, 1951.

_____. Personalidades cubanas. Gainesville, Fla.: Biblioteca del Bibliotecario, 1967.

Pérez, Emma. Historia de la Pedagogía en Cuba. La Habana, 1945.

Pérez Landa, Rufino. Bartolomé Masó. La Habana, 1947.

Pérez, Louis A., Jr. Army Politics in Cuba, 1898-1958. Pittsburgh: University of Pittsburgh Press, 1976.

Pérez Beato, Manuel. Habana antigua: apuntes históricos. La Habana, 1936.

Pérez Cabrera, José Manuel. Historiografía de Cuba. Mexico City: 1962.

Pérez de la Riva, Francisco. El café: historia de su cultivo y explotación en Cuba. La Habana: Jesús Montero, 1944.

Pericot y García, Luis. América indígena. Barcelona, 1962.

Pezuela, Jacobo de la. Diccionario geográfico, estadístico, histórico de la isla de Cuba. 4 vols. Madrid, 1863-1878.

_____. Ensayo histórico de la isla de Cuba. New York, 1842.

Pflaum, Irving P. Tragic Island: How Communism Came to Cuba. Englewood Cliffs, N.J.: Prentice-Hall, 1961.

Phillips, Ruby Hart. Cuban Sideshow. La Habana: Cuban Press, 1935.

_____. Island of Paradox. New York: McDowell, Obolensky, 1959.

_____. The Cuban Dilemma. New York: Ivan Obolensky, 1962.

Pichardo, Esteban. Caminos de la isla de Cuba. 3 vols. La Habana, 1865.

Pichardo Moya, Felipe. Los aborígenes de las Antillas. Mexico City: 1956.

Piedra, Carlos M. La inamovilidad de los trabajadores. La Habana: Cultural, 1945.

Piedra Martel, Manuel. Mis primeros treinta años: memorias. La Habana: Imp. Ucar, García y Cía., 1943.

Pincus, Arthur. Terror in Cuba. New York: Workers Defense League, 1936.

Pino-Santos, Oscar. El asalto a Cuba por la oligarquía financiera yanqui. La Habana: Casa de las Américas, 1973.

_____. El imperialismo norteamericano en la economía de Cuba. La Habana: Ed. Lex, 1960.

_____. Historia de Cuba. Aspectos fundamentales. La Habana: Editorial Nacional de Cuba, 1964.

Pío Elizalde, Leopoldo. Defamation. Mexico: Defensa Institucional Cubana, 1961.

Pittaluga, Gustavo. Diálogos sobre el destino. La Habana: Editorial Isla, S.A., 1960.

Plan Trienal de Cuba. La Habana: Cultural, 1938.

Plank, John N., ed. Cuba and the United States: Long Range Perspectives. Washington, D.C.: The Brookings Institution, 1967.

Playa Girón: derrota del imperialismo. La Habana: Ediciones R.,
1962.

Política internacional de la revolución cubana. La Habana:
Editora Política, 1966.

Ponte Domínguez, Francisco J. Historia de la Guerra de los Diez
Años. La Habana: Imprenta "El Siglo XX," 1958.

Portell Vilá, Herminio. Cuba y la conferencia de Montevideo. La
Habana: Imp. Heraldo Cristiano, 1934.

_____. Historia de Cuba en sus relaciones con los Estados Unidos
y España. 4 vols. La Habana: Jesús Montero, 1938-1941.

_____. Narciso López y su época (1848-1850). La Habana:
Compañía Editora de Libros y Folletos, 1952.

_____. Narciso López y su época (1850-1851). La Habana:
Compañía Editora de Libros y Folletos, 1952.

_____. The Nonintervention Pact of Montevideo and American
Interests in Cuba. La Habana: Molina y Cía., 1935.

_____. The Sugar Industry and Its Future. La Habana: Ucar,
García, 1943.

Porter, Robert P. Industrial Cuba. New York: G.P. Putnam,
1899.

Portuondo, José Antonio. Angustia y evasión de Julián del Casal.
La Habana: Molina y Cía., 1937.

_____. Crítica de la época. La Habana: Universidad Central
de Las Villas, 1965.

_____. "El contenido social de la literatura cubana," Jornadas.
Mexico: Fondo de Cultura Económica, 1944.

_____. Proceso de la cultura cubana. La Habana, 1938.

Portuondo del Prado, Fernando. Historia de Cuba. La Habana:
Molina y Cía. S.A., 1953.

Portuondo Linares, Serafín. Los independientes de color: historia
del Partido Independiente de Color. La Habana: Editorial
Librería Selecta, 1950.

Pratt, Julius. Expansionists of 1898. New York: P. Smith, 1964.

Primelles, León. Crónica cubana. 2 vols. La Habana: Ed. Lex,
1958.

Proclamas y leyes del gobierno provisional de la revolución, 1959.
La Habana: Editorial Lex, 1959.

Prohías, Rafael, and Casal, Lourdes. The Cuban Minority in the
United States: Preliminary Report on Need Identification and
Program Evaluation. Washington, D.C.: Cuban National
Planning Council, 1974.

Quesada, Gonzalo de. Archivo de Gonzalo de Quesada: documentos
históricos. Compiled by Gonzalo de Quesada y Miranda. La
Habana: Editorial de la Universidad de La Habana, 1965.

————. ¡En Cuba Libre! Historia documentada y anecdótica del
machadato. 2 vols. La Habana: Seoane, Fernández y Cía.,
1938.

Raggi Ageo, Carlos M. Condiciones económicas y sociales de la
República de Cuba. La Habana: Editorial Lex, 1944.

Ratliff, William E. Castroism and Communism in Latin America,
1959-1976. Stanford: Hoover, 1976.

Rauch, Basil. American Interest in Cuba, 1848-1855. New York,
1948.

Rauf, Mohammed A., Jr. Cuban Journal: Castro's Cuba as It
Really Is. New York: Crowell, 1964.

Ravelo Nariño, Agustín. El contrato de arrendamiento de finca
rústica en la legislación cubana. Santiago: Tipografía San
Román, 1956.

Reckford, Barry. Does Fidel Eat More Than Your Father?
London: André Deutsch, 1971.

Remos, Juan J. Proceso histórico de las letras cubanas. Madrid:
Ediciones Guadarrama, 1958.

Riera Hernández, Mario. Cuba libre, 1875-1958. Miami: Colonial
Press, 1968.

————. Cuba política, 1899-1955. La Habana: Impresora Modelo,
1955.

————. Historial obrero cubano, 1574-1965. Miami: Rema Press,
1965.

————. Un presidente constructivo. Miami: Colonial Press,
1965.

Ripoll, Carlos. José Martí, Thoughts/Pensamientos. A Bilingual
Anthology. New York: Las Américas Publishing Co., 1980.

_____. La generación del 23 en Cuba. New York: Las Américas Publishing Co., 1968.

Ritter, Archibald R.M. The Economic Development of Revolutionary Cuba: Strategy and Performance. New York: Praeger, 1974.

Riverand Brusone, Julio Le. Economic History of Cuba. La Habana: Instituto del Libro, 1967.

_____. La Habana: biografía de una provincia. La Habana, 1960.

_____. La república: dependencia y revolución. La Habana: Ed. Universal, 1966; Instituto del Libro, 1969.

_____. Reseña histórica de la economía cubana y sus problemas. Mexico: Litografía Machado, 1956.

Rivero, Nicolás. Castro's Cuba: An American Dilemma. Washington, D.C.: Luce, 1962.

Rivero Collado, Carlos. Los sobrinos del Tío Sam. La Habana: Instituto Cubano del Libro, 1976.

Rivero de la Calle, Manuel. Los aborígenes de Cuba: estudio histórico-etnográfico. La Habana, 1963.

Rivero Muñiz, José. El movimiento laboral cubano durante el período 1906-1911. La Habana: Empresa Consolidada de Artes Gráficas, 1962.

_____. El movimiento obrero durante la primera intervención (1899-1902). La Habana: Impresores Ucar, García, 1961.

_____. El primer partido socialista cubano. La Habana: Imprenta Nacional, 1962.

_____. Las tres sediciones de los vegueros en el siglo XVIII. La Habana: Impresora "El Siglo XX," 1951.

Roa, Ramón. Pluma y machete. La Habana, 1969.

Roa, Raúl. 15 años después. La Habana: Talleres Alfa, 1950.

_____. Retorno a la alborada. 2 vols. La Habana: Universidad de Las Villas, 1964.

Roberts, C. Paul, and Hamour, Mukhtar. Cuba, 1968: Supplement to the Statistical Abstract of Latin America. Los Angeles: University of California, 1970.

_____, eds. El triunfo popular en las elecciones. La Habana,
1946.

Robinson, Albert G. Cuba and the Intervention. New York:
Longman's Green and Co., 1905.

Roca, Blas. Los fundamentos de socialismo en Cuba. La Habana:
Ed. Popular, 1960.

_____. The Cuban Revolution: Report to the Eighth National
Congress of the Socialist Party of Cuba. New York: New
Century, 1961.

_____, Rodríguez, Carlos Rafael, and Luzardo, Manuel. En
defensa del pueblo. La Habana: Arrow Press, 1945.

Rodríguez Demorizi, Emilio. Papeles dominicanos de Máximo Gómez.
Montalvo, Dominican Republic, 1954.

Rodríguez-Embil, Luis. José Martí: el Santo de América. La
Habana: P. Fernández y Cía., 1941.

Rodríguez Ferrer, Miguel. Naturaleza y civilización de la grandiosa
isla de Cuba. 2 vols. Madrid, 1876.

Rodríguez Morejón, Gerardo. Fidel Castro, biografía. La Habana:
P. Fernández y Cía., 1959.

_____. Grau San Martín. La Habana: Ediciones Mirador, 1944.

_____. Menocal. La Habana: Cárdenas y Cía., 1941.

Roig, Enrique. El servicio militar obligatorio. La Habana:
Compañía Impresora y Papelería "La Universal," 1918.

Roig de Leuchsenring, Emilio. Curso de introducción a la historia
de Cuba. La Habana: Molina y Cía., 1938.

_____. Historia de la enmienda Platt: una interpretación de la
realidad cubana. Vol. 1. La Habana, 1935.

_____. Historia de La Habana. La Habana, 1938.

_____. "Homenaje a Martí en el Cincuentenario de la Fundación
del Partido Revolucionario Cubano, 1892-1942." Cuaderno de
Historia Habanera. La Habana: Cárdenas y Cía., 1942.

_____. Juan Gualberto Gómez, paladín de la independencia y la
libertad de Cuba. La Habana: Oficina del Historiador de la
Ciudad, 1954.

_____. La colonia superviva. La Habana: Imprenta "El Siglo XX," 1925.

_____. La lucha cubana por la república, contra la anexión y la Enmienda Platt, 1899-1902. La Habana: Oficina del Historiador de la Ciudad, 1959.

_____. Los problemas sociales de Cuba. La Habana, 1927.

_____. Males y vicios de Cuba republicana. La Habana: Oficina del Historiador de la Ciudad de La Habana, 1959.

_____. Revaloración de la historia de Cuba por los congresos nacionales de historia. La Habana: Oficina del Historiador de la Ciudad de La Habana, 1959.

_____. "Vida y pensamiento de Martí, 1892-1942, Vol. I." Colección histórica cubana y americana. La Habana: Municipio de La Habana, 1942.

Rojas R., Marta. La generación del centenario en el Moncada. La Habana: Ediciones R., 1964.

Rosell, Mirta, comp. and ed. Luchas obreras contra Machado. La Habana: Instituto Cubano del Libro, 1973.

Rosell Leyva, Florentino E. La verdad. Miami, 1960.

Rouse, Irving. The Ciboney. Washington, 1948.

Rousset, Ricardo V. Historial de Cuba. La Habana, 1918.

Royo Guardia, Fernando. Entierros aborígenes en Cuba. La Habana, 1940.

Rozitchner, León. Moral burguesa y revolución. Buenos Aires: Editorial Tiempo Contemporáneo, 1969.

Rubens, Horatio S. Liberty: the Story of Cuba. New York: Warren and Putnam, 1932.

Ruiz, Leovigildo. Diario de una traición. Miami: Florida Typesetting of Miami, 1965.

Ruiz, Ramón Eduardo. Cuba: the Making of a Revolution. Amherst: University of Massachusetts Press, 1968.

Saco, José Antonio. Colección de papeles científicos, históricos, políticos y de otros ramos sobre la isla de Cuba, ya publicados, ya inéditos. 3 vols. Paris, 1858-59.

_____. Contra la anexión. 5 vols. La Habana: Colección de Libros Cubanos, 1928.

_____. Historia de la esclavitud de la raza africana en el Nuevo Mundo. Barcelona, 1879.

Sagra, Ramón de la. Historia física, política y natural de la isla de Cuba. Paris, 1861.

Sainz de la Peña, Arturo. La revolución de agosto. La Habana: Imp. "La Prueba," 1909.

Saíz de la Mora, Santiago. El general Tacón en Cuba: consideraciones sobre su gobierno. La Habana, 1944.

San Martín, Rafael. El grito de la Sierra Maestra. Buenos Aires: Ediciones Gure, 1960.

Sánchez Amaya, Fernando. Diario del Granma. La Habana: Editorial Tierra Nueva, 1959.

Sánchez Roca, Mariano. Compilación ordenada y completa de la legislación cubana de 1951 a 1958. La Habana: Editorial Lex, 1960.

Sánchez Roig, Mario, and Gómez de la Maza, Federico. La pesca en Cuba. La Habana: Ministerio de Agricultura, 1952.

Sanguily, Manuel. Defensa de Cuba. La Habana: Oficina del Historiador de la Ciudad, 1948.

Sanjenís, A. Tiburón. La Habana: Librería Hispanoamericana, 1915.

Santaella Blanco, Antonio. La Masonería en la Revolución Cubana. Mexico: Editorial "Memphis," 1961.

Santovenia, Emeterio S. Armonía y conflictos en torno a Cuba. Mexico: Fondo de Cultura, 1956.

_____. "Correspondencia diplomática de la delegación cubana en Nueva York durante la guerra de independencia de 1895 a 1898." Publicaciones del Archivo Nacional de Cuba. Vol. 6. La Habana: Archivo Nacional de Cuba, 1945.

_____, and Shelton, Raúl M. Cuba y su historia. 2nd ed. Miami: Rema Press, 1966.

_____. El ABC ante la mediación. La Habana: Mazo, Caso y Cía., 1934.

_____. Historia de Cuba. La Habana, 1939.

_____, Santovenia, Antonio, and Pérez Cabrera, Manuel. La enseñanza de la historia en Cuba. Instituto Panamericano de Geografía e Historia. Mexico: Editorial Cultural, 1951.

Sapir, Boris. The Jewish Community in Cuba. New York: Jewish Teachers' Seminar and Peoples' University, 1948.

Sartre, Jean-Paul. Sartre on Cuba. New York: Ballantine Books, 1961.

Scheer, Robert, ed. The Diary of Che Guevara. New York: Bantam Books, 1968.

Secretaría de Educación. Homenaje a Enrique José Varona. La Habana: Molina y Cía., 1935.

Sedano y Cruzat, Carlos. Cuba de 1850 a 1873. Madrid, 1873.

Seers, Dudley. Cuba: Economic and Social Revolution. Durham, N.C.: Seeman Printery, 1964.

Seiglie y Llata, Oscar. El contrato de arrendamiento de finca rústica, el latifundio y la legislación azucarera. La Habana: Editorial Lex, 1953.

Selzer, Gregorio. Fidel Castro: la Revolución Cubana. Buenos Aires: Editorial Palestra, 1960.

Shelton Ovich, Raúl M., and Arredondo Gutiérrez, Alberto. Stages and Problems of Industrial Development in Cuba. Coral Gables, Fla.: Cuban Economic Research Project, University of Miami, 1965.

Silverman, Bertram. Man and Socialism in Cuba. New York: Atheneum, 1971.

Simons, William. Hands Off Cuba. New York: Workers Library Publishers, 1933.

Sklar, Barry A. "Cuba: Normalization of Relations." Issue Brief no. IB75030. Library of Congress, Congressional Research Service, March 3, 1976.

Smith, Earl E.T. The Fourth Floor. New York: Random House, 1962.

Smith, Robert F. The United States and Cuba: Business and Diplomacy. New York: Bookman Associates, 1961.

_____, ed. Background to Revolution: The Development of
Modern Cuba. New York: Alfred A. Knopf, 1966.

_____. What Happened in Cuba? A Documentary History.
New York: Twayne, 1963.

Smith, T. Lynn. Agrarian Reform in Latin America. New York:
Knopf, 1965.

Sobel, Lester A. Castro's Cuba in the 1970's. New York:
Facts on File, 1978.

Solano Alvarez, Luis. Mi actuación militar o apuntes para la
historia de la revolución de febrero de 1917. La Habana: Imp.
"El Siglo XX," 1920.

Sosa de Quesada, Arístides. El Consejo Corporativo de Educación,
Sanidad y Beneficiencia y sus instituciones filiales. La Habana:
Instituto Cívico Militar, 1937.

_____. Militarismo, anti-militarismo, seudo-militarismo. La
Habana: Instituto Cívico Militar, 1939.

_____. Por la democracia ... y por la libertad. La Habana:
P. Fernández, 1943.

Souchy, Agustín. Testimonios sobre la revolución cubana. Buenos
Aires: Editorial Reconstruir, 1960.

Souza, Benigno. Máximo Gómez. La Habana: Editorial Trópico,
1936.

Special Operations Research Office. Casebook on Insurgency and
Revolutionary Warfare: 23 Summary Accounts. Washington,
D.C.: The American University, 1963.

_____. Special Warfare Area Handbook for Cuba. Washington,
D.C.: The American University, 1961.

Stone, Elizabeth. Women and the Cuban Revolution. New York:
Pathfinder Press, 1981.

Strode, Hudson. The Pageant of Cuba. New York: Harrison
Smith and Robert Haas, 1934.

Suárez, Andrés. Cuba: Castroism and Communism, 1959-1966.
Translated by Joel Carmichael and Ernst Halperin. Cambridge:
MIT Press, 1967.

Suárez Núñez, José. El gran culpable. Caracas, 1963.

Suárez Rivas, Eduardo. Un pueblo crucificado. Miami: Service Offset Printers, 1964.

Suchlicki, Jaime. Cuba: From Columbus to Castro. New York: Scribner's, 1974.

_____. University Students and Revolution in Cuba, 1920-1968. Coral Gables, Fla.: University of Miami Press, 1969.

Suchlicki, Jaime, ed. Cuba, Castro, and Revolution. Coral Gables, Fla.: University of Miami Press, 1972.

_____. Cuba: Continuity and Change. Miami: North-South Center for the Institute of Interamerican Studies, University of Miami, 1985.

_____. Problems of Succession in Cuba. Miami: North-South Center for the Institute of Interamerican Studies, University of Miami, 1985.

Sutherland, Elizabeth. The Youngest Revolution: A Personal Report on Cuba. New York: Dial Press, 1969.

Szulc, Tad. Fidel: A Critical Portrait. New York: William Morrow and Co., 1986.

_____, and Meyer, Karl E. The Cuban Invasion. New York: Ballantine Books, Inc., 1962.

Tabares del Real, José A. Ensayo de interpretación de la revolución cubana. La Paz: Talleres Gráficos "Guthenberg," 1960.

_____. La revolución del 33: sus dos últimos años. La Habana, 1971.

Taber, Robert. M-26: Biography of a Revolution. New York: Lyle Stuart, 1961.

_____. The War of the Flea: A Study of Guerrilla Warfare, Theory and Practice. New York: Lyle Stuart, 1965.

Tang, Peter S.H., and Maloney, Juan. The Chinese Communist Impact on Cuba. Chestnut Hill, Mass.: Research Institute on the Sino-Soviet Bloc, 1962.

Tejeiro, Guillermo. Historia ilustrada de la colonia china en Cuba. 1947.

Testé, Ismael. Historia eclesiástica de Cuba. Burgos: Tipografía de la Editorial El Monte Carmelo, 1969.

Tetlow, Edwin. Eye on Cuba. New York: Harcourt, Brace, and World, 1966.

_____. The Cuban Crisis: A Documentary Record. New York: Foreign Policy Association, Headline series, No. 157, 1963.

Thomas, Hugh. Cuba: The Pursuit of Freedom, 1762-1969. New York: Harper and Row, 1971.

Torriente, Cosme de la. Cuarenta años de mi vida, 1898-1938. La Habana: Imp. "El Siglo," 1939.

_____. Cuba y los Estados Unidos. La Habana: Imprenta y Papelería de Rambla, Bouza y Cía., 1929.

_____. Fin de la dominación de España en Cuba (12 de agosto de 1898). La Habana: Imp. "El Siglo XX," 1948.

_____. Libertad y democracia. La Habana: Imp. "El Siglo XX," 1941.

_____. Libro Jubilar de Emeterio S. Santovenia. La Habana: Ucar, García, S.A., 1958.

_____. Por la amistad internacional. La Habana: Imprenta "El Siglo XX," 1951.

Torriente, Loló de la. Estudio de las artes plásticas en Cuba. La Habana, 1954.

Torriente Brau, Pablo de la. Realengo 18 y Mella, Rubén y Machado. La Habana: Ediciones Nuevo Mundo, 1962.

Torroella, Gustavo. Estudio de la juventud cubana. La Habana: Comisión Nacional de la UNESCO, 1963.

Trelles, Carlos M. Bibliografía cubana del siglo XIX. 8 vols. Matanzas, 1911-1915.

_____. Bibliografía cubana del siglo XX. 2 vols., 1916-1917.

_____. El progreso (1902 a 1905) y el retroceso (1906 a 1922) de la república de Cuba. Matanzas: Imp. de Tomás González, 1923.

_____. El sitio de La Habana y la dominación británica en Cuba. La Habana, 1925.

_____. Ensayo de bibliografía cubana de los siglos XVII y XVIII. Matanzas, 1907.

Tribunal de Cuentas. Recopilación y análisis de los ingresos presupuestales de Cuba. La Habana, 1953.

Truslow, Francis Adams. Report on Cuba. Baltimore: The Johns Hopkins Press, 1951.

Turosienski, Severin K., Education in Cuba. Washington, D.C.: U.S. Government Printing Office, 1943.

Ullivarri, Saturnino. Pirates y corsarios en Cuba. La Habana, 1931.

Unión Nacionalista. Nuestro propósito. La Habana, 1929.

Unión Social Económica de Cuba. Commercial Relations Between Cuba and the United States. La Habana, 1936.

United Fruit Company. Some Facts Regarding the Development and Operation of the United Fruit Company Sugar Properties in the Republic of Cuba. Preston, Cuba, 1944.

U.S. Congress, House Committee on Foreign Affairs, Subcommittee on Inter-American Affairs. "Soviet Activities in Cuba," In Hearings, 92nd Congress, 2nd session. Washington, D.C.: U.S. Government Printing Office, 1972.

_____. "Soviet Activities in Cuba." In Hearings, 93rd Congress. Washington, D.C.: U.S. Government Printing Office, 1974.

U.S. Congress, Senate Committee on the Judiciary, Subcommittee to Investigate the Administration of the Internal Security Act and Other Internal Security Laws. Hearings, 86th Congress, 2nd session, August-September 1960, and 87th Congress, 1st session, January-February 1961, February 1962. Washington, D.C.: U.S. Government Printing Office.

U.S. Congress, Senate Select Committee to Study Governmental Operations with Respect to Intelligence Activities. Alleged Assassination Plots Involving Foreign Leaders: An Interim Report. 94th Congress, 1st session. Washington, D.C.: U.S. Government Printing Office, 1975.

U.S. Department of Commerce. "United States Commercial Relations with Cuba: A Survey." In U.S., Congress, House Committee On International Relations, Subcommittees on International Trade and Commerce and on International Organizations. "U.S. Trade Embargo of Cuba." Hearings, 94th Congress, 2nd session. Washington, D.C.: U.S. Government Printing Office, 1976.

U.S. Department of Commerce and Bureau of Foreign and Domestic Commerce. Cuban Readjustment to Current Economic Forces. Trade Information Bulletin No. 725. Washington, D.C.: U.S. Government Printing Office, 1930.

U.S. Department of Commerce, Bureau of Foreign Commerce. Economic Development in Cuba, 1958. World Trade Information Service Economic Reports, April 1959. Washington, D.C.: 1959.

U.S. Department of Commerce, Bureau of Foreign and Domestic Commerce. Investment in Cuba. Washington, D.C.: U.S. Government Printing Office, 1956.

U.S. Department of State. Cuba, Latin America, and Communism. By Edwin M. Martin. Publication 7621. Washington, D.C., 1963.

_____. U.S. Policy Toward Cuba. Publication 7690. Washington, D.C., 1964.

U.S. Tariff Commission. Economic Controls and Commercial Policy in Cuba. Washington, D.C.: U.S. Government Printing Office, 1946.

_____. Mining and Manufacturing in Cuba. Washington, D.C.: U.S. Government Printing Office, 1947.

_____. The Effects of the Cuban Reciprocity Treaty of 1902. Washington, D.C.: U.S. Government Printing Office, 1929.

U.S. War Department. Census of Cuba, 1899. Washington, D.C.

Universidad Central Marta Abreu de Las Villas. La educación rural en Las Villas. La Habana: Impresores Ucar, García, 1959.

Universidad de La Habana. La Universidad de La Habana. La Habana, 1970.

_____. La Universidad de La Habana al Consejo Ejecutivo y a la Asamblea General de la Unión de Universidades de América Latina. La Habana, 1964.

Urrutia Lleó, Manuel. Fidel Castro and Company, Inc. New York: Frederick A. Praeger, Inc. 1964.

Valdés, Antonio José. Historia de la isla de Cuba. La Habana, 1813.

Valdés, Nelson P. Cuba: ¿socialismo democrático o burocratismo colectivista? Bogotá: Ediciones Tercer Mundo, 1973.

_____, and Bonachea, Rolando E. The Selected Works of Fidel Castro. 3 vols. Cambridge, Mass.: MIT Press, 1971.

Valdés Domínguez, Eusebio. Los antiguos diputados de Cuba y apuntes para la historia constitucional de esta Isla. La Habana, 1879.

Valdés-Miranda, Jorge. Cuba revolucionaria. La Habana.

Valdés y Aguirre, Fernando. Apuntes para la historia primitiva. Paris, 1859.

Valladares, Armando. Against All Hope. New York: Alfred A. Knopf, 1986.

Varela y Morales, Félix. Cartas a Elpidio. La Habana: Editorial de la Universidad de La Habana, 1945.

_____. El Habanero. La Habana: Editorial de la Universidad de La Habana, 1962.

_____. Ideario cubano. La Habana, 1953.

_____. Lecciones de Filosofía. 2 vols. La Habana: Editorial de la Universidad de La Habana, 1962.

_____. "Lógica," Instituciones de Filosofía Ecléctica. La Habana: Ediciones de la Universidad de La Habana, 1952.

_____. Miscelánea Filosófica. La Habana: Editorial de la Universidad de La Habana, 1944.

_____. Observaciones sobre la constitución política de la monarquía española. La Habana: Editorial de la Universidad de La Habana, 1944.

Varela Zaqueira, Eduardo. La política en 1905: episodios de una lucha electoral. La Habana: Imp. Rambla y Bouza, 1905.

Varona, Enrique. De la colonia a la República. La Habana: Editorial Cuba Contemporánea, 1919.

Varona, Manuel A. de. El drama de Cuba o la revolución traicionada. Buenos Aires: Editorial Marymar, 1960.

Vázquez, Eduardo I. Cuba independiente: primer período presidencial, 1902-1906. La Habana: Editorial Acosta, 1906.

Vázquez Rodríguez, Benigno. Precursores y fundadores. La Habana: Editorial Lex, 1958.

Vega Cobiellas, Ulpiano. El General Fulgencio Batista y la sucesión presidencial. La Habana: Editorial Colegio, 1957.

_____. La personalidad y la obra del General Fulgencio Batista Zaldívar. La Habana: Cultural, 1943.

_____. Los doctores Ramón Grau San Martín y Carlos Saladrigas Zayas. La Habana: Editorial Lex, 1944.

_____. Nuestra América y la evolución de Cuba. La Habana: Cultural, 1944.

Velasco, Carlos de. Estrada Palma: contribución histórica. La Habana: Imp. "La Universal," 1911.

Ventura Novo, Esteban. Memorias. Mexico City: Imp. M. León Sánchez, 1961.

Vilches de la Maza, Bartolomé. La tiranía de Machado. La Habana: Impresos "Martín," 1933.

Vilches González, Isidro A. Derecho cubano del trabajo. La Habana: Jesús Montero, 1948.

Vitier, Medardo. Las ideas en Cuba. La Habana: Seoane, Fernández y Cía., 1938.

Wallich, Henry C. Monetary Problems of an Export Economy: The Cuban Experience, 1914-1947. Cambridge, Mass.: Harvard University Press, 1950.

Weiss y Sánchez, Joaquín. Arquitectura cubana colonial. La Habana, 1936.

_____. La arquitectura cubana del siglo XIX. La Habana, 1960.

Welles, Benjamin Sumner. The Time for Decision. New York: Harper and Bros., 1944.

Weyl, Nathaniel. La estrella roja sobre Cuba. Buenos Aires: Editorial Freeland, 1961.

Williams, William Appleman. The United States, Cuba, and Castro. New York: Monthly Review Press, 1962.

Winocur, Marcos. Cuba a la hora de América. Buenos Aires: Ediciones Proyon, 1963.

Wood, Bryce. The Making of the Good Neighbor Policy. New York: The MacMillan Co., 1916.

Wright, Irene A. The Early History of Cuba, 1492-1586. New York: The MacMillan Co., 1916.

Wylie, Kathryn A. Survey of Agriculture in Cuba. Washington, D.C.: U.S. Department of Agriculture, Economic Research Service, 1969.

Yglesias, José. In the Fist of the Revolution: Life in a Cuban Country Town. New York: Pantheon, 1968.

Zamora, Juan Clemente. Derecho Constitucional: Cuba. Colección de documentos selectos. La Habana: Imp. "El Siglo XX," 1925.

_____. El proceso histórico. La Habana: Jesús Montero, Editor, 1938.

Zeitlin, Maurice. Revolutionary Politics and the Cuban Working Class. Princeton: Princeton University Press, 1967.

_____, and Scheer, Robert. Cuba: Tragedy in Our Hemisphere. New York: Grove Press, 1963.

Zéndegui, Guillermo de. Ambito de Martí. La Habana: P. Fernández y Compañía, 1953.

STATISTICAL PUBLICATIONS

Census Publications

Census of Cuba, Office of the Director. Report on the Census of Cuba, 1899. Washington, D.C.: U.S. Government Printing Office, 1900.

Dirección Central de Estadistica. Censo de población y viviendas, 6 de septiembre de 1970: datos preliminares. La Habana: Junta Central de Planificación, 1971.

_____. Memoria del censo ganadero (31 de agosto de 1967). La Habana: Junta Central de Planificación, 1969.

Oficina del Censo de los Estados Unidos. Censo de la República de Cuba, 1907. Washington, D.C.: U.S. Bureau of the Census, 1908.

República de Cuba. Censo de la República de Cuba, 1919. La Habana: Maza, Arroyo y Caso, 1920.

_____. Informe general del censo de 1943. La Habana: Fernández, 1945.

Ministerio de Agricultura. Memoria del censo agrícola nacional, 1946. La Habana: Fernández, 1951.

Oficina Nacional de los Censos Demográfico y Electoral. Censos de población, vivienda y electoral: informe general (enero 28 de 1953). La Habana: Fernández, 1955.

Statistical Yearbooks

Dirección Central de Estadística. Anuario Estadístico de Cuba, 1972.
La Habana: Junta Central de Planificación, 1974.

————. Anuario Estadístico de Cuba, 1973. La Habana: Junta
Central de Planificación, 1974.

————. Boletín Estadístico, 1964. La Habana: Junta Central de
Planificación, 1966.

————. Boletín Estadístico de Cuba, 1968. La Habana: Junta
Central de Planificación, 1970.

————. Boletín Estadístico de Cuba, 1970. La Habana: Junta
Central de Planificación, 1972.

————. Boletín Estadístico de Cuba, 1971. La Habana: Junta
Central de Planificación, 1973.

————. Compendio Estadístico de Cuba, 1966. La Habana, Junta
Central de Planificación, 1966.

Dirección General de Estadística. Anuario Estadístico de Cuba, 1957.
La Habana: Fernández, 1958.

SPECIAL HISTORIES

Aguilera Rojas, Eladio. Francisco V. Aguilera y la revolución de
Cuba de 1868. 2 vols. La Habana, 1909.

En el camino de la independencia: estudio histórico sobre la
rivalidad de los Estados Unidos y la Gran Bretaña en sus
relaciones con la independencia de Cuba. With an appendix,
"De Monroe a Platt," by Ramiro Guerra y Sánchez. La Habana,
1930.

González, Diego. Historia documentada de los movimientos
revolucionarios por la independencia de Cuba de 1852 a 1867.
2 vols. La Habana, 1939.

Harrington, M.R. Cuba antes de Colón. Translated by A. del
Valle and F. Ortiz. La Habana, 1935.

Irving, Washington. Vida y viajes de Cristóbal Colón. Translated
by José García de Villalta. Madrid, 1833.

Las insurrecciones en Cuba. 2 vols. Madrid, 1872-73.

Ortiz, Fernando. Historia de la arqueología Indocubana. La Habana, 1935.

Pirala, Antonio. Anales de la guerra de Cuba. 3 vols. Madrid, 1895-98.

Rodríguez, José Ignacio. Estudio histórico sobre el origen, desenvolvimiento y manifestaciones prácticas de la idea de la anexión de la isla de Cuba a los Estados Unidos de América. La Habana, 1900.

Valle, Adrián del. Historia documentada de la Conspiración de la Gran Legión del Aguila Negra. La Habana, 1930.

DOCUMENTS

Actas de las Asambleas de Representantes y del Consejo de Gobierno durante la Guerra de Independencia (1895-1899). Introduction by Joaquín Llaverías and Emeterio S. Santovenia. 6 vols. La Habana, 1933-38.

Arango y Parreño, Francisco de. Obras. 2 vols. La Habana, 1888.

Cabrales, Gonzalo. Epistolario de héroes, cartas y documentos históricos. La Habana, 1922.

Céspedes y Quesada, Carlos Manuel de. Carlos Manuel de Céspedes. Paris, 1895.

Chacón y Calvo, José M. Colección de documentos inéditos para la historia de Hispanoamérica. 6 vols. Cedulario cubano (los orígenes de la colonización). Vol. 2 (1493-1512). Madrid, 1929.

Colección de documentos inéditos relativos al descubrimiento, conquista y organización de las antiguas posesiones españolas de América y Oceanía, sacadas de los archivos del Reino y muy especialmente del de Indias. 42 vols. Madrid, 1864-84.

Colección de documentos inéditos relativos al descubrimiento, conquista y organización de las antiguas posesiones españolas de Ultramar. Second series, Isla de Cuba. Vols. 1, 4, 6. Madrid: Real Academia de la Historia, 1885-88-91.

Colección póstuma de papeles científicos, históricos, políticos y de otros ramos sobre la isla de Cuba. La Habana, 1881.

Cuevas, Mariano, S.J. Cartas y otros documentos de Hernán Cortés. Seville, 1915.

Diario de las Sesiones de la Cámara de Representantes de la isla de Cuba, 1898. La Habana, 1898.

Diario de sesiones de la Convención Constituyente. La Habana, 1928.

Diario de sesiones de la Convención Constituyente de la isla de Cuba. La Habana, 1900-01.

Diario de las Sesiones del Consejo de Administración de la isla de Cuba. La Habana, 1898.

Gayangos, Pascual de. Cartas y relaciones de Hernán Cortés al Emperador Carlos V. Madrid, 1861.

Giberga, Eliseo. Obras. 4 vols. La Habana, 1930-31.

Monte, Domingo del. Escritos. 2 vols. La Habana, 1929.

Montoro, Rafael. Obras. 4 vols. La Habana, 1930.

Obras completas de Martí: Cuba--discursos revolucionarios. 8 vols. La Habana: Editorial Trópico, 1937.

Obras completas de Martí: Cuba--política y revolución. 8 vols. La Habana: Editorial Trópico, 1936-37.

Obras de José Antonio Saco. 2 vols. New York, 1853.

Papeles existentes en el Archivo General de Indias relativos a Cuba y muy particularmente a La Habana. 2 vols. La Habana: Academia de Historia de Cuba, 1931.

Saco, José Antonio. Colección de papeles científicos, históricos, políticos y de otros ramos sobre la isla de Cuba. 3 vols. Paris, 1858-59.

Saco, José Antonio. Historia de la esclavitud de la raza africana en el Nuevo Mundo y en especial en los países américo-hispanos. Vol. 1, Barcelona, 1879; vol. 2, La Habana, 1893.

Saco, José Antonio. Historia de la esclavitud de los indios en el Nuevo Mundo. La Habana, 1883.

Sanguily, Manuel. Discursos y conferencias. 2 vols. La Habana, 1918-19.

Sanguily, Manuel. Obras. 9 vols. La Habana, 1925-30.

APPENDIX 1. COUNTRY BRIEF

GEOGRAPHY

Area: 114,478 square miles

Capital: Havana

Provinces and Capitals: Camagüey (Camagüey); Ciego de Avila
(Ciego de Avila); Cienfuegos (Cienfuegos); Ciudad de la Habana
(Havana); Granma (Bayamo); Guantánamo (Guantanamo); Habana
(Havana); Holguín (Holguín); Las Tunas (Victoria de las Tunas);
Matanzas (Matanzas); Pinar del Río (Pinar del Río); Sancti Spíritus
(Sancti Spíritus); Santiago de Cuba (Santiago de Cuba); Villa
Clara (Santa Clara).

Major ports: Havana, Santiago de Cuba, Matanzas, Nuevitas, Nipe,
Mariel.

PEOPLE

Population: 10,150,000 (estimate, close of 1985)

Language: Spanish

Literacy: 96 percent

LEADERSHIP

Fidel Castro Ruz: First secretary of Cuban Communist Party,
president of the councils of state and ministers, and commander
of the Armed Forces.

Raúl Castro Ruz: Second secretary of Cuban Communist Party, first vice president of the councils of state and ministers, and minister of Revolutionary Armed Forces.

ECONOMY

Estimated GNP: 13.3 billion US$

Agriculture: Main crops: sugar, tobacco, rice, vegetables, citrus fruits.

Major industries: Sugar milling, electricity, petroleum refining, nickel mining, food processing, cement.

Exports: $4.5 billion f.o.b. sugar, nickel, tobacco, shellfish.

Imports: $4.7 billion c.i.f. capital equipment, industrial raw materials, foodstuffs, petroleum.

Major trade partners: Exports: USSR, Eastern Europe, Canada, China, Japan, Spain; imports: USSR, Eastern Europe, China, Argentina, Japan, Spain.

Official monetary conversion rate: 1 peso equals US$1.33

APPENDIX 2. DIPLOMATIC RELATIONS[a]

MAJOR COMMUNIST COUNTRIES	Year Established
Albania	1960
Bulgaria	1960
Czechoslovakia	1960
East Germany	1963
Hungary	1960
Laos	1974
Mongolia	1960
North Korea	1960
People's Republic of China	1959
Poland	1960
Romania	1960
USSR	1960
Vietnam	1960

REMAINING COUNTRIES (arranged by continent)

Africa

Algeria	1961
Angola	1975
Benin	1974
Botswana	1977
Burundi	1974
Cameroon	1974
Cape Verde	1975
Chad	1976
Comoros	1976
Congo	1964
Egypt	1960
Equatorial Guinea	1972
Ethiopia	1975
Gambia	1979
Ghana	reestablished 1974
Guinea	1960

Africa (cont'd)	Year Established (cont'd)
Guinea-Bissau	1973
Lesotho	1979
Liberia	1974
Libya	1976
Madagascar	1974
Mali	1972
Mauritania	1972
Mauritius	1976
Mozambique	1975
Niger	1976
Nigeria	1974
Rwanda	1979
São Tomé and Principe	1976
Senegal	1974
Seychelles	1978
Sierra Leone	1972
Sudan	1979
Tanzania	1964
Togo	1979
Uganda	1974
Upper Volta (now Burkina Faso)	1975
Zaire[b]	reestablished 1979
Zambia	1972
Zimbabwe	1980

Asia-Oceania

Afghanistan	1975
Australia[c]	1973
Bangladesh	1972
Burma	1976
Hong Kong[c]	NA
India	prior to 1959
Indonesia	NA
Iran	reestablished 1979
Iraq	1960
Japan	prior to 1959
Jordan[d]	1979
Kampuchea	reestablished 1975
Kuwait	1974
Lebanon	1960
Malaysia	1975
Maldives	1977
Nepal	1975
Pakistan[d]	NA
Philippines	reestablished 1975
South Yemen	1972
Sri Lanka	1960

Asia-Oceania (cont'd)

Year Established (cont'd)

Syria	1965
Thailand	1963

Europe

Austria	prior to 1959
Belgium	prior to 1959
Cyprus	1960
Denmark	prior to 1959
Finland	prior to 1959
France	prior to 1959
Greece	prior to 1959
Iceland	prior to 1959
Italy	prior to 1959
Luxembourg	prior to 1959
Malta	1977
Netherlands	prior to 1959
Norway	reopened 1968
Portugal	prior to 1959
Spain	prior to 1959
Sweden	prior to 1959
Switzerland	prior to 1959
Turkey	prior to 1959
United Kingdom	prior to 1959
The Vatican	prior to 1959
West Germany	reestablished 1975
Yugoslavia	1960

Western Hemisphere

Argentina	reestablished 1973
The Bahamas[d]	1974
Barbados[d]	1972
Brazil	reestablished 1986
Canada	prior to 1959
Costa Rica[c]	1977
Dominica[d]	1980
Ecuador	reestablished 1979
Grenada	1979, suspended 1983
Guyana	1972
Jamaica	1972
Mexico	prior to 1959
Nicaragua	reestablished 1979
Panama	reestablished 1974
Peru	reestablished 1972
St. Lucia[d]	1979
Suriname[d]	1979

Western Hemisphere (cont'd)	Year Established (cont'd)
Trinidad and Tobago[d]	1972
United States[e]	1977
Venezuela	reestablished 1974

---

[a]Unless otherwise noted, there is an embassy in each of these countries.
[b]Established 1974, suspended 1977.
[c]Consular relations only.
[d]Relations at the ambassadorial level, but ambassador resides elsewhere.
[e]Special Interests Section

APPENDIX 3.

MEMBERSHIP IN INTERNATIONAL ORGANIZATIONS

United Nations (UN)

ECLA	Economic Commission for Latin America
FAO	Food and Agricultural Organization
GATT	General Agreement on Tariffs and Trade
IAEA	International Atomic Energy Agency
ICAO	International Civil Aviation Organization
ILO	International Labor Organization
IMCO	Intergovernment Maritime Consultative Organization
ITU	International Telecommunication Union
UNCTAD	UN Conference on Trade and Development
UNDP	UN Development Program
UNESCO	UN Educational, Scientific, and Cultural Organization
UNIDO	UN Industrial Development Organization
UPU	Universal Postal Union
WHO	World Health Organization
WIPO	World Intellectual Property Organization
WMO	World Meteorological Organization

Communist Organizations

CEMA	Council for Mutual Economic Assistance
IBEC	International Bank for Economic Cooperation
IIB	International Investment Bank

Regional Organizations

NAMUCAR	Caribbean Multinational Shipping Enterprise
PAHO	Pan American Health Organization
SELA	Latin America Economic System

Other

G77	Group of 77
IHO	International Hydrographic Organization

IRC	International Rice Commission
ISO	International Sugar Organization
IWC	International Wheat Commission
NAM	Nonaligned Movement
--	Permanent Court of Arbitration
--	Postal Union of the Americas and Spain
WTO	World Tourism Organization

APPENDIX 4. PRESIDENTS OF CUBA

Estrada Palma, Tomás (1902-1906)
Gómez, José Miguel (1909-1913)
Menocal, Mario G. (1913-1921)
Zayas, Alfredo (1921-1925)
Machado, Gerardo (1925-1933)
Céspedes, Carlos Manuel de (August-September 1933)
Grau San Martín, Ramón (1933-1934)
Hevia, Carlos (January 16-17, 1934)
Mendieta, Carlos (1934-1935)
Barnet, José A. (1935-1936)
Gómez, Miguel Mariano (May-December 1936)
Laredo Bru, Federico (1936-1940)
Batista, Fulgencio (1940-1944)
Grau San Martín, Ramón (1944-1948)
Prío Socarrás, Carlos (1948-1952)
Batista, Fulgencio (1952-1958)
Urrutia, Manuel (January-July 17, 1959)
Dorticós, Osvaldo (1959-1975)
Castro Ruz, Fidel (1975-   )

Schmitt